*Probability and Mathematical Statistics (Continued)*
PUKELSHEIM · Optimal Design of Experiments
PURI and SEN · Nonparametric Methods in General Linear Models
PURI, VILAPLANA, and WERTZ · New Perspectives in Theoretical and Applied
    Statistics
RAO · Asymptotic Theory of Statistical Inference
RAO · Linear Statistical Inference and Its Applications, *Second Edition*
ROBERTSON, WRIGHT, and DYKSTRA · Order Restricted Statistical Inference
ROGERS and WILLIAMS · Diffusions, Markov Processes, and Martingales,
    Volume II: Ito Calculus
ROHATGI · A Introduction to Probability Theory and Mathematical Statistics
ROSS · Stochastic Processes
RUBINSTEIN · Simulation and the Monte Carlo Method
RUBINSTEIN and SHAPIRO · Discrete Event Systems: Sensitivity Analysis and
    Stochastic Optimization by the Score Function Method
RUZSA and SZEKELY · Algebraic Probability Theory
SCHEFFE · The Analysis of Variance
SEBER · Linear Regression Analysis
SEBER · Multivariate Observations
SEBER and WILD · Nonlinear Regression
SERFLING · Approximation Theorems of Mathematical Statistics
SHORACK and WELLNER · Empirical Processes with Applications to Statistics
STAUDTE and SHEATHER · Robust Estimation and Testing
STOYANOV · Counterexamples in Probability
STYAN · The Collected Papers of T.W. Anderson: 1943–1985
WHITTAKER · Graphical Models in Applied Multivariate Statistics
YANG · The Construction Theory of Denumerable Markov Processes

*Applied Probability and Statistics*
ABRAHAM and LEDOLTER · Statistical Methods for Forecasting
AGRESTI · Analysis of Ordinal Categorical Data
AGRESTI · Categorical Data Analysis
ANDERSON and LOYNES · The Teaching of Practical Statistics
ANDERSON, AUQUIER, HAUCK, OAKES, VANDAELE, and WEISBERG ·
    Statistical Methods for Comparative Studies
ASMUSSEN · Applied Probability and Queues
* BAILEY · The Elements of Stochastic Processes with Applications to the Natural
    Sciences
BARNETT · Interpreting Multivariate Data
BARNETT and LEWIS · Outliers in Statistical Data, *Third Edition*
BARTHOLOMEW, FORBES, and McLEAN · Statistical Techniques for
    Manpower Planning, *Second Edition*
BATES and WATTS · Nonlinear Regression Analysis and Its Applications
BELSLEY · Conditioning Diagnostics: Collinearity and Weak Data in Regression
BELSLEY, KUH, and WELSCH · Regression Diagnostics: Identifying Influential
    Data and Sources of Collinearity
BHAT · Elements of Applied Stochastic Processes, *Second Edition*
BHATTACHARYYA and WAYMIRE · Stochastic Processes with Applications
BIEMER, GROVES, LYBERG, MATHIOWETZ, AND SUDMAN · Measurement
    Errors in Surveys
BIRKES and DODGE · Alternative Methods of Regression
BLOOMFIELD · Fourier Analysis of Time Series: An Introduction
BOLLEN · Structural Equations with Latent Variables
BOX · R.A. Fisher, the Life of a Scientist
BOX and DRAPER · Empirical Model-Building and Response Surfaces
BOX and DRAPER · Evolutionary Operation: A Statistical Method for Process
    Improvement
BOX, HUNTER, and HUNTER · Statistics for Experimenters: An Introduction to
    Design, Data Analysis and Model Building

*Continued on back end papers*

* Now available in a lower priced paperback edition in the Wiley Classics Library.

# Outliers in Statistical Data

## Third Edition

# Outliers in Statistical Data

## Third Edition

**VIC BARNETT**
*Rothamsted Experimental Station, UK*

and

**TOBY LEWIS**
*University of East Anglia, UK*

**JOHN WILEY & SONS**
Chichester • New York • Brisbane • Toronto • Singapore

*Other Wiley Editorial Offices*

John Wiley & Sons, Inc., 605 Third Avenue,
New York, NY 10158-0012, USA

Jacaranda Wiley Ltd, 33 Park Road, Milton,
Queensland 4064, Australia

John Wiley & Sons (Canada) Ltd, 22 Worcester Road,
Rexdale, Ontario M9W 1L1, Canada

John Wiley & Sons (SEA) Pte Ltd, 37 Jalan Pemimpin #05-04,
Block B, Union Industrial Building, Singapore 2057

*Library of Congress Cataloging-in-Publication Data*

Barnett, Vic.
    Outliers in statistical data / Vic Barnett and Toby Lewis. — 3rd ed.
    p.  cm.
    Includes bibliographical references and index.
    ISBN 0 471 93094 6
    1. Outliers (Statistics)  I. Lewis, Toby.  II. Title.
    QA276.B2849   1994                                     93-29289
    519.5′2—dc20                                           CIP

*British Library Cataloguing in Publication Data*

A catalogue record for this book is available from the British Library

ISBN 0 471 93094 6

Typeset in 10/12pt Times by Pure Tech Corporation, Pondicherry, India
Printed and bound in Great Britain by Biddles Ltd,
Guildford and King's Lynn

# Preface to First Edition

The concept of an outlier has fascinated experimentalists since the earliest attempts to interpret data. Even before the formal development of statistical method, argument raged over whether, and on what basis, we should discard observations from a set of data on the grounds that they are 'unrepresentative', 'spurious', or 'mavericks' or 'rogues'. The early emphasis stressed the contamination of the data by unanticipated and unwelcome errors or mistakes affecting some of the observations. Attitudes varied from one extreme to another: from the view that we should never sully the sanctity of the data by daring to adjudge its propriety, to an ultimate pragmatism expressing 'if in doubt, throw it out'.

The present views are more sophisticated. A wider variety of aims are recognized in the handling of outliers, outlier- generating models have been proposed, and there is now available a vast array of specific statistical techniques for processing outliers. The work is scattered throughout the literature of the present century, shows no sign of any abatement, but has not previously been drawn together in a comprehensive review. Our purpose in writing this book is to attempt to provide such a review, at two levels. On the one hand we seek to survey the existing state of knowledge in the outlier field and to present the details of selected procedures for different situations. On the other hand we attempt to categorize differences in attitude, aim, and model in the study of outliers, and to follow the implications of such distinctions for the development of new research approaches. In offering such a comprehensive overview of the principles and methods associated with outliers we hope that we may help the practitioner in the analysis of data and the researcher in opening up possible new avenues of enquiry.

Early work on outliers was (inevitably) characterized by lack of attention to the modelling of the outlier-generating mechanism, by informality of technique with no backing in terms of a study of the statistical properties of proposed procedures, and by a leaning towards the hardline view that outliers should be either rejected or retained with full import. Even today sufficient attention is not always paid to the form of the outlier model, or to the practical purpose of investigating outliers, in the presentation of methods for processing outliers. Many procedures have an *ad hoc*, intuitively justified, basis with little external reference in the sense of the relative statistical merits of different possibilities. In reviewing such techniques we will attempt to set them, as far as possible, within a wider framework of model, statistical principle, and practical aim, and we shall also consider the extent to which such basic considerations have begun to formally permeate outlier study over recent years.

Such an emphasis is reflected in the structure of the book.* The opening two chapters are designed respectively to motivate examination of outliers and to pose basic questions about the nature of an outlier. Chapter 1 gives a general survey of the field. In Chapter 2 we consider the various ways in which we can model the presence of outliers in a set of data. We examine the different interests (from *rejection* of unacceptable contamination, through the *accommodation* of outliers with reduced influence in robust procedures applied to the whole set of data, to specific *identification* of outliers as the facets of principal interest in the data). We discuss the statistical respectability of distinct methods of study, and the special problems that arise from the dimensionality of the data set or from the purpose of its analysis (single-sample estimation or testing, regression, analysis of data from designed experiments, examination of slippage in multisample data, and so on).

Chapter 3 examines at length the assessment of discordancy of outliers in single univariate samples. It discusses basic considerations and also presents a battery of techniques for practical use with comment on the circumstances supporting one method rather than another.

Chapter 4, on the accommodation of outliers in single univariate samples, deals with inference procedures which are robust in the sense of providing protection against the effect of outliers. Chapter 5 is concerned with processing several univariate samples both with regard to the relative slippage of the distributions from which they arise and (to a lesser extent) in relation to the accommodation of outliers in robust analysis of the whole set of data.

Chapters 6 and 7 extend the ideas and methods (in relation to the three interests: rejection, accommodation, identification) to single multivariate samples and to the analysis of data in regression, designed experiments, or time series situations. Chapter 8 gives fuller and more specific attention to the implications of adopting a Bayesian, or a non-parametric, approach to the study of outliers. The concluding Chapter 9 poses a few issues for further consideration or investigation.

The book aims to bring together in a logical framework the vast amount of work on outliers which has been scattered over the years in the various professional journals and texts, and which appears to have acquired a new

---

* Chapter numbers have been changed in the Second and Third Editions (see *Preface* and *Contents* to each Edition).

lease of life over the last decade or so. It is directed to more than one kind of reader: to the student (to inform him of the range of ideas and techniques), to the experimentalist (to assist him in the judicious choice of methods for handling outliers), and to the professional statistician (as a guide to the present state of knowledge and a springboard for further research).

The level of treatment assumes a knowledge of elementary probability theory and statistical method such as would be acquired in an introductory university-level course. The methodological exposition leans on an understanding of the principles and practical implications of testing and estimation. Where basic modelling and demonstration of statistical propriety are discussed, a more mathematical appreciation of basic principles is assumed, including some familiarity with optimality properties of methods of constructing tests and estimators and some knowledge of the properties of order statistics. Proofs of results are formally presented where appropriate, but at a heuristic rather than highly mathematical level.

Extensive tables of appropriate statistical functions are presented in an Appendix, to aid the practical worker in the use of the different procedures. Many of these tables are extracted from existing published tables; we are grateful to all the authors and publishers concerned, and have made individual acknowledgement at the appropriate places in our text. Other tables have been specially produced by us. The whole set of tables has been presented in as compact and consistent a style as possible. This has involved a good deal of selection and re-ordering of the previously published material; we have aimed as far as possible to standardize the ranges of tabulated values of sample size, percentage point, etc.

Copious references are given throughout the text to source material and to further work on the various topics. They are gathered together in the section entitled 'References and Bibliography' with appropriate page references to places of principal relevance in the text. Additional references augment those which have been discussed in the text. These will of course appear without any page reference, but will carry an indication of the main area to which they are relevant.

It is, of course, a privilege and pleasure to acknowledge the help of others. We thank Dave Collett, Nick Fieller, Agnes Herzberg, and David Kendall for helpful comments on early drafts of some of the material. We are particularly grateful to Kim Malafant who carried out the extensive calculations of the new statistical tables in Chapter 3. Our grateful thanks go also to Hazel Howard who coped nobly with the typing of a difficult manuscript.

We are solely responsible for any imperfections in the book and should be glad to be informed of them.

VIC BARNETT
TOBY LEWIS
*July, 1977*

# Preface to Second Edition

The first edition of this book took the opportunity of drawing together for the first time the vast and assorted literature of over a century on the topic of outliers. In offering a combination of specific methods for the experimentalist and a 'state-of-the-art' review for the professional statistician it was designed to provide a logical framework for, and comprehensive coverage of, the variety of topics and emphases in outlier study. About four hundred references were used as basic source material, in a climate where the subject seemed clearly 'to have acquired a new lease of life over the last decade or so'. What could not be anticipated was the sheer force and momentum of this revitalization of interest. In the six years since the appearance of the first edition about three hundred more published articles have appeared: in crude numerical terms, a seventy-five per cent expansion over the whole previous history of the subject. It is inevitable therefore that we should seek to reassess the current situation in the light of so many new ideas and refinements. The second edition aims to do this.

The modifications in the revised edition are of three types. Firstly, there are areas of enquiry which represent new topics of outlier study or which now need to be described within the context of outlier methodology. These are reflected in the new chapters (7 and 11) on outliers in directional data, and in time series, respectively. Secondly, there are the many contributions which refine, reassess, or extend our knowledge on specific aspects of outlier investigation. Such developments are incorporated by means of substantial expansion, and some judicious pruning, of the discussion of almost all the topics of the earlier edition of the book. Particular attention has been given to discordancy tests for univariate and multivariate samples and for data from structured models (linear models generally, and specific aspects of regression and designed experiment situations). At a more general level, there has been a welcome growth of emphasis on methods of accommodation (robust inference in the face of outliers) and on informal (often graphical) descriptive procedures. These latter areas of development have contributed to the stimulus for a final type of modification—a reordering or reemphasis of some of the basic ideas and principles. In particular this has prompted a separation of general approach from specific results for the study of univariate samples with regard both to tests of discordancy and methods of accommodation, and a reversal of order in the treatment of these two aspects. Thus we now have Chapter 3 and 4 on accommodation and Chapters 5 and 6 on discordancy testing, for univariate samples, distinguishing between general principle (Chapters 3 and 5) and specific method (Chapter 4 and 6) in each case. Accordingly the chapters dealing with slippage, multivariate outliers, outliers in linear models (regression, designed experiments), and Bayesian methods are now renumbered as Chapters 8, 9, 10, and 12, respectively.

There has been much activity in recent years on the theme of 'influential observations': as sample values which have disproportionate effect on estimates or tests of parameters in a model (particularly in the field of regression). Whilst not entirely coincident concepts, outliers and influential observations are clearly related in important aspects. Similarities and distinctions of aim and method are highlighted at appropriate stages in our revised study of outlier methods, especially in Chapter 10.

We are grateful to those who have pointed out misprints and ambiguities in the first edition; particularly to Dr N. A. Campbell. We have taken the opportunities of remedying such matters.

Whilst the new edition reflects the many developments and extensions mentioned above, it is important to stress that the basic emphasis and aim is unchanged. We continue to eschew the two extremes of pre-digested recipes for instant outlier management on the one hand and indulgence in mathematical formality or sophistication for its own sake on the other. Our aim remains one of explaining basic principle and developing associated method to a level where outlier techniques can be soundly and sanely applied (with relevant illustration and tabulation) and the researcher can be provided with a springboard for further exploration of a fascinating statistical topic.

VIC BARNETT
TOBY LEWIS
*December, 1983*

# Preface to Third Edition

Since the publication of the second edition of this book, work in outlier methodology has continued unabated. Indeed, over 1000 new refereed publications have appeared in the literature. These not only extend previous knowledge in existing fields, but have opened up whole new areas of enquiry. Clearly, a new edition was needed to do justice to these many developments and to maintain our objective of providing a comprehensive and up to date coverage of the subject.

We have thoroughly revised and updated the material on the range of topics that were covered in the earlier editions. At the same time, extra material has been introduced to reflect new themes and changing emphases.

In contrast, a small amount of material has been omitted, in particular the chapter on the peripheral topic of *outlying subsamples*.

Topics on which the coverage *is new or has been substantially changed or extended* include:

- **basic principles**; distribution theory under contamination models, measures of efficiency and performance for multiple outliers, assessment of masking and swamping, allocation of outliers.
- **univariate data**; new tests (including extreme value and Weibull), wider study of robustness and accommodation (including logistic and double exponential distributions), additional tables.
- **multivariate and structured data**; estimation of individual components of vector parameters, use of correlation estimators, deletion methods, elliptically symmetric distributions, graphical methods, least median of squares and $L_1$-norm methods, multivariate linear model, multiple outliers, non-linear regression (including logistic and generalized linear models).
- **special topics**; Bayesian methods, time series (ARIMA model, distinction of AO and IO outliers, model specification, new accommodation methods, diagnostics, multiple time series), directional data (methods for axial and vectorial spherical data, accommodation).

Further special topics are dealt with in *new chapters*—these include new methods for outliers in contingency tables, problems of sample surveys, statistical software and international standards and regulations. Practical illustrations remain important for reinforcement of ideas—new examples are included as also is discussion of data studies in the literature.

To cope with the vast expansion of material, it has been advantageous to restructure the book, which is now divided into the following four distinct parts: *Basic Principles, Univariate Data, Multivariate and Structured Data, Special Topics*. These reflect the headings used above to categorize the major areas of change. The *References and Bibliography* section has been substantially expanded to cover the new material.

We would like to thank Eileen Stoydin, Marian Joyce and Sharon Wilson for their most efficient assistance in the preparation of the typescript.

We hope that this new edition will prove to be valuable both to research statisticians and to practitioners in various fields.

VIC BARNETT
TOBY LEWIS
*April, 1993*

# Contents

# PART I

# Basic Principles

# CHAPTER 1

# Introduction

From the earliest times, there has been a concern for 'unrepresentative', 'rogue', or 'outlying' observations in sets of data. These are often seen as contaminating the data: reducing and distorting the information about the data source or generating mechanism. It is natural to seek means of interpreting or categorizing *outliers* and methods for handling them—sometimes perhaps rejecting them to restore the propriety of the data, or at least adopting methods of reducing their impact in any statistical analysis.

What are outliers and what is the outlier problem? An interesting answer is found in the following quotation.

In almost every true series of observations, some are found, which differ so much from the others as to indicate some abnormal source of error not contemplated in the theoretical discussions, and the introduction of which into the investigations can only serve . . . to perplex and mislead the inquirer.

The writer is referring of course to *outliers*, and the comments embody much of the basic nature of, and grounds for concern in, the presence of outliers in a data set. Interestingly, the remark was made over 130 years ago (Peirce, 1852) and referred to *previous* practice, making concern for outliers one of the earliest aspects of statistical method. Indeed, it predates to a substantial degree most of statistical methodology.

Inevitably much that is embodied in Peirce's remarks, and in proposals of the time, was primitive and restrictive when viewed from a present perspective. Outlying observations do not inevitably 'perplex' or 'mislead'; they are not necessarily 'bad' or 'erroneous', and the experimenter may be tempted in some situations not to reject an outlier but to welcome it as an indication of, say, some unexpectedly useful industrial treatment or surprisingly successful agricultural variety. Then again, it may not be necessary to adopt either of the extremes: of rejection (with the risk of loss of genuine information) or inclusion (with the risk of contamination). In appropriate circumstances we can use 'robust' methods of inference which employ all the data but minimize the influence of any outliers. Such methods will be said to *accommodate* the outliers.

We should note that the outlier problem, as well as being one of the earliest of statistical interests is also an *unavoidable* one. It is not enough to say, as some statisticians do, that one should not consider dealing with outlying observations unless furnished with information on their prior probabilities. Some will not even admit the concept of an outlier unless there is some obvious (or plausible) physical explanation of its presence! In fact, experimental scientists and others who have to deal with data and take decisions are forced to make judgements about outliers—whether or not to include them, whether to make allowances for them on some compromise basis, and so on.

Sometimes this is done in what appears, by modern standards, to be a particularly naive or inefficient way. For example, a chemistry textbook in current use (Calvin et al., 1949, reprinted in 1960) advises its readers to use Chauvenet's method: 'Any result of a series containing n . . . observations shall be rejected when the magnitude of its deviation from the mean of all measurements is such that the probability of occurrence of all deviations as large or larger is less than $1/2n$'. This proposal was one of the earliest for dealing with outliers, dating from the middle of the nineteenth century (Chauvenet, 1863). It turns out to be well-meaning but misguided! We shall say more about this, and about other early ideas for rejecting outliers, in Section 2.1.

Some basic characteristics of outliers can be illustrated by examples in English marital law: a rather unusual context in which to encounter the application of statistical method.

In 1949, in the case of *Hadlum* v. *Hadlum*, Mr Hadlum appealed against the failure of an earlier petition for divorce. His claim was based on alleged adultery by Mrs Hadlum, the evidence for which consisted of the fact that Mrs Hadlum gave birth to a child on 12 August 1945 which was 349 days later than when Mr Hadlum had left the country to serve King and nation!

The average gestation period for the human female is 280 days: 349 days seemed to Mr Hadlum to be surprisingly long. He adjudged it an *outlier*. He did not want his outlying observation to be *rejected* (discarded, ignored). On the contrary he wanted it to be identified (acknowledged) with appropriate consequences. (Such distinctions of aim in the handling of outliers will become a major factor in our later study).

Mr Hadlum was unsuccessful. The appeal judges agreed that some limit of credibility must be entertained but ruled that 349 days, whilst improbable, was not beyond the range of scientific possibility. In other similar cases conflicting views had prevailed. In *M.-T.* v. *M.-T.* also in 1949, the court had ruled that 340 days was *impossible* in the light of modern gynaecological experience, whilst for *Preston-Jones* v. *Preston-Jones* in 1951 the House of Lords had ruled that the 'limit' should be set at 360 days. *It is interesting that in none of these cases did the distribution of human gestation periods seem to figure in the evidence.* Figure 1.1 (see Barnett,

1978a, based on results in Chamberlain, 1975) shows a histogram of estimated gestation periods for a sample of 13 634 British births, with the above outliers superimposed. The vertical scale represents percentage of the sample born within a one-week period.

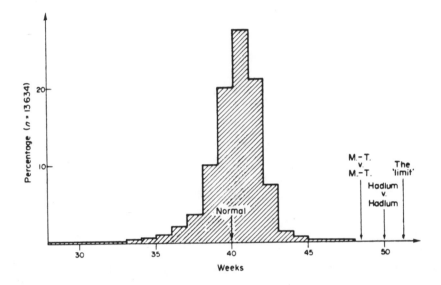

**Figure 1.1** Histogram of human gestation periods (reproduced from Barnett, 1978a, by permission of the Royal Statistical Society)

A much earlier case resurfaced at about the same time. In 1921, Mr Gaskill failed in a petition for divorce on the grounds of adultery based on an absence from the marital home of 331 days. In 1960, Mrs Gaskill succeeded in obtaining a divorce on grounds of separation for a period of 39 years: an outlier of quite another type!

Yet another example on the same theme highlights other aspects of outliers. Senn (1979) reported finding a 1916 medical diary in which the practitioner recorded details of pregnancies he had attended. From records of 98 births, Senn derived a frequency distribution of the doctor's 'prediction errors': discrepancies between 'date attended' and 'date expected'. These are shown as Figure 1.2 and exhibit some interesting features—perhaps again outliers—in the sense of some isolated prediction errors of a month or so in either direction (late and early). There are obvious possible explanations for such outliers!

To develop the basic ideas of outliers, let us note some particular features of the Hadlum case. The period of 349 days is *extreme*: surprisingly so, and we designate it an *outlier*. But this does not resolve the matter of the status of the observation, 349. It could be a genuine observation from the

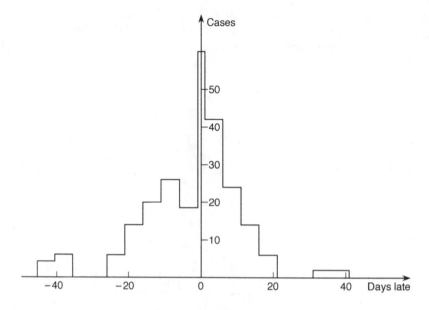

**Figure 1.2**   Histogram of birth prediction errors (in days)

distribution of gestation periods which just happens to be surprisingly extreme. This is essentially what the court ruled! On the other hand it might be a *contaminant*: an observation from some other distribution (in this case with later origin). This is precisely what Mr Hadlum was claiming!

Note yet a further prospect which would not be of interest to Mr Hadlum, but might be to someone else. A relatively unsurprising *apparent* gestation period of, say, 280 days could still be a contaminant. The *actual* gestation period could be, say, 260 days with a conception date 20 days later than had been assumed. This is just what is likely to be happening in the Senn data. Such crucial distinctions will be reiterated in a more formal manner later in this section.

With the development during this century of more detailed approaches to the statistical analysis of data, objectives have become clearer, principles more rigorously defined, and a vast array of sophisticated methodology has been constructed. Sample data may be analysed to assess the validity of some prescribed model, or to estimate or test hypotheses concerning relevant parameters. This greater sophistication in the design and use of statistical methods makes it even more important to be able to assess the integrity of a set of data.

What is known to be a good statistical procedure for estimating the mean of a normal distribution may be most inefficient if the distribution is not normal. The actual data being analysed can sound a warning for us! Perhaps one or more observations look suspicious when the data are considered as a sample from a normal distribution: they may have been incorrectly recorded (or measured), of course, or they may be a genuine reflection of a basic error in assuming an underlying normal distribution. Clearly such considerations are vitally important for proper statistical practice. We need a battery of techniques for assessing the integrity of a set of data with respect to an assumed model. We also should welcome the current avalanche of interest in statistical methods which are *robust* (i.e. retain reasonable validity) *over a range of possible probability models*. As particular aspects of these interests, we require methods for assessing, rejecting, making allowances for, or minimizing the influence of, outlying observations. The latter methods we will refer to as *outlier-robust*. The basic aim of this book is to bring together and to present a unified discussion of the rich variety of ways of handling outliers in statistical data, in relation to the nature of the outliers and to the aims of the investigation.

At this stage we must make clear what we mean by an **outlier**. (But note that, as explained on page 38, some writers use the word *outlier* in a different sense.) We shall define an outlier in a set of data to be *an observation (or subset of observations) which appears to be inconsistent with the remainder of that set of data*.

The phrase 'appears to be inconsistent' is crucial. It is a matter of subjective judgement on the part of the observer whether or not some observation (or set of observations) is picked out for scrutiny. What really matters is whether or not some observations are *genuine members* of the main population. If they are not, but are **contaminants** (arising from some other distribution), they may frustrate attempts to draw inferences about the original (basic) population.

Of course, any contaminants which occur in the midst of the data set will not be conspicuous; also, they are unlikely to seriously distort the inference process. What characterizes the 'outlier' is its impact on the observer (not only will it appear *extreme* but it will seem, in some sense, *surprisingly* extreme). Should *such* observations be foreign to the main population they may, by their very nature as contaminants, cause difficulties in the attempt to represent the population: they can grossly distort estimates (or tests) of parameters in the basic model for the population.

Accordingly the outlier problem progresses in the following way. We examine the data set. We decide that outliers exist (in the sense described above). Now we must ask: how should we react to the outliers, and what principles and methods can be used to support rejecting them, adjusting their values, or leaving them unaltered, prior to processing the principal mass of data? Clearly, *the answer depends on the form of the population*; techniques will be conditioned by, and specific to, the postulated (basic) model for the population. Thus, methods for the processing of outliers take on an entirely *relative* form: relative to the basic model.

One conceptual difficulty needs to be recognized at the outset. Opinion was, in the past, divided on precisely when it is justifiable to scrutinize outliers. There was little dispute that it is reasonable when outliers exist in the form of errors of observation, or mis-recording: that is, when they can be substantiated by practical considerations such as the sheer impossibility of a recorded value, or an obvious human error. It was sometimes claimed (as remarked above) that these are the only 'genuine' outliers and that if no such tangible explanation can be found for apparently unreasonable observations then their rejection, or accommodation by special treatment, is invalid.

However, two factors lead us to reject this nihilistic attitude. In the first place a variety of methods have been proposed for dealing with 'nontangible' outliers; these are widely used by statisticians and it is essential to present them in a classified manner for their better understanding and application.

Secondly, and more fundamentally, the concept of an outlier must be viewed in relative terms. Suppose we think that a sample arises from a normal distribution; one observation seems intuitively unreasonable (i.e. it is an outlier); an appropriate statistical test confirms its unacceptability. It begs the question to say that the unreasonable observation should not have been regarded as an outlier, on the grounds that it would not have appeared unreasonable if we had had in mind, say, a log-normal distribution, as a model for explaining the data. Be this as it may, the rogue observation *did appear as an outlier relative to our original model*, which presumably had some basis as an initial specification.

Examination of the outlier allows a more appropriate model to be formulated, or enables us to assess any dangers that may arise from basing inferences on the normality assumption. This is very much the way in which outliers have been discussed in the statistical literature, and will be considered here. We shall examine in subsequent chapters the various methods available for dealing with outliers in different situations, including some of the difficulties that inevitably arise.

To begin with we need to define, and explain, a little more formally, some basic considerations such as how we distinguish **extreme observations, outliers, contaminants**, etc.

Suppose we have a random univariate sample of size $n$

$$x_1, x_2, \ldots, x_n$$

from a distribution whose form we will denote by $F$.

We will write the ordered sample (from smallest to largest value) as

$$x_{(1)}, x_{(2)}, \ldots, x_{(n)}.$$

The observations $x_{(1)}$ and $x_{(n)}$ are the sample **extremes**. Whether we declare either of them to be an **outlier** depends on (at least an informal) consideration of how they appear in relation to the postulated model, $F$. In Figure 1.3(a), neither $x_{(1)}$ nor $x_{(n)}$ appears to be outlying. In Figure 1.3(b), $x_{(n)}$ is an *upper outlier* and $x_{(1)}$ also gives some cause for concern. We *might* declare $x_{(1)}$ to be a *lower outlier* (so that $x_{(1)}$, $x_{(n)}$ is an *outlier pair*). Note that we might declare $x_{(n-1)}$, $x_{(n)}$ as an upper outlier pair—see Figure 1.3(d)

So we see that *extreme values may or may not be outliers. Any outliers, however, are always extreme* (or relatively extreme) values in the sample! See Barnett (1988).

Suppose now that not all of the observations come from the distribution $F$, but one or two come from a distribution $G$ which has 'slipped' upward relative to $F$ (i.e. it has a larger mean). The observations from $G$ are termed **contaminants**. Such contaminants may appear as extremes, but *need* not do so. Figure 1.3(c) shows two contaminants (indicated ●), one of which is the upper extreme, the other is in the midst of the sample. But $x_{(n)}$, whilst an extreme and a contaminant, is not an outlier! In Figure 1.3(d), however, we see in contrast a *non-extreme* contaminant which is none the less outlying: as one of the outlier pair $x_{(n-1)}$, $x_{(n)}$.

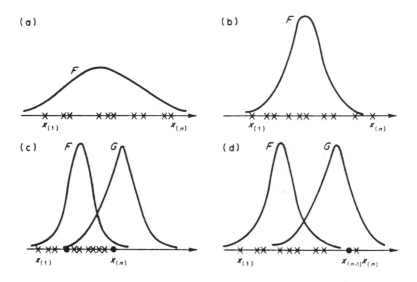

**Figure 1.3** Extremes, outliers, and contaminants

So we see that *outliers may or may not be contaminants; contaminants may or may not be outliers!* Of course, we have no way of knowing whether or not any observation is a contaminant. All we can do is concentrate attention on outliers as *the possible manifestation of contamination* and we shall see how statistical methods for examining outliers aim specifically at this prospect.

In the following sections of this chapter some practical examples are discussed briefly, to illustrate the ways in which the outlier problem may present itself. These also serve to motivate the different forms of statistical analysis considered in detail throughout the book.

## 1.1  HUMAN ERROR AND IGNORANCE

There are situations where outliers present little difficulty: where the manner of dealing with them is obvious and non-controversial. Such is the situation when human errors lead to clearly incorrect recording of data, or where lack of regard to practical factors results in serious misinterpretation.

In a study of low-temperature probabilities throughout the winter months in the UK, Barnett and Lewis (1967) analysed extensive data on hourly temperatures over several years and for various geographical sites. In the main, temperatures were recorded in degrees Fahrenheit. Among extensive data for Wick in northern Scotland the following hourly temperatures were found for the late evening of 31 December 1960 and early morning of 1 January 1961:

$$43, \ 43, \ 41, \ 41, \ 41, \ 42, \ 43, \ 58, \ 58, \ 41, \ 41$$

$$\uparrow$$

midnight

The values 58, 58 for midnight and 1.00 a.m. stand out in severe contrast to the others in this time-series section, and initially give one grounds for concern as to whether or not they are genuine—it seemed very warm at midnight for New Year in northern Scotland. On further enquiry, however, they were found to be perfectly reasonable! At midnight the Meteorological Office changed its measuring unit from degrees Fahrenheit to $\frac{1}{10}$ °C, so that in degrees Fahrenheit the values appear as follows (to the nearest degree):

$$43, \ 43, \ 41, \ 41, \ 41, \ 42, \ 43, \ 42, \ 42, \ 39, \ 39$$

$$\uparrow$$

midnight

These are much more satisfactory; so much for the 'outliers' 58 and 58! We have clearly *misinterpreted* them in the original form of the data.

Again, Finney (1974) gives an interesting example of the way in which a *recording error* may appear as an outlier in a set of data. He reports on measurements taken on the growth of poultry. For one bird, the weights (in kg) for successive weighings at regular intervals were shown as:

$$1.20, \ 1.60, \ 1.90, \ 1.55, \ 2.20, \ 2.25$$

From the manner in which the weights were determined and recorded it was clearly possible to commit recording errors of 0.50 kg or 1.00 kg. It seems highly likely that the fourth reading is a mis-recording of what should have been 2.05 kg. This conclusion is supported by biological considerations, and by the overall pattern of results for a large sample of birds.

Whilst one cannot be *certain* of the interpretation of 1.55 in this last example, some instances arise where recorded values are clearly impossible. In a student exercise, records were kept of the numbers of times a six occurred in ten throws of ten dice. One student returned the results:

$$2, 0, 3, 12, 2, 0, 1, 1, 3$$

Obviously the value 12 cannot be genuine. Since ten observations were asked for, it would seem merely that a comma has been omitted in a sequence of numbers which should have read:

$$2, 0, 3, 1, 2, 2, 0, 1, 1, 3$$

These examples all illustrate the effect of *non-statistical factors* attributable, to a greater or lesser extent, to lack of care in the recording or presentation of data. Processing of outliers (or spurious values of any sort) in such cases is not a matter of statistical analysis, but of native wit! Correct action also raises no difficulties in most cases. In the examples which follow, however, we will be very much concerned with *statistical* factors which affect the occurrence and treatment of outliers.

## 1.2 OUTLIERS IN RELATION TO PROBABILITY MODELS

Some very early data (Ryland, 1841) show the annual incomes of 91 scientific and literary societies in England in 1840. Figure 1.4 shows an extract of how the data appeared in the original publication. Omitting the organizations with incomes below £75, Table 1.1 presents the incomes of the remaining 69, with incomes rounded to the nearest pound and presented in increasing order. One value clearly stands out, the £7000 income of the Liverpool Mechanics' Institution. We have a clear *outlier* in the sense that this largest value is *surprisingly* high in relation to the others.

What are we to make of this outlier? Obviously there has to be a largest value and we might even expect to encounter a rather attenuated upper tail to the distribution for such a quantity as annual income. Note that we are immediately taking a rather different stance, arguing not in terms of a possible recording error but in terms of what we might reasonably and genuinely expect to encounter as the underlying distribution of values in such a situation. Thus we are directed to considering a *model* to explain the occurrence of the data.

Various possibilities immediately arise in relation to the outlying value 7000.

Table of the Income of Scientific and Literary Societies in England, and the Amount paid for Rates and Taxes in the Year 1840.
By ARTHUR RYLAND, Esq. (Laid before the Statistical Section of the British Association, 30th July, 1841.)

| Names of Societies. | Income. (£. s. d.) | Total Rates and Taxes. (£. s. d.) | Proportion to Income. | Window Tax. (£. s. d.) | Poor Rates. (£. s. d.) | City or Borough Rates. (£. s. d.) |
|---|---|---|---|---|---|---|
| Ashton-under-Lyne Mechanics' Institution | 140 0 0 | 1 5 8 | 1/110 | | 1 5 8 | |
| Bath Royal Literary and Scientific Society | 444 11 10 | 29 3 10 | 1/15 | | 29 3 10 | |
| ,, Mechanics' Institution | 138 6 6 | 6 18 6 | 1/20 | 6 18 6 | | |
| Beverley | 49 16 0 | 0 2 0 | | | | |
| Beccles Public Library | 90 0 0 | Gratuitous use of a Corporation building. | | | | |
| Belper Mechanics' Institution | 70 0 0 | Included in rent. | | | | |
| Bolton | 200 4 4 | None. | | | | |
| Brentford | | None; excused. | | | | |
| Bridport | | Excused. | | | | |
| Bristol Literary and Philosophical Society | 650 0 0 | 41 15 0 | 1/15 | | 15 15 0 | 26 0 0 |
| ,, Mechanics' Institution | 249 4 0 | 8 6 8 | 1/30 | | 3 10 0 | 4 16 0 |
| Birmingham Philosophical Institution | 431 10 0 | 68 9 0 | 1/6 | 8 19 0 | 34 5 0 | 25 5 0 |
| ,, Old Library | 700 0 0 | 17 11 0 | 1/40 | | 8 2 6 | 9 9 0 |
| ,, New Library | 290 0 0 | 20 2 6 | 1/14 | | 13 2 4 | 7 0 0 |
| ,, Mechanics' Institution | 400 0 0 | 34 13 9 | 1/11 | 6 3 5 | 16 12 0 | 11 18 0 |
| ,, Society of Arts | 335 0 0 | 68 8 0 | 1/5 | | 41 18 0 | 26 10 0 |
| Bury Mechanics' Institution | 200 10 0 | 1 4 3 | 1/160 | | 0 16 0 | 0 8 0 |
| Chichester Mechanics' Institution | 92 0 0 | 8 9 7 | 1/11 | 4 13 7 | 2 18 0 | 0 18 0 |
| Cambridge | 237 0 1 | 11 12 7 | 1/20 | 3 7 11 | 6 14 0 | 1 10 8 |
| Chertsey | 120 1 0 | 1 15 6 | 1/67 | | | 1 10 0 |
| Chippenham Literary and Scientific Institution | 45 0 0 | In rent. | | | | |
| Cockermouth Mechanics' Institution | 10 0 0 | In rent. | | | | |
| Cornwall Royal Polytechnic Society, Falmouth | 350 0 0 | Nominal. | | | | |
| Chester Mechanics' Institution | 77 10 0 | 3 2 0 | 1/25 | | 1 12 0 | 1 10 0 |
| Cheltenham Literary and Philosophical Society | 205 16 0 | 30 0 0 | 1/7 | | | |
| ,, Mechanics' Institution | | Exempted. | | | | |
| Colchester | 77 0 0 | 6 0 0 | 1/13 | | 5 0 0 | 1 0 0 |
| Darwen | 170 0 0 | 1 13 9 | 1/100 | | 1 0 3 | 0 13 6 |

| | | | | | | |
|---|---|---|---|---|---|---|
| Dover Philosophical Institution | 115 0 0 | Excused. | .. | .. | .. | .. |
| Devonport Mechanics' Institution | ... | Temporary buildings. | | .. | .. | .. |
| Evesham | 58 0 0 | 5 3 6 | 1/11 | .. | 3 16 4 | 1 6 7 |
| Gateshead | 79 7 11 | Included in rent. | .. | .. | .. | .. |
| Gloucester | 100 0 0 | 3 0 0 | 1/33 | .. | 3 0 0 | .. |
| Greenwich Society for Diffusion of Knowledge | ... | Gratuitous use of rooms. | | .. | .. | .. |
| Hastings and St. Leonard's Mechanics' Institution | 60 0 0 | 3 10 6 | 1/17 | .. | 2 14 0 | 0 16 6 |
| Hammersmith | 200 0 0 | 8 14 6 | 1/23 | 3 17 0 | 3 15 0 | 1 2 6 |
| Horncastle | 50 0 0 | Excused. | .. | .. | .. | .. |
| Horsham | 38 0 0 | None. | .. | .. | .. | .. |
| Hull Literary and Philosophical Society | ... | 5 0 0 | .. | 5 0 0 | .. | .. |
| ,, Subscription Library | ... | 12 18 4 | .. | .. | 8 10 0 | 4 8 4 |
| Halifax Mechanics' Institution | 135 15 11 | 3 1 7 | 1/44 | .. | 2 0 10 | 1 0 9 |
| Highgate Literary and Philosophical Institution | 230 0 0 | 14 0 0 | 1/16 | 8 10 0 | 5 10 0 | .. |
| Ipswich Mechanics' Institution | 309 5 1 | 7 14 0 | .. | .. | .. | .. |
| Keighley | 23 10 0 | 3 9 9 | 1/3 | .. | 2 8 3 | 1 1 6 |
| Kentish Town Mental Improvement | 9 10 0 | Gratuitous accommodation. | | | .. | .. |
| Leeds Mechanics' Institution | 150 0 0 | 1 13 0 | 1/90 | .. | .. | .. |
| ,, Literary Institution | ... | ... | .. | .. | .. | .. |
| ,, Literary and Philosophical Society | 456 5 6½ | 29 7 2¼ | 1/13 | 1 9 4½ | 15 8 0 | 12 9 9¾ |
| Lincoln Mechanics' Institution | 190 0 0 | 2 10 10 | 1/74 | .. | 1 2 6 | 1 8 4 |
| Liverpool Royal Institution | 500 0 0 | 29 9 4½ | 1/16 | .. | .. | .. |
| ,, Mechanics' Institution | 7,000 0 0 | None. | .. | .. | .. | .. |
| Louth | 90 0 0 | 3 0 0 | 1/30 | .. | 3 0 0 | .. |
| London Institution | 3,000 0 0 | 162 16 2 | 1/19 | .. | .. | .. |
| ,, Royal Society | 1,400 0 0 | 50 0 0 | 1/28 | .. | .. | .. |
| ,, ,, Institution, Albemarle Street | 2,363 0 0 | 126 5 9 | 1/18 | 53 3 4 | 60 10 5 | 12 12 0 |
| ,, Russell Institution | 900 0 0 | 53 0 0 | 1/17 | 13 2 2 | 16 12 0 | .. |
| ,, Eastern Literary and Philosophical Institution, Hackney | 200 0 0 | 19 1 10 | 1/10 | 6 3 6 | 8 6 8 | 4 11 8 |
| ,, Eastern Institution, Commercial Road | ... | 40 5 6 | .. | 11 10 6 | 25 0 0 | 3 15 0 |
| ,, City of, Literary and Scientific Institution | 1,877 15 0 | 46 17 4 | 1/40 | 15 17 2 | 13 14 0 | 17 6 2 |
| ,, Tower Hamlets' Chemical and Philosophical Society | ... | Exempted. | .. | .. | .. | .. |
| Lewes Mechanics' Institution | 134 5 1 | 14 15 7 | 1/9 | 5 15 8 | 5 17 7 | 3 2 4 |
| Manchester Literary and Philosophical Society | 207 19 3 | 17 3 8 | 1/12 | .. | .. | .. |
| ,, Royal Institution | 800 0 0 | 265 0 0 | 1/3 | .. | 130 0 0 | 135 0 0 |

**Figure 1.4** Incomes of Scientific and Literary Societies in England in 1840

(i) It might be that 7000 (whilst surprisingly large) is a reasonable value to encounter in relation to an *appropriate* choice of model for annual incomes: thus it does not reflect contamination, merely the inevitable wide variability of such incomes.

(ii) Perhaps 7000 really does reflect the fact that the Liverpool Mechanics' Institute is a different type of organization to the others: it is *contaminant* in the distribution of typical incomes of an otherwise homogeneous group of allied organizations.

(iii) The distinctions (i) and (ii) are highly relevant to what we want to do with the data. If we were interested in *identifying* aberrant organizations as a separate basis of study, demonstration of the correctness of the prospect (ii) is very important. If, notwithstanding the truth of (i) or (ii), we wanted to estimate the mean income for the main homogeneous group, we might wish to show good cause to reject the value 7000 before estimating the mean income for the homogeneous group, we might wish to show good cause to *reject* the value 7000 before estimating the mean, or (more subtly) find a different estimate of the mean which is relatively unaffected by (which *accommodates*, or is robust against) the outlying value.

We shall pursue all these possibilities in more detail later.

Whatever we decide to do, however, is likely to be highly dependent on what we assume about the underlying model for the bulk of the data (the 'homogeneous group').

Table 1.1 Incomes of the top 69 societies (in order of magnitude to the nearest pound)

| £77 | 112 | 169 | 249 | 700 |
|-----|-----|-----|-----|------|
| 77  | 115 | 170 | 290 | 800 |
| 79  | 120 | 170 | 300 | 844 |
| 80  | 120 | 190 | 309 | 900 |
| 80  | 120 | 200 | 335 | 900 |
| 84  | 125 | 200 | 350 | 1050 |
| 87  | 130 | 200 | 400 | 1300 |
| 90  | 135 | 200 | 404 | 1400 |
| 90  | 136 | 201 | 431 | 1878 |
| 90  | 138 | 206 | 445 | 2000 |
| 92  | 140 | 208 | 456 | 2363 |
| 100 | 147 | 230 | 500 | 3000 |
| 102 | 150 | 230 | 650 | 7000 |
| 110 | 150 | 237 | 650 |      |

A model that has been successfully applied in many situations involving incomes is the Pareto distribution (and the argument supporting such a model is unaffected by the fact that we have *truncated* the lower values in the original data set). Such a model has probability distribution function

$$F(x) = P(x \leqslant x) = 0 \qquad (x \leqslant a)$$
$$= 1 - (a/x)^r \qquad (x > a)$$

where $x$ is the income, $a$ is the minimum income (here 75) and $r$ is a shape parameter. The mean value of $X$ is $ra/(r - 1)$, $(r > 1)$.

We remarked above on the way in which outliers may influence the propriety of different methods of estimating parameters in the basic model. Let us consider another example. Suppose the following random observations were obtained for some variable of interest:

1.74, 1.46, − 1.28, − 0.02, − 0.40, 0.02, **3.89**, 1.35, − 0.10, 1.71

We wish to estimate the 'centre' of the parent population. Initial considerations suggest that the population may be normal, $N(\theta, 1)$, with mean $\theta$ and variance 1, so the *sample mean* would clearly be a sensible form of estimator. But the value 3.89 makes us suspicious of the $N(\theta, 1)$ assumption! *In fact*, these data were generated as a random sample from a Cauchy distribution, with probability density function

$$f(x) = \frac{1}{\pi}(1 + x^2)^{-1}.$$

The sample mean here is not even *consistent*, let alone of reasonable efficiency, and we should have made very poor use of our data in the estimation procedure had we used it as an estimator of location.

Observations far removed from the main body of the sample arise naturally in sampling both from a Pareto and a Cauchy distribution. This illustrates very clearly the *relative* nature of the element of 'surprise' in our declaration of an outlier.

An outlier is not only an extreme, but is surprisingly extreme. But what is surprising in relation to the $N(\theta, 1)$ model may not be so in relation to the Cauchy model. Observations would need to be an order of magnitude 'more extreme' for us to declare them outliers relative to a Cauchy model.

Such a phenomenon occurs not infrequently in biological contexts. For example, the distribution of the number of cones on a fir tree for trees in a given area of forest, or the distribution of the number of *lepidoptera* (butterflies or moths) of the same species present and observed in a particular location, are both characterized by high skewness. A typical sample from this latter type of distribution is given below; it refers to the number of individuals of a given species in a random sample of nocturnal *macrolepidoptera* caught in a light-trap at Rothamsted (Fisher, Corbet and Williams, 1943):

11, 54, 5, 7, 4, 15, **560**, 18, 120, 24, 3, 51, 3, 12, 84

Here we have a situation in which the outlying value (560) is, in fact, an inherent feature of the natural data pattern, and in no way anomalous!

There is also the fact that what one observer finds surprising, another may not.

An outlier is perceived to be surprisingly extreme relative to some null model, F. But some people are more easily surprised than others perhaps. Tingley and Field (1990) remark that whilst the sample

$$-3.0, \; -2.0, \; -1.1, \; 0.0, \; 6,0$$

'has no outliers', a small modification would produce an alternative sample

$$-3.0, \; -2.0, \; -1.2, \; 0.0, \; 6,0$$

in which 'the observation 6.0 is now an outlier'. We must confess that our surprise-sensor is far less acute than that of Tingley and Field!

## 1.3  OUTLIERS IN MORE STRUCTURED SITUATIONS

The examination of univariate samples for fitting models and estimating parameters, whilst an important part of statistical practice, has somewhat limited aims. More often, and more usefully, we need to consider more structured situations. For example, an interest in the way in which observations of a variable of principal interest vary with values of other variables, or vary with time, leads to the study of *regression* models, and *time-series* models, respectively. Or again, concern for the influence of different qualitative or quantitative factors on the principal variable leads to additive models analysed by *analysis of variance* techniques. These various models and techniques have, of course, their counterparts in the study of *multivariate* data.

In all of these more structured cases we must also expect to encounter, from time to time, unrepresentative data in the form of outliers. Here it is just as important as in the simple univariate sample to be able to recognize, interpret, and make allowance for outliers by using appropriately designed statistical techniques. Outliers may, as before, be of intrinsic interest in their own right, or may be indicative of incorrect specifications of the error structure, or of the basic model, with consequential implications for the choice of relevant inference procedures. With such more structured data two complications arise: *the outliers tend to be less intuitively apparent,* more hidden in the data mass, and *formal methods for their rejection or their 'accommodation' are less highly developed.*

The outlier problem is a complex one. Again one is looking for observations 'surprisingly far away from the main group'. But what is *surprising*? The *relative* nature of the element of surprise has already been stressed.

In what way does the person making the judgement assess the relationship of the outlying values to the main group?

For a univariate sample the process is straightforward. But the outlier may be a value in a designed experiment—one of the observed responses, say, in a Latin-square experiment. A simple visual inspection of the table of responses is unlikely to reveal the outlier. Its presence only comes to light when the parameters of the model are fitted and the deviations of the observed responses from the fitted values are tabulated.

Consider also the notion of outlying values in a regression analysis. It has become standard practice to carry out an examination of residuals as an essential part of any regression analysis. This gives a clue to the form of outliers in regression models: as outlying *residuals*. How does this differ from the nature of outliers in univariate (unstructured) samples? For one thing, the residuals are not independent; this may make outliers difficult to detect and also complicates the methodology.

It is useful to consider one or two practical examples of the existence of outliers in regression, time-series, and other contexts. No attempt will be made here to discuss implications in any detail (but see Chapters 7–12).

*Regression models*

The linear regression of one random variable, $Y$, on a controlled variable, $X$ (or conditional on the observed values of a *random* variable, $X$), is a model widely used for an initial study of the way in which $Y$ varies with $X$. Some data on the loads $x$ (in kg) applied to similar metal beams and the resulting extensions of length $y$ (in cm) for 9 such beams are shown (marked ●) in Figure 1.5. Clear evidence of a linearity of relationship can be observed with obvious sampling fluctuation in relation to a linear regression model

$$y = \alpha + \beta x + \varepsilon$$

with $E(\varepsilon) = 0$. If $\mathrm{Var}(\varepsilon) = \sigma^2$ (constant), however, what are we to make of the observation marked $A$? It seems to deviate in a more extreme manner from the linear model than do the other observations—perhaps to a surprising degree. Thus we might declare $A$ to be an *outlier*. Note that it is not an *extreme* value of $y$ (or of $x$). But the stimulus for declaring it to be an outlier is still its surprising discrepancy in relation to the model: *it breaks the expected pattern of results for the regression model.*

An interesting effect of the outlier $A$ is its *influence* on estimation of the parameters in the model. In particular, it greatly inflates the estimate of $\sigma^2$ if we do not remove it or employ an outlier-robust inference procedure (to *accommodate* it).

Suppose, however, that there had been another observation (shown as ○ and labelled $B$). This is *not an outlier* in the sense of disrupting the

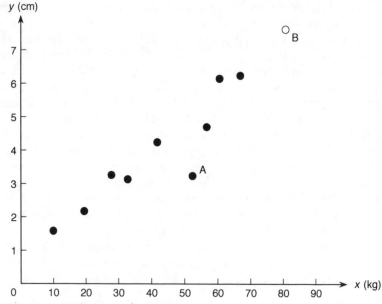

**Figure 1.5**  Relation between load and extension

model, but it is an extreme (in terms of both its $y$- and $x$-values). It is also highly influential in terms of estimation of $\alpha$ and $\beta$ (it makes the estimates more accurate, and *small* changes in its position can cause major changes in the estimates: particularly of $\alpha$).

An interesting further point is raised by this example. *Outlying behaviour* is assessed in terms of the extremeness of $y$ in relation to the regression model *for the relevant prescribed (or conditioned) value of $x$. Influential behaviour* can arise from an extreme $x$ value *per se* but this is a different concept to that of an outlier—we can hardly be *surprised* by the $x$-value since it is prescribed (or used conditionally). We shall return to this distinction between *outliers* and *influential* observations in Chapter 8 where we examine regression, and linear-model, outliers in detail.

*Time series*

Time series data are widely studied in commercial, industrial, meteorological, and sociological processes. Outliers can again arise and cause difficulties. Consider the following examples.

Chatfield (1975) presents some data on numbers of a particular product which are sold each quarter over a six-year period. These data are shown in Figure 1.6. Discussing the idea of outliers in the time series data, Chatfield suggests that there is reason to doubt the observation A. His reasoning again differs from our earlier examples of univariate, unstruc-

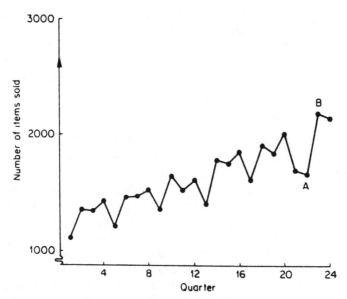

**Figure 1.6** Sales figures for a product over consecutive quarters for a period of six years (reproduced by permission of C. Chatfield and Chapman & Hall)

tured, data sets, where sample 'extremeness' was the stimulus. Once more, it is the break in the 'pattern' of results which makes observation A suspect.

For previous years a relatively low sales figure in the first quarter was followed by two intermediate values and a relatively high value. In the last year, the second quarter figure (A) breaks this pattern. Perhaps economic or accounting factors have produced a spurious result, possibly compensated for by the opposingly atypical value, B. Alternatively, the results for the final year may indicate a radical change in the cyclic pattern of sales over the year.

In Figure 1.7 we see (Fenton, 1975) a continuous trace of measured moisture content of Malaysian tobacco, being automatically monitored as it flows past the recording equipment on a conveyor belt at a particular

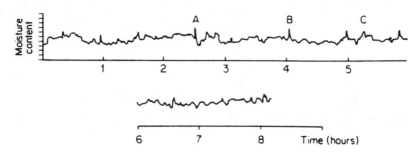

**Figure 1.7** Moisture content of Malaysian tobacco continuously monitored over eight hours

stage of the curing process. The equipment is known to suffer from occasional electronic 'hiccoughs' which appear as sharp spikes in the trace. An experienced observer comments that 'A and B are clearly outliers' in this sense, whilst 'C is unlikely to be an outlier'.

Chapter 10 is concerned with the problem of outliers in time series.


*Designed experiments*

Data from designed experiments can also contain outliers, though with so much variation possible even under a basic (uncontaminated) linear model it may not be an easy matter to distinguish them. Again the stimulus is one of *a surprising degree of departure from the expected pattern of results*—but mere sample inspection may not suffice to detect an outlier. In such situations (or indeed, for all data sets arising from linear models or even for unstructured multivariate data) it will be necessary to seek outliers through *derived* measures of the data rather than by mere observation of the sample values. As for regression models (illustrated above) the estimated *residuals* can again provide the appropriate measures.

John (1978) presents results for factorial experiments in which outliers are found. A $2^5$ experiment, with factors $A$, $B$, $C$, $D$ and $E$, is carried out to examine how the strength of sheet-metal which has been coated depends on various factors in the coating process. The results are shown in Table 1.2.

Can you see clearly *two outliers* at treatment combinations *ad* and *d*? Perhaps not, until we look at the estimated residuals after the fitting of a model with all main effects and first order interactions—where the yields for *ad* and *d* are now conspicuous outliers. See Figure 1.8. Note that *bd* and *bde* are not outliers!

**Table 1.2** Yields from a confounded $2^5$ experiment (reproduced from John 1978 by permission of the Royal Statistical Society)

| Treatment | Yield | Treatment | Yield | Treatment | Yield | Treatment | Yield |
|-----------|-------|-----------|-------|-----------|-------|-----------|-------|
| (1) | 1.4 | d | 5.0 | e | 1.7 | de | 9.5 |
| a | 1.2 | ad | 9.0 | ae | 2.0 | ade | 5.9 |
| b | 3.6 | bd | 12.0 | be | 3.1 | bde | 12.6 |
| ab | 1.2 | abd | 5.4 | abe | 1.2 | abde | 6.3 |
| c | 1.5 | cd | 4.2 | ce | 1.9 | cde | 8.0 |
| ac | 1.4 | acd | 4.4 | ace | 1.2 | acde | 4.2 |
| bc | 1.5 | bcd | 9.3 | bce | 1.0 | bcde | 7.7 |
| abc | 1.6 | abcd | 2.8 | abce | 1.8 | abcde | 6.0 |

Methods for handling outliers in designed experiment data will be discussed in Chapter 8.

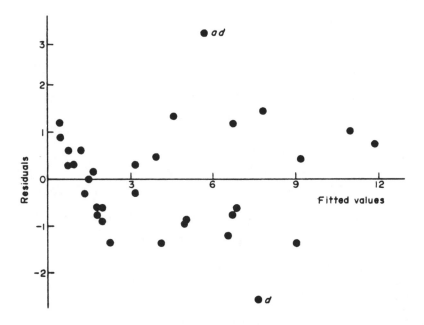

**Figure 1.8** $2^5$ experiment—residual plot (original observations) (reproduced from John, 1978, by permission of the Royal Statistical Society)

*Unstructured multivariate data*

Here also we can encounter outliers. As for univariate, or structured, data they may reflect deterministic factors (errors of measurement, misrecording, etc.) or be probabilistic in nature (causing us to question distributional assumptions). If we have a sample of multivariate observations, each consisting of a vector of measurements, there may be one particular observation for which none of the component measurements is 'surprising' in relation to its marginal distribution and yet *its assemblage of measurements* as a multivariate observation seems 'surprisingly far away from the main group of data'. *Perhaps it lies on the periphery of the data cloud.*

In this respect we have an intuitive notion of extremeness, but no overall formal ordering principle is available for multivariate data. Again we may have to process the data to some extent before we can readily detect an outlier. Afifi and Azen (1979, Section 5.1) present data on systolic and diastolic blood pressures for a group of 15 patients; a scatter diagram is shown as Figure 1.9, and we immediately note one outlier (observation *A*). But what about observations *B* and *C*?

Methods for detecting and statistically analysing outliers in multivariate data are considered in detail in Chapter 7.

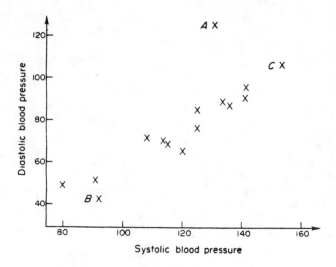

**Figure 1.9**  Systolic and diastolic blood pressures

## Outliers in directional data

The study of outliers has recently moved into many new areas. As an illustration of one such area we note the methods that have been developed to detect and analyse outliers in *directional data*: data in the form of directions in a plane or in space. Collett (1980) considers the former

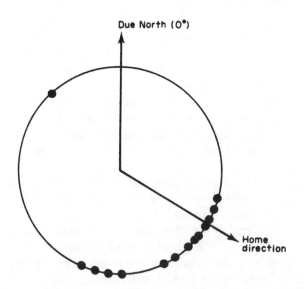

**Figure 1.10**  The orientations of 14 northern cricket frogs (reproduced from Collett, 1980, by permission of the Royal Statistical Society)

problem. One example he discusses is concerned with the homing ability of the northern cricket frog, *Acris crepitans*. Figure 1.10 shows the directions taken by 14 frogs when released from a dark environment. We seem to have one outlying 'lost soul'. Chapter 11 considers the outlier problem for directional data.

*Outliers in sample surveys*

In a *sample survey* or an *opinion poll* we draw a random sample from a fixed or predetermined finite population. The mechanisms of sampling and inference differ in a fundamental way from the infinite population (random variable) situation in that the sampler rather than 'nature' imposes the indeterminism (in the choice of the sample).

We could still, however, be *surprised* at the extremeness of some outcome. Suppose we are observing family size in a survey of a town and encounter a value 18. It is not an *impossible* value but prompts one to 'raise an eyebrow'! This is a relatively new and unexplored area for outlier study, but an interesting and important one, and it will be examined in Chapter 12.

## 1.4 BAYESIAN METHODS

During recent years there has been much interest in, and application of, *Bayesian* and *decision-theoretic* methods of statistical analysis. These differ from the more traditional methods both in terms of their interpretation of basic concepts and aims, and in their use of extra forms of relevant information (*prior probabilities* and *consequential costs*, additional to *sample data*). The sample data components of the information used in such approaches may well contain outlying observations. Again we must know how to deal with them.

These alternative methods of statistical analysis affect the outlier problem in two ways. On the one hand we might ask how a Bayesian, or decision-theoretic, analysis should try to cope with outliers. On the other hand we might ask if such forms of analysis can be used for the detection and processing of outliers. There are some conceptual difficulties here. A basic tenet of the Bayesian approach is that inferences, or decisions, are strictly *conditional* on the actual sample data that have been obtained. In the main, it is contrary to the Bayesian tradition to view the data within a framework of *alternative* sets of data *which might have arisen*. Thus the probability mechanism for generating the data appears to be irrelevant in the Bayesian approach (the data import is reflected by the likelihood function). It might appear to be similarly irrelevant to question the integrity of the realized data in the sense of examining the implications of outlying observations. However, since Bayesian methods revolve around the likelihood function which needs to be specified in any particular case and the likelihood depends on an assumed probability model, it would seem that, relative to that model, certain observations might be outliers. Thus, in the Bayesian context outliers might imply an incorrect specification of the likelihood.

One aspect of the Bayesian approach is its formalization of subjective impressions as an ingredient of statistical analysis. We have remarked above how subjective factors arise in judging whether or not outliers exist in a set of data. In the example on weights of birds in Section 1.1, for instance, the reading 1.55 looked suspicious and could well represent a recording error for 2.05. In the timeseries example on quarterly sales in Section 1.3, the pattern of sales over the last year appeared inconsistent with earlier experience and made us question the values A and B. In both these cases subjective judgement was involved. Might it not be better to use Bayesian methods directly in trying to reach a conclusion about the outliers, or in taking them into account in processing the data? For example, we might attempt to assign prior probabilities to different possible explanations of the outliers. Some efforts in this direction have been made and will be considered where appropriate throughout the book and particularly in Chapter 9.

## 1.5 STATISTICAL COMPUTING

It has become common practice to employ computers and computer packages (general statistical software systems) to conduct statistical analyses. Such facilities are becoming more and more sophisticated, with data-screening options, wide-ranging methodology and model-validation procedures often with 'smart' or 'semi-intelligent' (if not 'expert') intervention.

What of outlier detection, testing and accommodation? We might expect these to be incorporated in such computer packages. To an extent they are, but far less than might be expected. (The same is true of more general *robust inference methods*: the relationship between general robustness and accommodation (outlier-robustness) is explored in Chapter 3.)

Any up-to-date treatment of outlier methods must review (at least briefly) what current computer packages have to offer on this topic and some information is provided in Chapter 13. An interesting dilemma is evident, however, if we contrast the aim of statistical computing for routine repetitive data-screening and analysis, and the subjective 'one-off' stimulus of 'surprise' for the declaration of outliers.

## 1.6 SURVEY OF OUTLIER PROBLEMS

The informal discussion of ideas and examples in the earlier parts of this chapter enables us to draw up a broad classification of the types of enquiry we need to make in the study of outliers. No single-factor classification will do since we must consider (possibly jointly) the various distinctions:

(i) between deterministic and statistical causes of outliers,

(ii) between different specific probability models,

(iii) between univariate and multivariate (possibly structured, e.g. regression) data sets,

(iv) between different forms of statistical analysis in which the outliers are encountered,

(v) between single or multiple outliers, and

(vi) *most fundamentally*, between the different aims and purposes we may have in studying outliers.

Such a complex array of considerations does not lead to a particularly tidy subdivision of topics, but the arrangement of the succeeding chapters and their subsections has been chosen in what seems to be a natural progression from the simpler to the more complex considerations. Our main object has been to present a fairly full review of existing methodology in the treatment of outliers. We have aimed throughout both to provide sufficient theoretical detail to meet the interests of the mathematical statistician and to give a full enough description of practical methods to meet the needs of the data analyst. Application of methods is of paramount importance and we have included numerous relevant practical illustrations.

We shall start with a general discussion of the *different aims and purposes in* studying outliers and proceed to what is perhaps the most elementary aspect of this, the examination of outliers in single univariate samples from a given probability distribution. Special and detailed attention is given to the effect of outliers in estimation and test procedures and to specific tests of univariate outliers. Outliers in multivariate data are then examined. We progress to more structured models such as regression, designed experiments, time series, directional data, contingency tables and sample surveys. The notion of *influential observations* and the general topic of *robustness* are also considered where appropriate throughout the book so that their relationship to the outlier problem can be clearly seen. Bayesian methods are also considered in some detail and we conclude with comment on the future state of the art in the study of outliers.

CHAPTER 2

# Why Do Outlying Observations Arise and What Should One Do About Them?

## 2.1 EARLY INFORMAL APPROACHES

The existence of the problem of doubtful or anomalous values has been recognized for a very long time, certainly since the middle of the eighteenth century. Daniel Bernoulli, writing in 1777 about the combination of astronomical observations, asked:

is it right to hold that the several observations are of the same weight or moment, or equally prone to any and every error? . . . Is there everywhere the same probability? Such an assertion would be quite absurd, . . . I see no way of drawing a dividing line between those that are to be utterly rejected and those that are to be wholly retained; it may even happen that the rejected observation is the one that would have supplied the best correction to the others. Nevertheless, I do not condemn in every case the principle of rejecting one or other of the observations, indeed I approve it, whenever in the course of observation an accident occurs which in itself raises an immediate scruple in the mind of the observer, . . . If there is no such reason for dissatisfaction I think each and every observation should be admitted whatever its quality, as long as the observer is conscious that he has taken every care. (Allen, 1961).

We shall see that these remarks embody many of the diverse considerations in the handling of outliers: matters which were not formally distinguished until relatively recently (about 200 years, in fact, after the quoted views were expressed). They embody, *inter alia*, the distinct ideas of *rejection, accommodation* and *identification* which we shall be exploring in this chapter, as well as the possible *deterministic* explanation of outliers in some situations.

To take an even earlier example, Boscovich, attempting in the 1750s to determine the ellipticity of the earth by averaging ten measurements of excess of the polar degree over the equatorial, decided to discard the two extreme values of excess as outliers and recomputed the mean from the reduced sample of eight.

From this period until the middle of the nineteenth century, the main point of discussion in the literature with regard to outlying values is whether their **rejection** can be justified. Some writers took the view that observations should not be rejected purely on grounds of appearing inconsistent with the remaining data; Bessel and Baeuer, for example, wrote in 1838 that they had never rejected an observation merely because of its large residual, and that all observations ought to be allowed to contribute, with equal weight, to the result. Others, such as Boscovich, were quite prepared to reject any outliers. However, in these early days, rejection was not carried out according to any formal procedure, but was purely a matter of the observer's judgement. Legendre, for example, in 1805 was recommending the rejection of deviations 'adjudged too large to be admissible'. Indeed, a century later Saunder could write (1903):

I believe that the practice amongst computers of experience is to rely almost entirely on their individual judgement, taking into account the conditions of the observations, and drawing the line somewhere about those observations which give residuals of five times the probable error.

The first published objective test for anomalous observations was due to the American astronomer Peirce (1852). In Peirce's procedure, $k$ doubtful observations in a sample of $n$ should be rejected if

the probability of the system of errors obtained by retaining them is less than that of the system of errors obtained by their rejection multiplied by the probability of making so many, and no more, abnormal observations.

This last probability Peirce took to be $p^k(1 - p)^{n-k}$, where he began by defining $p$ as 'the probability, supposed to be unknown, of such an abnormal observation that it is rejected upon account of its magnitude', and then he assigned to $p$ the value $k/n$.

This bizarre test was followed in 1863 by the publication of the test for a single doubtful observation (already referred to in Chapter 1) by the American astronomer Chauvenet. Chauvenet's test has an attractive simplicity lacking in Peirce's; despite its now evident shortcomings, it persists in print to the present day, at any rate in a number of textbooks for students in engineering and the experimental sciences. The reasoning was as follows (see Stone, 1868). If $\theta(x)$ denotes the probability that an error is equal to or greater than $x$,

then the number of errors equal to or greater than $x$ which may fairly be expected in $n$ observations is $n\theta(x)$. If therefore we find $x$ such that $n\theta(x) = \frac{1}{2}$, any error greater than $x$ will have a greater probability against it than for it, and may, therefore, be rejected.

In effect, an observation is to be rejected if it lies outside the lower and upper $100/(4n)$ percentiles of the null distribution. So, in a sample of size $n$, the probability that an arbitrary observation is rejected is $1/(2n)$. Hence with this procedure the chance of *wrongly* rejecting a reasonable sample value is $1 - [1 - 1/(2n)]^n$, or, in a large sample, approximately $1 - e^{-1/2}$, i.e. about 40 per cent!

Such a policy is clearly unreasonable, and we shall reexamine it later to seek to understand its basic fallacy. Chauvenet's test, however, is not just an historical curiosity. It still attracts attention, as is evidenced in the recent works of Pagurova (1985) and Khalfina (1986, 1989).

Soon after Chauvenet, Stone (1868) introduced a rejection test based on the concept of a *modulus of carelessness, m*. This concept can be expressed in the following way: a given observer in a given sampling situation makes on average one mistake in every $m$ observations he takes. An observation is to be discarded if its deviation can be attributed with more probability to the observer's carelessness than to random variation. This means, in effect, that an observation is rejected if it lies outside the lower and upper $100/(2m)$ percentiles of the null distribution, so the test is essentially similar to Chauvenet's, becoming identical with it if $m = 2n$.

Some alternatives to *outright rejection* of extreme values were also being considered. Within a few years of Stone's rejection test, several methods were published—one by Stone himself—for the weighting of observations in calculating a sample mean. This can be regarded as a robust procedure for estimating a location parameter: it illustrates what we shall call the **accommodation** of outliers. In Rider's words (1933),

Since the object of combining observations is to obtain the best possible estimate of the true value of a magnitude, the principle underlying . . . [weighting] methods is that an observation which differs widely from the rest should be retained, but assigned a smaller weight than the others in computing a weighted average. Of course retention with an exceedingly small weight amounts to virtual rejection.

Glaisher (1872–73) was perhaps the first to publish such a weighting procedure, remarking, 'It will be seen that it supersedes the necessity for the *rejection* of anomalous observations'.

Glaisher's method was concerned with $n$ observations $x_i(i = 1, \ldots, n)$ from normal distributions, with a common mean $\mu$ required to be estimated, and with unknown and unequal variances. He proposed estimating the mean iteratively $\mu$ by a weighted combination of the $x_i$ with weights determined from the squared deviations of the values of the observations.

Stone (1873) followed a few months later with a criticism of Glaisher's method and a proposal for an alternative weighting procedure, based in effect on maximizing the likelihood. This leads to a weighted mean $\tilde{\mu}$ given by the $(n - 1)$th degree equation

$$\sum_{i=1}^{n}(x_i - \tilde{\mu})^{-1} = 0. \qquad (2.1.1)$$

The same method was published independently ten years later (Edgeworth, 1883), though Edgeworth subsequently (1887) acknowledged Stone's priority.

Another weighting method proposed at this time was by Newcomb (1886); see Stigler (1973b). His procedure assumes the $n$ observations to have come from a mixture of $r$ normal distributions, and evolves a final estimate of $\mu$ which is constructed as a weighted mean of $r^n$ different weighted means of the $x_i$. Rather interestingly, Newcomb refers in his paper to the 'evil' of a value; this turns out to be the mean squared error of an estimate, and is an interesting early use of the concept of a loss function.

It is also interesting to find at this period what we would now call *trimming*. In 1895 Mendeleev, the discoverer of the periodic table of elements (referring to the evaluation of the length of the Russian standard platinum–iridium metre from a set of eleven determinations) wrote:

I use . . . [the following] method to evaluate the harmony of a series of observations that must give identical numbers, namely I divide all the numbers into three, if possible equal, groups (if the number of observations is not divisible by three, the greatest number is left in the middle group): those of greatest magnitude, those of medium magnitude, and those of smallest magnitude: the mean of the middle group is considered the most probable . . . and if the mean of the remaining groups is close to it . . . the observations are considered harmonious. (See Harter, 1974–1976, Part I)

Following Chauvenet, Stone and others, a number of *ad hoc* rejection tests for outliers appeared in the literature up to the middle of the present century. In particular, we may note Wright's procedure (1884), which rejects any observation deviating from the mean by more than three times the standard deviation, or equivalently about four-and-a-half times the 'probable error'; the modified version by Wright and Hayford (1906), which adds to Wright's rule the further instruction:

Examine carefully each observation for which the residual exceeds 3.5 times the probable error, and reject it if any of the accompanying conditions are such as to produce lack of confidence

and Goodwin's procedure (1913), which rejects an outlying observation in a sample of $n$ if its deviation from the mean of the remaining $n - 1$ exceeds four times the average deviation of the $n - 1$.

One notes that Wright and Goodwin chose, for the critical ratios in their tests, values 4.5, 3.5, 4, and so on, which were independent of the sample size $n$. *This highlights two general defects of all the test procedures proposed up to this time*—they failed to distinguish between population variance and sample variance and, more importantly, they were erroneously based on the distributional behaviour of a random sample value rather than on that of an appropriate sample *extreme*. Thus, for example, Chauvenet's rule effectively leads to rejection of any observation $x_j$ for which $|x_j - m|/s$ is sufficiently large (where $m$ and $s$ estimate location and standard deviation, respectively)—but with no regard to the effect of sample size on distributional form of the statistic. More serious, however, is the failure to recognize that *it is an extreme* $x_{(1)}$ *or* $x_{(n)}$ which (by the very nature of outlier study) *should figure in the test statistic, rather than an arbitrary sample value* $x_j$. It is this failure which yields the earlier remarked 40% chance of rejecting at least one observation from any large, but *uncontaminated*, sample.

Perhaps the first writer to correct these two deficiencies explicitly with regard to outlier procedures was Irwin (1925), who pointed out the implications for outlier rejection of the unreliability of the sample standard deviation, $s$, as an estimate of its population analogue, $\sigma$, and also used extreme values in the test statistic. For the case where $\sigma$ is known, he proposed the test statistics

$$[x_{(n)} - x_{(n-1)}]/\sigma \quad \text{and} \quad [x_{(n-1)} - x_{(n-2)}]/\sigma$$

for testing upper outliers in a sample of size $n$. (As already indicated, $x_{(i)}$ denotes the $i$th *ordered* sample value). Ten years later an exact test based on a studentized criterion, $(x_{(n)} - \bar{x})/s$, was published by Thompson (1935). This was shortly followed by the classic paper by Pearson and Chandra Sekar (1936) entitled 'The efficiency of statistical tools and a criterion for the rejection of outlying observations'. A rationale for the treatment of outliers was beginning to take shape.

An encyclopaedic survey of outlier methods from the earliest times is included in Harter (1974–1976).

## 2.2 ORIGIN OF OUTLIERS, STATISTICAL METHODS AND AIMS

In the previous section we reviewed some of the early informal approaches which had been used for the study of outliers. In such work there was seldom any overt consideration of the basic nature of outliers, of any desirable (or even optimal) properties of the prevailing statistical methods, of the purpose of examining the outlying observations or of the manner in which outliers may reflect contamination of a basic probability model. Any detailed examination of the current state of theory and practice in relation to the study of outliers must consider such matters of origin, principle, aim and model. Thus in the remainder of this chapter we will present a general review of:

- *the various ways in which outliers arise,*
- *the nature and form of relevant statistical procedures,*
- *the different aims in examining outliers, and*
- *probabilistic models for contamination, which might account for the presence of outliers in a data set.*

Such matters are reconsidered in appropriate detail in later chapters where fuller treatments of specific topics are given. A general review of these matters of principle is given by Barnett (1978a).

### 2.2.1 The nature and origin of an outlier

We commence with what must be the most fundamental question: What is an outlier and how does it arise? The practical examples of Chapter 1 have shown some basic distinctions which need to be examined more systematically.

Let us recall what we mean by an 'outlier'. In observing a set of observations in some practical situation one (or more) of the observations 'jars' stands out in contrast to other observations, as an extreme value. As Grubbs (1969) remarks (italics inserted):

An outlying observation, or 'outlier', is one that *appears to deviate markedly* from other members of the sample in which it occurs.

Such outliers do not fit with the pattern we have in mind, at the outset of our enquiry, of what constitutes a *reasonable* set of data. We have subjective doubts about the propriety of the outlying values both in relation to the specific data set we have obtained and in relation to our initial views of an appropriate probability model to describe the generation of our data. *Note how our feelings about the data may, in this respect, differ quite widely with different possible basic probability models.*

If we anticipate a normal distribution we may react quite strongly to certain observations which would arouse no specific concern if the expected model is longer-tailed, say log-normal or Cauchy. The purpose of a body of statistical method for examining outliers is, in broad terms, to provide a means of assessing whether our *subjective* declaration of the presence of outliers in a particular set of data has important *objective* implications for the further analysis of the data.

Outliers may have arisen for purely *deterministic* reasons: a reading, recording, or calculating error in the data. When it is obvious that this is so the remedy is clear: the offending sample values should be removed from the sample or replaced by corrected values (when the method of 'correction' is unambiguously understood). In less clear-cut circumstances where we suspect, but cannot guarantee, such a tangible explanation for an outlier, no such obvious remedy is available to us, and we have no alternative but to *regard* the outlier as being of a *random*, or inexplicable, nature. It must then be assessed in terms of the variational properties of any random sample generated by the postulated basic probability model— as would be the case if we had no grounds for suspecting a deterministic origin.

Execution faults in assembling the data can also yield outliers of deterministic nature: e.g. we measure the weights of a sample of *judoka* but subsequently find that one of the sample members is a *sumo wrestler*. Again no statistical procedure is required—we remove, or replace, the alien observation, provided it can be clearly identified. Otherwise we must once more revert to a statistical approach relevant for an inexplicable (assumed random) manifestation of an outlier.

Such distinctions are apparent in the examples we have considered already in Section 1.1. They are further illustrated in practical situations described in Section 2.2.3 where we begin to see the need for a more sophisticated viewpoint which takes into account the way in which the postulated basic probability model may have to be replaced by one which specifically envisages *contamination* of the data.

For the moment, however, we restrict attention to the simple categorization of the different ways in which outliers may arise. Such a categorization has been discussed elsewhere in the literature: see, for example, the general review papers by Anscombe (1960a), Grubbs (1969), Barnett (1978a), Beckman and Cook (1983), and Barnett (1983b); also the monograph by Hawkins (1980a). The situation is described in the following way.

In taking observations, different sources of variability can be encountered. We can distinguish three of these.

INHERENT VARIABILITY. This is the expression of the way in which observations vary over the population; such variation is a natural feature of the population. It is uncontrollable and reflects the distributional properties of a correct basic model describing the generation of the data. Thus, for example, measurements of heights of men will reflect the amount of variability indigenous to the population (and this may well be reasonably modelled by a normal distribution).

MEASUREMENT ERROR. Often we take physical measurements on members of a population under study. Inadequacies in the measuring instrument superimpose a further degree of variability on the inherent factor. The rounding of obtained values, or mistakes in recording, compound the measurement error: they are part of it. Some control of this type of variability is possible.

EXECUTION ERROR. A further source of variability arises in the imperfect collection of our data. We may inadvertently choose a biased sample or include individuals not truly representative of the population we aimed to sample. Again, sensible precautions may reduce such variability but we may not be aware of the 'execution errors' and sometimes it may be appropriate to change our basic population model to encompass the prospect of anomalous sample members.

We can usefully attempt to classify outliers in relation to these *three types of variability*. An outlier in a set of data may in fact be a perfectly reasonable reflection of the natural inherent variation. If shown to be statistically unreasonable (in a manner described more formally in the next section) this reflects an inadequate *basic* model (unless of course it is merely a manifestation of Type I error in any statistical test we have applied to the data). We would hope to learn from the experience and adopt a more appropriate model; possibly incorporating the notion of *contamination*. But as Anscombe (1960a) points out:

In no field of observation can we entirely rule out the possibility that an observation is vitiated by a large measurement or execution error. . . . Several possible reasons . . . can usually be thought of without difficulty. In such cases, the reading will be checked or repeated if that is possible. If not, it may be rejected as spurious because of its big residual, even though there is no known reason for suspecting it. In sufficiently extreme cases, no one hesitates about such rejections. . . . If we could be sure that an outlier was caused by a large measurement or execution error which could not be rectified (and if we had no interest in studying such errors for their own sake), we should be justified in entirely discarding the observation and all memory of it. The act of observation would have failed; there would be nothing to report.

Anscombe (1960a) distinguishes in terminology between outliers arising from large variation of the *inherent* type, and those from large *measurement* or *execution* error. He calls the former 'outliers', the latter 'spurious observations'. *We shall make no such distinction*. The detailed study of statistical methods for outliers needs to encompass all derivative sources of variation; the only exceptions are outliers arising from clearly discernible deterministic mistakes of calculation, recording, etc. In this case rejection (or correction) is the only remedy; otherwise we need to clearly recognize many other possibilities, apart from outright rejection, for coping with outliers. These are discussed in Section 2.2.3 below.

### 2.2.2   Relevant statistical procedures for handling outliers

In the brief historical survey (in Section 2.1) of early approaches to the processing of outliers we noted two distinct statistical methods. We saw firstly the idea of 'testing' an outlier *with a view to determining whether it should be retained or* **rejected**. We then saw how effects were made to construct procedures for estimating (and why not for testing) values of parameters in the basic distribution *relatively free from the unreasonable influence of any outliers.*

This dichotomy of approach is a fundamental one, standing above the further refinements of aim considered in the next section, and we need to spend some time exploring its implications.

ACCOMMODATION OF OUTLIERS. Statistical methods designed to draw valid inferences about the population from which (the bulk of) a random sample has been obtained, and which will not be seriously distorted by the presence of outliers (or contaminants which they may reflect), will be termed **accommodation** procedures. They 'accommodate' the outliers at no serious inconvenience—or are *robust* against the presence of the outliers.

*Robustness* is a concept of much importance in the general problem of statistical inference and is by no means specific to outlier study. In recent years major efforts have been made to obtain statistical procedures which provide a measure of protection against various types of uncertainty of knowledge of the data generating mechanism. These include *robust* methods for estimating or testing summary measures of the underlying distribution or more general inference procedures: where the estimators tests or procedures retain desirable statistical properties over a range of different possible distributional forms. Informative and detailed texts on 'robust statistics' are provided by Huber (1981), Rousseeuw and Leroy (1987; on *Robust Regression and Outlier Detection*) and Hampel *et al.* (1986); other important references to the general field include Tukey (1960), Huber (1964, 1972), Bickel (1965, 1976), Jaeckel (1971a), Hampel (1974), Hogg (1974), Launer and Wilkinson (1979) and Ronchetti (1985).

One possible departure from the null model is where *contamination* has occurred, and when robust methods address themselves to this prospect *specifically* they are of course immediately relevant to the outlier problem. But the studies of robustness and of outliers are not coincident. Robust methods extend well beyond the prospect of contamination as the form of departure from the null model. Outlier procedures, in turn, are concerned with more than *accommodation* (robustness under a contaminated model) and include *tests of discordancy* for outlier *rejection, identification* and so on—as we shall see in the development below. We shall have cause to return to this distinction in more specific contexts later.

Thus in terms of our current interest, an obvious area in which we may wish to seek the protection of robust statistical methods is where we encounter, or anticipate, outliers in a set of data. As one example, extreme observations clearly have an extreme effect on the value of a sample variance! If we are interested in estimating a parameter in an initial model, but are concerned about the prospects of outliers, either arising from random execution errors of no specific relevance to our studies, or from random measurement error, we would want to use an estimator which is not likely to be highly sensitive to such outliers. A simple (if somewhat extreme) example is to be found in the use of the sample *median* as an estimator of location; see also Rousseeuw and Bassett (1990) on the use of the *'remedian'*. A more sophisticated example is the proposal (Hampel, 1974) that we should use as a sample measure of dispersion the *median deviation*,

$$s_m = \text{median} \{|x_i - \tilde{x}|\},$$

where $\tilde{x}$ is the sample median. In the same context, Shoemaker (1984) advances the case of the *midvariance* which effectively measures the variance of the middle of a distribution or sample. See also Simonoff (1987a) on robust estimation of scale.

We must, of course, recognize the importance of any impropriety in our basic model. If outliers arise because our initial model does not reflect the appropriate degree of inherent variation (we really need, say, a fatter-tailed distribution rather than the ubiquitous normal distribution initially adopted) then omission of extreme values to 'protect against outliers' is hardly a robust policy for estimating some measure of dispersion, say the variance. Rather than appropriately reducing the effect of extreme values it encourages underestimation!

If, on the other hand, a reasonable alternative hypothesis is one which expresses contamination of the initial model (perhaps expressing low-probability mixing, or slippage) the estimation or testing of parameters in the *initial model* may well be the matter of principal interest and it is sensible to employ appropriate robust procedures to protect against the occasional low-probability component or slipped value.

The idea that we may wish, in this spirit, to do more then *reject* outliers, that is, to devise statistically respectable means of *accommodating* them in a wider inferential scheme addressed to the initial model, implies that the outliers themselves are no longer of prime concern. We wish to proceed safely in spite of them! This is of the essence of the robustness concept.

As we have seen, some interest in this alternative view of the outlier problem begins to show itself in quite early work. The ideas of Glaisher (1872), Stone (1873), Edgeworth (1883, 1887), Newcomb (1886), Mendeleev (1895), Student (1927), and Jeffreys (1932) amount to reducing the weight attached to extreme values in estimation, the latter paper paying specific regard to outliers as the extreme members of the sample.

A review of later work which implicitly or explicitly attempts to accommodate outliers in the inference process conveniently divides itself into two components, and will be considered in detail in Chapters 3 and 5. The first component contains those methods of estimation which *implicitly* protect against outliers in placing less importance on extreme values than on other sample members. This emphasis is a feature of much of the extensive armoury of robust methods developed over the last 20 years or so and we shall briefly examine the relevance of some of this work to outlier study. The second component of the study of accommodation of outliers is *specifically* concerned with robustness in the face of outliers, and methods of estimation and testing are derived with particular regard to the nature of any model which might be needed to explain the presence of outliers (as manifestations of contamination). This field of study is expanding and such *specific accommodation techniques* will also be reviewed.

DISCORDANCY TESTS. The second type of statistical method for handling outliers, described in our review of early approaches, is that of '*testing*' an outlier with the prospect of rejecting it from the data set, or of '*identifying*' it as a feature of special interest.

We recall some earlier ideas. A sample $x_1, x_2, \ldots, x_n$ has *extremes*, $x_{(1)}$ and $x_{(n)}$. One of these might be declared an *outlier* (say $x_{(n)}$: an *upper outlier*) if its extremeness engenders surprise in relation to what we might informally expect of our basic model: $F$. The sample might contain a *contaminant*—an observation from some other distribution $G$, perhaps shifted upwards in relation to $F$. (See Figure 1.3.) The upper outlier $x_{(n)}$ *might* be the contaminant (or there might be another observation $x_j$ which is the contaminant, or there might be *no* contaminant). With these prospects we proceed as follows. Suppose the whole sample comes from $F$. Let us perform a statistical test (called a **discordancy test**) to examine whether $x_{(n)}$ (say) is not only an extreme (which we know) but is also *statistically unreasonable even when viewed as an extreme*, i.e. whether $x_{(n)}$ is significantly large in relation to the distribution of $X_{(n)}$ under $F$. (Note that lower and upper case symbols are used as customary to denote observations and random variables respectively.) If it is statistically unreasonable in this respect, we say that $x_{(n)}$ is a **discordant upper outlier** at the level of the test. Of course this procedure is not restricted to *upper* outliers: similarly we may demonstrate discordancy of a lower outlier $x_{(1)}$, or of an outlier pair $[x_{(1)}, x_{(n)}]$, etc.

What is achieved if $x_{(n)}$ is shown to be discordant? We will have established (at the level of the test) that it is not reasonable to believe that $x_{(n)}$ comes from $F$. If our only alternative is that a single observation comes from $G$ (is a *contaminant*) and the remainder of the sample comes from $F$ then the discordancy of $x_{(n)}$ demonstrated by the test directs us to the adoption of this alternative model (and to the implication that $x_{(n)}$ is the contaminant).

But there is a difficulty here! The test says: reject $F$ as a homogeneous source and adopt an alternative model $\bar{F}$ which declares that there is a contaminant. But why is $x_{(n)}$ the contaminant? At one level, because it is our only stimulus for belief that a contaminant is present. But much more study will be needed of how such a conclusion relates to the form of the alternative model, $\bar{F}$. Note also that, as with any pure significance test, our inference does not depend on the form of $\bar{F}$, only on the form of $F$. However, the form of $\bar{F}$ is crucial to any consideration of the properties of a particular test or to any comparison of rival tests. Thus we need to give some thought to just what sort of *alternative* (*outlier-generating, contamination*) *models* might be considered, and this will be the topic of Section 2.3 of the present chapter.

Let us consider an example of a test of discordancy. A well known set of data analysed by Peirce (1852) and by Chauvenet (1863) consisted of 15 observations made by Lt. Herndon in 1846 of the vertical semi-diameter of the planet Venus. They examined the residuals (about a simple model) to see if any outliers should be rejected. The residuals (in seconds) were as follows,

| | | | | |
|---|---|---|---|---|
| − 0.30 | + 0.48 | + 0.63 | − 0.22 | + 0.18 |
| − 0.44 | − 0.24 | − 0.13 | − 0.05 | + 0.39 |
| + 1.01 | + 0.06 | − 1.40 | + 0.20 | + 0.10 |

demonstrating a *lower outlier* $x_{(n)} = -1.40$. Suppose we were to adopt a normal basic model. Various tests of discordancy are possible here (see Chapter 6 below), including those with test statistics

$$t_1 = (\bar{x} - x_{(n)})/s$$

and

$$t_2 = [x_{(2)} - x_{(1)}]/[x_{(n)} - x_{(1)}],$$

where $\bar{x}$ and $s$ are sample mean and standard deviation, respectively. We have $t_1 = 2.58$ and $t_2 = 0.40$, and 5 per cent points for the appropriate null distributions are 2.71 and 0.44, respectively. (See Tables XIIIa and XIXa at the end of the book.)

Thus both tests coincide in their conclusion that the lower outlier is not (quite) discordant at the 5 per cent level. (Both Peirce and Chauvenet found it to be discordant on their primitive approaches!)

Armed with the idea of a test of discordancy, we can begin to appreciate differences in the way various authors define outliers.

Some writers, for example, use 'outlier' for an observation which is both surprising *and discordant*; a term such as 'suspect value' is then used to describe a surprising value (an outlier in our sense). For instance, we read (Grubbs, 1950):

both the largest and the smallest observations may appear to be 'different' from the remaining items in the sample. Here we are interested in testing the hypothesis that both . . . are truly 'outliers'.

See discussion of the meaning of outlier in Chapter 1.

### 2.2.3  Different aims in examining outliers

If our interest is specifically directed to inferring characteristics of the basic model $F$, irrespective of the presence and nature of any contaminants, then outliers have only nuisance value and we will wish to employ robust methods of analysis to minimize their impact. Here the aim is that of **accommodation** (per se) and any test of discordancy is likely to be irrelevant (except for the rare and extreme prospect of rejecting discordant outliers prior to the inference process).

On the other hand a test of discordancy is by no means reserved for situations where we wish to *reject* discordant outliers as manifestations of contamination.

Leaving aside the wider statistical analysis we intend to apply to the data (and for which purpose they were presumably assembled) an assessment of discordancy of some outliers must be viewed only as a first stage of study of the outliers themselves. What action are we to take if we adjudge one or more outliers to be discordant? This will depend on a variety of factors relating to our interest in the practical situation. Obvious possibilities arise including **rejection**: we may indeed decide to *reject* (or replace) the discordant outliers and proceed to analyse the residual (modified) data on the original model, but there are other prospects of possibly greater importance.

Alternatively, we may choose to modify the model for **incorporation** of the outliers in a non-discordant fashion, or we may concentrate attention on the discordant outliers as a welcome **identification** of unsuspected factors of practical importance.

Let us consider some numerical examples, where, at first sight, these separate possibilities seem reasonable. Suppose that in each case any declared outliers prove to be *discordant* on the basis of an appropriate discordancy test employing an 'initially reasonable' model.

A typical situation where rejection might be (and indeed was) the aim in mind is exemplified by the data on the vertical semi- diameter of Venus discussed at the end of Section 2.2.2.

If the outliers had proved to be discordant on an assumed normal distribution it is quite likely (bearing in mind the possibilities of inexplicable 'gross errors') that we would have chosen to *reject* them before proceeding to further study of the data. We cannot, of course, be sure that this action is entirely proper. Perhaps an appropriately more sophisticated non-normal model would *incorporate* them in a non-discordant fashion, or the purpose of the further analysis may support some form of partial *accommodation* short of complete incorporation (see comments below). Should we decide to reject the outliers then the stricture of Kruskal (1960b) makes sound sense.

As to practice, I suggest that it is of great importance to preach the doctrine that apparent outliers should *always* be reported, even when one feels that their causes are known or when one rejects them for whatever good rule or reason. The immediate pressures of practical statistical analysis are almost uniformly in the direction of suppressing announcements of observations that do not fit the pattern; we must maintain a strong sea-wall against these pressures.

The outright rejection of outliers has statistical consequences for the further analysis of the reduced sample. We would no longer have a random sample, but a censored one. The practice of replacing rejected (not deterministically explicable) outliers by statistical equivalents (further simulated random observations from the assumed underlying distribution) involves similar consequences. The practices of 'Winsorization' and trimming (see below) will also have distributional implications which must be allowed for.

The following data described by Karl Pearson (1931) give the capacities (in cc) of a sample of seventeen male Moriori skulls.

|      |      |      |      |      |      |
|------|------|------|------|------|------|
| 1230 | 1318 | 1380 | 1420 | *1630* | 1378 |
| 1348 | 1380 | 1470 | 1445 | 1360 | 1410 |
| 1540 | 1260 | 1364 | 1410 | 1545 |      |

The observation 1630 was suspected as being 'too large'; suppose it proves to be discordant. It may here be more appropriate to seek an alternative model which *incorporates* the value 1630 in a non-discordant way. Biological data often require skew distributions as models—see the example on *macrolepidoptera* in Section 1.2. There remains of course the possibility that *identification* of the outlier reflects the presence of a small number of another species in the population being studied, or (possibly less realistically) that it has arisen from a once-and-for-all error of measurement, or of recording.

Daniel (1959) reports the results of a $2^5$ factorial experiment where the 31 contrasts arranged in order of increasing absolute value are:

|         |          |          |          |          |        |         |
|---------|----------|----------|----------|----------|--------|---------|
| 0.0000  | 0.0281   | − 0.0561 | − 0.0842 | − 0.0982 | 0.1263 | 0.1684  |
| 0.1964  | 0.2245   | − 0.2526 | 0.2947   | − 0.3087 | 0.3929 | 0.4069  |
| 0.4209  | 0.4350   | 0.4630   | − 0.4771 | 0.5472   | 0.6595 | 0.7437  |
| − 0.7437 | − 0.7577 | − 0.8138 | − 0.8138 | − 0.8980 | 1.080  | − 1.305 |
| **2.147** | **− 2.666** | **− 3.143** |          |          |        |         |

The last three contrasts are discordant outliers on the normal model. But this is precisely what we are seeking: important effects of the experimental factors. Thus we *identify* the outliers as indications of features of practical importance rather than as tedious reflections of possible inadequacies in the model or measurement technique.

Let us return for a moment to the process of *accommodation* of outliers. The extreme prospect is that we decide to reject the outliers prior to estimation or testing, either because of clear tangible explanations of their presence or following a test of discordancy based on a confidently assumed model. But interest in robustness against outliers denies any great confidence in the appropriate model and preliminary tests of discordancy may not be feasible or desirable. Furthermore the severe act of *rejecting* outliers may be over-extravagant and over-specific.

There are possibilities of partial rejection. Although we may be suspicious of the actual values − 1.40 and 1.01 in the Lt. Herndon data for the planet Venus, we may none the less feel that the *direction* of the residuals carries information and wish to retain this information in some form. One possibility is to employ *Winsorization* where, for example, we replace the lower and upper extremes by their nearest neighbours. In this example, − 1.40 and 1.01 are replaced by − 0.44 and 0.63, respectively, thus making each of these latter values appear twice in the data.

Alternatively as an aid to robustness of estimation or testing we may choose to use an α-*trimmed* sample, in which a fixed fraction α of lower, and upper, extreme sample values are totally discarded before processing the sample. This 'old French custom' (Huber, 1972) is not specifically concerned with protecting against outliers, though these will clearly be candidates for trimming.

From the robustness standpoint, we are thus aiming to devise statistical procedures which do not directly examine the outliers, but in seeking to *accommodate* them and render them less serious in their influence on estimation or tests of summary measures of the underlying distribution, we may indeed partially or fully reject them—without recourse to any formal discordancy test.

Throughout this section we have identified different sources of variation: (*Inherent variation, measurement errors, execution errors*) yielding outliers of differing nature *random, deterministic*. Our aims may be *accommodation, incorporation, identification, rejection*. The complex structure and pattern of relationships is illustrated in Figure 2.1.

One further matter must be stressed at this stage. The study of outliers in a data set is often inevitably an informal screening process preceeding fuller more formal analysis of the data. As such we must expect to encounter (in the spirit of modern descriptive methods of data analysis) a plethora of *ad hoc* approaches. Amongst these, there are many *graphical methods* of highlighting outliers. These can have profound impact, by revealing outlying characteristics of the data much more clearly than are apparent from numerical values or simple scatter diagrams. Such techniques come into their own for multivariate or structured data and we shall be considering many examples at a later stage. But even for univariate data we can glimpse the potential advantages of a graphical approach. Consider

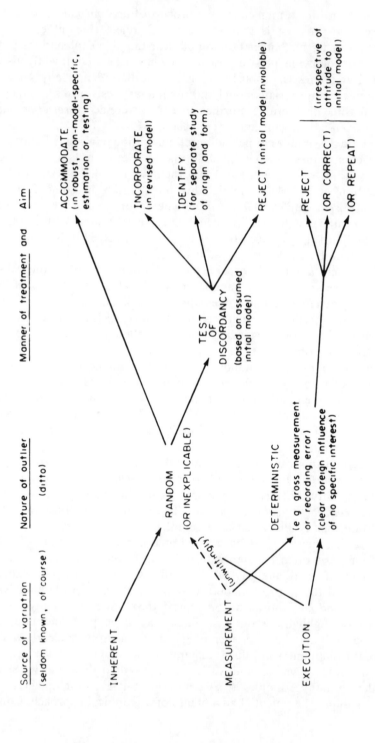

**Figure 2.1** Treatment of outliers

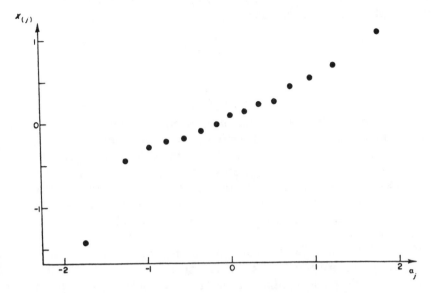

**Figure 2.2** Probability plot for the Venus vertical semi-diameter residuals

a simple probability plot (of ordered sample value $x_{(j)}$ against expected value $\alpha_j$ of standard normal order statistic—see Barnett, 1976a) for the Venus vertical semi-diameter residuals, shown in Figure 2.2. The outlier $-1.40$ is clearly discernible.

## 2.3 MODELS FOR CONTAMINATION

In the spirit of the previous section *we may regard the discordancy test as central to much of our interest in outliers*. Any statistical test must inevitably examine two hypothesis: a null hypothesis, or *working hypothesis*, which will be retained unless significant evidence is found to support its rejection, and an *alternative hypothesis* in favour of which the working hypothesis may need to be rejected.

For a test of discordancy of outliers, the working hypothesis will express some basic probability model for the generation of all the data with no contemplation of outliers; the alternative hypothesis expresses a way in which the model may be modified to incorporate or explain the outliers as reflections of contamination. We shall consider in some detail different forms of **contamination model**, or **outlier-generating model**, which have been proposed and studied.

Much of the early, or existing, work on tests of discordancy takes no specific regard for the nature of the working and alternative hypotheses. Thus, for example, when concerned with a single upper outlier in a set of independent observations $x_1, x_2, \ldots, x_n$ it is natural, and appealing, to contemplate statistics such as

$$[x_{(n)} - x_{(n-1)}]/D$$

or

$$[x_{(n)} - \bar{x}']/D$$

where $x_{(i)}$ is the $i$th *ordered* value in the sample (when all observations are placed in ascending order of magnitude), $\bar{x}'$ is the sample mean excluding the outlier $x_{(n)}$ and $D$ is some measure of the spread of the sample (again excluding consideration of $x_{(n)}$). Such statistics seem intuitively reasonable for assessing the discordancy of the outlier $x_{(n)}$, and were originally proposed purely on such an informal basis. (We shall return to the interesting question of whether the mean and the variance should indeed be calculated *after omission of the outlier*.)

Only at the more formal stage of determining the level or size of the test, or of constructing tables for assessing the statistical significance of the test statistic value, does it become unavoidable that the working hypothesis is specified. To go further and examine power characteristics of the test, or to construct tests with certain desirable statistical properties, *requires the additional specification of the alternative hypothesis*.

The historical development of tests of outliers mirrors such a progression in statistical sophistication. Whilst the declaration of the alternative hypothesis is the crux of the problem of defining just what we mean by contamination (and its possible manifestation as outliers), it still has not been very widely discussed. Perhaps this is inevitable: outliers are not easily defined, or incorporated in a generally acceptable form of model, and a degree of controversy still surrounds their study (see the opening pages of Chapter 1).

THE WORKING HYPOTHESIS

No fundamental difficulty exists in setting up the working hypothesis. We have repeatedly stressed the conditional nature of the outlier concept. An outlier is an observation which appears suspicious in the context of some provisional initial assignment of a probability model to explain the data-generating process. Thus the working hypothesis is merely a statement of the *initial* (basic) *probability model*.

In some discussion of the nature of tests of discordancy, or in attempts to accommodate outliers through robust statistical procedures, we may not need to be too specific about the probability model. It may suffice to declare, as a *working hypothesis*, $H$, that the data $x_j$ ($j = 1, 2, \ldots, n$) arise as independent observations from some common, but unspecified, distribution $F$. We can denote this as:

$$H : x_j \in F \qquad (j = 1, 2, \ldots, n).$$

But in the detailed analysis of real-world data we will often have much more specific model in mind: for example, that the data arise as a random sample from an *exponential distribution* with scale parameter $\theta$ (which may, or may not, be specified in value) or from a *homoscedastic linear regression* with *normal error structure*. Inevitably, much of the discussion of statistical tests of discordancy for outliers (as in so many areas of statistical enquiry) assumes a *normal* working hypothesis.

Chapter 6 will be concerned with the principles and criteria relevant to discordancy testing and Chapter 6 will present detailed forms of discordancy tests for individual outliers, or small groups of lower and upper outliers. Test statistics, relevant tables of percentage points and special advantages or disadvantages are described for a variety of tests. Extensive results are available for the normal and gamma (including exponential) distributions and these are presented along with information for other distributions such as the Pareto, Poisson, log-normal, uniform, binomial, and Poisson.

If, on the basis of a discordancy test, we adjudge an outlier (or small group of outliers) to be discordant we implicitly reject the working hypothesis in favour of some *alternative hypothesis*. Clearly, we must know what alternative hypothesis is being adopted! Any assessment of the power of the prevailing discordancy test depends on the form of the alternative hypothesis. All too frequently this aspect is ignored in the discussion of tests of outliers. We must now consider some of the possible forms of alternative hypothesis for discordancy tests and examine (briefly at this stage) the extent to which they figure in the construction or application of tests prescribed in the literature.

FORMS OF ALTERNATIVE HYPOTHESIS FOR CONTAMINATION

We can readily contemplate a variety of different forms. Some of these have been discussed in the literature; others have obvious appeal but do not appear to have been considered in any detail.

(i) **Deterministic alternative** The first type of alternative hypothesis which comes to mind covers the case of outliers caused by obvious identifiable gross errors of measurement, recording, and so on. Such an alternative hypothesis, which we term *deterministic*, is entirely specific to the actual data set, and the observed offending observations. Thus if our data $x_1, x_2, \ldots, x_n$ contain one observation, say $x_i$, which has clearly arisen from a mistaken reading or recording, we immediately reject any basic model, $F$, for the *whole* data set in favour of an alternative model which says that all $x_j$ ($j \neq i$) arise at random from $F$ whilst $x_i$ is quite different and requires rejection (or correction, or repeat reading). No discordancy test is needed. Rejection of the initial model in favour of the alternative is *deterministically* correct.

(ii) **Inherent alternative** In the terms of the discussion of sources of variability in the previous section, there must be the possibility that outliers have appeared in the data merely as a reflection of a greater degree of inherent variability than we initially anticipated. Perhaps what we thought was a sample from a normal distribution was really from a 'fatter-tailed' distribution. Upper outliers may reflect, say, that an initial assumption of a gamma distribution is best replaced with a log-normal distribution. Thus where outliers reflect a larger measure (or different form) of inherent variability than is encompassed in some basic model, $F$, it may be reasonable to express this by opposing the working hypothesis,

$$H : x_j \in F \qquad (j = 1, 2, \ldots, n)$$

that *all* observations arise from the distribution, $F$, with a suitably chosen alternative hypothesis,

$$\bar{H} : x_j \in G \qquad (j = 1, 2, \ldots, n),$$

that *all* observations arise from a distribution, $G$, under which the outliers no longer occasion the earlier degree of 'surprise'.

$F$ and $G$ may be different fully specified distributions, or may be distinct general parametric families of distributions. Note that here, and in the other alternative models described below, there are contraints on the form of $G$ in that it must have the potential to yield outliers. For example, its location may be shifted to one side or the other of that of $F$ or it may have *larger* dispersion or a *longer* tail in one direction. It can not have similar location and *smaller* dispersion, since we would not expect it to manifest outliers in this case.

Of course, we would hope to be able to distinguish between $F$ and $G$ by appropriate full-sample statistical procedures more powerful than the out-lier-specific discordancy test. On the other hand, in small sets of data our very motivation for considering a radically different form of model may stem entirely from the presence of outliers. Thus the *inherent* alternative hypothesis is relevant to tests of discordancy. Shapiro and Wilk have given such a test directed particularly to inherent alternatives, both for normal samples (1965) and for exponential samples (1972); detailed results on the power of their normal test against 45 different inherent alternatives are given by Shapiro, Wilk, and Chen (1968), and on the power of their exponential test against 15 different inherent alternatives by Shapiro and Wilk (1972).

(iii) **Mixture alternative** Rather than assume that outliers may reflect an unexpected degree or form of *inherent* variability, we might admit the possibility that the sample reflects low-level contamination from a population other than that represented by the basic model. We assume that such 'foreign' sample members, or contaminants, show themselves as outliers.

For example, in examining a sample of fossils from what is supposed to be a homogeneous population of the same species, we might inadvertently collect one or two fossils of a different species with different size characteristics. Whilst the presence of the different species might reasonably be ascribed to execution error when we are only interested in the predominant species, the term 'error' is not in general a good label for this type of manifestation. A particle physicist measuring characteristics of paths of radioactive particles is more likely to term it 'good fortune' than 'error' if he discovers in the form of outlying observations a basically new type of particle. (This illustrates again the distinction between *rejection* and *identification* in our response to a test of discordancy.)

In these terms a sensible alternative to the working hypothesis $H\!:\!F$ is an hypothesis of the form

$$\bar{H}\!:\!x_j \in (1 - \lambda)F + \lambda G \quad (j = 1, 2, \ldots, n)$$

which declares that outliers reflect the (small) chance $\lambda$ that observations arise from a distribution $G$, quite different from the initial model $F$. Such an alternative hypothesis will be called a *mixture* alternative, and it commonly figures in published work on outliers (see below). Note the constraints on possible forms of $G$ described in the discussion of the *inherent alternative* above.

As with the *inherent* type of alternative, we should again hope that the dichotomy

$$H\!:\!x_j \in F \quad (j = 1, 2, \ldots, n) \quad \text{versus} \quad \bar{H}\!:\!x_j \in (1 - \lambda)F + \lambda G \quad (j = 1, 2, \ldots, n)$$

is promoted, supported, and best analysed on a broader basis than the degree of 'surprise' engendered by the presence of one or two outliers in the data. But if the sample size, and the mixing parameter $\lambda$, are small, it may be that the outliers alone focus our attention on the possibility of a mixture model.

Box and Tiao (1968) and Guttman (1973b) discuss the implications of a mixture alternative for the study of outliers. Such a model is also used by Marks and Rao (1978, 1979) Prescott (1978) and Aitkin and Tunnicliffe Wilson (1980) and others. Box and Tiao (1968) remark that if the mixture prospect is revealed through outliers alone it will be necessary to assume that in a sample of at most 20 observations $\lambda$ is small: possibly less than 0.05, or at very most 0.10. Otherwise, occasions will arise where we encounter more observations from the distribution $G$ than we should be content to designate 'outliers'. (For larger samples, $\lambda$ will need to be even smaller on this argument.)

Guttman inverts this argument in support of the commonly made assumption that a set of data contains at most *one* contaminant! However, there seems to be a certain circularity in such arguments. If $\lambda$ is very small and $\bar{H}$ is a reasonable model we are unlikely to encounter more than one or two members of $G$ which (for appropriate $G$) may show up as outliers; if we encounter just one (or two) outliers it may be that $\bar{H}$ is an appropriate model to explain the outliers but $\lambda$ must perforce be small. But how are we to adjudge the propriety of the mixture alternative $(1 - \lambda)F + \lambda G$ with no evidence other than the one (or two) outliers whose discordancy we are to assess using this alternative model $(1 - \lambda)F + \lambda G$? There is no easy answer to this—but it is equally difficult to justify any other of the proposed forms of alternative hypothesis, although there seems to be greater intuitive appeal in the *slippage* type of alternative hypothesis we shall discuss next.

At this point (though not specifically concerned with solely the mixture alternative) it is relevant to point out an apparent philosophical inconsistency in the formulation of models to explain the presence of outliers. The model $(1 - \lambda)F + \lambda G$ merely declares that with a certain (small) probability observations might be generated by $G$. Yet the data have directed our attention to *specific* observations: the outliers, which appear typically as *extreme values* in the sample. Suppose we have just one (upper) outlier, $x_{(n)}$. The alternative hypothesis merely contemplates the possibility of *some* observations arising from $G$, not necessarily, specifically or solely $x_{(n)}$.

Statistical principles *may* lead to the conclusion that if there is only one observation from $G$ it is best (in some sense) adjudged to be $x_{(n)}$. But this seems to differ from our original subjective interest in the outlier $x_{(n)}$, *per se*, which would favour an alternative hypothesis specifically related to $x_{(n)}$. There appears to be no discussion of this matter in the literature. In any case, there is no obvious way of expressing such an interest through a mixture type of alternative hypothesis. But even for the slippage type of alternative discussed below, where such a facility can in fact be found, it does not seem to have been contemplated. We discuss this in some detail in Section 4.1.

As an example of the mixture model, Box and Tiao (1968) consider a Bayesian approach to outliers in which, under the working hypothesis, we have a random sample from a normal distribution $N(\mu, \sigma^2)$. The alternative hypothesis declares that, independently and with probabilities $1 - \lambda$, $\lambda$, respectively $(\lambda < 0.1)$, observations arise either from $N(\mu, \sigma^2)$ or from $N(\mu, b\sigma^2)$, with $b > 1$. Since $b > 1$ we might expect observations from the latter distribution $(G)$ to appear as extreme values, declared to be outliers. (We consider the results of Box and Tiao in more detail in Section 9.2. Their interest is in estimating $\mu$ rather than testing discordancy of outliers, so it comes under the heading of *accommodation*.)

The mixing distribution $G$ need not, of course, be restricted to a scale-shifted version of $F$; it could express a change of location or even a radically different form of distribution. As we have noted, only certain forms of mixture will be relevant to outliers: shifts of scale with $b > 1$ may give rise to upper and lower outliers in combination; shifts of location, upper or lower outliers separately depending on the direction of shift.

An early use of a mixture model for outliers is made by Dixon (1953). Tukey (1960) is interested in a mixture model with two normal distributions differing in *variance*.

**(iv) Slippage alternative** By far the most common type of alternative hypothesis as a model for contamination is what we shall refer to as the *slippage* alternative. It has been widely discussed and used, and figures (sometimes only implicitly) in work by Grubbs (1950), Dixon (1950), Anscombe (1960a), Ferguson (1961a), McMillan and David (1971), McMillan (1971), Guttman (1973b), Kimber (1979, 1982), Jain (1981a), Cook, Holschuh, and Weisberg (1982), Gather (1985), Stapanian *et al.* (1991) and many others. In its most usual form the slippage alternative states that all observations apart from some prescribed small number $k$ (1 or 2, say) arise independently from the initial model $F$ indexed by location and scale parameters, $\mu$ and $\sigma^2$, whilst the remaining $k$ are independent observations from a modified version of $F$ in which $\mu$ or $\sigma^2$ have been shifted in value ($\mu$ in either direction, $\sigma^2$ typically increased). In much published work $F$ is a *normal* distribution. The *models A* and *B* of Ferguson (1961a) are perhaps the most general expression of the *normal* slippage alternative, reflecting shifts of location and dispersion, respectively.

*Model A* $x_1, x_2, \ldots, x_n$ *arise independently from normal distributions with common variance,* $\sigma^2$. *(Under H they have common mean* $\mu$.) *There are known constants* $a_1, a_2, \ldots, a_n$ *(most of which will be zero), an unknown parameter* $\Delta$ *and an unknown permutation* $(v_1, v_2, \ldots, v_n)$ *of* $(1, 2, \ldots, n)$ *such that the normal distributions from which the* $x_i$ *arise have means*

$$\mu_i = \mu + \sigma\Delta a_{v_i} \qquad (i = 1, 2, \ldots, n) \tag{2.3.1}$$

$\bar{H}:\Delta \neq 0$ *(or one-sided analogues, e.g.* $\Delta > 0$ *when the* $a_i$ *have the same sign).*

*Model B* $x_1, x_2, \ldots, x_n$ *arise independently from normal distributions with common mean,* $\mu$. *(Under H they have common variance* $\sigma^2$.) *There are known positive constants* $a_1, a_2, \ldots, a_n$ *(most of which will be zero), an unknown parameter* $\Delta$ *and an unknown permutation* $(v_1, v_2, \ldots, v_n)$ *of* $(1, 2, \ldots, n)$ *such that the normal distributions from which the* $x_i$ *arise have variances*

$$\sigma^2_i = \sigma^2\exp(\Delta a_{v_i}) \qquad (i = 1, 2, \ldots, n) \tag{2.3.2}$$

$\bar{H}:\Delta > 0$ ($\Delta < 0$ *is irrelevant to the outlier problem*).

These models can encompass any number of contaminants in the data. Some particularizations, or modifications, are worth examining. For illustration we retain the normality assumption for $F$.

Anscombe (1960a) considers the case of a contaminant arising from a shift in the mean. He assumes that $\sigma^2$ is known, that $x_1, x_2, \ldots, x_{n-1}$ is a random sample from $N(\lambda, \sigma^2)$ and that under the alternative hypothesis $x_n$ arises independently from $N(\lambda + a\sigma, \sigma^2)$ with the value of $a$ unknown. Guttman (1973b) declares that under $\bar{H}$ one of the $x_i$ comes from $N(\mu + a, \sigma^2)$; which one is unknown. It may be any of the $x_i$ with equal probabilities, $1/n$. In a non-Bayesian framework this extension of the slippage alternative hypothesis (which is best described as an *exchangeable* alternative hypothesis—see below) can be expressed

$$\bar{H}: \text{one of } H_i \ (i = 1, 2, \ldots, n) \text{ holds},$$

where

$$H_i : x_i \sim N(\mu + a, \sigma^2). \tag{2.3.3}$$

If we are concerned with a contaminant arising from a shift in scale (or dispersion) rather than location, we would take as the corresponding slippage-type alternative hypothesis

$$\bar{H}': \text{one of } H_i' \ (i = 1, 2, \ldots, n) \text{ holds},$$

where

$$H_i' : x_i \sim N(\mu, b\sigma^2) \qquad (b > 1). \tag{2.3.4}$$

Both $\bar{H}$ and $\bar{H}'$ are immediate analogues of the alternative hypothesis in multisample *slippage tests* in the special case of samples each of just *one* observation (hence our terminology). If we wish to handle more than one outlier, $\bar{H}$ and $\bar{H}'$ need to be appropriately extended, of course, in the spirit of Ferguson's *models A* and *B*.

Such models have also been extensively applied to exponential distributions (see Section 5.5). Gather (1989) employs a broader generalization to non-normal (general location / scale) families of distributions.

Again we encounter the anomaly that this type of alternative hypothesis is non-specific with regard to which observation corresponds with the location-shifted (or scale-shifted) distribution. Suppose we encounter a single upper outlier, $x_{(n)}$, in the case of a location-shifted alternative hypothesis. As in the case of the mixture alternative, we might hope to test the working hypothesis against the *specific* analogous alternative that $x_{(n)}$ (specifically, rather than any single $x_i$) has arisen from $N(\mu + a, \sigma^2)$, with $a > 0$. As we have said, this prospect is not considered in the literature and it is not immediately obvious how such an alternative hypothesis should be best expressed, even in the slippage context; see Section 4.1 for further discussion.

(v) **Exchangeable alternative** A different approach to the form of the alternative hypothesis, extending the slippage formulation, is to be found in the work of Kale, Sinha, Veale, and others. Kale and Sinha (1971) and Veale and Kale (1972) were concerned respectively with estimating, and testing, the value of the mean, $\theta$, in an exponential distribution in a manner which is robust against contamination. The model for a single contaminant assumes, in its general form, that $x_1, x_2, \ldots, x_{i-1}, x_{i+1}, \ldots, x_n$ arise as independent observations from the distribution $F$ of the initial model, whereas $x_i$ is a random observation from a distribution $G$. It is further assumed that the index $i$ of the contaminant is equally likely to be any of $1, 2, \ldots, n$. The random variables $X_1, X_2, \ldots, X_n$ are, on this model, not independent, but they are exchangeable. We shall call such an alternative hypothesis an *exchangeable* alternative.

The likelihood under the alternative hypothesis of a single discordant observation is

$$L\{\mathbf{x}|F, G\} = \frac{1}{n} \sum_{i=1}^{n} g(x_i) \prod_{j \neq i} f(x_j) \qquad (2.3.5)$$

where $f(x)$, $g(x)$ are the probability (density) functions of the distributions $F$ and $G$. $F$ and $G$ might be taken from distinct families of distributions, or they may correspond merely with different parameter values within a parametric family. The latter is the case in the papers referred to above, where

$$f(x) = (1/\theta)e^{-(x/\theta)} \qquad (2.3.6)$$

and

$$g(x) = (b/\theta)e^{-(x/\theta)} \qquad (0 < b < 1). \qquad (2.3.7)$$

In its general form the exchangeable alternative is identical in form (if different in motivation) to that which would be employed in a *Bayesian* analysis of the slippage alternative with equal prior probabilities for the index of the observation which arises from the anomalous family, $G$, as described by Guttman (1973b).

A further extension of the exchangeable alternative is given by Joshi (1972b).

The exchangeable alternative model is of course readily extended to situations with more than one contaminant. See Mount and Kale (1973) for formal details of such an extension.

(vi) **Other alternative hypotheses** There are few other specific proposals for modelling contamination (either as an alternative hypothesis in a test of discordancy, or as a model for robust accommodation of outliers). Goldstein (1982) addresses himself to the basic theme of Bayesian modelling of 'contamination distributions'; Simar (1983) uses Bayesian methods in protecting against gross errors; Smith (1983) discusses Bayesian approaches to outliers and robustness, whilst Pettit and Smith (1983) consider Bayesian model comparisons when outliers are present in the data. O'Hagan (1988) also develops a Bayesian approach. Some ideas relating to the investigation of multivariate outliers implicitly employ other modelling concepts to those outlined above. See Chapter 7.

Remarks by Kruskal (1960b) are relevant to the general question of modelling. In particular, he points the need for models which allow the occurrence of (measurement error) outliers to depend on the value that would have occurred had an error not taken place—but he makes no specific proposals. Lovie (1986) employs the above classification of outlier models to discuss outlier 'identification'. Rauhut (1987) draws a distinction between such models, and heavy-tailed distributions; but the latter seem akin to the *inherent* alternative discussed above.

## 2.4 OUTLIER PRONENESS, OUTLIER RESISTANCE, OUTLIER LABELLING

Many concepts have been developed in recent years that are clearly relevant to the modelling of outliers but do not specifically address the idea of contamination, nor are they directed to accommodation or discordancy procedures, *per se*.

We start with the notion of **outlier proneness**. Recalling the subjective manner in which sample observations may be declared outliers we stress again the element of 'surprise' they engender. Such surprise should ideally be conditional on the initial model we have in mind for the data, so that, for example, extreme values in a Cauchy sample would need to be even more extreme than those in a normal sample if we were to declare them 'outliers'. A somewhat different attitude seems to be implied in work by Neyman and Scott (1971) and by Green (1974). Both works consider a method of distinguishing between families of distributions with regard to the differing extents to which they are likely to exhibit outliers. They define a concept of *outlier proneness*, and conclude, *inter alia*, that the families of log-normal and gamma distributions are outlier prone, whereas the family of Cauchy distributions is not.

The concept hinges on the probability $P(\kappa, n \,|\, F)$ that a random sample of size $n$ from a distribution $F$ (in a family $\mathscr{F}$) contains an extreme member $x_{(n)}$ which exceeds $x_{(n-1)}$ by more than an amount $\kappa(x_{(n-1)} - x_{(n)})$. If the supremum of $P(\kappa, n \,|\, F)$ over $\mathscr{F}$ is strictly less than unity, then they call $\mathscr{F}$ $(\kappa, n)$—*outlier-resistant*; otherwise it is $(\kappa, n)$—*outlier-prone*. If $\mathscr{F}$ is $(\kappa, n)$—outlier-prone for all $\kappa > 0$ and all $n > 2$ it is *outlier-prone completely*.

Green (1974) shows that this is so provided only that $\mathscr{F}$ is $(\kappa, n)$—outlier-prone for *some* $\kappa > 0, n > 2$. See also Kale (1975a, 1975b, 1977), Neyman (1979), O'Hagan (1979) and Goldstein (1983) for definitive contributions. Gather (1985) examines the effect of outlier proneness on the tail-behaviour of location estimators (with particular attention to trimmed means); Takahashi (1987) shows that the family of Weibull distributions is outlier-prone completely.

Gather and Rauhut (1990) review work on outlier-proneness and outlier-resistance and exhibit relationships with outlier-generating models. They discuss the role of the conditional distribution of $x_{(n)}$ given $x_{(n-1)}$, and of the **mean residual life function (MRLF)**:

$$E\{X - x \,|\, X > x\}.$$

Main (1987, 1988) proposes extensions of the concepts of *outlier-prone* and *outlier-resistant* distributions: in particular introducing the idea of **outlier neutral** families of distribution. Goldstein (1983) examines some Bayesian characteristics of an observation from an outlier-resistant distribution.

An *ad hoc* exploratory data proposal for outliers by Hoaglin, Iglewicz and Tukey (1986) suggests a 'resistant rule for identifying possible outliers' in the following form. If $F_L$ and $F_U$ are the lower and upper *fourths* or *hinges* (approximate quartiles), any observations are *labelled* as 'outside' if they are not within $F_L - k(F_U - F_L)$ and $F_U + k(F_U - F_L)$. A recommended choice for $k$ is the value 1.5. This is described as **outlier labelling** and is claimed to be 'resistant' in the sense that the procedure itself is felt to be unlikely to be affected by the 'possible outliers' or by masking. Simonoff (1987b) remarks that similar results can be obtained using 'standard' outlier detection procedures: referring to his own recent work (Simonoff, 1987a). Hoaglin and Iglewicz (1987) propose some 'fine-tuning' in respect of the manner in which the fourths are defined. See also Brant (1990).

Kimber (1990) extends the data analytic methods of Hoaglin, Iglewicz and Tukey (1986) to data from *skew* distributions, and suggests 'flagging' any observations 'labelled as outside' and calling them **outsiders**.

A concept of a so-called **straggler** (less extreme than a 'statistical outlier') appears in British Standards specifications; see Section 13.3 below.

Muñoz-Garcia, Moreno-Rebollo, and Pascual-Acosta (1990) propose a new definition of an outlier which seeks to avoid the subjective element of specification. Agreeing that an outlier should be 'a typical and/or erroneous' they add the condition that it should 'deviate decidedly from the general behaviour . . . with respect to the criteria [on which] it is to be analysed'. This appears to seek to combine *our* concept of outlying behaviour with that of discordancy or non-robustness.

A similar feature is evident in the extensive discussion of the identification of multiple outliers by Davies and Gather (1993). Outliers are *defined* as observations that lie in an *extremal region* of the null distribution $F$ characterized by $(\theta, \alpha_n)$ where $\theta$ is the true (typically unknown) value of the parameter specifying $F$ and $\alpha_n$ is some small value, specified by the observer, which is the probability of an observation falling in the extremal region. The dependence of $\alpha_n$ on $n$ might, for example, take the form

$$\alpha_n = 1 - (1 - \alpha)^{1/n}$$

for some specified small value $\alpha$: with obvious intuitive motivation. The extremal region has, of course, to be *estimated* since $\theta$ will be unknown. It is thus possible for a genuine observation (from $F$) to be designated an outlier. The concept is examined in relation to that of the finite sample breakdown point (see discussion of robustness in Section 3.1.2 below).

In this chapter we have introduced a large number of basic distinctions and critical concepts. They serve as the cornerstones of the whole study of outliers and are fundamental to the extended discussion of statistical method, and to the development of specific techniques, in the subsequent chapters.

# CHAPTER 3

# The Accommodation Approach: Robust Estimation and Testing

We outlined in Chapter 2 the need for accommodation of outliers by means of estimation or testing procedures which are *robust* against the presence of outliers—in the phrase of Anscombe and Barron (1966), 'desensitized to outliers'. We pointed out that outlier study is only one of a range of contexts in which the concept of robustness operates. 'What is a robust procedure?' Huber asks, in his 1972 Wald Lecture *Robust Statistics: A Review*, and goes on to say:

one never has a very accurate knowledge of the true underlying distribution; . . . the performance of some of the classical tests or estimates is very unstable under small changes of the underlying distribution; . . . some alternative tests or estimates . . . lose very little efficiency for an exactly normal law, but show a much better and more stable performance under deviations from it.

While for years one had been concerned mostly with what was later called 'robustness of validity' (that the actual confidence levels should be close to, or at least on the safe side of the nominal levels), one realized now that 'robustness of performance' (stability of power, or of the length of confidence intervals) was at least as important . . .

From the beginning, 'robustness' has been a rather vague concept; . . . if one wants to choose in a rational fashion between different robust competitors to a classical procedure, one has to make precise the goals one wants to achieve. (Huber, 1972)

Later, in his important expository book *Robust Statistics*, Huber says:

The word 'robust' is loaded with many . . . connotations. We use it in a relatively narrow sense: for our purposes, *robustness signifies insensitivity to small deviations from the assumptions*. (Huber, 1981)

Yes; but what *kind* of deviations? We must stress again the difference between *distributional robustness* (against *inherent alternatives*) and *robustness against contamination.* Consider for example a robust inference procedure for the parameters of a normal distribution based on data believed initially to come from such a distribution. Many such published procedures are robust against the possibility that the entire sample comes from some other distribution, possibly gamma or Cauchy, not too dissimilar to the normal but perhaps somewhat skew or fatter-tailed. This does not guarantee that it will provide good robustness against outliers arising from contamination. In this situation, the one which typically concerns us in outlier study, the sample may well have a high probability of conforming to the required normal model, but we need to accommodate the occasional extreme contaminant.

From our point of view, this implies that the procedures need to perform satisfactorily under alternative models of the kinds *which generate outliers.* (What is meant by 'satisfactory' performance is discussed in some detail in Section 3.1). Our attention is accordingly focused on alternative models of the categories discussed in Chapter 2, and specifically the *mixture, slippage,* and *exchangeable* models rather than the *inherent* type of alternative. Inherent alternatives, such as a Cauchy distribution for data normally distributed on the basic model, are covered by more general (non-outlier-specific) robustness procedures. For further discussion of links, and differences, between robustness and outlier study see Section 8.7 below.

In this chapter, we focus in the first place on principles of robust estimation and testing for a univariate sample of $n$ observations $x_1, \ldots, x_n$, all of which (on the basic model) belong to a distribution $F$. All the essential issues can be brought out in this context. Many of the detailed results published so far relate to normal or exponential $F$. In our general discussion (Section 3.1) of performance criteria for robust procedures, we shall frequently find it convenient to illustrate ideas on the assumption that $F$ is normal; our basic results apply, *mutatis mutandis,* to other distributions. In Section 3.2 we discuss methods for constructing such procedures, both general (non-outlier-specific) methods and methods designed to take specific account of outliers in the basic model. Detailed proposals for practical use in univariate samples are reviewed in Chapter 5. The accommodation issue in relation to more complex situations (e.g. multivariate data, linear models, regression, designed experiments, etc.) will be considered in detail where appropriate in the later chapters of the book.

Apart from technical methods for robust statistical analysis, we might also consider how 'robust' is the statistician's judgement; see Relles and Rogers (1977).

For examination of some real-life data sets, using robust methods, see Stigler (1977), Balasooriya and Chan (1981), Hill and Dixon (1982), and Rocke, Downs, and Rocke (1982). Rocke, Downs, and Rocke remark that, '[although] there is substantial literature on robust estimation, most scientists continue to employ traditional methods. They remain skeptical about the practical benefit of employing robust techniques . . . '. Accordingly the authors entitle their paper 'Are robust estimators really necessary?' and answer in the affirmative. They compare the performance of a number of estimators of location when applied both to the series of historical data sets in the physical sciences analysed by Stigler (1977) and to a collection of data sets published in 1976 in the *Journal of the Association of Official Analytical Chemists*. 'The results', they conclude, 'provide a strong testament to the practical utility of robust methods'.

An early (Victorian) interest in robustness is described by Stigler (1980).

Most of the existing work based on the *exchangeable* type of alternative model relates to exponential samples. This work is often further particularized by the assumption that there is just one contaminant. For details, see Section 5.5.

In the context of *slippage* alternatives, robust procedures have been discussed for normal samples by a number of writers, and we shall later review this work in detail.

Undoubtedly the type of alternative model most commonly envisaged, at any rate for normal samples, is a *mixture* model. Tukey (1960), in a seminal paper 'A survey of sampling from contaminated distributions', discusses robust estimation for samples where the basic model is normal,

$$H:x_j \in F \quad (j = 1, 2, \ldots, n),$$

and the alternative is a mixture of two normals,

$$\bar{H}:x_j \in (1 - \lambda)F + \lambda G \quad (0 < \lambda < 1; j = 1, 2, \ldots, n),$$

the normal distribution $G$ having either the same mean as $F$ but a larger variance, or the same variance but a shifted mean. Tukey calls the mixture $(1 - \lambda)F + \lambda G$ a *contaminated distribution*, the basic distribution $F$ being contaminated by the distribution $G$; the parameter $\lambda$, commonly a quite small fraction, is the amount of contamination, or contamination fraction, or just the *contamination*. Sample observations which come from $G$ are *contaminants*; under $\bar{H}$, the number of contaminants in a sample of $n$ observations will be a binomial random variable with parameters $n$, $\lambda$. We have already used the term 'contaminant' in Chapter 2. Slippage models can be regarded as contamination models in which the number of contaminants in a sample is fixed. We shall accordingly refer to the mixture model as one of 'random contamination' and the slippage model as one with a fixed number of contaminants.

In a random contamination model, the contaminating distribution $G$ need not necessarily be normal even if $F$ is normal. It may be symmetric about the mean of the basic normal distribution $F$, or it may not. We distinguish the cases of *symmetric contamination* and *asymmetric contamination* in the following manner for the case of a symmetric basic distribution

*F*. Contamination is symmetric if the contaminating distribution is symmetric about the centre of the distribution *F*. Thus if *F* is normal, *G* could be normal with the same mean as *F* but with a greater variance; or, more generally, it could be of arbitrary symmetric form centred at the mean of *F*. With asymmetric contamination, *G* may be symmetric about some value different from the centre of *F*, or it may be asymmetric. For example, again when *F* is normal, *G* may be normal (or non-normal symmetric) with a different mean from that of *F*, or it may be of arbitrary asymmetric form. We shall be mainly concerned with symmetric contamination of a basic symmetric distribution *F* mixed with only *one* contaminating distribution *G*. (One could envisage *F* being mixed with a set of contaminating distributions, e.g.

$$\bar{H}: x_j \in (1 - \lambda_1 - \lambda_2)F + \lambda_1 G_1 + \lambda_2 G_2 \qquad (j = 1, 2, \ldots, n)$$

or, still more generally,

$$\bar{H}: x_j \in \int G(\lambda, x) \, dK(\lambda) \qquad (j = 1, 2, \ldots, n)$$

where *F* is a particular distribution, say $F(x) = G(\lambda_0, x)$, from a one-parameter family $G(\lambda, x)$, and *K* is a mixing measure. For example, *F* might be $N(\mu, \sigma^2)$ and $G(\lambda, x)$ might be $N(\mu, \lambda\sigma^2)$ with $\lambda$ exponentially distributed and $\lambda_0 = 1$; the mixture specified by $\bar{H}$ is in this case a double-exponential distribution, illustrating that with such a mixture situation we have really moved over to an *inherent* alternative.)

If the basic distribution *F* is $N(\mu, \sigma^2)$, inferences about $\mu$ may assume $\sigma^2$ known or unknown; likewise, inferences about $\sigma^2$ may assume $\mu$ known or unknown. It might be thought that the cases $\sigma^2$ known or $\mu$ known would be only of academic interest, but this is not so; examples of practical situations involving knowledge of $\sigma^2$ or of $\mu$ do arise (cf. Sections 6.3 and 6.4).

What is the effect of random contamination? It may be considerable, even for very small values of $\lambda$. Suppose, for example, that $F:N(\mu, \sigma^2)$ is contaminated in the ratio $1 - \lambda:\lambda$ by another normal distribution $G:N(\mu, b\sigma^2)$ $(b > 1)$ having *b* times its variance. In a sample of size *n* from the mixture, there will be $R = 0, 1, \ldots$ contaminants, the random variable *R* having a binomial distribution with parameters $n, \lambda$. The sample mean $\bar{x}$ will be an unbiased estimator of $\mu$, with sampling variance

$$\text{var}(\bar{x}) = \frac{1}{n^2} E_R[(n - R)\sigma^2 + Rb\sigma^2] = \frac{\sigma^2}{n}[1 + (b - 1)\lambda]. \qquad (3.0.1)$$

The contamination has caused the sampling variance to increase, relative to $\sigma^2/n$, by a factor $1 + (b - 1)\lambda$. For $\lambda = 0.05$ and $b = 9$, i.e. for 5 per cent contamination by a distribution $G$ with three times the standard deviation of $F$—a not untoward situation—this factor is 1.4; for 10 per cent contamination by the same $G$ it is 1.8, a loss of efficiency of 44 per cent.

Now consider the effect of the contamination on the performance of the sample variance $s^2$ as an estimator of $\sigma^2$. A straightforward calculation gives, conditional on $R$ contaminants,

$$E(s^2 \mid R) = \frac{\sigma^2}{n}[(n - R) + Rb] \quad \text{as in (3.0.1),}$$

$$E(s^4 \mid R) = \frac{3\sigma^4}{n^2}(n - R + Rb^2).$$

$$+ \frac{(n^2 - 2n + 3)\sigma^4}{n^2(n - 1)^2}[(n - R)(n - R - 1) + 2(n - R)Rb + R(R - 1)b^2].$$

Hence, using $E(R) = n\lambda$, $E(R^2) = n^2\lambda^2 + n\lambda - n\lambda^2$, we get

$$\text{var}(s^2) = \frac{2\sigma^4}{n - 1}\left[1 + \tfrac{1}{2}\lambda(b - 1)(3b + 1) - \tfrac{1}{2}\lambda^2(b - 1)^2 - \frac{3\lambda(1 - \lambda)(b - 1)^2}{2n}\right].$$

$$(3.0.2)$$

The contamination has caused the sampling variance of $s^2$ to increase, relative to the sampling variance $2\sigma^4/(n - 1)$ under the basic model, by the factor in square brackets in (3.0.2). For $\lambda = 0.05$ and $b = 9$, the value of this factor is $6.52 - (4.56/n)$. Even with $\lambda$ as small as 0.01 and $b = 9$, the factor is $2.12 - (0.95/n)$; with a sample size as small as 7, a mere 1 per cent contamination by $N(\mu, 9\sigma^2)$ causes a loss of efficiency of 50 per cent in using the sample variance to estimate $\sigma^2$!

This striking effect must be due to the incidence, even if infrequent, of extreme values from the contaminating distribution. One might think that contaminants which have so pronounced an effect on the efficiency of estimation would show up unmistakably as outliers and could be rejected on the basis of some discordancy test. This is not so. To entertain such a hope with the sample of size 7 discussed above would be vain; that it would be equally so with large samples has been demonstrated cogently by Tukey (1960). He considers by way of example a sample of 1000 observations from a contaminated distribution $(1 - \lambda)F + \lambda G$, where $F$ is $N(\mu, \sigma^2)$, $G$ is $N(\mu, 9\sigma^2)$, and $\lambda = 0.01$ (the 1 per cent contamination of our earlier example). Some typical percentiles of the two distributions are as shown in Table 3.1. Thus the two cumulative distributions are indistinguishable for practical purposes for values of the variable between $\mu - 2\tfrac{1}{2}\sigma$ and $\mu + 2\tfrac{1}{2}\sigma$. Of the ten expected observations from $G$, only about 40 per cent (corresponding to standardized deviates $\pm 2.5/3 = \pm 0.833$)

will fall outside these bounds, and we may thus expect only two observations from the upper tail of the broader, rarer constituent [i.e. $G$], and another two from the lower tail. Beyond the same limits we will expect about six (in each tail) from the narrower constituent [i.e. $F$]. Unless one or both of the two are very extreme, the indication of non-normality will be very slight. . . . A sample of one thousand is likely to be of little help. (Tukey, 1960)

**Table 3.1**

| Cumulative probability | Percentile of $F$ | Percentile of $(1 - \lambda)F + \lambda G$ |
|---|---|---|
| 0.25 | $\mu - 0.67\sigma$ | $\mu - 0.68\sigma$ |
| 0.05 | $\mu - 1.64\sigma$ | $\mu - 1.67\sigma$ |
| 0.01 | $\mu - 2.33\sigma$ | $\mu - 2.41\sigma$ |
| 0.006 | $\mu - 2.51\sigma$ | $\mu - 2.64\sigma$ |

## 3.1  PERFORMANCE CRITERIA

### 3.1.1  Efficiency measures for estimators

Suppose we are estimating the location parameter $\mu$ of a symmetric population $F$ from a sample of size $n$. Our choice of estimator might be unrestricted, or we might decide on the other hand to confine ourselves to some restricted class of estimators, such as linear combinations $a_1 x_{(1)} + \cdots + a_n x_{(n)}$ of the sample order statistics. If the basic model holds good,

$$H:x_j \in F \quad (j = 1, 2, \ldots, n),$$

an 'optimal' estimator of $\mu$ can be defined among the estimators of the class we are considering; we will denote this by $\breve{\mu}$. For example, $\breve{\mu}$ might be the maximum likelihood estimator $\hat{\mu}$, or the linear unbiased estimator of form $\sum a_i x_{(i)}$ with minimum variance (the 'best linear unbiased estimator', or BLUE); while these estimators coincide for a normal distribution, they will be quite different for, say, a Cauchy distribution. However we define it, $\breve{\mu}$ serves as a yardstick for what an estimator can achieve in relation to the basic model. To fix ideas, suppose $F$ is $N(\mu, \sigma^2)$; $\breve{\mu}$ then would typically be the sample mean $\bar{x}$. We now propose some rival estimator, $T$, which we require to be robust in relation to some contamination model $\bar{H}$.

Starting with the simplest situation, we take $\bar{H}$ to be a simple and symmetric alternative, providing merely for the observations to belong to a specified mixture distribution with known symmetric contamination. That is,

$$\bar{H}:x_j \in (1 - \lambda)F + \lambda G \quad (j = 1, 2, \ldots, n) \quad\quad (3.1.1)$$

where typically $G$ could be $N(\mu, b\sigma^2)$, $b > 1$, and $\lambda$, $b$ are known. In view of the symmetry of the model we can reasonably assume $E(T | \bar{H}) = \mu$, and the question of bias in $T$ does not arise. Accordingly, if $\text{var}(\breve{\mu} | \bar{H})$ did not appreciably exceed $\text{var}(\breve{\mu} | H)$, $\breve{\mu}$ would itself be robust, and there would be no need to seek for a rival estimator $T$; the ratio

$$\frac{\text{var}(\breve{\mu} | \bar{H})}{\text{var}(\breve{\mu} | H)}$$

provides a quantitative indication of the need for a robust estimator alternative to $\breve{\mu}$. We can assume that this ratio exceeds unity substantially, as in the example discussed under equation (3.0.2).

For robustness of $T$ we require that $\text{var}(T | \bar{H})$ shall be substantially less than $\text{var}(\breve{\mu} | \bar{H})$; also, of course, $T$ must be a reasonable alternative to the optimal $\breve{\mu}$ when the data obey the basic model $F$, i.e. we require that $\text{var}(T | H)$ shall not be 'unduly' greater than $\text{var}(\breve{\mu} | H)$. So $\text{var}(T | \bar{H})/\text{var}(\breve{\mu} | \bar{H})$ is to be 'small', and $\text{var}(T | H)/\text{var}(\breve{\mu} | H)$ 'not much greater than unity'. Each of these ratios is, of course, a relative efficiency measure, and discussions of performance are often phrased in terms of efficiency.

An alternative terminology is that of *protection* and *premium*; these concepts, equivalent to the above relative efficiency measures, were introduced by Anscombe (1960a) in a classic paper. We quote the passage in which he introduces these terms as part of his exposition of a basic philosophy in the matter of accommodating outliers:

Rejection rules are not significance tests. . . . when a chemist doing routine analyses, or a surveyor making a triangulation, makes routine use of a rejection rule, he is not studying whether spurious readings occur (he may already be convinced they do sometimes), but guarding himself from their adverse effect. . . .
A rejection rule is like a householder's fire insurance policy. Three questions to be considered in choosing a policy are
(1) What is the premium?
(2) How much protection does the policy give in the event of fire?
(3) How much danger really is there of a fire?
Item (3) corresponds to the study of whether spurious readings occur in fact . . .
The householder, satisfied that fires *do* occur, does not bother much about (3), provided the premium seems moderate and the protection good. (Anscombe, 1960a)

The 'fire' here is the occurrence of outliers. The discussion is couched in terms of rejection rules, but applies to accommodation procedures in general. Anscombe goes on to ask:

In what currency can we express the premium charged and the protection afforded by a rejection rule? . . . variance will be considered here, although in principle any other measure of expected loss could be used. The premium payable may then be taken to be the percentage increase in the variance of estimation errors due to using the rejection rule, when in fact all the observations come from a homogeneous normal source; the protection given is the reduction in variance (or mean squared error) when spurious readings are present. (Anscombe, 1960a)

Clearly, in the case we have considered so far,

$$\text{premium} = \frac{\text{var}(T \mid H) - \text{var}(\check{\mu} \mid H)}{\text{var}(\check{\mu} \mid H)} \qquad (3.1.2)$$

and

$$\text{protection} = \frac{\text{var}(\check{\mu} \mid \bar{H}) - \text{var}(T \mid \bar{H})}{\text{var}(\check{\mu} \mid \bar{H})}. \qquad (3.1.3)$$

Whether the measurement of robustness is discussed in terms of variance ratios or relative efficiencies on the one hand, or premium and protection on the other, is a matter of taste. Papers which make explicit use of premium and protection include Anscombe and Barron (1966), Tiao and Guttman (1967), Guttman and Smith (1969, 1971), Guttman (1973a), Desu, Gehan, and Severo (1974), Marks and Rao (1978, 1979), Guttman and Kraft (1980), Rocke, Downs, and Rocke (1982, p.97), and Homan and Lachenbruch (1986).

From the simple situation just discussed, we may generalize in several directions.

*Asymmetric contamination*

If $F$ is $N(\mu, \sigma^2)$ and $G$ is, say, $N(\mu + a, \sigma^2)$, $E(T \mid \bar{H})$ will not in general be equal to $\mu$, and we have to take account of bias in our estimators. In this situation,

a natural criterion for judging them is their mean squared error, a measure which takes into account both their inherent variability and their distance from the estimand. (Jaeckel, 1971a)

This leads us to replacing $\text{var}(\check{\mu} \mid \bar{H}), \text{var}(T \mid \bar{H})$ in our discussion by the mean squared error values

$$MSE(\check{\mu} \mid \bar{H}) = E[(\check{\mu} - \mu)^2 \mid \bar{H}]; \qquad MSE(T \mid \bar{H}) = E[(T - \mu)^2 \mid \bar{H}].$$

There is also the question of how the bias affects performance. We have

$$E[(T - \mu)^2 \mid \bar{H}] = \text{var}(T \mid \bar{H})(1 + c^2)$$

where $c$ is the ratio of the bias to the standard deviation of the estimator.

Consider, for example, the performance of the sample mean $\bar{x}$ in estimating the mean $\mu$ of $F:N(\mu,\sigma^2)$ when there is contamination of amount $\lambda$ by $G:N(\mu + a, \sigma^2)$. With $R$ contaminants among the $n$ observations, we have

$$E(\bar{x}\mid R) = \mu + (a/n)R, \qquad E(\bar{x}^2\mid R) = [\mu + (a/n)R]^2 + (\sigma^2/n),$$

so that

$$E(\bar{x}) = \mu + \lambda a \tag{3.1.4}$$

and

$$\operatorname{var}(\bar{x}) = \frac{\sigma^2}{n} + \frac{\lambda(1 - \lambda)a^2}{n}. \tag{3.1.5}$$

Thus the estimator has bias $\lambda a$, and mean squared error

$$MSE(\bar{x}) = \frac{\sigma^2}{n} + a^2\left[\lambda^2 + \frac{\lambda(1 - \lambda)}{n}\right] \tag{3.1.6}$$

It follows from (3.1.4), (3.1.5) that if we want the bias of $\bar{x}$ to be less than, say, one-half the standard deviation, the sample size must satisfy

$$n < \frac{1}{4}\left(\frac{\sigma^2}{\lambda^2 a^2} + \frac{1 - \lambda}{\lambda}\right). \tag{3.1.7}$$

For example, with 1 per cent contamination, and $a$ equal to $5\sigma$, the sample size must not exceed 125. For further discussion see Huber (1964, p. 83; 1981, p. 104) and Jaeckel (1971a).

## Compound alternative $\bar{H}$

If in (3.1.1) we regard $\lambda$ as a parameter with a range of possible values (say $0 < \lambda \le \lambda_1$) rather than a single known quantity, our alternative model is a family of mixture distributions, indexed by $\lambda$. Consider any performance measure $M$, for example the protection measure defined in (3.1.3)

$$M = [\operatorname{var}(\breve{\mu}\mid\bar{H}) - \operatorname{var}(T\mid\bar{H})]/\operatorname{var}(\breve{\mu}\mid\bar{H}).$$

$M$ will now be a function of $\lambda$, and we can write it as $M(\lambda)$. (Similarly, $M$ could be a function of $b$, or of the two parameters $\lambda$ and $b$.) If our estimator $T$ is to possess robustness in relation to $\bar{H}$, it must perform 'satisfactorily' for every distribution which can arise under $\bar{H}$, and the value of $M(\lambda)$ must be 'satisfactory' for all values of $\lambda$ between zero and $\lambda_1$. For a *single* measure of performance this naturally suggests the value of the worst possible performance under $\bar{H}$, in other words the extreme value of $M(\lambda)$ (minimum or maximum as appropriate) over the range of possible values of $\lambda$. This measure could, for instance, take the form of the maximum variance of the estimator under $\bar{H}$, or on the other hand the minimum protection value, as $\lambda$ varies between zero and $\lambda_1$. Correspondingly in the two-parameter case we would use the extreme value of $M(\lambda, b)$.

*Estimation of scale and other parameters*

Robust estimation may of course be required, not only for location parameters, but for scale parameters (as with exponential distributions), dispersion or scale parameters (as with normal distributions), or shape parameters (as with Pareto distributions). Here we use the term 'scale parameter' for a dispersion parameter expressed in the same dimensions as the random variable, e.g. standard deviation (as opposed to variance). Our discussion of performance criteria has focused on the robust estimation of a location parameter, but obviously applies in essentials to any other kind of parameter. A scale parameter for a random variable $X$ can in any case be regarded as a location parameter for an appropriately transformed random variable $Y$: for example $Y = |X - \theta|$, where $\theta$ is some location parameter for $X$.

*Asymptotic measures*

The variances, relative efficiencies and mean squared errors we have used so far in our discussion have been actual (i.e. finite-sample) values. For example, the variance ratio $\mathrm{var}(s^2 | \bar{H})/\mathrm{var}(s^2 | H)$, indicating the non-robustness of the normal sample variance $s^2$ as an estimator of $\sigma^2$, was shown in (3.0.2) to have the value

$$M(\lambda, b) = 1 + \tfrac{1}{2}\lambda(b - 1)(3b + 1) - \tfrac{1}{2}\lambda^2(b - 1)^2 - \frac{3\lambda(1 - \lambda)(b - 1)^2}{2n} \quad (3.1.8)$$

for the contamination model $(1 - \lambda)N(\mu, \sigma^2) + \lambda N(\mu, b\sigma^2)$. For the typical numerical values $\lambda = 0.05$, $b = 9$, this gave

$$M(0.05, 9) = 6.52 - \frac{4.56}{n}. \quad (3.1.9)$$

Unless $n$ is small, this differs little in value from its limit as $n \to \infty$, i.e. 6.52. We can, if we wish, use this limiting value—the *asymptotic* variance ratio—as our measure, rather than the finite-sample variance-ratio (3.1.9). This choice between finite-sample and asymptotic values is obviously available for any measure involving variances, or efficiencies based on variance, or their premium and protection equivalents. With certain adjustments (see below), it is also available for measures based on mean squared errors.

As regards nomenclature, var $T$ will in general be of the form $(A/n) \times (1 + O(n^{-1}))$, so that $n$ var $T$ tends to a finite limit $A$ as $n \to \infty$; it is this limit which is conventionally called the asymptotic variance (and similarly for asymptotic mean squared error and asymptotic bias, see below). Huber (1964) argues in favour of using asymptotic measures:

Since ill effects from contamination are mainly felt for large sample sizes, it seems that one should primarily optimize large sample robustness properties.... the asymptotic variance is not only easier to handle, but . . . even for moderate values of $n$ it is a better measure of performance than the actual variance, because (i) the actual variance of an estimator depends very much on the behaviour of the tails of $H$ [$G$ in our notation].... (ii) If an estimator is asymptotically normal, then the important central part of its distribution and confidence intervals for moderate confidence levels can better be approximated in terms of the asymptotic variance than in terms of the actual variance. (Huber, 1964)

On these grounds he adopts the maximum asymptotic variance (over the family of alternative distributions in a compound $\bar{H}$) as a measure of performance. This measure has been widely used in the construction of robust estimators with optimal properties. The *minimax* robust estimator is that estimator (perhaps restricted to be of a particular type) whose maximum variance over the family of distributions in $\bar{H}$ is as small as possible. (It is possible that Huber's views on the usefulness of asymptotic measures may have become somewhat modified. See, for example, Huber, 1981, Sections 4.1, 7.4, and 7.5.)

A related prospect is to seek the *maximin* robust estimator. Here we consider the efficiencies of an estimator relative to other estimators each of which is known to perform well for individual distributions in the compound $\bar{H}$. The maximin estimator is one whose minimum efficiency relative to the individually satisfactory estimators is as large as possible. See, e.g. Huber (1964, 1981, 1983), Bickel (1965), Carroll (1979), Gastwirth (1966), Hogg (1967), Siddiqui and Raghunandanan (1967), Jaeckel (1971a, 1971b), Collins and Wiens (1985), and Martin and Zamar (1989), on use of the minimax and maximin criteria.

At the same time, finite-sample variance and efficiency measures are clearly the appropriate ones to use in many situations, such as Monte Carlo studies of robustness (cf. Andrews *et al.*, 1972), or the assessment of accommodation procedures for small samples, *per se*. (Extreme instances of the latter are studies relating to samples of size 3, such as those by Anscombe and Barron (1966), Willke (1966), Patil (1985), and Stefanski and Meredith (1986).) The choice resembles that between the finite-sample variance of an estimator $\hat{\theta}$ and the information-function reciprocal in the straightforward estimation of a parameter $\theta$. For robustness studies using finite-sample measures with sample sizes greater than 3 see, e.g. Dixon (1960), Birnbaum and Laska (1967), Birnbaum, Laska, and Meisner (1971), and Guttman (1973a). The finite-sample properties of various measures, in relation to their asymptotic values, have been investigated in detail by Gastwirth and Cohen (1970); see also Crow and Siddiqui (1967).

In the case of asymmetric contamination, variances under $\bar{H}$ are, as we have seen, replaced by mean squared errors. Now the result (3.1.6) illustrates that, while the variance of an estimator tends to zero as $n^{-1}$ when $n \to \infty$, its bias may be independent of $n$, or at any rate may tend to a non-zero limit. On this basis, comparisons between asymptotic mean squared errors of estimators would be meaningless. To meet this situation we might modify the contamination model (3.1.1) as follows. Instead of taking the amount of contamination to be a basic parameter, $\lambda$, we assume it to depend on sample size according to the relation

$$\lambda = \lambda(n) = \lambda_1 n^{-1/2}. \tag{3.1.10}$$

The alternative model is now

$$\bar{H}:x_j \in (1 - \lambda_1 n^{-1/2})F + \lambda_1 n^{-1/2}G \qquad (j = 1, 2, \ldots, n), \tag{3.1.11}$$

descriptive of a situation in which 'the amount of asymmetric contamination is large enough to affect the performance of the estimator, but is too small to be measured accurately at the given sample size' (Jaeckel, 1971a). In (3.1.4), (3.1.5), (3.1.6), the estimator $\bar{x}$ would now have bias $\lambda_1 a n^{-1/2}$, variance

$$\frac{\sigma^2}{n}[1 + 0(n^{-1/2})],$$

and mean squared error

$$\frac{\sigma^2 + \lambda_1^2 a^2}{n}[1 + 0(n^{-1/2})]$$

leading naturally to the definition of an *asymptotic bias* $\lambda_1 a$ and an *asymptotic mean squared error* $\sigma^2 + \lambda_1^2 a^2$.

### 3.1.2 Distributional properties of contaminated distributions

In investigating the performance of estimators, measures such as (3.1.3) require the calculation of various distributional properties under contamination. We have given examples of simple calculations under mixture alternatives, e.g. in (3.1.6) and (3.1.8). More often the calculation of the required properties, other than by simulation, is difficult or intractable, particularly under slippage alternatives. In recent years, however, a substantial body of work has been published on the distributional behaviour of order statistics and related quantities derived from contaminated samples. These results provide a basis for the theoretical evaluation of performance of estimation and testing procedures in the presence of outliers.

We refer first to a brief but fundamental early paper by Vaughan and Venables (1972). Suppose, very generally, that $n$ independent observations are taken, one from each of $n$ different continuous distributions $F_1, F_2, \ldots F_n$; and that

$$x_{(1)} < x_{(2)} < \ldots < x_{(n)}$$

are the ordered values in this sample of $n$, which we can regard as observed values of random variables $X_{(1)} < X_{(2)} \ldots < X_{(n)}$, respectively. If $r_1, r_2, \ldots, r_p$ is any chosen subset of size $p$ of the integers $1, 2, \ldots, n$, where $p$ can have any value from 1 to $n$, Vaughan and Venables (1972) give a general formula for the joint distribution of $X_{(r_1)}, X_{(r_2)}, \ldots, X_{(r_p)}$.

Particularizing, for purposes of outlier analysis, to the case

$$F_1 = F_2 = \ldots = F_{n-1} = F, \quad F_n = G \neq F, \tag{3.1.12}$$

we have (for appropriate forms of $G$) a contamination model for a slippage alternative in which $n - 1$ observations arise from a distribution $F$ and one from a distribution $G$. Basic formulae for the distribution of one ordered value $X_{(r)}$ in the contaminated sample and for the joint distribution of two ordered values $X_{(r)}, X_{(s)}$ have been given as follows by Shu (1978) and David and Shu (1978) (see also Arnold and Balakrishnan (1989), pp. 109–110):

$h_r(x) =$ probability density function of $X_{(r)}$ $(1 \leq r \leq n)$

$$= \binom{n-1}{r-1} [F(x)]^{r-2} [1 - F(x)]^{n-r-1} \{(r - 1)(1 - F(x))G(x)f(x)$$

$$+ F(x)(1 - F(x))g(x) + (n - r)F(x)(1 - G(x))f(x)\} \tag{3.1.13}$$

where the first term is omitted if $r = 1$ and the last term if $r = n$; here $f(x) = F'(x)$, $g(x) = G'(x)$ are the probability density functions of the distributions $F$ and $G$.

$h_{r,s}(x, y) =$ joint probability density function of $X_{(r)}, X_{(s)}$ $(1 \leq r < s \leq n)$

$$= \frac{(n - 1)!}{(r - 1)!(s - r - 1)!(n - s)!} [F(x)]^{r-2} [F(y) - F(x)]^{s-r-2} [1 - F(y)]^{n-s-1}$$

$$\times \{(r - 1)(F(y) - F(x))(1 - F(y))G(x)f(x)f(y)$$

$$+ F(x)(F(y) - F(x))(1 - F(y))(g(x)f(y) + f(x)g(y))$$

$$+ (s - r - 1)F(x)(1 - F(y))(G(y) - G(x))f(x)f(y)$$

$$+ (n - s)F(x)(F(y) - F(x))(1 - G(y))f(x)f(y)\} \tag{3.1.14}$$

where the first term is omitted if $r = 1$, the third term if $s = r + 1$, and the fourth and last term if $s = n$.

From these distributional formulae a large body of results has been produced on the moments and other properties of order statistics in such contaminated samples. Many of the results are due to Balakrishnan; important contributions have also been made by David, Joshi, Malik, Tiku, and others. For references to the extensive literature and authors, see the bibliographies in Arnold and Balakrishnan (1989, in particular chapter 5) and Balakrishnan (1992b).

These results have an evident application to robustness studies of estimation and testing procedures where the alternative hypothesis is of contamination by slippage of a single observation. Examples include studies by David and Shu (1978) and Arnold and Balakrishnan (1989) for normal samples, by Balakrishnan and Ambagaspitiya (1988) for double exponential samples, and by Balakrishnan (1992a) and Balakrishnan et al. (1992) for logistic samples. For exponential samples, where performance can be calculated without recourse to the results based on (3.1.13) and (3.1.14), there are a number of earlier studies by Kale and Sinha (1971), Veale and Kale (1972), Joshi (1972b), Kale (1975c), Veale (1975), Chikkagoudar and Kunchur (1980), and others. These robustness studies for various forms of the initial probability model (normal, logistic, etc.) will be reviewed in Chapter 5.

The above approach can be generalized to the multiple outlier situation. Consider the case of contamination by slippage of $k$ ($> 1$) observations. Replacing (3.1.12) by

$$F_1 = F_2 = \ldots = F_{n-k} = F, \ F_{n-k+1} = \ldots = F_{n-1} = F_n = G, \quad (3.1.15)$$

the general formula of Vaughan and Venables (1972) can be applied. This has recently been carried out by Balakrishnan (1992b), who gives a substantial body of results on the moments of order statistics in a sample with $k$ contaminants. He indicates that these results can in principle be applied to robustness studies in the multiple outlier situation, but refers to

the computational difficulties faced in the computation of moments of order statistics arising from a sample with more than one outlier [i.e. contaminant].

At the time of writing, we are not aware of any published application of his results to studies of robustness of accommodation procedures in the presence of multiple outliers. There is much work waiting to be done in this important area.

### 3.1.3 The qualitative approach: influence curves

So far in our discussion of performance criteria, we have confined ourselves to the question 'How does contamination affect the *precision* (and maybe the bias) of an estimator?' This prompted the various dispersion-based and efficiency-based criteria. But precision is only one aspect. In what way, one would like to know, does contamination *influence* a given estimator? For example, is the effect on the estimator proportional to the number of contaminants present? Supposing there is just one contaminant, how is its effect related to its magnitude? What is the worst possible effect that a single contaminant can have? In particular, is this effect bounded or not? As we show below, a contaminant in a sample of $n$ will, if large enough, shift the sample *mean* $\bar{x}$ beyond any bound; but two contaminants added to a sample of odd size $n = 2m - 1$ can at most shift the sample *median* from $x_{(m)}$ to $x_{(m-1)}$ or $x_{(m+1)}$, however far out these two contaminants may be. Aspects such as these underlie a powerful array of tools, which we will now describe, based on the *influence function* or *influence curve*. The approach is due to Hampel (1968, 1971); for a stimulating and highly readable exposition, see Hampel (1974). See also Section 9.2 for discussion of Bayesian aspects of the influence of a single large observation.

As usual, suppose we have a basic model $F$ and a random contamination model $(1 - \lambda)F + \lambda G$. If the contamination fraction $\lambda$ is small enough, the number of contaminants $R$ in a sample will effectively be either 0 or 1, so that for marginal comparisons of the performance of estimators in the neighbourhood of $\lambda = 0$ we need only consider the case of a single contaminant. Given $n$ 'good' (basic-model) observations $x_1, \ldots, x_n$ and an estimator $T(x_1, \ldots, x_n)$, we wish in principle to examine the effect on $T$ of substituting a contaminant for one of the $n$ observations. Denote the contaminant by $x_c$. The effect as defined would require averaging with respect to two random elements, first the random variation in $x_c$ as sampled from $G$, and second the random variation in the good value, $x_i$ say, which has been replaced by $x_c$. To sidestep these sources of variation we reformulate the problem more conveniently as follows. The contaminating distribution $G$ we will take to be atomic at $\xi$; that is to say, the contaminant $x_c$ has a fixed value $\xi$. And then we will work in terms of the effect on $T$ of *adding* the contaminant $\xi$ to the $n$ good observations, so that on the alternative model $T$ is based on an enlarged sample of size $n + 1$.

Suppose, for example, that $T(x_1, \ldots, x_n) = T$ is the sample mean $\bar{x}$. Write $T(x_1, \ldots, x_n, \xi) = T_c$ for the mean $\bar{x}_c$ based on the enlarged contaminated sample. Then the effect of adjoining $\xi$ is to change the value of the estimator by an amount

$$\bar{x}_c - \bar{x} = \frac{n\bar{x} + \xi}{n + 1} - \bar{x} = \frac{\xi - \bar{x}}{n + 1}. \qquad (3.1.16)$$

Naturally enough this is proportional to $1/(n + 1)$, that is to the amount of contamination in the sample; the effect standardized for the amount of contamination is

$$(n + 1)(\bar{x}_c - \bar{x}) = \xi - \bar{x}. \qquad (3.1.17)$$

This will, as we remarked above, exceed any bound for $\xi$ large enough. It is a linear function of the value of the contaminant.

Again, if $T$ is the sample variance $s^2$ for a distribution with unknown mean and variance, we have for the enlarged sample

$$ns_c^2 = (n - 1)s^2 + n\bar{x}^2 + \xi^2 - (n\bar{x} + \xi)^2/(n + 1),$$

giving

$$s_c^2 - s^2 = [(\xi - \bar{x})^2/(n + 1)] - (s^2/n).$$

The standardized effect is therefore

$$(n + 1)(s_c^2 - s^2) = (\xi - \bar{x})^2 - \frac{n + 1}{n} s^2. \tag{3.1.18}$$

The effect again exceeds any bound for $\xi$. large enough, but this time is a quadratic function of $\xi$.

Effects per unit of contamination, such as (3.1.17) and (3.1.18), are called *finite-sample influence functions* (or sometimes finite-sample influence curves). A finite-sample influence function depends on the argument $\xi$, on the estimator $T$, and in general (see, for example, the case of the sample median discussed below) on the basic distribution $F$; it may also depend explicitly, as (3.1.18) illustrates, on the sample size $n$. Accordingly we write it as $IC_{T, F; n}(\xi)$.

Equation (3.1.18) also suggests that, as with the variance and efficiency measures discussed earlier, we may wish to use the asymptotic equivalent

$$\lim_{n \to \infty} IC_{T, F; n}(\xi) = IC_{T, F}(\xi),$$

say. In fact, it is this asymptotic influence function, or simply the *influence function*, $IC_{T, F}(\xi)$, which is the really useful tool.

What is the influence function for $s^2$: $IC_{s^2, F}(\xi)$? If we let $n \to \infty$ in (3.1.18), we must not only replace $(n + 1)/n$ by 1, but also $\bar{x}$ and $s^2$ by $\mu$ and $\sigma^2$ respectively:

$$IC_{s^2, F}(\xi) = (\xi - \mu)^2 - \sigma^2. \tag{3.1.19}$$

This equation conveys the same information as the finite-sample version (3.1.18) regarding the unbounded quadratic influence of a contaminant on $s^2$. In deriving it, each estimator $T(x_1, \ldots, x_n)$ (e.g. $\bar{x}$ or $s^2$) on the right-hand side of (3.1.18) has been replaced by $\lim_{n \to \infty} T(x_1, \ldots, x_n)$ (e.g. $\mu$ or $\sigma^2$); this limiting form depends only on $F$ and we will write it $T(F)$. For example, if $T$ is the sample mean $\bar{x}$, $T(F) = \int x dF$; if $T$ is the sample variance $s^2$, $T(F) = \int (x - \mu)^2 dF$ where $\mu = \int x dF$.

Encompassing our procedure in a formal definition, we say that the *influence function of an estimator $T(x_1, \ldots, x_n)$ at the basic distribution $F$ is*

$$IC_{T,F}(\xi) = \lim_{\lambda \to 0}\{[T((1 - \lambda)F + \lambda G) - T(F)]/\lambda\} \qquad (3.1.20)$$

where $G$ is the atomic distribution

$$P(X = \xi) = 1. \qquad (3.1.21)$$

We may also write (3.1.20) as

$$IC_{T,F}(\xi) = \frac{\partial}{\partial \lambda}\{T[(1 - \lambda)F + \lambda G]\}|_{\lambda = 0} \qquad (3.1.22)$$

*Example 3.1 Sample mean. We have*

$$IC_{\bar{x},F}(\xi) = \lim_{\lambda \to 0}\{[(1 - \lambda)\mu + \lambda\xi - \mu]/\lambda\} = \xi - \mu, \qquad (3.1.23)$$

*which could also have been obtained by letting $n \to \infty$ in (3.1.17).*

*Example 3.2 Sample variance. Here*

$$IC_{s^2,F}(\xi) = \lim_{\lambda \to 0}\{[(1 - \lambda)(\mu^2 + \sigma^2) + \lambda\xi^2 - ((1 - \lambda)\mu + \lambda\xi)^2 - \sigma^2]/\lambda\}$$

$$= (\xi - \mu)^2 - \sigma^2$$

*as in* (3.1.19).

*Example 3.3 Sample median. It is not practicable to calculate the influence function of the sample median $\tilde{x}$ on a finite-sample basis, since the shift in sample median on moving from an odd to an even number of observations, or vice versa, is not defined. On an asymptotic basis, however, the calculation is straightforward. $T(F)$ is now the population median $m$, defined by*

$$\int_{-\infty}^{m} dF = \tfrac{1}{2}.$$

*$T((1 - \lambda)F + \lambda G)$, the median of the mixture distribution, is equal to $m + \Delta$, say, where $\Delta$ is positive or negative according as $\xi$ is greater or less than $m$. We assume $F$ to be continuous, with density $f$. To the first order of small quantities we have for $\Delta < 0$,*

$$\tfrac{1}{2} = (1 - \lambda)F(m + \Delta) + \lambda = (1 - \lambda)[\tfrac{1}{2} + \Delta f(m)] + \lambda = \tfrac{1}{2} + \tfrac{1}{2}\lambda + \Delta f(m),$$

*giving $\Delta = -\lambda/[2f(m)]$; similarly, for $\Delta > 0$, $\Delta = +\lambda/[2f(m)]$. Hence the influence function, $\lim_{\lambda \to 0}(\Delta/\lambda)$, is*

$$IC_{\tilde{x},F}(\xi) = \frac{\operatorname{sgn}(\xi - m)}{2f(m)}. \qquad (3.1.24)$$

*The influence of a contaminant on the sample median is thus seen to be bounded—an essential qualitative difference from, say, the sample mean!*

A readily calculated finite-sample representation of the influence function is the sensitivity function or *sensitivity curve* introduced by Tukey (1977). For this, the contaminant $\xi$ is added, not to a random sample $x_1, \ldots, x_n$ from $F$ as for the finite-sample influence function, but to a pseudo-sample $c_1, \ldots, c_n$ consisting of the order scores $c_j = E(X_{(j)})$ for a sample of size $n$ from $F$. This constructed sample may be thought of as smoothly representing the basic distribution. The sensitivity curve is given (as with the finite-sample $IC$) by $(n + 1)[T(c_1, \ldots, c_n, \xi) - T(c_1, \ldots, c_n)]$ regarded as a function of $\xi$. Other convenient order statistics could of course be used in place of order scores; for example, conditional centroids or medians. All of these have been extensively tabulated in the normal case (David *et al.*, 1968).

The sensitivity curve generalizes usefully to multiple outlier situations. In particular, *sensitivity contours* can readily be drawn to represent the influence of a pair of contaminants; see, for example, Prescott (1976b, 1978) and Kimber and Stevens (1981).

Reverting to the influence function, we note an important property. If we regard the argument $\xi$ as a random quantity distributed according to the basic model $F$, it can be shown (see, for example, Huber, 1981, p. 14) that the expectation of the influence function with respect to this variation in $\xi$ is zero:

$$\int IC_{T, F}(\xi)\,\mathrm{d}F(\xi) = 0, \qquad (3.1.25)$$

and that the mean squared value of the influence function,

$$\int [IC_{T, F}(\xi)]^2\,\mathrm{d}F(\xi),$$

is equal to the *asymptotic variance* of $T$. Thus we have a direct connection between the influence function and our earlier dispersion-based performance criteria.

We now describe some further parameters of the influence function which throw light on the robustness of an estimator.

*The gross-error sensitivity*

This is the supremum of the absolute value of the influence function,

$$\gamma_{T, F} = \sup_{\xi} | IC_{T, F}(\xi) | \qquad (3.1.26)$$

The gross-error sensitivity 'measures the worst approximate influence which a fixed amount of contamination can have on the value of the estimator (hence it may be regarded as an approximate bound for the bias of the estimator)' (Hampel, 1974).

*Example 3.4* Suppose $F$ is $N(\mu, \sigma^2)$; for $\bar{x}$ and for $s^2$, $\gamma_{T, F} = \infty$ (so that the effect that a contaminant can have on these estimators is unbounded), while for $\tilde{x}$ (the median)

$$\gamma_{T, F} = 1/[2f(\mu)], \text{ from (3.1.24); that is } \gamma_{T, F} = \sigma \sqrt{(2\pi)}/2 = 1.25\sigma.$$

*The local-shift sensitivity*

This is defined by

$$\beta_{T, F} = \sup_{\xi \neq \eta} |IC_{T, F}(\xi) - IC_{T, F}(\eta)| / |\xi - \eta|. \tag{3.1.27}$$

It measures the worst possible effect of 'adjusting' a contaminant by modifying its value, for example by Winsorizing (see Section 3.2.1).

*The rejection point*

Suppose that the influence function vanishes for all points $\xi$ outside some finite interval

$$|\xi - \mu| \leq \rho,$$

say, centred on $\mu$, the mean (or other appropriate location point) of $F$. This implies that observations outside $[\mu - \rho, \mu + \rho]$ have no influence on the estimator $T$—i.e. that the estimation procedure *rejects* such observations. $\rho = \rho_{T, F}$ is called the *rejection point* of the estimator. Examples of estimators with finite $\rho_{T, F}$ (and which therefore reject outliers beyond some particular distance) will be encountered in Chapter 5 (see, e.g., (5.1.1), (5.1.3), (5.1.19), (5.5.16)).

In some cases $IC_{T, F}$ may be very small, though not zero, for $|\xi - \mu|$ sufficiently large. The rejection point is then infinite, but outliers, though not explicitly rejected, have very little effect on the estimator.

*The breakdown aspect of performance*

We noted above that, in contrast to the sample mean, the influence of a contaminant on the sample median $\tilde{x}$ is bounded, and indeed that *two* contaminants, whatever their magnitudes, added to a sample of size $n = 2m - 1$ can at most shift the sample median from $x_{(m)}$ to $x_{(m-1)}$ or $x_{(m+1)}$. Why stop at two? Clearly the data can absorb a greater amount of contamination than this without the sample median becoming totally unreliable. With four contaminants added, the shift from $x_{(m)}$ is bounded (at most to $x_{(m-2)}$ or $x_{(m+2)}$); with $2m - 2$ contaminants added, it is still bounded (at most to $x_{(1)}$ or $x_{(2m-1)}$). But as soon as the number of added contaminants exceeds $2m - 2$, it is possible for $\tilde{x}$ to take any value whatsoever. From this point of view, the contamination becomes intolerable when its proportionate amount reaches $(2m - 1)/(4m - 2)$, i.e. one-half. We say that the sample median has *breakdown point* $\frac{1}{2}$. The breakdown point, $\pi_{T, F}$, for any estimator $T$ is the smallest proportion of contamination which can carry the value of the estimator over all bounds. It is an important measure of robustness; the idea is due to Hodges (1967) and Hampel (1968, 1971). A closely related measure called *test resistance* has been proposed by Ylvisaker (1977); see also Lambert (1982), Rieder (1982). For an illuminating discussion of the notion of breakdown point, see Donoho and Huber (1983).

We have now discussed in some detail what Huber calls the *stability* aspect of robustness,

in close analogy to the stability of a mechanical structure (say of a bridge): (i) the qualitative aspect: a small perturbation should have small effects; (ii) the breakdown aspect: how big can the perturbation be before everything breaks down; (iii) the infinitesimal aspect: the effects of infinitesimal perturbations. (Huber, 1972)

### 3.1.4 Robustness of confidence intervals

Suppose we have a sample $x_1, \ldots, x_n$ which comes, on some basic model $H$, from a distribution involving a location parameter $\mu$ and a scale parameter $\sigma$, both unknown; typically this distribution might be $N(\mu, \sigma^2)$. Consider the problem of constructing a confidence interval for $\mu$ at level $1 - \alpha$. Essentially this is built up from the following elements:

(i) an estimator $T$ of $\mu$;
(ii) an estimator $S_T$ of the standard deviation of $T$;
(iii) the distribution $\mathbf{D}$ of $(T - \mu)/S_T$, assuming the basic model.

If $u_1, u_2$ are lower and upper $\frac{1}{2}\alpha$-points of $\mathbf{D}$, the confidence interval is then

$$(T - u_2 S_T, T - u_1 S_T) \tag{3.1.28}$$

If the parent distribution is $N(\mu, \sigma^2)$, $T$, $S_T$ and $\mathbf{D}$ particularize to $\bar{x}$, $s/\sqrt{n}$ and the $t_{n-1}$ distribution in the classical procedure, and we get the familiar confidence interval

$$\left(\bar{x} - t_{n-1}(\alpha/2)\frac{s}{\sqrt{n}}, \bar{x} + t_{n-1}(\alpha/2)\frac{s}{\sqrt{n}}\right) \tag{3.1.29}$$

where $t_{n-1}(\alpha/2)$ is the upper $(\alpha/2)$-point of $t_{n-1}$.

If in fact the observations $x_1, \ldots, x_n$ come, not from $N(\mu, \sigma^2)$ as assumed, but from a contaminated distribution, the confidence interval (3.1.29) may be defective for two reasons. First, the distribution $\mathbf{D}$ of $(T - \mu)/S_T = (\bar{x} - \mu)/(s/\sqrt{n})$ may differ substantially from that of $t_{n-1}$, and the probability that the interval (3.1.29) covers the true value $\mu$ may thus differ substantially from $1 - \alpha$ (and may—a particularly undesirable occurrence—be substantially *less* than $1 - \alpha$). That is, the procedure may lack *robustness of validity*. Secondly, since $\bar{x}$ and s are sensitive to the presence of outliers in the sample, the confidence interval may be unnecessarily wide. It could well be preferable to use more robust estimators $T$ and $S_T$ in the construction (3.1.28), aiming at achieving satisfactory validity, and at the same time obtaining confidence intervals which tend to be shorter in the presence of outliers. From this point of view the procedure leading to (3.1.29) may lack *robustness of efficiency*, or 'robustness of performance' in Huber's words quoted at the beginning of this chapter. See Tukey and McLaughlin (1963), Dixon and Tukey (1968), Huber (1968, 1970).

The choice of $T$ and $S_T$ will be discussed in Chapter 5.

The following measures of performance of a confidence interval such as (3.1.28) have been proposed:

(i) The probability that the confidence interval fails to cover the true parameter value $\mu$ under a specified alternative model $\bar{H}$. This reflects the robustness of validity of the procedure. Takeuchi (1971) estimates it by the relative frequency of non-coverage of the parameter by the interval in a large number of simulations; he calls this the *error frequency* of the interval. Analogously, the measure itself may be called the *error probability*.

(ii) Suppose we have a compound alternative $\bar{H}$; a confidence interval $(T - a, T + a)$ of *given* length $2a$ will operate at different confidence levels for the different distributions that can arise under $\bar{H}$. For specified $a$, the minimum of these possible confidence levels gives a measure of 'guaranteed' performance. The idea is due to Huber (1968).

(iii) If the interval (3.1.28) has robustness of validity, a natural measure of its efficiency (for specified $\alpha$) is the ratio of the lengths of the intervals (3.1.29), (3.1.28); or, from another point of view, the ratio measures the relative efficiency of two procedures. Dixon and Tukey (1968, p. 86) call the *square* of this ratio the relative efficiency.

Geertsema (1987) investigates the robustness, under contamination, of a *sequential* procedure for calculating a confidence interval for the mean of a normal population. In his procedure, successive observations $x_1, x_2, x_3, \ldots$ are obtained, giving samples of successive sizes 1, 2, 3, ... with sample variances $0, s_2^2, s_3^2, \ldots$; sampling stops when $s_n^2/n \, (n > 1)$ first falls below some prescribed value.

### 3.1.5 Robustness of significance tests

This obviously bears a relation to the robustness of confidence intervals discussed above. There is an important difference, however, inasmuch as we now have a double family of alternative hypotheses. Suppose, to fix ideas, that a two-sided test of the hypothesis $\mu = \mu_0$ on the basis of an assumed normal sample is required. We can still think in terms of a basic model

$$H : x_j \in F \quad (j = 1, 2, \ldots, n) \quad \text{where } F \text{ is } N(\mu, \sigma^2), \quad (3.1.30)$$

and a contamination alternative $\bar{H}$ which might typically be

$$\bar{H} : x_j \in (1 - \lambda)F + \lambda G \quad (j = 1, 2, \ldots, n) \quad \text{where } G \text{ is } N(\mu, b\sigma^2). \quad (3.1.31)$$

$\bar{H}$ depends on the parameters $\lambda$, $b$, etc., which for present purposes we will denote by $\xi$.

The null hypothesis for the significance test is

$$H_0 : x_j \in F_0 \quad (j = 1, 2, \ldots, n) \quad \text{where } F_0 \text{ is } N(\mu_0, \sigma^2). \quad (3.1.32)$$

We are concerned with the behaviour of the test, both in relation to the usual *'Type II error' family of alternatives* (3.1.30) and in relation to the *family of alternatives*

$$\bar{H}_0 : x_j \in (1 - \lambda)F_0 + \lambda G_0 \quad (j = 1, 2, \ldots, n) \quad \text{where } G_0 \text{ is } N(\mu_0, b\sigma^2) \quad (3.1.33)$$

*expressing contamination* under the true value of $\mu$; more generally, we are concerned with the family (3.1.31), encompassing both types of departure from $H_0$.

For any choice of $T$ and $S_T$ we have, corresponding to the $(1 - \alpha)$-level confidence interval (3.1.28), a significance test at level $\alpha$ of $H_0$ against $\bar{H}_0$ with critical region

$$\mathscr{S} : (T - \mu_0)/S_T < u_1 \quad \text{or} \quad > u_2. \quad (3.1.34)$$

We can now formulate relevant measures of performance, as follows:

(i) The conventional power of the test, as a function of $\mu$; this is

$$\Pi_1(\mu) = P[(T - \mu_0)/S_T \in \mathscr{S} | H]. \quad (3.1.35)$$

(ii) The stability of the significance level under contamination, as a function of $\xi$; this is given by

$$\Pi_2(\xi) = \Pi_2(\xi, \alpha) = P[(T - \mu_0)/S_T \in \mathscr{S} | \bar{H}_0]. \quad (3.1.36)$$

It corresponds to the error probability of the equivalent confidence interval, as defined above.

(iii) For a compound alternative $\bar{H}$, the guaranteed significance level

$$\Pi_3 = \Pi_3(\alpha) = \min_{\xi} \Pi_2(\xi, \alpha). \quad (3.1.37)$$

This again corresponds to the guaranteed performance measure under contamination defined above for a confidence interval.

(iv) The stability of the power under contamination, as a function of both $\mu$, the argument of the power function, and $\xi$, the vector measure of contamination:

$$\Pi_4(\mu, \xi) = P[(T - \mu_0)/S_T \in \mathscr{S} \,|\, \bar{H}]. \tag{3.1.38}$$

(v) For a compound alternative $\bar{H}$, the guaranteed power at $\mu$:

$$\Pi_5(\mu) = \min_{\xi} \Pi_4(\mu, \xi). \tag{3.1.39}$$

These concepts apply to significance tests generally, though our discussion has been in the context of tests for the mean of a normal distribution. Veale and Kale (1972), for example, consider the testing of hypotheses for the parameter $\sigma$ of an exponential distribution with density $\sigma^{-1}\exp(-\sigma^{-1}x)$, and they develop a test (described in Chapter 5) robust against a contaminant arising from an exchangeable alternative model with contamination parameter $b$. Three measures of performance are tabulated, $p_m$, $p_t$, and $p_d$, each involving a comparison of the test with the test of the same size based on the sample sum, which is optimal under the basic model. In the notation of (3.1.35), (3.1.36), and (3.1.38), these measures can be written as

$$p_m = \check{\Pi}_1(\sigma) - \Pi_1(\sigma), \tag{3.1.40}$$

$$p_t = \check{\Pi}_2(b) - \Pi_2(b), \tag{3.1.41}$$

$$p_d = \check{\Pi}_4(\sigma, b) - \Pi_4(\sigma, b), \tag{3.1.42}$$

where $\Pi_1$, $\Pi_2$, $\Pi_4$ relate to the robust test and $\check{\Pi}_1$, $\check{\Pi}_2$, $\check{\Pi}_4$ to the optimal test. Interestingly, Veale and Kale call $p_m$ the *premium* and $p_t$ the *protection* involved in using the robust test, providing a natural extension of Anscombe's concepts of premium and protection in estimation discussed earlier; see (3.1.2) and (3.1.3). See also Iman and Conover (1977), Tiku (1980), Tiku and Singh (1981), Lambert (1982), and Wang (1981).

The qualitative approach to the performance of estimators, presented in Section 3.1.3, can also be applied to the performance of significance tests. Proceeding from the notion of the breakdown point of an estimator (page 000 above), He, Simpson, and Portnoy (1990) introduce the concepts of the *power* and *level breakdown functions of a test statistic*. The *power* breakdown function gives the amount of contamination of each alternative distribution that can carry the test statistic to a null value. The *level* breakdown function gives the amount of contamination of a null distribution that can carry the test statistic to each value in the alternative space.

We can illustrate these in terms of the particular test and models given in (3.1.30)–(3.1.34) above. The power breakdown function is the smallest proportion of contamination $\lambda$, as a function of $\mu$, needed under $\bar{H}$ to carry $(T - \mu_0)/S_T$ into the acceptance region (the complement of $\mathcal{S}$). The level breakdown function is the smallest proportion of contamination $\lambda$, as a function of $u$, needed under $\bar{H}_0$ to carry $(T - \mu_0)/S_T$ to the value $u$.

See also Donoho and Liu (1988), Hampel et al. (1986), and Simpson (1989).

## 3.2 METHODS OF ACCOMMODATION—GENERAL AND OUTLIER-SPECIFIC

We will now survey some of the general methods that exist for constructing robust estimators, tests, or confidence intervals. We shall also illustrate in this section how these methods can be made more specific to outliers, and so become genuine accommodation procedures. However, the detailed discussion of outlier-specific methods and techniques will form the subject-matter of Chapter 5.

### 3.2.1 Estimation of location

Suppose we are estimating a location parameter $\mu$. We consider here some standard methods which provide implicitly some protection against outliers, whilst not specifically constructed for this purpose. Since outliers manifest themselves as extreme observations this has the effect of protecting against their presence, even though the prime object may be to render less dramatic the effect of the tail behaviour of the generating distribution on the estimation of $\mu$.

We start with two familiar, simple, and intuitively appealing procedures, namely *trimming* and *Winsorizing*; in view of their key role we discuss them in some detail. The object is to control the variability due to the $r$ lowest sample values $x_{(1)}, \ldots, x_{(r)}$ and the $s$ highest ones $x_{(n-s+1)}, \ldots, x_{(n)}$. The choice of $r$ and $s$ is discussed later; for the moment we suppose they are pre-chosen parameters. If these $r + s$ observations are omitted, so that we confine ourselves to a censored sample of size $n - r - s$, we get the $(r, s)$-*fold trimmed mean*

$$\overset{\text{T}}{x}_{r,s} = (x_{(r+1)} + \cdots + x_{(n-s)})/(n - r - s). \tag{3.2.1}$$

If on the other hand the $r$ lowest sample values are each replaced by the value of the nearest observation to be retained unchanged, viz. $x_{(r+1)}$, and likewise the $s$ highest by $x_{(n-s)}$, so that we work with a transformed sample of size $n$, we get the $(r, s)$-*fold Winsorized mean*

$$\overset{\text{W}}{x}_{r,s} = (r x_{(r+1)} + x_{(r+1)} + \cdots + x_{(n-s)} + s x_{(n-s)})/n. \tag{3.2.2}$$

Often the amounts of lower-tail and upper-tail trimming or Winsorizing are the same, i.e. $r = s$, and we have the $r$-*fold symmetrically trimmed and Winsorized means*

$$\overset{\text{T}}{x}_{r, r} = (x_{(r + 1)} + \cdots + x_{(n - r)})/(n - 2r), \tag{3.2.3}$$

$$\overset{\text{W}}{x}_{r, r} = (rx_{(r + 1)} + x_{(r + 1)} + \cdots + x_{(n - r)} + rx_{(n - r)})/n. \tag{3.2.4}$$

However, when the basic distribution is asymmetrical it is natural to trim or Winsorize asymmetrically ($r \neq s$); for an application to exponential samples, see Kimber (1983a). See also Hertsgaard (1979).

Equivalently to (3.2.3) we can speak of the $\alpha$-*trimmed mean* $\overset{\text{T}}{x}(\alpha, \alpha)$, viz. the $r$-fold symmetrically trimmed mean in which the amount of trimming is, for convenience, specified by the proportion $2\alpha$ of the sample omitted rather than the number $2r$ of observations. With an $\alpha$-trimming procedure in which $\alpha$ has been specified beforehand, the number $\alpha n$ of observations supposed to be trimmed at each end may not be an integer; suppose its integer part is $r$, so that $\alpha n = r + f (0 < f < 1)$. We then omit $r$ observations at each end, and include the nearest retained observations, $x_{(r + 1)}$ and $x_{(n - r)}$, each with reduced weight $1 - f$:

$$\overset{\text{T}}{x}(\alpha, \alpha) = [(1 - f)x_{(r + 1)} + x_{(r + 2)} + \cdots + x_{(n - r - 1)} + (1 - f)x_{(n - r)}]/n(1 - 2\alpha).$$

$$\tag{3.2.5}$$

Similarly we can define $\alpha$-*Winsorized means* $\overset{\text{W}}{x}(\alpha, \alpha)$; there is now no need for any fractional weighting, since the number of lower-tail observations Winsorized into $x_{(r + 1)}$ is $r + f + 1 - f - 1 = r$. Thus

$$\overset{\text{W}}{x}(\alpha, \alpha) = (rx_{(r + 1)} + x_{(r + 1)} + \cdots + x_{(n - r)} + rx_{(n - r)})/n. \tag{3.2.6}$$

Clearly the 0-trimmed and 0-Winsorized means are both the same as the sample mean $\bar{x}$, and the $\frac{1}{2}$-Winsorized mean is the same as the sample median $\tilde{x}$; the $\frac{1}{2}$-trimmed mean, by a suitable limiting argument, can also be taken to be $\tilde{x}$. The $\frac{1}{4}$-trimmed mean, $\overset{\text{T}}{x}(\frac{1}{4}, \frac{1}{4})$, is called the *mid-mean*.

What is the influence function, $IC_{T, F}(\xi)$, for the $\alpha$-trimmed mean? If $F$ is continuous with density $f$ and mean $\mu$, and $0 \leqslant \alpha < \frac{1}{2}$, we have, in the notation of Section 3.1.3,

$$T(F) = \frac{1}{(1 - 2\alpha)} \int\limits_{x_\alpha}^{x_{1 - \alpha}} x \, dF \tag{3.2.7}$$

where $x_\alpha$ denotes the $\alpha$-quantile of $F : F(x_\alpha) = \alpha$. Hence

$$T[(1 - \lambda)F + \lambda G] = \left(\frac{1 - \lambda}{1 - 2\alpha}\right) \int\limits_{y_\alpha}^{y_{1 - \alpha}} x \, dF + \left(\frac{\lambda}{1 - 2\alpha}\right) \int\limits_{y_\alpha}^{y_{1 - \alpha}} x \, dG$$

where $y_\alpha$ is determined from

$$(1 - \lambda)F(y_\alpha) + \lambda = \alpha \quad (\xi < x_\alpha), \qquad (1 - \lambda)F(y_\alpha) = \alpha \quad (\xi > x_\alpha),$$

with a similar definition for $y_{1-\alpha}$.

Using (3.1.20) it follows that, for symmetric $F$, the influence function for the $\alpha$-trimmed mean is

$$IC_{T,F}(\xi) = \begin{cases} -(x_{1-\alpha} - \mu)/(1 - 2\alpha) & \text{for } \xi < x_\alpha \\ (\xi - \mu)/(1 - 2\alpha) & \text{for } x_\alpha \leq \xi \leq x_{1-\alpha} \\ (x_{1-\alpha} - \mu)/(1 - 2\alpha) & \text{for } \xi > x_{1-\alpha} \end{cases} \tag{3.2.8}$$

This is illustrated in Figure 3.1. It shows that

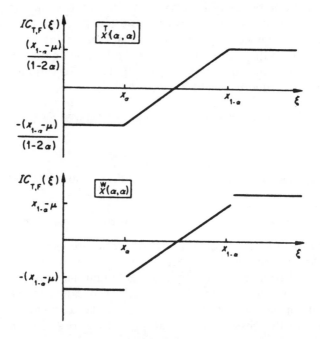

**Figure 3.1**   Influence functions for $\alpha$-trimmed and $\alpha$-Winsorized means

the 'influence' of an extreme outlier on the value of a trimmed mean is *not* zero, as one would naively expect (arguing that the outlier will be 'thrown out'); rather it is equal to the influence of an additional $x$ at $[x_\alpha$ resp. $x_{1-\alpha}] \ldots$ the $\alpha$-trimmed mean does not really 'throw out' outliers, in the sense of ignoring them completely, but in effect 'brings them in' towards the bulk of the sample. But what about the $\alpha$-Winsorized mean which had been designed specifically to 'bring in' outliers? (Hampel, 1974)

To answer this we have, for the $\alpha$-Winsorized mean $(0 \leq \alpha < \tfrac{1}{2})$,

$$T(F) = \alpha x_\alpha + \int\limits_{x_\alpha}^{x_{1-\alpha}} x\, dF + \alpha x_{1-\alpha}. \qquad (3.2.9)$$

Again assuming symmetric $F$, the influence function now comes out to have the following form, illustrated in Figure 3.1:

$$IC_{T,F}(\xi) = \begin{cases} -[(x_{1-\alpha} - \mu) + \alpha/f(x_\alpha)] & \text{for } \xi < x_\alpha \\ \xi - \mu & \text{for } x_\alpha \leqslant \xi \leqslant x_{1-\alpha} \\ +[(x_{1-\alpha} - \mu) + \alpha/f(x_\alpha)] & \text{for } \xi > x_{1-\alpha}. \end{cases} \qquad (3.2.10)$$

The IC is indeed bounded, the outliers 'brought in', but there is a jump at $[x_\alpha$ and $x_{1-\alpha}]$. . . . Furthermore both slope in the center and supremum differ from that of the $\alpha$-trimmed mean. . . . the . . . point is that the mass of the tails is put on single order statistics resp. single points in the limit, and shifting them . . . causes appreciable fluctuations of the Winsorized mean which are determined solely by the density in (and near) these points. A contamination in the central part, on the other hand, has the same influence as on the arithmetic mean, while the trimmed mean spreads the influence of outliers evenly over the central part, thus giving it a higher weight.

Thus . . . both the trimmed mean and the Winsorized mean restrict the influence of outliers, but in different ways. While the IC of the former is always continuous, the IC of the latter is discontinuous and very sensitive to the local behaviour of the true [sic] underlying distribution at two of its quantiles. (Hampel, 1974)

For distributional properties of trimmed means from a contaminated sample see Collins (1986), and for distributional properties of trimmed and Winsorized means from an uncontaminated sample (null model) see respectively Griffin and Pruitt (1987) and Griffin (1988). Martinsek (1988) discusses the use of trimmed and Winsorized means in a *sequential* procedure for estimating a population mean.

The use of trimming and Winsorization for robust estimation extends naturally to multivariate samples; for a bivariate application to robust estimation of the correlation coefficient, see Bebbington (1978). See also Chapter 7.

We now turn to three important general methods of robust estimation.

*L-estimators*

An obvious generalization of the trimming and Winsorization approach is to use estimators having the form of linear combinations of ordered sample values

$$\tilde{\mu} = \sum c_i x_{(i)} \qquad (3.2.11)$$

where the weights $c_i$ are lower in the extremes than in the body of the data set. The 'ultimate' example of such *linear order statistics estimators* (called '*L*-estimators' by Huber, 1972) is the sample median where $c_i = 0$ for *all* but the middle, or two middle, ordered observations. In contrast with the sample mean ($c_i = 1/n : i = 1, 2, \ldots, n$) this can on occasions effect considerable improvement: for example, for the Cauchy distribution where an even more drastic policy of assigning *negative* weight to extreme observations can yield further improvement still (Barnett, 1966). Other examples are found in trimmed means, in Winsorized means, and in the use of specific combinations of a few sample quantiles (for example see Gastwirth, 1966). A crucial set of tables for the construction of *L*-estimators in the presence of outliers is presented by David, Kennedy, and Knight (1977). Properties of *L*-estimators of location in symmetric distributions are discussed by David and Ghosh (1985).

Two other general principles of robust estimation yield what Huber (1972) calls maximum likelihood type estimators (*M*-estimators) and rank test estimators (*R*-estimators). See also the detailed discussion in Huber (1981).

*M-estimators*

*M*-estimators are obtained by solving an equation of the form

$$\sum_{i=1}^{n} \psi(x_i - \tilde{\mu}) = 0 \qquad (3.2.12)$$

to obtain an estimator $\tilde{\mu}$ where $x_1, x_2, \ldots, x_n$ is a random sample and $\psi(u)$ is some weight function with desirable features. For example if $|\psi(u)|$ is small for large $|u|$, $\tilde{\mu}$ will discount extreme sample values and protect against outliers. The particular choice $\psi(u) = f'(u)/f(u)$, where the distribution has probability density function of the form $f(x - \mu)$, will of course yield as $\tilde{\mu}$ a solution of the likelihood equation.

Joiner and Hall (1983) demonstrate an illuminating connection between *L*, *M* and *R* estimators by showing that efficient *L* and *R* estimators for $\mu$ can be defined by equations similar in form to (3.2.12) in terms of the same function $\psi(u) = f'(u)/f(u)$.

*R-estimators*

Hodges and Lehmann (1963) were the first to remark that estimators of $\mu$ could be obtained from certain rank test procedures, such as the Wilcoxon test. Such *R*-estimators often prove to be robust. An example is given by Huber (1972) who considers the two-sample rank test for location shift. If the samples are $x_1, x_2, \ldots, x_n$ and $y_1, y_2, \ldots, y_n$ the test statistic is

$$W(x_1, x_2, \ldots, x_n; y_1, y_2, \ldots, y_n) = \sum_1^n J[i/(2n + 1)] V_i \qquad (3.2.13)$$

where $J[i/(2n + 1)]$ is some suitably chosen function of the empirical distribution function for the combined sample, and $V_i = 1$ if the $i$th ordered value in the combined sample is one of the $x$ values (otherwise $V_i = 0$). We can derive an estimator $\tilde{\mu}$ as the solution of

$$W(x_1 - \tilde{\mu}, x_2 - \tilde{\mu}, \ldots, x_n - \tilde{\mu}; - x_1 + \tilde{\mu}, - x_2 + \tilde{\mu}, \ldots, - x_n + \tilde{\mu}) = 0 \qquad (3.2.14)$$

and the asymptotic behaviour of $\tilde{\mu}$ is obtained from the power function of the test. For symmetric distributions $R$-estimators can, for an appropriate choice of $J(\cdot)$, be asymptotically efficient and asymptotically normally distributed.

Jurečková (1985) discusses asymptotic relations between $L$, $M$ and $R$ estimators of location for a sample from a symmetric distribution, and shows that it is possible, under given conditions, to set up estimators of the three kinds which are asymptotically equivalent. See comments above on Joiner and Hall (1983).

### 3.2.2 Estimation of dispersion

A natural approach to the robust estimation of dispersion is to use a *location* estimation procedure to find some sort of typical *deviation* value, as follows. We have a sample of $n$ observations $x_1, \ldots, x_n$ of a random variable $X$. From these we get a location estimate $m$, and hence $n$ sample deviations

$$d_1 = x_1 - m, \ldots, d_n = x_n - m.$$

Choosing some suitable function $g(d)$ we average the $n$ values $g(d_1), \ldots, g(d_n)$ to obtain a dispersion estimator

$$S = \frac{1}{n^*} \sum_{i=1}^n g(d_i) \qquad (3.2.15)$$

with an appropriate divisor $n^*$ which is not necessarily the same as $n$; if we want an estimator in the same units as $X$, we can take

$$s = g^{-1}(S).$$

For example, the sample variance and sample standard deviation use $m = \bar{x}$, $g(d) = d^2$, $n^* = n - 1$; the mean absolute deviation uses $m = \bar{x}$, $g(d) = |d|$, $n^* = n$. For a *robust* estimator of similar structure, it is natural to choose for $m$ a robust estimate of location of $X$, such as those discussed in Section 3.2.1.

A further possibility is to calculate the estimator, not from the entire sample of $n$ values $x_1, \ldots, x_n$, but from a suitably trimmed or Winsorized sample. An early example of such an estimator is the $(r, r)$ *Winsorized variance*

$$\overset{W}{s}{}^2_{r,r} = \frac{1}{n-1}\left\{ r(x_{(r+1)} - m)^2 + \sum_{j=r+1}^{n-r}(x_{(j)} - m)^2 + r(x_{(n-r)} - m)^2\right\},$$

where $m$ is the $(r, r)$ Winsorized mean $\overset{W}{x}_{r,r}$; see Dixon and Tukey (1968).

In an extension of this approach, Simonoff (1987a) uses a reduced sample obtained from the entire sample of $n$ values, not by trimming or Winsorization, but by the application of discordancy tests (pp. 37–38 and Chapter 4) to identify suitable values for removal. His robust estimator of dispersion is the standard deviation of the reduced sample. This so-called 'rejection-plus-least-squares' procedure is shown to have good properties. For earlier examples of this type of approach, see Anscombe (1960a), Guttman and Smith (1969, 1971), Veale (1975), Marks and Rao (1978, 1979), Guttman and Kraft (1980), and Johnson, McGuire, and Milliken (1978).

In (3.2.15) we used the average of the $n$ values $g(x_i - m)$ as a location estimator for $g(X - m)$. But clearly we can generalize the method, and obtain a wide class of dispersion estimators, many of which should have good robustness properties, if we replace (3.2.15) by some other location estimator chosen for robustness. Specifically, we could use an $L$-estimator or an $M$-estimator or an $R$-estimator as discussed in Section 3.2.1. For a detailed discussion see Huber (1981).

Suppose, for example, that we choose $m$ to be the sample median $\tilde{x}$, and use as location estimator for $g(X - m)$ the sample *median* of $g(d_1), \ldots, g(d_n)$, which is a simple and robust $L$-estimator of location. Taking $g(d) = |d|$ we get

$$s_m = \text{median} \{ |x_1 - \tilde{x}|, \ldots, |x_n - \tilde{x}| \},$$

the *median deviation* (more precisely, the median absolute deviation from the median). This is a useful dispersion estimator, which goes back as far as Gauss; see Hampel (1974). Its properties will be discussed in Chapter 5, along with those of other estimators of dispersion for practical use.

$L$-estimation provides another general procedure for the robust estimation of dispersion. While linear combinations of the order statistics were introduced in Section 3.2.1 as estimators of location, such combinations

$$s_L = \sum_{i=1}^{n} b_i x_{(i)}$$

can clearly be used as estimators of dispersion if the weights $b_i$ are chosen suitably. An obvious example is the *sample range* $((b_1 = -1, b_2 = b_3 = \ldots = b_{n-1} = 0, b_n = 1)$. This estimator will clearly not be robust against outliers! For this purpose one would, as with $L$-estimators of location, seek weights

$b_i$ lower in the extremes than in the body of the data set. A simple and useful $L$-estimator of dispersion is the *semi-interquartile range* $Q$:

$$Q = \tfrac{1}{2}\{x_{(q_2)} - x_{(q_1)}\},$$

where $q_1$, $q_2$ are the nearest integers to $n/4$, $3n/4$ respectively. (Note that we can regard $Q$ as the *range* of the $\tfrac{1}{4}$-*trimmed sample*.)

Similar considerations to those outlined above apply to the robust estimation of *scale*; see Huber, 1981, Chapter 5. Note that a scale parameter for a random variable $X$ can be regarded as a location parameter for the random variable log $X$.

### 3.2.3 Hypothesis tests and confidence intervals

We first consider the standard method for making robust inferences about a parameter $\theta$, based on the deviation from $\theta$ of a robust estimate $T$, studentized by a robust estimate $S_T$ of the standard deviation of $T$. This was briefly outlined in Section 3.1.4.

To fix ideas, take the case where $x_1, x_2, \ldots, x_n$ is a random sample from $N(\mu, \sigma^2)$, $\mu$ and $\sigma^2$ unknown. We have the familiar test statistic

$$t = \frac{\bar{x} - \mu_0}{s/\sqrt{n}} \tag{3.2.16}$$

for testing the hypothesis $H : \mu = \mu_0$ against $\bar{H} : \mu \neq \mu_0$, or for constructing confidence intervals for $\mu$. The null-distribution of $t$ is of course Student's **t** with $n - 1$ degrees of freedom. If we were wrong in attributing the sample to the normal distribution, there might well be considerable errors in the significance probabilities indicated by the $t$-test, or in the confidence limits based on **t**. To produce a robust test or a robust confidence interval, various matters need to be more fully specified—in particular, we need to declare the types of departure from normality to be entertained and the criteria of robustness to be applied. As in all aspects of the study of robustness only certain types of alternative model are relevant to the outlier problem.

What we have just called the 'standard' approach is to replace $\bar{x}$ and $s$ in (3.2.16) by robust estimators $T$ and $S$ in such a way that the corresponding statistic is still distributed essentially as Student's **t** over the range of contemplated distributions, according to some criterion of the 'Student's t-ness' of the statistic. A typical proposal, due to Tukey and McLaughlin (1963) in a seminal paper, is to use a symmetrically trimmed mean $\overset{T}{\bar{x}}_{r,r}$ for $T$ and 'match' it with an estimated standard deviation $\overset{W}{s}_{r,r}$ based on the $(r, r)$-Winsorized sample; the appropriate degrees of freedom for $t$ are $n - 2r - 1$.

More generally we have a basic distribution $F$, not necessarily normal but typically symmetric. The range of alternative distributions contemplated often consists of a set of symmetrically contaminated versions of $F$ of a mixture type with symmetric (possibly normal) contaminating distributions. This, of course, bears on our interest in outliers—although sometimes wider families of distributions have been entertained with little direct outlier relevance. We wish to make inferences (in the form of hypothesis tests or interval estimates) regarding a location parameter $\theta$ of $F$, based on a studentized statistic of the type

$$(T - \theta)/S_T \qquad (3.2.17)$$

where $T$ is a robust estimator of $\theta$. Huber (1970) summarizes the basic aims, and examines various possibilities. In the main, we would hope to achieve high *robustness of performance* for $T$, and to then match $T$ with an estimated standard error $S(T)$ yielding high *robustness of validity* over a range of possible distributions for the sample. We can often expect asymptotic normality for (3.2.17), but when this occurs and how fast is the approach to normality depend to a considerable extent on the contemplated range of distributions. For a range including normal, Cauchy and various contaminated normal distributions Gross (1976) studies a wide variety of confidence interval procedures based on statistics of type (3.2.17). Other procedures of this nature have been proposed by Tukey and McLaughlin (1963), Huber (1970), Shorack (1976), and Gross (1977). See also Huber (1972).

Another method of robust inference has been put forward by Boos (1980), based on the use of $M$-estimators of location. Recalling (3.2.12), if we choose a suitable function $\psi$ and solve for $c$ the equation $\sum_{i=1}^{n} \psi(x_i - c) = 0$, the solution $c = \tilde{\theta}$ is an $M$-estimator of $\theta$; for example $\psi(x) = x$ and $\psi(x) = \mathrm{sgn}\, x$ give respectively as $M$-estimators of location the sample mean $\bar{x}$ and the sample median $\tilde{x}$. Boos considers the random variable

$$T = \left\{ - n^{-1/2} \sum_{i=1}^{n} \psi(x_i - \theta) \right\} \Big/ S$$

where $(n - 1)S^2 = \sum_{i=1}^{n} \psi^2(x_i - \tilde{\theta})$, and shows that, if the weight function $\psi(x)$ is chosen to be non-decreasing, with $\psi(-x) = \psi(x)$, then $T$ is distributed approximately as Student's t with $n - 1$ degrees of freedom. This enables robust confidence intervals for $\theta$ (and, by extension, hypothesis tests about $\theta$) to be set up. Robustness properties of these procedures are explored by Boos in relation to a variety of distributions including normal, triangular, logistic, double-exponential, and t on 3 degrees of freedom (but *not* including any contamination-type distributions with direct relevance to outliers).

The determination of robust confidence intervals for location or scale parameters can be approached in terms of other robust estimation methods. For example, although they are not specifically directed to outlier-type contamination, the range of non-parametric procedures do yield direct confidence intervals which may in their general robustness provide some basis for accommodating outliers. For a general discussion see Noether (1974). Further details are included in Lehmann (1975) and Noether (1967, 1973). For an outlier-specific proposal using non-parametric methods see Butler (1982). Robust permutation tests have been studied by Lambert (1985) and Keller-McNulty and Higgins (1987).

Finally, we turn to the robust version, due to Huber (1965; 1981, p. 264) of the classical *likelihood ratio test*. If we have two simple hypotheses $H_0$ and $H_1$, with likelihood ratio $f_1(x)/f_0(x) = r(x)$, say, for an observation $x$, then the classical test statistic based on $n$ observations $x_1, \ldots, x_n$ is

$$T = r_1 r_2 \ldots r_n,$$

where $r_i = r(x_i)$. Huber induces robustness by censoring (in effect, Winsorizing) any outliers among the $n$ values $r_i$. The procedure is to choose two constants $a, b$ ($0 < a < b$), to take

$$r_i' = r_i \quad \text{if} \quad a \leqslant r_i \leqslant b$$

$$r_i' = a \quad \text{if} \quad r_i < a$$

$$r_i' = b \quad \text{if} \quad r_i > b,$$

and to use as new test statistic

$$T' = r_1' r_2' \ldots r_n'.$$

This type of test has been investigated further by Rieder (1978), and has been generalized by Wang (1981) to the case where $H_0$ and $H_1$ involve nuisance parameters. The obvious application to robust sequential testing is discussed by Huber (1981, p. 273).

### 3.2.4 Adaptive robust procedures

Mention must be made of *adaptive methods* of statistical inference, in which the choice of inference procedures is allowed to depend in part on the actual sample to hand. Some such proposals have been made in the context of robust methods for estimation and hypothesis testing: see, for example, Beran (1974, 1978), Stone (1975), and De Wet and van Wyk (1979a,b); for a general account, see Hogg (1974). The concept can be illustrated by a simple but important problem, the choice of trimming parameter for a trimmed mean, as described by Jaeckel (1971b). Concerned with optimal choice of $\alpha$ in the $\alpha$-trimmed mean for estimating the location parameter of a symmetric distribution he proposes that we choose $\alpha$ in some permissible range $(\alpha_0, \alpha_1)$ so as to minimize the sample variance $s^2(\alpha)$ of $\overset{T}{x}(\alpha, \alpha)$. The resulting *optimal-trimmed mean* $\overset{T}{x}(\hat{\alpha}, \hat{\alpha})$ is shown to be asymptotically equivalent (in terms of variance) to the best estimator $\overset{T}{x}(\alpha, \alpha)$ (i.e. with minimum variance $\overset{T}{\sigma^2}(\alpha, \alpha)$) provided the truly best $\alpha$ happens to lie in the range $(\alpha_0, \alpha_1)$. Hogg (1974) stresses the difficulty in deciding what is an appropriate measure of location for an *asymmetric* distribution and follows up a suggestion by Huber (1972) that the measure might be *defined* in terms of the limiting form of some appealing estimator. He commends the trimmed mean $\overset{T}{x}(\alpha_1, \alpha_2)$ for this purpose with $\alpha_1$ and $\alpha_2$ (which may well differ in value) chosen adaptively to minimize an estimate $\overset{T}{s^2}(\alpha_1, \alpha_2)$ of the variance of $\overset{T}{x}(\alpha_1, \alpha_2)$. Other adaptive estimators and tests will be described where appropriate in Chapter 5.

The important topic of the *Bayesian* accommodation of outliers is discussed in Chapter 9.

# CHAPTER 4

# Testing for Discordancy: Principles and Criteria

We discussed in some detail in Chapter 2 the meaning of the term 'outlier', the nature of the outlier problem, and the variety of contexts in which outliers can arise. We drew the distinction between different types of action which may be called for in response to an outlier: rejection, and omission from the subsequent analysis; accommodation of its value by use of robust methods of estimation or testing which control its distorting effect on the result; using it as a clue to the existence of some previously unsuspected and possibly interesting factor; interpreting it as a signal to find a more appropriate model for the data. We considered the situation in which, as an alternative to accommodation and an indispensable prerequisite to the other types of action, a *detection* procedure was undertaken: a statistical test, which we have already termed in Section 2.2.2 a *test of discordancy*, to decide whether or not the outlier is to be regarded as a member of the main population. Such tests are often referred to in the literature as tests for the *rejection* of outliers, but, as we have stressed, rejection is not the only course open when an observation is detected as foreign to the main data set. In this chapter and later in the book we deal with tests of discordancy in different situations. We start in this chapter with the simplest situation: when the data, with the possible exception of any outliers, form a sample from a univariate distribution from a prescribed family (for example, gamma of unknown shape parameter; exponential; normal; normal with known variance).

We have already stressed in Section 2.2.2 the importance of the alternative model, by which the presence of possible contaminant outliers is accounted for, in assessing the properties of any particular discordancy test, and hence in constructing 'good' tests (tests with desirable performance characteristics). But quite apart from any consideration of alternative models it is not difficult in any particular situation to propose reasonable-looking test statistics. (It may be quite another matter, of course, to ascertain the critical values or percentage points against which the value of any such statistic should be judged, to determine the distribution of the statistic on the assumption that the outlier is consistent with the rest of the data, and to assess the advantages and disadvantages of the test procedure.)

*Example 4.1 Suppose that a civil engineer, wishing to find the mean crushing strength of concrete made from a particular mix, makes up a set of ten test cubes from the mix, allows a suitable hardening period, and then determines their strengths in kg/cm² to be as follows*:

$$280, \ 260, \ 330, \ 400, \ 370, \ 210, \ 520, \ 290, \ 310, \ 250.$$

*The value 520 seems to him to be out of line with the other nine values and he wishes to test it as an outlier. What test criteria might he use? The choice depends in the first place on the form of the distribution of crushing strengths of similar cubes from the same mix. Experience suggests that crushing strengths are normally distributed. However, the mean and variance will not be known.*

As regards the test criterion, we must surely expect that units of measurement are irrelevant: that any test is invariant with respect to changes of scale and origin in the data. For example, the ten values in the sample

$$6.28, \ 6.26, \ 6.33, \ 6.40, \ 6.37, \ 6.21, \ 6.52, \ 6.29, \ 6.31, \ 6.25$$

are linear transforms of the ten values in the first sample. Any test for 520 as an outlier in the first sample must give the same results when applied to 6.52 viewed as an outlier in the second sample. Had, say, an exponential distribution been assumed instead of a normal distribution, a test procedure would only need to remain unaltered under changes in scale, not shifts in origin, since practical support for a single (scale) parameter exponential model rests on a *natural* origin of measurement—see below.

Using the notation introduced in Section 2.2.2, let us arrange the ten values in ascending order and name them $x_{(1)}, x_{(2)}, \ldots, x_{(10)}$ respectively:

| $x_{(1)}$ | $x_{(2)}$ | $x_{(3)}$ | $x_{(4)}$ | $x_{(5)}$ | $x_{(6)}$ | $x_{(7)}$ | $x_{(8)}$ | $x_{(9)}$ | $x_{(10)}$ |
|------|------|------|------|------|------|------|------|------|------|
| 210 | 250 | 260 | 280 | 290 | 310 | 330 | 370 | 400 | 520 |

The figure shows these ten values as points on a line:

| 200 | 300 | 400 | 500 |

The reason the outlier $x_{(10)}$ appears aberrant is because it is 'widely separated' from the remainder of the sample *in relation to the spread of the sample*. This leads one to think of test statistics of the form $N/D$, where the numerator $N$ is a measure of the separation of $x_{(10)}$ from the remainder of the sample and the denominator $D$ is a measure of the spread of the sample. For the reason given above, $D$ must be of the same dimensions as $N$, i.e. in this example $D$ and $N$ would both be in kg/cm². For $N$ one might consider using the separation of $x_{(10)}$ from its nearest neighbour $x_{(9)}$, i.e. $x_{(10)} - x_{(9)} = 120$; or again the separation of $x_{(10)}$ from the other nine values considered as a group, say specifically from their mean $\bar{x}' = 300$. For $D$ one might use the range of this group, $x_{(9)} - x_{(1)} = 190$, or the spacing $x_{(9)} - x_{(8)} = 30$ which is markedly less than $x_{(10)} - x_{(9)}$, or perhaps the standard deviation $s' = 59.8$ of the nine values. These considerations suggest as possible test statistics such quantities as

$$y(9, 10; 1, 9) = \frac{x_{(10)} - x_{(9)}}{x_{(9)} - x_{(1)}} \qquad \left(\text{value here} = \frac{120}{190} = 0.63\right),$$

$$y(9, 10; 8, 9) = \frac{x_{(10)} - x_{(9)}}{x_{(9)} - x_{(8)}} \qquad \left(\text{value here} = \frac{120}{30} = 4.0\right),$$

$$T' = \frac{x_{(10)} - \bar{x}'}{s'} \qquad \left(\text{value here} = \frac{220}{59.8} = 3.68\right).$$

Statistics of the form

$$y(r, s; p, q) = \frac{x_{(s)} - x_{(r)}}{x_{(q)} - x_{(p)}} \tag{4.0.1}$$

—which we shall term *Dixon* statistics—have been investigated by Dixon (1950, 1951), Likeš (1966) and others; the $y$-notation is due to Likeš. An apparently attractive alternative is to judge the outlier by the ratio of the spacing $x_{(10)} - x_{(9)}$ to the range of all ten values including the outlier, giving

$$y(9, 10; 1, 10) = \frac{x_{(10)} - x_{(9)}}{x_{(10)} - x_{(1)}} \qquad \left(\text{value here} = \frac{120}{310} = 0.39\right),$$

but this is effectively the same statistic as $y(9, 10; 1, 9)$, since clearly

$$\frac{1}{y(9, 10; 1, 10)} - \frac{1}{y(9, 10; 1, 9)} \equiv 1.$$

In a similar way, the statistic

$$T = \frac{x_{(10)} - \bar{x}}{s} \qquad \left(\text{value here} = \frac{520 - 322}{89.5} = 2.21\right),$$

where $\bar{x}$, $s$ are the mean and standard deviation of all ten values including the outlier, is equivalent to $T'$ since the two quantities are functionally related; in fact (see Section 4.1.3)

$$\frac{(n-1)^2}{nT^2} - \frac{n(n-2)}{(n-1)T'^2} \equiv 1. \tag{4.0.2}$$

Properties of the test based on $T'$ (or $T$) have been discussed by Pearson and Chandra Sekar (1936), Grubbs (1950), and others.

As remarked earlier, it is easy to propose other test statistics for the above outlier example, for instance

$$\frac{x_{(10)} - \bar{x}}{x_{(10)} - x_{(1)}}.$$

To the best of our knowledge the properties of this statistic have not been studied and no percentage points are available, so no practical use can be made of it.

Consider now a different example. The table shows the lengths of stay (in days) of 92 patients in a hospital observation ward before they were transferred to a main ward (data by kind permission of J. Hoenig; the problem is described in Hoenig and Crotty, 1958).

| Length of stay in days | 1 | 2 | 3 | 4 | 5 | 6 | 7 | 8 | 9 | 10 | 11 | 21 | Total |
|---|---|---|---|---|---|---|---|---|---|---|---|---|---|
| Number of patients | | 11 | 18 | 28 | 8 | 12 | 5 | 5 | 1 | 1 | 1 | 1 | 1 | 92 |

Regarding $x_{(92)} = 21$ as an outlier, what criterion might be used for identifying it? The assumption of a normal distribution for the lengths of stay is not plausible, but there is some evidence to support the use of a gamma distribution with origin at zero. With such a distribution, a test criterion is required to be invariant under changes of scale. As before, we look for test statistics of the form $N/D$ where $N$ measures the separation of the outlier from the rest of the sample, and $D$, in the same units as $N$, measures the spread of the sample.

If the underlying gamma distribution (denoted $\Gamma(r, \lambda)$) has parameters $r$ and $\lambda$, i.e. if it has probability density function

$$f(x) = [\lambda^r \Gamma(r)]^{-1} x^{r-1} \exp(-x/\lambda), \tag{4.0.3}$$

then its mean is $r\lambda$ and its variance is $r\lambda^2$. If $r$ is known, but $\lambda$ is unknown, then the spread of the distribution can be measured not only by the sample standard deviation $s$ but also by the sample mean $\bar{x}$ or equivalently by the sample sum $\sum x_i$. This suggests that a useful statistic for identifying the upper outlier $x_{(92)}$ would be

$$\frac{x_{(92)} - \bar{x}}{\bar{x}}$$

or equivalently

$$\frac{x_{(92)}}{\sum x_i}$$

where $\sum x_i$ is the sum of all 92 observations.

As with $T$ and $T'$ in the normal case discussed above, the statistic is functionally related to, and hence equivalent to,

$$\frac{x_{(92)}}{\sum' x_i}$$

where $\sum' x_i$ is the sum of the 91 observations omitting the outlier. This statistic would of course be inappropriate for judging an upper outlier in a *normal* sample. On the other hand, statistics of Dixon's type such as

$$y(91, 92; 1, 91) = \frac{x_{(92)} - x_{(91)}}{x_{(91)} - x_{(1)}},$$

discussed above in the normal sample situation, are clearly applicable to gamma samples.

General considerations of this kind produce a wide choice of possible test statistics. We must now ask which tests are to be preferred for any particular situation, how can they be constructed, how can the requisite significance probabilities or levels be obtained, and how should the performance of the tests be assessed?

There are four main aspects to be considered.

• The basic model and the assessment of significance.

In any context, a statistical test might be constructed merely by setting up an intuitively appealing test statistic and rejecting or accepting some working hypothesis of no contamination on the basis of the value of the test statistic. This is true of tests of discordancy for outliers. Indeed, the subjective basis of the outlier concept and the long history of its study has tended in the past to encourage an informal attitude to proposals for 'outlier rejection'. At the very least, however, we need to be able to determine rejection criteria which enable significance to be assessed; and for this purpose it is essential to specify the working hypothesis, so that the distribution of the test statistic under the working hypothesis can either be known completely, or at least so that some bound on the tail probability of this *null* distribution can be obtained for any particular value of the test statistic. These issues are discussed in Section 4.1.

• $\begin{cases} \text{The alternative (outlier–generating) hypothesis.} \\ \text{The assessment of test performance, and the nature of the power concept.} \end{cases}$

Generally there is no *unique* discordancy test for any particular type of outlier manifestation. To choose between rival tests we need to have some useful measures of their relative performance, for example their power. Some possible measures and their properties are discussed in Section 4.2.1. The assessment of relative performance, e.g. by power calculations, depends of course on what alternative hypothesis is specified to explain the outliers, a topic discussed at some length in Chapter 2. It then requires knowledge of the distributional behaviour of the test statistic under this alternative hypothesis. This often presents difficult and complicated problems, and in the past many people have either ignored it or have confined themselves to simulated results. However, the body of theoretical work developed in recent years by Balakrishnan, David, Joshi, Tiku and others, which has been described in Section 3.1.2, is clearly of importance for test performance studies as well as for accommodation and robustness studies. Its application to test performance calculations remains largely to be exploited. Section 4.2.2 considers briefly the present position.

• What desirable properties should a discordancy test have?

Ideally we should wish to go further and construct tests which at their conception seek to express optimality properties or at least to satisfy certain useful practical constraints. Thus if we cannot (as is inevitably the case for most outlier tests) obtain tests which are globally *uniformly most powerful* we can at least strive for *local optimality* or *unbiasedness* or the satisfaction of certain *invariance* properties. Alternatively we might choose to construct tests by some accredited practical method, such as the *maximum likelihood ratio principle*, in the hope that the frequently encountered useful characteristics of such a method transfer to the outlier problem. These issues are treated in Section 4.1.2.

## 4.1  CONSTRUCTION OF DISCORDANCY TESTS

### 4.1.1  Test statistics

In further parts of this chapter we shall be considering in detail the construction of discordancy tests and their properties. We shall be answering some of the questions raised above, and in Chapter 6 we shall go on to give details of the precise form and application of the main tests now available. But it is first interesting to note that, in the earlier development of discordancy tests, test statistics tended to be chosen because of their intuitive appeal, irrespective of the form of the working and alternative hypotheses. While chosen on intuitive grounds they have in many cases been demonstrated (often *subsequent* to their original introduction and use) to be supported by statistical test principles applied to appropriate models. We begin with an account of the main categories of such test statistic. Augmenting the classification of Tietjen and Moore (1972) we can distinguish *seven* common types, considered in turn below. Some are more appropriate than others in different types of situation: for example, in examining a single upper outlier, or two lower outliers, or, perhaps, an upper and a lower outlier whilst safeguarding against the possibility of additional upper or lower outliers, and so on; see Chapter 6. For the question of what wider statistical support they may enjoy, see Section 4.1.2. The only associated concept we will discuss at this stage is that of *masking* (see below). The seven common types of test statistic are as follows.

### Excess/spread statistics

These are ratios of differences between an outlier and its nearest or next-nearest neighbour to the range, or some other measure of spread of the sample (possibly omitting the outlier and other extreme observations). Examples are

$$\frac{x_{(n)} - x_{(n-1)}}{x_{(n)} - x_{(2)}}$$

(Dixon, 1951, for examining an upper outlier $x_{(n)}$, avoiding $x_{(1)}$) or (Irwin, 1925)

$$\frac{x_{(n)} - x_{(n-1)}}{\sigma}$$

where $\sigma$ is the standard deviation in the basic model. Irwin's statistic assumes $\sigma$ is known and that the basic model is normal. Clearly we could replace $\sigma$ with an estimate which might possibly be based on a restricted sample which excludes observations we wish to protect against: such as the outlier $x_{(n)}$ or other extremes. If an independent estimate of $\sigma$ is available, this could also be used.

### Range/spread statistics

Here we replace the numerator with the sample range; for example (David, Hartley, and Pearson, 1954; Pearson and Stephens, 1964)

$$\frac{x_{(n)} - x_{(1)}}{s}.$$

Again $s$ might be replaced by a restricted sample analogue, independent estimate or known value of a measure of spread of the population. Using the range has the disadvantage that a significant result can represent discordancy of an upper outlier, a lower outlier, or both.

### Deviation/spread statistics

This latter difficulty is partly offset by using in the numerator a measure of the distance of an outlier from some measure of central tendency in the data. An example (Grubbs, 1950) for a lower outlier is

$$\frac{\bar{x} - x_{(1)}}{s}$$

As for $s$, $\bar{x}$ might be based on a restricted sample, or replaced with an independent estimate, or population value, of some convenient measure of location. A modification uses maximum deviation in the numerator; for example, $\max |x_i - \bar{x}|/s$ (Halperin *et al.*, 1955).

*Sums of squares statistics*

Somewhat different in form are test statistics expressed as ratios of sums of squares for the restricted and total samples: for example, the statistic

$$\sum_{i=1}^{n-2} (x_{(i)} - \bar{x}_{n,\,n-1})^2 \bigg/ \sum_{i=1}^{n} (x_i - \bar{x})^2$$

where $\bar{x}_{n,\,n-1} = \sum_{i=1}^{n-2} x_{(i)}/(n-2)$, proposed by Grubbs (1950) for testing two upper outliers $x_{n-1}$, $x_{(n)}$ in a normal sample; the same form of statistic proves useful for testing upper outliers in extreme-value samples (see Fung and Paul, 1985).

*Extreme/location statistics*

Another class of test statistics takes the form of ratios of extreme values to measures of *location*. These are particularly relevant to examining outliers where the initial model is from the *gamma* family of distributions. As an example, a discordancy test of an upper outlier may use

$$x_{(n)}/\bar{x}.$$

Such statistics have been examined by Epstein (1960a, b), Likeš (1966), Lewis and Fieller (1979), Hawkins (1972, 1980a), Kimber and Stevens (1981), and Kimber (1982).

*Higher-order moment statistics*

Statistics such as measures of skewness and kurtosis, not specifically designed for assessing outliers, can none the less be useful in this context; for example (Ferguson, 1961a)

$$\frac{n^{1/2} \sum (x_i - \bar{x})^3}{[\sum (x_i - \bar{x})^2]^{3/2}} \quad \text{and} \quad \frac{n \sum (x_i - \bar{x})^4}{[\sum (x_i - \bar{x})^2]^2}.$$

*W-statistics*

Another omnibus statistic of relevance to the testing of outliers is the *W*-statistic of Shapiro and Wilk (1965, 1972) and Shapiro, Wilk, and Chen

(1968) already referred to; see also Stephens (1978). This consists for normal data of the ratio of the square of a particular linear combination of all the ordered sample values to the sum of squares of the individual deviations about the sample mean; for exponential data it has the same form, but with the linear combination simplifying to $\bar{x} - x_{(1)}$.

In a review paper, Grubbs (1969) gives illustrative examples of the use of several different types of discordancy statistic, applying them to several sets of actual data.

One effect that many of these statistics must be prone to, to differing extents, is that of *masking*. This is an important concept and one which we shall have to spend some time investigating. It has been well described by Tietjen and Moore (1972):

Suspected observations sometimes form subgroups; i.e., several values are closer to each other than they are close to the bulk of the observations. This phenomenon makes [procedures for testing the most extreme observation] insensitive. It has been called the *masking effect*. The masking effect is the inability of a testing procedure to identify even a single outlier in the presence of several suspected values.

We shall discuss this phenomenon in some detail in Section 4.1.4, together with the related phenomenon of *swamping*.

Some test statistics have been proposed to provide tests for several outliers simultaneously. One example is the sums of squares statistic above due to Grubbs (1950) which is relevant to the case of *two* upper outliers. Such a procedure, in which several outliers are tested *en bloc* in a single test operation, is called a *block procedure*. We give a detailed discussion of block procedures in Section 4.3.1.

A different approach is to examine the outliers in *sequence*, using a hierarchical form of test. For instance, McMillan and David (1971) and McMillan (1971) describe what they call a 'sequential' test of several upper outliers in a normal sample (the terminology is confusing, since the test is not sequential in the usual sense of taking observations one at a time in order to draw conclusions as quickly as possible: it is in fact a fixed-sample-size test). They first examine the principal upper outlier $x_{(n)}$ by means of a deviation/spread statistic of form $[x_{(n)} - \bar{x}]/s$, where $s^2$ may be based on the sample data alone or combined with an independent estimate of variance. If $x_{(n)}$ proves to be discordant on this basis, they proceed to apply a similar test to $x_{(n-1)}$ in the reduced sample excluding $x_{(n)}$, using the statistic

$$\frac{x_{(n-1)} - \bar{x}'}{s'};$$

here $\bar{x}'$ and $s'$ are the sample mean, and an estimate of spread, obtained on omission of $x_{(n)}$. If the result is significant, $x_{(n-1)}$ is judged discordant and the procedure repeated for the next outlier, and so on until a non-discordant value is reached. The statistical properties of such a repeated procedure have been investigated by a number of authors, including McMillan and David (1971), McMillan (1971), Moran and McMillan (1973), Hawkins (1973, 1978, 1980a), Chikkagoudar and Kunchur (1987), Simonoff (1984b), and Sweeting (1983, 1986). As pointed out by Hawkins (1980a, p. 65), this approach can be modified by starting with the $k$th largest observation $x_{(n-k+1)}$ for some chosen $k(>1)$ and assessing its discordancy in relation to the $n - k$ smaller $x$-values; either it is judged discordant, and with it the $k - 1$ larger $x$-values, or the next observation $x_{(n-k+2)}$ is tested in relation to $x_{(1)}, \ldots, x_{(n-k+1)}$, and so consecutively. Kimber (1982) investigates this type of procedure for exponential samples.

We give a detailed discussion of such 'sequential' (or, as we shall call them, *consecutive*) tests of outliers in Section 4.3.2.

### 4.1.2  Statistical bases for construction of tests

Apart from intuitively based procedures, two widely applicable methods exist for setting up discordancy tests, as has been said earlier in this chapter. These are the *maximum likelihood ratio principle* and the *principle of local optimality* perhaps restricted to the classes of unbiased, or invariant, tests. Naturally the construction of the tests depends in the first instance on the alternative hypothesis employed to account for the outliers (Section 2.3).

Consider for example the testing of a single upper outlier $x_{(n)}$ in an exponential sample. Our working hypothesis is

$$H : x_j \in F \quad (j = 1, 2, \ldots, n)$$

declaring that all the observations $x_1, \ldots, x_n$ belong to the distribution $F$ with density $\theta e^{-\theta x}$ $(x > 0)$, $\theta$ being unknown. Suppose we have a slippage alternative $\bar{H}$ stating that $n - 1$ of the observations belong to $F$ and the remaining one, $x_n$ say, to the exponential distribution $G$ with density $\lambda\theta e^{-\lambda\theta x}(x > 0; \lambda < 1)$. We may write

$$H : \lambda = 1$$

$$\bar{H} : \lambda < 1.$$

The log likelihood of the observations on hypothesis $H$ is

$$L_H(\theta) = n \log \theta - n\theta\bar{x} \tag{4.1.1}$$

where $\bar{x}$ is the mean of $x_1, \ldots, x_n$. $L_H(\theta)$ is maximized by $\theta = 1/\bar{x}$, and its maximized value is

$$\hat{L}_H = -n \log \bar{x} - n \tag{4.1.2}$$

On hypothesis $\bar{H}$, the log likelihood of the observations is

$$L_{\bar{H}}(\theta, \lambda) = n \log \theta + \log \lambda - (n - 1)\theta\bar{x}' - \lambda\theta x_n \qquad (4.1.3)$$

where $\bar{x}'$ is the mean of $x_1, \ldots, x_{n-1}$. $L_{\bar{H}}(\theta, \lambda)$ is maximized when

$$n/\theta - (n - 1)\bar{x}' - \lambda x_n = 0$$

and

$$1/\lambda - \theta x_n = 0,$$

provided $\lambda \leq 1$, i.e. when $\theta = 1/\bar{x}'$ and $\lambda = \bar{x}'/x_n$, provided $x_n \geq \bar{x}'$. Its maximized value is accordingly (4.1.2) if $x_n < \bar{x}'$, otherwise it is

$$\hat{L}_{\bar{H}} = - (n - 1) \log \bar{x}' - \log x_n - n. \qquad (4.1.4)$$

The test statistic based on the maximum likelihood ratio is $\{\hat{L}_{\bar{H}} - \hat{L}_H\}$. This is equal to zero if $x_n < \bar{x}'$, while if $x_n \geq \bar{x}'$ it is

$$- \{(n - 1) \log \bar{x}' + \log x_n - n \log \bar{x}\}$$

$$= - (n - 1) \log \frac{n - T}{n - 1} - \log T, \quad \text{where } T = x_n/\bar{x}. \qquad (4.1.5)$$

It follows that the maximum likelihood ratio test is equivalent to rejecting $H$ when $T$ is large.

Strictly speaking, we are not in a position to use this test, because we do not know which of the observations is the discordant one belonging to $G$ if $\bar{H}$ is true. In practice this observation is assumed to be $x_{(n)}$, the outlier, and $T_{(n)} = x_{(n)}/\bar{x}$ is used as test statistic. If it were not for the presence of the outlier we would not be moved to query $H$ or to test the strength of the evidence for $\bar{H}$. This is an intuitive justification for the use of $T_{(n)}$, but clearly $T_{(n)}$ is not the maximum likelihood ratio test statistic for the $\bar{H}$ we have specified.

There are two ways in which we could legitimately establish $T_{(n)}$ as the appropriate test statistic. The first faces up squarely to the fact that our desire for a test of discordancy stems from our reaction to one specific observation; namely, the greatest observation, $x_{(n)}$. An alternative hypothesis can be set up which reflects this, in the form

$$\bar{H}' : x_{(1)}, x_{(2)}, \ldots, x_{(n-1)} \text{ belong to } F$$

$$x_{(n)} \text{ belongs to } G.$$

This hypothesis, which we may call the hypothesis of *labelled slippage*, identifies the *extreme* observation as the only possible discordant value.

If $y_1, \ldots, y_{n-1}$ is a random sample from $F$ and $y_n$ is a random observation from $G$, we can think of our ordered sample $x_{(1)}, \ldots, x_{(n)}$ as a particular realization $y_1, y_2, \ldots, y_n$ in which the observation $y_n$ turns out to be the largest. Thus the likelihood under $\bar{H}'$ is

$$P(x_{(1)}, \ldots, x_{(n)})$$

where $P(y_1, \ldots, y_n)$ is the likelihood of $y_1, \ldots, y_n$ *conditional on* $y_1 < \ldots < y_{n-1} < y_n$. Now each $y_j$ ($j = 1, \ldots, n-1$) may be regarded as the time to the first event in a Poisson process of rate $\theta$, and $y_n$ as the time to the first event in a Poisson process of rate $\lambda\theta$; by superposing these $n$ processes, assuming them independent, and considering which event occurs first, the probability that $y_1$ is the smallest of the $y$'s is seen to be $\theta/[(n-1)\theta + \lambda\theta] = 1/(n-1+\lambda)$. Continuing stepwise, we get

$$P(y_1 < \ldots < y_n) = 1/[(n-1+\lambda)(n-2+\lambda)\ldots(1+\lambda)]. \quad (4.1.6)$$

Hence the log likelihood of the observations on $\bar{H}'$ is

$$L_{\bar{H}'}(\theta, \lambda) = n\log\theta + \log\lambda - \theta(n\bar{x} - x_{(n)}) - \lambda\theta x_{(n)} + \sum_{j=1}^{n-1}\log(j + \lambda). \quad (4.1.7)$$

This is maximized when

$$n/\theta - n\bar{x} + x_{(n)} - \lambda x_{(n)} = 0$$

and

$$1/\lambda - \theta x_{(n)} + \sum_{j=1}^{n-1}(j+\lambda)^{-1} = 0.$$

The maximizing value of $\lambda$, $\hat{\lambda}$ say, must therefore satisfy

$$\sum_{j=0}^{n-1}(j + \hat{\lambda})^{-1} = nT_{(n)}/[n - (1 - \hat{\lambda})T_{(n)}] - (1/\hat{\lambda}). \quad (4.1.8)$$

The maximized value of $L_{H'}(\theta, \lambda)$ comes out to be

$$\hat{L}_{\bar{H}'} = -n\log\bar{x} - n - n\log[n - (1 - \hat{\lambda})T_{(n)}]$$
$$+ n\log n + \log\hat{\lambda} + \sum_{j=0}^{n-1}\log(j + \hat{\lambda}) \quad (4.1.9)$$

Under $H$, the log likelihood is

$$n\log\theta - n\theta\bar{x} - \log n!$$

with maximized value

$$\hat{L}_H = -n\log\bar{x} - n - \log n!$$

Hence $\hat{L}_H - \hat{L}_{\bar{H}'}$ depends on the observations in terms of $T_{(n)}$ and $\hat{\lambda}$ only, and is therefore a function of $T_{(n)}$ since in view of (4.1.8) $\hat{\lambda}$ is a function of $T_{(n)}$ only. In this example, therefore, the discordancy test statistic $T_{(n)}$

is equivalent to the maximum likelihood ratio test statistic in the labelled slippage formulation. (But see Section 4.3.1.)

The second way in which $T_{(n)}$ is established directly as the appropriate test statistic is by a multiple decision procedure applied to a set of $n$ alternative hypotheses

$$\bar{H}_i :x_i \text{ comes from } G \text{ (some particular } i) \atop x_j \text{ comes from } F \ (j \neq i)} \right\} \qquad (4.1.10)$$

for $i = 1, 2, \ldots, n$. This formulation is similar to the *model B* type of slippage alternative hypothesis considered by Ferguson (1961a) (see Section 2.3) specialized to the case of a single outlier and an exponential distribution. The decision criterion is that of maximizing the probability of adopting the correct $\bar{H}_i$ when slippage has occurred, subject to a prescribed probability of correct adoption of the basic hypothesis $H$ and to certain invariance conditions (index permutation, positive changes of scale, arbitrary changes of origin). In the present situation of an exponential basic model, changes in location are inappropriate and the procedure leads to adopting $\bar{H}_i$ if $T_j$ is maximized when $j = i$ and is sufficiently large. Thus the appropriate test statistic is precisely $T_{(n)}$.

Another way of handling (4.1.10) would be by means of a *two-stage maximum likelihood ratio* test: declaring as a discordant outlier that observation whose omission effects the greatest increase in maximized likelihood, provided that increase is significantly large. Thus we consider (see 4.1.5)

$$\hat{L}_{\bar{H}_i} - \hat{L}_H = \begin{cases} - (n - 1) \log \left( \dfrac{n - T_i}{n - 1} \right) - \log T_i & (T_i \geq 1) \\ 0 & (T_i \leq 1) \end{cases} \qquad (4.1.11)$$

Choosing $i$ to maximize (4.1.11) implies identifying the hypothesis $\bar{H}_i$ for which $x_i = x_{(n)}$; this $\bar{H}_i$ is adopted and $x_{(n)}$ declared a discordant outlier if $T_{(n)}$ is sufficiently large.

Still considering the case of a single upper outlier in an exponential sample, let us now construct a discordancy test on the basis of *one-sided local (invariant) optimality*. Assuming first the slippage alternative $\bar{H}$, with log likelihood

$$L_{\bar{H}} (\theta, \lambda) = n\log \theta + \log \lambda - (n - 1)\theta \bar{x}' - \lambda \theta x_n,$$

we have $\partial L_{\bar{H}} (\theta, \lambda)/\partial \lambda = (1/\lambda) - \theta x_n$.

Under the working hypothesis $H$, $\lambda = 1$, and $\partial L_{\bar{H}}/\partial \lambda$ is then equal to $1 - \theta x_n$. Replacing $\theta$ by its maximizing value $\hat{\theta} = 1/\bar{x}'$, we obtain $1 - (x_n/\bar{x}')$ as the locally optimal test statistic, which is equivalent to $T = x_n/\bar{x}$. If instead we take the labelled slippage alternative $\bar{H}'$, we get

$$\partial L_{\bar{H}'} (\theta, \lambda)/\partial \lambda = (1/\lambda) - \theta x_{(n)} + \sum_{j=1}^{n-1} (j + \lambda)^{-1}.$$

When $\lambda = 1$ this is equal to $\sum_{j=1}^{n} j^{-1} - \theta x_{(n)}$ ; substituting $\theta = \hat{\theta} = 1/\bar{x}$ we get $\sum_{j=1}^{n} j^{-1} - x_{(n)}/\bar{x}$ as the locally optimal test statistic, or in effect $T_{(n)}$.

But now consider the *block testing* of a lower and upper outlier pair $x_{(1)}, x_{(n)}$ in a gamma sample, where the working hypothesis is that the data $x_j$ ($j = 1, 2, \ldots, n$) belong to $\Gamma(r, \lambda)$, the gamma distribution with density $[\lambda^r \Gamma(r)]^{-1} x^{r-1} \exp(-x/\lambda)$ ($x > 0$; $r$ known, $\lambda$ unknown). Kimber (1988) shows that $R = [x_{(n)} - x_{(1)}]/\Sigma x_j$ is the locally most powerful scale-invariant test statistic against a labelled-slippage alternative in which $x_{(2)}, \ldots, x_{(n-1)}$ belong to $\Gamma(r, \lambda)$, $x_{(1)}$ belongs to $\Gamma(r, \lambda c)$ and $x_{(n)}$ belongs to $\Gamma(r, \lambda/c)$ ($c$ unknown, $0 < c < 1$). Kimber goes on to show that, on the other hand, the two-stage maximum likelihood ratio test statistic against the ordinary (unlabelled) alternative (in which two unspecified observations belong respectively to $\Gamma(r, \lambda c)$ and $\Gamma(r, \lambda/c)$) is **not** $R$ but

$$T = [\sqrt{x_{(n)}} - \sqrt{x_{(1)}}]^2 / \sum x_j.$$

Again, consider the testing of a single upper outlier in a *normal* sample, for which the statistic $T_{(n)} = \{x_{(n)} - \bar{x}\}/s$ is commonly used. As in the exponential case discussed above, it can be shown that $T_{(n)}$ is effectively the maximum likelihood ratio test statistic for a labelled slippage alternative, also for the corresponding multiple decision formulation, and the unidentifiable equivalent $T = \{x_n - \bar{x}\}/s$ is the corresponding statistic for the ordinary slippage alternative. The alternative hypotheses must be appropriately chosen; that $x_1, \ldots, x_{n-1}$ (in the case of $\bar{H}$) or $x_{(1)}, \ldots, x_{(n-1)}$ (in the cases of $\bar{H}'$ or $\bar{H}_n$) belong to $F : N(\mu, \sigma^2)$ and that $x_n$, or $x_{(n)}$, belongs to a normal distribution $G : N(\mu + a, \sigma^2)$ with a different mean $\mu + a$ ($a > 0$) but the same variance. If instead we make the alternative a normal distribution $N(\mu, b\sigma^2)$ ($b > 1$), i.e. with the *same mean* as $F$ but a *larger variance*, a different and less tractable statistic emerges. Likewise, in constructing tests of discordancy for normal samples based on the local optimality principle, Ferguson (1961a) obtains different tests for these two different alternatives. His tests, based respectively on sample skewness and kurtosis, are described in detail below.

These cases illustrate how crucially *the attribution of any optimality properties depends on the form adopted for the alternative hypothesis*. The choice of an alternative hypothesis is problematical; this uncertainty in turn may reduce the utility of any apparent 'optimum' properties of a test of discordancy. (For another example, see Sugiura and Sasamoto (1989), where it is shown that the locally best invariant test of a single outlier in an exponential sample is not everywhere 'best' in comparison with various other tests.) We must acknowledge this dilemma in reacting to the following results.

Continuing with the assumption of an initial normal distribution, and the commonly used *slippage* type of alternative hypothesis, Ferguson

(1961a) demonstrates that certain quasi-optimum tests of discordancy can be constructed in such situations. He considers, in the context of his *model A* (slippage of the mean), tests which are *invariant* with respect to the labelling of the observations and to changes of scale and location. (The change of scale must be effected by multiplication by a common *positive* quantity whilst the distribution for each individual observation may suffer a specific change of location.) We recall from (2.3.1) that *model A* declares

$$\mu_i = \mu + \sigma \Delta a_{v_i}.$$

Ferguson shows that a *locally best invariant* test of size $\alpha$ exists for testing $H : \Delta = 0$ against the one-sided alternative $\bar{H} : \Delta > 0$. It takes the following form: if $\mu_3(a) \gtrless 0$ reject $H$ whenever $\sqrt{b_1} \gtrless K_1$ where $K_1$ is chosen to yield a test of size $\alpha$ and

$$\mu_3(a) = \frac{1}{n} \sum_{i=1}^{n} (a_i - \bar{a})^3 \quad \left( \bar{a} = \frac{1}{n} \sum_{i=1}^{n} a_i \right) \tag{4.1.12}$$

$$b_1 = \frac{\sqrt{n} \sum (x_i - \bar{x})^3}{[\sum (x_i - \bar{x})^2]^{3/2}} \tag{4.1.13}$$

(this is just the coefficient of skewness statistic described above).

For a two-sided test, where the alternative hypothesis is $\bar{H}' : \Delta \neq 0$, there is a *locally best unbiased invariant test* of size $\alpha$ which takes the form: if $k_4(a) \gtrless 0$ reject $H$ whenever $b_2 \gtrless K_2$ where $K_2$ is chosen to yield a test of size $\alpha$, $k_4(a)$ is the fourth $k$-statistic of $a_1, a_2, \ldots, a_n$ and

$$b_2 = \frac{n \sum (x_i - \bar{x})^4}{[\sum (x_i - \bar{x})^2]^2} \tag{4.1.14}$$

(the sample coefficient of kurtosis).

With *model B*, the alternative hypothesis is, with

$$\sigma^2_i = \sigma^2 \exp(\Delta a_{v_i}),$$

$\bar{H}'' : \Delta > 0$. Under the same invariance requirements a *locally best invariant test* of size $\alpha$ exists and leads to rejection if $b_2 > K$, where $K$ is chosen to yield a test of size $\alpha$.

Still restricting attention to single outliers and an initial normal distribution, there are results for *slippage tests* (where we have samples from several normal distributions, one or more of which may have a mean, or variance, which differs from the others) which particularize, when samples are all of size 1, to tests of discordancy on a slippage type alternative hypothesis. Paulson (1952b) considers a multi-decision formulation where under $\mathcal{D}_0$ we decide that the observations all come from $N(\mu, \sigma^2)$ whereas under $\mathcal{D}_i$ ($i = 1, 2, \ldots, n$) we decide that $\mu_i$ has slipped to $\mu + a$ ($a > 0$). Under the restrictions that if all means are $\mu$ we accept $\mathcal{D}_0$ with probability $1 - \alpha$ and that the decision procedure is invariant with respect to the index of the slipped distribution and to positive change of scale and arbitrary change of origin, he shows there to be an optimum procedure in the sense of *maximizing the probability of making the correct decision* when slippage to the right has occurred. With the modification of proof described by Kudo (1956a) to cope with samples of size 1, this leads to the optimum decision rule: if

$$t = \frac{x_{(n)} - \bar{x}}{[\sum_{i=1}^{n} (x_i - \bar{x})^2]^{1/2}},$$

then when $t \leqslant h_\alpha$ conclude no discordant outlier, whilst if $t > h_\alpha$ conclude that $x_{(n)}$ is discordant, where $h_\alpha$ is chosen to ensure that $\mathcal{D}_0$ is adopted with probability $1 - \alpha$ in the null situation.

In Paulson (1952b) the procedure is shown to be the Bayes solution when equal prior probabilities are assigned to $\mathcal{D}_1, \mathcal{D}_2, \ldots, \mathcal{D}_n$. David (1981, p. 227) shows that it has corresponding support in non-Bayesian terms using the Neyman-Pearson lemma. Kudo (1956b) demonstrates that the optimality property remains when slippage in the mean is accompanied by decrease in variance for the slipped distribution, whilst Kapur (1957) adduces an *unbiasedness* property for the Paulson procedure in the sense that the probability of *incorrectly* taking any of the decisions $\mathcal{D}_0, \mathcal{D}_1, \ldots, \mathcal{D}_n$ never exceeds the probability of *correctly* taking any one of these decisions. David (1981, pp. 230–231) also remarks that an obvious modification to allow for an independent estimate of $\sigma^2$ remains optimum in the Paulson sense. Kudo (1956a) extends Paulson's results to the case of slippage either to the right or to the left. Truax (1953) presents the immediately parallel results for unidirectional slippage of the variance of one of the normal distributions.

A further extension, relating to simultaneous slippage of the means of two distributions by equal amounts, but in opposite directions, is given by Ramachandran and Khatri (1957).

In comparing the work of Ferguson with that of Paulson we note a basic distinction between the chosen measures of performance of the tests: power and maximum probability of correct action. This focuses attention on the complex issue, highlighted on page 93, of what constitutes an appropriate performance criterion for a test for discordancy. Some comments on the relative merits of *five* possible performance characteristics are made by David (1981, pp. 232–239). For a detailed discussion of these and related issues see Section 4.2.1.

Test statistics obtained by either the maximum likelihood ratio principle or the principle of local optimality have the required invariance properties. They are, in the first place, invariant under permutation of the subscripts of $x_1, \ldots, x_n$ and hence can be expressed as functions of the ordered values $x_{(1)}, \ldots, x_{(n)}$. (Any symmetric functions of the $x_{(i)}$ which figure here can of course be written as the corresponding functions of the unordered $x_i$.) Ferguson sets out to ensure scale and location invariance in his test criteria and asserts (for the normal case with unknown mean and variance which he is examining) that this implies that the observations will appear in the expressions for the criteria purely in terms of ratios of intervals between ordered values, of the type $\{x_{(a)} - x_{(b)}\}/\{x_{(c)} - x_{(d)}\}$. For example, the familiar $\{x_{(n)} - \bar{x}\}/s$ is a function of these ratios. (Here we have a justification for the use of Dixon statistics which are, of course, the ratios themselves.) Scale and location invariance will also hold in more specific circumstances, where certain parameter values are assumed known, although the condition that the test statistics are functions merely of ratios of differences between ordered values will not necessarily apply. For example, consider the testing of normal samples when the variance $\sigma^2$ is known. In this situation $\{x_{(n)} - \bar{x}\}/\sigma$, for instance, is scale- and location-invariant, and is a valid discordancy statistic. Similarly if $\mu$ but not $\sigma^2$ is known, $\{x_{(n)} - \mu\}/s$, again not expressible in terms of Dixon statistics, is a valid statistic. In some contexts dual invariance will not be appropriate, as we have remarked above in relation to exponential samples.

Reverting to the maximum likelihood ratio principle, this has a further application of great importance in the *detection* of outliers, quite apart from the assessment of their discordancy. In data situations such as regression or the results of designed experiments, outliers may be present, essentially as outlying *residuals*, but such values may not in general be immediately recognized as outliers in the same way that extreme values are in univariate samples. One way of defining the 'most outlying' observation or point in a sample is as the one whose omission produces the greatest increase in the maximized likelihood; this amounts, of course, to identifying it as the one out of the $n$ whose maximum likelihood ratio test statistic has the greatest value. The same principle can be used to detect the most outlying point in a multivariate sample. Once detected, in other words *identified*, it can of course be *tested* for discordancy (see Chapters 7 and 8). This principle is embodied in the use of the multiple decision formulation of the alternative hypothesis.

There are, of course, other methods of constructing discordancy tests for outliers.

For example, Kitagawa (1979) proposes the use of Akaike's information criterion (AIC), defined for a sample of observations and a parametric model as

$$AIC = -2\log \text{ (maximized likelihood of sample under model of best fit)}$$
$$+ 2 \text{ (number of fitted parameters).} \tag{4.1.15}$$

Suppose for example that we have a basic model under which $x_1, \ldots, x_n$ all belong to $N(\mu, \sigma^2)$ and a family of alternative models with slippage in location at both ends of the sample, under which the $r$ smallest values $x_{(1)}, \ldots, x_{(r)}$ belong to $N(\mu - a_1, \sigma^2)$ and the $s$ greatest values $x_{(n-s+1)}, \ldots, x_{(n)}$ belong to $N(\mu + a_2, \sigma^2)$ ($a_1 > 0$, $a_2 > 0$). We say a *family* of alternatives because in Kitagawa's procedure the numbers $r$ and $s$ of lower and upper outliers to be assessed are not pre-chosen, but are left to be determined by the test procedure itself. For successive values of $(r, s)$ starting with $(0, 0)$ (no contamination), the value, $\hat{I}_{r,s}$ say, of the AIC is calculated; the number of fitted parameters in (4.1.15) is 2 for $\hat{I}_{0,0}$, ($\mu$ and $\sigma^2$), 3 for $\hat{I}_{r,0}(\mu, \sigma^2, a_1)$, 3 for $\hat{I}_{0,s}(\mu, \sigma^2 a_2)$, and 4 for $I_{r,s}(\mu, \sigma^2, a_1, a_2)$. Inspection of the values

$$
\begin{array}{ccc}
\hat{I}_{0,0} & \hat{I}_{0,1} & \hat{I}_{0,2}\ldots \\
\hat{I}_{1,0} & \hat{I}_{1,1} & \hat{I}_{1,2}\ldots
\end{array}
$$
$$
\cdots\cdots\cdots\cdots\cdots
$$

in the matrix $\hat{I}_{r,s}$ as it builds up indicates which pair of values $(r,s)$ gives minimum $\hat{I}_{r,s}$. Suppose these values are $(3, 2)$, so that $\hat{I}_{3,2}$ is the least of the $\hat{I}$-values; then the observations $x_{(1)}, x_{(2)}, x_{(3)}, x_{(n-1)}$ and $x_{(n)}$ are declared to be discordant.

Kitagawa claims as advantages of this procedure that it does not involve reference to a significance level (or equivalent evaluation of a significance probability), and that the number of outliers is determined by the test itself. Be this as it may (and the latter feature abrogates any role of the statistician in reacting to outliers with *surprise*, while the former precludes the element of quantification of discordancy that one might have thought desirable), when the procedure is applied to the classical Herndon-Chauvenet data set on vertical semi-diameter residuals for Venus (Section 2.2.2 and Figure 2.2), using the above model for contamination with slippage in location at both ends of the sample, it declares discordant not only—1.40 and 1.01 but also the innocuous observation 0.63! This casts some doubt on the practical usefulness of the method, although Gather and Kale (1988) give an example of its satisfactory application to a data set on passage times of light.

For testing the discordancy of either a lower or an upper outlier, $x_{(1)}$ or $x_{(n)}$, in a sample from a symmetric distribution which is of exponential type (i.e. which satisfies

$$
\lim_{x \to \infty} \frac{d}{dx} \{[1 - F(x)]/f(x)\} = 0,
$$

e.g. a double-exponential distribution), George and Rousseau (1987) propose the use of the midrange $\frac{1}{2}(x_{(1)} + x_{(n)})$ as test statistic. They use the fact that on this type of working hypothesis the distribution of the midrange is approximately logistic. They state that it is well known that the midrange is sensitive to the presence of an outlier. Clearly, however, if both $x_{(1)}$ and $x_{(n)}$ are outliers the test is liable to be vitiated by masking.

Some other methods of constructing tests, for use in the consecutive testing of several outliers, are mentioned in Section 4.3.2.

Another matter which requires comment is the use of *nonparametric* methods. Nonparametric procedures for identifying outliers and testing their discordancy, or for rendering them uninfluential in a statistical analysis of the bulk of the data, have been presented in a variety of contexts. The original approach to the slippage problem (Mosteller, 1948) was nonparametric and generated a flood of refinements or modifications.

Various nonparametric proposals for identification and discordancy testing in a single unstructured univariate sample have been made by Walsh [1950 (and correction, 1953), 1959, 1965]; the accommodation issue is briefly considered by Walsh and Kelleher (1974).

It is a feature of nonparametric procedures that we adopt a minimum of distributional assumptions about the data-generating mechanism. Such procedures can have relatively low power in comparison with procedures specifically geared to a particular detailed parametric model. If such a model can be justified we would often wish to employ methods tailored to it. Otherwise, nonparametric methods are appealing in their lack of commitment to a model—their ubiquity or robustness. *But this type of appeal has an element of delusion* in the outlier context. Outliers are 'atypical' observations—they impress themselves on us by appearing to be unrepresentative of the overall sample data. With no knowledge (or assumptions) about the general distributional structure of the data-generation process we have no grounds for 'surprise', nothing 'typical' against which to ascribe 'atypicality'. The *macrolepidoptera* light-trap data at the end of Section 1.2 illustrate this well. It is only by considering (at least informally) the way in which the data might reasonably have been generated that we have any basis for examining the possibility of discordant outliers. There must always be the possibility of a homogeneous explanation of the values in any sample.

In its most extreme form a nonparametric approach makes no assumptions about the basic data-generating mechanism. At this level it seems a contradiction to seek to investigate outliers. Broad specifications such as symmetry of the basic distribution, or of location-slippage explanations of outliers, raise some prospect for outlier study but seem bound to be highly speculative in their conclusions. If such specifications are as much as we dare contemplate, then perhaps we have no alternative but to accept the highly limited assessment of outliers yielded by a nonparametric approach. But more than in almost any other area of statistical enquiry, the study of outliers hinges on as precise a model formulation as is feasible. To deliberately abandon the model, by seeking nonparametric (or distribution-free) methods in some broad aim of robustness, smacks of throwing out the bathwater before the baby has even been immersed.

*Bayesian methods* for examining outliers are discussed in detail in Chapter 9.

### 4.1.3  Inclusive and exclusive measures

In Section 4.1.1 we have listed seven common types of discordancy test statistic. Most of these are ratios of the form $N/D$, as discussed in the introduction to this chapter: here $N$ is a measure of the separation of the outlying value or values from the main mass of the sample and $D$ is a measure of the spread of the sample. Measures used for $N$ include the excess, the deviation and, in the gamma case, the extreme (as deviation from zero), all as defined in Section 4.1.1; measures used for $D$ include the standard deviation, the range, the sum of squares of observations corrected to the sample mean, and, again in the gamma case, the sample mean or sum. One ambiguity immediately presents itself: should the means and measures of spread which enter into these statistics be calculated from the complete data set including the outlier (or outliers), or from the reduced data set excluding the outlier? We would appear to have a double set of statistics, based respectively on *inclusive measures* and *exclusive measures*, and to be faced with a decision as to which is preferable. It turns out, however, that a statistic $N/D$ based on inclusive measures and its analogue based on exclusive measures are in many cases equivalent. The following examples will make this clear.

*Excess/range statistic for testing single upper outlier*

*Inclusive*
$$T = \frac{x_{(n)} - x_{(n-1)}}{x_{(n)} - x_{(1)}}$$
(4.1.16)

*Exclusive*
$$T' = \frac{x_{(n)} - x_{(n-1)}}{x_{(n-1)} - x_{(1)}}$$
(4.1.17)

Clearly

$$\frac{1}{T} - \frac{1}{T'} \equiv 1.$$

*Deviation/spread statistic for testing single lower outlier*

*Inclusive*
$$T = \frac{\bar{x} - x_{(1)}}{s}$$

*Exclusive*
$$T' = \frac{\bar{x}' - x_{(1)}}{s'}$$

where

$$n\bar{x} = \sum_{i=1}^{n} x_{(i)}, \qquad\qquad (n-1)\bar{x}' = \sum_{i=2}^{n} x_{(i)},$$

$$(n-1)s^2 = \sum_{i=1}^{n} (x_{(i)} - \bar{x})^2, \qquad\qquad (n-2)s'^2 = \sum_{i=2}^{n} (x_{(i)} - \bar{x}')^2.$$

We have $nsT = (n-1)s'T'$ and

$$(n-1)s^2 - (n-2)s'^2 = (\bar{x} - x_{(1)})^2 + (n-1)(\bar{x}' - \bar{x})^2$$

$$= s^2T^2 + \frac{1}{n-1}s^2T^2,$$

whence

$$\frac{(n-1)^2}{nT^2} - \frac{n(n-2)}{(n-1)T'^2} \equiv 1. \tag{4.1.18}$$

When the inclusive and exclusive forms of a discordancy statistic $N/D$ are functionally related, as above, the null distribution can be obtained by means of a very useful recursive argument. We shall discuss this in Section 4.1.5.

### 4.1.4  Masking and swamping

We have already referred to the importance of *masking* as a possible limitation of a discordancy test of a single outlier, or of a consecutive procedure making repeated use of such a single-outlier test. A converse danger, *swamping*, may affect a block discordancy test of a group of 2 or more outliers.

Suppose, to fix ideas, that we wish to test for discordancy two upper outliers $x_{(n-1)}$, $x_{(n)}$ in a normal sample.

One possibility is to proceed *consecutively* and first test $x_{(n)}$ in relation to the other $n-1$ values. If $x_{(n)}$ is adjudged discordant, we proceed to test $x_{(n-1)}$ in relation to $x_{(1)}, \ldots, x_{(n-2)}$; if $x_{(n)}$ is not adjudged discordant then neither is $x_{(n-1)}$. For example, with the following sample A:

$$3, 4, 7, 8, 10, 13, 951,$$

$x_{(7)} = 951$ will be declared discordant and $x_{(6)} = 13$ will not: a correct judgement.

On the other hand, with sample B:

$$3, 4, 7, 8, 10, 949, 951,$$

$x_{(7)} = 951$ will *not* be adjudged discordant because of the proximity of $x_{(6)} = 949$; so the test fails at the first stage. In a phrase due to Murphy

(1951), $x_{(6)}$ has had a *'masking effect'* on the identification of $x_{(7)}$ in the consecutive test; more briefly, $x_{(6)}$ has masked $x_{(7)}$. The masking phenomenon was discussed as early as 1936 by Pearson and Chandra Sekar.

Note that, in another form of consecutive procedure, discussed in Section 4.3, $x_{(n-1)}$ is tested *first* in relation to $x_{(1)}, \ldots, x_{(n-2)}$. If $x_{(n-1)}$ is not adjudged discordant (as with Sample *A*), $x_{(n)}$ is then tested in relation to $x_{(1)}, \ldots, x_{(n-2)}, x_{(n-1)}$. If $x_{(n-1)}$ *is* adjudged discordant (as with Sample B), this automatically implies the discordancy of $x_{(n)}$ *as well, and the outlier-pair* $x_{(n-1)}, x_{(n)}$ are jointly declared discordant. In this approach, the risk of masking is removed. If there were more than 2 outliers (*k*, say) we would of course have started with $x_{(n-k+1)}$.

The other approach is to use a *block* test of $x_{(n-1)}, x_{(n)}$ as a pair. With sample B, such a block test will rightly declare the pair to be discordant. With sample A the block test will again declare the pair $x_{(6)} = 13$, $x_{(7)} = 951$ to be discordant; the truly discordant value $x_{(7)}$ has 'carried' the value $x_{(6)} = 13$ with it, resulting in a false judgement as regards $x_{(6)}$. This danger affecting block procedures was first described by Fieller (1976) who named it the *'swamping effect'*; $x_{(7)}$ has swamped $x_{(6)}$ in the block test.

A block procedure may also, of course, be vulnerable to masking; e.g. a high value of $x_{(n-2)}$ may have a masking effect on a block test of the upper outlier-pair $x_{(n)}, x_{(n-1)}$.

In considering whether any particular test procedure is vulnerable to masking or swamping, judgements have usually been made on purely qualitative grounds. For instance, a discordancy test of a single upper outlier $x_{(n)}$ in a normal sample using the statistic

$$\frac{x_{(n)} - x_{(n-1)}}{x_{(n)} - x_{(2)}}$$

is clearly vulnerable to a possible masking effect from $x_{(n-1)}$ (if its value should be unduly near to $x_{(n)}$) but immune from any masking effect from $x_{(1)}$ (even if it is an extreme lower outlier). But vulnerable *to what extent?* If two different tests of $x_{(n)}$ are both seen to be vulnerable to masking by $x_{(n-1)}$, can one make a quantitative comparison of their vulnerability? This question has been addressed by Bendre and Kale (1985 [correction 1986], 1987) and Bendre (1989), who have proposed measures of the masking and swamping effects which enable these effects to be assessed quantitatively. We now consider their proposals in some detail. To fix ideas, suppose that we have:

a sample $x_j$ ( $j = 1, 2, \ldots, n$);

a statistic $T_k$ for testing a block of $k$ upper outliers $x_{(n-k+1)}, \ldots, x_{(n-1)}, x_{(n)}$ for discordancy; with $k = 1$ this covers the case of a statistic $T_1$ for testing a single upper outlier $x_{(n)}$;

a working hypothesis $H : x_j \in F( j = 1, 2, \ldots, n)$;

a test procedure '*If $T_k$ exceeds some critical value $t_k(n, \alpha)$, declare the k outliers* $x_{(n-k+1)}, \ldots, x_{(n-1)}, x_{(n)}$ *discordant*', the value of $t_k$ $(n, \alpha)$ being given by

$$P(T_k > t_k(n, \alpha) \mid H) = \alpha,$$

where $\alpha$ is the significance probability or level. (This includes the case of a test procedure for a single outlier '*If $T_1$ exceeds a critical value $t_1(n, \alpha)$, declare $x_{(n)}$ discordant*', with $P(T > t_1(n, \alpha) \mid H) = \alpha$.

Finally, let us assume a *labelled slippage* alternative model under which any discordant values belong to a contaminating distribution $G$, related to $F$ in terms of a slippage parameter $\Delta$. For example, if $F$ is the exponential distribution $E(\lambda)$ with density

$$\lambda^{-1} \exp(- x/\lambda) \qquad (x > 0), \tag{4.1.19}$$

$G$ could be the scale-slipped exponential distribution $E(\Delta \lambda)$ $(\Delta > 1)$; if $F$ is the normal distribution $N(\mu, \sigma^2)$, $G$ could be the location-slipped normal distribution $N(\mu + \Delta\sigma, \sigma^2)$ where without loss of generality $\Delta > 0$, or again the scale-slipped normal distribution $N(\mu, \Delta^2 \sigma^2)$ $(\Delta > 1)$. Absence of slippage (i.e. the basic model) in these cases is represented by $\Delta = 1, 0, 1$ respectively; extreme slippage corresponds to $\Delta \to \infty$.

Since we are testing $k$ outliers *en bloc*, the number of contaminants envisaged is either 0 (hypothesis $H$), or $k$; the alternative hypothesis then being

$$\bar{H}_k : x_{(1)}, x_{(2)}, \ldots, x_{(n-k)} \in F,$$

$$x_{(n-k+1)}, \ldots, x_{(n-1)}, x_{(n)} \in G.$$

We want the test to have good *power* (see Section 4.2.1 below), the power $\pi_k$ $(\Delta)$ of the test being the probability that $x_{(n-k+1)}, \ldots, x_{(n-1)}, x_{(n)}$ are declared discordant when $\bar{H}_k$ is true:

$$\pi_k (\Delta) = P(T_k > t_k(n, \alpha) \mid \bar{H}_k).$$

In fact, however, the number of contaminants could be some number $c$ which is neither 0 (hypothesis $H$) nor $k$ (hypothesis $\bar{H}_k$). If $c > k$, the possibility of masking arises; if $0 < c < k$, the possibility of swamping arises. The alternative hypothesis when there are $c$ contaminants is

$$\bar{H}_c : x_{(1)}, x_{(2)}, \ldots, x_{(n-c)} \in F,$$

$$x_{(n-c+1)}, \ldots, x_{(n-1)}, x_{(n)} \in G.$$

With $c$ contaminants the power of the test is

$$P_k (\Delta, c) = P(T_k > t_k(n, \alpha) \mid \bar{H}_c);$$

clearly $P_k(\Delta, k) = \pi_k(\Delta)$. Note that any critical $T_k$-value $t_k(n, \alpha)$ depends on $n$ and $\alpha$ but not on $c$.

Consider first the situation $c > k$ and the consequent possibility of masking. Following Bendre and Kale (1985, 1987) (but using our own notation), define the *(finite $\Delta$) masking effect* as

$$M_k(\Delta, c) = P_k(\Delta, k) - P_k(\Delta, c) \qquad (c \geq k). \qquad (4.1.20)$$

Clearly, $0 \leq M_k(\Delta, c) \leq \pi_k(\Delta)$. $M_k(\Delta, c)$ gives the percentage of samples in which the test detects the $k$ discordant observations if these are the only discordant observations but fails to detect them in the presence of a greater number, $c$, of discordant observations. When the slippage is $\Delta$ the test suffers from masking effect if $M_k(\Delta, c) > 0$, and is free from masking effect if $M_k(\Delta, c) = 0$. Obviously, $M_k(\Delta, k) = 0$.

Under extreme slippage we get the *limiting masking effect*

$$M_k(c) = \lim_{\Delta \to \infty} M_k(\Delta, c).$$

This is useful as a single overall measure of masking; it is essentially an upper bound to the masking effect. Note that the range of values of $c$ for which the limiting masking effect is *zero* can often be obtained directly by simple consideration of an 'extreme sample'; see the following example.

*Example 4.2 Suppose F is the exponential distribution* $\mathbf{E}(\lambda)$ *given by* (4.1.19), *G is* $\mathbf{E}(\Delta \lambda)$, $k = 1$, *and we are testing a single upper outlier* $x_{(n)}$ *for discordancy using the statistic*

$$T_1 = x_{(n)} / \sum x_j.$$

*(This is the statistic denoted by* $T_{\mathrm{Ea1}}$ *and treated in detail on pages.* 197–8) *What is the vulnerability to masking in this situation?*

With $c > 1$ discordant observations, the greatest possible masking effect would occur with the extreme sample consisting of $n - c$ zeros and $c$ equal positive values which without loss of generality can be taken as 1's; the value of $T_1$ is then $1/c$. If this is equal to or greater than the critical value $t_1(n, \alpha)$, $x_{(n)}$ is declared discordant as a single upper outlier; thus there can be no masking effect with this test if $c \leq h(n, \alpha)$ where $h(n, \alpha) = [1/t_1(n, \alpha)]$, the integer part of $1/t_1(n, \alpha)$. Values of $t_1(n, \alpha)$ are tabulated, for $\alpha = 0.05$ and 0.01 and various values of $n$, in Table III pages 473–4 (entries for $r = 1$). Using this table, the values of $h(n, 0.05)$ and $h(n, 0.01)$ are easily derived and are presented for a range of values of $n$ in Table 4.1.

Table 4.1

| $n$ | 15 | 20 | 24 | 30 | 40 | 60 | 120 |
|---|---|---|---|---|---|---|---|
| $h(n, 0.05)$ | 2 | 3 | 4 | 5 | 6 | 8 | 15 |
| $h(n, 0.01)$ | 2 | 3 | 3 | 4 | 5 | 7 | 13 |

As shown above, $h(n, \alpha)$ is the maximum number of discordant observations which the test at level $\alpha$ will accommodate without risk of masking. Table 4.1 indicates that, broadly speaking, the test is free from the limiting masking effect at the 5 per cent level of significance when the number of discordant values $c$ is not more than about one-seventh of the sample size $n$, and at the 1 per cent level when $c$ is not more than about one-eighth of the sample size.

These results are, however, very conservative. Bendre and Kale (1985) show that the limiting masking effect $M_1(c)$, which is of course 0 when $t_1(n, \alpha) \leqslant 1/c$, is given by

$$M_1(c) = 1 - \sum_{i=1}^{h(n, \alpha)} (-1)^{i-1} \binom{c}{i} (1 - it_1(n, \alpha))^{c-1} \qquad (4.1.21)$$

when $1/c < t_1(n, \alpha) < 1$. Using this formula, we find that the limiting masking effect is negligible for values of $c$ substantially higher than those in Table 4.1. For example, with $\alpha = 0.05$ the limiting masking effect $M_1(c)$ is less than 1 per cent for values of $c$ up to the following:

| $n$ | 15 | 20 | 24 | 30 | 40 | 60 | 120 |
|---|---|---|---|---|---|---|---|
| Max $c$ for $M_1(c) < 0.01$ | 3 | 4 | 5 | 7 | 11 | 18 | 41 |

Let us turn now to the situation $k > 1, 0 < c < k$ and the consequent possibility of *swamping*. Bendre (1989) defines (in her own notation) the *finite $\Delta$ swamping effect*, analogously to (4.1.20), as

$$S_k (\Delta, c) = P_k(\Delta, k) - P_k(\Delta, c) \quad (0 < c \leqslant k), \qquad (4.1.22)$$

with, under extreme slippage, the *limiting swamping effect*

$$S_k(c) = \lim_{\Delta \to \infty} S_k(\Delta, c)$$

which can be used as a single overall measure of swamping. In fact, the greater the loss of power under the alternative hypothesis the *less* the swamping effect, so $1 - S_k(\Delta, c)$ and $1 - S_k(c)$ would be more directly appropriate as measures of swamping. But this is a trivial distinction. More to the point, the analogy with (4.1.20) is in our view not a good one.

For, consider the three scenarios

(a) $P_k(\Delta, k)$ high, $P_k(\Delta, c)$ high
(b) $P_k(\Delta, k)$ high, $P_k(\Delta, c)$ low
(c) $P_k(\Delta, k)$ low, $P_k(\Delta, c)$ low.

If $c > k$, (4.1.20) indicates a low value of $M_k(\Delta, c)$ in cases (a) and (c) and a high value in case (b). And clearly the test's vulnerability to masking is indeed low or zero in case (a) and high in case (b). In case (c) we might have, say $P_k(\Delta, k) = 0.1$, $P_k(\Delta, k + 1) = 0$; the extra discordant value would cause complete masking, but on the other hand the measure $M_k(\Delta, k + 1)$ would be low. However, the test is an inefficient one of low power, so we can regard its vulnerability to masking as of little relevance. All in all, $M_k(\Delta, c)$ can be taken as a satisfactory measure of masking effect.

The situation is different, however, if $0 < c < k$. From (4.1.22) we have a high value of $1 - S_k(\Delta, c)$ in cases (a) and (c) and a low value in case (b). The test's vulnerability to swamping is indeed high in case (a) and low in case (b). But in case (c) the measure of swamping $1 - S_k(\Delta, c)$ is high even if $P_k(\Delta, c) = 0$—no swamping at all!

According we suggest $P_k(\Delta, c)$ $(0 < c \leq k)$ as a measure of the finite $\Delta$ swamping effect, rather than $S_k(\Delta, c)$ or $1 - S_k(\Delta, c)$, with

$$P_k(c) = \lim_{\Delta \to \infty} P_k(\Delta, c) \quad (0 < c \leq k)$$

as the limiting swamping effect.

*Example 4.3 Suppose F is the normal distribution* $N(\mu, \sigma^2)$ *and G is the location-slipped normal distribution* $N(\mu + \Delta\sigma, \sigma^2)$. *Consider the following block tests of two upper outliers* $(k = 2)$:

$$T_2 = [x_{(n-1)} + x_{(n)} - 2\bar{x}]/s$$

$$U_2 = S^2_{n-1, n}/S^2$$

*where* $\bar{x} = \sum_{j=1}^{n} x_j /n$; $s^2 = \sum_{j=1}^{n} (x_j - \bar{x})^2/(n - 1)$;

$S^2 = (n - 1)s^2$, *the sum of squares of the deviations from* $\bar{x}$ *of the n observations* $x_1, \ldots, x_n$; *and* $S^2_{n-1, n}$ *is the sum of squares of the deviations of* $x_{(1)}, x_{(2)}, \ldots, x_{(n-2)}$ *from their own mean* $\sum_{j=1}^{n-2} x_{(j)}/(n - 2)$.

$T_2$, $U_2$ *are respectively the statistics denoted by* $T_{N3}$, $T_{N4}$ *treated on pages 244-5. Do these tests differ in regard to their vulnerability to swamping?*

With just one discordant observation $(c = 1)$, the greatest possible swamping effect would occur with the extreme sample consisting of $n - 1$ zeros and one positive value; the values of $T_2$ and $U_2$ are then $(n - 2)/n^{1/2}$ and 0. So there is no swamping effect with the $T_2$-test if $(n - 2)/n^{1/2} \leq t_2(n, \alpha)$. Values of $t_2(n, 0.05)$ and $t_2(n, 0.01)$ are tabulated in Table XVa page 491 (entries for $k = 2$); using these, we find that with just one discordant value the test at level 5 per cent is free from any risk of swamping when the sample size $n$ does not exceed 21, and at level 1 per cent when $n$ does not exceed 26. The $U_2$-test, on the other hand, is always vulnerable to the limiting swamping effect, since the extreme value 0 is always significant.

Note that *sensitivity contours* can often be used to investigate the properties of tests with regard to masking and swamping when at most two outliers are present; for examples, see Kimber (1982, 1983b).

### 4.1.5  Assessment of significance

We have now considered at some length the construction of statistics for testing discordancy. Suppose we decide to use a particular statistic $T$ to test an outlying value or configuration $x$ for discordancy in some particular data set. The value of $T$ for the data is found to be $t$: should $x$ be declared discordant or not? To judge this, we refer $t$ to the distribution of $T$ on the working hypothesis of no contamination—the *null distribution* of $T$. The weight of the evidence for judging $x$ to be discordant is measured by the *significance probability* attaching to $t$, i.e. the probability that, on the working hypothesis, $T$ takes values more indicative of discordancy than $t$. We will denote this significance probability, $P(T > t)$ or $P(T < t)$ as the case may be, by $SP(t)$. We can either evaluate $SP(t)$, and declare $x$ discordant if $SP(t)$ is sufficiently small; or we can compare $t$ with known significance levels (or critical values) $t_\alpha$ given by $SP(t_\alpha) = \alpha$ with $\alpha$ equal to say 0.05 to 0.01.

Should $SP(t)$ be calculated on a one-sided or two-sided basis? At the end of the day this does not greatly matter. Changing the value of a significance probability by a factor of, say, 2 does not greatly affect the weight of evidence represented by that value. To the extent that this issue does matter, we make the following comments.

Some proposed test statistics are clearly suited to one-sided tests, e.g.

$$\frac{x_{(n)} - \bar{x}}{s}$$

whilst others relate to two-sided tests, e.g.

$$\frac{\max |x_i - \bar{x}|}{s}$$

David (1981, p. 223) discusses a possible conflict of choice in relation to which of these types of test should be used in practice. He remarks as follows:

Indeed, the question may be raised whether we should not always use a two-sided test, since applying a one-sided test in the direction indicated as most promising by the sample at hand is clearly not playing fair. This criticism of what is often done in practice is valid enough, except that sticking to a precise level of significance may not be crucial in exploratory work, frequently the purpose of tests for outliers. Strictly speaking, one-sided tests should be confined to the detection of outliers in cases where only those in a specified direction are of interest, or to situations such as the repeated determination of the melting point of a substance, where outliers due to impurities must be on the low side since impurities depress the melting point. Similar arguments show that it is equally incorrect to pick one's outlier test after inspection of the data.

This is a commonly expressed viewpoint but it is in opposition to the attitude we have adopted in this book. We define outliers in subjective terms relative to a particular set of data—the data themselves initiate our interest in outliers. If we declare there to be an upper outlier, $x_{(n)}$, it seems natural, therefore, to use the appropriate one-sided criterion, and to ascribe discordancy if, say,

$$(x_{(n)} - \bar{x})/s > C$$

for some suitable value of $C$. Of course, if we wish to protect ourselves against other outliers we should reflect this in the choice of test statistic. Also, if our suspicions about $x_{(n)}$ are well founded then use of max $|x_i - \bar{x}|/s$ rather than $(x_{(n)} - \bar{x})/s$ is not going to be materially important. What matters is

(i) our declaration of the outliers;
(ii) the alternative model we employ.

    Suppose for the moment that the values of $SP(t)$ or $t_\alpha$ are not available whether in a formula or a tabulation; in principle, we then need to find the null distribution of $T$. This may be difficult in many cases, and algebraically intractable in others so that simulation is required. However, there are two useful methods of wide application which we now describe. One is the recursive method mentioned at the end of Section 4.1.3, for calculating the null distributions of certain types of discordancy statistic. The other is the calculation of approximations to $SP(t)$ by the use of Bonferroni-type inequalities.

*A recursive algorithm for the null distribution of a test statistic*

This method is applicable to discordancy test statistics which have inclusive and exclusive forms which are functionally related, as in Section 4.1.3. While it is not our intention in this book to give detailed proofs of all results used, we feel it is worthwhile setting out this recursive argument in some detail for a simple particular case, namely that of an upper outlier in an exponential sample. It will then be sufficient to state other results obtainable by the same type of argument as they arise, without giving the argument in detail.

*The distribution of the ratio of the greatest observation to the sum of the observations for an exponential sample*

In the introduction to this chapter we remarked on the usefulness of $x_{(n)}/\sum x_i$ (or some simple function of it) for judging a single upper outlier in a gamma sample. For its detailed application we need to know its distribution in the null case, i.e. when there is *no discordant value* present. Let us consider this specifically for the exponential distribution, $\Gamma(1, \lambda)$.

Suppose $X_1, \ldots, X_n$ are $n$ independent identically distributed (i.i.d.) random variables each exponentially distributed with probability density function

$$f(x) = \lambda e^{-\lambda x} \quad (x > 0). \tag{4.1.23}$$

Write

$$A_X = \sum_{j=1}^{n} X_j, \quad T_j = X_j / A_X \quad (j = 1, \ldots, n).$$

Then $\mathbf{T} = (T_1, \ldots, T_n)'$ is a vector random variable of dimension $n - 1$, since

$$\sum T_j = 1;$$

and $A_X$, $\mathbf{T}$ are statistically independent.

When $X_n$ is omitted, $X_1, \ldots, X_{n-1}$ are $n - 1$ i.i.d. random variables with the distribution (4.1.23) and we may rename these as $Y_j (j = 1, \ldots, n - 1)$, with

$$A_Y = \sum_{j=1}^{n-1} Y_j.$$

Also write $T_j^* = Y_j / A_Y (j = 1, \ldots, n - 1)$, and $\mathbf{T}^* = (T_1^*, \ldots, T_{n-1}^*)'$, of dimension $n - 2$. Clearly $X_n$, $A_Y$, $\mathbf{T}^*$ are independent, hence $X_n / A_Y$ and $\mathbf{T}^*$ are independent.

Corresponding with $X_n / A_X = T_n$, write $X_n / A_Y = T_n'$. Clearly

$$\frac{1}{T_n} - \frac{1}{T_n'} = 1. \tag{4.1.24}$$

Hence when $X_n / A_X = t$, it follows that

$$X_n / A_Y = t / (1 - t).$$

Suppose now that $g_n(t)$ is the probability density function, and $G_n(t)$ the distribution function, of the random variable $X_{(n)} / A_X$, where $X_{(n)}$ is the greatest of the $X_i$. Then

$$g_n(t) \delta t = P[X_{(n)} / A_X \in (t, t + \delta t)]$$

$$= nP[X_n / A_X \in (t, t + \delta t), X_n = X_{(n)}]$$

$$= nP[X_n / A_X \in (t, t + \delta t), X_1 < X_n, \ldots, X_{n-1} < X_n]$$

$$= nP[X_n / A_X \in (t, t + \delta t), X_j / A_Y < t / (1 - t) \quad \text{for} \quad j = 1, \ldots, n - 1]$$

$$= nP[T_n \in (t, t + \delta t), T_j^* < t / (1 - t) \quad \text{for} \quad j = 1, \ldots, n - 1].$$

From (4.1.24), $T_n$ and $\mathbf{T}^*$ are independent. Hence

$$g_n(t)\delta t = nP[T_n \in (t, t + \delta t)]P[T_j^* < t/(1 - t) \quad \text{for} \quad j = 1, 2, \ldots, n - 1].$$

The second of these probabilities is

$$P\left(\max_j T_j^* < \frac{t}{1 - t}\right) = G_{n-1}\left(\frac{t}{1 - t}\right).$$

The first probability is equal to

$$P\left[T_n' \in \left(t', t' + \frac{dt'}{dt}\delta t\right)\right]$$

where $t' = t/(1 - t)$, so that we conclude

$$g_n(t)\delta t = nP\left[T_n' \in \left(t', t' + \frac{dt'}{dt}\delta t\right)\right]G_{n-1}\left(\frac{t}{1 - t}\right). \tag{4.1.25}$$

Since, from (4.1.23), $(n - 1)T_n'$ has the F-distribution on 2 and $2(n - 1)$ degrees of freedom, we have in the exponential case the recurrence relationship

$$g_n(t) = n(n - 1)(1 - t)^{n-2}G_{n-1}\left(\frac{t}{1 - t}\right) \tag{4.1.26}$$

which gives the null distribution of the outlier statistic $X_{(n)}/A_X$ for a sample of size $n$, in terms of the corresponding distribution for a sample of size $n - 1$.

The range of possible values of $X_{(n)}/A_X$ is from $1/n$ to 1, hence $G_n(t) = 1$ for $t \geq 1$. The following recursive calculation arises from (4.1.26).

| Range for $t(1 - t)^{-1}$ | $G_{n-1}\left(\dfrac{t}{1-t}\right)$ | Range for $t$ | $g_n(t)$ | $G_n(t)$ |
|---|---|---|---|---|
| $[1, \infty]$ | $1$ | $[\frac{1}{2}, 1]$ | $n(n - 1)(1 - t)^{n-2}$ | $1 - n(1 - t)^{n-1}$ |
| $[\frac{1}{2}, 1]$ | $1 - (n - 1)\left(\dfrac{1 - 2t}{1 - t}\right)^{n-2}$ | $[\frac{1}{3}, \frac{1}{2}]$ | $n(n - 1)(1 - t)^{n-2}$ $- n(n - 1)^2(1 - 2t)^{n-2}$ | $1 - n(1 - t)^{n-1}$ $+ \dfrac{n(n - 1)}{2!}(1 - 2t)^{n-1}$ |
| $[\frac{1}{3}, \frac{1}{2}]$ | $1 - (n - 1)\left(\dfrac{1 - 2t}{1 - t}\right)^{n-2}$ $+ \dfrac{(n - 1)(n - 2)}{2!}$ $\times \left(\dfrac{1 - 3t}{1 - t}\right)^{n-2}$ | $[\frac{1}{4}, \frac{1}{3}]$ | $n(n - 1)(1 - t)^{n-2}$ $- n(n - 1)^2(1 - 2t)^{n-2}$ $+ \dfrac{n(n - 1)^2(n - 2)}{2!}$ $\times (1 - 3t)^{n-2}$ | $1 - n(1 - t)^{n-1}$ $+ \dfrac{n(n - 1)}{2!}(1 - 2t)^{n-1}$ $- \dfrac{n(n - 1)(n - 2)}{3!}(1 - 3t)^{n-1}$ |

and so on. The density function consists of a succession of smoothly connected arcs in the intervals $[\frac{1}{2}, 1], [\frac{1}{3}, \frac{1}{2}], [\frac{1}{4}, \frac{1}{3}], \ldots, [1/n, 1/(n-1)]$. This well known result was first given by Fisher (1929).

There is, of course, nothing in the method of derivation of (4.1.26) which is specific to the exponential distribution and it can be applied in other circumstances. Additionally, it is easily modified for handling, say, $X_{(1)}/A_X$, which will also be of interest.

*Bonferroni-type approximations to the values of significance probabilities*

We first illustrate this method by an example. In order to test for discordancy a single upper outlier $x_{(n)}$ in a normal sample with unknown mean $\mu$ and *known* variance $\sigma^2$, suppose we use the test statistic

$$T = \frac{x_{(n)} - \bar{x}}{\sigma} \tag{4.1.27}$$

where $\bar{x} = (1/n) \sum_{i=1}^{n} x_i$ . What can we say about $SP(t)$?

$x_1, \ldots, x_n$ are observations of random variables $X_1, \ldots, X_n$ which on the working hypothesis are independent and identically distributed, as $N(\mu, \sigma^2)$. Write $T_i = (X_i - \bar{X})/\sigma (i = 1, \ldots, n)$, $E_i$ for the event $T_i > t$, and $E$ for the event $T > t$, so that $SP(t) = P(E)$. Clearly $T$ is the greatest of the $n$ quantities $T_i$, and so $E$ is the union of the $n$ events $E_i$. Bounds on $P(E)$ in terms of the component events $E_i$ are therefore given by Bonferroni's inequalities. The first two inequalities are

$$P(E) \leqslant nP(E_1) \tag{4.1.28}$$

$$P(E) \geqslant nP(E_1) - \binom{n}{2} P(E_1 \cap E_2) \tag{4.1.29}$$

since the $n$ events $E_i$ are equiprobable, and likewise the $\binom{n}{2}$ events $E_i \cap E_j$. The random variable $T_1$ is distributed $N(0, (n-1)/n)$, hence

$$P(E_1) = \Phi(-tn^{1/2}/(n-1)^{1/2}). \tag{4.1.30}$$

The events $E_1$ and $E_2$ are not independent, and there is no simple expression for $P(E_1 \cap E_2)$ analogous to (4.1.30). However, it can be shown (Doornbos, 1976) that

$$P(E_1 \cap E_2) < \{P(E_1)\}^2.$$

Thus, from (4.1.29)

$$P(E) > nP(E_1) - \frac{n-1}{n}\{nP(E_1)\}^2/2 \tag{4.1.31}$$

$$> nP(E_1) - \tfrac{1}{2}\{nP(E_1)\}^2 \qquad\qquad (4.1.32)$$

In discordancy testing situations we may either have an observed value $t$ and wish to find an approximate value for its significance probability $SP(t)$, or we may wish to find an approximate value for the critical value $t_\alpha$ corresponding to a chosen significance level $\alpha$ such as 0.05. For example, take our statistic $T$ of (4.1.27) with $n = 10$. The 5 per cent and 1 per cent critical values of $T$ are in fact given in Table XIIIe, page 486 (see also Test Nσ1 Worksheet, pages 244–5):

$$t_{0.05} = 2.44, \qquad t_{0.01} = 2.93.$$

However, in the absence of tables we could approximate these critical values by using (4.1.28) and (4.1.30):

$$0.05 = SP(t_{0.05}) \simeq 10P(E_1),$$

$$= 10\Phi(-1.0541 t_{0.05}),$$

giving $-1.0541 t_{0.05} \simeq \Phi^{-1}(0.005) = -2.576,$

$$t_{0.05} \simeq 2.444;$$

and similarly

$$t_{0.01} \simeq 2.932.$$

Conversely if we had an observed value $t = 2.444$ and wished to test it for discordancy, (4.1.30) would give

$$P(E_1) = \Phi(-1.0541 \times 2.444) = \Phi(-2.576) = 0.00500,$$

$$10P(E_1) = 0.0500\,;$$

whence, from the inequalities (4.1.28) and (4.1.31),

$$0.0500 - \tfrac{9}{20}(0.0500)^2 < SP(2.444) \leqslant 0.0500,$$

$$0.0489 < SP(2.444) \leqslant 0.0500.$$

The simpler lower bound (4.1.32) is commonly used in preference to (4.1.31); in the above example it gives 0.04875 instead of 0.0489, a trivial difference.

Clearly, Bonferroni-type inequalities can be set up for any discordancy test statistic expressible as the greatest (or least) of a set of identically (though not independently) distributed random variables $T_i$. Many statistics of types described in Section 4.1.1, particularly the deviation/spread, sums of squares, and extreme/location types, are in this category. Providing that the common marginal distribution of the $T_i$'s is known, an upper bound $nP(E_1) = \alpha$, say can be calculated for $SP(t)$ from (4.1.28). If the test relates to a single outlier, (4.1.32) or (4.1.31) can be used to give a lower bound for $SP(t)$:

$$\alpha - \tfrac{1}{2}\alpha^2$$

or, marginally better,

$$\alpha - \frac{n-1}{2n}\alpha^2.$$

With the test statistic (4.1.27) in the above example, the common distribution of the $T_i$'s is normal, and the tail probabilities $P(E_1)$, which are often very small in practice, can readily be found from tables. More usually $\sigma^2$ is unknown when testing outliers in normal samples, and $P(E_1)$ is then defined in terms of a $t$-distribution. Conventional tables of $t$ do not give the large deviate-values corresponding to the very small tail probabilities arising in the present context; tables giving these large deviate-values for $t$ have been published by Bailey (1977).

For a more detailed discussion of the use of Bonferroni inequalities in discordancy testing, see Hawkins (1980a), where a variety of specific applications are described. See also Worsley (1982) for an improved Bonferroni inequality and its application to outlier testing in linear regression.

## 4.2  THE ASSESSMENT OF TEST PERFORMANCE

### 4.2.1  Measures of performance

We raised earlier the question of what constitutes an appropriate performance criterion for a test of discordancy. A key measure is the significance level, although its interpretation is to some extent problematical (see Collett and Lewis, 1976). Comparison of tests of the same significance level must of course depend on the alternative hypothesis we have in mind for explaining the outliers (Section 2.3).

Consider first the *slippage alternative*. To fix ideas, suppose we are testing an upper outlier $x_{(n)}$ in a univariate sample $x_1, \ldots, x_n$. The null hypothesis is

$$H : x_j \in F \quad (j = 1, 2, \ldots, n),$$

i.e. all the observations arise from a distribution $F$ which is, say, $N(\mu, \sigma^2)$ with $\mu$, $\sigma^2$ unknown. We envisage a slippage alternative $\bar{H}$ which states that $n - 1$ of the observations belong to $F$ and the $n$th observation $x_n$, which we will now rename $x_c$ and call the *contaminant*, belongs to a different distribution $G$. If $F$ is $N(\mu, \sigma^2)$, $G$ may be $N(\mu + \sigma\Delta, \sigma^2)$ (for slippage in location) or $N(\mu, \sigma^2 \exp \Delta)$ (for slippage in dispersion); the hypotheses can then be written

$$H : \Delta = 0$$

versus

$$\bar{H} : \Delta > 0.$$

We wish to test $x_{(n)}$ for discordancy. For the moment, let us distinguish between two kinds of test statistic, 'general' and 'specific' say. (We will not need to maintain this distinction for long.) To construct a *general* statistic $Z_{(n)}$, we start with a measure $Z_i$ of the positioning of any observation $x_i$ in relation to the rest of the sample; for example, $Z_i$ could be

$$(x_i - \bar{x})/s \text{ or } (x_i - \bar{x})/(x_{(n)} - x_{(1)}).$$

By particularizing to $x_{(n)}$ we get the corresponding discordancy statistic $Z_{(n)}$, e.g.

$$Z_{(n)} = (x_{(n)} - \bar{x})/s \text{ or } (x_{(n)} - \bar{x})/(x_{(n)} - x_{(1)}).$$

A *specific* statistic, on the other hand, is sensibly defined only in relation to the outlier $x_{(n)}$, and cannot be meaningfully embedded in some set of statistics of like form ranging over the $n$ sample members; e.g.

$$Z = (x_{(n)} - x_{(n-1)})/s.$$

Suppose our discordancy statistic is of the general type. The test takes the form: 'Adjudge $x_{(n)}$ discordant if $Z_{(n)} > z_\alpha$, where $z_\alpha$ is the critical value for preassigned significance level $\alpha$ defined by

$$P(Z_{(n)} > z_\alpha \mid H) = \alpha.' \tag{4.2.1}$$

Since we assume a slippage alternative, one of the sample observations under $\bar{H}$ will be the contaminant $x_c$; and since $Z_{(n)}$ is a general statistic, a corresponding measure $Z_c$ exists for the contaminant. In the context of this particular set of assumptions, David (1981) suggested the following five probabilities as 'reasonable measures' of the performance of $Z_{(n)}$.

(i)                     $$P_1 = P(Z_{(n)} > z_\alpha \mid \bar{H}) \tag{4.2.2}$$

This is the *probability under $\bar{H}$ that the outlier is identified as discordant*, in other words the *power function*.

(ii)                    $$P_2 = P(Z_c > z_\alpha \mid \bar{H}). \tag{4.2.3}$$

(iii)                   $$P_3 = P(Z_c = Z_{(n)}, Z_{(n)} > z_\alpha \mid \bar{H}) \tag{4.2.4}$$

This is the *probability that the contaminant is the outlier and is identified as discordant*.

(iv)                    $$P_4 = P(Z_c = Z_{(n)} > z_\alpha, Z_{(n-1)} < z_\alpha \mid \bar{H}). \tag{4.2.5}$$

(v)                     $$P_5 = P(Z_c > z_\alpha \mid Z_c = Z_{(n)}; \bar{H}) \quad . \tag{4.2.6}$$

This is the *probability that, when the contaminant is the outlier, it is identified as discordant.*
David observes that:

$P_1$ measures the probability of significance [i.e. adjudged discordancy] for any reason whatever and is thus especially suitable for sounding a general alarm. . . . $P_2$, $P_3$ and $P_4$ focus with increasing severity on the correct detection of the outlier . . .; only $P_4$ specifically excludes the possibility that good observations might be significant [adjudged discordant] in addition to . . . [the contaminant]. We see that

$$P_1 \geqslant P_2 \geqslant P_3 \geqslant P_4.$$

Hawkins (1980a), who uses the notation $\beta_1$, $\beta_2$, $\beta_3$, $\beta_4$, $\beta_5$ for $P_3$, $P_1$, $P_5$, $P_2$, $P_4$ respectively, gives some further inequalities for these probabilities. He discusses the various measures of performance, and recommends the use of $\beta_1$ and $\beta_2$, i.e. $P_3$ and $P_1$. In another paper (Hawkins, 1978) he points to the merits of $P_3$.

In point of fact there is reason to think that $P_1$, $P_3$ and $P_5$ are all useful measures; on the other hand, the information conveyed by $P_2$ and $P_4$ would seem to be rather limited. Suppose for example that the contaminant $x_c$ is the second greatest observation, $x_{(n-1)}$. For a 'good' test we want a high probability of identifying $x_c$ as discordant but this should appropriately be done by reference to the null distribution of $Z_{(n-1)}$ and not of $Z_{(n)}$, so that $z_\alpha$ is not the appropriate critical value and $P_2$ not the appropriate measure. Similar considerations apply to $P_4$, based as it is on the inequality $Z_{(n-1)} < z_\alpha$. In fact, David (1981) labels $P_2$, $P_4$ respectively as the 'probability that $X_1$ [$x_c$ in our notation] is significantly large' and the 'probability that only $X_1$ [i.e. $x_c$] is significant'. While these definitions appear attractive, the concept of 'significance' on which they rest is ill-defined.

We can therefore discard $P_2$ and $P_4$ as performance criteria. Once we do this, the need to distinguish between general and specific tests for discordancy disappears, since $P_1$, $P_3$, and $P_5$ do not depend for their definition on the test being of one or the other type. (For this generalization the event '$Z_c = Z_{(n)}$' in the definitions of $P_3$ and $P_5$ needs to be rewritten as '$x_c = x_{(n)}$'.)

$P_1$, $P_3$ and $P_5$ may be considered to contain between them all the relevant information about the performance of a discordancy test against a slippage alternative. $P_1$ is a convenient general measure, for the reason indicated by David (1981) (see above). There are strong arguments, however, for preferring $P_3$ as a measure. Dixon (1950), discussing the assessment of a number of discordancy criteria for outliers in normal samples, says:

The performance of the . . . criteria is measured by computing the proportion of the time the contaminating distribution provides an extreme value and the test discovers the value [i.e. $P_3$]. Of course, performance could be measured by the proportion of the time the test gives a significant value when a member of the contaminating population is present in the sample, even though not at an extreme [i.e. $P_1$]. However, since it is assumed that discovery of an outlier will frequently be followed by the rejection of an extreme we shall consider discovery a success only when the extreme value is from the contaminating distribution.

For a 'good' test, we require $P_3$ and $P_5$ to be high. We also want $P_1 - P_3$ to be low; this is the probability that the test wrongly identifies a good observation as discordant. $P_3/P_5$ is the probability that the contaminant shows up as the outlier, and it might appear that one would like this ratio to be as large as possible, in conflict with the requirement for a high value of $P_5$. Consider, however, two hypothetical tests with performance measures as follows:

$$\text{test } A: P_3 = \tfrac{1}{2}, P_5 = 1, P_3/P_5 = \tfrac{1}{2}.$$

$$\text{test } B: P_3 = \tfrac{1}{2}, P_5 = \tfrac{1}{2}, P_3/P_5 = 1.$$

In test $A$, the contaminant has only 50 per cent chance of showing up as the extreme value, indicating that the degree of contamination is not severe on average (the value of $\Delta$ in $\bar{H}$ is not large); however, when the contaminant does appear as the outlier, it is certain to be detected. In test $B$, on the other hand, contamination is more severe and the contaminant is always the extreme value. But it is only detected as discordant 50 per cent of times. What we require of a test is that it should identify contamination when this is sufficiently manifest, so test $A$ is preferable to test $B$, and a high value of $P_5$ is desirable rather than a high value of $P_3/P_5$. The same line of argument indicates that $P_5$ takes precedence over $P_3$ as a measure of test performance. To sum up, a good test may be characterized by high $P_5$, high $P_3$ (implying high $P_1$), and low $P_1 - P_3$.

So far we have been talking in terms of a slippage alternative. Consider now the assessment of performance of a discordancy test against a *mixture* alternative. As before, take the particular situation of an upper outlier $x_{(n)}$ under test in a univariate sample. Under $\bar{H}$ the number of contaminants in the sample is no longer fixed as in the slippage case, but is a binomially distributed random number which may be 0, 1 or more. The following events are clearly relevant to the assessment of performance:

$D$: that the test identifies $x_{(n)}$ as discordant.
$E$: that $\bar{H}$ holds and the sample contains one or more contaminants.
$F$: that $\bar{H}$ holds and $x_{(n)}$ is a contaminant.

The direct analogue of the performance measure $P_1$ defined in (4.2.2), i.e. the *power*, is $P(D \mid \bar{H})$. However $P(D \mid E)$, which is a power function conditional on the actual presence of contamination, is also a useful measure; denote it by $P_6$.

Analogously to the measures $P_3$, $P_5$ of (4.2.4) and (4.2.6) we can define measures $P(F \cap D \mid E)$ and $P(D \mid F)$ (which is of course the same as $P(D \mid F \cap E)$).

Characteristics of a good test are, by the previous line of argument, a high value of $P(D \mid F)$, a high value of $P(D \mid E)$, and a low value of $P(D \mid \bar{H}) - P(D \cap F \mid \bar{H})$, the probability that the test identifies as discordant an observation actually generated by the basic model; i.e., high $P_5$, high $P_6$, and low $P_1 - P_3$.

In the case of a discordancy test against an *inherent* alternative, the situation simplifies. There is now no specifiable contaminant observation, and the probabilities $P_3$ and $P_5$ (and the events $E$ and $F$) are undefined. The appropriate measure of performance of the test is the power $P(D \mid \bar{H})$.

### 4.2.2 Distributional properties under the alternative hypothesis

An account has been given in Section 3.1.2 of the considerable body of work on the distributional properties of order statistics in samples from contaminated distributions. Much of this work is concerned with moments of order statistics and their recurrence relations under the contamination model. However, basic formulae such as (3.1.13), (3.1.14) for the *densities* of order statistics in contaminated samples are clearly of direct relevance to the calculation of performance measures such as (4.2.2), (4.2.4), (4.2.6). So far, these formulae do not appear to have been exploited to any noticeable extent for test performance calculations, and reliance has continued to be placed on the use of simulation; for a recent example of an extensive simulation-based empirical power study of various tests against contamination alternatives, see Balakrishnan and Ambagaspitiya (1989).

## 4.3  THE MULTIPLE OUTLIER PROBLEM

We have discussed the testing of a single outlier for discordancy, and have also referred to situations when the number of observations which appear aberrant in relation to the main data mass is more than one. We may have, for instance, a normal sample of size $n$, with two upper outliers $x_{(n-1)}$ and $x_{(n)}$ both of which are unusually far to the right of the other $n - 2$, or a normal sample with two lower outliers $x_{(2)}$ and $x_{(1)}$ unusually far to the left, or again a sample with a lower and an upper outlier-pair, $x_{(1)}$ and $x_{(n)}$, widely bracketing the main data mass. Again, a normal sample may contain three extreme values which appear to be outlying in relation to the main $n - 3$, perhaps all three upper, perhaps two upper and one lower, and so on.

Similar situations may arise with gamma samples, though in the particular case of the exponential distribution its $J$-shape makes it likely that all outliers presenting themselves will be upper ones, not lower. Two or more outliers may in fact be encountered in samples from most univariate distributions or from distributions of higher dimensionality; and it is possible to find two or more outlying points in a regression, two or more outlying residuals underlying the observations from a designed experiment, two or more outlying directions in a spherical sample, or two or more outlying values in a time-series. In all these *multiple outlier* situations there are $k(> 1)$ outliers or outlying points in a data set of size $n$, and the analyst envisages the possibility of up to $k$ contaminants. Appropriate tests of discordancy will therefore be required. We defer for a short while the question of how the analyst might decide on the value of $k$.

Faced with a multiple outlier situation there is, as we have said earlier, a basic choice between two types of procedure, *block procedures* and *consecutive procedures*. The latter are sometimes referred to as 'sequential', 'stepwise', or 'recursive' procedures—see e.g. Dixon (1950), Prescott (1979), Hawkins (1980a), and Jain (1981a); see also Section 4.3.2. We prefer the word *consecutive* to *sequential*. Sequential testing, in common statistical parlance, implies that the sample size is not fixed but is determined in each realization in relation to the values of the earlier observations. In the usual context of successive testing of multiple outliers this sequential property applies to the number of times the test is used, not to the sample size, which is *fixed*.

(A true *sequential* situation of discordancy testing is discussed by Quesenberry (1986): namely the screening of outliers in process quality control by variables. Typically, observations $x_1, x_2, \ldots, x_n$ are observed one at a time in chronological sequence. Each observation as it appears is tested for discordancy in relation to the already screened set of previous observations, using one of the available single-outlier tests, and deleted if appropriate. As an example, Quesenberry analyses a set of data on metal machining measurements in a manufacturing process. Screening is required here because, as he explains,

[this] process, as with many metal machining processes, is prone to produce occasional outlying units. One possible cause of these outliers is the occurrence of chips. Even though the machine has a chip guard, occasionally a machined chip will fall against the cutting blade and cause one or two outlying units before it burns or falls off. If these outlying units are used in computing automated adjustments for the process, then serious overadjustments can result.)

Suppose to fix ideas that, specifying $k = 2$, we wish to test for discordancy two upper outliers $x_{(n-1)}, x_{(n)}$ in an exponential sample of size $n$. Take $x_{(1)}, \ldots, x_{(n-2)}$ as belonging to the distribution $F$ with density $\theta e^{-\theta x}(x > 0)$, and $x_{(n-1)}$ and $x_{(n)}$ as belonging to exponential distributions $G_1, G_2$ with respective densities $\lambda \theta e^{-\lambda \theta x}$, $\mu \theta e^{-\mu \theta x}(x > 0)$. The working hypothesis is

$$H : \lambda = \mu = 1.$$

If we take as the single alternative hypothesis

$$\bar{H} : \lambda = \mu < 1,$$

we are led to a single discordancy test, as the result of which we either accept both outliers as consistent with the rest of the sample, or adjudge them both discordant. A possible test statistic in this context would be

$$[x_{(n-1)} + x_{(n)}] \Big/ \sum_{i=1}^{n} x_i.$$

This exemplifies a block procedure—we would be testing the multiple outliers *en bloc*.

On the other hand, we could go for a pair of consecutive alternatives to $H$, which might typically be as follows:

$$\bar{H}' : \lambda = 1, \mu < 1$$

$$\bar{H}'' : \lambda < 1.$$

A familiar procedure is first to test $H$ against $\bar{H}'$ using a test for a *single* upper outlier; if $H$ is accepted, both outliers are declared consistent with the remainder of the sample, and the discordancy test terminates. If $H$ is rejected, $\bar{H}''$ is tested against a revised working hypothesis confined to $x_{(1)}, \ldots, x_{(n-1)}$,

$$H'' : \lambda = 1.$$

Again we would use a test for a single upper outlier. We thus have a *consecutive* procedure, with three possible paths:

Accept $H \rightarrow$ adjudge neither $x_{(n)}$ nor $x_{(n-1)}$ discordant.
Reject $H$, accept $H'' \rightarrow$ adjudge $x_{(n)}$ discordant, but not $x_{(n-1)}$.
Reject $H$, reject $H'' \rightarrow$ adjudge both $x_{(n)}$ and $x_{(n-1)}$ discordant.

For many years, consecutive procedures in the literature were performed in this way 'from the outside inwards': the most extreme outlier was tested first, then the second most extreme, and so on. But the now widely used alternative of doing it the other way round is often preferable. Suppose that, in our above example, we *first* test $H''$ against $\bar{H}''$, using a test for $x_{(n-1)}$ as a single upper outlier in the sample $x_{(1)}, \ldots, x_{(n-1)}$ *omitting* $x_{(n)}$. If $H''$ is *rejected*, the outlier $x_{(n-1)}$ is declared discordant *and with it the more extreme outlier* $x_{(n)}$. If $H''$ is *accepted*, we *then* test the working hypothesis $H$ against $\bar{H}'$. The three possible paths with this 'inside-out' consecutive procedure are:

Reject $H''$ → adjudge both $x_{(n-1)}$ and $x_{(n)}$ discordant.

Accept $H''$, reject $H$ → adjudge $x_{(n)}$ discordant, but not $x_{(n-1)}$.

Accept $H''$, accept $H$ → adjudge neither $x_{(n-1)}$ nor $x_{(n)}$ discordant.

We will refer to these two types of procedure as *inward* and *outward* consecutive procedures respectively. Hawkins (1980a) terms them *forward selection* and *backward elimination stepwise procedures*, and considers in detail their relative advantages and disadvantages; see below.

In the above discussion, the choice between block and consecutive procedures has been illustrated. An apparently similar kind of choice is, of course, familiar in the testing of the relevance of a subset of the regressor variables $x_1, x_2, \ldots$ in a regression analysis. In the regression situation the data context sometimes gives a guide as to when a block procedure is appropriate, in preference to the consecutive testing of variables one by one which is the norm. For example, we might be studying the effect of nine factors $x_1, \ldots, x_9$ on the efficiency, $y$, of a domestic heating device tested *in situ*. Suppose $x_1, x_2$ are properties of the fuel used, $x_3, \ldots, x_7$ are different features of the internal construction of the house, and $x_8, x_9$ are measurements of ambient temperature at ground and roof level outside the house. Then it is reasonable to test the significance of $x_8$ and $x_9$ jointly—a block procedure—on the basis that if $y$ is affected at all by the outside temperature we should clearly take both of the outside temperature measurements into account.

In the usual multiple outlier situation, all the observations have the same status on the working hypothesis, and discordancy tests are only invoked when particular values show up as outliers; thus no guidance is available from the data context as to whether a block procedure should be used. The exception is when we have prior information which leads us to focus on some particular subset of the data—for example, if $x_{(1)}, x_{(n)}$ were observations made by experimenter A and the other $n - 2$ by experimenter B.

*Example 4.4   An interesting example, in which the 'prior' information only came to hand after the data had been analysed, relates to sex ratios in the light-brown apple moth. If the ith adult female moth produces a progeny of size $n_i$ consisting of $m_i$ males and $f_i$ females $(m_i + f_i = n_i)$, $x_i = m_i/n_i$ is the observed sex ratio; $m_i$ can be regarded as a value from a binomial distribution $\mathbf{B}(n_i, p_i)$, where $p_i$ is the sex ratio parameter associated with parent female i. Geier and Briese (1977) showed that there were two kinds of female moths, as characterized by their progenies: 'normal type', for which $p_i$ is approximately one-half, and 'Q-type', with few or no males in the progeny $(p_i \ll \frac{1}{2})$. In an investigation by Geier and Briese, reported in Geier, Briese, and Lewis (1978), 439 progenies were collected, their x-values being illustrated in Figure 4.1. The progeny sizes $n_i$ were mostly in the range 20–70. There were 413 progenies with sex ratios (shown as a histogram in Figure 4.1) well fitted by assuming a common value $p_i = p \simeq \frac{1}{2}$ (in fact, estimated from the data as*

*0.506). These progenies can all be regarded as of normal type; their m-values conform to the model*

$$m_i \sim \mathbf{B}(n_i, p). \tag{4.3.1}$$

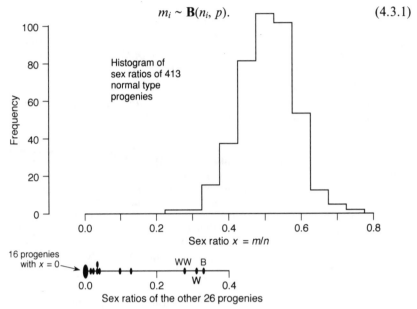

**Figure 4.1**  Sex ratios of 439 apple moth progenies

*The remaining 26 progenies had low sex ratios, 16 actually zero (progeny all female) and 10 shown in Figure 4.1 as points on the horizontal line below the histogram. Seven of these 10 were in the range 0.02–0.13 and three, marked WW, W, B on the diagram, had higher values as follows:*

**Table 4.2**

| Locality | $m_i$ | $n_i$ | Sex ratio $x_i = m_i/n_i$ |
|---|---|---|---|
| Wangi Wangi (WW) | 13 | 47 | 0.277 |
| Waikerie (W) | 18 | 58 | 0.310 |
| Bicheno (B) | 14 | 43 | 0.326 |

*The 16 + 7 sex ratios not exceeding 0.13 could be satisfactorily modelled by assuming that their parameters $p_i$ were random values from a beta distribution with two unknown parameters r, s (say), implying that $m_i$ for a progeny of size $n_i$ had a mixed-binomial distribution given by*

$$P(m_i = m) = \frac{\Gamma(r + s)}{\Gamma(r)\,\Gamma(s)} \binom{n_i}{m} \frac{\Gamma(r + m)\,\Gamma(s + n_i - m)}{\Gamma(r + s + n_i)} \tag{4.3.2}$$

*The corresponding 23 progenies, with low m-values all consistent with the model (4.3.2), could all be assigned to Q-type.*

*However, the three sex ratios in Table 4.2 were significantly too large to be fitted by this model (significance probability 0.0025). They were thus upper outliers with respect to model (4.3.2) for the Q-type data. At the same time, their values had been perceived as low in relation to $\frac{1}{2}$, and for this reason they had not in the first place been included with the 413 cases clearly assigned to normal-type; they were thus lower outliers with respect to model (4.3.1) for the normal-type data. Entomologically, there were only two kinds of progeny, normal and Q-type, and the problem was to **allocate** each of the three outliers to one or other group; to postulate a third group for them would make no entomological sense.*

Note the difference from the usual problem of outlier identification, in which we have an outlying value (or set of outlying values) which are extreme with respect to some main population; we wish to judge by a discordancy test whether or not it is statistically reasonable to regard the outlier (or outliers) as belonging, though extreme, to the main population. If found not reasonable then the outlier is judged *discordant* and requires a separate probability model from the main population. Here we have a different problem. Our outlying values lie *between two main populations* and are extreme with respect to *both*; we wish to *allocate* each outlier to one or other of the populations and thus avoid setting up separate models for them. Is this statistically reasonable, and if so how should the allocation be done? (Lewis, 1987)

We can thus add **allocation** to the catalogue of aims listed in Figure 2.1 on page 42. Reverting to the apple moth data, the three $x$-values in Table 4.2 had been judged discordant with respect to model (4.3.2); further consideration suggested that, though they were lower outliers with respect to model (4.3.1), they might be statistically reasonable in relation to this model. The use of a *block test* for the three outliers was therefore indicated. On the working hypothesis, each $x_i = m_i/n_i$ was effectively a value from a normal distribution $N(\frac{1}{2}, 1/4n_i)$. There were three different $n$-values, 47, 58 and 43, but using the greatest of these, 58, ensured a conservative test, i.e. one somewhat overvaluing the weight of evidence in favour of discordancy. Accordingly a block test was carried out of the three $x$-values as lower outliers in a sample of size $413 + 3 = 416$ from a normal distribution $N(\frac{1}{2}, 1/4 \times 58)$, using the test statistic (page 247)

$$T_{N\mu\sigma_3} = (3 \times 0.5 - 0.277 - 0.310 - 0.326)/(1/232)^{1/2} = 8.94.$$

Clearly from Table XVg (page 493) this value is not significant, so the analysis indicated that the three parent females in Table 4.2 were all of normal type. (Note the interesting use of a normal sample discordancy test with $\mu$ and $\sigma$ both known.)

Actually, Geier and Briese later found from observations on second-generation progenies descended from these parent females that the Bicheno parent female in Table 4.2 was Q-type, not normal type. The rare (but not incredibly rare) event associated with the above-mentioned P-value of 0.0025 had in fact occurred!

The choice between a block procedure and a consecutive procedure in a multiple outlier situation depends on the relative performances of the test procedures in relation to an alternative hypothesis $\bar{H}$. The first direct quantitative comparisons of block and consecutive procedures appear to be those of McMillan and David (1971) and McMillan (1971). McMillan and David consider a normal sample with known variance, unity say, containing two contaminants from a normal distribution also having unit variance but with mean slipped to the right. They evaluate $P_3$ (Section 4.2.1) for a block discordancy test based on the sum of the two largest deviations from the sample mean; and they also consider a consecutive procedure based at each stage on the largest deviation from the mean, evaluating the probabilities that at least one contaminant is identified as discordant and that both contaminants are so identified; see Section 4.3.2 below. McMillan (1971) gives corresponding results in terms of studentized deviations for the case when the underlying variance is unknown. Hawkins (1973) extends McMillan's results and gives values of the power function for the consecutive test in various cases. Other comparative performance studies of multiple outlier tests include those by Hawkins (1978) and Prescott (1978) in relation to normal samples and Kimber (1982) in relation to exponential samples, both Prescott and Kimber making extensive use of sensitivity contours. However, it is still largely the case that, as Hawkins (1980a) observes,

relatively little is known about the performance of [the various available] procedures in the presence of multiple outliers

and much research remains to be done in this area.

Consecutive procedures have an obvious appeal, but, as has long been recognized, the *inward* form of consecutive procedure suffers from an inherent limitation, the possible effect of *masking*. On the other hand, *outward* procedures are largely immune to masking, providing that the actual number of contaminants in the sample does not exceed the number of outliers $k$ assumed in the test procedure. On the whole the outward structure is preferable to the inward; see Prescott (1978), Hawkins (1980a) and Kimber (1982). See also Hawkins (1980b) and page 236 for comments on the case where the number of contaminants is *less than* the prescribed $k$.

This brings us back to the question of deciding on a value for $k$, which we raised at the beginning of Section 4.3. For a *block* procedure, and again for an *outward consecutive* procedure, $k$ needs to be specified. With an *inward* consecutive procedure, there is a choice. On the one hand, a number of outliers up to a specified maximum number $k$ can be tested consecutively; or, alternatively, a sequence of single-outlier tests can be carried out until the first non-discordant result is reached, say on the $(m + 1)$th test, in which case $k$ is determined by the test process itself as equal to $m$.

The value of $k$ chosen for the operation of a block test or consecutive test on a given sample is in effect the *maximum number of contaminants which the sample is assumed to contain*. It may well be dictated by straightforward inspection of the data, as the number of observations which 'obviously' stand out from the rest of the data set. Often, however, it is preferable to decide the value of $k$ by some process of calculation from the data, rather than by visual impression. The problem of deciding on the number of outliers in a sample has been considered by various authors, including Tietjen and Moore (1972), Tiku (1975), Rosner (1975, 1977), Kale (1976), and Jain and Pingel (1981a).

For given sample size $n$ there must be an effective upper limit to the number of outliers $k$. To take an extreme case, one cannot consider $k = n - 1$ observations as outliers in relation to the remaining 1! Indeed the concept of the 'main data mass' is hardly meaningful if $k \geqslant n/2$, say. Intuitively, an upper limit of the form $k_{max} = Cn^\alpha$ suggests itself for $k$, where $C$ is a positive constant (perhaps 1) and $\alpha$ a constant between 0 and 1. Taking $\alpha = \frac{1}{2}$, $C = 1$ for example, we would get $k_{max} = \sqrt{n}$, so that one would not deal with more than 10 outliers in a sample of 100. As Chhikara and Feiveson (1980) observe,

it is reasonable to consider three potential outliers in a data set of 10 observations, but it is unrealistic to expect 30 outliers out of a data set of 100 observations. In the latter case, the outlier detection problem becomes one of discrimination between two or more classes of data.

An early approach to the determination of $k$ is due to Tietjen and Moore (1972). To fix ideas, suppose that the number of upper outliers is to be estimated; they propose locating the largest gap between adjacent observations in the sample arranged in ascending order to the right of the sample mean, and declaring $k$ to be the number of observations to the right of this largest gap. If, for example, $x_{(n)} - x_{(n-1)}$ is the largest gap, $k = 1$. Other procedures based on the largest gap, in weighted or generalized form, are due to Tiku (1975) and Jain and Pingel (1981a).

Another approach is due to Kale (1976). As before, suppose that the number, $k$, of upper outliers is to be determined. For any $m = 0, 1, 2, \ldots$, the ordered observations $x_{(1)}, \ldots, x_{(n-m)}$ in the reduced sample obtained by omitting the $m$ most extreme values may be compared with the order scores $E(X_{(1)}), \ldots, E(X_{(n-m)})$ for a sample of size $n - m$ from the null distribution. The essence of Kale's procedure is to calculate, for each $m$, a measure of agreement between the ordered observations and the order scores. This measure might, for example, be the mean of the squared differences, or some other suitable statistic. The value of $m$ giving the greatest measure of agreement is declared to be the number of outliers $k$.

For further discussion of this problem, see Rosner (1975).

### 4.3.1 Block procedures for multiple outliers in univariate samples

The considerations governing discordancy tests for single outliers, discussed earlier in this chapter, extend to block-type discordancy tests for multiple outliers—in the construction of tests, the existence in some cases of inclusive and exclusive measures and of recursive relations for null distributions, and in the setting up of performance criteria.

As in the single outlier case (Section 4.1), appealing test statistics can be set up on an intuitive basis; and 'best' tests can be constructed on the maximum likelihood ratio principle or again on the principle of local optimality. See, for example, Gather and Kale (1988), Gather (1989).

Consider again the testing of a pair of upper outliers $x_{(n-1)}$, $x_{(n)}$ in an exponential sample. On intuitive grounds—maybe by generalization from the case of a single upper outlier—one could propose

$$T_{(n-1, n)} = \{x_{(n-1)} + x_{(n)}\}/2\bar{x} \qquad (4.3.3)$$

as a sensible statistic. Again, Dixon-type statistics such as $\{x_{(n)} - x_{(n-2)}\}/\{x_{(n-2)} - x_{(1)}\}$ have a natural appeal. On the other hand, let us see where the maximum likelihood ratio principle leads. Our working hypothesis is

$$H : x_j \in F \qquad (j = 1, 2, \ldots, n)$$

declaring that all the observations $x_1, \ldots, x_n$ belong to the distribution $F$ with density $\theta e^{-\theta x}(x > 0)$, $\theta$ unknown.

Suppose first that we have a slippage alternative $\bar{H}$ stating that $n - 2$ of the observations belong to $F$ and the remaining two, $x_{n-1}$ and $x_n$ say, come from the exponential distribution $G_\lambda$ with density $\lambda\theta e^{-\lambda\theta x}(x > 0 ; \lambda < 1)$. Calculations similar to those of equations (4.1.1) to (4.1.5) lead to

$$T = (x_{n-1} + x_n)/2\bar{x} \qquad (4.3.4)$$

as test statistic, providing that $(x_{n-1} + x_n)/2 \geqslant \bar{x}''$, the mean of $x_1, \ldots, x_{n-2}$. If instead of $\bar{H}$ we adopt the corresponding *labelled* slippage alternative

$$\bar{H}' : x_{(1)}, \ldots, x_{(n-2)} \text{ belong to } F$$

$$x_{(n-1)}, x_n \text{ belong to } G_\lambda,$$

calculations similar to those of equations (4.1.6) to (4.1.9) lead to $T_{(n-1, n)}$ as test statistic. Alternatively, $T_{(n-1, n)}$ can be set up on the basis of a multiple decision argument as in (4.1.10).

But $\bar{H}$, $\bar{H}'$ are not the only pair of slippage alternatives. Consider instead the slippage alternative $\bar{H}''$ stating that $x_1, \ldots, x_{n-2}$ belong to $F$, $x_{n-1}$ belongs to $G_\lambda$ and $x_n$ belongs to the exponential distribution $G_\mu$ with density $\mu\theta e^{-\mu\theta x}(x > 0; \mu < \lambda)$. The maximum likelihood ratio statistic is now not $T$ but

$$T'' = \left(\frac{x_{n-1} \, x_n}{\bar{x}^2}\right)^{1/n} \left(n - \frac{x_{n-1} + x_n}{\bar{x}}\right)^{(n-2)/n} \tag{4.3.5}$$

The multiple decision argument leads to the statistic

$$T''_{(n-1,\, n)} = \left(\frac{x_{(n-1)} \, x_{(n)}}{\bar{x}^2}\right)^{1/n} \left(n - \frac{x_{(n-1)} + x_{(n)}}{\bar{x}}\right)^{(n-2)/n} \tag{4.3.6}$$

On the other hand, the labelled slippage alternative corresponding to $\bar{H}''$, viz.

$$\bar{H}''' : x_{(1)}, \, \ldots, \, x_{(n-2)} \text{ belong to } F$$

$$x_{(n-1)} \text{ belongs to } G_\lambda$$

$$x_{(n)} \text{ belongs to } G_\mu$$

does *not* now lead to $T''_{(n-1,\, n)}$, but gives a maximum likelihood ratio test statistic which cannot be expressed in closed form.

As regards performance criteria for tests of multiple outliers, we begin by generalizing the discussion of Section 4.2.1 to the block test situation. Consider, as before, a univariate sample $x_1, x_2, \ldots, x_n$, with null hypothesis $H : x_j \in F$ ($j = 1, 2, \ldots, n$). Instead of a single outlier, say $x_{(n)}$, we have a group of $k$ outliers ($k > 1$), say $x_{(n-k+1)}, \ldots, x_{(n-1)}, x_{(n)}$. Using a *block discordancy test statistic*, $Z$ say, we have a test which takes the form: 'Adjudge $x_{(n-k+1)}, \ldots, x_{(n-1)}, x_{(n)}$ discordant if $Z > z_\alpha$, where $z_\alpha$ is the critical value for preassigned significance level $\alpha$ defined by

$$P(Z > z_\alpha \,|\, H) = \alpha. \text{ '} \tag{4.3.7}$$

Consider first a slippage alternative (*unlabelled*) $\bar{H}$ under which $n - k$ of the observations belong to $F$ and $k$ observations, the *contaminants*, belong to some different distribution $G$. Analogously to (4.2.2), (4.2.4) and (4.2.6), three useful measures of the performance of the test are $P_1$, $P_3$ and $P_5$, where $P_1$ is the *power function*

$$P_1 = P(Z > z_\alpha \,|\, \bar{H}); \tag{4.3.8}$$

$P_3$ is the probability that the $k$ contaminants are the outliers and are identified as discordant; and $P_5$ is the probability that, when the contaminants are the outliers, they are identified as discordant.

Beyond the discussion of Section 4.2.1 further measures have been proposed, using the following variables $C$ and $W$ defined under $\bar{H}$:

$C$ = the number of contaminants correctly identified, i.e. declared discordant

$W$ = the number of 'false positives', i.e. non-contaminants wrongly declared discordant.

The proposed measures originate with Beckman and Cook (1983) in their impressive review of outlier methodology (written by invitation to mark the occasion of *Technometrics*'s 25th year of publication—what better topic than outliers for such an occasion!). Fung and Paul (1985), in a study of outlier detection procedures for extreme-value samples with slippage models for contamination, use the six measures listed below. As Fung and Paul state, their six criteria are similar to those suggested earlier by Beckman and Cook; however, we shall use their list in the following discussion rather than Beckman and Cook's, for two reasons. First, there appears to be some discrepancy in Beckman and Cook's paper between their definition of the word 'outlier' on their page 121 (a collective to refer to *either* an observation that appears surprising [an outlier in the terminology of this book] *or* a contaminant) and their use of the word 'outlier' in their proposed performance measures on page 132 (evidently a contaminant). Then, their second criterion, $P(C > 0, W = 0)$, is not directly appropriate to a block test since, if $W = 0$, $C$ can only equal 0 or $k$.

Clearly, under $\bar{H}$, if $Z > z_\alpha$ then the block test declares $k$ observations discordant, and $C + W = k$; if $Z \leq z_\alpha$ then $C = W = 0$. Write $P(C = c, W = k - c) = \psi_c$, $c = 0, 1, \ldots, k$. Then

$$P_1 = P(Z > z_\alpha) = P(C + W = k) = \psi_0 + \psi_1 + \ldots + \psi_k; \qquad (4.3.9)$$

and

$$P_3 = P(W = 0, Z > z_\alpha) = \psi_k. \qquad (4.3.10)$$

The six measures used by Fung and Paul (1985) are as follows:

(i) $P(C = k)$; this $= \psi_k = P_3$ from (4.3.10).
(ii) $P(W = k) = \psi_0$.
(iii) $P(0 < C < k)$; this $= \psi_1 + \ldots + \psi_{k-1} = P_1 - P_3 - \psi_0$.
(iv) $E(C)$, the expected number of detected contaminants; this $= \sum_{i=1}^{k-1} i\,\psi_i$.
(v) $E(W)$, the expected number of false positives.
(vi) $E(C)/[E(C) + E(W)]$, the ratio of the expected number of real contaminants declared discordant to the expected number of observations declared discordant.

Since $E(C + W) = kP(Z > z_\alpha) = kP_1$, this list, together with $P_1$ and $P_3$, only amounts to four independent measures, viz. $P_3$; $P_1$ (or equivalently, as recommended in Section 4.2.1, $P_1 - P_3$); $\psi_0$, the probability that the test declares $k$ discordant observations none of which are in fact contaminants; and $E(C)$, the expected number of contaminants declared discordant.

Suppose now that the performance of the test is being assessed against a *mixture* alternative. Under $\bar{H}$, the number of contaminants is now a binomially distributed random variable whose possible values include 0; denote it by $B$.

If $Z > z_\alpha$, $C + W = k$ and $C$ takes values $0, 1, \ldots, \min(k, B)$; if $Z \leqslant z_\alpha$, $C = W = 0$ whatever the value of $B$. The performance of the test on any particular occasion obviously depends on the realized value of $B$. If $B$ happens to be zero the power $P_1$ has the small value $\alpha$, and generally when $B < k$ a block test of $k$ outliers is to a lesser or greater extent inappropriate. The measures $P_1$, $P_3$, $P_5$, $P_6$ in Section 4.2.1 can reasonably be carried over to the block situation by redefining the events $D$, $E$, $F$ as follows.

$D$: that the test declares discordant the $k$ outliers
$x_{(n-k+1)}, \ldots, x_{(n-1)}, x_{(n)}$.
$E_b$: that $\bar{H}$ holds and the sample contains at least $b$ contaminants $(b = 1, 2, \ldots, k, k+1, \ldots)$.
$F$: that $\bar{H}$ holds and all of the $k$ outliers are contaminants.

Clearly $P(F \mid B) = 0$ for $B < k$.
We can now write

$$P(D \mid \bar{H}) = P_1 = P(Z > z_\alpha);$$

$$P(D \cap F \mid E_k) = P_3 = P(C = k)/P(E_k);$$

$$P(D \mid F) = P_5 = P(C = k)/P(F);$$

$$P(D \mid E_k) = P_6 = P(B \geqslant k, Z > z_\alpha)/P(B \geqslant k).$$

Characteristics of a good test in this situation are high $P_5$, high $P_6$, low $P_1 - P_3$; also, as indicated above, low $P(B < k)$, i.e. high $P(E_k)$. Expected values such as $E(C)$ and $E(W)$, which Fung and Paul (1985) employed as criteria in the slippage context (see above), could also be used to measure test performance against a mixture alternative. Such measures would involve averaging over the binomial $B$-distribution, and should therefore be particularly well-suited to the mixture context.

Published work on block procedures includes Grubbs (1950), Tietjen and Moore (1972), Tiku (1975), Prescott (1978) and Hawkins (1977, 1978, 1979, 1980a) for normal samples, Likeš (1966, 1987) and Kimber and Stevens (1981) for exponential samples, Tiku (1975) and Lewis and Fieller (1979) for gamma samples, and Fung and Paul (1985) for extreme-value samples.

### 4.3.2 Consecutive procedures for multiple outliers in univariate samples

Procedures for testing multiple outliers consecutively for discordancy have been investigated by a number of authors. Notable among these are the following whose work we have already referenced: Beckman and Cook, David, Hawkins, Jain, Jain and Pingel, Kimber, Prescott, Rosner (1975, 1977), Tiku; together with Rosner (1983), Simonoff (1984 a,b, 1987 a,b,c), and Sweeting (1983, 1986). For early work, see Pearson and Chandra Sekar

(1936), Dixon (1953), Ferguson (1961b), Tietjen and Moore (1972), McMillan and David (1971), and McMillan (1971).

Sweeting (1983, 1986) presents a versatile outward consecutive procedure, based on what he calls 'scale-free spacings'. In his method, outlined below, the null distribution of the test statistic is known *approximately* for samples from a wide variety of distributions and *exactly* for exponential and uniform samples. He gives examples of its application to exponential, normal, Weibull, logistic and gamma samples.

Given an ordered sample $x_{(1)} < x_{(2)} < \ldots < x_{(n)}$ from a continuous distribution $F$ with density $f(x)$, the spacings are the quantities

$$x_{(i)} - x_{(i-1)} \quad (i = 2, \ldots, n).$$

From these, Sweeting (1983, 1986) defines the statistics

$$T_i = D_i / \sum_{j=2}^{i} D_j \quad (i = 2, \ldots, n) \tag{4.3.11}$$

where

$$D_i = h_i(x_{(i)} - x_{(i-1)}), \ h_i = nf(q_i), \ E(q_i) = (i-1)/n. \tag{4.3.12}$$

The $T_i$ are Sweeting's *scale-free spacings*. For exponential $F$, and again for uniform $F$, Sweeting (1983) shows that the $T_i$ are independently distributed with distribution functions

$$P(T_i < t) = 1 - (1 - t)^{i-2} \quad (0 \leq t \leq 1) \tag{4.3.13}$$

Similar results hold asymptotically for a wide range of other distributions (Sweeting, 1986).

Applying these results to, for instance, the testing of $k$ upper outliers, the quantities $T_{n-k+1}, T_{n-k+2}, \ldots$ are tested consecutively outwards, with test distributions (4.3.13), until significance (or a final non-significant $T_n$) is reached. For a level-$\alpha$ test, positive constants $c_{n-k+1}, c_{n-k+2}, \ldots, c_n$ with $\Sigma c = 1$ are chosen such that the level of the test based on $T_i$ is $1 - (1 - \alpha)^{c_i}$, $i = n - k + 1, \ldots, n$. Sweeting offers no guidance on the question of 'optimal' choice of the $c_i$, remarking only that 'Monte Carlo studies may throw some light on this' (Sweeting, 1983). He shows how to use his procedure for the testing of a combination of lower and upper outliers.

In the usual consecutive procedures, successively changing values of sample size are used in the sequence of stages: $n, n - 1, n - 2, \ldots$ (inward) or $n - k, n - k + 1, n - k + 2, \ldots$ (outward). An interesting variant is proposed by Balasooriya (1989), in which each stage involves the full sample size $n$. He gives a method for the detection of discordant outliers in an exponential sample, using the idea of successive 'prediction' of ordered sample values; his assumed basic model is the exponential distribution with density $(1/\lambda) \exp(-x/\lambda)$ $(x > 0)$, but the method could in principle be applied to other models. Suppose we knew only the $r$ lowest ordered values $x_{(1)}, x_{(2)}, \ldots, x_{(r)}$ in the sample of $n$; based on this information, the maximum likelihood estimate of $\lambda$ is

$$\hat{\lambda} = \left[ \sum_{i=1}^{r} x_{(i)} + (n - r)x_{(r)} \right]/r,$$

and the best linear unbiased predictor of $x_{(r+1)}$ is

$$\hat{x}_{(r+1)} = x_{(r)} + \hat{\lambda}/(n - r).$$

In fact $x_{(r+1)}$ is of course known, but it can be compared with $\hat{x}_{(r+1)}$ to provide a single-outlier test for consecutive use; the test statistic used by Balasooriya is $[x_{(r+1)} - x_{(r)}] / [\hat{x}_{(r+1)} - x_{(r)}]$.

In principle, consecutive procedures present no fresh problem of test construction, since they merely involve repeated use of single outlier tests from the available repertoire. However, there are important choices to be made. Which single outlier test should be used in a given data situation? What value of significance level should be used at each stage? Some recommended procedures for normal samples, which illustrate the issues involved, are presented in Chapter 6 (Test **N19**, pages 235–6).

Criteria for assessing the performance of a consecutive procedure are required, and here we are in a fresh situation. The measures described in Section 4.2.1 will need generalizing since there may be, by definition, not one contaminant but several. Suppose for example that the alternative hypothesis $\bar{H}$ envisages two discordant values, liable to appear as upper outliers, in a sample of $n$. The following events defined under $\bar{H}$ would seem to be relevant in the context of an *inward* procedure (for an outward procedure the definitions of $D_0$, $D_1$ and $D_2$ below would need changing in an obvious way):

$E_1$: that $x_{(n)}$ is one of the two contaminants
$E_2$: that $x_{(n-1)}$ is one of the two contaminants
$E = E_1 \cap E_2$: that the two contaminants are the two outliers
$D_0$: that $x_{(n)}$ is not adjudged discordant (on the first test)
$D_1$: that $x_{(n)}$ is adjudged discordant (on the first test) but $x_{(n-1)}$ is not adjudged discordant (on the second test)
$D_2$: that $x_{(n)}$, $x_{(n-1)}$ are both adjudged discordant (requiring two tests).

'Total' measures corresponding to $P_1$, $P_3$, $P_5$ in Section 4.2 will be respectively $P(D_2)$, $P(D_2 \cap E)$, $P(D_2 | E)$. But 'partial' measures are also of interest, such as $P(D_1 | E)$ and $P(D_0 | E)$ ( $= 1 - P(D_1 | E) - P(D_2 | E)$).

The measures of performance used by McMillan and David (1971) and McMillan (1971) in their pioneering papers on consecutive testing are not in fact any of those we have listed above, but are

$$P(C_1), \ P(C_2), \ P(C_3)$$

defined as follows. Denoting by $R$ the discordancy region for the first-stage test based on all $n$ observations, $C_1$ is the event that at least one of the two contaminants is in $R$ (and so adjudged discordant), and $C_2$ the event that both contaminants are in $R$. $C_3$ is the event that at least one of the two contaminants is in $R$ and the other is in the discordancy region for a second test based on the reduced sample omitting the first contaminant. The values of $P(C_1)$ and $P(C_3)$ should correspond reasonably closely with $P(D_1 \mid E) + P(D_2 \mid E)$ and $P(D_2 \mid E)$ respectively, at any rate when the degree of contamination is marked enough to make it virtually certain that the contaminants appear as the outliers. However, in general the use of $P(C_1)$, $P(C_2)$, and $P(C_3)$ seems rather arbitrary. $P(C_2)$ in particular seems difficult to interpret.

An event which can usefully be defined is

$D_3$: that $x_{(n)}$ is not adjudged discordant (on the first test), but $x_{(n-1)}$ would be adjudged discordant on a hypothetical second test omitting $x_{(n)}$ from the sample.

$P(D_3)$ is clearly a measure of the masking effect.

The alternative approach to the assessment of performance *via sensitivity contours* is well illustrated by Prescott (1978) and Kimber (1982).

# PART II

## *Univariate Data*

# CHAPTER 5

# Accommodation Procedures for Univariate Samples

## 5.1 ESTIMATION OF LOCATION

### 5.1.1 Estimators based on trimming or Winsorization

A basic problem in the use of trimmed or Winsorized means is choosing the extent of trimming or Winsorization. Should we employ an asymmetric $(r \neq s)$, or symmetric $(r = s)$, scheme; how should $(r, s)$ be chosen (or $\alpha$ in the symmetric proportionate schemes)? There are no simple answers. The range (and degree of specification) of possible basic and alternative models and in particular the presumed direction of contamination, the variety of performance criteria which may be adopted, the dependence on sample size, and so on, all affect this choice. Symmetrically trimmed estimators $\overset{\text{T}}{x}(\alpha, \alpha)$ whose properties have been studied include the cases $\alpha = 0.05, 0.10, 0.15, 0.25$ (the mid-mean) and of course $\alpha = 0.5$ (the median); see Table 5.1 below. With an asymmetric basic model such as a gamma distribution, asymmetric trimming is appropriate; details are discussed in Section 5.5. See also Hertsgaard (1979) for some specific results.

Adaptive choice of $(r, s)$ or $\alpha$ can be advantageous, and some recommended schemes for adaptive trimming are discussed below.

Some modifications of trimmed or Winsorized means change the nature of the problem of choosing the degree of trimming or Winsorization. We might, for example, contemplate eliminating (or transferring) sample values in terms of some quantitative measure of their extremeness, rather than merely on the basis of their rank order. Suppose we consider sample residuals $z_j (j = 1, 2, \ldots, n)$, defined on some appropriate basis. For example, if the basic model is $N(\mu, \sigma^2)$ we might use $z_j = x_j - \bar{x}$. *Modified trimming* was first introduced by Anscombe (1960a) with his 'rejection rule' according to which, in a situation where $\sigma$ is known, $\mu$ is estimated by

$$\left.\begin{array}{ll} \bar{x} & \text{if } |z_j| < c\sigma \text{ (all } j) \\ \overset{\text{T}}{x}_{1,0} & \text{if } |z_{(1)}| \geqslant c\sigma \text{ and } |z_{(1)}| > |z_{(n)}| \\ \overset{\text{T}}{x}_{0,1} & \text{if } |z_{(n)}| \geqslant c\sigma \text{ and } |z_{(n)}| > |z_{(1)}| \end{array}\right\} \qquad (5.1.1)$$

for a suitable choice of $c$. If $\sigma$ is unknown it is replaced by $s$.

This means that for an alternative model of the slippage type, incorporating precisely one contaminant, the observation with maximum absolute residual is trimmed if it is *sufficiently* extreme. As a generalization of this idea we have the class of *Anscombe estimators* $\hat{\mu}_k(c)$. An Anscombe estimator is

a mean of those observations which remain after certain of the observations, which 'look' as if they were generated from $N(\theta + A, \sigma^2)$, are discarded. (Guttman and Kraft, 1980)

Specifically, an integer $k$ is chosen to reflect the amount of contamination to be guarded against. Of the $k$ largest absolute residuals $|z_j|$ there will be some number $\hat{k}$ say $(0 \leqslant \hat{k} \leqslant k)$ of 'unduly large' absolute residuals for which

$$|z_j| > cs \qquad (5.1.2)$$

where $c$ is a suitably chosen constant. The corresponding $\hat{k}$ observations are then trimmed and $\hat{\mu}_k(c)$ is calculated as the mean of the remaining $n - \hat{k}$ observations. Anscombe's original estimator (5.1.1), with $s$ replacing the rather artificial choice of $\sigma$, is $\hat{\mu}_1(c)$ in this notation.

Guttman and Kraft (1980) investigate the performance of Anscombe estimators in terms of premium and protection for various sample sizes $n$ between 16 and 48, and compare them with a useful class of $R$-estimators: the Hodges–Lehmann linearized estimators (see below). They conclude that $\hat{\mu}_k(c)$ is vulnerable to the choice of $k$ and that the Hodges–Lehmann type estimator is on the whole to be preferred.

The premium and protection of the original Anscombe estimator (5.1.1) has also been investigated by Guttman and Tiao (1978) in relation to the prospect of correlation between the observations, and they show that the estimator lacks robustness to departures from independence. This is a different kind of alternative from the main ones we have been considering in our study of robust procedures, but it is of practical importance, and as we shall see has relevance in the time-series context. See also Portnoy (1977).

As with modified trimming, *modified Winsorization* can be used: for example, when a single observation is replaced by its nearest neighbour in the ordered sample if its maximum absolute residual is *sufficiently extreme*. Thus Guttman and Smith (1969) suggest estimating $\mu$ in $N(\mu, \sigma^2)$ by

$$\left.\begin{array}{ll} \bar{x} & \text{if } |z_j| < c\sigma \text{ (all } j) \\ \overset{\text{W}}{x}_{1,0} & \text{if } |z_{(1)}| \geqslant c\sigma \text{ and } |z_{(1)}| > |z_{(n)}| \\ \overset{\text{W}}{x}_{0,1} & \text{if } |z_{(n)}| \geqslant c\sigma \text{ and } |z_{(n)}| > |z_{(1)}| \end{array}\right\} \qquad (5.1.3)$$

for a suitable choice of $c$ (again $s$ replaces $\sigma$ if $\sigma$ is unknown).

Another possibility, termed *semi-Winsorization* (Guttman and Smith, 1969) replaces the sufficiently extreme observation (that with largest absolute residual if this exceeds $c\sigma$) with the appropriate cut-off point, $\bar{x} - c\sigma$ or $\bar{x} + c\sigma$, rather than with its nearest neighbour. Again $\mu$ is estimated by the mean of the treated sample, or by $\bar{x}$ if $|z_j| < c\sigma$ for all $j$, with $s$ used in place of $\sigma$ if $\sigma$ is unknown.

A number of proposals have been made for *adaptive* trimming. In relation, for example, to the optimal choice of $\alpha$ in the $\alpha$-trimmed mean for estimating the location parameter of a symmetric distribution Tukey and McLaughlin (1963) and later Jaeckel (1971b) propose that we choose $\alpha$ in some permissible range $(\alpha_0, \alpha_1)$ to minimize the sample variance $s^2(\alpha)$ of $\overset{T}{x}(\alpha, \alpha)$. The resulting *optimal-trimmed mean* $\overset{}{x}(\hat{\alpha}, \hat{\alpha})$ is shown to be asymptotically equivalent (in terms of variance) to the best estimator $\overset{T}{x}(\alpha, \alpha)$ (i.e. with minimum variance $\sigma^2(\alpha, \alpha)$) provided the truly best $\alpha$ happens to lie in the range $(\alpha_0, \alpha_1)$. In a valuable exposition of adaptive robust procedures Hogg (1974) proposes basing the choice of $\alpha$ on the tail length of the distribution sampled, as measured by the statistic

$$Q_n(\beta_1, \beta_2) = \frac{\bar{U}_n(\beta_1) - \bar{L}_n(\beta_1)}{\bar{U}_n(\beta_2) - \bar{L}_n(\beta_2)}. \qquad (5.1.4)$$

Here $\bar{U}_n(\beta)$, $\bar{L}_n(\beta)$ denote the respective averages of the $n\beta$ largest and $n\beta$ smallest observations in the sample (using fractions of the borderline observations if $n\beta$ is not an integer). If the value of $Q_n(\beta_1, \beta_2)$ indicates heavy tails then a higher percentage is trimmed, and vice versa. Typically, the value of $Q_\infty$ (0.2, 0.5) is 1.75 for a normal distribution and 1.92 for a double-exponential distribution. Clearly the greater the value of $Q_n$, the greater the value of $\alpha$ that should be chosen. Hogg (1974) suggests a discrete choice of $\alpha$ corresponding to division of the range of values of $Q_n$ (0.05, 0.5) or of $Q_n$ (0.2, 0.5) into 'bands'. De Wet and van Wyk (1979b) modify Hogg's scheme by providing a continuous choice of $\alpha$ as a function of $Q = Q_n$ (0.2, 0.5) according to the following type of piecewise linear relationship:

$$\alpha = \begin{cases} A_1 & \text{for } Q \leqslant C_1 \\ A_1 + \dfrac{Q - C_1}{C_2 - C_1}(A_2 - A_1) & \text{for } C_1 \leqslant Q \leqslant C_2 \\ A_2 & \text{for } Q \geqslant C_2 \end{cases} \qquad (5.1.5)$$

The parameters $A_1$, $A_2$, $C_1$, $C_2$ are at choice. De Wet and van Wyk recommend two particular choices, viz.

$$A_1 = 0.05, A_2 = 0.40, C_1 = 1.75, C_2 = 1.95$$

and

$$A_1 = 0.10, \ A_2 = 0.40, \ C_1 = 1.80, \ C_2 = 1.95,$$

leading to adaptive trimmed means which they denote by HG1 and HG2 respectively. They investigate the efficiency of these estimators and their robustness properties in relation to various alternatives including 5 per cent and 10 per cent contamination of $N(0, 1)$ with $N(0, 9)$. It emerges that HG1 and HG2, which are easy to compute and appealing from a practical point of view, compare favourably in performance with a number of other recommended estimators (H15, 25A, JAE, JOH) considered below (see Section 5.4). This assumes a symmetrical basic model. For a highly asymmetric basic model such as an exponential distribution, this type of adaptive trimmed mean may perform poorly (Carroll, 1979).

### 5.1.2 L-estimators (linear order statistics estimators)

Suppose $x_{(1)} < x_{(2)} \ldots < x_{(n)}$ denotes the ordered sample. An $L$-estimator of the location parameter $\mu$ is a linear function

$$T_n = \sum_{j=1}^{n} a_j x_{(j)} \tag{5.1.6}$$

of the $x_{(j)}$ ($j = 1, 2, \ldots, n$). Such linear order statistics estimators have been widely studied for specific uncontaminated samples (see, for example, the lengthy review in David, 1981). Much of this work is directed to censored samples. Whilst not specifically concerned with the problem of outliers, in that the reason for censoring is seldom considered and no outlier-specific alternative model employed, some linear order statistics estimators for censored samples will possess general robustness properties which carry over to the outlier problem. There is good reason, however, to consider estimators of the form (5.1.6) specifically in the context of robust estimation from possibly contaminated samples. Indeed, we have already considered examples of such estimators including the sample median and trimmed and Winsorized means—each yielded by a particular choice of the $a_j$ in (5.1.6). The choice of the $a_j$ for an $L$-estimator may also be made *adaptively*, as instanced by an adaptive trimmed mean. When, as is often the case, the object is to estimate the centre of a symmetric distribution we frequently encounter the further (natural) assumption that the weights are symmetrically valued; that is, $a_j = a_{n+1-j}$. Under appropriate conditions the corresponding $T_n$ is consistent and asymptotically normal (Chernoff, Gastwirth, and Johns, 1967; Bickel, 1967; Jaeckel, 1971a).

Let us consider some further examples of $L$-estimators which have been proposed and investigated.

Gastwirth and Cohen (1970) consider, primarily for a normal basic model with symmetric contamination, the estimator

$$T_n = \gamma(x_{([pn])+1} + x_{(n-[pn])} + (1 - 2\gamma)\tilde{x} \tag{5.1.7}$$

$(0 < p < 1, 0 < \gamma < 1)$. This is a weighted combination of the lower and upper $p$th sample fractiles, each with weight $\gamma$, and the sample median, with weight $1 - 2\gamma$. They compare it with a number of other estimators, all but one (the Hodges–Lehmann estimator) of $L$-type.

Two special cases of (5.1.7) are *Gastwirth's estimator*

$$T_n = 0.3x_{([n/3]+1)} + 0.4\tilde{x} + 0.3x_{(n-[n/3])}, \qquad (5.1.8)$$

proposed and investigated by Gastwirth (1966), and the *trimean*

$$T_n = (h_1 + 2\tilde{x} + h_2)/4 \qquad (5.1.9)$$

where $h_1$ and $h_2$ are the sample quartiles.

Patel, Mudholkar, and Fernando (1988) comment that

(the) trimmed means, trimean and Gastwirth's estimator are perhaps the simplest among reasonably good robust estimators proposed in the literature.

They give approximations for studentized versions of these estimators in terms of t-distributions, which can be used in a convenient way for obtaining confidence intervals for a location parameter $\mu$ and for calculating significance probabilities when testing hypotheses about $\mu$. Details are given in Section 5.4 below.

David and Shu (1978) have studied the robustness of a number of different $L$-estimators of location against contamination by slippage of a single observation in a sample of 10 when the basic model is normal. The estimators include the trimmed means $\overset{T}{\bar{x}}_{1,1}$ and $\overset{T}{\bar{x}}_{2,2}$, the Winsorized means $\overset{W}{\bar{x}}_{1,1}$ and $\overset{W}{\bar{x}}_{2,2}$, the median $\tilde{x}$, the Gastwirth mean (5.1.8), and the linearly weighted means

$$L_1 = (x_{(2)} + 3x_{(3)} + 5x_{(4)} + 7x_{(5)} + 7x_{(6)} + 5x_{(7)} + 3x_{(8)} + x_{(9)})/32 \qquad (5.1.10)$$

and

$$L_2 = (x_{(3)} + 3x_{(4)} + 5x_{(5)} + 5x_{(6)} + 3x_{(7)} + x_{(8)})/18. \qquad (5.1.11)$$

For slippage in location, the bias is uniformly least for the median, followed by the Gastwirth mean, $L_2$, $\overset{T}{\bar{x}}_{2,2}$ and $L_1$ in that order; but the mean square error (MSE) is greatest for the median. The relative ranking of the estimators as regards MSE depends somewhat on the amount of slippage, but the median, the Gastwirth mean, $L_2$, $L_1$ and $\overset{T}{\bar{x}}_{2,2}$ turn out to be (uniformly) in decreasing order of MSE: a largely reverse ranking to that for bias! $L_2$ and $\overset{T}{\bar{x}}_{2,2}$ appear to be reasonable compromise 'good choices'. For slippage in variance the same MSE ordering applies.

A useful *adaptive* form of $L$-estimation employs the notion of *skipping* (Tukey, 1977) and examples are investigated by Andrews et al. (1972). The lower and upper sample quartiles are taken as *hinges* $h_1$ and $h_2$. An interval $(c_1(\eta), c_2(\eta))$ is defined in terms of $h_1$ and $h_2$ and a parameter $\eta$ such that the interval becomes wider as $\eta$ increases. The skipping process involves deleting observations in the tails of the sample (outside the interval $(c_1, c_2)$) preliminary to calculation of the *trimean* of the *retained* observations. In *iterative skipping* the process is repeated with recalculated hinges at each stage until the retained data set remains constant: *multiple skipping* repeats this process by skipping applied to the retained data set with different values of $\eta$ at each stage.

Other adaptive *L*-estimators have been considered by Birnbaum and Miké (1970), Takeuchi (1971), and Jaeckel (1971a). We also mention the '*shorth*' which is the sample mean of the shortest half of the sample (defined as $x_{(l)}, \ldots, x_{(l+[n/2])}$ where $l$ minimizes $x_{(l+[n/2])} - x_{(l)}$). See Andrews *et al.* (1972), Hampel *et al.* (1986).

### 5.1.3  M-estimators (maximum likelihood type estimators)

As we saw in Section 3.2.1, *M*-estimators for a location parameter $\mu$ are based on a generalization of the least squares principle. Suppose, on the basic model, that the sample comes from a continuous distribution with distribution function $F(x)$ and density function $f(x)$. The principle is to estimate $\mu$ by $T_n = T_n(x_1, \ldots, x_n)$ chosen to minimize

$$\sum_{j=1}^{n} \rho(x_j - T_n)$$

where $\rho(\ )$ is some real-valued non-constant function. As special cases we note that $\rho(t) = t^2$ yields the sample mean, $\rho(t) = |t|$ yields the sample median, whilst $\rho(t) = -\log f(t)$ yields the maximum likelihood estimator. If $\rho(\ )$ is continuous with derivative $\psi(\ )$, equivalently we estimate $\mu$ by $T_n$ satisfying

$$\sum_{j=1}^{n} \psi(x_j - T_n) = 0, \qquad (5.1.12)$$

as formulated in (3.2.12).

Usually we restrict attention to convex $\rho(\ )$, so that $\psi(\ )$ is monotone and $T_n$ unique. Under quite general conditions $T_n$ can be shown to have desirable properties as an estimator. If $\rho(\ )$ is convex $T_n$ is unique, translation invariant, consistent, and asymptotically normal (Huber 1964, 1967). We now discuss particular choices for $\rho(\ )$ or $\psi(\ )$.

One particular estimator with desirable properties of robustness arises from putting

$$\rho(t) = \begin{cases} \frac{1}{2}t^2 & |t| \leq \kappa \\ \kappa|t| - \frac{1}{2}\kappa^2 & |t| > \kappa \end{cases} \qquad (5.1.13)$$

for a suitable choice of $\kappa$. (This estimator is in fact related to the *Winsorized mean*; it can be shown to be equivalent to the sample mean of a sample in which all observations $x_j$ such that $|x_j - T_n| > \kappa$ are replaced by $T_n - \kappa$ or $T_n + \kappa$, whichever is the closer. We have multiple semi-Winsorization operating at both ends of the ordered sample.)

Another $M$-estimator, with

$$\rho(t) = \begin{cases} \frac{1}{2}t^2 & |t| \leq \eta \\ \frac{1}{2}\eta^2 & |t| > \eta \end{cases} \tag{5.1.14}$$

can be similarly interpreted as a *trimmed mean*; $T_n$ is now the sample mean of those observations $x_j$ satisfying $|x_j - T_n| < \eta$. This extends the modified trimming above from rejection of a single extreme value to rejection of all sample values whose residuals about $T_n$ are sufficiently large in absolute value. See Huber (1964) for details.

When the basic model involves a scale parameter (the distribution function is of the form $F[(x - \mu)/\sigma]$), modified forms of $M$-estimator have been proposed. The estimator of $\mu$ is now a solution $T_n$ of an equation of the type

$$\sum_{j=1}^{n} \psi[(x_j - T_n)/S] = 0 \tag{5.1.15}$$

where the scale parameter estimator $S$ is robust for $\sigma$ and is estimated either independently by some suitable scheme or simultaneously with $\mu$ by joint solution of (5.1.15) and an equation of the form

$$\sum_{j=1}^{n} \chi[(x_j - T_n)/S] = 0. \tag{5.1.16}$$

Different choices for $\psi(\ )$ [and for $\chi(\ )$] yield a large assortment of $M$-estimators which have been discussed in the literature. One example due to Hampel (see Andrews *et al.*, 1972, or Hogg, 1974) employs Huber's $\rho(t)$ as given by (5.1.13), i.e.

$$\psi(t) = \begin{cases} t & |t| \leq \kappa \\ \kappa \operatorname{sgn} t & |t| > \kappa \end{cases}, \tag{5.1.17}$$

with $S$ taken as

$$\operatorname{median}\{|x_j - \tilde{x}|\}/(0.6745) \tag{5.1.18}$$

where $\tilde{x}$ is the sample median.

Hampel (1974) terms

$$s_m = \operatorname{median}\{|x_j - \tilde{x}|\}$$

the *median deviation* (by analogy with the mean deviation). He outlines its earlier but limited usage, going back as far as Gauss, and recommends it as a quick robust scale parameter estimator and 'as a basis for the rejection of outliers'. In the present context he also advocates its use in (5.1.15) for developing *three-part descending M-estimators* where

$$\psi(t) = \begin{cases} t & |t| \leq a \\ a \operatorname{sgn} t & a < |t| \leq b \\ a(c \operatorname{sgn} t - t)/(c - b) & b < |t| \leq c \\ 0 & |t| > c. \end{cases} \qquad (5.1.19)$$

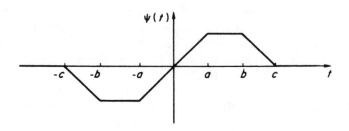

A somewhat similar proposal in Andrews *et al.* (1972) employs

$$\psi(t) = \begin{cases} \sin(t/d) & |t| \leq d\pi \\ 0 & |t| > d\pi \end{cases} \qquad (5.1.20)$$

with the properties of the estimator investigated for the specific choice $d = 2.1$.

Related estimators based on preliminary modified Winsorization of observations whose residuals exceed $cs$ (for some choice of $c$) in absolute value have also been proposed and examined (see for example, Andrews *et al.*, 1972, where they are referred to as *one-step Huber estimators*).

Cheng (1991) generalizes to $M$-estimation, say of a location parameter $\mu$, the well-known use (see, e.g. Section 5.4 below) of jackknife pseudo-values

$$\xi_j = \hat{\mu} + (n-1)(\hat{\mu} - \hat{\mu}_{-j}) \qquad (j = 1, \ldots, n),$$

where $\hat{\mu}_{-j}$ is the estimator of the same form as an estimator $\hat{\mu}$, but based on the reduced sample obtained by omitting the observation $x_j$. His estimator $T_n$ is chosen to satisfy

$$\sum_{j=1}^{n} \psi (\xi_i - T_n) = 0$$

for suitable $\psi$. He gives an example in which his proposed estimator has smaller asymptotic variance than both the ordinary $M$-estimator and the ordinary jackknife estimator.

There is a vast range of possible $M$-estimators. Even in the cases described above a great deal of choice remains in terms of how to estimate $\sigma$ and what values to take for cut-off points such as $\kappa$, $\eta$, $a$, $b$, $c$ and $d$. Many theoretical, numerical and simulation studies have been made and we shall review some of the results later (Sections 5.2, 5.3). Some key references are the large-scale empirical study by Andrews *et al.* (1972), also Hampel (1974), Hogg (1974), Huber (1964), Jaeckel (1971a), and Leone, Jayachandran, and Eisenstat (1967). An important general treatment is given in the book by Hampel *et al.* (1986).

There is of course scope for an adaptive approach with $M$-estimation. It is reasonable to contemplate choosing, for example, relevant cut-off points in the light of the sample data. The actual function $\psi(\ )$ in (5.1.12) is chosen adaptively in a proposal by Moberg, Ramberg, and Randles (1978). Given a sample, they calculate the symmetric tailweight statistic $Q_n(0 \cdot 05, 0 \cdot 5)$ defined in (5.1.4), together with the analogous skewness statistic $R_n(0 \cdot 05, 0 \cdot 5)$ where

$$R_n(\beta_1, \beta_2) = \frac{\bar{U}_n(\beta_1) - \bar{M}_n(\beta_2)}{\bar{M}_n(\beta_2) - \bar{L}_n(\beta_1)} \tag{5.1.21}$$

and $\bar{M}_n(\beta)$ denotes the average of the middle $n\beta$ observations. They then select one of five different functions $\psi(\ )$ according to the values of the $Q$- and $R$-statistics. The robustness of the estimator is assessed by the authors on the basis of its performance when applied to 'the $k$-population selection problem' with intrinsic alternatives, so its specific robustness properties against outliers are not evaluated.

Rather than employing separate estimates of $\sigma$ in the case where $\mu$ and $\sigma$ are unknown, we can pursue a joint estimation process which consists of simultaneous solution of (5.1.15) and (5.1.16). The example which has received most attention is known as *Huber's proposal* 2. Huber (1964) suggested (primarily for a *normal* basic model) that we employ $\psi(\ )$ as defined in (5.1.17) with some preliminary choice of value for $\kappa$ and take

$$\chi(t) = \psi^2(t) - \beta(\kappa) \tag{5.1.22}$$

where

$$\beta(\kappa) = \int \psi^2(t) \, dt. \tag{5.1.23}$$

This form of $\chi(t)$ was motivated by consideration of reasonable $M$-estimators of $\sigma$ (see Section 5.2). The corresponding (5.1.15) and (5.1.16) need to be solved iteratively for suitably chosen starting values. The resulting estimators of $\mu$ (and of $\sigma$) have received much attention (see, for example, Andrews *et al.*, 1972; Bickel, 1965; Huber, 1964) and we shall consider them further below.

**5.1.4  R-estimators (rank test estimators)**

As already explained in Section 3.2.1 an estimator $T_n$ of $\mu$ based on the pooled ranks of the $n$ quantities $x_j - T_n$ and the $n$ quantities $T_n - x_j$ ($j = 1, \ldots, n$) is an *R-estimator*.

An important *R*-estimator, based on the one-sample Wilcoxon test, which has received much attention, is the *Hodges-Lehmann estimator* (Hodges and Lehmann, 1963). This is the median of the set of $n(n + 1)/2$ pairwise means $(x_j + x_l)/2$ ($j < l; j = 1, 2, \ldots, n; l = 1, 2, \ldots, n$). Whilst simple in form its calculation can be tedious if $n$ is at all large. More easily calculable versions have been proposed, based on means of symmetrically placed ordered sample values—there are only $[(n + 1)/2]$ such means. For example we have the *folded-median* type estimators. The sample is folded by replacing $x_1, \ldots, x_n$ with $[x_{(1)} + x_{(n)}]/2, [x_{(2)} + x_{(n-1)}]/2. \, .$, and the median of the folded sample is chosen as the estimator (the *Bickel–Hodges estimator*). Reordering and further folding (with or without trimming) is also contemplated. See Andrews *et al.* (1972).

A simple estimator for the location parameter of a large sample is the *ninther*, proposed by Tukey (1978). Taking first the case $n = 9$, this is the median of the three medians of sub-samples of 3 into which the sample is randomly divided. For $n = 9k$ ($k > 1$), Tukey proposes dividing the sample into $k$ sets of 9 and estimating $\mu$ by the median of the $k$ ninthers. While the efficiency for a normal sample of 9 is only 55 per cent, the procedure appears to be robust against outliers and is in Tukey's phrase 'low-effort' from the application point of view.

The *remedian*, proposed by Rousseeuw and Bassett (1990), is an extension of this idea, providing a robust estimator of location which is economical in computing when the size $n$ of the data set is large. Assuming first that $n$ is of the form $b^k$ where $b$, the *base*, and $k$, the *exponent*, are integers, the data are divided into $b^{k-1}$ groups each of size $b$ and the medians of these groups calculated; the process is repeated for these $b^{k-1}$ medians, then for the resulting $b^{k-2}$ medians, and so on until a single estimate, the remedian, is obtained. The total computing storage required is $bk$, i.e. $O(\log n)$ for fixed $b$, in contrast to the storage requirement $O(n)$ for most robust estimators. As regards robustness, the breakdown point is $[b/2]^k/n$. While $b = 3$ gives the smallest total storage, $b = 9$ is recommended for practical purposes as giving a satisfactory trade-off between robustness and storage (interestingly in agreement with Tukey's *ninther*). In the general case when $n$ is not of the form $b^k$, it can be written

$$n = n_1 + n_2 b + \ldots n_k \, b^{k-1},$$

so that the data can be divided into groups of sizes $n_1, n_2 b, \ldots, n_k b^{k-1}$. From the group of size $n_j \, b^{j-1}$ ($j = 1, \ldots, k$), $n_j$ remedians are calculated; and a final 'weighted remedian' estimator is then derived by combining these $(n_1, n_2, \ldots, n_k)$ values with weights $1, b, \ldots, b^{k-1}$.

For a related approach, see Bradley (1984) with an application to time series.

### 5.1.5 Other estimators

*MD estimators (minimum distance estimators)*

Suppose we have a sample of size $n$ from a distribution with distribution function $F(x, \mu)$ involving an unknown parameter $\mu$: a location parameter. If the distribution is continuous, say with density $f(x, \mu)$, the $n$ ordered values $x_{(1)}, \ldots, x_{(n)}$ in the sample can be taken as distinct; the empirical distribution function $\hat{F}(x)$ is then given by

$$\hat{F}(x) = \begin{cases} 0 & (x < x_{(1)}) \\ \dfrac{i}{n} & (x_{(i)} \leq x < x_{(i+1)}), \quad i = 1, \ldots, n-1 \\ 1 & (x_{(n)} \leq x). \end{cases} \tag{5.1.24}$$

We can estimate $\mu$ by minimizing some measure of the distance or discrepancy between $F(x, \mu)$ and $\hat{F}(x)$, for example the Kolmogorov distance

$$\sup_x |\hat{F}(x) - F(x, \mu)|$$

or the Cramér–von Mises distance

$$\int_{-\infty}^{\infty} (\hat{F}(x) - F(x, \mu))^2 f(x, \mu) dx.$$

We could weight the distance to be minimized by some suitably chosen weighting function $\psi(F(x, \mu))$, giving, for example, a weighted Kolmogorov distance

$$\sup |\hat{F}(x) - F(x, \mu)| \psi(F(x, \mu)).$$

Some of the *minimum distance estimators* obtained by such a procedure have good robustness properties against outliers.

It may well be asked why an estimator obtained by minimization of a discrepancy measure that is useful for goodness-of-fit purposes (and, hence, in many cases extremely sensitive to outliers or general discrepancies from the model) should be hoped to possess any desirable 'robustness' properties. It turns out that, in most cases . . . , while the discrepancy measure itself may be fairly sensitive to the presence of outliers, the value [of the parameter] which minimizes the discrepancy . . . is much less so. (Parr and Schucany, 1980)

Parr and Schucany (1980) give Monte Carlo estimates of the variances of a wide range of estimators of the means of samples of sizes 10 and 20 with a basic $N(0, 1)$ model and alternatives including contamination by 10 per cent $N(0, 9)$ and 10 per cent $N(0, 100)$. Included among their estimators are *MD* estimators based on the Kolmogorov and Cramér–von Mises distances, Kuiper's $V$ and Watson's $U^2$, as well as a number of recommended estimators from Andrews *et al.* (1972). They show that, while the *MD* estimator based on Kuiper's $V$ performs poorly, those based on Kolomogorov or Cramér–von Mises distances have 98 per cent efficiency or better compared with the best of the recommended estimators for the relevant alternative when this is the 10 per cent $N(0, 9)$ mixture; this reduces to 87 per cent for the 10 per cent $N(0, 100)$ mixture alternative.

If the sample comes from a discrete distribution, say with probability function $p(x, \mu)$, it can be taken to consist of $r_1$ values equal to $x_1$, $r_2$ values equal to $x_2, \ldots, r_k$ values equal to $x_k$, where $x_1, \ldots, x_k$ are distinct and $r_1 + \ldots + r_k = n$. The empirical probability function $\hat{p}(x)$ is given by

$$\hat{p}(x) = r_i/n \ (x = x_i, \ i = 1, \ldots, k)$$

$$0 \text{ otherwise.} \tag{5.1.25}$$

$\mu$ may then be estimated by minimizing the Hellinger distance, which in this context is

$$\sum_x (\sqrt{\hat{p}(x)} - \sqrt{p(x, \mu)})^2,$$

or equivalently by maximizing

$$\sum_x \sqrt{\hat{p}_x(x)} \sqrt{p(x, \mu)}.$$

Simpson (1987) studies the minimum Hellinger distance estimator (*MHD*) in the context of discrete data, in particular count data. He shows that the *MHD* performs well in the presence of outliers, and gives little weight to count data that are improbable relative to the basic model. In the Poisson case, the asymptotic breakdown point for the *MHD* estimator is in fact 1/2. At the same time the *MHD* estimator is asymptotically equivalent to the maximum likelihood estimator when there is no contamination. As an example, Simpson applies the method to a set of mutagenicity test data, noting that anomalously large counts are not unusual in this context (see also Simpson, 1989). In relation to *MHD* estimation see also Beran (1977) and Tamura and Boos (1986).

For other robust minimum distance estimators see, for example, Boos (1981), Parr and De Wet (1981), Hulsmann (1987), and Donoho and Liu (1988). Beran (1978) proposes an *adaptive MD* estimation procedure. Heathcote (1977) suggests an estimator based on the distance between the theoretical and empirical characteristic functions. An extensive bibliography of minimum distance estimation is given by Parr (1981).

*P-estimators (Pitman-type estimators)*

Johns (1979) proposes estimating $\mu$ by a *P*-estimator of the form

$$T_n = \frac{\displaystyle\int \mu \prod_{i=1}^{n} \gamma(x_i - \mu)\,d\mu}{\displaystyle\int \prod_{i=1}^{n} \gamma(x_i - \mu)\,d\mu},$$

where $\gamma$ is a suitably chosen function. If $\gamma$ is taken to be the density $f$, we get the Pitman estimator. Johns discusses possible choices of $\gamma$ to achieve robustness, in particular the form

$$\gamma(x) = 1/(1 + x^2) + (a + bx^2)/(1 + x^2)^3,$$

and shows that the corresponding $T_n$ performs well for a normal basic model in the presence of variance slippage contamination. *P*-estimators have the advantage of not requiring iterative solution methods for their computation.

For another class of robust estimators with good robustness properties against mixture alternatives, see Pollak (1979).

An extensive account of the topic is given in the book by Hampel *et al.* (1986).

## 5.2  ESTIMATION OF SCALE OR DISPERSION

The general problem here is to estimate robustly a dispersion parameter $\sigma$ for the distribution of a random variable $X$; for example, $\sigma$ might be the standard deviation of $X$. If the distribution function is of the form $F[(x - \mu)/\sigma]$ (as with a Weibull distribution or a normal distribution), $\sigma$ is a *scale parameter*. If $X$ is a non-negative variable with distribution function of the form $F(x/\sigma)$ (as with an exponential distribution), $\sigma$ can be called a *pure scale parameter*.

The robust estimation of a scale parameter $\sigma$ may be approached by the general method of (3.2.15); in the case of a pure scale parameter, $d_i = x_i$ rather than $x_i - m$. Alternatively, the methods of *L*-, *M*-, or *R*-estimation can be used; a detailed discussion is given in Huber (1981, Chapters 5 and 6).

We have already indicated how, when the location $\mu$ is unknown, simultaneous *M*-estimators of $\mu$ and $\sigma$ can be obtained as solutions of equations of type (5.1.15), (5.1.16). If, for example, $\psi(u)$ is taken as sgn $u$ and $\chi(a)$ as sgn $(|u| - 1)$, we get the median jointly with the *median deviation* of (5.1.18),

$$s_m = \text{median}\,\{|x_j - \tilde{x}|\}.$$

This estimator of scale might by its nature be expected to provide reasonable protection against the influence of discordant values in the sample—see Hampel (1974). It arises 'as the $M$-estimate of scale with the smallest possible gross-error sensitivity at the normal (and many other) models'. Hampel regards it as the counterpart of the median as location estimator, and he discusses the form of its influence function: in particular the gross-error sensitivity and breakdown point. In these respects the median deviation is seen to be more desirable than the *semi-interquartile range, Q* (although an equivalence exists for symmetric samples and, asymptotically, for a symmetric basic model). $Q$ might also be expected to protect against outliers, but it may well be over-protective.

In spite of the relatively low efficiency of $s_m$ for uncontaminated data ($c$. 40 per cent in the normal case) it appears to possess a robustness absent from estimators which are more efficient for homogeneous data. See Tukey (1960) and Stigler (1973b).

Andrews *et al.* (1972), in studying the characteristics of some rather attractive robust location estimators (such as the Huber 'proposal 2' estimate with $\kappa = 1.5$, the Bickel one-step modification with Winsorization of residuals at $\pm \kappa s$, where $s$ is a robust scale estimate, and Hampel's three-part descending $M$-estimator with one of the cut-off points determined from a robust scale estimator), provide evidence to suggest that using a scale estimate based on $s_m$ (their 'robust scale estimate') is preferable to using one based on $Q$. In the simpler case of robust estimation of a *pure* scale parameter for a non-negative random variable $X$ whose density is of the form $(1/\sigma) f(x/\sigma)$, joint estimation with a separate location parameter is not involved. The techniques of robust estimation of location could be applied in this situation since $\log X (= Y$, say) has $\log \sigma (= \phi$, say) as a location parameter, though properties depending on symmetry in the $X$-distribution are lost under the log transformation. As pointed out by Thall (1979), an equation

$$\sum_{j=1}^{n} \psi(y_j - T_n) = 0,$$

of form (5.1.12) with some suitably chosen $\psi(\ )$, for setting up an $M$-estimator of location $\phi$ for the sample $y_1, \ldots, y_n$, leads to an equation of form

$$\sum_{j=1}^{n} \zeta\left(\frac{x_j}{S_n}\right) = 0 \qquad (5.2.1)$$

where $\zeta(u) \equiv \psi(\log u)$. This is clearly the procedure of (5.1.15) adapted to the situation where there is no separate estimation of location.

The choice of function

$$\zeta(u) = \frac{u f'(u)}{f(u)} + 1$$

gives the maximum likelihood estimator of $\sigma$ as $S_n$ in (5.2.1). With $\zeta(u) =$ sgn($|u| - 1$), we get the median deviation from zero, i.e. median $\{|x_j|\}$. Thall (1979) discusses other choices of $\zeta(\ )$ designed for an exponential basic model; these are, however, non-outlier-specific.

As regards $L$-estimation of a scale parameter $\sigma$, it is natural to consider estimators of the form

$$S_n = \sum_{j=1}^{n} a_j x_{(j)} \tag{5.2.2}$$

analogously to (5.1.6). When the distribution function of $X$ is of the form $F[(x - \mu)/\sigma]$ with unknown location $\mu$, the weights $a_j$ must satisfy

$$\sum_{j=1}^{n} a_j = 0. \tag{5.2.3}$$

Important particular cases include the *range* $x_{(n)} - x_{(1)}$, as mentioned in Section 3.2.2, and the *quasi-ranges* $x_{(n-1)} - x_{(2)}$, $x_{(n-2)} - x_{(3)}$, . . . .

There have been a number of studies of scale estimators of type (5.2.2), including censored (trimmed) or Winsorized equivalents (see, for example, Sarhan and Greenberg, 1962, pp. 218–251; David, 1981, pp. 136–158). The interest in this work lies predominantly in what loss occurs relative to the full sample equivalent for a *prescribed distribution*. What reference there is to robustness is not specific to accommodating outliers. $L$-estimation of the standard deviation $\sigma$ of a normal distribution *specifically in relation to the accommodation of outliers* has been studied by David (1979); details of his results are discussed in Section 5.4.

## 5.3  HYPOTHESIS TESTS AND CONFIDENCE INTERVALS

There are many one-sample location tests and equivalent confidence interval procedures based on statistics of type (3.2.17) and possessing robustness properties of some kind. Some tests and confidence intervals giving specific protection against outliers in normal samples are described in Section 5.4.

An interesting generalization of (3.2.17) is proposed by Johnson (1978). To test the null hypothesis $\mu = \mu_0$ for a sample $x_1, \ldots, x_n$ with mean $\bar{x}$, variance $s^2$ and sample third moment $m_3$, he uses the following statistic based on the Cornish-Fisher expansion of $\bar{x}$:

$$\{\bar{x} - \mu_0 + a + b(\bar{x} - \mu_0)^2\}/(s/\sqrt{n}),$$

where $a = m_3/(6ns^2)$ and $b = m_3/(3s^4)$. This is tested as a value of Student's **t** on $n - 1$ degrees of freedom.

Turning to nonparametric tests, these could be expected from their very nature to possess robustness properties of some kind. For the one-sample location problem, Lambert (1981) investigates the influence function of the test statistic (which she describes as the 'influence function of the test') for a number of tests including the sign test and the Wilcoxon test; comparison of the influence functions shows the Wilcoxon test to have better robustness properties than the others when the basic model is normal. A similar situation obtains for the two-sample location problem, where a study of non-parametric tests by Eplett (1980) shows the Wilcoxon test to be fairly insensitive to contamination by slippage in location, and to compare particularly favourably in this respect with the test based on normal scores; see also Iman and Conover (1977), Rousseeuw and Ronchetti (1981), and Hampel *et al.* (1986, Section 3.6).

Adaptive nonparametric test procedures are also available, though not designed primarily to afford protection against outliers. For the one-sample location problem, an efficient test of this kind is described by Jones (1979). An adaptive non-parametric test procedure for the two-sample location problem, involving selection of test (Wilcoxon, median or modified rank) according to the values of statistics $Q$, $R$ of types (5.1.4) and (5.1.21) calculated from the pooled observations, is proposed by Hogg, Fisher, and Randles (1975). See also Wei (1981).

## 5.4 ACCOMMODATION OF OUTLIERS IN UNIVARIATE NORMAL SAMPLES

From the vast amount of material on general robustness reviewed above we now select for more detailed comment some results which relate most closely to our theme: the accommodation of outliers. We shall be concerned with robustness in relation to families of distributions which correspond with the alternative models for outliers previously discussed. The information below sometimes represents specific study of outlier-type families of distributions—more frequently it is obtained by judicious selection from more embracing robustness studies. It will sometimes prove convenient to use abbreviated notation to describe certain estimators and in such cases we adopt that used in the famous 'Princeton study' by Andrews *et al.* (1972). This *tour de force* continues to be a key reference in any account of robust procedures. It consists of a study of 68 different location estimators in terms of various asymptotic characteristics as well as a variety of finite sample characteristics for samples of sizes 5, 10, 20, and 40 for a large range (*c.* 20) of different possible data-generating models. Consideration is also given to such matters as the relative ease of computation of the estimators.

We give as Table 5.1 a list of the 68 estimators with brief descriptions and the notation used for them in the original publication (Andrews *et al.*, 1972).

In the present section we concentrate on the case of a basic *normal* model $F$, with either a *mixture type* alternative

$$(1 - \lambda)F + \lambda G$$

($G$ is also usually normal with the same mean as $F$ but larger variance) or a *slippage type* alternative (where the slippage occurs either in the mean or in the variance). In Section 5.5 we review results for the exponential distribution, mostly assuming an *exchangeable* type of alternative model; and in Section 5.6 we review results for the logistic distribution, and the double exponential distribution, with slippage type alternatives.

Suppose our sample arises from a mixture distribution $(1 - \lambda)F + \lambda G$ where $F$ is $N(\mu, \sigma^2)$, and we wish to estimate $\mu$. Huber (1964) shows that the $M$-estimator with

$$\psi(t) = \begin{cases} t & |t| \leq \kappa \\ \kappa \, \text{sgn} \, t & |t| > \kappa \end{cases} \tag{5.4.1}$$

is minimax among translation invariant estimators for symmetric $G$. Tabulated values suggest that choice of $\kappa$ is not highly critical—performance being fairly insensitive and reasonable over the range $1 \leq \kappa \leq 2$ for $\lambda < 0.2$. For asymmetric $G$ the above estimator proves to be biased. Huber discusses the extent of the bias and concludes that attempts at substantially reducing it may be quite costly in terms of asymptotic variance. The straightforward $M$-estimators implicitly depend on knowing the values of $\sigma^2$ and of $\lambda$. Scale invariant versions (allowing more realistically for unknown $\sigma^2$) have been discussed in Section 5.1.3. These include estimators

**Table 5.1**   The 68 estimators used in the 'Princeton study' with notation and description

| | |
|---|---|
| A15 | Huber $M$-estimator, $\kappa = 1.5$, robust scaling |
| A20 | Huber $M$-estimator, $\kappa = 2.0$, robust scaling |
| ADA | Adaptive $M$-estimator, $\psi$ bends at ADA, 4.5, 8.0 |
| AMT | $M$-estimator, $\psi$ is sin function |
| BH | Bickel–Hodges estimator |
| | |
| BIC | Bickel modified adaptive trimmed mean |
| CML | Cauchy maximum likelihood |
| CPL | Cauchy–Pitman (location only) |
| CST | Iteratively C-skipped trimean |
| CTS | CTS-skipped trimean |
| | |
| D07 | One-step Huber, $c = 0.7$, start = median |
| D10 | One-step Huber, $c = 1.0$, start = median |
| D15 | One-step Huber, $c = 1.5$, start = median |
| D20 | One-step Huber, $c = 2.0$, start = median |
| DFA | Maximum estimated likelihood |
| | |
| GAS | Gastwirth's estimator |
| H07 | Huber proposal 2, $\kappa = 0.7$ |

**Table 5.1** (*Contd.*)

| | |
|---|---|
| H10 | Huber proposal 2, $\kappa = 1.0$ |
| H12 | Huber proposal 2, $\kappa = 1.2$ |
| H15 | Huber proposal 2, $\kappa = 1.5$ |
| | |
| H17 | Huber proposal 2, $\kappa = 1.7$ |
| H20 | Huber proposal 2, $\kappa = 2.0$ |
| H/L | Hodges–Lehmann estimator |
| HGL | Hogg 69, based on kurtosis |
| HGP | Hogg 67, based on kurtosis |
| | |
| HMD | $M$-estimator, $\psi$ bends at 2.0, 5.5 |
| JAE | Adaptive trimmed mean (Jaeckel) |
| JBT | Restricted adaptive trimmed mean |
| JLJ | Adaptive linear combination of trimmed means |
| JØH | John's adaptive estimator |
| | |
| JWT | Adaptive form based on skipping |
| LJS | Least favourable distribution |
| M | Mean |
| M15 | One-step Huber, $c = 1.5$, start = mean |
| MEL | Mean likelihood |
| | |
| OLS | 'Olshen's estimator' |
| P15 | One-step Huber, $c = 1.5$, start = median, robust scale |
| SHO | Shorth (for shortest half) |
| SJA | Symmetrized adaptive trimmed mean |
| SST | Iteratively s-skipped trimean |
| | |
| TAK | Takeuchi's adaptive estimator |
| THL | T-skipped Hogg 69 |
| THP | T-skipped Hogg 67 |
| TOL | T-skipped Olshen |
| TRI | Trimean |
| | |
| 5% | 5% symmetrically trimmed mean |
| 10% | 10% symmetrically trimmed mean |
| 12A | $M$-estimator, $\psi$ bends at 1.2, 3.5, 8.0 |
| 15% | 15% symmetrically trimmed mean |
| 17A | $M$-estimator, $\psi$ bends at 1.7, 3.4, 8.5 |
| | |
| 21A | $M$-estimator, $\psi$ bends at 2.1, 4.0, 8.2 |
| 22A | $M$-estimator, $\psi$ bends at 2.2, 3.7, 5.9 |
| 25% | 25% symmetrically trimmed mean |
| 25A | $M$-estimator, $\psi$ bends at 2.5, 4.5, 9.5 |
| 2RM | 2-times folded median |

**Table 5.1** (*Contd.*)

| | |
|---|---|
| 33T | Multiply-skipped trimean |
| 3R1 | 3-times folded median trimming 1 |
| 3R2 | 3-times folded median trimming 2 |
| 3RM | 3-times folded median |
| 3TO | Multiply-skipped mean, $3k$ deleted |
| | |
| 3T1 | Multiply-skipped mean, max $(3k, 2)$ deleted |
| 4RM | 4-times folded median |
| 50% | Median (or 50% symmetrically trimmed mean) |
| 5T1 | Multiply-skipped mean, max $(5k, 2)$ deleted |
| 5T4 | Multiply-skipped mean, max $(5k, 2)$ but $\leq 0.6N$ deleted |
| | |
| A | 0.75(5T4) + 0.25(D20) |
| B | 0.50(5T4) + 0.50(D20) |
| C | 0.25(5T4) + 0.75(D20) |

where $\sigma^2$ is estimated robustly, perhaps in terms of median deviation or interquartile range and possibly involving associated Winsorization or having multi-part form.

Alternatively, $\mu$ may be estimated simultaneously with $\sigma^2$, as for example in Huber's 'proposal 2' (that is, solving (5.1.15) and (5.1.16) with $\chi(t)$ given by (5.1.22) and (5.1.23) and with $\psi(t)$ as in (5.4.1)). Some quantitative features of this approach, for symmetric $G$, are given by Huber (1964) for a range of values of $\kappa$ and $\lambda$. Minimax optimality, and asymptotic variances, are highly limited criteria, but it is of some relevance (if not specific to outliers) to note that various $M$-type estimators find broad support on various bases in the study by Andrews *et al.* (1972). Although no simple prescription of 'best estimator' is feasible, the different contributors tend to include among their recommendations $M$-estimators and *one-step Huber estimators* (both using (5.4.1), median deviation to estimate $\sigma$ and with $\kappa$ in the vicinity of 1.5), *Huber's proposal* 2 (with somewhat smaller $\kappa$) and the *three-part descending estimators* (with $a$, $b$, $c$ in the regions of 2, 4, and 8, or with $a$ chosen adaptively and $b$, $c$ in the regions of 4 and 8). From Siddiqui and Raghunandanan (1967) we can make a limited asymptotic comparison of the *Hodges–Lehmann estimator, trimmed and Winsorized means*, and *Gastwirth and Cohen's estimator* (5.1.7).

With the mixture model where $F$ and $G$ are $N(\mu, \sigma^2)$ and $N(\mu, 9\sigma^2)$, respectively, and the mixing parameter $\lambda$ is restricted to at most 0.05, there is little to choose between the first three estimators (the best trimming factor has the value in the region of $\alpha = 0.20$) with minimum efficiency about 95 per cent, almost 10 per cent higher than that of the best form of (5.1.7). (It is worth noting that Gastwirth's version (5.1.8) achieves an efficiency of about 80 per cent or more for the set of *inherent* alternatives: normal, Cauchy, double-exponential, logistic. See Gastwirth, 1966.)

Asymptotic properties need however to be augmented with information on finite sample behaviour.

In Gastwirth and Cohen (1970) we find tables of means, variances, and covariances of order statistics for samples of sizes up to 20 from contaminated normal distributions

$$(1 - \lambda)F + \lambda G$$

where $F$ is $N(0, 1)$, $G$ is $N(0, 9)$ and $\lambda = 0.01, 0.05, 0.10$. These are useful for comparing the performance of order-statistics-based robust linear estimators of the mean in the corresponding range of contaminated normal distributions. The authors tabulate some results showing that over the types of estimator they consider (including mean, median, trimmed means, Winsorized means, combinations of the median with equally weighted fractiles of the form (5.1.7), and the Hodges–Lehmann estimator), it is again the trimmed means which perform well in terms of minimax variance both asymptotically and at the different finite sample sizes: typically (for $n = 20$) needing trimming factor $\alpha$ in the regions of

$$0.12\text{–}0.20 \quad (0.05 \leqslant \lambda \leqslant 0.1),$$

$$0.05\text{–}0.15 \quad (0.01 \leqslant \lambda \leqslant 0.05)$$

(Note that published support for $\alpha$-values as high as 0.25–0.30 is based on minimax performance over wide-ranging families of *distinct* distributions— normal to Cauchy. Such a catholic situation does not accord with the outlier models we have been considering.)

Some Monte Carlo results reported by Huber (1972), from the Andrews *et al.* (1972) work, are illuminating. For samples of size 20 he compares 20 estimators in terms of their estimated variance when $(20 - k)$ observations come from $N(0, 1)$ and $k$ come from $N(0, 9)$. This amounts to a scale-slippage type model. When $k = 1$ some estimators show up better than others: the $\alpha$-trimmed mean with $\alpha = 0.05$ or 0.10, H20 and H15 (Huber's 'proposal 2' with $\kappa = 2.0$ or 1.5), A15 and P15 (Huber $M$-estimates, with $\kappa = 1.5$ and median dispersion scale estimate, in direct form, and in one-step form starting with the median), and 25A (Hampel's three-part descending estimator with $a = 2.5, b = 4.5, c = 9.5$)). For $k = 2, 3$ the $\alpha$-trimmed means remain impressive with $\alpha$ advancing to 0.1 or 0.15, and 0.15, respectively, as do H15 and H10, respectively, and, for $k = 2$ alone, A15, P15, 25A. With 18 observations from $N(0, 1)$ and two from $N(0, 100)$, 25A stands out as better than most other estimators. Often the estimated variances show only small differences, and to put the above recommendations in perspective it is useful to examine Table 5.2 extracted from the tabulated results in Huber (1972).

**Table 5.2**  Monte Carlo variances of $n^{1/2}T_n$ for selected estimators and distributions; sample size $n = 20$

(*Italics indicate the estimator with the smallest variance for each distribution*)

| | | N(0, 1) | | $(n-k)$N(0, 1) plus $k$N(0, 9), $n = 20$ | | | 18N(0, 1) plus 2N(0, 100) |
|---|---|---|---|---|---|---|---|
| | | $n = \infty$ | $n = 20$ | $k = 1$ | $k = 2$ | $k = 3$ | |
| | Mean | *1.00* | *1.00* | 1.40 | 1.80 | 2.20 | 10.90 |
| | $\alpha = 0.05$ | 1.026 | 1.02 | *1.16* | 1.39 | 1.64 | 2.90 |
| Trimmed | $\alpha = 0.10$ | 1.060 | 1.06 | 1.17 | *1.31* | 1.47 | 1.46 |
| mean | $\alpha = 0.15$ | 1.100 | 1.10 | 1.19 | 1.32 | *1.44* | 1.43 |
| | $\alpha = 0.25$ | 1.195 | 1.20 | 1.27 | 1.41 | 1.50 | 1.47 |
| | median | 1.571 | 1.50 | 1.52 | 1.70 | 1.75 | 1.80 |
| Huber | $k = 2.0$ | 1.010 | 1.01 | 1.17 | 1.41 | 1.66 | 1.78 |
| (1964) | $k = 1.5$ | 1.037 | 1.04 | *1.16* | 1.32 | 1.49 | 1.50 |
| prop. 2 | $k = 1.0$ | 1.107 | 1.11 | 1.21 | 1.34 | *1.44* | 1.43 |
| | $k = 1.7$ | 1.187 | 1.20 | 1.27 | 1.42 | 1.49 | 1.47 |
| Hodges–Lehmann | | 1.047 | 1.06 | 1.18 | 1.35 | 1.50 | 1.52 |
| Gastwirth (1966) | | 1.28 | 1.23 | 1.30 | 1.45 | 1.52 | 1.50 |
| Jaeckel (1969) | | 1.000 | 1.10 | 1.21 | 1.37 | 1.47 | 1.45 |
| Hogg (1967) | | 1.000 | 1.06 | 1.28 | 1.56 | 1.79 | 1.79 |
| Takeuchi (1971) | | 1.000 | 1.05 | 1.19 | 1.38 | 1.53 | 1.32 |
| A15 | | 1.037 | 1.05 | 1.17 | 1.33 | 1.47 | 1.49 |
| P15 | | 1.037 | 1.05 | 1.17 | 1.33 | 1.47 | 1.49 |
| Hampel 25A | | 1.025 | 1.05 | *1.16* | 1.32 | 1.49 | *1.26* |
| Hampel 12A | | 1.166 | 1.20 | 1.26 | 1.40 | 1.47 | 1.32 |

We should also bear in mind the computational effort involved in constructing the estimators. Trimmed means are fairly easily determined, and whilst possibly needing some iteration the various Huber-type estimators (such as H15, A15, P15) are not unreasonable. In comparison, the Hodges–Lehmann estimator can be most time-consuming.

Hodges (1967) uses Monte Carlo methods to examine the extent to which some simple location estimators are efficient with respect to estimating the mean of a normal distribution and are able to 'tolerate extreme values' in the sense of not being influenced by the $r$ lowest and $r$ highest extremes. Thus $\overset{\text{T}}{x}_{r,r}$ and $\overset{\text{W}}{x}_{r,r}$ have 'tolerance' $r$, the median $\tilde{x}$ has tolerance $[(n-1)/2]$, $\bar{x}$ has tolerance 0. A modified more easily calculated type of Hodges–Lehmann estimator, BH, is the median of the means of symmetrically chosen pairs of ordered observations (the *Bickel–Hodges folded median*; see also Bickel and Hodges, 1967). BH has tolerance $[(n-1)/4]$ and is shown by sampling experiments with $n = 18$ to have efficiency about 95 per cent relative to $\bar{x}$. (But this needs careful interpretation—there is no contemplation of an alternative outlier generating model, we are merely estimating $\mu$ with reduced consideration of extreme values whether or not they are discordant. We do not learn how BH compares with other estimators for a prescribed mixture- or slippage-type model.)

Other performance characteristics are also important, for example those based on the influence curve including such features as gross-error sensitivity, local-shift sensitivity, and rejection point. Hampel (1974) tabulates such quantities for many of the estimators we have discussed assuming an uncontaminated normal distribution. Though we would be better armed for current purposes if the distribution were of a mixture or slippage type, the results are interesting. One comment, in particular, adds to the summary above:

. . . three-part descending $M$-estimators pay a small premium in asymptotic variance or gross-error-sensitivity, as compared with Huber-estimators, in order to be able to reject outliers completely. (Hampel, 1974)

While earlier comparative studies such as those of Andrews et al. (1972) and Wegman and Carroll (1977) are based on Monte Carlo methods, the body of theoretical results obtained by Balakrishnan and others on the distributional properties of contaminated distributions enable performance measures of robust procedures to be calculated exactly in many cases (see Section 3.1.2 above). Arnold and Balakrishnan (1989) use this direct approach to consider robustness properties of the following 11 estimators of location for normal samples of size 10:

Sample mean $\bar{x}$

(1,1) trimmed mean $\bar{x}^{T}_{1,1}$

(2,2) trimmed mean $\bar{x}^{T}_{2,2}$

(1,1) Winsorized mean $\bar{x}^{W}_{1,1}$

(2,2) Winsorized mean $\bar{x}^{W}_{2,2}$

Sample median $\tilde{x}$

Modified maximum likelihood estimator

$$M_{1,1} = [n\,\bar{x}^{W}_{1,1} - (1 - \beta_1)(x_{(2)} + x_{(n-1)})]/((n - 2 + 2\beta_1)$$

Modified maximum likelihood estimator

$$M_{2,2} = [n\bar{x}^{W}_{2,2} - 2(1 - \beta_2)(x_{(3)} + x_{(n-2)})]/(n - 4 + 4\beta_2)$$

(where $\beta_1$, $\beta_2$ are defined in Tiku (1980) and tabulated in Tiku, Tan, and Balakrishnan (1986)).

$$L\text{-estimator } L_1 = \sum_{i=1}^{(n/2)-1} (2i - 1)(x_{(i+1)} + x_{(n-i)})/2\left(\frac{n}{2} - 1\right)^2$$

$$L\text{-estimator } L_2 = \sum_{i=1}^{(n/2)-2} (2i-1)(x_{(i+2)} + x_{(n-i-1)})/2 \left(\frac{n}{2} - 2\right)^2$$

Gastwirth's estimator $G$: see (5.1.8) above.

They evaluate the bias and mean square error (MSE) for each estimator when the basic model is $N(\mu, 1)$ and the alternative specifies location-slippage by a single observation arising from $N(\mu + a, 1)$, with $a = 0(0.5)2(1)4, \infty$. On the criterion of bias, the ordering of the estimators from greatest to least bias (i.e. best performance) is

$$\bar{x}, \overset{W}{x}_{1,1}, M_{1,1}, \overset{T}{x}_{1,1}, \overset{W}{x}_{2,2}, M_{2,2}, L_1, \overset{T}{x}_{2,2}, L_2, G, \tilde{x} \qquad (5.4.2)$$

for all the $a$-values tabulated.

On the MSE criterion the ordering of the estimators depends to some extent on the value of $a$, but the following four sets of partial orderings (in the direction of increasingly good performance) are common to all the $a$-values:

$$\left.\begin{array}{l} \tilde{x}, G, L_2, L_1, \overset{T}{x}_{2,2}; \\[4pt] \overset{W}{x}_{1,1}, M_{1,1}; \\[4pt] \overset{W}{x}_{2,2}, M_{2,2}; \\[4pt] \overset{W}{x}_{2,2}, \overset{T}{x}_{1,1}. \end{array}\right\} \qquad (5.4.3)$$

These orderings are almost the reverse of the ordering (5.4.2). Yet again, no uniform conclusion is possible!

Arnold and Balakrishnan (1989) also consider the case when the alternative model is of scale slippage with a single observation arising from $N(\mu, b)$, for $\sqrt{b} = 0.5, 1(1)4, \infty$. All the estimators are unbiased for all $b$. Evaluation of the variances gives the same partial orderings as in (5.4.3) for $b \geq 1$ (their $b = 0.25$ is not of course relevant to the outlier situation; it would tend to produce an 'inlying' contaminant).

A detailed study of the accommodation of outliers in slippage models is presented by Guttman and Smith (1969, 1971) and Guttman (1973a). They consider three specific methods based on modified trimming, modified Winsorization, and semi-Winsorization (the 'A-rule', after Anscombe, 1960a; the 'W-rule' and the 'S-rule') for estimating the mean (Guttman and Smith, 1969; Guttman, 1973a) and the variance (Guttman and Smith, 1971) for a normal basic model $N(\mu, \sigma^2)$ with a location, or scale, slippage alternative model to explain the behaviour of one or two observations. Performance characteristics are restricted to the premium-protection ideas of Anscombe (1960a), that is, variance ratios or relative efficiencies (see Section 3.1.1).

Consider first the case where we wish to estimate $\mu$ robustly under Ferguson's *model A* for slippage of the mean or *model B* for slippage of scale. Here we assume that $x_1, x_2, \ldots, x_n$ arise from $N(\mu, \sigma^2)$, but entertain the prospect that at most one observation may have arisen from $N(\mu + a, \sigma^2)$ (*model A*) or from $N(\mu, b\sigma^2)$ with $b > 1$ (*model B*). In either case the three robust estimators considered are those described in Section 5.1.1, namely the *modified trimmed mean*, the *modified Winsorized mean* and the *semi-Winsorized mean* which we denote $T_A$, $T_W$, and $T_S$. Guttman and Smith (1969) determine and compare the finite-sample premium and protection measures for these estimators. Detailed results are presented for the case where $\sigma^2$ is known. When $\sigma^2$ is unknown (and replaced by the full-sample unbiased variance estimate $s^2$) computational difficulties restrict the amount of information readily obtainable.

Under the basic model $\bar{x}$ is optimal for $\mu$. Putting $\check{\mu} = \bar{x}$ in (3.1.2) and (3.1.3) (with obvious modification of the latter to allow for any bias in the typical candidate estimator $T$) we can determine the premium and protection measures for $T_A$, $T_W$, and $T_S$. The premiums have the general form:

$$\text{Premium} = nE(U^2)/\sigma^2 \qquad (5.4.4)$$

when $T$ is re-expressed as $\bar{x} + U$. The protection is

$$\{E[(\bar{x} - \mu)^2] - E[(T - \mu)^2]\}/E[(\bar{x} - \mu)^2]$$

evaluated under the alternative hypothesis, and for *model A* and *model B*, respectively, we have:

$$\text{Protection} = \begin{cases} - n^2 E[U(U + 2a\sigma)/n]/[\sigma^2(n + a^2)]. & (5.4.5) \\ - n^2 E[U(U + 2\bar{x} - 2\mu)]/[\sigma^2(n + b - 1)]. & (5.4.6) \end{cases}$$

To determine (5.4.4) and (5.4.5) or (5.4.6) we have to investigate the first two moments of the incremental estimators $U_A$, $U_W$, $U_S$ under the basic and alternative models. Simple closed form expressions are not available, but Guttman and Smith (1969) develop an appropriate computational (Monte Carlo type) procedure and provide graphs and tables for comparing $T_A$, $T_S$ and $T_W$ for sample sizes up to 10 at premium levels of 5 per cent and 1 per cent and for different values of the slippage parameters $a$ and $b$.

The general conclusions are, for these sample sizes, that under *model A* $T_S$ is best for small $a$, $T_W$ for intermediate $a$, and $T_A$ for large $a$. Under *model B*, $T_A$ is not a contender; $T_S$ is best for small $b$, $T_W$ for large $b$. The estimators are defined in terms of the cut-off values $c$ in (5.1.1), (5.1.3), which depend on the chosen premium, the sample size and the type of estimator; Guttman and Smith (1969) tabulate approximate values of $c$ for premiums of 5 per cent and 1 per cent and $n = 3, 4(2)10$.

Extensions to larger sample sizes (encompassing the prospect of one or two discordant values) are considered by Guttman (1973a), again principally for the case of known $\sigma^2$. An interesting feature of this work is the replacement of the residuals by adjusted (independent) residuals to facilitate the calculation of premium and protection for larger sample sizes. See also Tiao and Guttman (1967) and Marks and Rao (1978). Under the basic model the residuals $x_j - \bar{x}$ ($j = 1, 2, \ldots, n$) have common variance $(n - 1)\sigma^2/n$, and covariance $- \sigma^2/n$. Thus if $u$ is an observation from $N(0,1)$, independent of the $x_j$, the *adjusted residuals*

$$z_j = x_j - \bar{x} + \sigma u/\sqrt{n} \qquad (5.4.7)$$

are *independent* observations from $N(0, \sigma^2)$. For reasonable sample sizes the $z_j$ will differ little from the true residuals $x_j - \bar{x}$; the induced independence, however, renders the determination of performance measures of the corresponding $T_A$, $T_S$, $T_W$ more tractable and enables some quantitative comparisons to be made.

For a modified procedure designed to respond to a mixture rather than a slippage contamination model, see Marks and Rao (1979).

We next consider some specific procedures for estimation of dispersion *per se*; after this we will review specific procedures for joint estimation of location and dispersion or scale.

Johnson, McGuire, and Milliken (1978) propose estimators of $\sigma^2$ for a normal distribution, specifically robust against contamination by location slippage of a given number of observations $k$. Taking $\sum_{i=1}^{n} (x_i - \bar{x})^2$ in the well-known alternative form $(1/n) \sum_{i < j} d_{ij}^2$ where $d_{ij} = |x_i - x_j|$, their method is essentially to remove large values from among the $n(n - 1)/2$ terms $d_{ij}^2$ and to use the trimmed sum with appropriate divisor as estimator of $\sigma^2$. Their first estimator, $s_1^2$ say, is given by

$$n(n - 1)s_1^2 = \sum d_{ij}^2$$

where $\sum d_{ij}^2$ is calculated from only those values of $d_{ij}$ which are less than some constant $c$. For their second estimator, $s_2^2$ say, $k$ extreme values are identified among the $n$ observations $x_1, \ldots, x_n$, leaving $n - k$ 'non-extreme' values, and the $k(n - k)$ differences $d_{ij}$ between an extreme and a non-extreme observation are excluded in calculating $\sum d_{ij}^2$; $s_2^2$ is then given by

$$\{k(k - 1) + (n - k)(n - k - 1)\}s_2^2 = \sum d_{ij}^2.$$

Their third estimator, $s_3^2$ say, is given by

$$s_3^2 = s_2^2/u(k, n),$$

where $u(k, n)$ is a tabulated function giving the expected value of $s_2^2/\sigma^2$ under the basic model of no contamination, i.e. $s_3^2$ is constructed to be unbiased for an uncontaminated normal sample. The authors show by a Monte Carlo study that the best of these estimators are those of type $s_3^2$, providing the chosen value of $k$ is not less than the actual number of contaminants in the sample; for practical purposes it is reasonable to take

$k = [n/2]$, the greatest possible value for $k$. This gives a fair measure of protection against outliers. For example, with 10 per cent contamination the bias is approximately $0.6\sigma^2$ even when the extent of slippage in the mean is as high as $9\sigma$. The loss of efficiency under the basic model is not great; the $s_3^2$ estimators have efficiencies of 95 per cent and upwards, depending on $n$, when $k = 1$, and their lowest efficiencies, reached when $k = [n/2]$, are still of the order of 80 per cent. However, the estimators may well be less robust against contamination by slippage in the *variance*.

We have already referred in Section 5.2 to the study by David (1979) of $L$-estimators of the standard deviation $\sigma$ of a normal distribution. He considers twelve different $L$-estimators and investigates their robustness against contamination by location slippage of a single observation in samples of sizes 5, 10 and 20. This work is a companion study to the one by David and Shu (1978) on robustness of $L$-estimators of location (see Section 5.1.2).

We denote the 12 chosen $L$-estimators by $MVU, MVU_{1,1}, MVU_{2,2}$; $G, G_{1,1}, G_{2,2}$; $k_0W_0, k_1W_1, k_2W_2$; $k_{01}(W_0 + W_1), k_{012}(W_0 + W_1 + W_2)$; and GMD. $MVU, MVU_{1,1}$ and $MVU_{2,2}$ are the minimum-variance unbiased $L$-estimators of form (5.1.6) based respectively on the order statistics of the complete sample, the (1,1) trimmed sample and the (2,2) trimmed sample, $G, G_{1,1}$, and $G_{2,2}$ are the related, but often more convenient, Gupta estimators (for details see, for example, Gupta (1952), Sarhan and Greenberg (1962, p. 208), or David (1981, pp. 131–133)). Next denote the sample range and $q$th quasi-range ($q = 1, 2, \ldots$) by $W_0, W_q$ respectively; thus

$$W_0 = x_{(n)} - x_{(1)}$$
$$W_1 = x_{(n-1)} - x_{(2)}$$
$$W_2 = x_{(n-2)} - x_{(3)}, \text{ etc.}$$

Applying scaling factors $k_0, k_1, k_2$ chosen to give unbiasedness for $\sigma$ in the null case, we get the estimators $k_0W_0, k_1W_1, k_2W_2$. Likewise $k_{01}(W_0 + W_1)$ and $k_{012}(W_0 + W_1 + W_2)$ are scaled versions of the 'thickened ranges' $W_0 + W_1$ and $W_0 + W_1 + W_2$, unbiased when there is no contamination. Finally GMD is Gini's mean difference

$$GMD = k_G \sum_{i=1}^{n} \left(i - \frac{n+1}{2}\right) x_{(i)}$$

where the scaling factor $k_G$ secures unbiasedness in the null case.

Some of these estimators such as $MVU_{2,2}$ and $k_2W_2$ are of course inapplicable for the smallest sample size $n = 5$ considered by David. For his next sample size, $n = 10$, the estimators are given by the formulae in Table 5.3 (adapted from David, 1979).

**Table 5.3** Selected $L$-estimators for a normal standard deviation $\sigma$; sample size $n = 10$

| | |
|---|---|
| MVU | $0.2044W_0 + 0.1172W_1 + 0.0763W_2 + 0.0436W_3 + 0.0142W_4$ |
| $\text{MVU}_{1,1}$ | $0.4034W_1 + 0.1074W_2 + 0.0616W_3 + 0.0201W_4$ |
| $\text{MVU}_{2,2}$ | $0.7021W_2 + 0.0947W_3 + 0.0310W_4$ |
| G | $0.1944W_0 + 0.1265W_1 + 0.0829W_2 + 0.0475W_3 + 0.0155W_4$ |
| $G_{1,1}$ | $0.3150W_1 + 0.2064W_2 + 0.1182W_3 + 0.0386W_4$ |
| $G_{2,2}$ | $0.5592W_2 + 0.3202W_3 + 0.1046W_4$ |
| $k_0W_0$ | $0.3249W_0$ |
| $k_1W_1$ | $0.4993W_1$ |
| $k_2W_2$ | $0.7621W_2$ |
| $k_{01}(W_0 + W_1)$ | $0.1968W_0 + 0.1968W_1$ |
| $k_{012}(W_0 + W_1 + W_2)$ | $0.1564W_0 + 0.1564W_1 + 0.1564W_2$ |
| GMD | $0.1772W_0 + 0.1379W_1 + 0.0985W_2 + 0.0591W_3 + 0.0197W_4$ |

David tabulates the bias and MSE of the various estimators of $\sigma$ when the basic model is $N(\mu, \sigma^2)$ and there is one contaminant from $N(\mu + a, \sigma^2)$. Not surprisingly, the influence function is unbounded for the six estimators which involve $W_0$. The other six estimators, which are based essentially on trimmed samples, give good protection against the type of contamination specified. Maximum bias is of the order of $0.19\sigma$ for $\text{MVU}_{1,1}$, $G_{1,1}$ and $k_1W_1$, and $0.14\sigma$ for $\text{MVU}_{2,2}$, $G_{2,2}$ and $k_2W_2$; these maximum levels are effectively reached when $a = 4\sigma$. Conversely, maximum MSE is of the order of $0.15\sigma^2$ for $\text{MVU}_{1,1}$ and $G_{1,1}$ (though $0.17\sigma^2$ for $k_1W_1$), and $0.19\sigma^2$ for $\text{MVU}_{2,2}$, $G_{2,2}$ and $k_2W_2$; as with bias, these levels are effectively reached when $a = 4\sigma$ or even $3\sigma$. For best accommodation properties in the presence of an outlier $\text{MVU}_{1,1}$ and $G_{1,1}$ are recommended. With $\text{MVU}_{2,2}$, $G_{2,2}$, and $k_2W_2$ there is a substantial loss of efficiency under the basic model ($a = 0$). David states that similar results hold in the case of dispersion slippage, i.e. when the contaminant comes from $N(\mu, b\sigma^2)$, and he points out that this is to be expected since the distribution $N(\mu, b\sigma^2)$ with slipped variance can be regarded as a mixture of distributions $N(\mu + a, \sigma^2)$ with slipped mean when $a$ is allowed to vary according to a normal distribution $N(0, (b - 1)\sigma^2)$.

In passing we should note the useful tables of means, variances, and covariances of normal order statistics in the presence of contamination, presented by David, Kennedy, and Knight (1977).

As regards other dispersion estimators giving specific protection against outliers, we have referred to the use of the interquartile range, $Q$, the median deviation (e.g. Hampel, 1974)

$$s_m = \text{median}\{|x_{(j)} - \tilde{x}|\},$$

the various quasi-ranges, and quadratic measures using trimmed or Winsorized samples, such as $S^{W2}_{r,r}$ (e.g. Dixon and Tukey, 1968) or, with an analogous notational interpretation, $S^{T2}_{r,r}$.

Dispersion estimators based on samples subjected to modified trimming, modified Winsorization, or semi-Winsorization have also been designed. Guttman and Smith (1971) define robust dispersion estimators $\tilde{S}^2_A$, $\tilde{S}^2_W$, and $\tilde{S}^2_S$ of $\sigma^2$ analogous to their location estimators $T_A$, $T_W$, and $T_S$ for a normal slippage model where all but possibly one observation arise from $N(\mu, \sigma^2)$ and at most one discordant value arises either from $N(\mu + a, \sigma^2)$ or from $N(\mu, b\sigma^2)$ with $b > 1$. The same principle applies of rejecting or modifying the observation with largest absolute residual, should this be sufficiently large. The proposed estimators take the forms shown in Table 5.4.

**Table 5.4**  Forms of $\tilde{S}^2_A$, $\tilde{S}^2_W$, $\tilde{S}^2_S$

| $\tilde{S}^2_A$ | $\tilde{S}^2_W$ | $\tilde{S}^2_S$ | Condition |
|---|---|---|---|
| $ds^2$ | $ds^2$ | $ds^2$ | $z^2_{(1)} < \kappa s^2$ and $z^2_{(n)} < \kappa s^2$ |
| $ds^2_{(1)}$ | $d \max [s^2_{(2,1)}, s^2_{(n,1)}]$ | $\dfrac{d}{n-1}[(n-2)s^2_{(1)} + \kappa s^2]$ | $z^2_{(1)} \geqslant \kappa s^2$ and $z^2_{(1)} > z^2_{(n)}$ |
| $ds^2_{(n)}$ | $d \max [s^2_{(1,n)}, s^2_{(n-1,n)}]$ | $\dfrac{d}{n-1}[(n-2)s^2_{(n)} + \kappa s^2]$ | $z^2_{(n)} \geqslant \kappa s^2$ and $z^2_{(n)} > z^2_{(1)}$ |

In each estimator $\kappa$ must be prescribed, and $d$ is then chosen to ensure unbiasedness in the null case (no discordant value). Forms given in Table 5.4 apply to the more usual case where $\mu$ is unknown. If $\mu$ were known we would merely replace $\bar{x}, \bar{x}_{(1)}$ etc. in $s^2, s^2_{(1)}$ etc. by the true mean $\mu$. (Subscript indices in brackets refer to omitted ordered observations.)

For an outlier-robust method of estimating a variance by use of the influence function, illustrated by an application to electrical power plant data, see Chernick and Murthy (1983). For robust estimation of a variance *function* (e.g. the relationship between the variance of a response in an immunoassay to its mean value), by a method affording protection against outliers, see Raab (1981).

We now review some outlier-specific accommodation procedures involving the joint estimation of $\mu$ and $\sigma$ in the normal case. The object may simply be to estimate these parameters *per se*, or it may be to set up significance tests or confidence intervals based on the estimators.

We refer first to some earlier numerical studies of studentized location estimators of the form $\sqrt{n}(T - \mu)/S$ for a normal basic model. Dixon and Tukey (1968) study by qualitative arguments and Monte Carlo methods the sampling behaviour of

$$t_W = (\bar{x}^W_{r,r} - \mu)/\sqrt{\{S^{W2}_{r,r}/[n(n-1)]\}}. \tag{5.4.8}$$

They conclude that

$$(h - 1)t_W(n - 1)$$

has a distribution which is well approximated by Student's t distribution with $h - 1$ degrees of freedom when $h = n - 2r$. No consideration is given to how the distribution of $t_W$ changes for, say, a mixture model $(1 - \lambda)F + \lambda G$ where $F$ and $G$ are similarly centred but differently scaled normal distributions, which would be germane to the outlier problem.

Just such a mixture model is examined, however, by Leone, Jayachandran, and Eisenstat (1967). Again by Monte Carlo methods, they examine the sampling behaviour of studentized forms, $\sqrt{n}(T - \mu)/S$, of robust location estimators, for the mixture model $(1 - \lambda)F + \lambda G$ with $\lambda = 0.05$ and $0.10$ and $F$ and $G$ both normal. Symmetric and asymmetric contamination are considered: specifically $F$ is $N(0, 1)$ and $G$ is $N(a, b)$ with $a = 0, \frac{1}{2}, 1$ and $b = 1, 9, 25$. The estimators considered are various joint estimators $(T, S)$ obtained under Huber's proposal 2.

The goodness-of-fit of $\sqrt{n}(T - \mu)/S$ to a Student's t distribution is examined for samples of size $n = 20$ (also studied is the extent to which their $T$-estimators, and the Hodges–Lehmann estimator, have approximate normal distributions).

Broad conclusions from a mass of empirical results include the following.

(i) For proximity of $\sqrt{n}(T - \mu)/S$ to Student's t distribution over the contemplated range of models, reasonable choice of $\kappa$ in Huber's proposal 2 is in the region of 1.8 or 1.9. The fit is reasonable except for extreme cases such as $\lambda = 0.1, b = 25$.

(ii) The Huber ($\kappa = 1, 1.5, 2$) and Hodges–Lehmann location estimators are reasonably normal, except again in extreme cases, e.g. $\lambda = 0.1$, $a = 1, b = 25$.

A much more extensive study is reported in the important paper by Gross (1976), who considers from the point of view of confidence interval robustness the 25 joint estimators $(T, S)$ listed in Table 5.5. Using Monte Carlo techniques on simulated samples of sizes 10 and 20 from a spectrum of distributions, he compares the 95 per cent confidence intervals of form

$$(T - t^*S/\sqrt{n}, \ T + t^*S/\sqrt{n})$$

based on each $(T, S)$. The distributions studied include the normal as basic model, the Cauchy, and four contamination models, two of mixture type and two of dispersion slippage type. *Robustness of performance* is then measured by the expected confidence interval length, and *robustness of validity* by the true level of the conservative confidence interval based on the greatest of the values of $t^*$ over the spectrum of distributions examined.

**Table 5.5** A list of 25 estimators $T$, $S$ of Normal $\mu$, $\sigma$ (Gross, 1976)

| Code (Gross, 1976) | Brief description of $T$; $S$ | Location estimator $T$ in notation of Andrews et al. (1972) and Table 5.1 |
|---|---|---|
| 1 MEAN | mean; standard deviation | M |
| 2 TM10 | trimmed mean; Winsorized standard deviation, $\alpha = 0.10$ | 10% |
| 3 TM25 | trimmed mean; Winsorized standard deviation, $\alpha = 0.25$ | 25% |
| 4 TM35 | trimmed mean; Winsorized standard deviation, $\alpha = 0.35$ | — |
| 5 TMMV | trimmed mean with minimum estimated standard deviation | JAE |
| 6 TMZ | adaptive trimmed mean with Winsorized standard deviation | — |
| 7 MED | median; median absolute deviation | 50% |
| 8 W18 | wave estimator; estimated asymptotic standard deviation, $k = 1.8$ | AMT with $k = 1.8$ |
| 9 W24 | wave estimator; estimated asymptotic standard deviation, $k = 2.4$ | AMT with $k = 2.4$ |
| 10 W18J | jackknifed wave estimator, $k = 1.8$ | — |
| 11 W24J | jackknifed wave estimator, $k = 2.4$ | — |
| 12 WMIN | wave estimator with minimum estimated standard deviation | — |
| 13 W18A | asymmetric wave estimator, $k = 1.8$ | — |
| 14 W24A | asymmetric wave estimator, $k = 2.4$ | — |
| 15 H12 | Hampel estimator; estimated asymptotic std. deviation, corners at (1.71, 2.85, 11.4) | Like 12 A, 17A, 21A, 22A and 25A, but with parameters as stated |
| 16 H15 | Hampel estimator; estimated asymptotic std. deviation, corners at (1.71, 2.28, 5.7) | |
| 17 H22 | Hampel estimator; estimated asymptotic std. deviation, corners at (2.25, 3.75, 15.0) | |
| 18 H25 | Hampel estimator; estimated asymptotic std. deviation, corners at (2.25, 3.0, 7.5) | |
| 19 H12J | jackknifed Hampel estimator H12 | — |
| 20 H15J | jackknifed Hampel estimator H15 | — |
| 21 H21J | jackknifed Hampel estimator H21 | — |
| 22 H25J | jackknifed Hampel estimator H25 | — |
| 23 BS74 | bisquare estimator; estimated asymptotic standard deviation, $k = 7.4$ | Like P15, but with parameter $k$ as stated |
| 24 BS82 | bisquare estimator; estimated asymptotic standard deviation, $k = 8.2$ | |
| 25 BS90 | bisquare estimator; estimated asymptotic standard deviation, $k = 9.0$ | |

From the detailed results he recommends the following three estimators $(T, S)$ as giving good robustness of performance while maintaining an error rate close to the specified 5 per cent across the spectrum of distributions tested:

the wave estimator W24
the Hampel estimator H22
the bisquare estimator BS82.

We give in full the equations for these estimators in terms of the median $\tilde{x}$ and the median deviation $s_m$:

W24

$$T = \tilde{x} + ks_m \tan^{-1}\{\textstyle\sum \sin z_i / \textstyle\sum \cos z_i\}$$

$$S = ks_m\{n \textstyle\sum \sin^2 z_i\}^{1/2}/\textstyle\sum \cos z_i$$

where

$$z_i = (x_i - \tilde{x})/ks_m, \quad k = 2.4, \text{ and}$$

the summations $\sum$ include only those terms for which $|z_i| < \pi$.

H22

$$T = \tilde{x} + s_m \textstyle\sum \psi(z_i)/\textstyle\sum \psi'(z_i)$$

$$S = s_m\{n \textstyle\sum \psi^2(z_i)\}^{1/2}/\textstyle\sum \psi'(z_i)$$

where $z_i = (x_i - \tilde{x})/s_m$,

$$\psi(z) = \begin{array}{ll} z & (|z| < a) \\ a & (a \le |z| < b) \\ a(c - |z|)/(c - b)\,\mathrm{sgn}\,z & (b \le |z| < c) \\ 0 & (c \le |z|), \end{array}$$

and $a = 2.25, b = 3.75, c = 15.0$.

BS82

$$T = \tilde{x} + ks_m \textstyle\sum \psi(z_i)/\textstyle\sum \psi'(z_i)$$

$$S = ks_m\{n \textstyle\sum \psi^2(z_i)\}^{1/2}/\textstyle\sum \psi'(z_i)$$

where $z_i = (x_i - \tilde{x})/ks_m$, $k = 8.2$, and

$$\psi(z) = \begin{array}{ll} z(1 - z^2)^2 & (|z| < 1) \\ 0 & (|z| \ge 1). \end{array}$$

Among the many points of interest in Gross's results, one notes that the various jackknifed estimators all perform poorly. See Miller (1974, p. 13); his observation 'The jackknife is not a device for correcting outliers' is prophetic!

Patel, Mudholkar, and Fernando (1988) propose the use of the following joint estimators $(T, S)$:

(A)  $T_A$ = trimean (5.1.9), $S_A$ = interquartile range $x_{(d)} - x_{(a)}$
(B)  $T_B$ = Gastwirth estimator (5.1.8), $S_B$ = intertertile range $x_{(c)} - x_{(b)}$

where $a, b, c, d$ are the nearest integers to $n/4, n/3, 2n/3, 3n/4$, respectively; (C) $T_C$ = $\alpha$-trimmed mean $\overset{T}{x}(\alpha, \alpha)$, $S_C$ = $\alpha$-Winsorized standard deviation $\overset{W}{s}(\alpha, \alpha)$, with possible values $\alpha = 0.05(0.05)0.25$; $(T_C, S_C)$ has previously been considered by Gross (1976), see Table 5.5.

They show that the null distributions of $(T - \mu)/S$ are well approximated by $k$ $t_\nu$ distributions and give the following values, based on Monte Carlo calculations, for the scale factor $k$ and the degrees of freedom $\nu$ of Student's $t$:

| Studentized location estimator | $k$ | $\nu$ |
|---|---|---|
| $(T_A - \mu)/S_A$ | $1 - 5/(3n)$ | $2.00 + 0.372\,n$ |
| $(T_B - \mu)S_B$ | $1 - 4/(3n)$ | $2.25 + 0.202\,n$ |
| $(T_C - \mu)/S_C$ | $1 + 16[n\,\alpha]^{1/2}\exp(2[n\,\alpha] - n)$ where $r_\alpha = 2.92, 2.68, 2.60, 2.48$ for $\alpha = 0.05, 0.10, 0.15, 0.25.$ | $n - r_\alpha[n\alpha] - 1$ |

Monte Carlo power comparisons of the various statistics are presented, for various mixture models of contamination. In the light of these it is recommended that estimators based on a high proportion of trimming, such as the 25 per cent trimmed mean, the trimean, or the Gastwirth estimator, should be used when the sample is expected to come from a normal distribution with substantial contamination.

## 5.5  ACCOMMODATION OF OUTLIERS IN GAMMA (INCLUDING EXPONENTIAL) SAMPLES

Most of the published work on specific procedures for accommodation of outliers in gamma samples relates to the exponential case, and we will consider this first. Suppose our sample $x_1, x_2, \ldots, x_n$ comes from an exponential distribution with density

$$f(x, \theta) = \frac{1}{\theta}\exp(-x/\theta) \qquad (5.5.1)$$

apart from the prospect that one observation may be a contaminant: arising from a distribution with density $f(x, b\theta)$ for some $b > 1$. In the first instance we assume this observation is unspecified or *unlabelled*. We have a basic model $H$:$f(x, \theta)$ and a scale-slippage alternative model $\bar{H}$ where the slippage relates to just one observation. Basic early contributions were by Kale and Sinha (1971), Joshi (1972b), and Veale and Kale (1972). Assuming the *exchangeable* version of the slippage model to account for a possible contaminant, Kale and Sinha (1971) study the performance of 'restricted' $L$-estimators

$$S(\mathbf{l}) = \sum_{j=1}^{m} l_j x_{(j)} \qquad (5.5.2)$$

which ignore the largest $n - m$ observations. They show that the one-sided Winsorized mean

$$S_{m,n} = \frac{1}{m+1}\left[\sum_{j=1}^{m-1} x_{(j)} + (n - m + 1)x_{(m)}\right], \qquad (5.5.3)$$

with divisor $m + 1$ instead of $n$, is 'optimal' in the sense of minimizing $MSE(S(\mathbf{l})\,|\,b = 1)$. Note that there is no suggestion of optimality under the alternative model where $b > 1$. In the null case, (5.5.3) has efficiency $(m + 1)/(n + 1)$ relative to the optimal full-sample version: $S_{n,n} = \sum x_j/(n + 1)$. It is argued, however, that this loss of efficiency under the basic model (where we do not need to protect against a contaminant) may be offset by a corresponding gain under $\bar{H}$ (where we do need to do so). Accordingly $MSE(S_{m,n}\,|\,b > 1)$ is investigated. The cases $n = 3, 4$ are studied in detail and this gain (relative to $S_{n,n}$) is confirmed for sufficiently large $b$. Typically if $n = 4$, $m = 3$ the relative efficiency rises from 0.8 at $b = 1.1$ to 1.0 at $b = 2$, 4 at $b = 5$, 15 at $b = 10$, and ultimately becomes infinite.

The questions of choice of $m$, and performance for larger $n$, are taken up by Joshi (1972b). When $b < 2$ it turns out that no $m \neq n$ improves on $S_{n,n}$. But for more extreme contamination (larger $b$) substantial gains in relative efficiency are available. Table II on page 472 (extracted from Joshi, 1972b) presents the optimal choice $m^*$ for $m$ and associated relative efficiency $e_{m^*}$ for values of $b$ in the range 2–20. Specifically, if $b = 1/h$ the table presents results for $h = 0.05(0.05)0.50$.

Of course, $b$ will not be known and Joshi suggests an *ad hoc* procedure which consists of first calculating $S_{n-1,n}$ as a provisional estimate $\tilde{\theta}$, then estimating $b$ from

$$nS_{n,n} = (n + b - 1)\tilde{\theta} \qquad (5.5.4)$$

for the purpose of determining $m^*$ from Table II. The corresponding $S_{m^*,n}$ is used for $\tilde{\theta}$ in (5.5.4), and a new $m^*$ is determined from the table. The process is repeated until $m^*$ becomes stable, at which stage $\theta$ is estimated by the corresponding $S_{m^*,n}$.

*Example 5.1 Failures of a critical electronic component occur from time to time in a navigational aid. On failure the component is replaced by a new one. Records show the ordered values of lifetime for 9 components to be*

1.6 2.8 2.9 4.1 9.8 14.1 16.7 22.1 54.3

*Here* $n = 9$ *and we have* $S_{9,9} = 12.84$ *and* $S_{8,9} = 10.689$. *From* (5.5.4) *we get* $b = 2.811$. *Thus from Table II,* $m^* = 8$. *We do not need to proceed further. We estimate* $\theta$ *by* 10.689 *for an efficiency gain (if b is truly* 2.811*) of about* 28 *per cent.*

Other aspects of this approach are discussed by Sinha (1973a; moment properties and limiting form of the MSE), Sinha (1973c; refinements for the two-parameter, location shifted, case), Sinha (1973d; some exact distributional results, including lengths of confidence intervals for $n = 4$, $m = 3$).

Veale and Kale (1972) consider a corresponding hypothesis test. Under $H$ the UMP size-$\alpha$ test of $H_0$: $\theta = 1$ versus $H_1$: $\theta > 1$ has critical region of the form:

$$S_{n,n} > C_{\alpha,n}$$

The robustness of this test is examined by considering its performance under the contamination model $\bar{H}$. An expression for the power function $\beta(b, \theta)$ is obtained. Veale and Kale proceed to examine tests based on $S_{m,n}(m < n)$. For any $m$, a UMP size-$\alpha$ test again exists with rejection for sufficiently large $S_{m,n}$. However, consideration of power shows not surprisingly that for the basic (uncontaminated) model we are best to take $m = n - 1$ if we cannot take $m = n$. Robustness properties of this test are discussed in terms of 'premium' and 'protection' measures (see Section 3.1.5), and some tabulated values are presented.

In subsequent work on outlier-robust estimation of $\theta$, we will see that a variety of estimators have been studied, particularly in the following categories:

$L$-estimators, including various forms of trimmed and Winsorized mean;
Anscombe-type estimators (see Section 5.1.1);
moment estimators.

Kale (1975c) presents a wider study of robust estimation of scale parameters under an exchangeable model consisting of two components in the exponential family, with $(n - k)$ observations from one and $k$ ($\geq 1$) from the other. Employing a maximum likelihood approach he obtains as estimator in the case of an exponential distribution the $(0, k)$ *trimmed mean*

$$\overset{\text{T}}{x}_{0,k} = \sum_{j=1}^{n-k} x_{(j)}/(n - k) \tag{5.5.5}$$

rather than the Winsorized mean of the Kale and Sinha (1971) approach. (Trimmed means also arise for location estimates in the normal case with known, common, $\sigma^2$.)

The trimmed, and Winsorized, means are compared for $k = 1$ using premium-protection measures. $\overset{T}{x}_{0,1}$ provides greater protection (lower MSE as $b \rightarrow \infty$) than $S_{n-1,n}$ but at a higher premium (higher MSE as $b \searrow 1$). We should recall, however, that $S_{n-1,n}$ is not necessarily the optimal form of $S_{m,n}$.

These comparisons are taken up by Chikkagoudar and Kunchur (1980) who, in an extension of Joshi's approach described above, propose the full-sample $L$-estimator

$$S_{CK} = \sum_{j=1}^{n} a_j x_{(j)} \qquad (5.5.6)$$

where

$$a_j = \frac{1}{n}\left(1 - \frac{2j}{n(n+1)}\right). \qquad (5.5.7)$$

They construct this estimator as a linear combination of the $n$ estimators

$$\tilde{\theta}_i = \frac{1}{n}\left(\sum_{j=1}^{n} x_j - x_i\right),$$

each $\tilde{\theta}_i$ being weighted proportionally to the rank of the omitted observation $x_i$. A detailed comparative study is made of the efficiencies, in terms of MSE, of $S_{CK}$ and three other estimators: $S_{n,n}$, the full-sample version of $S_{m,n}$ in (5.5.3); $S_{m^*,n}$, where Joshi's optimal value $m^*$ is used in (5.5.3); and the (0,1) trimmed mean

$$\overset{T}{x}_{0,1} = \sum_{j=1}^{n-1} x_{(j)}/(n-1).$$

As regards the efficiency of $S_{CK}$ relative to $S_{n,n}$, which they tabulate for $n = 2(1)5(5)20$, 30 and for $b = 1/0.75 = 1.33$, 10 and 20, typical values are as follows:

|          | $n = 2$ | 3 | 4 | 15 | 30 |
|----------|---------|-------|-------|-------|-------|
| $b = 1.33$ | _0.921_ | _0.986_ | 1.001 | 1.005 | 1.003 |
| 10       | 4.71    | 2.48  | 1.92  | 1.20  | 1.11  |
| 20       | 4.39    | 2.38  | 1.86  | 1.18  | 1.10  |

With the two italicized exceptions, $S_{CK}$ is more efficient than $S_{n,n}$. The relative efficiency is systematically higher when $b = 10$ than when $b = 20$ or 1.33, suggesting that $S_{CK}$ has maximum advantage over $S_{n,n}$ at some value of $b$ between 1.33 and 10.

Compared with the trimmed mean $\overset{T}{x}_{0,1}$, $S_{CK}$ is more efficient for all $n$ when $b \leqslant 1/0.45 = 2.22$, but for high values of $b$ the situation is dramatically reversed; for example, when $n = 15$ the efficiency of $S_{CK}$ relative to $\overset{T}{x}_{0,1}$ is 1.18 when $b = 1.33$, but only 0.12 when $b = 10$ and 0.029 when $b = 20$! As regards $S_{m^*,n}$, $S_{CK}$ is more efficient for values of $b$ not greatly in excess of 1, specifically for $n \geq 6$ and $b \leq 2.5$; but $S_{m^*,n}$ is otherwise more efficient, certainly so when $n \geq 6$ and $b \geq 1/0.35 = 2.86$. 'However', Chikkagoudar and Kunchur (1980) claim:

in practical situations moderate values of . . . [$b$] are very important. Because, if . . . [$b$] is quite . . . [large] the outlying observation will be quite large as compared to other observations . . . and so it can be easily identified; if . . . [$b$] is quite . . . [small], then the outlying observation will be so similar to others that the whole sample can be considered as homogeneous.

They also discuss the case when the slippage parameter $b$ is *known*, but they do not suggest a practical application.

Veale (1975) proposes to estimate $\theta$ in (5.5.1) from a sample $x_1$, $x_2, \ldots, x_n$ with, as before, the prospect that one observation may arise from a scale-slipped exponential distribution with density $f(x, b\theta)$, but with the assumption that this observation, $x_n$ say, is *specified*. This model is called by Gather (1986) the *identified-outlier* model. (It is not the same as our *labelled slippage* model, in which it is the *greatest* observation, $x_{(n)}$, which may arise from a contaminating distribution.) The model, Veale observes,

may be appropriate in life testing situations where the timing mechanism for one of the items fails, yielding an overestimated lifetime . . . [(case $b > 1$)], or where one of the items is accidently [*sic*] subjected to an excessive (and imprecisely measured) stress, yielding an underestimated lifetime . . . [(case $b < 1$)]. *Per contra*, all of the data may be 'trustworthy', but prior information may enable one to stratify the data on some characteristic, e.g. component age or priority ('primary' and 'stand-by' units) or test conditions ('simulated' and 'real world'). In such stratified samples (with $n - 1$ observations in one strata and one observation in the other), the . . . model may be appropriate. (Veale, 1975)

(The 'stratified' situation, where $x_n$ may or may not be a *contaminant*, does indeed invite a robust procedure accommodating $x_n$. In the first type of situation described, however, one might think that $x_n$ is *known* to be a contaminant, and that it would therefore be best to estimate $\theta$ from the reduced sample $x_1, \ldots, x_{n-1}$.)

For the case $b > 1$ Veale considers seven different estimators based on the statistics

$$A = \sum_{j=1}^{n} x_j, \qquad A' = \sum_{j=1}^{n-1} x_j = A - x_n, \qquad (5.5.8)$$

including $A/n = \bar{x}$, the sample mean; $A/(n + 1) = S_{n,n}$ as defined in (5.5.3); and $A'/n$ which may correspondingly be written $S_{n-1,n-1}$. From a comparative study of mean squared errors and biases he recommends the following estimator from among the seven:

$$S_V = \min (A'/n, A/(n + 1))$$
$$\min (S_{n-1,n-1}, S_{n,n}).$$

(5.5.9)

With this procedure, $\theta$ will be estimated by $S_{n-1,n-1}$ except when $x_n$ is less than $A'/n$, i.e. less than $(n - 1)/n$ times the mean of the other $n - 1$ observations—hardly an upper outlier.

Continuing with the estimation of $\theta$ in (5.5.1) subject to the possibility that one of the $n$ observations arises from $f(x, b\theta)$, but reverting to the assumption that this observation is unspecified, Rauhut (1982) proposes two estimators of a rather similar structure to those considered by Veale (1975). Since the possible contaminant is not identified, we work once again in terms of the *ordered* observations. As in (5.5.8), write

$$A = \sum_{j=1}^{n} x_j = \sum_{j=1}^{n} x_{(j)}$$

(5.5.10)

but now

$$A^* = \sum_{j=1}^{n-1} x_{(j)} = A - x_{(n)}.$$

(5.5.11)

Rauhut's procedure is to select as estimator either $A/(n + 1)$ (based on the full sample) or $A^*/n$ (based on the reduced sample omitting $x_{(n)}$) according to how 'extreme' $x_{(n)}$ is, as judged by some criterion. Two alternative criteria are proposed, viz.

$$(x_{(n)} - x_{(n-1)})/(x_{(n-1)} - x_{(1)}) \quad \text{and} \quad x_{(n)}/\bar{x}.$$

(5.5.12)

As will be seen in Chapter 6, these are in fact discordancy test statistics for an upper outlier in an exponential sample, being respectively equivalent to the statistic (see page 199)

$$T_{E2} = (x_{(n)} - x_{(n-1)})/(x_{(n)} - x_{(1)})$$

(5.5.13)

and the statistic (see pages 197–8)

$$T_{Ea1} = x_{(n)}/A.$$

(5.5.14)

Rauhut's two estimators based on these criteria are

$$S_{R1} = A/(n + 1) \quad \text{if} \quad T_{E2} < c$$
$$A^*/n \quad \text{if} \quad T_{E2} \geq c$$

(5.5.15)

and

$$S_{R2} = A/(n + 1) \quad \text{if} \quad T_{Ea1} < c'$$
$$A^*/n \quad \text{if} \quad T_{Ea1} \geq c'$$

(5.5.16)

where $c$, $c'$ are suitably chosen positive constants.

These resemble Veale's estimator $S_V$ in (5.5.9). In spirit they are Anscombe-type estimators as defined in Section 5.1.1, the inequalities in (5.5.15) and (5.5.16) being analogous to those in (5.1.1), (5.1.2) and (5.1.3).

The estimators $S_{R1}$ and $S_{R2}$ are assessed in terms of premium and protection, and compared with $A/(n + 1) = S_{n,n}$ and $A^*/n = [(n - 1)/n]\overset{\mathrm{T}}{x}_{0,1}$; the constants $c$, $c'$ are chosen so as to give prescribed levels of premium of 5 per cent and 1 per cent. The broad conclusion is that $S_{R2}$ should be used in preference to any of the other estimators when there is the possibility of an outlier to be accommodated.

Gather (1986) considers ten Anscombe-type estimators of the exponential parameter $\theta$ in relation to the prospect of a single contaminant. She specifies four different models for this contaminant: $H_I$ (identified-outlier, see page 178), $H_S$ (unlabelled slippage), $H_E$ (exchangeable), and $H_L$ (labelled slippage). Her procedure, like that of Rauhut (1982), is to select as estimator either $S_{n,n} = n\bar{x}/(n + 1)$ or some other statistic $T$, the choice being made according to some criterion; she uses the same two criteria (5.5.12) as Rauhut. The ten estimators $S_{G1}, S_{G2}, \ldots, S_{G10}$ with their associated statistics $T_1, T_2, \ldots, T_{10}$ are as follows:

'Use $T_r$ as estimator if $x_n > c\bar{x}$; use $S_{n,n}$ otherwise.'
Assumed contamination model $H_I$.

$$S_{G1}: T_1 = A'/n \quad \text{(equation (5.5.8))}$$

$$S_{G2}: T_2 = A'/(n - 1)$$

'Use $T_r$ as estimator if $x_{(n)} > c\bar{x}$; use $S_{n,n}$ otherwise.'
Assumed contamination model $H_S$, $H_E$ or $H_L$.

$$S_{G3}: T_3 = \sum_{j=1}^{n-1} x_{(j)}/n$$

$$S_{G4}: T_4 = \sum_{j=1}^{n-1} x_{(j)}/(n - 1) = \overset{\mathrm{T}}{x}_{0,1}$$

$$S_{G5}: T_5 = \left[\sum_{j=1}^{n-1} x_{(j)} + x_{(n-1)}\right]/(n + 1)$$

$$S_{G6}: T_6 = \left[\sum_{j=1}^{n-1} x_{(j)} + x_{(n-1)}\right]/n = \overset{\mathrm{W}}{x}_{0,1}$$

'Use $T_r$ as estimator if $x_{(n)} - x_{(n-1)} > c(x_{(n-1)} - x_{(1)})$; use $S_{n,n}$ otherwise.'
Assumed contamination model $H_S$, $H_E$, or $H_L$.

$$S_{G7}: T_7 = T_3$$

$$S_{G8}: \quad T_8 = T_4$$

$$S_{G9}: \quad T_9 = T_5$$

$$S_{G10}: \quad T_{10} = T_6$$

Gather reports the results of a study of premium and protection values for the various estimators (using exact calculation and not simulation). Her main conclusion is that $S_{G6}$ *(using either the (0,1) Winsorized mean or $n\bar{x}/(n + 1)$ according to the value of $x_{(m)}/\bar{x}$) is definitely the best* of the eight estimators $S_{G3}$ to $S_{G10}$ for any of the models $H_S$, $H_E$, $H_L$.

Motivated by life-testing problems, Homan and Lachenbruch (1986) consider methods for the robust estimation of $\theta$, both for uncensored samples from (5.5.1) and for randomly censored samples; in the latter case the likelihood of a censored observation $x_j$ is $\exp(- x_j/\theta)$.

Eight methods are reviewed. As well as estimators already discussed here, such as the Winsorized mean (5.5.3), the trimmed mean (5.5.5) and the Chikkagoudar–Kunchur estimator $S_{CK}$ (5.5.6, 5.5.7), these include two simple $L$-estimators based respectively on just one, or two, of the order statistics. They are

$$S_{HL1} = c_k \, x_{(k)} \tag{5.5.17}$$

where

$$1/c_k = \sum_{i=1}^{k} 1/(n - i + 1) \tag{5.5.18}$$

and

$$S_{HL2} = c_l x_{(l)} + c_m x_{(m)}, \tag{5.5.19}$$

where optimal (variance-minimizing) values of $k$, $l$, $m$ and $c_l$, $c_m$ are given in Harter (1969b).

Homan and Lachenbruch report extensive performance comparisons based on simulations of size 500, for samples of sizes 10 and 20, both uncensored and censored, and again both uncontaminated and contaminated. Contamination is according to a mixture model with $f(x, 1)$ as the basic model (see (5.5.1)) and a random number of contaminants from $f(x, a)$, $a = 5, 10$. See Homan and Lachenbruch's paper for details of their results.

Robust estimators of $\theta$ in the presence of a single contaminant are again considered by Joshi (1988) and by From (1991), in each case with an exchangeable model for the contaminant which is assumed to arise from the scale-slipped exponential distribution $f(x, a\theta)$ (see (5.5.1)). Joshi's approach is to estimate $\theta$ and the slippage parameter $a$ jointly. He obtains the joint *moment estimators*, $\tilde{a}$ and $S_{J1} = \hat{\theta}$ given by the equations

$$\sum_{j=1}^{n} x_j = (n - 1 + a)\theta \tag{5.5.20}$$

$$\sum_{j=1}^{n} x_j^2 = 2(n - 1 + a^2)\theta^2; \tag{5.5.21}$$

and, since these equations do not give real values for $\tilde{a}$ and $\tilde{\theta}$ if $\sum x_j^2 < 2(\sum x_j)^2/n$, he suggests using a *modified moment estimator* $S_{J2}$ given by solving (5.5.20) and

$$x_{(n)} = [aB(n, 1/a) + \sum_{j=1}^{n-1} (1/j)]\theta, \tag{5.5.22}$$

where $B$ is the beta function.

Maximum likelihood estimators $\hat{a}$ and $S_{J3} = \hat{\theta}$ are given by (5.5.20) together with

$$\sum_{j=1}^{n} (x_j - a\theta) \exp[(a - 1)x_j/a\theta] = 0. \tag{5.5.23}$$

The estimators $S_{J2}$ and $S_{J3}$ require iterative calculation. Joshi reviews them (without any detailed performance calculations) in comparison with $S_{m,n}$, $S_{CK}$ and $S_{R1}$ and observes that

a clear cut choice between these estimators is difficult. . . . In general, we recommend the use of $[S_{CK}]$ over other estimators mainly due to its simple form and robust nature. (Joshi, 1988).

From (1991) breaks new ground in relation to this particular problem. Like previous authors, he considers $L$-estimators of $\theta$ in (5.5.1) of the general form

$$T(c) = \sum_{j=1}^{n} c_j x_{(j)},$$

with weights $\mathbf{c} = (c_1, \ldots, c_n)$. He presents expressions for MSE $(T(c))$ under the same alternative (contamination) model $H$ as in Joshi (1988), and for the values $\mathbf{c}^* = (c_1^*, \ldots, c_n^*)$ of the weights which minimize MSE $(T(c))$. In a pilot calculation of the values of $\mathbf{c}^*$, for $n = 3$ and $n = 5$ and for a range of values of the slippage parameter $a$ between $10/9 = 1\cdot11$ and 20, the optimal weights $(c_1^*, \ldots, c_n^*)$ turn out to be very nearly of the form

$$\begin{aligned} c_1^* &= c_2^* = \ldots = c_M^* = D_1, \text{ say} \\ c_{M+1}^* &= c_{M+2}^* = \ldots = c_n^* = D_2, \text{ say} \end{aligned} \tag{5.5.24}$$

for some constants $D_1 > D_2$ (depending on $n$ and $a$) and some integer $M$ in the range $1 \leqslant M \leqslant n - 1$. This is illustrated by the following values of $\mathbf{c}$ taken from Table I of From (1991).

| $n = 3$ | $a$ | $c_1^*$ | $c_2^*$ | $c_3^*$ | | |
|---|---|---|---|---|---|---|
| | 2 | 0.259 | 0.263 | 0.140 | | |
| | 5 | 0.276 | 0.275 | 0.048 | | |
| | 10 | 0.270 | 0.265 | 0.024 | | |
| $n = 5$ | $a$ | $c_1^*$ | $c_2^*$ | $c_3^*$ | $c_4^*$ | $c_5^*$ |
| | 2 | 0.174 | 0.174 | 0.173 | 0.177 | 0.096 |
| | 5 | 0.186 | 0.185 | 0.185 | 0.184 | 0.030 |
| | 10 | 0.181 | 0.181 | 0.181 | 0.176 | 0.015 |

From confirms by further calculations that the pattern (5.5.24) holds good for values of $n$ up to 50 and for $a \geq 1$. Accordingly he proposes the simple $L$-estimator of $\theta$

$$S_F = D_1 \sum_{j=1}^{n} x_{(j)} + D_2 \sum_{j=M+1}^{n} x_{(j)} \qquad (5.5.25)$$

where, for given $n$ and $a$, $D_1, D_2$ and $M$ are chosen to minimize $MSE(S_F \mid \bar{H})$. It is found that $MSE(S_F \mid \bar{H})$ is no more than 0.1 per cent greater than $MSE(T(\mathbf{c}^*))$ for $n \leq 50$ and $a \geq 1$.

Table I, abridged from From (1991) Table III, gives the optimal values of $D_1, D_2$ and $M$ required in (5.5.25), for $n = 4(2)12, 15(5)30, 40, 50$ and $1/a = 0.1(0.1)0.9$. Since $D_2 < D_1$ and $M/n$ is nearly 1, $S_F$ can be regarded as a 'partially trimmed mean' with a small amount of one-sided 'partial trimming'.

The calculation of $S_F$ for a given data set involves the slippage parameter $a$; this is unknown, and so must be estimated. From recommends the use of $\tilde{a}$ given by

$$n S_{n,n} = (n - 1 + \tilde{a}) S_{n-1,n} \qquad (5.5.26)$$

as in (5.5.4).

The author compares, in terms of MSE, the performances of $S_F$, Kale and Sinha's estimator $S_{m,n}$ with optimal $m$, and $S_{CK}$. $S_F$ is found to have the smallest MSE in every case.

Finally, he comments that

Chikkagoudar and Kunchur (1980) argue that the weights $[c_i]$ should be monotonic decreasing in $i$ for a robust estimator. This paper shows that the overall change in the decreasing nature of the weights should be abrupt and sudden (at $[M]$) and not gradual, and that the optimal weights do not always decrease as the index increases. (From, 1991)

Most of the work reviewed above relates to the prospect of a single contaminant. Maximum likelihood estimation of parameters in the presence of $k > 1$ contaminants is considered, both for exponential and normal samples, by Gather and Kale (1988). They also present likelihood ratio tests for testing the null hypothesis of no contaminants against a slippage alternative declaring $k$ contaminants.

The problem of *tolerance intervals* for exponential distributions in the presence of a contaminant is considered by Ranganathan and Kale (1983). We now turn from the particular case of exponential samples to gamma samples in general, and assume that our sample $x_1, \ldots, x_n$ comes from a gamma distribution with density

$$f(x, \rho, \sigma) = x^{\rho - 1} e^{-x/\sigma} / [\sigma^\rho \Gamma(\rho)] \tag{5.5.27}$$

Here $\sigma$ is the scale parameter and $\rho$ the shape parameter; $\rho = 1$ gives the exponential case (5.5.1). We assume that $\sigma$ is unknown. Some useful robust procedures are proposed by Kimber (1983a) both for $\rho$ known and for $\rho$ unknown. A typical situation when $\rho$ is known arises

when a measurement is replicated $m$ times for each of $n$ subjects. Each of the $n$ within-subjects sample variances has, assuming Normality, a gamma distribution with shape parameter $(m - 1)/2$. From this gamma sample we may wish to estimate the within-subjects variance robustly. (Kimber, 1983a)

For estimating $\sigma$ in this situation, Kimber proposes the use of a trimmed mean with an unbiasing factor: namely,

$$S_K = \overset{\text{T}}{x}_{r, s} / h \tag{5.5.28}$$

where the trimming parameters $r$, $s$ are at choice, and the unbiasing factor $h = h(\rho, r, s, n)$ is chosen so that $E(S_K) = \sigma$ when all the observations conform to the basic model (5.5.27). Writing $r = n\alpha$, $s = n\beta$, a discussion is given of the choice of trimming proportions $\alpha$, $\beta$, and of how it depends on the data situation. It is suggested that a reasonable choice, in the absence of any advance information about the distribution of outliers, is to take $\alpha = \beta$ whatever the value of $\rho$. The total proportion trimmed, $\alpha + \beta$, should be chosen to correspond to the greatest level of contamination one expects to be present, or is prepared to cope with, in the data.

As regards the unbiasing factor $h(\rho, r, s, n)$, which we rewrite for convenience as $g(\rho, \alpha, \beta, n)$, Kimber indicates methods for calculating it exactly, but shows that in practice it is adequate to use the following asymptotic formula, which gives a satisfactory approximation:

$$g(\rho, \alpha, \beta, n) \simeq \left(1 + \frac{1}{2n\rho}\right) g(\rho, \alpha, \beta, \infty)$$

$$= \left(\rho + \frac{1}{2n}\right) \left\{1 + \frac{A^\rho e^{-A} - B^\rho e^{-B}}{(1 - \alpha - \beta)\Gamma(\rho + 1)}\right\}. \tag{5.5.29}$$

Here $A$ and $B$ denote respectively the lower $\alpha$-quantile and upper $\beta$-quantile of the gamma distribution (5.5.27) with $\sigma = 1$, in other words $2A$ and $2B$

are the lower $100\alpha$ per cent and upper $100\beta$ per cent points of $\chi^2$ on $2\rho$ degrees of freedom.

The properties of $S_K$ are investigated in relation to the influence function, the efficiency and the gross-error sensitivity, and the estimator is shown to compare favourably, not only with $\bar{x}/h$ (the mean with unbiasing factor—not robust) and $\tilde{x}/h$ (the median with unbiasing factor—robust but inefficient), but also with the maximum likelihood estimator based on the censored sample, and with selected $M$-estimators. $S_K$ is to be recommended, providing only that in choosing the values of $\alpha$ and $\beta$ the actual proportion of contamination in the sample has not been underestimated.

It is further shown that the distribution of $S_K$ on the basic model (5.5.27) can be well approximated by a gamma distribution with scale parameter $\sigma/n\rho E$ and shape parameter $n\rho E$, where $E$ is the efficiency of $S_K$. The following empirical approximation for $E$ is given for the case when $0 \leqslant \alpha \leqslant \beta \leqslant 0.15$:

$$E \simeq 1 - \beta(0.6 + \rho^{-0.8}). \tag{5.5.30}$$

These results enable robust significance tests for $\sigma$ to be carried out and robust approximate confidence intervals for $\sigma$ to be constructed. See also Kimber (1983c).

Dixit (1989) considers parameter estimation for a sample from the gamma distribution with density $f(x, \rho, \sigma)$ (5.5.27), which is contaminated by a known number, $k$, of contaminants arising from a shape-slipped gamma distribution with density $f(x, \rho + b, \sigma)$ ($b \neq 0$). He presents procedures for obtaining both moment estimators and maximum likelihood estimators, (i) jointly for $\rho, \sigma, b$; (ii) for $\rho$ and $b$ when $\sigma$ is known; (iii) for $\sigma$ and $b$ when $\rho$ is known; and (iv) for $\rho$ and $\sigma$ when $b$ is known. Most of these procedures involve substantial computation. Some simulated values of bias and MSE are presented. Dixit (1991) again considers estimation for a gamma sample with density $f(x, \rho, \sigma)$ contaminated by a known number of contaminants, but this time assumes that they come from a scale-slipped gamma distribution with density $f(x, \rho, \sigma/\alpha)$.

## 5.6 ACCOMMODATION OF OUTLIERS IN LOGISTIC AND DOUBLE EXPONENTIAL SAMPLES

This section deals with the robust estimation of location and scale parameters $\mu$ and $\sigma$ in contaminated samples from the **logistic distribution** with distribution function

$$F_L(x; \mu, \sigma) = 1/\{1 + \exp(-\pi(x - \mu)/\sigma \sqrt{3})\}$$
$$(-\infty < x < \infty) \tag{5.6.1}$$

and the **double exponential distribution** with density

$$f_{DE}(x; \mu \; \sigma) = (1/2\sigma) \exp\left(- \left| x - \mu \right| /\sigma\right)$$

$$(- \infty < x < \infty)$$

$$(5.6.2)$$

In considering these two distributions we will be reviewing principally the work of Balakrishnan (1992a) and Balakrishnan and Ambagaspitiya (1988). They explore the behaviour of a range of estimators of $\mu$ and $\sigma$, taking the forms listed below. (Note that, for the logistic distribution (5.6.1), $\sigma$ is the standard deviation, while for the double exponential distribution (5.6.2), $\sigma\sqrt{2}$ is the standard deviation; this is the notation of the two papers cited, which we retain for convenience of reference.)

### 5.6.1  Estimators of the mean $\mu$

The estimators considered are as follows.

1.      The sample mean $\bar{x}$.
2.      The sample median $\tilde{x}$.

3, 4, 5.  The best linear unbiased estimators $BLUE\mu(0,0)$, $BLUE\mu(\cdot1,\cdot1)$, $BLUE\mu(\cdot2,\cdot2)$, based respectively on the full sample and on $(\cdot1,\cdot1)$ and $(\cdot2,\cdot2)$ symmetrically trimmed samples. For the weights $c_j$ of these $L$-estimators $\Sigma c_j x_{(j)}$, see Balakrishnan (1992a) for the logistic distribution and Govindarajulu (1966) for the double exponential distribution.

6, 7.   The trimmed means $\overset{T}{x}(\cdot1,\cdot1)$ and $\overset{T}{x}(\cdot2,\cdot2)$.
8, 9.   The Winsorized means $\overset{W}{x}(\cdot1,\cdot1)$ and $\overset{W}{x}(\cdot2,\cdot2)$.
10.     The Gastwirth mean (5.1.8).
11, 12. The linearly weighted means $L(\cdot1,\cdot1)$, $L(\cdot2,\cdot2)$ given, for even $n$, by

$$L(\alpha,\alpha) = \sum_{j=1}^{h} (2j - 1)(x_{(\alpha n + j)} + x_{(n - \alpha n - j + 1)})/2h^2 \qquad (5.6.3)$$

where $h = (0.5 - \alpha)n$.

13.     The estimator $RS\mu$ proposed by Raghunandanan and Srinivasan (1970, 1971). Define, for $m = 1, 2, \ldots, [n/2]$, the *quasi-ranges*

$$w_{(m)} = x_{(n - m + 1)} - x_{(m)} \qquad (5.6.4)$$

and the *quasi-midranges*

$$v_{(m)} = (x_{(m)} + x_{(n - m + 1)})/2 \qquad (5.6.5)$$

The $RS$ estimator of $\mu$ is the optimal quasi-midrange, i.e. the $v_{(m)}$ with the smallest variance.

14, 15. The modified maximum likelihood estimators $MML\mu(\cdot1,\cdot1)$, $MML\mu(\cdot2,\cdot2)$ based on $(\cdot1,\cdot1)$ and $(\cdot2,\cdot2)$ symmetrically trimmed samples. These are of the form

$$MML\mu(\alpha,\alpha) = \{(1 - 2\alpha)\overset{T}{\bar{x}}(\alpha, \alpha) + \alpha\beta(x_{(\alpha n + 1)}$$

$$+ x_{(n - \alpha n)})\}/(1 - 2\alpha + 2\alpha\beta) \quad (5.6.6)$$

For required values of $\beta$, see Tiku (1980) and Tiku, Tan, and Balakrishnan (1986).

16, 17, 18. (Used for logistic but not for double exponential samples.) Approximate maximum likelihood estimators $AML\mu(\cdot 1,\cdot 1)$, $AML\mu(\cdot 2,\cdot 2)$, $AML\mu(\cdot 3,\cdot 3)$, based on the respective symmetrically trimmed samples. Writing $\alpha n = r$, they are given by

$$AML\mu(\alpha, \alpha) = B/[m(n + 1)^2]$$

where

$$B = 2 \sum_{i = r + 1}^{n - r} i(n + 1 - i)x_{(i)} + r(r + 1)(n - r)(x_{(r + 1)} + x_{(n - r)}) \quad (5.6.7)$$

and

$$m = [n(n + 1)(n + 2) - 2r(r + 1)(r + 2)]/3(n + 1)^2 \quad (5.6.8)$$

See Balakrishnan (1992a).

## 5.6.2   Estimators of the scale parameter $\sigma$

The estimators considered are as follows.

*For logistic samples*

1.         The sample standard deviation $s$.
2, 3, 4.   The best linear unbiased estimators $BLUE\sigma(0,0)$, $BLUE\sigma(\cdot 1,\cdot 1)$, $BLUE\sigma(\cdot 2,\cdot 2)$, based respectively on the full sample and on $(\cdot 1,\cdot 1)$ and $(\cdot 2,\cdot 2)$ symmetrically trimmed samples. For the weights in these $L$-estimators, see Balakrishnan (1992a).
5, 6, 7.   The estimators $RS\sigma(0, 0)$, $RS\sigma(\cdot 1,\cdot 1)$, $RS\sigma(\cdot 2,\cdot 2)$ proposed by Raghunandanan and Srinivasan (1970). $RS\sigma(\alpha,\alpha)$ is based on the untrimmed quasi-ranges $w_{(m)}$ in (5.6.4), i.e. those for which $n\alpha < m \leq [n/2]$, and is the optimal (i.e. minimum variance) unbiased linear combination $\Sigma c_i w_{(i)}$ of those quasi-ranges, where each $c_i$ is either 0 or a constant value $c$ not dependent on $i$. Values of the constants required for calculating $RS\sigma(\alpha,\alpha)$ are given in Raghunandanan and Srinivasan (1970).
8, 9, 10.  The trimmed standard deviations

$$\overset{T}{s}(\cdot 1,\cdot 1), \quad \overset{T}{s}(\cdot 2,\cdot 2), \quad \overset{T}{s}(\cdot 3,\cdot 3).$$

11, 12, 13. The modified maximum likelihood estimators $MML\sigma(\cdot1,\cdot1)$, $MML\sigma(\cdot2,\cdot2)$, $MML\sigma(\cdot3,\cdot3)$, based on $(\cdot1,\cdot1)$, $(\cdot2,\cdot2)$, and $(\cdot3,\cdot3)$ symmetrically trimmed samples. For details, see Tiku (1980), Tiku, Tan and Balakrishnan (1986).

14, 15, 16. Approximate maximum likelihood estimators $AML\sigma(\cdot1,\cdot1)$, $AML\sigma(\cdot2,\cdot2)$, $AML\sigma(\cdot3,\cdot3)$, based on $(\cdot1,\cdot1)$, $(\cdot2,\cdot2)$ and $(\cdot3,\cdot3)$ symmetrically trimmed samples. They are readily calculated as follows.

Write

$$r = \alpha n$$

$$\gamma_i = [i/(n+1)] - [i(n+1-i)/(n+1)^2] \log[i/(n+1-i)]$$

$$(i = 1, \ldots, n) \quad (5.6.9)$$

$$C = 2 \sum_{i=r+1}^{n-r} i(n+1-i)x_{(i)}^2 + r(r+1)(n-r)(x_{(r+1)}^2 + x_{(n-r)}^2) \quad (5.6.10)$$

and

$$D = \sum_{i=r+1}^{n-r} (2\gamma_i - 1)x_{(i)} - r\gamma_{(n-r)}(x_{(r+1)} - x_{(n-r)}) \quad . \quad (5.6.11)$$

Then the estimator of $\sigma$ is given by

$$AML\sigma(\alpha,\alpha) = [\pi/2(n-2r)\sqrt{3}]\{D + [D^2 + 4(n-2r)$$

$$\times (C - B^2/m)/(n+1)^4]^{1/2}\} \quad (5.6.12)$$

where $B$ and $m$ are given by (5.6.7) and (5.6.8) respectively. See Balakrishnan (1992a).

*For double exponential samples*

Balakrishnan and Ambagaspitiya (1988) consider seven different estimators of $\sigma$ in (5.6.2) for each of two sample sizes, $n = 10$ and 20. Analogously to scale estimators 2, 3, 4 for logistic samples, they study the best linear unbiased estimators $BLUE\sigma(\cdot1,\cdot1)$, $BLUE\sigma(\cdot2,\cdot2)$. For the required weights, see Govindarajulu (1966). Analogously to Raghunandanan and Srinivasan's scale estimators 5, 6, 7 for logistic samples, they examine the estimators $RS\sigma(\alpha, \alpha)$, where $\alpha = 0, \cdot1, \cdot2, \cdot3$ for $n = 10$, and $\alpha = 0, \cdot05, \cdot1, \cdot2$ for $n = 20$. For parameter values, see Raghunandanan and Srinivasan (1971).

### 5.6.3  Robust estimation for logistic samples

Balakrishnan (1992a) assumes that any contamination in the sample $x_1, \ldots, x_n$ from the logistic distribution (5.6.1) is by slippage of a single observation, either slippage in location with $n - 1$ observations arising from $F_L(x; \mu, \sigma)$ and one from a logistic distribution $F_L(x; \mu + \alpha\sigma, \sigma)$, or by slippage in scale with $n - 1$ observations arising from $F_L(x; \mu, \sigma)$ and one from $F_L(x; \mu, b\sigma)$.

As regards estimation of location, the 18 estimators of $\mu$ considered by Balakrishnan (1992a) are listed above (Section 5.6.1). Using theoretical results referred to in Section 3.1.2, he tabulates the bias and MSE of these estimators under the location slippage alternative for $n = 10, 20$ and $a = 0.5(0.5)3, 4, \infty$; he also tabulates their variances (the biases being zero) under the scale slippage alternative for $n = 10, 20$ and $b = 0.5(0.5)2(1)4, \infty$.

The main conclusions emerging from these tables are as follows. *The sample mean and the BLUE$\mu$ (0,0) based on the complete sample are quite non-robust against single-outlier contamination. The sample median also performs badly in regard to its MSE under location slippage and its variance under scale slippage.* The approximate maximum likelihood estimators $AML\mu(\cdot1,\cdot1)$ and $AML\mu(\cdot2,\cdot2)$ perform well under both alternative models. *The $(\cdot2,\cdot2)$ trimmed mean $\overset{T}{x}(\cdot2,\cdot2)$ performs similarly well, and can be recommended as an efficient and robust estimator of* $\mu$.

Turning now to the estimation of scale, Balakrishnan (1992a) studies 16 estimators of $\sigma$ as listed above (Section 5.6.2). He tabulates the bias and MSE of these estimators under the location slippage alternative for $n = 10, 20$ and $a = 0.5(0.5)3, 4$, and under the scale slippage alternative for $n = 10, 20$ and $b = 0.5(0.5)2(1)4$. *The sample standard deviation, and the estimators BLUE$\sigma$(0,0) and RS$\sigma$(0,0) based on the complete sample, are all shown to be quite non-robust.* The trimmed and *MML* estimators have a substantial bias, and also a larger mean square error than the corresponding approximate maximum likelihood estimators. *The recommended estimators are these latter estimators with $(\cdot1,\cdot1)$ and $(\cdot2,\cdot2)$ symmetric trimming,* viz. $AML\sigma(\cdot1,\cdot1)$ and $AML\sigma(\cdot2,\cdot2)$; they are efficient and robust, and do not require the use of any special tables for their evaluation.

### 5.6.4    Robust estimation for double exponential samples

If $F$ denotes a continuous distribution symmetric about zero with density $f(x)$, denote by $\tilde{F}$ the distribution obtained by folding the density at zero; its density $\tilde{f}(x)$ will equal $f(x) + f(-x)$ $(x > 0)$, 0 otherwise. Suppose now that we have a contaminated sample with $n - 1$ values from $F$ and one contaminant from a scale-slipped distribution $G$ also symmetric about zero. Among the results on the distributional properties of contaminated distributions referred to in Section 3.1.2, Balakrishnan (1988) has expressed the moments of order statistics in such samples in terms of the moments of order statistics in samples with $n - 1$ values from $\tilde{F}$ and one contaminant from $\tilde{G}$.

These results are used by Balakrishnan and Ambagaspitiya (1988) to study the outlier-robustness properties of the estimators of $\mu$ and $\sigma$ listed in Sections 5.6.1 and 5.6.2 for samples from the double exponential distribution with density $f_{DE}(x; \mu, \sigma)$ in (5.6.2). A slippage alternative is assumed, in which $n - 1$ observations arise from $f_{DE}(x; \mu, \sigma)$ and one from the scale-slipped double exponential distribution $f_{DE}(x; \mu, b\sigma)$.

As regards estimation of location, the variances of the 16 listed estimators of $\mu$ are tabulated for $n = 10$ and 20 and for $b = 0.1$, 0.5, 0.75, 1(1)4(2)10. The tables indicate that the *best estimators of location* $\mu$ *in these conditions are the median, the Gastwirth mean, and the linearly weighted mean* $L(\cdot 2, \cdot 2)$ *based on* $(\cdot 2, \cdot 2)$ *trimming (see (5.6.3))*. (The $BLUE\mu$ estimators and the $RS\mu$ estimator also perform well as regards variance, but they all require special tables, unlike the three estimators just recommended.)

Balakrishnan and Ambagaspitiya (1988) also compare the seven $BLUE\sigma$ and $RS\sigma$ estimators of scale listed in Section 5.6.2. The values of bias and MSE for these estimators are tabulated by them for the above values of $n$ and $b$. In the light of these results, they *recommend the use of* $BLUE\sigma(\cdot 1, \cdot 1)$ *or* $RS\sigma(\cdot 1, \cdot 1)$ based on the $(\cdot 1, \cdot 1)$ symmetrically trimmed sample when estimating $\sigma$ *for small sample sizes*, and the use of $RS\sigma(\cdot 1, \cdot 1)$ *for larger sample sizes*.

# CHAPTER 6

# Specific Discordancy Tests for Outliers in Univariate Samples

Having considered in Chapter 4 general principles and procedures for discordancy tests of outliers in univariate samples, we now present detailed information on a wide range of useful tests. 'Useful' means two things here, firstly that the test performs reasonably well, even if not optimally, in relation to some meaningful alternative hypothesis; secondly that some information on percentage points is available—at least an inequality, if not an extensive tabulation. The main types of distribution for which useful tests are available are gamma and normal, and these are dealt with in Section 6.2, Section 6.3 respectively. Some tests for samples from other distributions, including log-normal, truncated exponential, uniform, extreme-value, Pareto, binomial, and Poisson, are described in Section 6.4.

## 6.1 GUIDE TO USE OF THE TESTS

We will be presenting information on a large number of tests, including descriptions of their forms, indications of properties, details of evaluations or tabulations of percentage points, relevant references, and examples of practical application. To keep such a large amount of information within bounds a unified method of presentation is adopted with separate sections (or subsections) devoted to different distributions.

In each case (gamma, normal, uniform, etc.) we commence with some general discussion of the types of outlier situation where the distribution might be appropriate. We then present for each distribution a **contents list** of tests (pp. 195–6, 218–21, 251); we have labelled the tests **G1**, **G2**, . . . for gamma samples, **N1**, **N2**, . . . for normal samples, and so on, as is explained in detail later. For each individual test we then give a **worksheet**, which presents systematically the following information where available:

*Label and purpose of test*

*Test statistic*, denoted by $T$ with the test label as subscript. For example, we denote the statistic for test **N1** by $T_{N1}$.

*Test distribution*, i.e. the distribution of the statistic on the working hypothesis of no contamination. The probability density function and the distribution function for this distribution are denoted respectively by $f_n(t)$, $F_n(t)$, where $n$ is the sample size.

*Recurrence relationship* for the test distribution (where appropriate).

*Simple inequality for the significance probability* (where appropriate). The significance probability attaching to an observed value $t$ of a discordancy statistic $T$ is denoted here by $SP(t)$. That is to say, $SP(t)$ is the probability that, on the working hypothesis, $T$ takes values more discordant than $t$ (for most tests this means $T > t$).

*Tabulated significance levels* in the form of references to where these will be found in the set of tables in the Appendix at the back of the book, together with source attribution.

*Further tables*: references to books and journals extending our tabulated significance levels, or tabulating other quantities of interest, for example, power.

*References* to other published material on the test such as derivation of the test distribution, optimality properties, power considerations, etc.

*Properties of test*: advantages and disadvantages, including statements of any secondary features of the data against which the test provides a particular safeguard, such as a suspicious *least* value when testing for an *upper* outlier; whether the test has a theoretical validation, such as being a maximum likelihood ratio test for some alternative; information on power or other performance measures, if available.

Illustrative *examples* are also given for some of the tests.

The following notation is used for standard distributions, random variables and functions:

*Notation for distributions or random variables*

$N(\mu, \sigma^2)$     normal with mean $\mu$ and variance $\sigma^2$

$t(\nu)$     Student's $t$ with $\nu$ degrees of freedom

$F(\nu_1, \nu_2)$     variance-ratio (or $F$) with $\nu_1$ and $\nu_2$ degrees of freedom

$\Gamma(r, \lambda)$     gamma with scale parameter $\lambda$ and shape parameter $r$, i.e. with density $f(x) = [\lambda^r \Gamma(r)]^{-1}(x^{r-1})\exp(-x/\lambda)$ $(x > 0)$

$E(\lambda)$     exponential with mean $\lambda$, i.e. with density $f(x) = \lambda^{-1}\exp(-x/\lambda)$ $(x > 0)$—same as $\Gamma(1, \lambda)$

$E(\lambda; a)$     exponential with scale parameter $\lambda$ and origin at $a$, i.e. with density $f(x) = \lambda^{-1}\exp[-(x-a)/\lambda]$ $(x > a)$ and $0(x < a)$

$P(\mu)$     Poisson with mean $\mu$

$B(n, p)$     binomial with parameters $n$, $p$

$H(N; n, r)$    hypergeometric with parameters $N$; $n$, $r$

$\phi(t)$    probability density function of $N(0,1)$, i.e. $(2\pi)^{-1/2} \exp(-\frac{1}{2}t^2)$

$\Phi(t)$    distribution function of $N(0, 1)$, i.e. $\int_{-\infty}^{t} \phi(u)du$

$B(r, s)$    beta function with parameters $r$ and $s$, i.e. $\Gamma(r)\Gamma(s)/\Gamma(r + s)$

$b_{r,s}(t)$    beta density with parameters $r$ and $s$, i.e. $[B(r, s)]^{-1}t^{r-1}(1 - t)^{s-1}$
$(0 \leq t \leq 1)$

Note. The tests described in detail below contain many *block tests* for *several outliers* (e.g. **Ga5 (Ea5)**, **N11**). These are, of course, vulnerable to *swamping*: the phenomenon whereby a block of outliers adjudged discordant *in toto* may contain a non-discordant outlier 'carried along' by the import of other members of the block. As discussed in Section 4.3, a natural response to this prospect is to consider *consecutive tests* (sometimes called *sequential* or *recursive tests*) for several outliers, where we examine up to $k$ outliers, whether in order of their 'aberrance' or in reverse order, before reaching a conclusion on a discordant outlying subset. Several consecutive tests of several outliers are described in detail below. In the cases where these are constructed by multiple application of a specific test of a single outlier this is mentioned under the test in question and appropriate references are given.

## 6.2 DISCORDANCY TESTS FOR GAMMA (INCLUDING EXPONENTIAL) SAMPLES

Until the 1960s, most of the published work on outliers in univariate samples was in the context of normal distributions. However, problems of outliers in samples from gamma distributions, and in particular from exponential distributions, are of considerable practical importance. Outlier situations in exponential samples arise naturally in such contexts as life testing; outliers in $\chi^2$ samples arise in analysis of variance; outliers in gamma samples of arbitrary shape parameter arise with skew-distributed data, for which a gamma distribution is often a useful pragmatic model; and outliers in both gamma and, specifically, exponential samples arise in any contexts where Poisson processes are appropriate basic models, e.g. in studying traffic flow, failures of electronic equipment, or biological aggregation. Attention to such problems in the literature has developed considerably from the 1970s onwards, with early contributions by Epstein, 1960a, 1960b; Laurent, 1963; Basu, 1965; Likeš, 1966; Kabe, 1970; Kale and Sinha, 1971. Full details of relevant references will appear in appropriate places throughout the chapter.

Further applications arise through transformation of the data. Procedures for outliers in exponential samples can sometimes be applied to outliers in samples from other distributions, such as the extreme-value distribution and the Weibull, by transforming the observations. For example, if the $n$ values $x_1, \ldots, x_n$ are (on the working hypothesis) a sample from the extreme-value distribution, with distribution function

$$P(X \leq x) = \exp\{- \exp[- (x - a)/b]\},$$

then the $n$ transformed values $\exp(- x_1/b), \ldots, \exp(- x_n/b)$ are a sample from the exponential distribution with mean $\exp(- a/b)$. Thus if $b$ is known but $a$ unknown, an outlier in the extreme-value sample can be tested by applying to the transformed values a suitable discordancy test for an exponential sample. See Section 6.4.4 for details.

In contrast, outlier tests for some other distribution can sometimes be employed for data from exponential or gamma distributions, by appropriately transforming the data to the form of the other distribution. For example, Kimber (1979) describes tests for a single outlier in a gamma sample (with *unknown* shape *and* scale parameters) which use square root, and cube root, transformations to normality and the normal tests, **N2** and **N15**, presented in Section 6.3.1 below. See test **Ga16** on pages 212–3 for further details of this approach.

Outlier situations can also arise in the context of *shifted* exponential or gamma distributions. If the origin of the exponential distribution $E(\lambda)$ with density $\lambda^{-1}\exp(- x/\lambda)(x > 0)$ is shifted to $x = a$, say, we get the distribution $E(\lambda; a)$ with density $\lambda^{-1}\exp[- (x - a)/\lambda](x > a)$. (Similar remarks apply of course to the gamma distribution $\Gamma(r, \lambda)$.)

Now some discordancy tests for exponential or gamma samples do require the assumption that the origin of the distribution is at zero, or at any rate is known; for example, the test based on the statistic $x_{(n)}/\sum x_i$ assumes that the origin is zero, and it can obviously be adapted to *known* non-zero origin $a$ by using $(x_{(n)} - a)/(\sum x_i - na)$ as statistic.

In contrast there are other tests which do not depend on knowledge of the origin, for example the Dixon-type test based on the statistic $(x_{(n)} - x_{(n-1)})/(x_{(n)} - x_{(1)})$; such tests are useful for two reasons. First, they are needed in the data contexts, sometimes encountered, where a shifted gamma or exponential distribution is the appropriate model. For instance, the development times of diapausing pupae of the cotton bollworm under conditions of constant temperature may be regarded as exponentially distributed with non-zero origin $t_{\min}$, this parameter being a minimum development time. Secondly, they are useful for testing outliers in samples from Pareto distributions, since again a transformation technique can be applied. Specifically, if the $n$ values $x_1, \ldots, x_n$ are a sample from a Pareto distribution with origin $a(> 0)$ and shape parameter $r$, then the $n$ transformed values $\log x_1, \ldots, \log x_n$ are a sample from an exponential distribution with origin $\log a$ and scale parameter $1/r$, i.e. from $E(1/r; \log a)$ in our notation. See Section 6.4.5.

### 6.2.1   Gamma samples: contents list and details of tests

The gamma distribution with scale parameter $\lambda$ and shape parameter $r$, i.e. the distribution with density

$$f(x) = [\lambda^r \Gamma(r)]^{-1} x^{r-1} \exp(- x/\lambda) \quad (x > 0),$$

is denoted by $\Gamma(r, \lambda)$. If the origin is shifted to $a$, the density is $[\lambda^r \Gamma(r)]^{-1} \times (x - a)^{r-1} \exp[- (x - a)/\lambda]$ $(x > a)$. $\Gamma(\nu/2, 2)$ is the $\chi^2$-distribution with $\nu$ degrees of freedom, $\chi_\nu^2$. $\Gamma(1, \lambda)$ is the exponential distribution with mean $\lambda$, denoted here by $E(\lambda)$. The corresponding distribution with origin shifted to $x = a$ is denoted by $E(\lambda; a)$.

In all the tests given here, $\lambda$ is assumed unknown. Except in tests **Ga16** and **Ga17**, the shape parameter $r$ is assumed *known*. The tests are classified as follows:

| Code | Distribution under the working hypothesis |
|---|---|
| **G** | gamma with unknown origin |
| **E** | exponential with unknown origin |
| **Ga** | gamma with known origin 0 (or more generally $a$) |
| **Ea** | exponential with known origin 0 (or more generally $a$) |

Needless to say, any **G**- or **Ga**-test can be applied to an exponential sample as a special case ($r = 1$); **E**- and **Ea**-tests are specific to the exponential case (generally because tables are only available for this case).

| Label | Worksheet page no. | Description of test | Statistic |
|---|---|---|---|
| **Ga1(Ea1)** | 197 | Test of a single upper outlier $x_{(n)}$ in a gamma sample | $x_{(n)}/\sum x_j$ |
| **Ea2** | 198 | Test of a single upper outlier $x_{(n)}$ in an exponential sample | $\dfrac{x_{(n)} - x_{(n-1)}}{x_{(n)}}$ |
| **E2** | 199 | Test of a single upper outlier $x_{(n)}$ in an exponential sample irrespective of origin | $\dfrac{x_{(n)} - x_{(n-1)}}{x_{(n)} - x_{(1)}}$ |
| **Ga3(Ea3)** | 199 | Test of a single lower outlier $x_{(1)}$ in a gamma sample | $x_{(1)}/\sum x_j$ |
| **E4** | 200 | Test of a single lower outlier $x_{(1)}$ in an exponential sample with origin unknown | $\dfrac{x_{(2)} - x_{(1)}}{x_{(n)} - x_{(1)}}$ |
| **Ga5(Ea5)** | 201 | Test of $k(\geqslant 2)$ upper outliers $x_{(n-k+1)}, \ldots, x_{(n)}$ in a gamma sample | $\dfrac{x_{(n)} + \ldots + x_{(n-k+1)}}{\sum x_j}$ |
| **Ea6** | 203 | Test of $k(\geqslant 2)$ upper outliers $x_{(n-k+1)}, \ldots, x_{(n)}$ in an exponential sample | $\dfrac{x_{(n)} - x_{(n-k)}}{x_{(n)}}$ |

*(Contd.)*

| Label | Worksheet page no. | Description of test | Statistic |
|-------|------|---------------------|-----------|
| E6 | 204 | Test of $k(\geqslant 2)$ upper outliers $x_{(n-k+1)}, \ldots, x_{(n)}$ in an exponential sample irrespective of origin | $\dfrac{x_{(n)} - x_{(n-k)}}{x_{(n)} - x_{(1)}}$ |
| Ga7(Ea7) | 204 | Test of a lower and upper outlier-pair $x_{(1)}, x_{(n)}$ in a gamma sample | $\dfrac{x_{(n)}}{x_{(1)}}$ |
| Ga8(Ea8) | 205 | Test of a lower and upper outlier-pair $x_{(1)}, x_{(n)}$ in a gamma sample | $\dfrac{x_{(n)} - x_{(1)}}{\sum x_j}$ |
| Ga9(Ea9) | 206 | Test of a lower and upper outlier-pair $x_{(1)}, x_{(n)}$ in a gamma sample | $\dfrac{(\sqrt{x_{(n)}} - \sqrt{x_{(1)}})^2}{\sum x_j}$ |
| E10 | 207 | Test of a lower and upper outlier-pair $x_{(1)}, x_{(n)}$ in an exponential sample with origin unknown | $\dfrac{x_{(n-1)} - x_{(2)}}{x_{(n)} - x_{(1)}}$ |
| Ga11(Ea11) | 207 | Test of $k(\geqslant 2)$ lower outliers $x_{(1)}, \ldots, x_{(k)}$ in a gamma sample | $\dfrac{x_{(1)} + \ldots + x_{(k)}}{\sum x_j}$ |
| E12 | 208 | Test of a lower outlier-pair $x_{(1)}, x_{(2)}$ in an exponential sample with origin unknown | $\dfrac{x_{(3)} - x_{(1)}}{x_{(n)} - x_{(1)}}$ |
| Ea13 | 209 | General Dixon-type test for an exponential sample, using knowledge of origin $a$ | $\dfrac{x_{(s)} - x_{(r)}}{x_{(q)} - a}$ |
| E13 | 210 | General Dixon-type test for an exponential sample, irrespective of origin | $\dfrac{x_{(s)} - x_{(r)}}{x_{(q)} - x_{(p)}}$ |
| E14 | 210 | Test for contamination in an exponential sample | Shapiro and Wilk's W-statistic. See worksheet |
| Ea15 | 211 | Test for contamination in an exponential sample | $\dfrac{\sum x_j^2}{(\sum x_j)^2}$ |
| Ga16 | 212 | Testing for discordancy in a gamma sample of unknown shape parameter $r$ by transformation of the variables | See worksheet |
| Ga17 | 213 | Testing for discordancy in a gamma sample of unknown shape parameter $r$ | Kimber (1979); see worksheet |
| Ea18 | 213 | Consecutive test of $k$ upper outliers in an exponential sample | $x_{(n-j+1)} / \sum_1^{n-j+1} x_{(i)}$ $(j = k, k-1, \ldots, 1)$ |
| Ea19 | 214 | Consecutive test of $k$ lower outliers in an exponential sample | $x_{(j+1)} / \sum_1^{j+1} x_{(i)}$ $(j = k, k-1, \ldots, 1)$ |
| Ea20 | 215 | Dixon-type consecutive test of $k$ upper outliers in an exponential sample | $\dfrac{x_{(n-j+1)} - x_{(n-j)}}{x_{(n-j+1)}}$ $(j = k, k-1, \ldots, 1)$ |

# WORKSHEETS

**Ga1(Ea1)** *Discordancy test of a single upper outlier $x_{(n)}$ in a gamma (or exponential) sample*

*Test statistic:*

$$T_{\text{Ga1}} = \frac{\text{outlier}}{\text{sum of observations}} = \frac{x_{(n)}}{\sum x_j}.$$

*Test distribution:*

For $\Gamma(r, \lambda)$,    $f_n(t) = n b_{r,(n-1)r}(t)$   if $t \geq \frac{1}{2}$.

For $E(\lambda)$,    $F_n(t) = \sum_{j=0}^{[1/t]} (-)^j \binom{n}{j} (1 - jt)^{n-1}, \quad 0 \leq t \leq 1,$

where $[1/t]$ denotes the integer part of $1/t$.

*Recurrence relationship:*

$$f_n(t) = n b_{r,(n-1)r}(t) F_{n-1}\{t/(1 - t)\}.$$

*Inequality:*

$$SP(t) \leq n P[\mathbf{F}_{2r, 2(n-1)r} > (n - 1)t/(1 - t)]; \quad \text{equality for } t \geq \frac{1}{2}.$$

**Table 6.1**

| Excess cycle time $X$ | Frequency | $X$ | Frequency |
|:---:|:---:|:---:|:---:|
| 1 | 18 | 11 | 6 |
| 2 | 12 | 12 | 7 |
| 3 | 18 | 13 | 2 |
| 4 | 16 | 14 | 1 |
| 5 | 10 | 15 | 3 |
| 6 | 4 | 21 | 3 |
| 7 | 9 | 32 | 2 |
| 8 | 9 | 35 | 1 |
| 9 | 2 | 92 | 1 |
| 10 | 7 | | 131 |

*Tabulated significance levels:* Table III, pp. 473–4; reproduced (with appropriate change of notation) from Eisenhart, Hastay, and Wallis (1947), Tables 15.1 and 15.2, pages 390–391.

*References:* Fisher (1929), Cochran (1941). Consecutive testing of a number of exponential upper outliers using test **Ea1** is discussed by Kimber (1982) and Chikkagoudar and Kunchur (1987). See test **Ea18**.

*Properties of test*: No special features. All purpose, maximum likelihood ratio test for labelled slippage alternative.

*Example 6.1.    Table* 6.1 *shows a sample of* 131 *excess cycle times in steel manufacture.*

*The sample of size* 130 *obtained by omitting the outlier* $x_{(131)} = 92$ *has mean* $\bar{x} = 6.44$, *variance* $s^2 = 38.14$, *standard deviation* $s = 6.18$, *and third and fourth moments about the mean* $m_3 = 493.4$, $m_4 = 13444$. *Hence* $\bar{x}/s = 1.04$, $m_3/s^3 = 2.09$, $m_4/s^4 = 9.24$, *suggesting that the distribution may reasonably be assumed to be exponential* $((\mu/\sigma = 1, \mu_3/\sigma^3 = 2, \mu_4/\sigma^4 = 9$ *for an exponential distribution). On this assumption we can test the outlier* 92 *for consistency with the other* 130 *values using test* **Ea1**. *The value of* $T_{Ea1}$ *is* $t = 92/929 = 0.0990$, *so*

$$SP(t) \leqslant 131P\left(F_{2,260} > \frac{130 \times 0.0990}{0.9010}\right)$$

$$= 131P(F_{2,260} > 14.28)$$

$$= 131\left(1 + \frac{14.28}{130}\right)^{-130} = 0.00017,$$

*i.e. the evidence for regarding the value* 92 *as being too large to have arisen from the same distribution as the other* 130 *values is very strong.*

**Ea2 Discordancy test of a single upper outlier $x_{(n)}$ in an exponential sample**

*Test statistic:*

$$T_{Ea2} = \frac{\text{excess}}{\text{outlier}} = \frac{x_{(n)} - x_{(n-1)}}{x_{(n)}}.$$

*Test distribution:*

$$F_n(t) = 1 - n(n-1)B\left(\frac{2-t}{1-t}, n-1\right) \quad (0 \leqslant t \leqslant 1).$$

*Tabulated significance levels:* Table IV, page 475; abridged from Likeš (1966), Table 1, page 49, where 10 per cent, 5 per cent, and 1 per cent points are given for $n = 2(1)20$.

*References:* Likeš (1966), Chikkagoundar and Kunchur (1987). The latter paper discusses consecutive testing of a number of upper outliers using test **Ea2**. See test **Ea20**.

*Properties of test:* Vulnerable to masking effect from $x_{(n-1)}$.

*Example 6.1, continued: Applying test* **Ea2** *to the example discussed in Worksheet* **Ga1(Ea1)**, *the value of* $T_{Ea2}$ *with* $n = 131$ *is* $t = (92 - 35)/92 = 0.6196$.

*Hence*

$$SP(t) = 1 - F_n(t) = 131 \times 130B(3.629, 130)$$

$$= 131 \times 130\Gamma(3.629)\Gamma(130)/\Gamma(133.629)$$

$$= 131!(2.629)(1.629)(0.897)/[(132.63)^{133.13}e^{-132.63}\sqrt{(2\pi)}]$$

$$= 0.0013$$

*Compare $SP(t) \leq 0.000\ 17$ for test* **Ea1**.

**E2** *Discordancy test of a single upper outlier* $x_{(n)}$ *in an exponential sample with unknown origin*

*Test statistic:*

$$T_{E2} = \frac{\text{excess}}{\text{range}} = \frac{x_{(n)} - x_{(n-1)}}{x_{(n)} - x_{(1)}}.$$

$T_{E2}$ for a sample of size $n$ has the same test distribution as $T_{Ea2}$ for a sample of size $n - 1$; use Worksheet **Ea2**.

*References:* Likeš (1966), Kabe (1970).

*Properties of test:* Dixon-type test. Vulnerable to masking effect from $x_{(n-1)}$.

**Ga3(Ea3)** *Discordancy test of a single lower outlier* $x_{(1)}$ *in a gamma (or exponential) sample*

*Test statistic:*

$$T_{Ga3} = \frac{\text{outlier}}{\text{sum of observations}} = \frac{x_{(1)}}{\sum x_j}.$$

*Test distribution:*

For $E(\lambda)$,

$$f_n(t) = n(n-1)(1 - nt)^{n-2} \qquad \left(0 \leq t \leq \frac{1}{n}\right),$$

$$= 0 \qquad \left(t \geq \frac{1}{n}\right).$$

For $\Gamma(2, \lambda)$ and $\Gamma(3, \lambda)$, the following expressions are available for small $n$:

$\Gamma(2, \lambda)$: $f_2(t) = 12t(1 - t) \quad (0 \leq t \leq \frac{1}{2})$

$\quad f_3(t) = 60t(1 - 3t)(1 - 3t^2) \quad (0 \leq t \leq \frac{1}{3})$

$\quad f_4(t) = 168t(1 - 4t)^2(1 + 3t - 12t^2 - 4t^3) \quad (0 \leq t \leq \frac{1}{4})$

$$f_5(t) = 360t(1 - 5t)^3(1 + 8t - 18t^2 - 80t^3 + 64t^4) \quad (0 \leq t \leq \tfrac{1}{5})$$

$$f_6(t) = 660t(1 - 6t)^4(1 + 15t - 360t^3 + 864t^5) \quad (0 \leq t \leq \tfrac{1}{6})$$

$\Gamma(3,\lambda)$: $\quad f_2(t) = 60t^2(1 - t)^2 \quad (0 \leq t \leq \tfrac{1}{2})$

$$f_3(t) = 504t^2(1 - 3t)(1 - 2t + 4t^2 - 18t^3 + 21t^4) \quad (0 \leq t \leq \tfrac{1}{3})$$

$$f_4(t) = 1980t^2(1 - 4t)^2(1 - 8t + 28t^2 - 224t^3 + 1540t^4 - 5266t^5$$
$$+ 11032t^6 - 16832t^7 + 13696t^8) \quad (0 \leq t \leq \tfrac{1}{4})$$

*Recurrence relationship:*

$$f_n(t) = nb_{r,(n-1)r}(t)(1 - F_{n-1}\{t/(1 - t)\}).$$

*Inequality:*

$$SP(t) < nP(\mathbf{F}_{2r, 2(n-1)r} < (n - 1)t/(1 - t)).$$

*Tabulated significance levels:* Table V, page 476; specially compiled.

*Reference:* Lewis and Fieller (1979).

*Properties of test:* All purpose, maximum likelihood ratio test.

**E4 Discordancy test of a single lower outlier $x_{(1)}$ in an exponential sample with unknown origin**

*Test statistic:*

$$T_{E4} = \frac{\text{excess}}{\text{range}} = \frac{x_{(2)} - x_{(1)}}{x_{(n)} - x_{(1)}}.$$

*Test distribution:*

$$F_n(t) = (n - 2)B\left(\frac{1 + (n - 2)t}{1 - t}, n - 2\right) \quad (0 \leq t \leq 1).$$

*Tabulated significance levels:* Table VI, page 477; abridged from Likeš (1966), Table 2, page 51, where 10 per cent, 5 per cent, and 1 per cent points are given for $n = 3(1)20$.

*References:* Likeš (1966), Kabe (1970).

*Properties of test:* Dixon-type test. Note a practical difficulty in applying it: the smallest values $x_{(1)}$, $x_{(2)}$ need to be given to a sufficient degree of accuracy, which frequently will not be the case in practice (e.g. excess cycle times data, Table 6.1).

**Ga5(Ea5)** *Discordancy test of $k(\geqslant 2)$ upper outliers in a gamma (or exponential) sample*

*Test statistic:*

$$T_{\text{Ga5}} = \frac{\text{sum of outliers}}{\text{sum of observations}} = \frac{x_{(n-k+1)} + \ldots + x_{(n)}}{\sum x_j}.$$

*Test distribution:*

$$k = 2 \quad f_n(t) = n(n-1)^2 \left\{ \frac{1}{2}(nt-2)(1-t)^{n-3} - (1-t)^{n-2} \sum_{j=1}^{n-3} (1+j)^{-1} \right.$$

$$\left. - 2^{2-n} \sum_{j=1}^{r-1} (-1)^j \binom{n-2}{j} j^{-1}[2 - (j+2)t]^{n-2} \right\}$$

$$((r+1)^{-1} \leqslant t/2 < r^{-1})$$

for $r = 2, 3, \ldots, n-1$, and

$$= 0 \qquad (t < 2/n \text{ or } t \geqslant 1)$$

$$F_n(t) = 1 - n(n-1) \left\{ \frac{1}{2}(nt-1)(1-t)^{n-2} - (1-t)^{n-1} \sum_{j=1}^{n-3} (1+j)^{-1} \right.$$

$$\left. - 2^{2-n} \sum_{j=1}^{r-2} (-1)^j \binom{n-2}{j} j^{-1}(j+2)^{-1}[2 - (j+2)t]^{n-1} \right\}$$

$$((r+1)^{-1} \leqslant t/2 < r^{-1})$$

for $r = 2, 3, \ldots, n-1$.

For the exponential case ($r = 1$) and general $k$,

$$f_n(t) = \frac{n-1}{B(k+1, n-k)} \left[ \frac{1}{k} \left( \frac{1}{n-k} \right)^{n-2} \Sigma_1 + \left( \frac{1}{k} \right)^{n-k-1} \Sigma_2 \right] \left( \frac{k}{n} < t < 1 \right)$$

where

$$\Sigma_1 = \sum_{i=0}^{n-k-2} \sum_{j=0}^{n-k-2} (-1)^{n-k+i+j} \binom{n-k-1}{i} \binom{n-2}{j}$$

$$(nt-k)^{n-2-j}(1-t)^j(n-k-i-1)^j \left( \frac{i+1}{k} \right)^{n-k-2-j}$$

and

$$\Sigma_2 = \sum_{i=k/(t-1)}^{n-k-2} (-1)^{n-k-1-i} \binom{n-k-1}{i} [(k+i+1)t - k]^{n-2} \left(\frac{1}{i+1}\right)^k$$

*Inequality:*

$$SP(t) \le \binom{n}{k} P\left(F_{2kr, 2(n-k)r} > \frac{(n-k)t}{k(1-t)}\right). \qquad (6.2.1)$$

*For* $k = 2$

$$SP(t) < \binom{n}{2}(1-t)^{n-2}[(n-2)t + 1] \qquad (2/n \le t < 1) \qquad (6.2.2)$$

and

$$SP(t) < \binom{n}{2}(1-t)^{n-2}\{(n-2)t + 1 - 2(1-t)[\log(n-2) + \gamma]\} + \delta \qquad (6.2.3)$$

where

$$\delta = 0 \qquad (\tfrac{2}{3} \le t < 1)$$

$$= \binom{n}{3} 2^{3-n}(2-3t)^{n-1} \qquad (2/n \le t < \tfrac{2}{3})$$

and                    $\gamma = 0.577\,2157.$

*Tabulated significance levels: For the exponential case* $(r = 1)$ with two upper outliers $(k = 2)$, Kimber and Stevens (1981) give 5 per cent and 1 per cent points for $n = 5(1)20(2)30(5)50(10)100$. These are reproduced in Table VII on page 477. Likewise for the exponential case, Chikkagoudar and Kunchur (1983) give the following 5 per cent and 1 per cent points for $n = 6(1)12$ and $k(\le n/2) = 3(1)6$ (Table 6.2):

**Table 6.2**

| | 5 per cent points | | | | 1 per cent points | | | |
|---|---|---|---|---|---|---|---|---|
| *k* <br> *n* | 3 | 4 | 5 | 6 | *k* | 3 | 4 | 5 | 6 |
| 6 | 0.9299 | | | | | 0.9610 | | | |
| 7 | 0.8875 | | | | | 0.9286 | | | |
| 8 | 0.8459 | 0.9212 | | | | 0.8937 | 0.9505 | | |
| 9 | 0.8064 | 0.8884 | | | | 0.8589 | 0.9242 | | |
| 10 | 0.7697 | 0.8560 | 0.9152 | | | 0.8251 | 0.8964 | 0.9426 | |
| 11 | 0.7358 | 0.8248 | 0.8883 | | | 0.7930 | 0.8686 | 0.9203 | |
| 12 | 0.7046 | 0.7951 | 0.8617 | 0.9107 | | 0.7628 | 0.8414 | 0.8972 | 0.9368 |

*References:* Fieller (1976), Kimber and Stevens (1981), Chikkagoudar and Kunchur (1983).

*Properties of test:* Maximum likelihood ratio test. Inequality (6.2.3) is better than (6.2.2) (which is based on (6.2.1)) for all $n \geq 4$ when $\frac{2}{3} \leq t < 1$. Inequality (6.2.1) is unlikely to be useful unless $k$ is small. Kimber and Stevens (1981) compare the performance of this test with the test **E14** using sensitivity contours and conclude that it is better if two upper outliers are present. For more than two upper outliers, the omnibus test **E14** is superior. Chikkagoudar and Kunchur (1983) compare the test with test **Ea6** (see under **Ea6** for details).

**Ea6** *Discordancy test of* $k (\geq 2)$ *upper outliers* $x_{(n-k+1)}, \ldots, x_{(n-1)}, x_{(n)}$ *in an exponential sample*

*Test statistic:*

$$T_{Ea6} = \frac{\text{excess}}{\text{outlier}} = \frac{x_{(n)} - x_{(n-k)}}{x_{(n)}}$$

*Test distribution:*

$$F_n(t) = \frac{1}{B(k+1, n-k)} \sum_{j=0}^{k} (-1)^j \binom{k}{j} B\left(\frac{jt}{1-t} + k + 1, n - k\right).$$

*Tabulated significance levels:* Chikkagoudar and Kunchur (1983) give the following 5 per cent and 1 per cent points for $n = 4(1)12$ and $k(\leq \frac{1}{2}n) = 2(1)6$ (Table 6.3).

**Table 6.3**

| | | Critical values* of $T_{Ea6}$ for $\alpha = 0.05$ and $\alpha = 0.01$ | | | |
|---|---|---|---|---|---|
| $k$ $n$ | 2 | 3 | 4 | 5 | 6 |
| 4 | 0.9435 | | | | |
| | 0.9768 | | | | |
| 5 | 0.8994 | | | | |
| | 0.9481 | | | | |
| 6 | 0.8628 | 0.9293 | | | |
| | 0.9211 | 0.9635 | | | |
| 7 | 0.8329 | 0.9001 | | | |
| | 0.8977 | 0.9242 | | | |
| 8 | 0.8081 | 0.8750 | 0.9219 | | |
| | 0.8775 | 0.9232 | 0.9549 | | |
| 9 | 0.7871 | 0.8535 | 0.9005 | | |
| | 0.8600 | 0.9061 | 0.9387 | | |
| 10 | 0.7691 | 0.8348 | 0.8816 | 0.9176 | |
| | 0.8447 | 0.8909 | 0.9242 | 0.9500 | |
| 11 | 0.7534 | 0.8184 | 0.8648 | 0.9008 | |
| | 0.8312 | 0.8773 | 0.9104 | 0.9361 | |
| 12 | 0.7396 | 0.8039 | 0.8499 | 0.8857 | 0.9148 |
| | 0.8192 | 0.8651 | 0.8983 | 0.9241 | 0.9450 |

*Upper value in each cell refers to $\alpha = 0.05$ and lower value to $\alpha = 0.01$.

*References:* Likeš (1966), Chikkagoudar and Kunchur (1983).

*Properties of test:* The power of the test is inherently less than that of the competitor test **Ea5**, but only by 7 per cent at most when $k = 1$ or 2; see Chikkagoudar and Kunchur (1983). Unlike test **Ea5**, however, it can be used as a discordancy test for $x_{(n)}$ if it is desired to insure against masking by $k - 1$ upper values below $x_{(n)}$.

**E6 *Discordancy test of* $k (\geqslant 2)$ *upper outliers* $x_{(n - k + 1)}, \ldots, x_{(n - 1)}, x_{(n)}$ *in an exponential sample with unknown origin***

*Test statistic:*

$$T_{E6} = \frac{\text{excess}}{\text{range}} = \frac{x_{(n)} - x_{(n - k)}}{x_{(n)} - x_{(1)}}$$

$T_{E6}$ for a sample of size $n$ has the same test distribution as $T_{Ea6}$ for a sample of size $n - 1$; use Worksheet **Ea6**.

**Ga7(Ea7) *Discordancy test of a lower and upper outlier-pair* $x_{(1)}, x_{(n)}$ *in a gamma (or exponential) sample***

*Test statistic:*

$$T_{Ga7} = \frac{\text{upper outlier}}{\text{lower outlier}} = \frac{x_{(n)}}{x_{(1)}}.$$

*Test distribution:*

$$F_n(t) = \frac{n}{2^{nr}[\Gamma(r)]^n} \int_0^\infty u^{r - 1} e^{-u/2} I^{n - 1} \, du \quad \text{where} \quad I = \int_u^{tu} v^{r - 1} e^{-v/2} dv.$$

*Recurrence relationship:*

$$f_n(t) = \frac{n(n - 1)rt^{r - 1}}{(1 + t)^{2r}} F_{n - 2}(t).$$

*Inequality:*

$$SP(t) \leqslant n(n - 1) P(F_{2r, 2r} > t).$$

*Tabulated significance levels:* Table VIII, pp. 478–9; reproduced (with appropriate change of notation) from Pearson and Hartley (1966), Table 31, page 202.

*References:* Hartley (1950), David (1952), Rosado (1987), Kimber (1989).

*Properties of test:* As with test **E4**, not suitable where rounding makes value of $x_{(1)}$ imprecise.

Hartley (1950) gives some values for the power of test **Ga7(Ea7)** in comparison with Bartlett's global test for heterogeneity of variances, the alternative hypothesis being that the $n$ population variances are a random sample from a log-normal distribution. The relative power of test **Ga7** is 100 per cent when $n = 2$, and takes values in the range 90–100 per cent for larger sample sizes (up to twelve). Hartley's figures must be treated with caution, in view of inaccuracies in his tables of percentage points of $T_{Ga7}$ (later corrected by David, 1952).

With particular reference to an *exponential* basic model, Rosado (1987) proves that, for a scale-slippage alternative in which $n - 2$ observations come from a common $E(\delta)$ distribution and two from $E(\delta')$ where $\delta' \neq \delta$ and $\delta, \delta'$ are both unknown, a maximum likelihood ratio test for an outlier-pair can only select $x_{(1)}, x_{(2)}$ or $x_{(n-1)}, x_{(n)}$ for discordancy testing. In this situation the pair $x_{(1)}, x_{(n)}$ is not a candidate for testing, contrary, Rosado states, 'to what is done in Barnett and Lewis (1978)' (the reference is to the test **Ea7**). In our view a test such as **Ea7** is designed for a different situation, in which there is a prospect of multiple outliers including members from each end of the sample. An illuminating discussion of the underlying issues is given by Kimber (1989).

*Example 6.2. The times at which every fourth vehicle travelling westward along a main road in Hull passed an observer were recorded as follows (min:sec):*

$$19:57, \quad 20:14, \quad 20:20, \quad 20:38, \quad 20:50,$$
$$21:30, \quad 21:38, \quad 21:46, \quad 22:07.$$

*There are eight time intervals, viz. 17, 6, 18, 12, 40, 8, 8, and 21 seconds; if the traffic flow is assumed to be random (i.e. in accord with a Poisson process), these will be independent values from a gamma distribution with shape parameter $r = 4$. Taking the values 40 and 6 as upper and lower outliers, their ratio is $40/6 = 6.7$. The 5 per cent significance point for $T_{Ga7}$ with $n = 8, r = 4$ is 10.5, so on the basis of this test there is no reason to believe that the traffic flow was not random. (This conclusion is unaffected if we make maximal allowance for rounding error and use values 40.5, 5.5 instead of 40, 6, giving a ratio 7.4.)*

**Ga8 (Ea8)** *Discordancy test of a lower and upper outlier-pair $x_{(1)}, x_{(n)}$ in a gamma (or exponential) sample*

*Test statistic:*

$$T_{Ga8} = \frac{\text{range}}{\text{sum of observations}} = \frac{x_{(n)} - x_{(1)}}{\sum x_j}$$

*Test distribution:*

For $E(\lambda)$, $F_n(t) = (n - 1)!\ t^{n-1}\ (0 \leqslant t \leqslant 1/(n - 1))$

$$(n - 1)!t^{n-1} + \sum_{k=1}^{n-j-1} (-1)^k \binom{n-1}{k-1}[(n - k)t - 1]^{n-1}$$

$$(1/(j + 1) < t \leqslant 1/j, j = 1, 2, \ldots, n - 2).$$

For large $r$ and large $n$,

$$F_n(t) \simeq 1 - P(T_{N\sigma6} > nt/\sqrt{r}),$$

see p. 248 and tabulated significance levels of $T_{N\sigma6}$ in Table XVIIb p. 494.

*Inequality:*

$$SP(t) \leqslant nP[F(2r, 2(n - 1)r) > (n - 1)t/(1 - t)].$$

*Reference:* Kimber (1988).

*Properties of test:* Locally best test for a labelled scale-slippage alternative in which $x_{(2)}, x_{(3)}, \ldots, x_{(n-1)}$ arise from $\Gamma(r, \lambda)$, $x_{(1)}$ arises from $\Gamma(r, \lambda\phi)$ and $x_{(n)}$ arises from $\Gamma(r, \lambda/\phi)$ where $\phi(0 < \phi < 1)$ and $\lambda$ are unknown. The test is similar in structure and behaviour to the next listed test **Ga9(Ea9)**, but of the two **Ga9(Ea9)** performs slightly better.

**Ga9(Ea9)** *Discordancy test of a lower and upper outlier-pair* $x_{(1)}$, $x_{(n)}$ *in a gamma (or exponential) sample*

*Test statistic:*

$$T_{Ga9} = \frac{(\sqrt{x_{(n)}} - \sqrt{x_{(1)}})^2}{\sum x_j}$$

*Test distribution:* For large $r$ and large $n$,

$$F_n(t) \simeq 1 - P(T_{N\sigma6} > 2(nrt)^2),$$

see p. 248 and tabulated significance levels of $T_{N\sigma6}$ in Table XVIIb p. 494.

*Inequality:*

$$SP(t) \leqslant nP[F(2r, 2(n - 1)r) > (n - 1)t/(1 - t)].$$

*Tabulated significance levels:* Table IX, page 480, obtained from Kimber (1988), Table I, page 1060; these 5 per cent and 1 per cent points have been estimated on the basis of simulations of sizes 10 000.

*Further tables:* Kimber (1988) gives values of power for $n = 10$ and $r = 1, r = 4$, for situations with differing combinations of at most two upper and lower outliers.

*Reference:* Kimber (1988).

*Properties of test:* Maximum likelihood ratio test for a scale-slippage alternative (unlabelled) in which $n - 2$ observations arise from a common $\Gamma(r, \lambda)$ distribution, one further observation from $\Gamma(r, \lambda\phi)$ and one from $\Gamma(r, \lambda/\phi)$ where $\phi(0 < \phi < 1)$ and one from $\lambda$ are unknown. Limited studies of power (see above) suggest that the test generally performs either as well as or better than **Ga7(Ea7)** on samples actually containing a lower and an upper outlier. It is also preferable to **Ga8(Ea8)**, see above.

**E10  *Discordancy test of a lower and upper outlier-pair* $x_{(1)}, x_{(n)}$ *in an exponential sample with unknown origin***

*Test statistic:*

$$T_{E10} = \frac{\text{reduced range}}{\text{range}} = \frac{x_{(n-1)} - x_{(2)}}{x_{(n)} - x_{(1)}}.$$

*Test distribution:*

$$F_n(t) = 1 - (n - 1)!(1 - t)^2 \sum_{j=1}^{n-3} \frac{(-)^{j+1} j}{(j + 1)!(n - 3 - j)!(1 + jt)\{n - 1 - (n - j - 2)t\}}.$$

*Reference:* Kabe (1970).

**Ga11(Ea11)  *Discordancy test of* $k(\geqslant 2)$ *lower outliers in a gamma (or exponential) sample***

*Test statistic:*

$$T_{Ga11} = \frac{\text{sum of outliers}}{\text{sum of observations}} = \frac{x_{(1)} + \ldots + x_{(k)}}{\sum x_j}$$

*Test distribution:*
For $E(\lambda)$ and $k = 2$,

$$f_n(t) = \frac{n(n-1)^2}{n-2} [(1 - \tfrac{1}{2}nt)^{n-2} - (1 - (n - 1)t)^{n-2}] \quad \text{for} \quad 0 < t < \frac{1}{n-1},$$

$$= \frac{n(n-1)^2}{n-2} (1 - \tfrac{1}{2}nt)^{n-2} \qquad \qquad \text{for} \quad \frac{1}{n-1} < t < \frac{2}{n},$$

$$= 0 \quad \text{otherwise.}$$

*Inequality:*

$$SP(t) < \binom{n}{k} P\left( \mathbf{F}(2kr, 2(n-k)r) < \frac{(n-k)t}{k(1-t)} \right).$$

*Reference:* Fieller (1976), Lewis and Fieller (1979).

*Properties of test:* Maximum likelihood ratio test.

*Example 6.3. Epstein (1960b, p. 171) considers a life test in which the failure times of ten items are observed, totalling $\sum x_j = 600$ units. The failure times of the first two items to fail are the shortest of the ten, and total 24 units, so we can write $x_{(1)} + x_{(2)} = 24$. It is assumed that failure times under the given conditions are exponentially distributed. Epstein tests whether the first two items to fail can be regarded as having failed abnormally early; making use of a straightforward F-test, he concludes in favour of this hypothesis. Suppose however that the ten items were placed on test at different starting times and that the two shortest failure times occurred, not necessarily first, but randomly in chronological sequence, so that there is no a priori reason to consider these two items as different from the rest. How strong is the evidence for regarding them as inconsistent with the rest in view of their failure times? The value of $T_{Ea11}$ is*

$$t = \frac{24}{600} = 0.04, \text{ which is in the range } 0 < t < \frac{1}{9}.$$

*Hence*

$$SP(t) = \int_0^{0.04} \frac{810}{8} \{(1 - 5u)^8 - (1 - 9u)^8\} \, du.$$

$$= \frac{90}{8} \left\{ \frac{(0.64)^9}{9} - \frac{(0.80)^9}{5} - \frac{1}{9} + \frac{1}{5} \right\} = 0.721$$

*This is the significance probability attaching to the observed ratio $t = 0.04$, i.e. there is no real evidence for regarding it as abnormally low—a contrary conclusion to Epstein's, on our modified premise.*

**E12 Discordancy test of a lower outlier-pair $x_{(1)}, x_{(2)}$ in an exponential sample with unknown origin**

*Test statistic:*

$$T_{E12} = \frac{\text{excess}}{\text{range}} = \frac{x_{(3)} - x_{(1)}}{x_{(n)} - x_{(1)}}.$$

*Test distribution:*

$$F_n(t) = 1 - (n - 3)\left[(n - 1)B\left(\frac{1 + (n - 3)t}{1 - t}, n - 3\right) - (n - 2)B\left(\frac{1 + (n - 2)t}{1 - t}, n - 3\right)\right].$$

*References:* Likeš (1966), Kabe (1970).

*Properties of test:* As for test **E4**, page 200.

**Ea13** **General Dixon-type discordancy test for an exponential sample, using knowledge of the origin** $a$

*Test statistic:*

$$T_{\text{Ea}13} = \frac{x_{(s)} - x_{(r)}}{x_{(q)} - a}, \quad 0 \leqslant r < s < q \leqslant n,$$

where $x_{(0)}$ is to be interpreted as $a$, thus including the case

$$T_{\text{Ea}13} = \frac{x_{(s)} - a}{x_{(q)} - a}.$$

*Test distribution:*

$$1 - F_n(t) = \frac{n!}{(n - q)!}(1 - t)$$

$$\times \left\{ \sum_{i=1}^{q-s}\sum_{k=1}^{s-r} \frac{(-)^{i+k}(q - r - i)![(n - s + k)t + (n - q + i)(1 - t)]^{-1}}{(i - 1)!(k - 1)!(q - s - i)!(s - r - k)!(q - i)!(n - s + k)} \right.$$

$$\left. + \sum_{j=1}^{r}\sum_{k=1}^{s-r} \frac{(-)^{q-s+j+k}(s - r + j - 1)![(n - s + k)t + (n - r + j)(1 - t)]^{-1}}{(j - 1)!(k - 1)!(r - j)!(s - r - k)!(q - r + j - 1)!(n - s + k)} \right\}$$

where the first of the double sums is omitted if $q = s$, and the second is omitted if $r = 0$.

*References:* Likeš (1966), Kabe (1970).

*Properties of test:* Applicable to any combination of lower and/or upper outliers. For example, $[x_{(n-2)} - x_{(4)}]/[x_{(n)} - a]$ would be a suitable statistic for a block test of discordancy of three lower and two upper outliers. **Ea2**, **Ea6**, **Ea7** are important particular cases.

**E13** *General Dixon-type discordancy test for an exponential sample irrespective of origin*

*Test statistic:*

$$T_{E13} = \frac{x_{(s)} - x_{(r)}}{x_{(q)} - x_{(p)}}, \quad 1 \leqslant p \leqslant r < s \leqslant q \leqslant n, \quad q - p > s - r.$$

$T_{E13}$ for sample size $n$ and observations $x_{(p)}$, $x_{(q)}$, $x_{(r)}$, $x_{(s)}$ has the same test distribution as $T_{Ea13}$ for sample size $n - p$, origin $a = 0$, and observations $x_{(q-p)}$, $x_{(r-p)}$, $x_{(s-p)}$ instead of $x_{(q)}$, $x_{(r)}$, $x_{(s)}$; use Worksheet **Ea13**.

*References:* Dixon (1950, 1951), Likeš (1966), Kabe (1970). Note that a factor $(q - p - i)!$ needs inserting in the denominator of the first double sum in Kabe's equation (13), page 17. Patil, Kovner and King (1977) present detailed tables of percentage points for generalized forms of Dixon-type statistic including that of the outlier test of Basu (1965): namely,

$$[x_{(n)} - x_{(1)}] / \sum_{j=2}^{n} [x_{(j)} - x_{(1)}].$$

*Properties of test:* Applicable to any combination of lower and/or upper outliers. For example, $[x_{(n-2)} - x_{(4)}]/[x_{(n)} - x_{(1)}]$ would be a suitable statistic for a block test of discordancy for three lower and two upper outliers. **E2**, **E4**, **E6** are important particular cases.

**E14** *Two-sided test for contamination in an exponential sample irrespective of origin*

*Test statistic:*

$$T_{E14} = \text{Shapiro and Wilk's 'W-Exponential' statistic}$$

$$= \frac{n(\bar{x} - x_{(1)})^2}{(n - 1) \sum_{j=1}^{n} (x_j - \bar{x})^2}.$$

*Tabulated significance levels:* Table X, page 481; abridged from Shapiro and Wilk (1972). Table 1, pages 361–362, where lower and upper 0.5, 1, 2.5, 5, and 10 per cent points and the 50 per cent points are given for $n = 3(1)100$.

*Further tables:* Shapiro and Wilk (1972) give values for the power of the test against 15 different inherent alternatives (see Chapter 2, page 46).

*Reference:* Shapiro and Wilk (1972). Stephens (1978) describes a modification for the case of known origin.

*Properties of test:* A useful omnibus test against inherent alternatives. In the outlier context, significantly *high* values of $T_{E14}$ indicate the presence of one or more high discordant values $x_{(n)}$, $x_{(n-1)}$, . . . and/or one *low* discordant value $x_{(1)}$; significantly *low* values of $T_{E14}$ indicate the presence of a number of low discordant values $x_{(1)}$, $x_{(2)}$, . . . . See also the comments on the properties of test **Ga5(Ea5)** on page 201.

**Ea15 Test for contamination in an exponential sample**

*Test statistic:* $T_{Ea15}$ = Greenwood's statistic

$$= \frac{\text{sum of squares of observations}}{\text{square of sum of observations}} = \frac{\sum x_j^2}{(\sum x_j)^2}$$

*Test distribution:* See Currie (1981).

*Tabulated significance levels:* We give below a table of 5 per cent and 1 per cent critical values (Table 6.4). This is abridged from Burrows (1979), Table 1, page 257, and Currie (1981), Table 1, page 361, where lower and upper 1, 5, 10(10)40 per cent points and the 50 per cent points are given for $n = 3(1)11$ (Burrows) and $n = 12(1)21$ (Currie), and from Stephens (1981), Table 1, page 365, where lower and upper 1, 2.5, 5 and 10 per cent points are given for $n = 13(2)21(5)31(10)61(20)101, 201, 501$. Note that 'sample size' in the above three authors' papers, which they denote by $n$, is $n + 1$ in our notation.

**Table 6.4**  Critical values for 5 per cent and 1 per cent tests for the presence of an undefined number of discordant values in an exponential sample, using Greenwood's statistic $T_{Ea15}$

| $n$ | 5% | 1% |
|---|---|---|
| 3 | 0.770 | 0.890 |
| 4 | 0.617 | 0.756 |
| 5 | 0.509 | 0.640 |
| 6 | 0.432 | 0.547 |
| 7 | 0.374 | 0.475 |
| 8 | 0.329 | 0.414 |
| 9 | 0.294 | 0.371 |
| 10 | 0.265 | 0.333 |
| 12 | 0.221 | 0.274 |
| 14 | 0.188 | 0.232 |
| 16 | 0.164 | 0.201 |
| 18 | 0.145 | 0.177 |
| 20 | 0.130 | 0.157 |
| 26 | 0.0982 | 0.117 |
| 31 | 0.0814 | 0.0958 |

**Table 6.4** (*Contd.*)

| n | 5% | 1% |
|---|-----|-----|
| 41 | 0.0604 | 0.0697 |
| 51 | 0.0478 | 0.0545 |
| 61 | 0.0394 | 0.0445 |
| 81 | 0.0291 | 0.0323 |
| 101 | 0.0230 | 0.0253 |
| 201 | 0.0111 | 0.0119 |
| 501 | 0.0043 | 0.0045 |

$n$ = number of observations.

*References:* Greenwood (1946), Moran (1953), Bartholomew (1957), Burrows (1979), Currie (1981), Stephens (1981), Sugiura and Sasamoto (1989).

*Properties of test:* A useful omnibus test against inherent alternatives. In the outlier context, significantly *high* values of $T_{\text{Ea}15}$ indicate the presence of discordant values which may be low or high or both. Sugiura and Sasamoto (1989) show that it is the locally best invariant discordancy test for outliers in exponential samples and more generally in gamma samples of any shape parameter. In the exponential case they compare its performance with 8 other tests, including **Ea1, Ea3** and **Ea7**; the tests using statistics **N1, N2** and **N15** (see pages 221, 223 and 231) as applied to exponential samples; and a test based on Bartlett's 'homogeneity of variance' statistic. Tables of simulated power comparisons are presented for $k = 1, 2$ and 3 contaminants of various magnitudes: the cases $k = 2$ and $k = 3$ include all the possible lower and upper combinations. (In the cases $k = 2$, $k = 3$ the single-outlier tests such as **Ea1** are presumably applied as inward consecutive procedures, though this is not stated in the paper.) Broadly speaking, **Ea15** and **Ea1** have similar power, which is higher than that of the other tests; and **Ea15** is less vulnerable than **Ea1** to masking. An exception is the relatively unimportant situation when there are one or more extreme lower outliers, i.e. values very near zero, in which case the Bartlett test has the highest power.

**Ga16 *Procedure for testing one or more outliers for discordancy in a gamma sample of unknown shape parameter* r**

Transform the values $x_1, \ldots, x_n$ in the gamma sample to $y_1 = \sqrt{x_1}$, $\ldots, y_n = \sqrt{x_n}$ and apply to the values $y_1, \ldots, y_n$ (all taken positively) a discordancy test for a sample from a normal distribution with unknown mean and variance (tests **N1–N19**, Section 6.3.1). (For if $X$ is distributed as $\Gamma(r, \lambda)$, then $\sqrt{X}$ is distributed approximately as $N(\sqrt{[\lambda(r - \frac{1}{4})]}, \frac{1}{4}\lambda)$; and when $r$ and $\lambda$ are both unknown, the mean and variance of this approximating normal distribution are both unknown.)

For example, if three upper outliers $x_{(n)}$, $x_{(n-1)}$, $x_{(n-2)}$ in the gamma sample are to be tested for discordancy, $\sqrt{x_{(n)}}$, $\sqrt{x_{(n-1)}}$, and $\sqrt{x_{(n-2)}}$ will be the three greatest values in the $y$-sample, and test **N3** could appropriately be used, with $[\sqrt{x_{(n)}} + \sqrt{x_{(n-1)}} + \sqrt{x_{(n-2)}} - 3\bar{y}]/s_y$ as test statistic, where $\bar{y}$ and $s_y$ are the mean and standard deviation of the $y$-values.

A specific proposal of this type is examined by Kimber (1979) who suggests that we should use the more accurate *cube root* transformation, under which $\sqrt[3]{x}$ is distributed approximately as $N\{(r\lambda)^{1/3}[1 - 1/(9r)], (\lambda^2/r)^{1/3}/9\}$, and employ the normal tests **N2** or **N15** of Section 6.3.1. Power comparisons revealed little difference in performance of the two tests. Both behaved well except for small sample sizes ($n \leq 10$) where they were not satisfactory when *only* lower outliers were encountered.

**Ga17** *Discordancy test of one or more outliers in a gamma sample of unknown shape parameter r*

*Test statistic:*

$$T_{Ga17} = D_{(n)}/\sum D_j$$

where $D_j = -\log(x_j/\bar{x}) - (n - 1)\log[(n - x_j/\bar{x})/(n - 1)]$.

*Tabulated significance levels:* Kimber (1979) presents 5 per cent and 1 per cent critical values, determined by simulation, for a conservative version of the test with sample sizes $n = 5(1)20$.

*Reference:* Kimber (1979).

*Properties of the test:* Maximum likelihood ratio test for slippage in either direction of the scale parameter for a single observation. Performance characteristics, and comparisons with tests of type **Ga16**, are discussed by Kimber (1979). The conclusions about the test are not too felicitous particularly for small sample sizes ($n \leq 15$).

**Ea18** *Consecutive test of k upper outliers in an exponential sample*

*Test procedure:*

$$\text{Put } S_j = x_{(n-j+1)}/\sum_{i=1}^{n-j+1} x_{(i)} \qquad \text{for } j = 1, 2, \ldots, n - 1.$$

Choose a greatest possible number, $k$, of contaminants. For a test at level $\alpha$, determine critical values $s_j$ ($j = 1, 2, \ldots, k$) to satisfy

$$P\left[\bigcap_1^k (S_j < s_j)\right] = 1 - \alpha \quad \text{and} \quad P(S_1 > s_1) = \ldots = P(S_k > s_k),$$

under the working (no contamination) model.

If $S_k > s_k$, declare the $k$ largest observations discordant; if $S_j < s_j$ ($j = k, k - 1, \ldots, l + 1$) but $S_l > s_l$, declare the $l$ largest observations discordant ($l = k - 1, k - 2, \ldots, 1$); if $S_j < s_j$ ($j = k, k - 1, \ldots, 1$), declare no observations discordant.

*Test distributions:* For $j = 1, 2, \ldots, n - 1$,

$$P(S_j > s) = \sum_{i=1}^{[1/s]} (-1)^{i+1} \binom{n}{i+j-1} \binom{i+j-2}{j-1} \{(1 - is)/(1 + js - s)\}^{n-j}$$
$$(1/(n - j + 1) < s < 1)$$

Formulae from which

$$P\left[ \bigcap_{1}^{k} (S_j < s_j) \right]$$

can be calculated are given in Kimber (1982). However, an approximation which avoids the necessity for this calculation is to use $\alpha/k$ for the common value of $P(S_j > s_j)$ ($j = 1, \ldots, k$).

*Tabulated significance levels:* Table XI, pages 482–4, gives for $k = 2, 3$, and 4 the values of $s_k, s_{k-1}, \ldots, s_1$ for $\alpha = 0.05$ and 0.01 and $n = 5k(1)20(2)30(5)$ 50(10)100(20)140, reproduced from Kimber (1982), Tables 1–3 pp. 266–267.

*References:* Kimber (1982), Chikkagoudar and Kunchur (1987).

*Properties of test:* Chikkagoudar and Kunchur (1987) compare **Ea18** with the corresponding consecutive test, **Ea20** below, based on the Dixon-type statistics $D_j = [x_{(n-j+1)} - x_{(n-j)}]/x_{(n-j+1)}$ ($j = 1, \ldots, k$). Their calculations, which relate to the particular case $k = 2$, $\alpha = 0.05$, indicate that **Ea18** is superior to **Ea20** as regards power but that **Ea20** performs better than **Ea18** as regards vulnerability to swamping and masking; accordingly neither test is uniformly preferable to the other. In view of the greater availability of significance levels, **Ea18** has the practical advantage. Kimber (1982) examines sensitivity contours for **Ea18** and describes applications to some practical problems.

### Ea19 *Consecutive test of k lower outliers in an exponential sample*

*Test procedure:* Define $S_j$ as in Worksheet **Ea18**, and put

$$T_j = S_{n-j} = x_{(j+1)} / \sum_{i=1}^{j+1} x_{(i)} \qquad \text{for} \quad j = 1, 2, \ldots, n - 1.$$

Choose a greatest possible number, $k$, of contaminants. For a test at level $\alpha$, determine critical values $t_j$ ($j = 1, 2, \ldots, k$) to satisfy

$$P\left[\bigcap_1^k (T_j < t_j)\right] = 1 - \alpha \text{ and } P(T_1 > t_1) = \ldots = P(T_k > t_k),$$

under the working (no contamination) model.

If $T_k > t_k$, declare the $k$ smallest observations discordant; if $T_j < t_j$ ($j = k, k - 1, \ldots, l + 1$) but $T_l > t_l$, declare the $l$ smallest observations discordant ($l = k - 1, k - 2, \ldots, 1$); if $T_j < t_j$ ($j = k, k - 1, \ldots, 1$), declare no observations discordant.

*Test distributions:* See Worksheet **Ea18**, nothing that $T_j = S_{n-j}$.

*Tabulated significance levels:* Table XII, pages 482–4, gives for $k = 2, 3$, and 4 the critical values of $t_k, t_{k-1}, \ldots, t_1$, for $\alpha = 0.05$ and 0.01 and $n = 5k(5)20, 50, 100, 200$, reproduced from Kimber (1982), Table 4 p. 270.

*Reference:* Kimber (1982).

*Properties of test:* Preferable to a test based on test statistic $T_{Ea11}$, because uninfluenced by the behaviour of the largest observations in the sample. As with **E4** (*q.v.*), **Ga7(Ea7)**, etc., the smallest values need to be given to a sufficient degree of accuracy.

**Ea20 *Dixon-type consecutive test of $k$ upper outliers in an exponential sample***

*Test procedure:*

Put $D_j = [x_{(n-j+1)} - x_{(n-j)}]/x_{(n-j+1)}$   for   $j = 1, 2, \ldots, n - 1$.

Choose a greatest possible number, $k$, of contaminants. For a test at level $\alpha$, determine critical values $d_j$ ($j = 1, 2, \ldots, k$) to satisfy

$$P\left[\bigcap_1^k (D_j < d_j)\right] = 1 - \alpha \text{ and } P(D_1 > d_1) = \ldots = P(D_k > d_k),$$

under the working (no contamination) model.

If $D_k > d_k$, declare the $k$ largest observations discordant; if $D_j < d_j$ ($j = k, k - 1, \ldots, l + 1$) but $D_l > d_l$, declare the $l$ largest observations discordant ($l = k - 1, k - 2, \ldots, 1$); if $D_j < d_j$ ($j = k, k - 1, \ldots, 1$), declare no observations discordant.

*Test distributions:* See Chikkagoudar and Kunchur (1987).

*Tabulated significance levels:* For the particular case $k = 2$, Chikkagoudar and Kunchur (1987) give (Table 6.5) the following critical values of

$d_2$, $d_1$ for $\alpha = 0.05$ and $n = 4(1)20$ (actually, with $k = 2$ the case $n = 4$ would not seem to be realistic):

Table 6.5

| n | $d_2$ | $d_1$ | n | $d_2$ | $d_1$ |
|----|-------|-------|----|-------|-------|
| 4 | 0.902 | 0.873 | 13 | 0.574 | 0.676 |
| 5 | 0.824 | 0.830 | 14 | 0.561 | 0.667 |
| 6 | 0.765 | 0.797 | 15 | 0.548 | 0.657 |
| 7 | 0.718 | 0.770 | 16 | 0.538 | 0.650 |
| 8 | 0.683 | 0.748 | 17 | 0.527 | 0.642 |
| 9 | 0.653 | 0.730 | 18 | 0.519 | 0.636 |
| 10 | 0.629 | 0.714 | 19 | 0.510 | 0.629 |
| 11 | 0.608 | 0.700 | 20 | 0.503 | 0.624 |
| 12 | 0.590 | 0.688 | | | |

*References:* Chikkagoudar and Kunchur (1983, 1987).

*Properties of test:* See Worksheet **Ea18**.

## 6.3 DISCORDANCY TESTS FOR NORMAL SAMPLES

Historically, the motivation for a statistical treatment of outliers came first from the problems of combining astronomical observations, and *repeated measurements* or *determinations* must always be one of the main contexts in which discordancy problems arise. In very many cases errors of measurement may plausibly be assumed to follow a normal distribution, whether through the operation of the central limit theorem on contributory error components, or purely as an empirical fact. It is not surprising, therefore, that the vast body of published methodology on outliers from the eighteenth century to the present day rests on the working hypothesis of a normal distribution. Indeed, it is only in the last twenty-five years or so that outliers in exponential and other non-normal models have been specifically considered.

When the normal distribution is being used in this way as a kind of all-purpose probability model, the mean $\mu$ and variance $\sigma^2$ will both in general be unknown, and any discordancy test for outliers will reflect this. However, discordancy tests also arise in situations when information is available concerning $\mu$ or $\sigma^2$ or both.

The value of $\mu$ may be known. The variance $\sigma^2$ may be known exactly, or again some information on its value may be available in the form of an estimate independent of the particular sample of observations under study for discordancy. This estimate may perhaps be an item of background information 'from the files'. A quite different context giving rise to such as estimate is in analysis of variance, when we may find a surprising value among a set of treatment means, and have available the residual mean square to assist in judging its discordancy.

Outlier situations with $\mu$ unknown but $\sigma^2$ known may arise in quality control, where past experience provides reasonably accurate knowledge of the process variance.

Outlier situations with $\sigma^2$ unknown but $\mu$ known may arise, for example, in paired comparison situations where the sample values we are considering for discordancy are differences between corresponding responses and so have mean $\mu = 0$ on the working hypothesis.

The case with $\mu$, $\sigma^2$ both known is of limited methodological interest, any discordancy test being based simply on the appropriate extreme-value distribution. However, we have included this case as it has practical interest. It could arise, for example, in the validation of tables of random normal deviates; or, less esoterically, in reaching decisions on classification for taxonomic, anthropological, or even legal purposes. Consider, for example, the analysis of paired comparisons of Poisson counts to which a normalising transformation has been applied. See also Example 4.4, pages 128–30.

Finally, the technique of transformation of the observations leads to a further range of applications of normal-sample discordancy tests, as we have already seen in the discussion of test **Ga 16** on page 212–3 where tests designed for normal samples with $\mu$, $\sigma^2$ both unknown were applied to outliers in samples from gamma distributions with unknown shape parameter. Tests designed for normal samples *with known variance* $\sigma^2$ are particularly useful, since they can be applied to outliers in Poisson samples and binomial samples. For example, if the $n$ values $x_1, \ldots, x_n$ are on the working hypothesis a sample from a Poisson distribution $\mathbf{P}(\mu)$, then the $n$ transformed values $\sqrt{(x_1 + \frac{3}{8})}, \ldots, \sqrt{(x_n + \frac{3}{8})}$ are (provided the mean $\mu$ is not too small) a sample from a distribution which is approximately $\mathbf{N}(\sqrt{\mu} + \frac{3}{8}, \frac{1}{4})$. For details of discordancy testing in the Poisson and binomial cases, see Section 6.4 below.

### 6.3.1 Normal samples: contents list and details of tests

The tests are classified as follows, according to the information available regarding the mean and variance of the normal distribution $\mathbf{N}(\mu, \sigma^2)$ assumed in the working hypothesis.

| Code | Information |
| --- | --- |
| **N** | $\mu$ and $\sigma^2$ both unknown |
| **N$\nu$** | $\mu$ unknown. Information available on $\sigma^2$ independent of the sample in the form of an estimate $v = s_\nu^2$ such that $vs_\nu^2/\sigma^2$ is distributed as $\chi_\nu^2$. |
| **N$\mu$** | $\mu$ known, $\sigma^2$ unknown |
| **N$\sigma$** | $\sigma^2$ known, $\mu$ unknown |
| **N$\mu\sigma$** | $\mu$ and $\sigma^2$ both known |

In the case **N**, $\mu$ is estimated on the working hypothesis by $\bar{x} = \sum x_j/n$, and $\sigma^2$ by $s^2 = \sum(x_j - \bar{x})^2/(n - 1)$. In the case **Nv**, $\sigma^2$ can be estimated by the independent estimators $s^2$, $s_v^2$, or by the pooled estimator

$$\tilde{s}^2 = [\sum(x_j - \bar{x})^2 + v s_v^2]/(n - 1 + v).$$

In the case **Nμ**, $\sigma^2$ is estimated by $s^2(\mu) = \sum(x_j - \mu)^2/n$.

The sum of squares of the deviations from $\bar{x}$ of the $n$ observations $x_1, x_2, \ldots, x_n$, $\sum_{j=1}^{n}(x_j - \bar{x})^2$, is denoted by $S^2$. Thus $s^2 = S^2/(n - 1)$. If $x_{(n)}$ is omitted, the sum of squares of the deviations of the remaining $n - 1$ observations from their own mean is denoted by $S_n^2$. $S_{n-1,n}^2$ is the corresponding sum of squares when $x_{(n-1)}$, $x_{(n)}$ are both omitted, and so on. The quantity $\sum_{j=1}^{n}(x_j - \mu)^2$, which is of relevance to case **Nμ**, is denoted by $S^2(\mu)$ with $s^2(\mu) = S^2(\mu)/n$, and with similar definitions for $S_n^2(\mu)$, $S_{n-1,n}^2(\mu)$ etc. In the case **Nv**, the sum of squares $\sum(x_j - \bar{x})^2 + v s_v^2$ is denoted by $\tilde{S}^2$.

Tests not involving the sample mean or population mean are coded both as **N** and **Nμ** (or, where $\sigma^2$ is known, both as **Nσ** and **Nμσ**). Examples of such tests are **N7(Nμ7)**, with test statistic $(x_{(n)} - x_{(n-1)})/(x_{(n)} - x_{(1)})$ and **Nσ6(Nμσ6)**, with test statistic $(x_{(n)} - x_{(1)})/\sigma$.

In a situation where $\mu$ is known but no appropriate test is listed under code **Nμ** because no significance levels are available, or similarly where $\sigma^2$ is known but no appropriate test is listed under code **Nσ**, there may be an appropriate N-test which can be used though with some loss of efficiency. For example, to test for a lower and upper outlier-pair $x_{(1)}$, $x_{(n)}$ with $\mu$ known, $\sigma^2$ unknown, **N6** could be used: significance levels for a test based on $(x_{(n)} - x_{(1)})/s(\mu)$ are not available. Where both $\mu$ and $\sigma^2$ are known an **Nσ**-test (or possibly an **Nμ**-test) can be used if necessary.

In view of the symmetry of the normal distribution, any test for an upper outlier, upper outlier-pair etc., can be used for a lower outlier, lower outlier-pair etc., with the obvious modifications. For example, two lower outliers $x_{(1)}$, $x_{(2)}$ can be tested for discordancy by test **N3** using the statistic $(2\bar{x} - x_{(2)} - x_{(1)})/s$. To save space, such tests are only given here in terms of the upper outlier situation.

| Label | Worksheet page no. | Description of test | Statistic |
|-------|--------------------|--------------------|-----------|
| **N1** | 221 | Test of upper outlier $x_{(n)}$ | $\dfrac{x_{(n)} - \bar{x}}{s}$ or equivalently $S_n^2/S^2$ |
| **N2** | 223 | Test of extreme outlier (two-sided form of **N1**) | $\max\left(\dfrac{x_{(n)} - \bar{x}}{s}, \dfrac{\bar{x} - x_{(1)}}{s}\right)$ or equivalently $\min(S_n^2/S^2, S_1^2/S^2)$ |
| **N3** | 224 | Test of $k(\geqslant 2)$ upper outliers $x_{(n-k+1)}, \ldots, x_{(n)}$ | $\dfrac{x_{(n-k+1)} + \ldots + x_{(n)} - k\bar{x}}{s}$ |

(*Contd.*)

| Label | Worksheet page no. | Description of test | Statistic |
|---|---|---|---|
| N4 | 225 | Test of $k(\geqslant 2)$ upper outliers $x_{(n-k+1)}, \ldots, x_{(n)}$ | $S^2_{n-k+1, \ldots, n-1, n}/S^2$ |
| N5 | 225 | Test of lower and upper outlier-pair $x_{(1)}, x_{(n)}$ | $S^2_{1,n}/S^2$ |
| N6 | 226 | Test of lower and upper outlier-pair $x_{(1)}, x_{(n)}$ | $\dfrac{x_{(n)} - x_{(1)}}{s}$ |
| N7(N$\mu$7) | 226 | Dixon-type test of upper outlier $x_{(n)}$ | $\dfrac{x_{(n)} - x_{(n-1)}}{x_{(n)} - x_{(1)}}$ |
| N8(N$\mu$8) | 227 | Dixon-type test of extreme outlier (two-sided form of N7(N$\mu$7) ) | $\max\left[\dfrac{x_{(n)} - x_{(n-1)}}{x_{(n)} - x_{(1)}}, \dfrac{x_{(2)} - x_{(1)}}{x_{(n)} - x_{(1)}}\right]$ |
| N9(N$\mu$9) | 228 | Dixon-type test of upper outlier $x_{(n)}$ | $\dfrac{x_{(n)} - x_{(n-1)}}{x_{(n)} - x_{(2)}}$ |
| N10(N$\mu$10) | 228 | Dixon-type test of upper outlier $x_{(n)}$ | $\dfrac{x_{(n)} - x_{(n-1)}}{x_{(n)} - x_{(3)}}$ |
| N11(N$\mu$11) | 229 | Dixon-type test of two upper outlier $x_{(n-1)}, x_{(n)}$ | $\dfrac{x_{(n)} - x_{(n-2)}}{x_{(n)} - x_{(1)}}$ |
| N12(N$\mu$12) | 229 | Dixon-type test of two upper outliers $x_{(n-1)}, x_{(n)}$ | $\dfrac{x_{(n)} - x_{(n-2)}}{x_{(n)} - x_{(2)}}$ |
| N13(N$\mu$13) | 229 | Dixon-type test of two upper outliers $x_{(n-1)}, x_{(n)}$ | $\dfrac{x_{(n)} - x_{(n-2)}}{x_{(n)} - x_{(3)}}$ |
| N14 | 230 | Test of one or more upper outliers | Sample skewness $g_1 = \sqrt{b_1}$ |
| N15 | 231 | Two-sided test of one or more outliers, irrespective of their directions | Sample kurtosis $b_2$ |
| N16 | 232 | Block test of $k$ outliers irrespective of directions (i.e. of how many upper and how many lower) | Tietjen and Moore's $E_k$-statistic. See worksheet |
| N17 | 232 | Test for contamination in a normal sample | Shapiro and Wilk's W-statistic. See worksheet |
| N18 | 234 | Test of extreme outlier (two-sided): as N2 but with robust estimate $s_b$ | $\max\left(\dfrac{x_{(n)} - \bar{x}}{s_b}, \dfrac{\bar{x} - x_{(1)}}{s_b}\right)$ |
| N19 | 235 | 'Recursive' (consecutive) tests of several outliers | Based on Rosner (1975), etc. See worksheet |
| Nv1 | 237 | Test of upper outlier $x_{(n)}$ | $\dfrac{x_{(n)} - \bar{x}}{s_v}$ |
| Nv2 | 237 | Test of upper outlier $x_{(n)}$ | $\dfrac{x_{(n)} - \bar{x}}{\tilde{s}}$ |
| Nv3 | 238 | Test of extreme outlier (two-sided form of Nv1) | $\max\left(\dfrac{x_{(n)} - \bar{x}}{s_v}, \dfrac{\bar{x} - x_{(1)}}{s_v}\right)$ |

*(Contd.)*

| Label | Worksheet page no. | Description of test | Statistic |
|---|---|---|---|
| **Nv4** | 239 | Test of extreme outlier (two-sided form of **Nv2**) | $\max\left(\dfrac{x_{(n)} - \bar{x}}{\tilde{s}}, \dfrac{\bar{x} - x_{(1)}}{\tilde{s}}\right)$ |
| **Nv5** | 240 | Test of $k(\geqslant 2)$ upper outliers $x_{(n-k+1)}, \ldots, x_{(n)}$ | $\dfrac{x_{(n-k+1)} + \ldots + x_{(n)} - k\bar{x}}{\tilde{s}}$ |
| **Nv6** | 240 | Test of lower and upper outlier-pair $x_{(1)}, x_{(n)}$ | $\dfrac{x_{(n)} - x_{(1)}}{s_v}$ |
| **Nv7** | 241 | Test of lower and upper outlier-pair $x_{(1)}, x_{(n)}$ | $\dfrac{x_{(n)} - x_{(1)}}{\tilde{s}}$ |
| **Nv8** | 241 | Two-sided discordancy test of $k$ outliers (irrespective of directions) | Modification of Tietjen and Moore's $E_k$-statistic. See worksheet |
| **Nμ1** | 241 | Test of upper outlier $x_{(n)}$ | $\dfrac{x_{(n)} - \mu}{s(\mu)}$ |
| **Nμ2** | 242 | Test of extreme outlier (two-sided form of **Nμ1**) | $\max\left(\dfrac{\|x_{(n)} - \mu\|}{s(\mu)}, \dfrac{\|\mu - x_{(1)}\|}{s(\mu)}\right)$ or equivalently $\min(S_n^2(\mu)/S^2(\mu), S_1^2(\mu)/S^2(\mu))$ |
| **Nμ3** | 242 | Test of $k(\geqslant 2)$ upper outliers $x_{(n-k+1)}, \ldots, x_{(n)}$ | $\dfrac{x_{(n-k+1)} + \ldots + x_{(n)} - k\mu}{s(\mu)}$ |
| **Nμ4** | 243 | Test of two upper outliers $x_{(n-1)}, x_{(n)}$ | $S_{n-1,n}^2(\mu)/S^2(\mu)$ |
| **Nμ5** | 243 | Test of lower and upper outlier-pair $x_{(1)}, x_{(n)}$ | $S_{1,n}^2(\mu)/S^2(\mu)$ |
| **Nμ6** | 243 | Two-sided test of one or more extreme outliers | $\dfrac{\sum_{j=1}^{n}(x_j - \mu)^4}{ns^4(\mu)}$ |
| **Nμ14** | 244 | Block test of $k$ outliers irrespective of directions | See worksheet |
| **Nσ1** | 244 | Test of upper outlier $x_{(n)}$ | $\dfrac{x_{(n)} - \bar{x}}{\sigma}$ |
| **Nμσ1** | 245 | Test of upper outlier $x_{(n)}$ | $\dfrac{x_{(n)} - \mu}{\sigma}$ |
| **Nσ2** | 246 | Test of extreme outlier (two-sided form of **Nσ1**) | $\max\left(\dfrac{x_{(n)} - \bar{x}}{\sigma}, \dfrac{\bar{x} - x_{(1)}}{\sigma}\right)$ |
| **Nσ3** | 246 | Test of $k(\geqslant 2)$ upper outliers $x_{(n-k+1)}, \ldots, x_{(n)}$ | $\dfrac{x_{(n-k+1)} + \ldots + x_{(n)} - k\bar{x}}{\sigma}$ |
| **Nμσ3** | 247 | Test of $k(\geqslant 2)$ upper outliers $x_{(n-k+1)}, \ldots, x_{(n)}$ | $\dfrac{x_{(n-k+1)} + \ldots + x_{(n)} - k\mu}{\sigma}$ |
| **Nσ4** | 247 | Test of two upper outliers $x_{(n-1)}, x_{(n)}$ | $S_{n-1,n}^2/\sigma^2$ |
| **Nμσ4** | 247 | Test of two upper outliers $x_{(n-1)}, x_{(n)}$ | $S_{n-1,n}^2/(\mu)/\sigma^2$ |

(*Contd.*)

| Label | Worksheet page no. | Description of test | Statistic |
|-------|-----|---------------------|-----------|
| N$\sigma$5 | 248 | Test of lower and upper outlier-pair $x_{(1)}$, $x_{(n)}$ | $S^2_{1,n}/\sigma^2$ |
| N$\mu\sigma$5 | 248 | Test of lower and upper outlier-pair $x_{(1)}$, $x_{(n)}$ | $S^2_{1,n}(\mu)/\sigma^2$ |
| N$\sigma$6(N$\mu\sigma$6) | 248 | Test of lower and upper outlier-pair $x_{(1)}$, $x_{(n)}$ | $\dfrac{x_{(n)} - x_{(1)}}{\sigma}$ |
| N$\sigma$7(N$\mu\sigma$7) | 249 | Test of upper outlier $x_{(n)}$ | $\dfrac{x_{(n)} - x_{(n-1)}}{\sigma}$ |
| N$\sigma$8(N$\mu\sigma$8) | 249 | Test of two upper outliers $x_{(n-1)}$, $x_{(n)}$ | $\dfrac{x_{(n-1)} - x_{(n-2)}}{\sigma}$ |
| N$\sigma$9(N$\mu\sigma$9) | 250 | Test of $k$ lower and $k$ upper outliers | $\dfrac{x_{(n-k+1)} - x_{(k)}}{\sigma}$ |

**N1** *Discordancy test of a single outlier $x_{(n)}$ in a normal sample with $\mu$ and $\sigma$ unknown*

*Test statistic:*

$$T_{N1} = \text{internally studentized extreme deviation from mean} = \frac{x_{(n)} - \bar{x}}{s}.$$

An equivalent statistic is:

$$\frac{\text{reduced sum of squares}}{\text{total sum of squares}} = \frac{S^2_n}{S^2} = 1 - \frac{n}{(n-1)^2}\, T^2_{N1}.$$

*Recurrence relationship:*

$$f_n(t) = \frac{n}{n-1}\left(\frac{n}{\pi}\right)^{1/2} \frac{\Gamma\!\left(\dfrac{n-1}{2}\right)}{\Gamma\!\left(\dfrac{n-2}{2}\right)}\left(1 - \frac{nt^2}{(n-1)^2}\right)^{(n-4)/2}$$

$$\times F_{n-1}\left[\left(\frac{n^2(n-2)t^2}{(n-1)((n-1)^2 - nt^2)}\right)^{1/2}\right] \left(\frac{1}{\sqrt{n}} \le t \le \frac{n-1}{\sqrt{n}}\right)$$

with

$$F_2(t) = 0\left(t < \frac{1}{\sqrt{2}}\right),\ 1\left(t > \frac{1}{\sqrt{2}}\right).$$

*Inequality:*

$$SP(t) \le nP\left(\mathbf{t}(n-2) > \left[\frac{n(n-2)t^2}{(n-1)^2 - nt^2}\right]^{1/2}\right).$$

This is an equality when $t \geq [(n - 1)(n - 2)/2n]^{1/2}$.

*Tabulated significance levels:* Table XIIIa, page 485; abridged from Grubbs and Beck (1972), Table I, pages 848–850, where 0.1, 0.5, 1, 2.5, 5, and 10 per cent points are given for $n = 3(1)147$.

*Further tables:* Dixon (1950) gives graphs of the performance measure $P_3$ (see page 122) for $n = 5, 15$, and for alternatives of slippage in location and slippage in dispersion by one and two observations; the figures are derived from sampling experiments of size 200 at most. Ferguson (1961a) gives tables of power $P_1$ (page 122) for the alternative of slippage in location by a single observation. David and Paulson (1965) give graphs of performance measure $P_2$ (page 122) in relation to the same alternative, for $n = 4(2)10$. McMillan (1971) gives graphs of the performance measures $P(C_1)$, $P(C_2)$, $P(C_3)$ (page 139) when **N1** is used consecutively for the testing of two upper outliers; some corrections to these results are given by Moran and McMillan (1973). See also Hawkins (1978) (further details below) and **N19**.

*References:* Pearson and Chandra Sekar (1936), Dixon (1950, 1962), Grubbs (1950, 1969), Kudo (1956a), Ferguson (1961a, 1961b), Quesenberry and David (1961), David and Paulson (1965), Stefansky (1971), McMillan (1971), Moran and McMillan (1973), Hawkins (1978).

*Properties of test:* **N1** is the maximum likelihood ratio test for a location-slippage alternative in which one observation arises from a normal distribution $N(\mu + a, \sigma^2)$, $a > 0$. For this alternative, it has the optimal property of being the scale-and location-invariant test of given size which maximizes the probability $P_3$ of identifying the contaminant as discordant. Vulnerable to masking effect when there is more than one contaminant, but less so than **N7(Nμ7)**. Not very suitable for *consecutive* use when testing several outliers; preferable in this case to use a block procedure or to use **N15** consecutively. See also Hawkins (1978) for properties of a *consecutive* version of the test for two upper outliers, also of **Nv2**, with tabulated 10 per cent, 5 per cent, 1 per cent and 0.1 per cent points for $n = 5(1)15(5)30$. Hawkins calculates the power of the test for $n = 10$ and compares it with the $k = 2$ cases of **N3** (and **Nv5**) and **N4**, concluding that (when two outliers are present), 'the recursive test is never best, but is never poor'. The study by Prescott (1978) of behaviour of outlier tests assessed by means of sensitivity surfaces includes the present test (as well as **N2**, **N9**, **N10**, **N11**, **N6**, **N3**, and recursive forms) and highlights the conflict of optimal behaviour when one outlier is present and vulnerability to masking if *more* than one outlier is present. The specific form of the general test of Tiku (1975) for the normal distribution has a somewhat similar structure to that of the present test. It is compared empirically and by power calculations with this test (**N1**) and with **N4** and **N14**; see also Hawkins (1977), Tiku (1977) and Srivastava (1980). The latter reference deals with equicorrelated observations. See also **N19**.

Table XIIIa prompts a general comment. As Shiffler (1988) observes,

Outliers are sometimes defined as those values for which $[(x_{(n)} - \bar{x})/s]$ exceeds 3 ... or 4 ...

and he gives references to books which exemplify this. He is right; a naive tendency does still persist in places for people to think 'If $x_{(n)}$ is more than (say) 3 (or 4) standard deviations above the mean it must be suspect, discard it.' As we have already remarked, in Chapter 2 and elsewhere, it is obvious that the critical number of standard deviations above the mean for judging $x_{(n)}$ to be discordant depends on the sample size $n$. This is highlighted by the well-known result that, whatever the sample, $(x_{(n)} - \bar{x})/s$ cannot exceed $(n - 1)/\sqrt{n}$ (as can be seen, e.g., from (4.1.18) ). So in the normal sample case, which underlies the use of fixed figures such as 3 and 4, Table XIIIa and the above inequality show, for instance, that a value of $x_{(n)}$ three standard deviations above the mean is not significant (at 5%) in a sample of size 60, *is* significant (at 1%) in a sample of size 15, and can never occur in a sample of size 10. So much for the three standard deviations rule!

**N2** *Two-sided discordancy test of an extreme outlier in a normal sample with* $\mu$ *and* $\sigma^2$ *unknown*

*Test statistic:*

$$T_{N2} = \max\left(\frac{x_{(n)} - \bar{x}}{s}, \frac{\bar{x} - x_{(1)}}{s}\right).$$

An equivalent statistic is:

$$\min\left(\frac{S_n^2}{S^2}, \frac{S_1^2}{S^2}\right) = 1 - \frac{n}{(n - 1)^2} T_{N2}^2.$$

*Inequality:*

$$SP(t) \leq 2P(T_{N1} > t). \text{ Equality holds for } t > \sqrt{[(n - 1)(n - 2)/2n]}.$$

*Tabulated significance levels:* Table XIIIb, page 485; derived from Pearson and Hartley (1966), Table 26b, page 188.

*Further tables:* Ferguson (1961a) gives tables of power $P_1$ for alternatives of slippage in location by a single observation and by two observations. See comment on Prescott (1979) below.

*References:* Kudo (1956a), Ferguson (1961a, 1961b), Quesenberry and David (1961), Tietjen and Moore (1972). Tietjen and Moore, working in terms of the equivalent statistic $\min(S_n^2/S^2, S_1^2/S^2)$, present the inequality for $SP(t)$ as an equality. Prescott (1979) presents a *consecutive* form of this test, with tabulated percentage points for up to three outliers obtained by simulation. See **N19** for further details.

*Properties of test:* Maximum likelihood ratio test for a location-slippage alternative in which one observation arises from a normal distribution $N(\mu + a, \sigma^2)$, $a \neq 0$. For this alternative, has the optimal property of being the scale- and location-invariant test of given size which maximizes the probability ($P_3$, page 122) of identifying the contaminant as a discordant outlier. Vulnerable to masking effect in small samples when there are two outliers in the same direction. See also comments under **N1**.

**N3** *Discordancy test of $k$ upper outliers* $x_{(n-k+1)}, \ldots, x_{(n-1)}, x_{(n)}$ *in a normal sample with $\mu$ and $\sigma^2$ unknown*

*Test statistic:*

$T_{N3}$ = sum of internally studentized deviations from the mean

$$= \frac{x_{(n-k+1)} + \ldots + x_{(n-1)} + x_{(n)} - k\bar{x}}{s}$$

*Inequality:*

$$SP(t) \leq \binom{n}{k} P\left( \mathbf{t}(n-2) > \left[ \frac{n(n-2)t^2}{k(n-k)(n-1) - nt^2} \right]^{1/2} \right).$$

This is an equality when $t \geq [k^2(n-1)(n-k-1)/(nk+n)]^{1/2}$.

*Tabulated significance levels:* Table XVa, page 491, specially compiled on the basis of simulations of sizes 10 000.

*References:* Murphy (1951), Kudo (1956a), Ferguson (1961b), McMillan (1971), Fieller (1976). McMillan gives results for the comparative performance of tests **N3** and **N4** as applied to two upper outliers ($k = 2$); see **N4** Worksheet.

*Properties of test:* **N3** is the maximum likelihood ratio test for a location-slippage alternative in which $k$ observations arise from a common normal distribution $(\mu + a, \sigma^2)$, $a > 0$. For this alternative, it has the optimal property of being the scale- and location-invariant test of given size which maximizes the probability of identifying the $k$ contaminants as discordant. See also comments under **N1**.

**N4 Discordancy test of $k(\geqslant 2)$ upper outliers** $x_{(n-k+1)}, \ldots, x_{(n-1)}, x_{(n)}$ **in a normal sample with $\mu$ and $\sigma^2$ unknown**

*Test statistic:*

$$T_{N4} = \frac{\text{reduced sum of squares}}{\text{total sum of squares}} = \frac{S^2_{n-k+1,\ldots,n-1,n}}{S^2}.$$

*Tabulated significance levels:* Table XVb, page 491; values for $k = 2$ abridged from Grubbs and Beck (1972), Table II, pages 851–853, where 0.1, 0.5, 1, 2.5, 5, and 10 per cent points are given for $n = 4(1)149$; values for $k = 3$ and $k = 4$ abridged from Tietjen and Moore (1972), Table I, pages 587–590, where 1, 2.5, 5, and 10 per cent points are given for $k = 1(1)10$ and $n = \max[3, 2k]\,(1)20(5)50$. Note that values of $T_{N4}$ *smaller* than the tabulated level are significant.

*References:* Grubbs (1950, 1969), Dixon (1950), McMillan (1971), Tietjen and Moore (1972), Fieller (1976).

*Properties of test:* **N4** is the maximum likelihood ratio test for a location-slippage alternative in which $k$ observations arise from separate normal distributions each with variance $\sigma^2$ but with distinct means all exceeding $\mu$.

A study by McMillan (1971) of the performance measure $P_2$ (see page 122) for tests **N3** and **N4** in the case $k = 2$ indicates that **N4** is more robust than **N3** against departures from the relevant alternative.

Commonly the number, $k$, of outliers to be tested will have been chosen, either as being the number of manifest outliers, or as a parameter in a data-processing procedure. As an alternative Tietjen and Moore (1972) suggest proceeding in the following way: find the 'largest gap', i.e. the largest of the intervals $x_{(n)} - x_{(n-1)}, x_{(n-1)} - x_{(n-2)}, \ldots$, to the right of the mean $\bar{x}$, and fix upon the observations to the right of this gap ($k$ in number, say) for testing as upper outliers. Tietjen and Moore show that **N4** applied as a block test to these $k$ outliers has rather better performance, in terms of proportion of contaminants correctly identified as discordant, than consecutive test procedures using either **N14** or **N17**, and much better performance than consecutive procedures using either **N1** or (an unspecified) one of the Dixon-type tests.

See also comments under **N1**.

**N5 Discordancy test of a lower and upper outlier-pair** $x_{(1)}, x_{(n)}$ **in a normal sample with $\mu$ and $\sigma^2$ unknown**

*Test statistic:*

$$T_{N5} = \frac{\text{reduced sum of squares}}{\text{total sum of squares}} = \frac{S^2_{1,n}}{S^2}.$$

*Tabulated significance levels:* Table XVIa, page 493; specially compiled on the basis of simulations of sizes 10 000. Values of $T_{N5}$ *smaller* than the tabulated level are significant.

*References:* Grubbs (1950), Ferguson (1961b), Fieller (1976).

*Properties of test:* Maximum likelihood ratio test for a location-slippage alternative in which two observations arise from separate normal distributions $N(\mu + a_1, \sigma^2)$, $N(\mu + a_2, \sigma^2)$, $a_1 < 0 < a_2$.

**N6** *Discordancy test of a lower and upper outlier-pair $x_{(1)}$, $x_{(n)}$ in a normal sample with $\mu$ and $\sigma^2$ unknown*

*Test statistic:*

$$T_{N6} = \text{internally studentized range} = \frac{x_{(n)} - x_{(1)}}{s}.$$

*Inequality:*

$$SP(t) \leq n(n-1)P\left(\mathbf{t}(n-2) > \left[\frac{(n-2)t^2}{2n-2-t^2}\right]^{1/2}\right).$$

This is an equality when $t \geq [\frac{3}{2}(n-1)]^{1/2}$.

*Tabulated significance levels:* Table XVIIa, page 494, abridged from Pearson and Hartley (1966), Table 29c, page 200, where lower and upper limits and lower and upper 0.5, 1, 2.5, 5, and 10 per cent points are given for $n = 3(1)20(5)100, 150, 200, 500, 1000$.

*Further tables:* Shapiro, Wilk, and Chen (1968) give values for the power of the test against 45 different inherent alternatives (see Chapter 2, page 46).

*References:* David, Hartley, and Pearson (1954), Pearson and Stephens (1964), Shapiro, Wilk, and Chen (1968).

*Properties:* As a test against an *inherent* alternative, **N6** has good power properties against a variety of symmetric distributions alternative to the normal, but performs poorly with respect to asymmetric alternatives. See also comments under **N1** and **N19**.

**N7(Nμ7)** *Discordancy test of a single upper outlier $x_{(n)}$ in a normal sample with $\sigma^2$ unknown*

*Test statistic:*

$$T_{N7} = \frac{\text{excess}}{\text{range}} = \frac{x_{(n)} - x_{(n-1)}}{x_{(n)} - x_{(1)}} \quad \text{(Dixon's } r_{10} \text{ statistic)}.$$

*Test distribution:* For small $n$, we have:

$$f_3(t) = \frac{3\sqrt{3}}{2\pi} (t^2 - t + 1)^{-1}$$

$$f_4(t) = \frac{2}{\sqrt{3}} f_3(t)[(1 - 2t)(4t^2 - 4t + 3)^{-1/2} - (t - 2)(3t^2 - 4t + 4)^{-1/2}]$$

*Tabulated significance levels:* Table XIXa, page 498, abridged from Dixon (1951), Table I, page 73, where upper 0.5, 1, 2 per cent points, upper and lower 5, 10, 20, 30, 40 per cent points, and the 50 per cent point are given for $n = 3(1)30$.

*Further tables:* Dixon (1950) gives graphs of the performance measure $P_3$, based on sampling experiments of comparatively small size (66–200 replications). Ferguson (1961a) gives tables of power $P_1$ for the alternative of slippage in location by one observation.

*References:* Dixon (1950, 1951), Ferguson (1961a, 1961b).

*Properties of test:* Mainly effective when there is at most one contaminant, otherwise vulnerable to possible masking effect of $x_{(n-1)}$ and/or $x_{(1)}$; see properties of test N1. The performances of tests N7(Nμ7) as measured both by $P_3$ and $P_1$, against the alternative of slippage in location by a single observation, are effectively the same for sample sizes up to 15.

**N8(Nμ8)** *Two-sided discordancy test of an extreme outlier in a normal sample with* $\sigma^2$ *unknown*

*Test statistic:*

$$T_{N8} = \max\left[\frac{x_{(n)} - x_{(n-1)}}{x_{(n)} - x_{(1)}}, \frac{x_{(2)} - x_{(1)}}{x_{(n)} - x_{(1)}}\right].$$

*Inequality:*

$$SP(t) \leqslant 2P(T_{N7} > t)$$

This is an equality when $t \geqslant \frac{1}{2}$.

*Tabulated significance levels:* Table XIXb, page 498; specially compiled on the basis of simulations of sizes 10000.

*Reference:* King (1953).

*Properties of test:* Two-sided form of **N7(Nμ7)**.

**N9(Nμ9)** *Discordancy test of a single upper outlier* $x_{(n)}$ *in a normal sample with* $\sigma^2$ *unknown*

*Test statistic:*

$$T_{N9} = \frac{x_{(n)} - x_{(n-1)}}{x_{(n)} - x_{(2)}} \quad \text{(Dixon's } r_{11} \text{ statistic)}.$$

*Test distribution:* For $n = 4$ we have:

$$f_4(t) = \frac{3\sqrt{3}}{\pi}(t^2 - t + 1)^{-1/2}[1 + (t - 2)[3(4 - 4t + 3t^2)]^{-1/2}].$$

*Tabulated significance levels:* Table XIXc, page 498; abridged from Dixon (1951), Table II, page 74, where more extensive values are given, for the parameter values detailed in Worksheet **N7(Nμ7)**.

*Further tables:* Dixon (1950) gives graphs of $P_3$; see Worksheet **N7(Nμ7)**.

*References:* Dixon (1950, 1951).

*Properties of test:* Advantage: avoids any possible masking effect of lowest sample value $x_{(1)}$ (by inflation of the denominator). Disadvantage: vulnerable to masking effect of $x_{(n-1)}$. See also Worksheet **N12(Nμ12)** in relation to performance, and comments under **N1**.

**N10(Nμ10)** *Discordancy test of a single upper outlier* $x_{(n)}$ *in a normal sample with* $\sigma^2$ *unknown*

*Test statistic:*

$$T_{N10} = \frac{x_{(n)} - x_{(n-1)}}{x_{(n)} - x_{(3)}} \quad \text{(Dixon's } r_{12} \text{ statistic)}.$$

*Tabulated significance levels:* Table XIXd, page 498; abridged from Dixon (1951), Table III, page 75, where more extensive values are given, for the parameter values detailed in Worksheet **N7(Nμ7)**.

*Further tables and references:* As for **N9(Nμ9)**.

*Properties of test:* Avoids any possible masking effect of the two lowest observations $x_{(1)}$, $x_{(2)}$ on the testing of $x_{(n)}$, but is vulnerable to any masking effect of $x_{(n-1)}$. See also Worksheet **N13(Nμ13)** in relation to performance, and comments under **N1**.

**N11(Nμ11)** *Discordancy test of an upper outlier-pair* $x_{(n-1)}$, $x_{(n)}$ *in a normal sample with* $\sigma^2$ *unknown*

*Test statistic:*

$$T_{N11} = \frac{x_{(n)} - x_{(n-2)}}{x_{(n)} - x_{(1)}} \quad \text{(Dixon's } r_{20} \text{ statistic)}.$$

*Tabulated significance levels:* Table XIXe, page 498; abridged from Dixon (1951), Table IV, page 76, where more extensive values are given, for the parameter values detailed in Worksheet **N7(Nμ7)**.

*Further tables and references:* As for **N9(Nμ9)**.

*Properties of test:* **N11(Nμ11)** can also be used as a discordancy test for a single upper outlier $x_{(n)}$ which avoids the risk of masking by $x_{(n-1)}$. See also comments under **N1**.

**N12(Nμ12)** *Discordancy test of a single upper outlier* $x_{(n-1)}$, $x_{(n)}$ *in a normal sample with* $\sigma^2$ *unknown*

*Test statistic:*

$$T_{N12} = \frac{x_{(n)} - x_{(n-2)}}{x_{(n)} - x_{(2)}} \quad \text{(Dixon's } r_{21} \text{ statistic)}.$$

*Tabulated significance levels:* Table XIXf, page 498; abridged from Dixon (1951), Table V, page 77, where more extensive values are given, for the parameter values detailed in Worksheet **N7(Nμ7)**.

*Further tables and references:* As for **N9(Nμ9)**.

*Properties of test:* Avoids any possible masking effect from $x_{(1)}$. Can be used as a discordancy test for a single upper outlier $x_{(n)}$, and for this purpose is to be preferred to test **N9(Nμ9)**, since its performance is similar against a single contaminant and it avoids the risk of masking from $x_{(n-1)}$ if there is more than one contaminant.

**N13(Nμ13)** *Discordancy test of an upper outlier-pair* $x_{(n-1)}$, $x_{(n)}$ *in a normal sample with* $\sigma^2$ *unknown*

*Test statistic:*

$$T_{N13} = \frac{x_{(n)} - x_{(n-2)}}{x_{(n)} - x_{(3)}} \quad \text{(Dixon's } r_{22} \text{ statistic)}.$$

*Tabulated significance levels:* Table XIXg, page 498; abridged from Dixon (1951), Table VI, page 78, where more extensive values are given, for the parameter values detailed in Worksheet **N7(Nμ7)**.

*Further tables:* As for **N9(Nμ9)**.

*References:* Dixon (1950, 1951), Ferguson (1961b).

*Properties of test:* Avoids any possible masking effect from the two lowest observations $x_{(1)}$, $x_{(2)}$. Can be used as a discordancy test for a single upper outlier $x_{(n)}$, and for this purpose is superior to test **N10(Nμ10)**, having a similar performance against a single contaminant but being more robust against the presence of a second upper outlier at $x_{(n-1)}$.

**N14** *Discordancy test of one or more upper (or lower) outliers in a normal sample with* $\mu$ *and* $\sigma^2$ *unknown*

*Test statistic:*

$$T_{N14} = \text{sample skewness} = \frac{n^{1/2} \sum_{j=1}^{n} (x_j - \bar{x})^3}{\left[ \sum_{j=1}^{n} (x_j - \bar{x})^2 \right]^{3/2}}.$$

The value tested for discordancy is $x_{(n)}$ or $x_{(1)}$ according as the sign of $\sum (x_j - \bar{x})^3$ is + or −. For more than one outlier, apply test *consecutively.*

*Tabulated significance levels:* Table XXa, page 499; abridged from Pearson and Hartley (1966), Table 34B, page 207, where 5 per cent and 1 per cent points are given for $n = 25(5)50(10)100(25)200(50)1000(200)2000(500)5000$, and from Ferguson (1961a), Table I, page 281, where estimated 10, 5, and 1 per cent points are given for $n = 5(5)25$.

*Further tables:* Ferguson (1961a) gives tables of power $P_1$ against the alternative of slippage in location by a single observation. Shapiro, Wilk, and Chen (1968) give values for power against 45 different *inherent* alternatives (see Chapter 2, page 46).

*References:* Ferguson (1961a, 1961b), Shapiro, Wilk, and Chen (1968).

*Properties of test:* **N14** is the locally best invariant test of given size against a location-slippage alternative in which $k$ of the $n$ observations arise from separate normal distributions $N(\mu + a_1, \sigma^2), \ldots, N(\mu + a_k, \sigma^2)$, $a_1 > 0$, $a_2 > 0, \ldots, a_k > 0$, whatever the values of the $a$'s, and whatever the value of $k$ provided only that the contamination proportion $k/n$ under the alternative hypothesis is less than $\frac{1}{2}$.

Its power is nearly as good as that of **N1** against slippage in location for a single observation by medium or large amounts. It also has good power against inherent Cauchy and log-normal alternatives. See also comments under **N1**.

**N15** *Discordancy test of one or more outliers (irrespective of their directions) in a normal sample with* $\mu$ *and* $\sigma^2$ *unknown*

*Test statistic:*

$$T_{N15} = \text{sample kurtosis} = \frac{n \sum_{j=1}^{n} (x_j - \bar{x})^4}{\left[ \sum_{j=1}^{n} (x_j - \bar{x})^2 \right]^2}.$$

The value tested for discordancy is whichever of $x_{(n)}$ or $x_{(1)}$ is further from $\bar{x}$. Discordancy is indicated by *high* values of the statistic. For more than one outlier, apply test *consecutively*.

*Tabulated significance levels:* Table XXb, page 499; abridged from Pearson and Hartley (1966), Table 34C, page 208, where lower and upper 1 per cent and 5 per cent points are given for $n = 50(25)150(50)700(100)1000$ $(200)2000(500)5000$, from Ferguson (1961a), Table II, page 282, where estimated 1, 5, and 10 per cent points are given for $n = 5(5)25$, and from D'Agostino and Tietjen (1971) where simulated percentiles (including the upper 10, 5, $2\frac{1}{2}$, 2, and 1 per cent points) are given for $n = 7(1)10, 12, 15(5)50$.

*Further tables:* Ferguson (1961a) gives tables of power against alternatives of slippage in location by a single observation and by two observations. Shapiro, Wilk, and Chen (1968) give values for power against 45 different inherent alternatives (see Chapter 2, page 46).

*References:* Ferguson (1961a, 1961b), Shapiro, Wilk, and Chen (1968).

*Properties of test:* **N15** is the locally best unbiased invariant test of given size against a location-slippage alternative in which $k$ of the $n$ observations arise from separate normal distributions $N(\mu + a_1, \sigma^2), \ldots, N(\mu + a_k, \sigma^2)$, where $a_1, \ldots, a_k$ differ from zero but are otherwise arbitrary, provided that the contamination proportion $k/n$ under the alternative hypothesis is less than 0.21. **N15** is also the locally best invariant test of given size against a dispersion-slippage alternative in which $k$ of the observations arise from separate normal distributions $N(\mu, b_1\sigma^2), \ldots, N(\mu, b_k\sigma^2)$, $b_1 > 1, \ldots, b_k > 1$, irrespective of the proportion $k/n$.

Its power is nearly as good as that of **N2** against slippage in location for a single observation by medium or large amounts. Against slippage in location by two observations it is superior to **N2** in power, greatly so when the sample size is less than, say, 20.

**N15** has the advantage of being robust against possible masking effects. It is suitable for *consecutive* use in the possible presence of more than one outlier. See also **N19**.

**N16** *Two-sided discordancy test of k outliers (irrespective of directions) in a normal sample with μ and σ² unknown*

*Test statistic:*

$$T_{N16} = \text{Tietjen and Moore's } E_k\text{-statistic} = \frac{\sum_{j=1}^{n-k} (r_{(j)} - \bar{r}_{n-k})^2}{\sum_{j=1}^{n} (r_{(j)} - \bar{r})^2}$$

where $r_j = |x_j - \bar{x}|$, the absolute deviation of $x_j$ from the sample mean; $\{r_{(j)}\}$ are the values of the $r_j$ in ascending order, $r_{(1)} < r_{(2)} < \ldots < r_{(n)}$; $\bar{r}$ is the mean of all the $r$'s; and $\bar{r}_{n-k}$ is the mean of the $(n - k)$ lowest $r$'s, i.e.

$$\bar{r}_{n-k} = (r_{(1)} + \ldots + r_{(n-k)})/(n - k).$$

*Tabulated significance levels:* Table XXI, page 500, extracted from Tietjen and Moore (1972), Table II, pages 591–593, where 1, 5, and 10 per cent points are given for $k = 1(1)10$ and $n = [\max(3, 2k)](1)20(5)50$. Erroneous entries in Tietjen and Moore's tables have been amended; see Tietjen and Moore (1979). See also comments on Hawkins (1979) below.

*References:* Tietjen and Moore (1972), Hawkins (1979).

*Properties of test:* A pragmatic test procedure. Hawkins (1979) describes a possible shortcoming of the test, 'apparently noticed' by Rosner (1975), and proposes a straightforward remedy. The problem is that in using absolute deviations from the overall mean, $\bar{x}$, highly extreme values can mask less extreme (but still possible discordant) values near to $\bar{x}$, with the effect that they do not feature in the outlying subset identified by the test. To overcome this he suggests that $r_{(n)}$ should be determined; then the observation in $r_{(n)}$ is omitted before $r_{(n-1)}$ is determined, and so on. Modified tables of percentage points are now needed. Hawkins (1979) provides these and also shows how to incorporate an independent external variance estimate, $s_v^2$. He tabulates simulation estimates of the 5, 1 and 0.1 per cent points for $v = 0$, 10, 20, $k = 2(1)10$ and $n = h(5)30(10)50(25)100$ where $h$ is the multiple of 5 closest to $2k$. See also Hawkins and Perold (1977).

**N17** *Two-sided test for contamination in a normal sample with μ and σ² unknown*

*Test statistic:*

$$T_{N17} = \text{Shapiro and Wilk's } W\text{-statistic}$$

$$= \left( \sum_{i=1}^{[n/2]} a_{n,n-i+1} [x_{(n-i+1)} - x_{(i)}] \right)^2 \Big/ S^2$$

where $[n/2]$ denotes the integer part of $n/2$, and the $a_{n,j}$ are tabulated constants (Table XXIIb, page 502; extracted from Shapiro and Wilk (1965), Table 5, pages 603–604, where values of these constants are given for $n = 2(1)50$).

*Test distribution:* For $n = 3$, we have:

$$f_3(t) = \frac{3}{\pi} (t - t^2)^{-1/2} \quad (\tfrac{3}{4} \leqslant t \leqslant 1).$$

An approximate normalizing transformation of $W = T_{N17}$ is given by Royston (1982a):

$$Z = [(1 - W)^\lambda - a]/b \approx N(0, 1)$$

where $\lambda$, $a$ and $b$ are functions of $n$ given below; *large positive* values of $Z$ are significant.

$\lambda$, $\log a$ and $\log b$ are calculated as polynomials $\sum c_r u^r$ in $u$, where

$$u = \log n - 3 \text{ for } 7 \leqslant n \leqslant 20, \quad u = \log n - 5 \text{ for } 21 \leqslant n \leqslant 2000.$$

The polynomial coefficients $c_0, c_1, \ldots$ are given in Table 6.6.

**Table 6.6**

|  | $c_0$ | $c_1$ | $c_2$ | $c_3$ | $c_4$ | $c_5$ | $c_6$ |
|---|---|---|---|---|---|---|---|
| $\lambda (7 \leqslant n \leqslant 20)$ | 0.1189 | 0.1334 | 0.3279 | | | | |
| $\lambda (21 \leqslant n \leqslant 2000)$ | 0.4804 | 0.3188 | 0 | $-0.0242$ | 0.0088 | 0.0030 | |
| $\log a (7 \leqslant n \leqslant 20)$ | $-0.3754$ | $-0.4921$ | $-1.1243$ | $-0.1994$ | | | |
| $\log a (21 \leqslant n \leqslant 2000)$ | $-1.9149$ | $-1.3789$ | $-0.0418$ | 0.1066 | $-0.0351$ | $-0.0150$ | |
| $\log b (7 \leqslant n \leqslant 20)$ | $-3.1580$ | 0.7294 | 3.0185 | 1.5588 | | | |
| $\log b (21 \leqslant n \leqslant 2000)$ | $-3.7354$ | $-1.0158$ | $-0.3319$ | 0.1774 | $-0.0164$ | $-0.0322$ | 0.0038 |

*Tabulated significance levels:* Table XXIIa, page 501; abridged from Shapiro and Wilk (1965), Table 6, page 605, where lower and upper 1,2,5, and 10 per cent points, and the 50 per cent point, are given for $n = 3(1)50$. Values of $T_{N17}$ *smaller* than the tabulated level are significant. For $n > 50$, use Royston's normalizing transformation given above.

*Further tables:* Shapiro, Wilk, and Chen (1968) give values for the power of the test against 45 different inherent alternatives (see Chapter 2, page 46). Chen (1971) gives values for the power against contamination by a given number of observations (either 1 or 2) in small samples ($n \leqslant 10$), or with a given contamination probability per observation (0.05, 0.10 or 0.20) in larger samples (up to $n = 50$); he deals with shifts both in location and in dispersion.

*References:* Shapiro and Wilk (1965), Shapiro, Wilk, and Chen (1968), Chen (1971), Royston (1982a, 1982b, 1986, 1989).

*Properties:* A useful omnibus test, both against inherent alternatives and against slippage alternatives. It is, however, sensitive to the presence of ties when the raw data are grouped or rounded. In this situation a modified test procedure given by Royston (1989) should be used. Effectively, $W = T_{N17}$ is recalculated with $S^2$ in the denominator replaced by $S^2 - [(n - 1)/(12n]\sum h_j^2$, where $h_j$ is the length of the grouping interval containing $x_j (j = 1, \ldots, n)$. New parameters $a^*$, $b^*$ in the approximate normalizing transformation $[(1 - W)^\lambda - a^*]/b^* \approx N(0, 1)$ are calculated from polynomial formulae given in Royston (1989).

**N18** *Test of extreme outlier (two-sided) in a normal sample with* $\mu$ *and* $\sigma^2$ *unknown, using a robust estimator of* $\sigma$

*Test statistic:*

$$T_{N17} = \max\left(\frac{x_{(n)} - \bar{x}}{s_b}, \frac{\bar{x} - x_{(1)}}{s_b}\right)$$

where

$$s_b^2 = (V/W)\sqrt{n}$$

with

$$V = \left[\sum_{|u_i| < 1} (x_i - m)(1 - u_i^2)^4\right]^{1/2}$$

$$W = \max\left\{1, -1 + \left|\sum_{|u_i| < 1} (1 - u_i^2)(1 - 5u_i^2)\right|\right\}$$

and

$$u_i = (x_i - m)/(9 \text{ median } |x_i - m|)$$

with $m$ the sample median.

*Tabulated significance levels:* No tabulation, but extensive simulation supports $2.064n^{0.5}/(n - 1)^{0.387}$ as upper 5 per cent point (with maximum error of $0.02\sqrt{n}$).

*Reference:* Iglewicz and Martinez (1982).

*Properties of test:* A pragmatic test procedure. The form of $s_b^2$ is motivated by a proposed scale estimator of Mosteller and Tukey (1977, p. 208) based on an estimate of the asymptotic variance of the biweight location estimator. Effectively, observations more than $6\sigma$ from $m$ are ignored. If the $u_i$ are small, $s_b^2$ is approximately $\sum(x_i - m)^2/(n - 1)$. Some empirical comparisons are made by lglewicz and Martinez (1982) with **N2**, **N14** and **N15** as single-outlier procedures and, in consecutive forms, as multiple outlier procedures. Although the limited power comparisons reveal no serious defects, the method is laborious and requires more detailed investigation.

**N19** *Consecutive ('recursive') test of up to $k$ outliers (irrespective of directions) in a normal sample with $\mu$ and $\sigma^2$ unknow*n

*Test procedure:* Suppose $T$ is a test statistic for a (two-sided) discordancy test of a single outlier in a normal sample. Let $x_1, x_2, \ldots, x_k$ (for prescribed $k$) be the observations yielding the maximum value of $T$ in subsamples $s_1, s_2, \ldots, s_k$ where $s_j$ ($j = 1, 2, \ldots, k$) is the set of observations excluding $x_1, x_2, \ldots, x_{j-1}$. (That is, $x_1$ produces the maximum value of $T$ in the complete sample; $x_2$ produces the next largest value of $T$ *calculated for the sample of $n - 1$ observations on omission of $x_1$*, and so on.) Suppose the successive values of $T$ so obtained are $T_1, T_2, \ldots, T_k$.

We determine $\lambda_i(\beta)$ where $P[T_i > \lambda_i(\beta)] = \beta$ for $i = 1, 2, \ldots, k$ and

$$P\left\{\bigcup_1^k [T_i > \lambda_i(\beta)]\right\} = \alpha.$$

Then a level-$\alpha$ test operates as follows. If $T_k > \lambda_k(\beta)$ then $x_1, x_2, \ldots, x_k$ are discordant. Otherwise we proceed by examining $T_l$ for $l = k - 1, k - 2, \ldots, 1$ until $T_l > \lambda_l(\beta)$, at which stage $x_1, x_2, \ldots, x_l$ are adjudged discordant at level $\alpha$. (If $T_l \leq \lambda_l(\beta)$ for all $l = 1, 2, \ldots, k$, we conclude, of course, that there are no discordant outliers).

Specific forms of such a consecutive test employ $T_{N1}$, $T_{N6}$ and $T_{N15}$ and are termed respectively ESD, STR and KUR procedures by Rosner (1975). His RST procedure (Rosner, 1977) uses

$$T = \max |x_i - a|/b$$

where $a$ and $b$ are trimmed mean and standard deviation calculated on omission of the $k$ smallest, and $k$ largest, sample values. The JST procedure replaces $a$ and $b$ with $m$ and $d$: the median and interquartile range of the trimmed sample.

*Tabulated significance levels:* Various tables of simulation estimates of percentage points have been presented, as follows:

ESD     Rosner (1975)      5, 1, 0.5 per cent; $n = 10(1)20(5)50$; $k = 2$
        Jain (1981b)       10, 5, 1 per cent; $n = 20(10)60(20)100$; $k = 2(1)5$
        Chhikara and
        Feiveson (1980)    Charts of 5 per cent; $n \leq 100$, $k = 1(1)10$
KUR     Jain (1981b)       10, 5, 1 per cent; $n = 20(10)60(20)100$; $k = 2(1)5$
RST     Rosner (1977)      10, 5, 1 per cent; $n = 10(5)20(10)50(25)100$; $k = 2$
                                                          $n = 20(10)50(25)100$; $k = 3, 4$
        Jain (1981b)       10, 5, 1 per cent; $n = 20(10)60(20)100$; $k = 2(1)5$

Table XXIII on pages 503–5 presents critical values of $\lambda_i(\beta)$ for 5 and 1 per cent tests of types ESD, KUR and RST for $n = 20(10)60(20)100$ and $k = 2(1)5$ extracted from Jain (1981b).

*Further tables:* Tables and charts for power comparisons and robustness studies are presented (from simulation studies) by Rosner (1975), Jain (1981a) and Jain and Pingel (1981b).

*References:* Rosner (1975, 1977), Prescott (1979), Chhikara and Feiveson (1980), Jain (1981a, 1981b), Jain and Pingel (1981b). Simonoff (1984a) presents a computer program (in FORTRAN) which simultaneously calculates critical values for the ESD, KUR and KST procedures and for the consecutive $E_k$ procedure described in Worksheet **N16** under *'Properties of test'* (details in Hawkins, 1979). The ESD, KUR and KST procedures and an analogous consecutive procedure based on $T_{N8}$ are compared for performance in Simonoff (1984b).

*Properties of test:* These procedures embody the estimation of the *number* of contaminants in the sample. The policy of examining samples of successively reduced size in *reverse order* protects against masking effects in the more usual forms of consecutive test. The inconvenience of having to recalculate summary statistics at each of the $k$ stages is eliminated in the trimmed sample methods RST and JST described above. (Earlier versions used progressive degrees of trimming.) Prescott (1979) shows that use of the alternative representation of **N1** in terms of ratios of variances provides similar reduction of effort for the ESD procedure. Hawkins (1980b) points out a problem which arises from having to specify at the outset the number, $k$, of outliers. If in fact there are $l < k$ contaminants, it would be desirable that the probability of declaring *more* than $l$ discordant outliers should be low (in the region of the overall significance level we are aiming at). He comments that this feature is lacking in Prescott's (1979) proposal, and suggests a means of restoring it. It seems that none of the procedures is consistently more powerful than the others, and that there is danger of swamping, particularly if most outliers are at one end of the sample (Jain, 1981a; Jain and Pingel, 1981b). **Ea18**, **Ea19** and **Ea20** (pages 213, 214 and 215) are one-sided versions of such an approach for exponential samples.

**Nv1** *Discordancy test of a single upper outlier* $x_{(n)}$ *in a normal sample with* $\mu$ *unknown and an independent estimate of* $\sigma^2$ *known*

*Test statistic:*

$T_{Nv1}$ = externally studentized extreme deviation from the mean = $\dfrac{x_{(n)} - \bar{x}}{s_v}$.

*Recurrence relationship:*

$$f_n(t) = \left(\frac{n^3}{(n-1)\pi v}\right)^{1/2} \frac{\Gamma\left(\dfrac{v+1}{2}\right)}{\Gamma\left(\dfrac{v}{2}\right)} \left(1 + \frac{nt^2}{(n-1)v}\right)^{-(v+1)/2} F_{n-1}\left(\frac{nt}{n-1}\right)$$

with $F_1(t) = 0 (t < 0)$, $1 (t > 0)$.

*Inequality:*

$$SP(t) < nP(\text{t}(v) > [n/(n-1)]^{1/2}t).$$

*Tabulated significance levels*: Table XIVa, page 487, abridged from Pearson and Hartley (1966), Table 26, pages 185–186, where 10, 5, 2.5, 1, 0.5, and 0.1 per cent points are given for $n = 3(1)10$, 12 and $v = 10(1)20$, 24, 30, 40, 60, 120, $\infty$, and further 5 and 1 per cent points for the additional $v$-values 5(1)9.

*Further tables:* David and Paulson (1965) give graphs of performance measure $P_2$ (see page 122) for an alternative model of slippage in location by one observation. McMillan (1971), subsequently amended by Moran and McMillan (1973), gives graphs of performance measures $P(C_1)$, $P(C_2)$, $P(C_3)$ (see page 139) for the consecutive testing of two upper outliers using **Nv1**.

*References:* Nair (1948, 1952), David (1956a, 1956b), David and Paulson (1965), McMillan (1971), Moran and McMillan (1973).

*Properties of test*: Makes no use of the *internal* estimate of variance; if there is at most one contaminant this wastes information, and test **Nv2** is preferable; on the other hand it offers a safeguard against the risk of masking if there is more than one contaminant.

**Nv2** *Discordancy test of a single upper outlier* $x_{(n)}$ *in a normal sample with* $\mu$ *unknown and an independent estimate of* $\sigma^2$ *known*

*Test statistic:*

$T_{Nv2}$ = externally and internally studentized extreme deviation from the

mean = $\dfrac{x_{(n)} - \bar{x}}{\tilde{s}}$.

*Recurrence relationship:*

$$f_n(t) = \left(\frac{n^3}{\pi(n-1)(n-1+v)}\right)^{1/2} \frac{\Gamma\left(\frac{n-1+v}{2}\right)}{\Gamma\left(\frac{n-2+v}{2}\right)} \left(1 - \frac{nt^2}{(n-1)(n-1+v)}\right)^{(n-4+v)/2}$$

$$\times F_{n-1}\left[\left(\frac{n^2(n-2+v)t^2}{(n-1)^2(n-1+v) - n(n-1)t^2}\right)^{1/2}\right],$$

$$t \le [(n-1)(n-1+v)/n]^{1/2},$$

with $F_1(t) = 0(t < 0), 1(t > 0)$.

*Inequality:*

$$SP(t) < nP\left(\mathbf{t}(n-2+v) > \left[\frac{n(n-2+v)t^2}{(n-1)(n-1+v) - nt^2}\right]^{1/2}\right).$$

*Tabulated significance levels:* Table XIVb, page 488; derived from Quesenberry and David (1961) Tables 1 and 2, page 388, where 5 and 1 per cent points of $(n-1+v)^{-1/2}T_{Nv2}$ are given for $n = 3(1)10, 12, 15, 20$ and $v = 0(1)10, 12, 15, 20, 24, 30, 40, 50$.

*Further tables:* David and Paulson (1965) give graphs of performance measure $P_2$ (see page 122) for an alternative model of slippage in location by one observation. McMillan (1971), with subsequent amendments by Moran and McMillan (1973), gives graphs of performance measures $P(C_1), P(C_2), P(C_3)$ (see page 139) when **Nv2** is used consecutively for testing two upper outliers.

*References:* Kudo (1956a), Quesenberry and David (1961), David and Paulson (1965), McMillan (1971), Moran and McMillan (1973).

*Properties of test:* For a location-slippage alternative in which one observation arises from a normal distribution $N(\mu + a, \sigma^2), a > 0$, **Nv2** has the optimal property of being the scale- and location-invariant test of given size which maximizes the probability of identifying the contaminant as discordant.

### Nv3 Two-sided discordancy test of an extreme outlier in a normal sample with μ unknown and an independent estimate of $\sigma^2$ known

*Test statistic:*

$T_{Nv3}$ = externally studentized extreme absolute deviation from the mean

$$= \max\left(\frac{x_{(n)} - \bar{x}}{s_v}, \frac{\bar{x} - x_{(1)}}{s_v}\right).$$

*Inequality:*

$$SP(t) < 2P(T_{Nv1} > t) < 2nP(t(v)) > [n/(n - 1)]^{1/2}t).$$

*Tabulated significance levels:* Table XIVc, page 489; derived from Halperin *et al.* (1955), Tables 1 and 2, pages 187–188, where bounds on the 5 and 1 per cent points are given for $n = 3(1)10$, 15, 20, 30, 40, 60 and $v = 3(1)10$, 15, 20, 30, 40, 60, 120, $\infty$ subject to $v \geqslant n$.

*Reference:* Halperin *et al.* (1955).

*Properties of test:* An appropriate test for comparing treatment means in analysis of variance.

## Nv4 Two-sided discordancy test of an extreme outlier in a normal sample with μ unknown and an independent estimate of $\sigma^2$ known

*Test statistic:*

$T_{Nv4}$ = externally and internally studentized extreme absolute deviation from the mean

$$= \max\left(\frac{x_{(n)} - \bar{x}}{\tilde{s}}, \frac{\bar{x} - x_{(1)}}{\tilde{s}}\right).$$

*Inequality:*

$$SP(t) < 2P(T_{Nv2} > t) < 2nP\left(t(n - 2 + v) > \left[\frac{n(n - 2 + v)t^2}{(n - 1)(n - 1 + v) - nt^2}\right]^{1/2}\right).$$

*Tabulated significance levels:* Table XIVd, page 489; derived from Quesenberry and David (1961), Tables 3 and 4, pages 389–390, where bounds on the 5 and 1 per cent points of $(n - 1 + v)^{-1/2}T_{Nv4}$ are given for $n = 3(1)10$, 12, 15, 20 and $v = 0(1)10$, 12, 15, 20, 24, 30, 40, 50. An error in Quesenberry and David's Table 3 (the entry for $n = 7$, $v = 4$) has been corrected.

*References:* Kudo (1956a), Quesenberry and David (1961).

*Properties of test:* For a location-slippage alternative in which one observation arises from a normal distribution $N(\mu + a, \sigma^2)$, $a \neq 0$, **Nv4** has the optimal property of being the scale- and location-invariant test of given size which maximizes the probability of identifying the contaminant as discordant.

**Nv5** *Discordancy test of* $k(\geqslant 2)$ *upper outliers* $x_{(n-k+1)}, \ldots, x_{(n-1)}, x_{(n)}$ *in a normal sample with* $\mu$ *unknown and an independent estimate of* $\sigma^2$ *known*

*Test statistic:*

$T_{\text{Nv5}}$ = sum of jointly (externally and internally) studentized deviations from the mean

$$= \frac{x_{(n-k+1)} + \ldots + x_{(n-1)} + x_{(n)} - k\bar{x}}{\tilde{s}}.$$

*Inequality:*

$$SP(t) \leqslant \binom{n}{k} P\left(\mathbf{t}(n-2+v) > \left[\frac{n(n-2+v)t^2}{k(n-k)(n-1+v) - nt^2}\right]^{1/2}\right).$$

This is an equality when $t \geqslant [k^2(n-1+v)(n-k-1)/(nk+n)]^{1/2}$.

*References*: McMillan (1971), Fieller (1976).

*Properties of test*: As for **N3**.

**Nv6** *Discordancy test of a lower and upper outlier-pair* $x_{(1)}, x_{(n)}$ *in a normal sample with* $\mu$ *unknown and an independent estimate of* $\sigma^2$ *known*

*Test statistic:*

$$T_{\text{Nv6}} = \text{externally studentized range} = \frac{x_{(n)} - x_{(1)}}{s_v}.$$

*Inequality:*

$$SP(t) < n(n-1)P(\mathbf{t}(v)) > t\sqrt{2}).$$

*Tabulated significance levels:* Table XVIIc, pages 495–6; abridged from Pearson and Hartley (1966), Table 29, pages 191–193, where 10, 5, and 1 per cent points are given for $n = 2(1)20$ and $v = 1(1)20$, 24, 30, 40, 60, 120, $\infty$.

*Further tables:* Harter (1969a) gives upper and lower 0.1, 0.5, 1, 2.5, 5, 10(10)40 per cent points and the median, and extends the sample sizes to $n = 22(2)40(10)100$.

*References:* Dixon (1950), Thompson (1955), Moore (1957), David (1962), Harter (1969a), Fieller (1976).

**Nv7** *Discordancy test of a lower and upper outlier-pair* $x_{(1)}$, $x_{(n)}$ *in a normal sample with* $\mu$ *unknown and an independent estimate of* $\sigma^2$ *known*

*Test statistic:*

$$T_{\text{Nv}7} = \text{externally and internally studentized range} = \frac{x_{(n)} - x_{(1)}}{\tilde{s}}.$$

*Inequality:*

$$SP(t) \leq n(n - 1)P(\mathbf{t}(n - 2 + v)) > [(n - 2 + v)t^2/(2n - 2 + 2v - t^2)]^{1/2}).$$

This is an equality when $t \geq [\frac{3}{2}(n - 1 + v)]^{1/2}$.

*Reference:* Fieller (1976).

**Nv8** *Two-sided discordancy test of $k$ outliers (irrespective of directions) in a normal sample with $\mu$ unknown and an independent estimate of $\sigma^2$ known*

*See:* worksheet for **N16** under the heading *Properties of test*.

*Reference:* Hawkins (1979).

**N$\mu$1** *Discordancy test of a single upper outlier* $x_{(n)}$ *in a normal sample with* $\mu$ *known and* $\sigma^2$ *unknown*

*Test statistic:*

$$T_{\text{N}\mu1} = \frac{x_{(n)} - \mu}{s(\mu)}.$$

*Recurrence relationship:*

$$f_n(t) = \left(\frac{n}{\pi}\right)^{1/2} \frac{\Gamma\left(\dfrac{n}{2}\right)}{\Gamma\left(\dfrac{n-1}{2}\right)} \left(1 - \frac{t^2}{n}\right)^{(n-3)/2}$$

$$\times F_{n-1}\left(\left[\frac{(n-1)t^2}{n - t^2}\right]^{1/2}\right) \quad (-\sqrt{n} \leq t \leq \sqrt{n})$$

with $F_1(t) = 0 \, (t < -1)$, $\frac{1}{2}(-1 < t < 1)$, $1 \, (t > 1)$.

*Tabulated significance levels:* Table XIIIc, page 485; specially compiled on the basis of simulations of sizes 10 000.

*Properties of test:* Note that $T_{\text{N}\mu1}$ can take negative values. The statistic $S_n^2(\mu)/S^2(\mu) = 1 - (1/n)T_{\text{N}\mu1}^2$ is therefore not equivalent to $T_{\text{N}\mu1}$, in contradistinction to the one-one relationship between $S_n^2/S^2$ and $T_{\text{N}1}$. The occurrence of a negative value for $T_{N\mu1}$ has probability $1/2^n$ on the working hypothesis and is therefore rare except for very small samples.

**N$\mu$2** *Two-sided discordancy test of an extreme outlier in a normal sample with $\mu$ known and $\sigma^2$ unknown*

*Test statistic:*

$$T_{N\mu2} = \max\left(\frac{|x_{(n)} - \mu|}{s(\mu)}, \frac{|\mu - x_{(1)}|}{s(\mu)}\right).$$

*Recurrence relationship:*

$$f_n(t) = nb_{1/2,(n-1)/2}(t)F_{n-1}[t/(1-t)] \quad (0 \leqslant t \leqslant 1).$$

*Inequality:*

$$SP(t) \leqslant 2nP(\mathbf{t}(n-1)) > [(n-1)t/(1-t)]^{1/2}).$$

This is an equality when $t \geqslant \frac{1}{2}$.

*Tabulated significance levels:* Table XIIId, page 485; derived from Eisenhart, Hastay, and Wallis (1947), Tables 15.1 and 15.2, pages 390–391, where 5 and 1 per cent points of $(T_{N\mu2})^2/n$ are given (as part of a larger table) for $n = 2(1)10, 12, 15, 20, 24, 30, 40, 60, 120$.

*References:* Cochran (1941), Fieller (1976), Lewis and Fieller (1979).

*Properties of test:* Maximum likelihood ratio test when the alternative is that one observation arises from a normal distribution $N(\mu, b\sigma^2)$, $b > 1$. Note that $(T_{N\mu2})^2/n$ has the same distribution as $T_{Ga1}$ with $r = \frac{1}{2}$.

**N$\mu$3** *Discordancy test of $k(\geqslant 2)$ upper outliers $x_{(n-k+1)}, \ldots, x_{(n)}$ in a normal sample with $\mu$ known and $\sigma^2$ unknown*

*Test statistic:*

$$T_{N\mu3} = \frac{x_{(n-k+1)} + \ldots + x_{(n-1)} + x_{(n)} - k\mu}{s(\mu)}.$$

*Inequality:*

$$SP(t) \leqslant \binom{n}{k}P(\mathbf{t}(n-1)) > [(n-1)t^2/(nk-t^2)]^{1/2}).$$

*Tabulated significance levels:* Table XVc, page 492, specially compiled on the basis of simulations of sizes 10000.

**N$\mu$4 Discordancy test of two upper outliers $x_{(n-1)}$, $x_{(n)}$ in a normal sample with $\mu$ known and $\sigma^2$ unknown**

*Test statistic:*

$$T_{N\mu4} = S_{n-1,n}^2(\mu)/S^2(\mu).$$

*Tabulated significance levels:* Table XVd, page 492; specially compiled on the basis of simulations of sizes 10 000. Values of $T_{N\mu4}$ *smaller* than the tabulated level are significant.

**N$\mu$5 Discordancy test of a lower and upper outlier-pair $x_{(1)}$, $x_{(n)}$ in a normal sample with $\mu$ known and $\sigma^2$ unknown**

*Test statistic:*

$$T_{N\mu5} = S_{1,n}^2(\mu)/S^2(\mu).$$

*Tabulated significance levels:* Table XVIb, page 493, specially compiled on the basis of simulations of sizes 10 000. Values of $T_{N\mu5}$ *smaller* than the tabulated level are significant.

**N$\mu$6 Discordancy test of one or more outliers (irrespective of their directions) in a normal sample with $\mu$ known and $\sigma^2$ unknown**

*Test statistic:*

$T_{N\mu6}$ = sample kurtosis based on deviations from $\mu$

$$= \frac{\sum_{j=1}^{n}(x_j - \mu)^4}{ns^4(\mu)}.$$

The value tested for discordancy is $x_{(n)}$ or $x_{(1)}$, according as $|x_{(n)} - \mu|$ is greater or less than $|\mu - x_{(1)}|$. The presence of a discordant value is indicated by a *high* value of the test statistic. For more than one outlier, apply test consecutively.

*Tabulated signficance levels:* Table XXc, page 500, specially compiled on the basis of simulations of sizes 10000.

*Reference:* Ferguson (1961a).

*Properties of test:* **N$\mu$6** is the locally best invariant test of given size against a location-slippage alternative in which $k$ of the $n$ observations arise from separate normal distributions $N(\mu + a_1, \sigma^2), \ldots, N(\mu + a_k, \sigma^2)$, $a_1 \neq 0, \ldots, a_k \neq 0$, provided that the contamination proportion $k/n$ under the alternative hypothesis is less than $\frac{1}{3}$ (strictly, provided that $k < \frac{1}{3}(n + 2)$).

The test is suitable for consecutive use in the possible presence of more than one outlier.

**N μ 14** *Discordancy test of k outliers (irrespective of directions) in a normal sample with μ known and $\sigma^2$ unknown*

*Test statistic:*

$$T_{N\mu14} = \frac{\sum_{j=n-k+1}^{n} d_{(j)}^2}{S^2(\mu)}$$

where $d_j = |x_j - \mu|$, the absolute deviation of $x_j$ from the population mean; and $\{d_{(j)}\}$ are the values of the $d_j$ in ascending order, $d_{(1)} < d_{(2)} < \ldots < d_{(n)}$.

*Inequality:*

$$SP(t) < \binom{n}{k} P\left( \mathbf{F}(k, n-k) > \frac{(n-k)t}{k(1-t)} \right).$$

*References:* Fieller (1976), Lewis and Fieller (1979).

*Properties of test:* **N μ 14** is the maximum likelihood ratio test for an alternative in which $k$ of the $n$ observations arise from a common normal distribution $N(\mu, b\sigma^2)$, $b > 1$.

The test statistic $T_{N\mu14}$ has the same distribution as $T_{Ga5}$ with $r = \frac{1}{2}$.

**N σ 1** *Discordancy test of a single upper outlier $x_{(n)}$ in a normal sample with $\sigma^2$ known and μ unknown*

*Test statistic:*

$$T_{N\sigma1} = \text{standardized extreme deviation from the mean} = \frac{x_{(n)} - \bar{x}}{\sigma}.$$

*Test distribution:*

$$F_n(t) = \exp\left\{ -\frac{1}{2n} \frac{d^2}{dt^2} \right\} [\Phi(t)]^n.$$

*Recurrence relationship:*

$$f_n(t) = \left( \frac{n^3}{2\pi(n-1)} \right)^{1/2} \exp\left( -\frac{1}{2} \frac{nt^2}{(n-1)} \right) F_{n-1}\left( \frac{nt}{n-1} \right),$$

with $f_2(t) = 2\exp(-t^2)/\sqrt{\pi}$.

*Inequality:*

$$SP(t) < n\Phi[-n^{1/2}t/(n-1)^{1/2}].$$

*Tabulated significance levels:* Table XIIIe, page 486; abridged from Grubbs (1950), Table III, page 45, where 10, 5, 1, and 0.5 per cent points are given for $n = 2(1)25$; values for $n = 30, 40, 50, 100$ specially calculated.

*Further tables:* A table in Nair (1948), also in Pearson and Hartley (1966), gives lower and upper 10, 5, 2.5, 1, 0.5, and 0.1 per cent points for $n = 3(1)9$. Dixon (1950) gives graphs of performance measure $P_3$. David (1956b) gives a table of values of power $P_1$ against the alternative that one observation arises from a normal distribution $N(\mu + a, \sigma^2)$, $a > 0$, for $a/\sigma = 1(1)4$ and $n = 3(1)10, 12, 15, 20, 25$. David and Paulson (1965) correct some errors in this table. McMillan and David (1971) give graphs of performance measures $P(C_1)$, $P(C_2)$, $P(C_3)$ (page 139) for the consecutive use of $N\sigma 1$ in testing two upper outliers.

*References:* McKay (1935), Nair (1948), Grubbs (1950, 1969), Dixon (1950, 1962), David (1956b), Kudo (1956a), Ferguson (1961b), McMillan and David (1971), Fieller (1976).

*Properties of test:* $N\sigma 1$ is the maximum likelihood ratio test for the above-stated alternative of slippage in location by $a > 0$ for one observation. For this alternative, it has the optimal property of being the scale-and location-invariant test of given size which maximizes the probability $P_3$ of identifying the contaminant as discordant. For the same alternative, some typical values of power $P_1$ are as shown in Table 6.7 (David and Paulson, 1965).

**Table 6.7**

| | $a/\sigma$ $n$ | 2 | 3 | 4 | 5 |
|---|---|---|---|---|---|
| Test at 5% | 3 | 0.31 | 0.63 | 0.87 | 0.98 |
| significance | 10 | 0.27 | 0.62 | 0.89 | 0.99 |
| level | 25 | 0.21 | 0.54 | 0.86 | 0.98 |
| Test at 1% | 3 | 0.14 | 0.40 | 0.71 | 0.92 |
| significance | 10 | 0.12 | 0.41 | 0.76 | 0.95 |
| level | 25 | 0.09 | 0.35 | 0.72 | 0.94 |

Unlike test **N1**, **N$\sigma$1** can be used effectively when there is more than one contaminant, being relatively unaffected by the risk of masking.

**N$\mu\sigma$1** *Discordancy test of a single upper outlier* $x_{(n)}$ *in a normal sample with both* $\mu$ *and* $\sigma^2$ *known*

*Test statistic:*

$T_{N\mu\sigma 1}$ = standardized extreme deviation from the population mean

$$= \frac{x_{(n)} - \mu}{\sigma}.$$

*Test distribution:*

$$F_n(t) = [\Phi(t)]^n.$$

*Tabulated significance levels:* Table XIIIg, page 486; the entries for sample sizes up to $n = 30$ are extracted from Pearson and Hartley (1966), Table 24, page 184, where lower and upper 10, 5, 2.5, 1, 0.5, and 0.1 per cent points are given for $n = 1(1)30$; the entries for $n > 30$ have been specially compiled.

*Reference:* Dixon (1962).

*Properties of test:* Maximum likelihood ratio test when the alternative is that one observation arises from a normal distribution $N(\mu + a, \sigma^2)$, $a > 0$.

**N$\sigma$2 Two-sided discordancy test of an extreme outlier in a normal sample with $\sigma^2$ known and $\mu$ unknown**

*Test statistic:*

$$T_{N\sigma2} = \max\left(\frac{x_{(n)} - \bar{x}}{\sigma}, \frac{\bar{x} - x_{(1)}}{\sigma}\right).$$

*Inequality:*

$$SP(t) < 2P(T_{N\sigma1} > t).$$

*Tabulated significance levels:* Table XIIIf, page 486; extracted from Halperin et al. (1955), Tables 1 and 2, pages 187–188 ($v = \infty$; see Worksheet Nv3)

*Reference:* Kudo (1956a).

*Properties of test:* As for **N2**, except that **N$\sigma$2** (unlike **N2**) is relatively unaffected by masking from other outliers.

**N$\sigma$3 Discordancy test of $k(\geqslant 2)$ upper outliers $x_{(n-k+1)}, \ldots, x_{(n-1)}, x_{(n)}$ in a normal sample with $\sigma^2$ known and $\mu$ unknown**

*Test statistic:*

$T_{N\sigma3} = $ sum of standardized deviations from the mean

$$= \frac{x_{(n-k+1)} + \ldots + x_{(n-1)} + x_{(n)} - k\bar{x}}{\sigma}.$$

*Inequality:*

$$SP(t) < \binom{n}{k}\Phi[-n^{1/2}t/(kn - k^2)^{1/2}].$$

*Tabulated significance levels:* Table XVe, page 492; values for $k = 2$, $n \leqslant 20$ extracted from McMillan and David (1971), Table I, page 82, where 5 and 1 per cent points are given for $n = 4(1)27$; values for $k = 2$, $n \geqslant 30$ and for $k = 3$ and $k = 4$ specially compiled on the basis of simulations of sizes 10 000.

*Further tables:* McMillan and David (1971) give graphs of the performance measure $P_2$ for the case $k = 2$.

*References:* Kudo (1956a), McMillan and David (1971), Fieller (1976).

*Properties of test:* As for **N3**.

**N$\mu\sigma$3** *Discordancy test of* $k (\geqslant 2)$ *upper outliers* $x_{(n-k+1)}, \ldots, x_{(n-1)}, x_{(n)}$ *in a normal sample with both* $\mu$ *and* $\sigma^2$ *known*

*Test statistic:*

$T_{\mathrm{N}\mu\sigma3}$ = sum of standardized deviations from the population mean

$$= \frac{x_{(n-k+1)} + \ldots + x_{(n-1)} + x_{(n)} - k\mu}{\sigma}$$

*Tabulated significance levels:* Table XVg, page 493; specially compiled on the basis of simulations of sizes 10 000.

**N$\sigma$4** *Discordancy test of two upper outliers* $x_{(n-1)}, x_{(n)}$ *in a normal sample with* $\sigma^2$ *known and* $\mu$ *unknown*

*Test statistic:*

$$T_{\mathrm{N}\sigma4} = S^2_{n-1,n}/\sigma^2.$$

*Tabulated significance levels:* Table XVf, page 492, specially compiled on the basis of simulations of sizes 10000. Values of $T_{\mathrm{N}\sigma4}$ *smaller* than the tabulated level are significant.

**N$\mu\sigma$4** *Discordancy test of two upper outliers* $x_{(n-1)}, x_{(n)}$ *in a normal sample with both* $\mu$ *and* $\sigma^2$ *known*

$$T_{\mathrm{N}\mu\sigma4} = S^2_{n-1,n}(\mu)/\sigma^2.$$

*Tabulated significance levels:* Table XVh, page 493; specially compiled on the basis of simulations of sizes 10 000. Values of $T_{\mathrm{N}\mu\sigma4}$ *smaller* than the tabulated level are significant.

**Nσ5** *Discordancy test of a lower and upper outlier-pair* $x_{(1)}$, $x_{(n)}$ *in a normal sample with* $\sigma^2$ *known and* $\mu$ *unknown*

*Test statistic:*

$$T_{N\sigma5} = S_{1,n}^2/\sigma^2.$$

*Tabulated significance levels:* Table XVIc, page 493; specially compiled on the basis of simulations of sizes 10 000. Values of $T_{N\sigma5}$ *smaller* than the tabulated level are significant.

**Nμσ5** *Discordancy test of a lower and upper outlier-pair* $x_{(1)}$, $x_{(n)}$ *in a normal sample with both* $\mu$ *and* $\sigma^2$ *known*

*Test statistic:*

$$T_{N\mu\sigma5} = S_{1,n}^2(\mu)/\sigma^2.$$

*Tabulated significance levels:* Table XVId, page 493; specially compiled on the basis of simulations of sizes 10 000. Values of $T_{N\mu\sigma5}$ *smaller* than the tabulated level are significant.

**Nσ6(Nμσ6)** *Discordancy test of a lower and upper outlier-pair* $x_{(1)}$, $x_{(n)}$ *in a normal sample with* $\sigma^2$ *known*

*Test statistic:*

$$T_{N\sigma6} = \text{standardized range} = \frac{x_{(n)} - x_{(1)}}{\sigma}.$$

*Test distribution:*

$$F_n(t) = n \int_{-\infty}^{\infty} \phi(x)[\Phi(x) - \Phi(x - t)]^{n-1}\,dx.$$

*Tabulated significance levels:* Table XVIIb, page 494; abridged from Harter (1969a), Table A7, pages 372–374, where lower and upper 0.01, 0.05, 0.1, 0.5, 1, 2.5, 5, 10(10)40 per cent points and the 50 per cent point are given for $n = 2(1)20(2)40(10)100$.

*Further tables:* Dixon (1950) gives graphs of performance measure $P_3$.

*References:* Tippett (1925), Pearson (1926, 1932), Pearson and Hartley (1942), Dixon (1950, 1962), Harter (1969a).

**N$\sigma$7(N$\mu\sigma$7)** *Discordancy test of a single upper outlier* $x_{(n)}$ *in a normal sample with* $\sigma^2$ *known*

*Test statistic:*

$$T_{N\sigma7} = \frac{x_{(n)} - x_{(n-1)}}{\sigma}.$$

*Test distribution:*

$$F_n(t) = 1 - n \int_{-\infty}^{\infty} \phi(x + t)[\Phi(x)]^{n-1}dx.$$

*Tabulated significance levels:* Table XIXh, page 499; derived from Irwin (1925), Table II, page 239, where values of $F_n(t)$ are given for $t = 0.1(0.1)5.0$ and $n = 2$, 3, 10(10)100(100)1000.

*Further tables:* Dixon (1950) gives graphs of performance measure $P_3$.

*References:* Irwin (1925), Dixon (1950).

*Properties of test:* Analogous in concept to a Dixon-type test. Performance $P_3$ is comparable to that of test **N$\sigma$1** if there is just one contaminant, but compares unfavourably with **N$\sigma$1** if there is more than one contaminant, owing to incidence of masking by $x_{(n-1)}$. However, **N$\sigma$7** could be a useful test for a *lower* outlier (with test statistic $[x_{(2)} - x_{(1)}]/\sigma$) in a life-test data situation in which, for practical reasons, only the shortest lifetimes were actually observed.

**N$\sigma$8(N$\mu\sigma$8)** *Discordancy test of an upper outlier-pair* $x_{(n-1)}$, $x_{(n)}$ *in a normal sample with* $\sigma^2$ *known*

*Test statistic:*

$$T_{N\mu\sigma8} = \frac{x_{(n-1)} - x_{(n-2)}}{\sigma}.$$

*Test distribution:*

$$F_n(t) = 1 - n(n-1) \int_{-\infty}^{\infty} \phi(x + t)[1 - \Phi(x + t)][\Phi(x)]^{n-2}dx.$$

*Tabulated significance levels:* Table XIXi, page 499; derived from Irwin (1925), Table III, page 242, where values of $F_n(t)$ are given for $t = 0.1(0.1)2.0$ and $n = 3$, 10(10)100(100)1000, and also for $t = 2.1(0.1)4.0$ in the case $n = 3$ (a sample size obviously irrelevant in the present context).

*Reference:* Irwin (1925).

*Properties of test:* Analogous to a Dixon-type test. As with **Nσ7**, could be an appropriate test in a life-testing context. **Nσ7** and **Nσ8** have a historical standing as being two of the earliest published tests in modern outlier methodology.

**Nσ9(Nμσ9)** *Discordancy test of k lower and k upper outliers* ($k \geqslant 2$) *in a normal sample with* $\sigma^2$ *known*

*Test statistic:*

$$T_{N\sigma9} = (k - 1)\text{th standardized quasi-range} = \frac{x_{(n-k+1)} - x_{(k)}}{\sigma}.$$

*Tabulated significance levels:* Table XVIII, page 497; abridged from Harter (1969b), Table A7, pages 295–319, where lower and upper 0.01, 0.05, 0.1, 0.5, 1, 2.5, 5, 10(10)40 per cent points and the 50 per cent point are given for $k = 1(1)9$ and $n = 2k(1)20(2)40(10)100$.

*Reference:* Harter (1969b).

*Properties of test:* Advantage is taken here of Harter's very extensive tables of quasi-ranges to provide a test for discordancy in the tails of a large sample whose main central mass can be assumed normal.

### 6.4 DISCORDANCY TESTS FOR SAMPLES FROM OTHER DISTRIBUTIONS

Many of the discordancy tests given in Section 6.2 for exponential samples can be used for samples from Pareto distributions and from distributions of asymptotic extreme-value type (i.e. Gumbel, Fréchet, and Weibull distributions), by simple transformation of the data. Likewise, various discordancy tests given in Section 6.3 for normal samples can be used for samples from log-normal distributions. Since a square root transformation converts a Poisson random variable into a variable distributed approximately normally with variance $\frac{1}{4}$, whatever the value of the Poisson mean provided it is not too small, the **Nσ** tests in Section 6.3 can be used for samples from Poisson distributions. Similarly, the **Nσ** tests can be used for samples from binomial distributions under appropriate conditions.

Apart from the use of transformations, a few specific discordancy tests are available for Poisson and binomial samples and for samples from other distributions.

Details of discordancy tests for the various distributions are given below in the following sections:

### 6.4.1 Log-normal samples

If $X$ is a log-normal random variable with parameters $\mu$ and $\sigma$, $\log X$ is $N(\mu, \sigma^2)$. Thus as a discordancy test for any outlier or outliers in a log-normal sample, we need only take the logarithms of the observations and apply an appropriate normal sample test (from those listed in Section 6.3) to the transformed sample.

No such facility is available for the generalized three- parameter log-normal distribution, in which $\log(X - \xi)$ is $N(\mu, \sigma^2)$, $\xi$ being an unknown location parameter.

### 6.4.2 Truncated exponential samples

If the exponential distribution with density $(1/\lambda)\exp(-x/\lambda)\,(x > 0)$ is truncated at $x = a$ we get the distribution with density $(1/\lambda) \times [1 - \exp(-a/\lambda)]^{-1} \exp(-x/\lambda)$ for $0 < x < a$ and $0$ for $x > a$. Such a distribution might arise, for example, in life testing data where for practical reasons the test is not allowed to continue for longer than some preassigned duration $a$. Upper outlier problems are perhaps of limited interest with such data, in view of the truncation, but two Dixon-type discordancy tests are available in the literature (Wani and Kabe, 1971), and we give details of these below. The first is for a single upper outlier, the second (more useful one) for a single *lower* outlier.

**TE1** *Discordancy test of a single upper outlier* $x_{(n)}$ *in a truncated exponential sample*

*Test statistic:*

$$T_{\text{TE1}} = \frac{\text{excess}}{\text{range}} = \frac{x_{(n)} - x_{(n-1)}}{x_{(n)} - x_{(1)}}.$$

*Test distribution:*

$$f_n(t) = \frac{(n-1)!}{(1 - e^{-a})^n} \sum_{j=2}^{n-1} \frac{(-)^{n+j}}{(j-2)!(n-j)!} \left\{ \frac{1}{u_j^2} - \frac{1}{(n-u_j)^2} e^{-na} \right.$$

$$\left. - \left[ \frac{1}{u_j^2} - \frac{1}{(n-u_j)^2} + \frac{na}{u_j(n-u_j)} \right] e^{-au_j} \right\}$$

where $u_j = n - j + 1 - (n - j)t$, $(0 \leq t \leq 1)$.

*Reference:* Wani and Kabe (1971).

**TE2** *Discordancy test of a single lower outlier* $x_{(1)}$ *in a truncated exponential sample*

*Test statistic:*

$$T_{TE2} = \frac{\text{excess}}{\text{range}} = \frac{x_{(2)} - x_{(1)}}{x_{(n)} - x_{(1)}}.$$

*Test distribution:*

$$f_n(t) = \frac{(n-1)!}{(1 - e^{-a})^n} \sum_{j=3}^{n} \frac{(-)^{n+j}}{(j-3)!(n-j)!} \left\{ \frac{1}{u_j^2} - \frac{1}{(n-u_j)^2} \right\} e^{-na}$$

$$- \left[ \frac{1}{u_j^2} - \frac{1}{(n-u_j)^2} + \frac{na}{u_j(n-u_j)} \right] e^{-au_j} \right\}$$

where $u_j = n - j + 1 + (j-2)t$, $(0 \leqslant t \leqslant 1)$.

*Reference:* Wani and Kabe (1971).

### 6.4.3  Uniform samples

Denote the lower and upper bounds of the uniform distribution by $a$, $b$ respectively, so that its density is $1/(b - a)$ for $a < x < b$, and 0 otherwise. Given the ordered sample $x_{(1)} < x_{(2)} < \ldots < x_{(n)}$, it is well known that the $n + 1$ intervals

$$x_{(1)} - a, \ x_{(2)} - x_{(1)}, \ \ldots, \ x_{(n)} - x_{(n-1)}, \ b - x_{(n)}$$

are distributed as $n + 1$ independent exponential random variables with a common (arbitrary) scale parameter, conditional upon their sum having the constant value $b - a$. For testing outliers, therefore, we can use the fact that the ratio of the means of any two non-intersecting combinations of these intervals will have an F-distribution on the working hypothesis. This leads at once to a general Dixon-type discordancy test (labelled **U**) applicable to any combination of lower and upper outliers. If $a$ and $b$ are unknown, the statistic for testing $h$ lower and $k$ upper outliers ($h \geqslant 0$, $k \geqslant 0$, $h + k > 0$) is

$$T_U = \frac{x_{(n)} - x_{(n-k)} + x_{(h+1)} - x_{(1)}}{x_{(n-k)} - x_{(h+1)}} \cdot \frac{n - k - h - 1}{k + h}.$$

In principle, no tabulated significance levels are required, other than those provided in standard F-tables, since

$$SP(t) = P(\mathbf{F}(2(k + h), 2(n - k - h - 1)) > t).$$

For example, if $k$ upper outliers are to be tested for discordancy ($k \geqslant 1$), $T_U$ would take the form

$$\frac{x_{(n)} - x_{(n-k)}}{x_{(n-k)} - x_{(1)}} \cdot \frac{n - k - 1}{k}$$

with test distribution $\mathbf{F}(2k, 2(n - k - 1))$.

However, for convenience we provide in Table XXIV, page 506, 5 per cent and 1 per cent critical values for testing the following frequently used forms of $T_U$:

$$\frac{x_{(n)} - x_{(n-1)}}{x_{(n)} - x_{(1)}}; \quad \frac{x_{(n)} - x_{(n-2)}}{x_{(n)} - x_{(1)}}; \quad \frac{x_{(n)} - x_{(n-2)}}{x_{(n)} - a} \quad (a \text{ known}).$$

These are reproduced from Barnett and Roberts (1993), Table II.

If $a$ is known in this case (but not $b$), it is preferable to use this extra information and employ the statistic

$$\frac{x_{(n)} - x_{(n-k)}}{x_{(n-k)} - a} \cdot \frac{n - k}{k}$$

with test distribution $\mathbf{F}(2k, 2(n - k))$.

Similarly, if $b$ is known (but not $a$) and $h$ lower outliers are to be tested, the statistic

$$\frac{x_{(h+1)} - x_{(1)}}{b - x_{(h+1)}} \cdot \frac{n - h}{h}$$

should be used, with test distribution $\mathbf{F}(2h, 2(n - h))$.

If both lower and upper outliers are to be tested ($h > 0$, $k > 0$), the above statistic $T_U$ should be used even if $a$ and/or $b$ are known.

*Example 6.4. Crushed rock is widely used for road and rail bases. In assessing the usefulness of any particular type and source of rock for this purpose, one of the measures of durability is the Sulphate Soundness Test (SST) value, which can be determined for any individual specimen of rock. Denote the SST value of a specimen by X. Table 6.8 gives the ordered values $x_{(1)}, x_{(2)}, \ldots, x_{(12)}$ of X for 12 specimens of rock of a particular type from a particular quarry. The data in this example are due to the Department of Main Roads, New South Wales, Australia, who gave kind permission for their use.*

**Table 6.8**

| $j$ | 1 | 2 | 3 | 4 | 5 | 6 | 7 | 8 | 9 | 10 | 11 | 12 |
|-----|-----|-----|-----|-----|-----|-----|-----|-----|-----|-----|-----|-----|
| $x_{(j)}$ | 0.8 | 1.0 | 1.0 | 2.1 | 2.6 | 2.6 | 6.5 | 6.7 | 7.7 | 8.6 | 9.6 | 52.8 |

$x_{(12)} = 52.8$ *is evidently an outlier.*

*The distribution of X-values for specimens from the same type of rock and the same quarry is skew, and analysis of a number of data sets shows that the corresponding distribution of the transformed random variable $U = 1/\sqrt{X}$ can reasonably be assumed to be* uniform. *As a typical example, 23 specimens of rock of a certain type from a certain quarry gave the X-values in Table 6.9 and the U-values shown diagrammatically in Figure 6.1. The sample skewness and kurtosis of the 23 U-values are respectively $0.15, -1.15$, in good agreement with the population values $\gamma_1 = 0$, $\gamma_2 = -1.2$ for a uniform distribution.*

**Table 6.9**

| 2.5 | 2.7 | 2.9 | 3.3 | 3.6 | 4.4 | 4.8 | 4.8 | 5.2 | 5.5 | 5.5 | 5.7 |
|-----|-----|-----|-----|-----|-----|-----|-----|-----|-----|-----|-----|
| 6.5 | 7.7 | 8.7 | 10.7 | 13.0 | 13.1 | 15.7 | 16.7 | 18.0 | 19.0 | 26.8 | |

**Figure 6.1** Distribution of values of $1/\sqrt{(SST)}$ for 23 specimens of rock

*The values $x_{(1)} = 0.8$, $x_{(11)} = 9.6$, $x_{(12)} = 52.8$ in Table 6.8 transform respectively to $u_{(12)} = 1.118$, $u_{(2)} = 0.323$, $u_{(1)} = 0.138$. On the working hypothesis, the 12 values $u_{(j)}$ constitute an ordered sample from a uniform distribution with unknown bounds $a < 0.138$, $b > 1.118$. Let us test $u_{(1)}$ as a lower outlier, using the test statistic $T_U$. The observed value of $T_U$ is*

$$t = \frac{u_{(2)} - u_{(1)}}{u_{(12)} - u_{(2)}} \cdot \frac{20}{2} = 2.33,$$

*and the significance probability of this value is*

$$SP(2.33) = P(\mathbf{F}(2, 20) > 2.33) = 0.12.$$

*So $u_{(1)}$ is not adjudged discordant, and the x-value 52.8 is quite consistent with the other eleven x-values in Table 6.8!*

This illustrates once again that the existence of an outlier is always *relative to a particular model*, and an observation may be outlying in relation to one model but consistent with the main data set in relation to another model.

### 6.4.4 Gumbel, Fréchet, and Weibull samples

The asymptotic extreme-value distributions of the first, second, and third types, in other words the Gumbel, Fréchet, and Weibull distributions, are well known as models for extreme observations such as annual maximum wind speeds, floods (as greatest-value phenomena), endurance limits in fatigue testing (as smallest-value phenomena), annual minimum temperatures, oldest ages of individuals in a population, and shortest lives of manufactured items. In analysing extreme-value data it is obviously important to remove where possible the biasing effect of any contaminants which may be present, and the testing of outliers for discordancy is of particular relevance.

The Gumbel distribution, often called '*the* extreme-value distribution', depends on a location parameter $a$ and a positive scale parameter $b$; the Fréchet and Weibull distributions depend on a location parameter $\alpha$ and a positive scale parameter $\beta$ and also on a positive shape parameter $r$. In terms of these parameters their distribution functions $P(X < x)$ are given in Table 6.10. Each distribution has two forms, according as it relates to greatest-value or least-value extremes. We will denote the Gumbel greatest-value and least-value distributions by $\mathbf{GU}_{max}(a, b)$ and $\mathbf{GU}_{min}(a, b)$ respectively. It can be shown that if the shape parameter is reparametrized as $\lambda = 1/r$ (Weibull), $\lambda = -1/r$ (Fréchet), the Gumbel distribution corresponds to the limiting case $\lambda \to 0$.

**Table 6.10**

| | Distribution of greatest values | Distribution of least values |
|---|---|---|
| Gumbel | $\exp[-e^{-(x-a)/b}], \; -\infty < x < \infty$ | $1 - \exp[-e^{-(a-x)/b}], \; -\infty < x < \infty$ |
| Fréchet | $\exp\left[-\left(\dfrac{x-\alpha}{\beta}\right)^{-r}\right], \; x > \alpha$ | $1 - \exp\left[-\left(\dfrac{\alpha-x}{\beta}\right)^{-r}\right], \; x < \alpha$ |
| Weibull | $\exp\left[-\left(\dfrac{\alpha-x}{\beta}\right)^{r}\right], x < \alpha$ | $1 - \exp\left[-\left(\dfrac{x-\alpha}{\beta}\right)^{r}\right], x > \alpha$ |

If $X$ has a $\mathbf{GU}_{min}(a, b)$ distribution, the transformed random variable $Y = \exp(X/b)$ has an exponential distribution with origin 0 and mean $\exp(a/b)$. If we know the value of $b$ we can test the discordancy of an outlier or a set of outliers in a sample from the $X$-distribution by transforming each observed value $x_i$ to $y_i = \exp(x_i/b)$ and using on the $y$'s a discordancy test for an exponential sample with origin 0.

Correspondingly, if $X$ has a $\mathbf{GU}_{max}(a, b)$ distribution, $Y = \exp(-X/b)$ has an exponential distribution with origin 0 and mean $\exp(-a/b)$, so if $b$ is known we can test the discordancy of one or more outliers in a sample from the $X$-distribution by transforming each $x_i$ to $y_i = \exp(-x_i/b)$ and using an exponential discordancy test on the $y$'s. In this case an upper outlier $x_{(n)}$ in the $x$-sample converts to a lower outlier $y_{(1)}$ in the $y$-sample, so the test on the $y$-values must be chosen accordingly.

If neither $a$ nor $b$ are known, various discordancy test criteria with location and scale invariance can be constructed, such as Dixon statistics.

Thirteen such tests, each for $k = 1$, 2 or 3 outliers, have been studied by Fung and Paul (1985), significance levels calculated for them where not already known, and their performances compared; worksheets for the best of these tests are given below, coded as **GU1–GU6**. Tests **GU3** and **GU6** can in fact be used for more than three outliers, though their properties for such values of $k$ have not been studied. To save space, all tests given here are for the least-value Gumbel distribution $GU_{min}$; to test $k$ *upper* (lower) outliers for discordancy in a greatest-value sample one uses a test of $k$ *lower* (upper) outliers in a least-value sample.

*Weibull and Fréchet samples*

The test procedures for Gumbel samples can be used for Weibull and Fréchet samples with parameters $\alpha$. $\beta$ $r$, *provided that the location parameter $\alpha$ is known*. If $W$ has a Weibull distribution and $F$ a Fréchet distribution (whether greatest-value or least-value), then

$$\log |W - \alpha| \text{ is distributed } GU_{min}\left(\log \beta, \frac{1}{r}\right) \qquad (6.4.1)$$

and $\qquad \log |F - \alpha| \text{ is distributed } GU_{max}\left(\log \beta, \frac{1}{r}\right) \qquad (6.4.2)$

So if the values of $\beta$ and $r$ are unknown, we can test the discordancy of one or more outliers in a Weibull sample by transforming each observed value $w_i$ to $x_i = \log |w_i - \alpha|$ and using on the $x$'s a **GU** discordancy test; similarly for a Fréchet sample.

Equivalently to (6.4.1) and (6.4.2),

$$|W - \alpha|^r \text{ is distributed exponentially with mean } \beta^r \qquad (6.4.3)$$

and $\qquad |F - \alpha|^{-r} \text{ is distributed exponentially with mean } \beta^{-r} \qquad (6.4.4)$

So in testing the discordancy of one or more outliers in a Weibull sample when $\beta$ is unknown but $\alpha$ and $r$ both known, we can bypass the use of **GU** discordancy tests and go straight to the wide range of available exponential tests by employing the transformation

$$y_i = |w_i - \alpha|^r; \text{ similarly, for a Fréchet sample}$$

$$y_i = |f_i - \alpha|^{-r}.$$

**GU1** *Discordancy test of $k$ = 1, 2, or 3 upper outliers in a Gumbel least-value sample with a and b unknown*

*Test statistic:*

$$T_{GU1} = \frac{\text{excess}}{\text{range}} = \frac{x_{(n)} - x_{(n-k)}}{x_{(n)} - x_{(1)}}.$$

*Tabulated significance levels:* Table XXVa, page 507, abridged from Fung and Paul (1985), Table I, pages 900–901, where 10, 5 and 1 per cent points, evaluated by simulation, are given for $n = 5(1)20$ when $k = 1$, $n = 8(1)20$ when $k = 2$ and $n = 10(2)20$ when $k = 3$.

*Further tables:* Fung and Paul (1985) give tables, based on simulations of size 10 000, of various performance measures for 13 tests including **GU1–GU6**, both for a location-slippage alternative and for an alternative with slippage in both location and scale. The performance measures are basically $P_3$ and $P_1$–$P_3$ (see pages 122–4) when $k = 1$ and extensions of these measures when $k = 2, 3$.

*References:* Fung and Paul (1985), Paul and Fung (1986).

*Properties of test:* **GU1**, **GU2** and **GU3** appear to be the best overall among the available tests for upper outliers, and have broadly comparable performances for each value of $k$ (1, 2, 3) and for the two slippage alternatives considered. **GU3** performs well consistently throughout the cases studied, but has the disadvantage of requiring a table of special constants for calculation of the test statistic; on the other hand, it has the advantage of being applicable for values of $k$ exceeding 3 (see Worksheet **GU3**). **GU2** has marginally the best performance when $k = 3$, and by and large the use of **GU2** can be recommended for values of $k$ and $n$ for which significance levels are available.

**GU2 *Discordancy test of $k$ = 1, 2, or 3 upper outliers in a Gumbel least-value sample with a and b unknown***

*Test statistic:*

$$T_{\mathrm{GU2}} = \frac{\text{reduced sum of squares}}{\text{total sum of squares}} = \frac{S_{n-k+1,\ldots,n-1,n}^2}{S^2}$$

(notation as on page 218).

*Tabulated significance levels:* Table XXVb, page 507, abridged from Fung and Paul (1985), Table I, page 900, where 10, 5 and 1 per cent points, evaluated by simulation, are given for $n = 5(1)20$ when $k = 1$, $n = 8(1)20$ when $k = 2$ and $n = 10(2)20$ when $k = 3$.

*Further tables:* See Worksheet **GU1**.

*Reference:* Fung and Paul (1985).

*Properties of test*: See Worksheet **GU1**.

**GU3** *Discordancy test of* $k$ = 1, 2, *or* 3 *upper outliers in a Gumbel least-value sample with a and b unknown*

*Test statistic:*

$$T_{GU3} = \left(\frac{n - k - 1}{k}\right)\left(\sum_{i=n-k}^{n-1} c_{n,i}[x_{(i+1)} - x_{(i)}]\right) \bigg/ \left(\sum_{i=1}^{n-k-1} c_{n,i}[x_{(i+1)} - x_{(i)}]\right)$$

where the $c_{n,i}$ are tabulated constants (Table XXVI, pages 508–9; derived from Mann, Scheuer, and Fertig (1973), Table I, pages 393–398, where values of $1/c_{n,i}$ are given for $n = 3(1)25$). In fact $1/c_{n,i} = E(x_{(i+1)} - x_{(i)})$ for a sample of size $n$ from $GU_{min}(0, 1)$.

*Test distribution:*

$$F_n(t) \simeq P(F(2k, 2n - 2k - 1) < t). \tag{6.4.5}$$

Theoretical considerations (see, e.g., Pyke (1965), Mann, Scheuer, and Fertig (1973), Fung and Paul (1985)) lead to the approximation

$$F_n(t) \simeq P(F(2k, 2n - 2k - 2) < t). \tag{6.4.6}$$

Fung and Paul (1985) calculate, by simulation, 10, 5 and 1 per cent points of the following statistic equivalent to $T_{GU3}$:

$$L = \left(\sum_{i=n-k}^{n-1} c_{n,i}[x_{(i+1)} - x_{(i)}]\right) \bigg/ \left(\sum_{i=1}^{n-1} c_{n,i}[x_{(i+1)} - x_{(i)}]\right), \tag{6.4.7}$$

for $n = 5(1)20$ when $k = 1$, $n = 8(1)20$ when $k = 2$ and $n = 10(2)20$ when $k = 3$. Their results indicate that our approximation (6.4.5) is better than (6.4.6) and gives good results throughout the parameter ranges covered by their tables. Incidentally, the following small corrections need to be made to Fung and Paul's paper: (1) page 899, definition of $L$, summation in numerator should start at $i = n - k$ not $n - k - 1$; (2) page 904 line 3, the degrees of freedom of the second $\chi^2$ should be $2(n - 1)$ not $2(n - 2)$; (3) page 904 line 5, the second Beta parameter should be $n - k - 1$ not $n - 2$.

*Further tables:* See above, and Worksheet **GU1**.

*References:* Mann, Scheuer, and Fertig (1973), Fung and Paul (1985).

*Properties of test:* See Worksheet **GU1**.
    The fact that (6.4.5) is a good approximation when $k = 1, 2, 3$ suggests that it can be used to extend the test to values of $k$ exceeding 3. However, no calculations have so far been done to check on this.

**GU4** *Discordancy test of k* = 1, 2, *or* 3 *lower outliers in a Gumbel least-value sample with a and b unknown*

*Test statistic:*

$$T_{GU4} = \frac{\text{excess}}{\text{range}} = \frac{x_{(k+1)} - x_{(1)}}{x_{(n)} - x_{(1)}}$$

*Tabulated significance levels:* Table XXVc, page 507, abridged from Fung and Paul (1985), Table I, pages 900–901, where 10, 5 and 1 per cent points, evaluated by simulation, are given for $n = 5(1)20$ when $k = 1$, $n = 8(1)20$ when $k = 2$ and $n = 10(2)20$ when $k = 3$.

*Further tables:* See Worksheet **GU1**.

*References:* Fung and Paul (1985), Paul and Fung (1986).

*Properties of test:* **GU5** and **GU6** appear to be the best overall among the available tests for lower outliers, and have broadly comparable performances for each value of $k$ (1, 2, 3) and for the two slippage alternatives considered. **GU4** performs well in the case of a single lower outlier ($k = 1$), and is comparable with **GU5** and **GU6** in this case.

**GU5 Discordancy test of** $k$ = 1, 2, **or** 3 **lower outliers in a Gumbel least-value sample with** *a* **and** *b* **unknown**

*Test statistic:*

$$T_{GU5} = \frac{\text{reduced sum of squares}}{\text{total sum of squares}} = \frac{S^2_{1,2,\ldots,k}}{S^2}$$

(notation as on page 218).

*Tabulated significance levels:* Table XXVd, page 507, abridged from Fung and Paul (1985), Table I, page 900, where 10, 5 and 1 per cent points, evaluated by simulation, are given for $n = 5(1)20$ when $k = 1$, $n = 8(1)20$ when $k = 2$ and $n = 10(2)20$ when $k = 3$.

*Further tables:* See Worksheet **GU1**.

*Reference:* Fung and Paul (1985).

*Properties of test:* See Worksheet **GU4**. In the sense detailed above, this test appears to be the 'best buy' for testing 1, 2 or 3 lower outliers.

**GU6 Discordancy test of $k$ = 1, 2, or 3 lower outliers in a Gumbel least-value sample with $a$ and $b$ unknown**

*Test statistic:*

$$T_{GU6} = \left(\frac{n - k - 1}{k}\right)\left(\sum_{i=1}^{k} c_{n,i}[x_{(i+1)} - x_{(i)}]\right)\Bigg/\left(\sum_{k+1}^{n-1} c_{n,i}[x_{(i+1)} - x_{(i)}]\right)$$

where the $c_{n,i}$, which are the same constants as in Worksheet **GU3**, are given in Table XXVI, pages 508–9.

*Test distribution:* As in Worksheet **GU3**.

Fung and Paul's simulated percentage points for $L$ in (6.4.7) apply equally to the statistic

$$L' = \left(\sum_{i=1}^{k} c_{n,i}[x_{(i+1)} - x_{(i)}]\right)\Bigg/\left(\sum_{i=1}^{n-1} c_{n,i}[x_{(i+1)} - x_{(i)}]\right)$$

equivalent to $T_{GU6}$, so the comments in Worksheet **GU3** regarding the use of (6.4.5) apply equally to **GU6**.

*Further tables and References:* See Worksheet **GU3**.

*Properties of test:* See Worksheet **GU4**.

The fact that (6.4.5) is a good approximation when $k = 1, 2, 3$ suggests that it can be used to extend the test to values of $k$ exceeding 3. However, no calculations have so far been done to check on this.

### 6.4.5 Pareto samples

In addition to its well known role in economics as a model for the distribution of incomes, the Pareto distribution can be used as a pragmatic model for other skew-distributed data characterized by a main mass of low values at one end and a gradation to a long tail of infrequently occurring high values at the other. In data of this nature, high-valued outliers requiring test may well arise.

A Pareto random variable $X$ is characterized by two parameters, the minimum value $a(a > 0)$, and the shape parameter $r(r > 0)$. Its distribution function can be written

$$P(X < x) = 0 \qquad (x \leqslant a)$$
$$= 1 - (a/x)^r \quad (x \geqslant a)$$

If $Y = \log X$, the distribution function of $Y$ is

$$P(Y < y) = 0 \qquad (y \leqslant \log a)$$

$$= 1 - \exp[- r(y - \log a)] \quad (y \geq \log a)$$

i.e. $Y$ has an exponential distribution with origin $\log a$ and scale parameter $1/r$. Suppose then that we have, on the working hypothesis, a Pareto sample $x_{(1)}, \ldots, x_{(n)}$, containing one or more outliers. The transformed quantities $\log x_{(1)}, \ldots, \log x_{(n)}$ will also be in ascending order, and so can be written $y_{(1)}, \ldots, y_{(n)}$, an ordered sample from the exponential $Y$-distribution. If $a$ and $r$ are both unknown, the outlying value or values in the $x$-sample can be tested for discordancy by applying to the corresponding $y$-values the appropriate test for an exponential sample with unknown origin. The available tests given in Section 6.2 are the Dixon-type tests **E2, E4, E6, E10, E12, E13**.

If $a$ is known, the transformation $Z = \log (X/a)$ should be used. On the working hypothesis the transformed quantities $z_{(1)} = \log [x_{(1)}/a], \ldots, z_{(n)} = \log [x_{(n)}/a]$ belong to an exponential distribution with origin at 0 (and density $re^{-rz}(z > 0)$). All the **G** and **E** tests listed in Section 6.2 are applicable to the transformed sample.

### 6.4.6 Poisson samples

Outliers in Poisson samples may arise in any of the numerous practical contexts giving rise to Poisson-distributed data, in particular where the data are counts of events occurring randomly in a given time, or counts of individuals scattered randomly over a given length, area, or volume.

Suppose, to fix ideas, that we wish to test for discordancy an upper outlier $x_{(n)}$ in a sample from a Poisson distribution whose mean $\mu$ is unknown. The argument used earlier (page 92) in the case of a gamma sample might suggest $x_{(n)}/\sum x_j$ as a possible statistic, being of the form $N/D$ and invariant under change of scale. This statistic cannot be used as it stands, since its null distribution depends on $\mu$. However, the null distribution of $x_{(n)}$ *conditional on the observed value of* $\sum x_j$ does *not* depend on $\mu$, since the distribution of $x_{(1)}, \ldots, x_{(n)}$ conditional on the observed value of $\sum x_j$ is multinomial with parameters $(\sum x_j, 1/n, \ldots, 1/n)$. Using this fact, Doornbos (1976) has shown how to set up a discordancy test for (say) $x_{(n)}$, based on its null distribution conditional on $\sum x_j$. The table of critical values of $x_{(n)}$ (for example, for a *nominal* 5 per cent test), will thus have an entry corresponding to each *pair of values* $n$, $\sum x_j$. (As always with discrete distributions, the significance probability attaching to each critical integer value will not be exactly 5 per cent.)

Discordancy tests of this *conditional* type are given below for four outlier situations, namely a single upper outlier, a single lower outlier, an upper outlier-pair, and a lower outlier-pair (tests **P1–P4**).

As an alternative to these specific and conditional tests we may, as already mentioned, use the fact that the transform $\sqrt{(X + \frac{3}{8})}$ of a Poisson random variable $X$ with mean $\mu$ is approximately $N(\sqrt{(\mu + \frac{3}{8})}, \frac{1}{4})$, providing that $\mu$ is not too small, say at least 4 or 5. This will give us tests which are approximate but not conditional on the observed value of $\sum x_j$. Thus to test for discordancy any outlier or outliers in a Poisson sample of unknown but not too small mean, we can take the transformed values $\sqrt{(x_j + \frac{3}{8})}$ $(j = 1, \ldots, n)$ and apply any appropriate $N\sigma$ test with $\sigma = \frac{1}{2}$ to the transformed sample. Paul and Barnwal (1987) have investigated the properties of a number of such *unconditional* approximate procedures, and recommend two tests as having good power properties. These are listed below as **P5** and **P6**.

**P1** *Discordancy test of a single upper outlier $x_{(n)}$ in a Poisson sample of unknown mean*

*Test statistic:* $T_{P1} =$ outlier $x_{(n)}$, tested conditionally on the observed sum of observations $\sum x_j$.

*Inequality:*

$$np_1 - \frac{n(n-1)}{2} p_1^2 < SP(t) < np_1$$

where $p_1 = P[\mathbf{B}(\sum x_j, 1/n) \geq t]$.

*Tabulated significance levels:* Table XXVIIa, page 510; specially compiled.

*Further tables:* Doornbos (1976), Table I, gives nominal 5 per cent and 1 per cent critical values of $x_{(n)}$ for $n = 2(1)10$ and $\sum x_j = 2(1)25$, together with the actual levels of significance attaching to each (necessarily discrete) entry.

*Reference:* Doornbos (1976).

**P2** *Discordancy test of a single lower outlier $x_{(1)}$ in a Poisson sample of unknown mean*

*Test statistic:* $T_{P2} =$ outlier $x_{(1)}$, tested conditionally on the observed sum of observations $\sum x_j$.

*Inequality:*

$$np_2 - \frac{n(n-1)}{2} p_2^2 < SP(t) < np_2$$

where $p_2 = P[\mathbf{B}(\sum x_j, 1/n) \leq t]$.

*Tabulated significance levels:* Table XXVIIb, page 511; specially compiled.

*Reference:* Doornbos (1976).

**P3 *Discordancy test of an upper outlier-pair*** $x_{(n-1)}$, $x_{(n)}$ ***in a Poisson sample of unknown mean***

*Test statistic:* $T_{P3}$ = sum of outliers $x_{(n-1)} + x_{(n)}$, tested conditionally on the observed sum of observations $\sum x_j$.

*Inequality:*

$$SP(t) < \binom{n}{2} P\left[ \mathbf{B}\left( \sum x_j, \frac{2}{n} \right) \geqslant t \right].$$

*Tabulated significance levels:* Table XXVIIIa, page 512; specially compiled.

*Further tables:* Doornbos (1976), Table III, gives nominal 5 per cent and 1 per cent critical values of $x_{(n-1)} + x_{(n)}$ for $n = 4(1)10$ and $\sum x_j = 5(1)25$.

*Reference:* Doornbos (1976).

**P4 *Discordancy test of a lower outlier-pair*** $x_{(1)}$, $x_{(2)}$ ***in a Poisson sample of unknown mean***

*Test statistic:* $T_{P4}$ = sum of outliers $x_{(1)} + x_{(2)}$, tested conditionally on the observed sum of observations $\sum x_j$.

*Inequality:*

$$SP(t) < \binom{n}{2} P\left[ \mathbf{B}\left( \sum x_j, \frac{2}{n} \right) \leqslant t \right].$$

*Tabulated significance levels:* Table XXVIIIb, page 513; specially compiled.

*Further tables:* Doornbos (1976), Table IV, gives nominal 5 per cent and 1 per cent critical values of $x_{(1)} + x_{(2)}$ for the following cases: $n = 4$ and $n = 5$, $\sum x_j = 4(1)25$; $n = 6$, $\sum x_j = 14(1)25$; $n = 7$ and $n = 8$, $\sum x_j = 17(1)25$.

*Reference:* Doornbos (1976).

**P5 *Discordancy test of a single upper outlier*** $x_{(n)}$ ***in a Poisson sample of unknown mean***

*Test statistic:*    $T_{P5} = (y_{(n)} - y_{(n-1)})/\sigma = 2(y_{(n)} - y_{(n-1)})$,

where $y_{(j)} = \sqrt{(x_{(j)} + \frac{3}{8})}$.

*Test distribution* and *Tabulated significance levels:* As for test **N$\sigma$7**, page 249.

*Further tables:* Paul and Barnwal (1987) give a table of 10 per cent, 5 per cent and 1 per cent points for samples of sizes 5, 10, 20, 30, 50, 100 from Poisson distributions with means $\mu = 5, 10, 15, 25, 50, 100$, estimated on the basis of simulations of sizes 15 000; these are, as one would expect, nearly independent of $\mu$. The authors also give a table of estimates of power for $T_{P5}$ and a rival test statistic

$$(y_{(n)} - y_{(n-1)})/(y_{(n)} - y_{(1)}).$$

*Reference:* Paul and Barnwal (1987).

*Properties of test:* Has better power properties than other tests based on the normalizing transformation.

**P6 Discordancy test of a single lower outlier $x_{(1)}$ in a Poisson sample of unknown mean**

*Test statistic:*   $T_{P6} = (y_{(2)} - y_{(1)})/\sigma = 2(y_{(2)} - y_{(1)})$,

see **P5**.

*Other information* as for **P5**. Under *Further tables* the rival test statistic is $(y_{(2)} - y_{(1)})/(y_{(n)} - y_{(1)})$.

### 6.4.7 Binomial samples

Suppose we have a sample $x_1, \ldots, x_n$ in which each observation $x_j$ is a value from a binomial distribution $\mathbf{B}(m, p_j)$ with $m$ known, $p_j$ unknown. On the working hypothesis the $p_j$ are all equal, say to $p$ (unknown). Data of this kind could arise, for instance, in sampling inspection of mass-produced items, where successive samples each of $m$ items are inspected and $x_j$ is the number of defectives found on the $j$th occasion; or again in experiments to compare, say, the germination rates of seeds under different conditions, $m$ seeds being tested under each condition, with replication of the control.

The null distribution of a statistic such as $x_{(n)}/\sum x_j$ depends on the unknown $p$, but the null distribution of $x_1, \ldots, x_n$ conditional on the observed value of $\sum x_j$ does not depend on $p$, being multihypergeometric with parameters $(nm; \sum x_j; m, \ldots, m)$. Therefore, as with Poisson samples discussed above (see page 261), discordancy tests for outliers are carried out conditionally on the observed value of $\sum x_j$. In principle, the *three* quantities $n$, $m$, and $\sum x_j$ are needed for entering any table of critical values of (say) $x_{(n)}$.

Discordancy tests are given below for an upper outlier and a lower outlier; these are in effect the same test. For convenience, the table of critical values that we give for this test is entered with $n$, $m$, and $m - x_{(n)}$, the tabulated quantity being $\sum x_j$.

Once again, transformation is available as an alternative to these specific discordancy tests. For a binomial random variable $X$ with parameters $m$, $p$ the transform $\sin^{-1}[(X/m)^{1/2}]$ is distributed approximately $N(\sin^{-1}(p^{1/2})$, $1/(4m))$. Thus to test for discordancy any outlier or outliers in a binomial sample with known $m$ but unknown $p$, we can apply an appropriate $N\sigma$ test with $\sigma = 1/(2m^{1/2})$ to the sample of transformed values $\sin^{-1}[(x_j/m)^{1/2}]$.

**B1** *Discordancy test of a single upper outlier $x_{(n)}$ in a binomial sample of unknown probability parameter*

*Test statistic:* $T_{B1}$ = outlier $x_{(n)}$, tested conditionally on the observed sum of observations $\sum x_j$.

*Inequality:*

$$np_1 - \frac{n(n-1)}{2} p_1^2 < SP(t) < np_1$$

where $p_1 = P[\mathbf{H}(nm; \sum x_j, m) \geq t]$.

*Tabulated significance levels:* Table XXIX, pages 514–5; specially compiled.

*Reference:* Doornbos (1976).

*Example:* See the example in Worksheet **B2**.

**B2** *Discordancy test of a single lower outlier $x_{(1)}$ in a binomial sample of unknown probability parameter*

*Test statistic:* $T_{B2}$ = outlier $x_{(1)}$, tested conditionally on the observed sum of observations $\sum x_j$.

*Inequality:*

$$np_2 - \frac{n(n-1)}{2} p_2^2 < SP(t) < np_2$$

where $p_2 = P[\mathbf{H}(nm; \sum x_j, m) \leq t]$.

*Tabulated significance levels:* Table XXIX, pages 514–5. This is the table for test **B1** (upper outlier); to use it for test **B2**, replace the given binomial sample $x_1, \ldots, x_n$ with lower outlier $x_{(1)}$ by the complementary binomial sample $y_1 = m - x_1, \ldots, y_n = m - x_n$ and apply test **B1** to the *upper* outlier $y_{(n)} = m - x_{(1)}$.

*Reference:* Doornbos (1976).

*Example 6.5. Suppose that, in the inspection of quality of a manufactured item, five specimens are selected randomly from each of ten batches and tested to destruction, with the following results:*

| Batch | A | B | C | D | E | F | G | H | I | J |
|---|---|---|---|---|---|---|---|---|---|---|
| Number of good items out of five | 5 | 4 | 4 | 5 | 4 | 1 | 4 | 5 | 3 | 4 |

*Can one take it as obvious that batch F is out of line with the others?*

*On the basic model we have a sample of ten from a distribution* $B(m, p)$ *where* $m = 5$ *and* $p$ *is unknown. We wish to test the lower outlier* $x_6 = x_{(1)} = 1$ *for discordancy. To convert into an upper outlier test, we consider instead the numbers of failures* $y_1, \ldots, y_{10} = 0, 1, 1, 0, 1, 4, 1, 0, 2, 1$. *The upper outlier* $y_{(10)} = 4 = m - 1$. *Entering Table XXIX with* $n = 10$ *and* $m = 5$, *we see that this outlier would be significant at 1 per cent if* $\sum y_j$ *were 7 or less, and would be significant at 5 per cent if* $\sum y_j$ *were 10 or less. In fact* $\sum y_j = 11$, *so the evidence for regarding batch F as discordant is weak.*

**B3** **Test of discordancy of a single outlier in a binomial sample of unknown probability parameter**

A pragmatic test for an outlier in a binomial sample is proposed by Collings, Margolin, and Oehlert (1981). It consists of examining the 'residuals' $x_i - \bar{x}$, taking their studentized forms, detecting an outlier as that observation with maximum absolute studentized residual, and declaring it discordant if the absolute value of its studentized residual is large enough. The null distribution of the studentized residuals is approximated by a folded standard normal distribution.

# PART III

# *Multivariate and Structured Data*

# CHAPTER 7

# Outliers in Multivariate Data

The study of outliers is as important for multivariate data as it is for univariate samples. Many factors carry over immediately. The basic notions, of an outlier as an observation which engenders surprise owing to its extremeness, and of its discordancy in the sense of that 'extremeness' being statistically unreasonable in terms of some basic model, are not constrained by the dimensionality of the data. Nor is the importance of seeking inference procedures that are relatively unaffected by contamination of the data source, i.e. outlier-robust (accommodation) methods. But the expression of these ideas and aims is by no means as straightforward when we operate in more than one dimension.

## 7.1 PRINCIPLES FOR OUTLIER DETECTION IN MULTIVARIATE SAMPLES

As Gnanadesikan and Kettenring (1972) remark, a multivariate outlier no longer has a simple manifestation as an observation which 'sticks out at the end' of the sample. The sample has no 'end'! But, notably in bivariate data, we may still perceive an observation as suspiciously aberrant from the data mass, particularly so if the data are represented in the form of a scatter diagram. See Figure 7.1(a), and the observation labelled $A$. This observation happens to be an extreme in the $x_2$-direction, but it is not extreme in the $x_1$-direction. A multivariate outlier need not be an extreme in any of its components. Someone who is short and fat need not be the shortest, or the fattest, person around. But that person can still be an 'outlier'.

Kempthorne and Mendel (1990) offer some interesting comments on the intuitive stimulus to the declaration of a bivariate outlier.

The idea of extremeness still inevitably arises from some form of 'ordering' of the data. What is now lacking is any notion of order, and hence of extremeness, as a *formal* stimulus to the declaration of an outlier. Although no unique unambiguous form of total ordering is possible for multivariate data, different types of sub-(less than total) ordering principle may be defined and employed. Barnett (1976b) surveys the role of sub-ordering in multivariate analysis. In attempting to express clearly the subjective stimulus to the declaration of an outlier in a multivariate sample it is useful to employ some such *sub-ordering* principle.

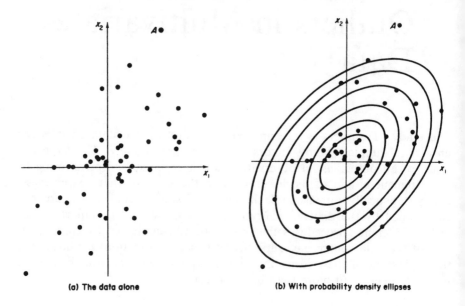

(a) The data alone                    (b) With probability density ellipses

**Figure 7.1**  A bivariate sample. (a) The data alone. (b) With probability density ellipses

This imposes a crucial further stage in the process of investigating outliers. For a univariate sample we needed to

- detect any outliers in the data in terms of their extremeness relative to the basic model $F$

and either

- test the discordancy of the outliers with respect to the basic model $F$

or

- employ (robust) accommodation methods of inference.

The detection stage was straightforward for a univariate sample; *it was obvious what was meant by extremeness.* But for multivariate data this is not so and we need an earlier step in a process:

- *adopt an appropriate sub-ordering principle as a basis for expressing extremeness of observations.*

Barnett (1976b) categorizes sub-ordering principles in four types: *marginal, reduced* (or *aggregate*), *partial* and *conditional*. For outlier study, *reduced subordering* is almost the only principle that has been employed and we shall consider this in some detail. (Other contributions to multivariate ordering are included in Oja, 1983, Brown and Hettmanspeiger, 1989, and Small, 1990.)

With reduced sub-ordering we transform any multivariate observation $x$, of dimension $p$, to a scalar quantity $R(x)$. We can then order a sample $x_1, x_2, \ldots, x_n$ in terms of the values $R_j = R(x_j)$ $(j = 1, 2, \ldots, n)$. That observation $x_i$ which yields the maximum value $R_{(n)}$ is then a candidate for declaration as an outlier—provided its extremeness is surprising relative to the basic model $F$. Furthermore, an outlier $x_i$ will be adjudged discordant if $R_{(n)}$ is unreasonably large in relation to the distribution of $R_{(n)}$ under $F$. Thus the principle of a *test of discordancy* is the same as it was for a univariate outlier.

However, many problems now arise which are specific to the multivariate case. Clearly, we may lose useful information on multivariate structure by employing reduced (or any other form of) sub-ordering. The special cases where $R(x)$ singles out some particular marginal component of $x$ (this is marginal sub-ordering), or indentifies, say, the first principal component of the multivariate sample, intuitively demonstrate this risk.

So how are we to choose the reduction measure $R(x)$? Subjective choice is fraught with danger; multivariate data do not reveal their structure readily (or reliably) to casual observation. We might hope that an appropriate choice of $R(x)$ (or use of some broader sub-ordering method) would emerge from applying some relevant statistical principle in the context of a suitable specification of a basic model $F$ and alternative (contamination) model $\bar{F}$. This is seldom so! Finally, even if we have chosen $R(x)$, the distributional form of $R_{(n)}$ under $F$ may not be easy to determine or employ as a basis for applying a test of discordancy.

We have to acknowledge the degree of arbitrariness which may result from having to make some ad hoc choice of $R(x)$. We may be fortunate in that the chosen $R(x)$ turns out to be reasonable in terms of some sensible criteria; it may even prove to be tractable. But the form of $F$ is of course critical. What is reasonable for one $F$ will not be so for another.

Consider an example where we choose to represent a multivariate observation $x$ by means of a **distance measure**,

$$R(x; x_0, \Gamma) = (x - x_0)'\Gamma^{-1}(x - x_0)$$

where $x_0$ reflects the location of the data set or underlying distribution and $\Gamma^{-1}$ applies a differential weighting to the components of the multivariate observation inversely related to their scatter or to the population variability. For example, $x_0$ might be the zero vector $0$, or the true mean $\mu$, or the sample mean $\bar{x}$, and $\Gamma$ might be the variance-covariance matrix $V$ or its sample equivalent $S$, depending on the state of our knowledge about $\mu$ and $V$.

If the basic model $F$ were multivariate normal, $N(\boldsymbol{\mu}, V)$, the corresponding form

$$R(\mathbf{x}; \boldsymbol{\mu}, V) = (\mathbf{x} - \boldsymbol{\mu})' V^{-1} (\mathbf{x} - \boldsymbol{\mu}) \qquad (7.1.1)$$

has substantial practical appeal in terms of probability density ellipsoids and we shall see (Section 7.3.1) that it also has much broader statistical support. In fact, the sample shown in Figure 7.1(a) is from a bivariate normal distribution and the appropriate elliptic density contours in Figure 7.1(b) highlight any intuitive concern we had for the observation $A$. For other distributions $R(\mathbf{x}; \mathbf{x}_0, \Gamma)$, may or may not be appropriate as we shall find later but it is none the less widely used often in a form employing *robust* versions of $\mathbf{x}_0$ and $\Gamma$. We shall explore this further in Sections 7.3.2 and 7.3.3 (alternative forms for specific non-normal distributions) and 7.4.4 (wider use of 'robust' forms of $R(\mathbf{x}; \mathbf{x}_0, \Gamma)$).

A Bayesian approach to outlier detection ('detection of spuriosity') arises from the work of Guttman (1973b) referred to in Section 2.3. The approach can be immediately extended to multivariate data $\mathbf{x}_1, \mathbf{x}_2, \ldots, \mathbf{x}_n$ arising from $N(\boldsymbol{\mu}, V)$ if there is no contamination, and with contamination (if present) manifest in a single observation from $N(\boldsymbol{\mu} + \mathbf{a}, V)$. Again it is through the posterior distribution of $\mathbf{a}$ that we seek to detect a discordant outlier. The detection criteria again revolve around the values taken by a set of weights attached, in the posterior distribution of $\mathbf{a}$, to the separate observations. It is interesting to note that the implicit concept of extremeness used to detect the outlier is again expressible in terms of the distance metric, $R(\mathbf{x}; \boldsymbol{\mu}, V)$.

Barnett (1979) has considered general principles for the detection of multivariate outliers. The simplest one is:

**Principle A.** *The most extreme observation is that one, $\mathbf{x}_i$, whose omission from the sample $\mathbf{x}_1, \mathbf{x}_2, \ldots, \mathbf{x}_n$ yields the largest incremental increase in the maximized likelihood under $F$ for the remaining data. If this increase is surprisingly large, declare $\mathbf{x}_i$ to be an outlier.*

This principle requires only the basic model $F$ to be specified. If we are prepared to adopt an alternative (contamination) model $\bar{F}$, e.g. of slippage type, we can set up a more sophisticated principle. For example, suppose that $\bar{F}$ is of a slippage type under which one observation is a contaminant. Then we could set up an alternative principle as follows:

**Principle B.** *The most extreme observation is that one, $\mathbf{x}_i$, whose assignment as the contaminant in the sense of $\bar{F}$ maximizes the difference between the loglikelihoods of the sample under $\bar{F}$ and $F$. If this difference is surprisingly large, declare $\mathbf{x}_i$ to be an outlier.*

If $F$ and $\bar{F}$ are fully specified (have no parameters of unknown value) then of course we use joint probability, rather than likelihood, in **Principles**

A and B. The ideas extend in a natural way to outlying subsets of the data, rather than single outliers. We shall see later (Sections 7.3.1, 7.3.2 and 7.3.3) how **Principle A** (at least) can be employed for constructing discordancy tests in specific cases.

Mathar (1985) extends univariate concepts of *outlier proneness* and of *outlier-resistance* to the multivariate case.

As in our study of methods for univariate outliers we proceed to consider accommodation procedures (Section 7.2) and then tests of discordancy (Section 7.3). With the greater complexity of the multivariate case, however, we will need to examine a wide range of informal proposals for outlier detection and processing—these are discussed in Section 7.4.

## 7.2 ACCOMMODATION OF MULTIVARIATE OUTLIERS

In the analysis of multivariate data we frequently need to employ statistical methods which will be relatively unaffected by the possible presence of contaminants in the sample under investigation. That is, we need methods for the *accommodation of outliers*. For multivariate data such methods are less well developed than for the univariate case and furthermore they tend to be directed to general robustness considerations rather than to be designed specifically with outliers in mind.

Some outlier-specific proposals do, however, appear in the literature. Often they concentrate on a basic normal distribution or have an informal structure with intuitive, rather than theoretically justified, appeal. We shall review briefly some of the techniques available for estimating parameters (in multivariate distributions), and for conducting various forms of multivariate analysis (e.g. principal component analysis and discriminant analysis), in ways which are likely to be robust against the presence of outliers.

### 7.2.1 Estimation of parameters in multivariate distributions

Suppose under the basic model $F$ we have a location (or mean) vector $\mu$ and a covariance matrix $V$. We shall be interested in robust estimation of $\mu$ and of $V$ (and of derived quantities such as the correlation matrix $P$). Consider first the estimation of $\mu$.

We have referred in Chapters 3 and 5 to the 'premium-protection' rules of Anscombe (1960a) which take the form of joint rejection/estimation procedures. For a univariate sample $x_1, x_2, \ldots, x_n$ from $N(\mu, \sigma^2)$, with a location-slippage alternative model $N(\mu + a, \sigma^2)$ for the generation of at most one of the $x_j$, we examine the maximum absolute residual

$$\max_{j = 1, 2, \ldots, n} |x_j - \bar{x}|.$$

*If this is sufficiently large* we omit the observation $x_i$ yielding the maximum absolute residual and estimate $\mu$ by the sample mean of the remaining $n - 1$ observations; otherwise we estimate $\mu$ merely by $\bar{x}$. There is an obvious multivariate generalization in which we consider $R_{(n)}(\mathbf{x}; \bar{\mathbf{x}}, S)$ [or $R_{(n)}(\mathbf{x}; \bar{\mathbf{x}}, V)$, depending on our state of knowledge about $V$] and if it is sufficiently large omit the observation $\mathbf{x}_i$ yielding $R_{(n)}$ before estimating $\mu$ from the residual sample; if $R_{(n)}$ is not sufficiently large we estimate $\mu$ from the total sample by means of $\bar{\mathbf{x}}$. Such an approach is implicitly taken up by Golub, Guttman, and Dutter (1973) in greater detail and generality. Their work is broader in that they consider a general normal linear model and augment the Anscombe-type rule by corresponding rules based on Winsorization and 'semi-Winsorization' of residuals. The greater detail is reflected in their discussion of the problems of determining the premium and protection. To ease the task of such determinations they also propose the use of orthogonalized ('adjusted') residuals as a basis for approximating the premium and protection. This aspect is discussed more fully in our study of the linear model in Section 8.3.

Note that this type of approach is an *adaptive* one, in that the action taken depends on the sample values actually observed. Also, it need not be restricted to a single outlier, nor to a normal basic model. We could consider $R_{(n)}$, $R_{(n-1)}$, . . . , removing all those which are sufficiently large, using a reduction measure $R(\mathbf{x})$ *appropriate* to any contemplated basic model, $F$.

Of course, there are immediate generalizations of the principles of trimming and of Winsorization that can be applied to the estimation of multivariate location parameters (or any other summary measures). Prior to estimating relevant parameters, we could trim or Winsorize *a specific number*, or *proportion*, of the 'most extreme' values in the sample, assessing extremeness in terms of some *appropriate* reduction measure, $R(\mathbf{x})$. (This differs from the Anscombe-type rule in removing the requirement that $R_{(n)}$, $R_{(n-1)}$. . . need to be sufficiently large before trimming or Winsorization takes place.)

A different approach to multivariate outlier accommodation can be based on the Bayesian analysis by Guttman (1973b) outlined in Section 7.1. Guttman is concerned with the posterior distribution of **a** for a basic normal model $N(\mu, V)$, with a mean-slippage alternative model $N(\mu + \mathbf{a}, V)$ for at most one of the observations. Examination of the posterior distribution of $\mu$ (and $V$) is germane to the accommodation issue.

Gnanadesikan and Kettenring (1972) are concerned with robust estimation of multivariate location and dispersion. Whilst not preoccupied with the outlier problem some of their proposed robust estimators will *en passant* provide protection against outliers. Various robust estimators of $\mu$ are reviewed (and have their performance characteristics examined by simulation in the bivariate normal case). The estimators mostly take the form of vectors of robust univariate estimators for the distinct marginal components of $\mu$, such as the vector of sample medians, or of trimmed means. A number of contributions in the literature have also been concerned, *inter alia*, with robust estimation of the covariance matrix $V$ either in its own right or as an ingredient in some variety of multivariate analytic method. Outlier-robust techniques for conventional forms of multivariate analysis will be considered in the next sub-section (7.2.2). Here we consider some results on the accommodation of outliers in estimating $V$ itself, the generalized variance $|V|$, the correlation matrix $P$, or individual correlation coefficients.

A large body of attention has focused on use of Huber's $M$-estimators (see Section 5.1.3) for estimating $V$ (often simultaneously with $\mu$, since we are unlikely to know the location parameter of the distribution.) Hampel (1973) proposes an invariant iterative method for estimating $V$ which is equivalent to use of an $M$-estimator. Maronna (1976) specifically examines robust $M$-estimators for $V$ and $\mu$, setting the tone for others by concentrating on affine-invariance (e.g. invariance under changes of scale and origin) and a basic model $F$ which is assumed to have an elliptically symmetric density. Maronna considers problems of existence and uniqueness of $M$-estimators, their consistency and asymptotic normality, and their robustness properties as expressed by the influence function and breakdown bounds (see Section 3.1.2). The breakdown point is shown, under reasonable assumptions, to be at most $(p + 1)^{-1}$ (where $p$ is the dimensionality of the data) and this is seen to cast some doubt on the suitability of $M$-estimators for high-dimensional data. We shall return to this matter shortly.

Of course, measures of robustness have to be obtained on the basis of some assumed contamination model. Maronna (and many others whose work is reported here) assumes a mixture model with a degenerate (one- or two-point) distribution for any contaminant. We should recall that the breakdown point essentially expresses in this context the maximum proportionate contamination that can be tolerated before an estimator under the contamination model is capable of suffering its maximum possible discrepancy from the value it would take under the basic model.

Huber (1977a) presents a systematic study of the affine-invariant situation considered by Maronna, with the aim of determining the most general form of $M$-estimator for $V$. Modifying the conditions employed by Maronna only effects a slight increase in the breakdown point to $p^{-1}$ and again doubt is expressed about use of $M$-estimators of $V$ for high-dimensional data.

Huber (1981, Chapter 8) provides a clear review of robust estimation of $V$ and $P$. He identifies different possible interests in robust estimation of $V$; concern for the individual elements of $V$ or of $P$, interest in estimating variances for a selected subset of the marginal components and estimation of the 'shape matrix' for some relevant elliptical distribution with density of the form

$$f(\mathbf{x}) = |V| h[(\mathbf{x} - \boldsymbol{\mu})' V^{-1}(\mathbf{x} - \boldsymbol{\mu})]$$

where $h(\mathbf{y})$ is a spherically symmetric density in $p$-dimensional space. In the latter case $V$ is not uniquely identifiable, although $V'V$ will be, and this means that some suitable condition must be placed on $V$ to make it estimable. It is such considerations which lead to an interest in affine-invariant estimation procedures. For the other interests (in the elements of $V$ or of $P$ *per se*, or in subsets of variances or covariances) coordinate-dependent rather than invariant procedures are more relevant—certainly so, for example, if we wish to estimate the correlation matrix $P$.

The effect of outliers on estimation of $V$ can take various forms, leading to serious bias, reduction of precision or even 'break-down'. (Even one or two outliers in a large data set can wreak havoc!) One possibility is that the uncontaminated data lie closely in a tight ellipsoidal region of effectively fairly low dimension, whilst contaminants are diffuse and essentially spherically distributed. The reverse prospect is also feasible; it may indeed be more likely. Invariant procedures will not distinguish these alternatives and it is this fact which underlies an interpretation of some $M$-estimators in terms of a rather unusual form of Winsorization (see below).

The relationship between the uncontaminated data mass and a small number of contaminants is expressed, of course, through an appropriate contamination model, $\bar{F}$. In work on robust estimation of $V$, $\bar{F}$ is nearly always taken to have a mixture formulation but only a limited (and rather artificial) range of contaminating distributions is considered. Some qualitative effects of outliers on estimation of $V$, and corresponding attitudes to robust estimation, are considered by Devlin, Gnanadesikan, and Kettenring (1981) and by Campbell (1980) who claims that outliers 'tend to deflate correlations and possibly inflate variances'.

A tangible form for outlier-robust $M$-estimators, relevant to an assumed elliptically symmetric basic model and an associated normal contamination distribution, is exhibited by Maronna (1976) and by Campbell (1980). The estimators of $\boldsymbol{\mu}$ and $V$ are obtained as iteratively derived simultaneous solutions to the equations

$$\boldsymbol{\mu} = \sum_{j=1}^{n} w_j \mathbf{x}_j \bigg/ \sum_{j=1}^{n} w_j \tag{7.2.1}$$

and

$$V = \sum_{j=1}^{n} w_j^2 (\mathbf{x}_j - \boldsymbol{\mu})(\mathbf{x}_j - \boldsymbol{\mu})' \bigg/ \left( \sum_{j=1}^{n} w_j^2 - 1 \right) \tag{7.2.2}$$

where

$$w_j = w(R_j)/R_j \tag{7.2.3}$$

and $R_j$ is the sample value of the reduced measure (7.1.1) relevant to the normal distribution: that is,

$$R_j = R(\mathbf{x}_j; \boldsymbol{\mu}, V) = (\mathbf{x}_j - \boldsymbol{\mu})' V^{-1} (\mathbf{x}_j - \boldsymbol{\mu}).$$

We need to specify the weight function $w(t)$ and this is commonly taken as a continuous function of bounded influence (Hampel, 1974). Campbell (1980) makes the specific proposal

$$w(t) = \begin{cases} t & (t < t_0) \\ t_0 \exp[-\dfrac{1}{2}(t - t_0)^2/a] & (t > t_0) \end{cases} \tag{7.2.4}$$

with $t_0 = \sqrt{p} + b$, for suitable choice of positive values for $a$ and $b$.

The effect of (7.2.3) and of (7.2.4) is to give full weight to all observations $\mathbf{x}_j$ not too far (in terms of the $R_j$ value) from $\boldsymbol{\mu}$ and to downweight more extreme values. Note that, in view of the implicit form of (7.2.1) and (7.2.2) such a differential weighting scheme does not have immediate interpretation in respect of action for specific $\mathbf{x}_j$ in isolation. In fact, it can be interpreted (as with the Anscombe-type rules) as an adaptive form of Winsorization. Some forms of robust $M$-estimators (see, for example, Huber, 1977a) have a superficially rather odd form: $\mathbf{x}_j$ with sufficiently large $R_j$ are pulled inwards and other $\mathbf{x}_j$ with sufficiently small $R_j$ are *pushed outwards*. But this is not so strange a policy when we consider the underlying affine-invariance condition and the type of model where low-dimensional data clouds have superimposed higher-dimensional contaminants, or vice versa.

An algorithm for calculating robust $M$-estimates of $\boldsymbol{\mu}$ and $V$ is given by Maronna (1976) who also presents some small-sample Monte Carlo results on their properties, Campbell (1980) also exhibits some Monte Carlo based results; as do Devlin, Gnanadesikan, and Kettenring (1981) within a wider-ranging review of various methods of robust estimation of covariance and correlation matrices (but with major attention to the latter) where they consider three approaches: separate bivariate analyses, and estimation of the full $P$ (or $V$) matrix using either (ellipsoidal) multivariate trimming or $M$-estimators. We shall return below to the estimates based on bivariate sub-components. The multivariate trimming method takes the following form. Start with provisional full-sample estimates $\tilde{\boldsymbol{\mu}} = \bar{\mathbf{x}}$ and $\tilde{V}_0 = S$ for $\boldsymbol{\mu}$ and $V$. Reject a fixed percentage (10 per cent is chosen) of the observations with largest $R(\mathbf{x}_j; \tilde{\boldsymbol{\mu}}_0, \tilde{V}_0)$ values. Recalculate the sample mean and covariance matrix on the reduced sample as second stage estimates $\tilde{\boldsymbol{\mu}}_1$ and $\tilde{V}_1$, and repeat the process until some convergence criterion for the elements of $P$ is satisfied. The $M$-estimation methods are similar to those described above.

The broad conclusions are that the multivariate methods are better than the use of separate bivariate estimates (and trimming has the advantage that the breakdown point is limited to the trimming percentage irrespective of dimensionality; compare this with $(p + 1)^{-1}$ for the $M$-estimators). The 'separate-estimates' approach has some support in terms of cost effectiveness and safeguards against missing data, and very high dimensionality: some 'tailoring' is needed of course to ensure positive definite matrix estimates (see below).

An early proposal for *direct* robust estimation of $V$ in positive definite form was made by Gnanadesikan and Kettenring (1972) who suggest ranking the multivariate sample in terms of some reduction measure $R(\mathbf{x}; \mathbf{x}^*, I)$ where $\mathbf{x}^*$ is a robust estimator of $\boldsymbol{\mu}$, omitting a small proportion of the sample having the largest $R(\mathbf{x}; \mathbf{x}^*, I)$ values, and computing a matrix

$$A_0 = \sum (\mathbf{x}_i - \mathbf{x}^*)(\mathbf{x}_i - \mathbf{x}^*)'$$

where the summation extends over the *retained* sample members. The whole sample is then ranked in terms of $R(\mathbf{x}; \mathbf{x}^*, A_0)$, again a small proportion of the observations having the largest $R(\mathbf{x}; \mathbf{x}^*, A_0)$ are omitted, and $V$ is estimated as the appropriate multiple of the matrix of sums of squares and cross-products of the *finally retained* sample members. The procedure is intuitively appealing, but only limited empirical investigation is reported. Another ingenious method of constructing a robust estimator of $V$, without recourse to estimation of $\boldsymbol{\mu}$, is also presented.

Nyquist (1984) presents an alternative form of $M$-estimator for $\boldsymbol{\mu}$ and for $V$ for outlier accommodation.

Rousseeuw and van Zomeren (1990) are also concerned with robust estimation of $\boldsymbol{\mu}$ and $V$. Rejecting the $M$-estimators of Campbell (1980) in view of the low breakdown point, they suggest alternative robust estimators with a higher breakdown point. Stahel (1981) and Donoho (1982) proposed a form for such estimators, but Rousseeuw and van Zomeren prefer the *minimum volume ellipsoid (MVE) estimator* introduced by Rousseeuw (1985). On this approach, $\boldsymbol{\mu}$ is estimated by the centre of the minimum volume ellipsoid covering half of the observations and $\Gamma$ is also estimated for this same construct. A warning is given against using the method unless there are *at least five observations per dimension*, to avoid the 'curse of dimensionality' which may lead to an unrepresentative MVE. The purpose of this work is not accommodation, however; it is detection with the avoidance of the 'masking' that the authors claim arises if we use estimates of $\boldsymbol{\mu}$ and $V$ which include any outliers (contaminants). The illustrative examples look impressive, but we need to recall the discussion in Section 4.1.3 of the frequent irrelevance (at least in the univariate case) of whether we use *'inclusive'* or *'exclusive'* measures. See also the rather sceptical comments of Cook and Hawkins (1990). (The Rousseeuw and van Zomeren results effectively assume a normal basic model, $F$; see also Kempthorne and Mendel, 1990.)

The low breakdown point for $M$-estimators also underlies the work of Lopuhaä (1989) on robust estimation of $\mu$ and $V$, where $M$-estimators are compared with the corresponding $S$-estimators (of Rousseeuw and Yohai, 1984) and found to have, asymptotically, similar efficiency but higher breakdown point.

Little (1988) uses the EM algorithm for maximum likelihood estimation of $\mu$, $V$ in $N(\mu, V)$ (and multivariate $t$ and contaminated normal distributions) when data are incomplete in the sense that values are missing for some of the marginal variables for some of the sample observations. The aim is robustness against outliers whilst retaining high efficiency in the null case. The success of the approach is demonstrated by simulation studies.

Tiku and Balakrishnan (1988a and 1988b) present outlier-robust tests of a mean, or of equality of two means, in the bivariate and p-variate normal cases, respectively.

### 7.2.2 Estimation of individual components of vector parameters

Gnanadesikan (1977, Section 5.2.3) presents a comprehensive review of the robust estimation of $\mu$ and $V$ with detailed proposals for constructing estimators from robust estimators of the *individual elements* of $\mu$ and $V$ as well as for directly 'multivariate' estimators of $\mu$ and $V$.

We shall pursue this distinction, by considering specific proposals for outlier-robust estimation of the component members of $\mu$, $V$ (or the correlation matrix, $P$).

We start with the robust estimation of individual covariances $v_{ij}$ or correlation coefficients $\rho_{ij}$. Huber (1981, Section 8.2) considers this topic. We note that for given $a$, $b$,

$$v_{ij} = \frac{1}{4ab} [\text{Var}(aX_i + bX_j) - \text{Var}(aX_i - bX_j)], \qquad (7.2.5)$$

so that robust estimators of $v_{ij}$ or $\rho_{ij}$ might be based on robust estimators of variances. Gnanadesikan and Kettenring (1972) suggest using (7.2.5) with

$$a = [S(X_i)]^{-1} \quad \text{and} \quad b = [S(X_j)]^{-1},$$

where $S(X)$ is a robust estimate of $\text{Var}(X)$ (utilizing trimming or Winsorization), to obtain a 'robust' estimate of $v_{ij}$. A corresponding robust estimate of $\rho_{ij}$ has the form $\frac{1}{4}[S(aX_i + bX_j) - S(aX_i - bX_j)]$. The latter form is not restricted to the interval $[-1, 1]$ and Huber (1981) reports on a modified form

$$\tilde{\rho}_{ij} = \frac{S(aX_i + bX_j) - S(aX_i - bX_j)}{S(aX_i + bX_j) + S(aX_i - bX_j)}$$

which restores this restriction. Correspondingly, we can consider a derived covariance estimator

$$\tilde{v}_{ij} = \tilde{\rho}_{ij} S(X_i) S(X_j).$$

One difficulty with such an approach is that, if we form estimators of $V$ or of $P$ from separate robust estimators of their elements, the resulting $\tilde{V}$ and $\tilde{P}$ may not be positive definite. Devlin, Gnanadesikan, and Kettenring (1975) propose a remedy involving 'shrinking' $\tilde{P}$ until it is positive definite, and rescaling it if an estimate of $V$ is required. (They also review various *ad hoc* estimators of the correlation coefficient, for a bivariate normal distribution, based on partitioning the sample space, on transformations of Kendall's $\tau$ or on normal scores. See also Huber (1981, Section 8.3) for transformation methods based on the sample product–moment correlation coefficient.)

Mosteller and Tukey (1977, Chapter 10) propose a robust estimator of $V$ based on robust regression estimation: regressing $x_j$ on $x_1, x_2, \ldots, x_{j-1}$ for $j = 2, 3, \ldots, p$ (where $x_j$ is the $j$th component of the typical observation vector **x**.)

Pasman and Shevlyakov (1987) consider robust estimation of the correlation coefficient, $\rho$, in a bivariate normal distribution, comparing a range of existing estimators with newly proposed ones. Gideon and Hollister (1987) define a new rank correlation coefficient with claimed advantages in contaminated samples over the conventional estimators.

A number of studies have recently appeared of the robustness of tests of (and confidence intervals for) the correlation coefficient $\rho$ in a bivariate normal distribution under the prospect of contamination. Srivastava and Lee (1983) make numerical comparisons of confidence intervals based on five transformations of the sample correlation coefficient, $r$ (including Fisher's $z$ and Samiuddin's $t$). They show that these two transformations yield confidence intervals which are relatively robust against (a particular form of contamination) provided $\rho \leq 0.3$, but are seriously affected if $\rho > 0.3$.

Srivastava and Lee (1984) show that tests of $\rho = 0$ based on Fisher's $z$ (and other transformations) are *not* robust in the face of contamination under a mixture model (with scale inflation). Extensions of this study under modified conditions are described in Srivastava and Lee (1985).

Robust tests for the correlation coefficient are also described in Tiku and Balakrishnan (1986) and Tiku (1987 and 1988).

Radhakrishnan and Kshirsagar (1981) examine influence functions for various multivariate parameteric functions including correlation coefficients, and the generalized variance $|V|$. Cleveland (1979) is concerned with accommodation of outliers in his proposed method for 'smoothing' scatter diagrams.

Another approach to outlier-robust estimation of correlation uses the ideas of convex hull 'peeling' or ellipsoidal peeling (see Bebbington, 1978; Titterington, 1978). Barnett (1976b) had suggested that the most extreme group of observations in a multivariate sample are those lying on the convex hull (with those on the convex hull of the remaining sample, the second most extreme group, etc). Figure 7.2 shows the successive convex hulls for the bivariate sample presented in Figure 7.1. Bebbington (1978) proposes trimming off the observations on the convex hull prior to estimating the correlation coefficient (in bivariate data) by means of the sample product-moment correlation coefficient. Titterington (1978) attempts to overcome computational and conceptual difficulties, particularly with higher-dimensional data, through trimming in terms of 'ellipsoids of minimal content' rather than convex hulls (cf. discussion of $MVE$ above).

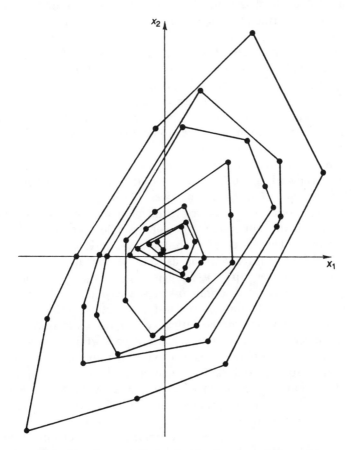

**Figure 7.2** Convex hulls for the bivariate data of Figure 7.1

### 7.2.3 Outliers in multivariate analyses

Any form of multivariate analysis is, of course, likely to be susceptible to departures from the assumed structure and error model. This is particularly so when the data contain outliers as the manifestation of contamination. There have been some proposals for modified forms of multivariate analysis which give protection against outliers.

Campbell (1980) presents a method of outlier-robust principal component analysis in which $V$ is estimated (for an elliptically symmetric base model) by the $M$-estimator described by (7.2.1) to (7.2.3) above. The method embodies a means of detecting outliers in the course of such a principal component analysis. Delvin, Gnanadesikan, and Ketternring (1981) also extend their study of the relative merists of different methods of robust estimation of $P$ to estimation of the principal components of $P$. Jolliffe (1989) provides an interesting example (with $p = 5$ and $n = 34$) of how a contaminant can affect the outcomes of a principal component analysis, and of how the principal components can reveal an outlier. See also Critchley (1985).

Another important area of multivariate analysis is canonical variate analysis. Campbell (1982) adopts a similar approach to the one he advocates for principal components for the development of an outlier-robust canonical variates analysis; again an elliptically symmetric basic model is assumed. A general conclusion is that canonical roots are much more susceptible to outliers than are the canonical vectors. A biological example on whelks illustrates the method.

The problem of accommodating outliers in a discriminant analysis has also been considered: again by Campbell (1978). A normal basic model is assumed, with equal covariance matrices $V$ for the two groups, and the contamination model is of mixture form incorporating the prospect of a small proportion of contaminants at some single specific value $\mathbf{x}$, in the first group only. The effect of a contaminant is examined by means of the influence function for various aspects of the analysis: namely, the Mahalanobis distance $(\boldsymbol{\mu}_1 - \boldsymbol{\mu}_2)' V^{-1}(\boldsymbol{\mu}_1 - \boldsymbol{\mu}_2)$, the coefficient vector $\boldsymbol{\alpha} = V^{-1}(\boldsymbol{\mu}_1 - \boldsymbol{\mu}_2)$ and the discriminant means $\boldsymbol{\alpha}'\boldsymbol{\mu}_i$ ($i = 1, 2$). The influence functions are used to determine criteria for detecting outliers. Approximations to the distributions of the influence functions are obtained and an example on discrimination of two species of rock crab illustrates how the relative influence of different observations might be assessed in practice. In effect, this approach does not yield a method of analysis which is outlier-robust, but provides a method of assessing the effect that outliers might have on the analysis. Critchley and Vitiello (1991) further examine Campbell's approach to the influence of outliers in linear discriminant analysis with particular regard to estimates of misclassification probabilities. (See also Radhakrishnan and Kshirsagar, 1981, on the form of influence functions for a range of quantities involved in different types of multivariate analysis.) Radhakrishnan (1983) applies these ideas in discriminant analysis *with more than two groups*.

An approach to linear discriminant analysis by Ganeshanandam and Krzanowski (1989) extends the so-called *'leave-one-out'* approach of Lachenbruch and Mickey (1968) (and others) in which 'each observation is removed from the training data in turn, and the removed individual is classified on the basis of the allocation rule computed from the remaining individuals'. Although not pursued by the authors, this could clearly be relevant to outlier detection or accommodation. An outlier-robust Bayesian decision procedure for classifying observations into one of two classes is presented and assessed by Medvedev and Kharin (1988).

A more specific accommodation procedure is presented by Dallal and Hartigan (1980) who develop a non-parametric measure of 'monotone association' for a bivariate sample. Using the notion of *monotone cover* (being the difference between the maximum lengths of data subsequence for which both components are non-decreasing, and one is non-decreasing and the other non-increasing) a simple test of association is constructed. It is shown to be more robust against outliers than the rank correlation test.

Feigin and Cohen (1978) investigate a model representing the similarity of assessment (*concordance*) of a group of investigators required to rank a common set of items according to a declared characteristic. In illustrating their method on some sociological data they highlight its facility for exhibiting outliers, and examine the effect of outliers on the overall analysis.

A robust approach to the analysis of covariance is described by Birch and Myers (1982).

Escoffier and LeRoux (1976) discuss the influence of outliers on the stability of factors in correspondence analysis. For factor analysis, Tanaka and Odaka (1989) propose an iterative method to detect 'influential observations', whilst Hirsch, Wu, and Tway (1987) examine the 'reliability' of factor analysis in the presence of outlying data.

A multidimensional scaling procedure claimed to be highly resistant to outliers is described (and examined using simulation) by Spence and Lewandowski (1989).

## 7.3 DISCORDANCY TESTS

The notion of a test of discordancy is as relevant to multivariate data as it is to univariate samples, although conceptual and manipulative difficulties have limited the number of formal and specific proposals. It will come as no surprise that most work centres on the normal distribution. If not as ubiquitously relevant as some might claim, the normal distribution proves amenable to the construction of tests of discordancy with desirable statistical properties and a useful degree of unity of form. Section 7.3.1 considers tests of discordancy for multivariate normal samples, whilst the succeeding sub-sections present the more sparse information available for specific non-normal situations.

### 7.3.1 Multivariate normal samples

Suppose $x_1, x_2, \ldots, x_n$ is a sample of $n$ observations of a $p$-component random variable $X$ assumed under the basic model to have a $p$-dimensional **normal distribution**, $N(\mu, V)$, where $\mu$ is the $p$-vector of means, and $V$ the $p \times p$ variance-covariance matrix. A possible alternative model which would account for a single contaminant is the slippage alternative, obtained as a multivariate adaptation of the univariate *models A* (slippage of the mean) and *B* (slippage of the variance) discussed by Ferguson (1961a)—see Section 2.3. Specifically, the alternative hypotheses are:

*Model A*    $E(X_i) = \mu + a$ (some $i$) $(a \neq 0)$

$E(X_j) = \mu$    $(j \neq i)$

with variance-covariance matrix $V(X_j) = V$ $(j = 1, 2, \ldots, n)$.

*Model B*    $V(X_i) = bV$    (some $i$) $(b > 1)$

$V(X_j) = V$    $(j \neq i)$

with mean vector $E(X_j) = \mu (j = 1, 2, \ldots, n)$.

Initially, we shall examine a test of discordancy based on the *two-stage maximum likelihood ratio principle* (that is, *Principle B* of Section 7.1). We consider *models A* and *B* separately, with various assumptions about what parameter values are known.

*Model A, V known*

On the basic model the likelihood of the sample $x_1, x_2, \ldots, x_n$ is proportional to

$$P_\mu(x \mid V) = \frac{1}{|V|^{n/2}} \exp\left\{ -\frac{1}{2} \sum_{j=1}^{n} (x_j - \mu)' V^{-1}(x_j - \mu) \right\}. \qquad (7.3.1)$$

The maximized log-likelihood is (apart from the constant term)

$$L(x \mid V) = -\frac{1}{2} \sum_{j=1}^{n} (x_j - \bar{x})' V^{-1}(x_j - \bar{x}). \qquad (7.3.2)$$

Under the alternative (*model A*) hypothesis of a single contaminant the corresponding maximized log-likelihood is

$$L_A(\mathbf{x} \mid V) = -\frac{1}{2} \sum_{j \neq i} (\mathbf{x}_j - \bar{\mathbf{x}}')' V^{-1}(\mathbf{x}_j - \bar{\mathbf{x}}') \qquad (7.3.3)$$

where $\bar{\mathbf{x}}'$ is the sample mean of the $(n-1)$ observations excluding $\mathbf{x}_i$ and $i$ is chosen to maximize

$$L_A(\mathbf{x} \mid V) - L(\mathbf{x} \mid V).$$

Thus we are led to declare as the outlier $\mathbf{x}_{(n)}$ that observation $\mathbf{x}_i$ for which $R_i(\bar{\mathbf{x}}, V) = (\mathbf{x}_i - \bar{\mathbf{x}})' V^{-1}(\mathbf{x}_i - \bar{\mathbf{x}})$ is a maximum, so that implicitly the observations have been ordered in terms of the reduced form of sub-ordering based on the distance measure $R(\mathbf{x}; \bar{\mathbf{x}}, V)$ already introduced with intuitive support in Section 7.1. It also arises directly from the application of **Principle A** of Section 7.1. Furthermore we will declare $\mathbf{x}_{(n)}$ a *discordant* outlier if

$$R_{(n)}(\bar{\mathbf{x}}, V) \equiv (\mathbf{x}_{(n)} - \bar{\mathbf{x}})' V^{-1}(\mathbf{x}_{(n)} - \bar{\mathbf{x}}) = \max_{j=1,\dots,n} R_j(\bar{\mathbf{x}}, V)$$

is significantly large.

The null distribution of $R_{(n)}(\bar{\mathbf{x}}, V)$ is not readily determined in exact form nor very tractable.

However, it has been studied by Siotani (1959) who discusses the problems associated with determining percentage points of $R_{(n)}(\mathbf{x}_0, \Gamma)$ when $\Gamma = V$ and $\mathbf{x}_0$ is either $\mathbf{0}$, $\boldsymbol{\mu}$ or $\bar{\mathbf{x}}$. For the latter case, and of immediate relevance to us at the present stage of the discussion, he presents approximate upper 5, $2\frac{1}{2}$, and 1 per cent points of $R_{(n)}(\bar{\mathbf{x}}, V)$ for $p = 2(1)4$ and $n = 3(1)10(2)20(5)30$. The critical values for 5 per cent and 1 per cent tests of discordancy of a single outlier in a multivariate normal sample where $V$ is known are reproduced as Table XXX on page 516.

Marco, Young, and Turner (1987) show that this test is robust (identical in its properties) if the observations $\mathbf{x}_i$ are not independent, but each pair $\mathbf{x}_i, \mathbf{x}_j$ has the same correlation matrix. See also Young, Pavur, and Marco (1989) and Rasmussen (1988).

If $\boldsymbol{\mu}$ were known, $R(\mathbf{x}; \boldsymbol{\mu}, V)$ would be the appropriate reduction measure and the $R_j(\boldsymbol{\mu}, V)$ would be independent $\chi_p^2$ variates. We would then have to relate their maximum $R_{(n)}(\boldsymbol{\mu}, V)$ to the distribution of the maximum observation in a random sample of size $n$ from a $\chi_p^2$ distribution.

Gupta (1960) has considered the distribution of the order statistics from gamma samples and has tabulated percentage points for a range of sample sizes, and values of the two parameters. (Note however, that the last six lines of some of the tables in Gupta (1960) are incorrect; they are revised in the 'Errata' section of the journal *Technometrics*, 1960, **2**, 523.) Suitably extracted and modified values serve for the outlier problem with $\boldsymbol{\mu}$ and $V$ known, and Table XXXI on page 516 presents upper 5 per cent and 1 per cent points of $R_{(n)}(\mathbf{\mu}, V)$ for $p = 2(2)10$ and $n = 3(1)20$, 25, 50, 100, 200, 500, 1000. Note that only *even* values of $p(\geq 2)$ are accessible from this table.

In the particular case of a bivariate sample ($p = 2$), $R_{(n)}(\mathbf{\mu}, V)/2$ has the distribution of the maximum of $n$ independent exponential variates (mean 1) and its percentage points are easily determined. For a level-$\alpha$ test we would conclude that $\mathbf{x}_{(n)}$ (the observation $\mathbf{x}_i$ yielding $R_{(n)}(\mathbf{\mu}, V)$) is a discordant outlier if $R_{(n)}(\mathbf{\mu}, V) > \xi_\alpha$ where

$$\alpha = P\{R_{(n)}(\mathbf{\mu}, V) > \xi_\alpha\} = 1 - \{F(\xi_\alpha/2)\}^n \qquad (7.3.4)$$

with

$$F(x) = 1 - e^{-x}.$$

Thus

$$\xi_\alpha = -2\ln[1 - (1 - \alpha)^{1/n}] \qquad (7.3.5)$$

provides an explicit value for use in the test.

Some care is needed in using some of the published proposals for testing outliers in multivariate normal samples. Barnett (1983b) refers to one particular proposal where there is probability approaching one (as the sample size increases) of detecting at least one discordant outlier in any *non-contaminated* sample!

The assumption of known $V$ is in general unrealistic. We therefore proceed to examine the two-stage maximum likelihood ratio test for *model A* when both $\mathbf{\mu}$ and $V$ are unknown.

*Model A, V unknown*

With $V$ unknown (as well as $\mathbf{\mu}$) the maximized log-likelihood under the basic model is (apart from the constant factor)

$$L(\mathbf{x}) = -\frac{n}{2}\log|A| \qquad (7.3.6)$$

where $A$ is the matrix of sums of squares and cross-products of the observations about the component sample means: that is

$$A = \sum_{j=1}^{n} (\mathbf{x}_j - \bar{\mathbf{x}})(\mathbf{x}_j - \bar{\mathbf{x}})'. \qquad (7.3.7)$$

Under the *model A* alternative the maximized log-likelihood is

$$L_A(\mathbf{x}) = -\frac{n}{2}\log|A^{(i)}| \qquad (7.3.8)$$

where $A^{(i)}$ is the restricted matrix obtained on omission of $\mathbf{x}_i$ and $i$ is chosen to maximize

$$L_A(\mathbf{x}) - L(\mathbf{x}).$$

Thus when $V$ is unknown it seems at first sight that quite a different principle is advanced for the declaration of an outlier $\mathbf{x}_i$ and for the assessment of its discordancy. Here we are implicitly ordering the multivariate observations in terms of reduced sub-ordering based on the values of $|A^{(j)}|$. The $|A^{(j)}|$ are ordered, and the observation corresponding with the smallest value of $|A^{(j)}|$ is declared an outlier. Equivalently, if we denote

$$\mathscr{R}_j = \frac{|A^{(j)}|}{|A|}$$

the sample points are 'ordered' in accord with the ordered $\mathscr{R}_j$ and the outlier is that observation corresponding with the smallest $\mathscr{R}_j$, $\mathscr{R}_{(1)}$. If $\mathscr{R}_{(1)}$ is significantly low in value the outlier is adjudged discordant. Thus the outlier is that observation whose removal from the sample effects the greatest reduction in the 'internal scatter' of the data set. But the distinction of principle for declaring an outlier in the case of unknown $V$, compared with the case where $V$ is known, turns out to be less profound than might appear at first sight. Clearly we can rewrite

$$\mathscr{R}_j = \frac{\left| A - \left(\dfrac{n}{n-1}\right)(\mathbf{x}_j - \bar{\mathbf{x}})(\mathbf{x}_j - \bar{\mathbf{x}})' \right|}{|A|} = 1 - \left(\frac{n}{n-1}\right) R_j(\bar{\mathbf{x}}, A) \quad (7.3.9)$$

and minimization of $\mathscr{R}_j$ becomes equivalent to maximization of

$$R_j(\bar{\mathbf{x}}, A) = R_j(\bar{\mathbf{x}}, S)/(n-1). \quad (7.3.10)$$

Thus the outlier is *again* that observation whose 'distance' from the body of the data set measured in terms of the familiar quantity (7.1.1) is a maximum, provided we replace $\mu$ and $V$ by $\bar{\mathbf{x}}$ and $S$. (This supports informal proposals by Healy, 1968; see Section 7.4.3 for more details.)

Once more the distribution of the test statistic is highly complicated. Little is known in detail of the joint distribution of the $\mathscr{R}_j$ or more particularly of the distribution of the minimum, $\mathscr{R}_{(1)}$.

However, there is a deal of useful tabulated material on approximate percentage points for $\mathscr{R}_{(1)}$, and on the corresponding statistic for assessing the discordancy of a *pair* of outliers in a multivariate normal sample. The tables appear in work by Wilks (1963) which was the first detailed applications-oriented study of outlier detection in multivariate normal data. We shall consider this work in some detail and apply it to a problem concerned with yields of grass in a long-term agricultural experiment.

Interested in testing outlying observations in a sample from a multivariate normal distribution with unknown mean vector and variance-covariance matrix, Wilks (1963) proposes an intuitively based representation of the sample in terms of the sum of squares of the volumes of all simplexes that can be found from $p$ of the sample points augmented by the sample mean $\bar{x}$. He shows (Wilks, 1962) that this is just $(p!)^{-2}|A|$, where $A$ is the matrix defined above. He calls $|A|$ the *internal scatter* of the sample and suggests that a sensible criterion for the declaration of an outlier is to choose that sample member whose omission leads to the least value for the so-called *one-outlier scatter ratio*

$$\mathcal{R}_j = \frac{|A^{(j)}|}{|A|}.$$

But this is precisely the likelihood ratio criterion and corresponding test statistic. Wilks shows that the $\mathcal{R}_j$ are identically distributed *Beta* variates $\mathcal{B}((n - p - 1)/2, p/2)$ with a joint distribution symmetric over $R^n$ subject to

$$\sum \mathcal{R}_j = n\left(1 - \frac{p}{n - 1}\right)$$

$$0 \leqslant \mathcal{R}_j \leqslant 1 \qquad (j = 1, 2, \ldots, n).$$

The joint distribution is intractable, but Wilks uses Bonferroni inequalities to obtain an upper bound for the distribution function of $\mathcal{R}_{(1)}$ (which he denotes $r_1$), and hence lower bounds for the lower percentage points of $\mathcal{R}_{(1)}$ thus enabling conservative tests of significance for a single outlier to be conducted. In comparison with exact results due to Grubbs (1950) for the case $p = 1$, the approximate values seem reasonable, though it must be stressed that their accuracy for $p > 1$ has not been thoroughly assessed since there is at present no yardstick (in terms of *exact* probabilities) for comparison. However, experience elsewhere with Bonferroni inequalities (e.g. for outliers in linear model data) suggests that the approximations should be good.

Wilks (1963) tabulates lower bounds to the lower 10, 5, $2\frac{1}{2}$, and 1 per cent points of $\mathcal{R}_{(1)}$ for $p = 1(1)5$ and $n = 5(1)30(5)100(100)500$. For closer comparability with Tables XXX and XXXI, selected values from Wilks's tables have been transformed via (7.3.9) and (7.3.10) into approximate upper percentage points for $R_{(n)}(\bar{x}, S)$. Table XXXII on page 517 presents critical values for 5 per cent and 1 per cent tests of discordancy for a single outlier in a multivariate normal sample where $\mu$ and $V$ are estimated by $\bar{x}$ and $S$, respectively. The table covers the ranges $p = 2(1)5$, $n = 5(1)10(2)20(5)50$, 100, 200, 500. *Note that* the elements of $S$ are *unbiased* sample variance and covariance estimates (with divisors $n - 1$). Extended tables (with 20, 10, 5, 2.5, 1 and 0.5 per cent critical values for $p = 2(1)10$,

12, 15, 20 and a similar range of sample sizes to that of Table XXXII appear in Jennings and Young (1988). These were obtained by simulation. Wilks adopts a similar approach to the testing *en bloc* of 2, 3, or 4 outliers in a multivariate sample, by considering for the *s*-outlier case (*s* = 2, 3, 4) the *s-outlier scatter ratios*

$$\mathscr{R}_{j_1, j_2, \ldots, j_s} = \frac{|A^{(j_1, j_2, \ldots, j_s)}|}{|A|}$$

where $|A^{(j_1, j_2, \ldots, j_s)}|$ is the internal scatter when $x_{j_1}, x_{j_2}, \ldots, x_{j_s}$ are omitted from the sample. Again it is the subset of observations that minimizes $\mathscr{R}_{j_1, j_2, \ldots, j_s}$, which is declared the outlying subset and their discordancy must be assessed in terms of how small is

$$r_s = \min \mathscr{R}_{j_1, j_2, \ldots, j_s}.$$

For $s = 2$, an outlying subset of two observations, Wilks tabulates square roots of lower bounds for the lower percentage points of $r_2$. Extracted values are reproduced as Table XXXIII on page 518 for precisely the same set of significance levels and values of $p$ and $n$ as are used in Table XXXII. Thus Table XXXIII gives critical values for 5 per cent and 1 per cent tests of discordancy (based on $\sqrt{r_2}$) of outlier pairs in multivariate normal samples where $\mu$ and $V$ are unknown for $p = 2(1)5$ and $n = 5(1)10(2)20(5)50, 100, 200, 500$.

Fung (1988) considers Wilks 2 and 3 outlier tests in more detail. Critical values are presented for block tests for two and three outliers for significance levels 0.1, 0.05, and 0.01 and sample sizes $n = 10(1)20(2)30(5)50$ [for two outliers] and $n = 15(1)20(2)30(5)50$ [for three outliers] for dimensions $p = 2, 3, 4,$ and 5. The critical values are obtained from approximations of Andrews (1971).

*Example 7.1   Rothamsted Experimental Station is one of the oldest agricultural research centres. One of the 'classical experiments' is Park Grass in which the growth of grasses has been monitored under various fixed treatment regimes for about* 150 *years. Table 7.1 presents the yields of grass (in t/ha dry matter) on two totally untreated plots for the 50 years from 1941 to 1990 (reproduced by permission of Rothamsted Experimental Station). Figure 7.3 is a scatter diagram of these yields. Several observations appear as outliers (notably those for* 1965, 1971, 1969, 1982 *and* 1942 *marked as A, B, C, D and E on Figure 7.3). It is interesting to test some of them for discordancy. For illustrative purposes we assume a bivariate normal distribution for yields, although no detailed study has been made of the underlying distribution.*

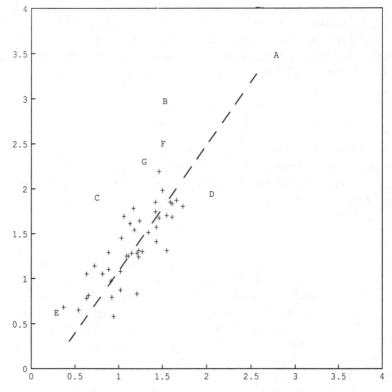

**Figure 7.3**  Rothamsted Park Grass plots 3 and 12, 1941–1990

*Calculations show that A, B, C, D, and E produce the five largest values of $R_i(\bar{x}, S)$ (in decreasing order). We have*

$$A: \quad R_{(50)}(\bar{x}, S) = 14.07$$
$$B: \quad R_{(49)}(\bar{x}, S) = 10.87$$
$$C: \quad R_{(48)}(\bar{x}, S) = \phantom{0}7.53$$
$$D: \quad R_{(47)}(\bar{x}, S) = \phantom{0}5.49$$
$$E: \quad R_{(46)}(\bar{x}, S) = \phantom{0}4.55$$

*From Table XXXII we find that A is a discordant outlier; the 5 per cent critical value (for $n = 50$, $p = 2$) is 12.23.*

*If we test the outlier pair (A, B) on Wilks's test we have $\sqrt{r_2} = 0.727$ which is significant almost at the 1 per cent level (the 1% critical value is 0.733). Of course, the two tests (of a single outlier, and of an outlier pair) are not independent.*

*An alternative approach would be to conduct a sequential series of tests, omitting at each stage the observation yielding the largest R-value and reducing the sample size by one (with appropriate recalculation at each stage). The overall size of such a series of tests is not easily determined but it is interesting to observe what happens in this example. We have the following results:*

| Stage | Largest R-value | Year | Observation |
|-------|-----------------|------|-------------|
| 1 | 14.07 | 1965 | A |
| 2 | 11.91 | 1971 | B |
| 3 | 9.00 | 1969 | C |
| 4 | 8.04 | 1964 | F |
| 5 | 9.39 | 1979 | G |

*We note that for the first three stages we obtain A, B, and C: the three 'most extreme' observations in terms of $R_i(\bar{x}, S)$ for the full sample. At stages 4 and 5, however, two new candidates enter (observations F and G in Figure 7.3 for 1964 and 1979).*

**Table 7.1** Yields (in $t/ha$ dry matter) for plots 3 and 12 from 1941 to 1990

| Year | Plot 3 | Plot 12 | Year | Plot 3 | Plot 12 |
|------|--------|---------|------|--------|---------|
| 1941 | 0.85 | 1.26 | 1966 | 1.43 | 2.16 |
| 1942 | 0.26 | 0.59 | 1967 | 1.31 | 1.48 |
| 1943 | 1.03 | 1.66 | 1968 | 1.52 | 1.28 |
| 1944 | 0.34 | 0.65 | 1969 | 0.72 | 1.87 |
| 1945 | 1.14 | 1.75 | 1970 | 1.15 | 1.51 |
| 1946 | 1.18 | 0.80 | 1971 | 1.50 | 2.94 |
| 1947 | 1.52 | 1.67 | 1972 | 1.40 | 1.54 |
| 1948 | 1.12 | 1.25 | 1973 | 1.24 | 1.27 |
| 1949 | 0.62 | 0.78 | 1974 | 1.18 | 1.25 |
| 1950 | 0.89 | 0.76 | 1975 | 0.91 | 0.55 |
| 1951 | 1.00 | 1.42 | 1976 | 1.06 | 1.22 |
| 1952 | 1.58 | 1.80 | 1977 | 1.20 | 1.21 |
| 1953 | 1.63 | 1.84 | 1978 | 1.70 | 1.77 |
| 1954 | 0.99 | 1.05 | 1979 | 1.26 | 2.27 |
| 1955 | 1.10 | 1.58 | 1980 | 0.85 | 1.07 |
| 1956 | 0.69 | 1.11 | 1981 | 1.47 | 1.95 |
| 1957 | 0.60 | 1.02 | 1982 | 2.03 | 1.91 |
| 1958 | 1.21 | 1.61 | 1983 | 0.99 | 0.84 |
| 1959 | 0.51 | 0.62 | 1984 | 1.08 | 1.22 |
| 1960 | 1.56 | 1.82 | 1985 | 1.58 | 1.65 |
| 1961 | 1.39 | 1.82 | 1986 | 0.78 | 1.02 |
| 1962 | 1.20 | 1.28 | 1987 | 1.39 | 1.71 |
| 1963 | 1.43 | 1.64 | 1988 | 1.40 | 1.38 |
| 1964 | 1.48 | 2.47 | 1989 | 0.60 | 0.75 |
| 1965 | 2.75 | 3.45 | 1990 | 0.88 | 0.94 |

Caroni and Prescott (1992) present a detailed sequential application of the Wilks's single outlier test for testing up to $k$ outliers with controlled probability of type 1 error overall. It uses the same principles as Rosner (1975, 1977) and presents tabulated material to implement the procedure.

We should remark that the various formal tests of discordancy for multivariate normal outliers discussed in this section have the desirable property of being invariant with respect to the location and scale of the measurement basis of the observations.

For *model B*, having a single contaminant in a normal sample, Ferguson (1961a) has derived a multi-decision procedure (see Section 4.1.2) with certain optimal properties. Here $\mu$ and $V$ are still unknown, and contamination is reflected in inflation of $V$ for a single observation. See also Kudo (1957).

Denoting by $\mathscr{D}_i$ the decision to regard $\mathbf{x}_i$ as the contaminant $(i = 1, 2, \ldots, n)$ with $\mathscr{D}_0$ the decision to declare no contamination of the sample, Ferguson considers those decision rules which satisfy four conditions:

(a) each is invariant under the addition to $\mathbf{X}_i$ of a constant vector;
(b) each is invariant under the multiplication of $\mathbf{X}_i$ by a common non-singular matrix;
(c) the probability $p_i(\mathscr{D}_i)$ of declaring $\mathbf{x}_i$ the contaminant when this is true is independent of $i$;
(d) the probability of correctly declaring no contamination is $1 - \alpha$, for a preassigned $\alpha$ in $(0, 1)$; that is, the procedure has size $\alpha$.

He seeks that decision rule which maximizes $p_i(\mathscr{D}_i)$. It turns out to have a familiar form: the optimum rule is to reject $\mathbf{x}_i$ as the discordant value if $\mathbf{x}_i$ yields the maximum value $R_{(n)}(\bar{\mathbf{x}}, S)$ and

$$R_{(n)}(\bar{\mathbf{x}}, S) > K$$

where $K$ is chosen to satisfy the test size condition $(d)$.

Thus, as for *model A*, we again declare the outlier to be the observation with maximum generalized distance $R_j(\bar{\mathbf{x}}, S)$, and assess it as discordant if that maximum, $R_{(n)}(\bar{\mathbf{x}}, S)$, is sufficiently large.

Additionally, however, Ferguson demonstrates that this procedure is *uniformly best over all values of $b > 1$*.

Thus when $\mu$ and $V$ are unknown it is immaterial whether we adopt the *model A* or *model B* formulation of the alternative hypothesis describing the occurrence of a single outlier. In either case the test has the same form, and can be implemented by using the Table XXXII on page 517.

Summarizing the status of the (identical) *model A*, and *model B*, tests of discordancy when $\mu$ and $V$ are unknown, we have that, in relation to *model A*, the test is the likelihood ratio test, whereas, relative to *model B*, it has the uniform optimality property of maximizing $p_i(\mathcal{D}_i)$ irrespective of the value of $b$.

Barnett (1978b) summarizes the properties of discordancy tests for multivariate normal data.

Siotani (1959) tabulates approximate percentage points for a studentized form of $R_{(n)}(\bar{\mathbf{x}}, S)$ where $S$ is replaced by an external unbiased estimate $S_v$ of $V$ having a Wishart distribution with $v$ degrees of freedom. These are of value for an informal test of discordancy of a single outlier in a multivariate sample, where $V$ is *not* estimated from the sample itself but by means of such an external estimate.

Table XXXIV on pages 519–20, extracted from Siotani (1959), presents approximate 5 per cent an 1 per cent critical values for a test of a single outlier in a bivariate normal sample ($p = 2$) where $\boldsymbol{\mu}$ and $V$ are unknown, and $V$ is estimated by $S_v$. The table covers the values $n = 3(1)14$ and $v = 20(2)40(5)60$, 100, 150, 200.

Schwager and Margolin (1982) further consider the mean-slippage alternative model and seek discordancy tests which are invariant under a natural set of transformations. They show that the locally best invariant test is essentially based on a *multivariate sample kurtosis coefficient*

$$b_{2,p} = n \sum_{j=1}^{n} [(\mathbf{x}_j - \bar{\mathbf{x}})' S^{-1}(\mathbf{x}_j - \bar{\mathbf{x}})]^2$$

(see Mardia, 1970, 1974, 1975a) with rejection of the no-contamination model under typical practical conditions for sufficiently large values of the kurtosis coefficient.

Mardia and Kanazawal (1983) discuss the null distribution of $b_{2,p}$ and Stapanian *et al.* (1991) present simulation-based critical values for selected $p$ and $n$ up to 25 and 500, respectively.

We return to the use of $b_{2,p}$ in a more general context in Section 7.3.4 after we have considered the specific cases of exponential and Pareto distributions.

### 7.3.2 Multivariate exponential samples

Frequently we encounter multivariate data for which the normal distribution is quite unsuitable, in view of manifest *skewness*. Thus we need to consider models expressing skewness and we shall consider in this section, and the next, two such prospects: a *multivariate exponential* model and a *multivariate Pareto* model.

Many forms of multivariate exponential distribution have been proposed. One of these, due to Gumbel (1960), has for the bivariate case a probability density function

$$f(x_1, x_2) = [(1 + \theta x_1)(1 + \theta x_2) - \theta] \exp(- x_1 - x_2 - \theta x_1 x_2) \qquad (7.3.11)$$

with $x_1$, $x_2 > 0$ and the association parameter $\theta$ in the range $(0, 1)$. The product moment correlation coefficient is

$$\rho = \frac{1}{\theta} e^{-1/\theta} \int\limits_{1/\theta}^{\infty} \frac{e^{-z}}{z} \, dz - 1$$

which varies monotonically from 0 to $-0.4037$ as $\theta$ goes from 0 to 1 (although such a measure has dubious relevance for such a highly skew distribution).

Since the bivariate exponential random variable $\mathbf{X}$ is restricted to the upper quadrant it is possible to distinguish a directional concept of extremeness and Barnett (1979) has considered a discordancy test for an 'upper' outlier. Applying a directional form of **Principle A** of Section 7.1 an appropriate reduction measure

$$R(\mathbf{X}) = X_1 + X_2 + \theta X_1 X_2 \qquad (7.3.12)$$

is obtained. Thus an upper outlier is detected as that observation $(x_{1i}, x_{2i})$ which yields the largest value of $R(\mathbf{x})$ over the sample of $n$ observations. It is adjudged discordant if the corresponding $R_{(n)} = \max_i R(\mathbf{x}_i)$ is sufficiently large.

The distribution of $R_{(n)}$ is tractable. The random variable $T = 1 + \theta R$ has distribution function

$$H(t) = 1 - e^{-(t-1)/\theta} \left( \frac{t \ln t}{\theta} + 1 \right).$$

Thus $R_{(n)}$ has distribution function

$$G_{(n)}(r) = \left\{ 1 - \frac{e^{-r}}{\theta} [(1 + \theta r)\ln(1 + \theta r) + \theta] \right\}^n$$

and the critical value $\beta_\alpha$ for a level-$\alpha$ test of discordancy satisfies

$$\beta_\alpha = \ln\left[ 1 + \frac{(1 + \theta\beta_\alpha)\ln(1 + \theta\beta_\alpha)}{\theta} \right] - \ln[1 - (1 - \alpha)^{1/n}]. \qquad (7.3.13)$$

We can easily solve the implicit equation (7.3.13) numerically and Table XXXV (freshly compiled) on page 520 gives 5 per cent and 1 per cent critical values for $\theta = 0.1(0.1)1.0$ and $n = 4(1)10, 12, 15, 20(10)40, 60, 100$.

Of course, the above results assume that $\theta$ is known. Estimation of $\theta$ is difficult for the distribution (7.3.11), but consideration of the range of values of $\beta_\alpha$ in Table XXXV for different values of $\theta$ at a given sample size shows relative stability and a conservative assessment of discordancy is feasible. Alternatively, we could face up directly to any lack of knowledge of the value of $\theta$, and employ an estimator $\tilde{\theta}$ and an associated reduction measure $\tilde{R}(\mathbf{x})$. Some simulation results for a test of discordancy using a particular form of $\hat{\theta}$ are given by Barnett (1979)—maximum discrepancies of 5 to 10 per cent (for given $n$ and $\alpha$) are encountered in comparison with the results of Table XXXV.

Generalizations of this test from bivariate to higher-dimensional data are straightforward in principle.

### 7.3.3 Multivariate Pareto samples

Barnett (1979) also considers a discordancy test for an 'upper' outlier in another skew bivariate distribution: namely, that one of the two Pareto distributions considered by Mardia (1962) which has probability density function

$$f(x_1, x_2) = a(a + 1)(\theta_1\theta_2)^{a + 1}(\theta_2 x_1 + \theta_1 x_2 - \theta_1\theta_2)^{-(a + 2)} \qquad (7.3.14)$$

for $x_1 \geq \theta_1 \geq 0$, $x_2 \geq \theta_2 \geq 0$ and $a > 0$. The correlation coefficient is $\rho = a^{-1}(a > 2)$.

This time the appropriate restricted form of **Principle A** (assuming $\theta_1$, $\theta_2$ and $a$ known) yields a reduction measure

$$R(\mathbf{X}) = \frac{X_1}{\theta_1} + \frac{X_2}{\theta_2} - 1 \qquad (7.3.15)$$

in place of (7.3.12). We now find that $R$ has distribution function

$$G(r) = 1 - r^{-a}(1 + a + a/r)$$

and the sample maximum $R_{(n)}$ has distribution function $H(r_{(n)}) = [G(r_{(n)})]^n$, so that the critical value $\gamma_\alpha$ for a level-$\alpha$ discordancy test for an upper outlier is obtained from

$$\delta\gamma_\alpha^{(a + 1)} - (a + 1)\gamma_\alpha + a = 0, \qquad (7.3.16)$$

with $\delta = 1 - (1 - \alpha)^{1/n}$.

Again, iterative numerical solution readily yields $\gamma_\alpha$ for given $a$, and Table XXXVI (from Barnett, 1979) on page 521 presents 5 per cent and 1 per cent critical values for $a = 2.5(0.5)4$, 5, 7, 10, 20, 40 and $n = 4(1)6(2)12$, 15, 20, 30, 40. Once more the need to estimate $\theta_1$, $\theta_2$ (and $a$) presents serious difficulties. Replacing $\theta_1$ and $\theta_2$ by the intuitively appealing (if not fully efficient) estimators $x_{1(1)}$ and $x_{2(1)}$ (the marginal-sample minima) shows that Table XXXVI yields a conservative test, on the observation that

$$\frac{x_{1j}}{x_{1(1)}} + \frac{x_{2j}}{x_{2(1)}} \leq \frac{x_{1j}}{\theta_1} + \frac{x_{2j}}{\theta_2} \quad (\text{all } j = 1, 2, \ldots, n).$$

The accuracy of this conservative test has also been examined by simulation in Barnett (1979). Critical values $\gamma_\alpha$ are highly dependent on the value of $a$ and more work is needed to deal with the case where $a$ is unknown.

### 7.3.4 Discordancy tests for more general distributions

There have been several studies of outlier tests when the basic model $F$ takes a more general form. Specifically, $F$ is assumed to be an **elliptically symmetric distribution** with probability density function,

$$f(x) \propto |V|^{-1/2} g\{(x - \mu)' V^{-1}(x - \mu)\}$$

for some non-negative function $g(\cdot)$ and $p \times p$ positive definite matrix, $V$.

Sinha (1984) introduced this model and showed that in this more general case a locally optimum test of a single outlier can be produced (for a location-slippage alternative model) using the Mardia (1970) *multivariate kurtosis statistic*, $b_{2,p}$, described in Section 7.3.1 above, thus extending the results of Schwager and Margolin (1982).

Das and Sinha (1986) show that the same test based on the Mardia kurtosis statistic is also locally optimum when the alternative model has a dispersion-slippage form. The mean-slippage, and dispersion-slippage, alternative models are immediate multivariate generalizations of the Ferguson (1961a) models, described at the beginning of Section 7.3.1 above.

Further work characterizing multivariate elliptical distributions in terms of the multivariate coefficient of kurtosis, with applications to outlier testing appears in Berkane and Bentler (1987, 1988). Hara (1988) describes an optimum decision rule for multiple outliers in the context of (*inter alia*) elliptically contoured distributions with location slippage or scale inflation.

### 7.3.5 A transformation approach

We have seen that discordancy tests for outliers in multivariate normal samples have simple forms, desirable properties and extensively tabulated critical values. It is tempting to enquire whether we could handle outliers in samples from specific non-normal multivariate distributions *by initially transforming the data to normality*. This prospect has been considered by Barnett (1983a) and some proposals have been made.

Suppose we are sampling a bivariate random variable $(X_1, X_2)$ with marginal and conditional distributional functions denoted by $F_{X_1}(x_1)$, $F_{X_2}(x_2)$ and $F_{X_1|x_2}(x_1)$ and $F_{X_2|x_1}(x_2)$, respectively. (Extensions to higher-dimensional data sets are possible.) Consider the transformation $(X_1, X_2) \rightarrow (U_1, U_2)$ effected by solving

$$\left. \begin{array}{l} F_{X_1}(x_1) \quad = \Phi(u_1) \\ F_{X_2|x_1}(x_2) = \Phi(u_2) \end{array} \right\} \qquad (7.3.17)$$

where $\Phi(\ )$ is the distribution function of the standard normal distribution, $N(0, 1)$. It is readily confirmed that $U_1$ and $U_2$ are independent $N(0, 1)$.

The idea is that we use (7.3.17) to transform our random sample $(x_{11}, x_{21}), (x_{12}, x_{22}), \ldots, (x_{1n}, x_{2n})$ to $(u_{11}, u_{21})(u_{12}, u_{22}), \ldots, (u_{1n}, u_{2n})$, detecting an outlier as that observation with the maximum value of the appropriate (normal-distribution-based) reduction measure: which is, of course, $R(U) = U_1^2 + U_2^2$. We can then test the discordancy of the outlier by reference to the appropriate null distribution. In fact, $R_{(n)}/2$ is distributed as the largest order statistic from a random *exponential* sample of size $n$, so that the critical value $\xi_\alpha$ for a level-$\alpha$ test is just

$$\xi_\alpha = -2\ln[1 - (1 - \alpha)^{1/n}].$$

It all sounds too simple and of course there are some difficulties and some limitations.

The reduction measure $R(U)$ is a natural one for the *transformed* data set. It implies, of course, that *some other* measure has effectively been used in $(X_1, X_2)$ space. It would be interesting to enquire what appeal the implied measure has in the *original* space. In particular, the implied measure cannot be directional (cf. study of 'upper' outliers in Sections 7.3.2 and 7.3.3). Use of the transformation (7.3.17) can involve some complicated algebraic manipulation in particular cases—we should also note that (7.3.17) is somewhat arbitrary in form: for example, we could have used the marginal distribution of $X_2$ and the conditional distribution of $X_1 | x_2$. These points are discussed, and illustrated for the exponential and Pareto distributions (of Sections 7.3.2 and 7.3.3) by Barnett (1983a). The most serious restriction, however, is that the (uncontaminated) distribution of $(X_1, X_2)$ is assumed to be fully specified for the above procedure to be operable. If any parameters of the distribution are unknown in value they would have to be estimated. The method would of course need to take account of this matter, but no general proposals have yet been made for handling this extended prospect.

## 7.4 INFORMAL METHODS FOR MULTIVARIATE OUTLIERS

A host of informal proposals have been made for detecting outliers in multivariate data by quantitative or graphical methods. These cannot be regarded as tests of discordancy; they may be based on derived reduction measures (but with no supporting distribution theory) or, more commonly, they are presented simply as aids to intuition in picking out multivariate observations which are suspiciously aberrant from the bulk of the sample.

In univariate samples the 'suspicious' observation which is to be declared an outlier is obvious on simple inspection. It is an extreme observation in the sample. In multivariate data the extremeness concept is a nebulous one, as we have remarked above. Various forms of initial processing of the data, involving transformation, study of individual marginal components of the observations, judicious reduction of the multivariate observations to scalar quantities in the forms of reduction measures or linear combinations of components, changes in the coordinate bases of the observations, and appropriate methods of graphical representation, can all help to identify or highlight a suspicious observation. If several such procedures are applied simultaneously (or individually) to a set of data they can help to overcome the difficulty caused by the absence of a natural overall ordering of the sample members. An observation which clearly stands out on one, or preferably more, processed re-representations of the sample becomes a firm candidate for identification as an outlier.

Gnanadesikan and Kettenring (1972) remark:

The consequences of having . . . [outliers] in a multivariate sample are intrinsically more complex than in the much discussed univariate case. One reason for this is that a multivariate outlier can distort not only measures of location and scale but also those of orientation (i.e. correlation). A second reason is that it is much more difficult to characterize the multivariate outlier. A single univariate outlier may be typically thought of as 'the one which sticks out on the end', but no such simple idea suffices in higher dimensions. A third reason is the variety of types of multivariate outliers which may arise: a vector response may be faulty because of a gross error in one of its components or because of systematic mild errors in all of its components.

The complexity of the multivariate case suggests that it would be fruitless to search for a truly omnibus outlier protection procedure. A more reasonable approach seems to be to tailor detection procedures to protect against specific types of situations, e.g., correlation distortion, thus building up an arsenal of techniques with different sensitivities. This approach recognizes that an outlier for one purpose may not necessarily be one for another purpose! However, if several analyses are to be performed on the same sample, the result of selective segregation of outliers should be more efficient and effective use of the available data.

An early example of an informal graphical procedure is described by Healy (1968) who proposes plotting the ordered $R_j(\bar{\mathbf{x}}, S)$ against the expected values of the order statistics of a sample of size $n$ from $\chi_p^2$. He is principally concerned with bivariate data ($p = 2$). Outliers are detected as observations yielding values $R_{(n)}(\bar{\mathbf{x}}, S)$, $R_{(n-1)}(\bar{\mathbf{x}}, S)$ . . . lying above the expected straight line (if $X$ were normal and $\mu$, $V$ known). Healy considers $R_j(\bar{\mathbf{x}}, S)$ on intuitive grounds, rather than from any formal (alternative-model, or test-construction) principle. It is of interest to consider an example which Healy describes.

*Example 7.2   A bivariate sample of 39 observations of the logarithms of daily fat intake and serum cholesterol level for a group of hospital patients (data taken from Begg, Preston, and Healy, 1966) have values of $R_j(\mathbf{x}, S)$ calculated, where $S$ is the sample variance–covariance matrix. Ignoring the* *inaccuracies introduced by estimating* $\mu$ *and* $V$, *we might consider plotting the ordered 'distances'* $R_j$ *against the expected values*

$$\frac{2}{n}, \frac{2}{n} + \frac{2}{n-1}, \frac{2}{n} + \frac{2}{n-1} + \frac{2}{n-2}, \ldots, \frac{2}{n} + \frac{2}{n-1} \ldots + \frac{2}{1}$$

*(with* $n = 39$*) of the order statistics for a sample from* $\chi_2^2$. *Instead, Healy introduces a further level of approximation, based on the approximate*

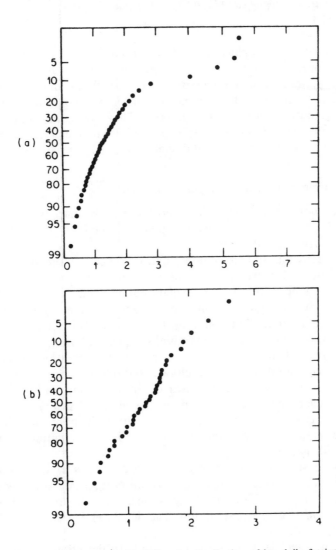

**Figure 7.4** Normal plots of $\sqrt{R_j}(\overline{\mathbf{x}}, S)$ for the distribution of log daily fat intake and log serum cholesterol in a sample of 39 hospital patients. (a) Complete sample: (b) omitting four extreme values (reproduced by permission of the Royal Statistical Society)

*normality of $\sqrt{\chi^2}$, on the grounds that this simplifies the graphical process and even enhances the prospects of distinguishing outliers (in view of the reduced coefficient of variation of upper extreme sample values). Thus the square roots of the ordered distances for the sample of 39 observations are plotted on normal probability paper, with the results shown in Figure 7.4(a).*

*He concludes that the bivariate normal model is not unreasonable, but that there seem to be four outliers. Figure 7.4(b) is the corresponding plot with the four outliers removed. (Figures 7.4(a) and 7.4(b) are reproduced from Healy, 1968.)*

We shall consider graphical procedures in more detail later (Section 7.4.2).

Such informal methods do not (in general) lead to any formal test of discordancy; they seldom even adopt any specific assumptions about the distribution from which the sample has arisen or about the nature of an alternative (outlier generating) hypothesis. They are to be viewed as initial data screening procedures, in the spirit of the current interest in 'data analysis' methods for the representation and summary of large-scale sets of data.

### 7.4.1   Marginal outliers and linear constraints

We should not underestimate the importance of the marginal samples (that is, the univariate samples of each component value in the multivariate data) in the occurrence of outliers.

It is perfectly plausible for contamination to occur in one of the marginal variables alone: for example by mis-reading or an error of recording. It could even happen that a single marginal variable is intrinsically more liable to contamination. We must be careful, however, not to adopt too simplistic an approach in examining this prospect. It is tempting to restrict attention to the corresponding marginal sample, exploiting univariate methods with their advantages of straightforward representation of the outliers (as sample extremes) and well-developed methods of testing or accommodation. But apart from obvious cases of gross errors, readily and unambiguously identified, concentration on the *marginal sample* to investigate the possibility of *marginal contamination* will be inappropriate. It ignores the correlation structure within the multivariate observation, resulting in loss of information.

This is illustrated in some work by Barnett (1983c) who considers a sample from a bivariate normal distribution $N(\mu_1, \mu_2, \sigma_1^2, \sigma_2^2, \rho)$ where contamination may have occurred by slippage of the mean of the *first component only* for one observation. The detection of such an outlier is by no means simple.

Adopting **Principle A** of Section 7.1 leads to detection of the outlier as that observation which takes the minimum value [over the sample $(x_{11}, x_{21}), (x_{12}, x_{22}), \ldots, (x_{1n}, x_{2n})$] of the quantity

$$\left\{1 - \frac{n}{(n-1)^2}\begin{pmatrix} x_{1j} - \bar{x}_1 \\ x_{2j} - \bar{x}_2 \end{pmatrix}' S^{-1}\begin{pmatrix} x_{1j} - \bar{x}_1 \\ x_{2j} - \bar{x}_2 \end{pmatrix}\right\}\left\{1 - \frac{n}{(n-1)^2}\frac{(x_{2j} - \bar{x}_2)^2}{S_{22}}\right\}^{-1}$$

where the $\bar{x}_i (i = 1, 2)$ are the marginal sample means, and $S_{22}$ is the second-component unbiased variance estimate. This is readily interpreted. The outlier is 'as far out as possible' in terms of the familiar reduction measure $R(\mathbf{x}; \bar{\mathbf{x}}, S)$ but also 'as close as possible to the second-component mean'. This is, of course, quite a different principle from consideration of the first-component extremes.

Another situation in which we may have easy access to the detection of outliers in where we anticipate a simple (usually linear) relationship between the components of the multivariate observation or between the expected values of the components. For example, our multivariate observation might consist of proportionate measurements, such as the proportions of the total body length of a reptile corresponding with biologically distinct sections of the body, or it may be the three angles of a triangle in a geographic survey. In the first case consistency of representation demands that the proportions should have unit sum; in the second the sum of the components should, apart from errors of measurement, add to 180°. In either case, marked departures of the sum of the component values from their expected sum can highlight gross errors of measurement or of recording as an indication of outliers. Fellegi (1975) comments on the presence of outliers in the editing of multivariate data where just such 'pre-identified relation-ships' hold. He includes consideration of less specific forms of relationship where, for example, we have information on what constitutes a reason-able range of values which may be taken by some ratio of marginal components.

Note that outliers identified in this way need not (indeed are unlikely to) show up merely on consideration of the marginal samples.

See also Katz and Brown (1988)

### 7.4.2 Graphical and pictorial methods

In relation to multivariate outliers, Rohlf (1975) remarks as follows:

Despite the apparent complexity of the problem, one can still characterize outliers by the fact that they are somewhat isolated from the main cloud of points. They may not 'stick out on the end' of the distribution as univariate outliers must, but they must 'stick out' somewhere. Points which are not internal to the cloud of points (i.e. which are somewhere on the surface of the cloud) are potentially outliers. Techniques which determine the position of a point relative to the others would seem to be useful. A second important consideration is that outliers must be separated from the other points by distinct gaps.

We saw this principle in operation in *Example 7.1*.

With this emphasis it is natural to consider different ways in which we can merely *look at the data* to see if they seem to contain outliers. A variety of methods employing different forms of pictorial or graphical representation have been proposed with varying degrees of sophistication in terms of the preprocessing of the data prior to its display and of the subsequent fuller analysis of the data.

For obvious reasons, bivariate data are the most amenable to informative display, although it will be apparent that some of the approaches do not depend vitally on the dimensionality of the data. Review of such methods of 'informal inference' applied to general problems of analysis of multivariate data (not solely the detection of outliers) are presented by Gnanadesikan (1977) and by Barnett (1981). Gnanadesikan (1977) includes a lengthy discussion (in Section 6.4) of the multivariate outlier problem, with copious practical illustration of proposed methods, which provides a natural source for fuller study. Accordingly we shall consider only some of the more basic ideas here, and in the subsequent sections of this Chapter. Other relevant references are Wilk and Gnanadesikan (1964), Gnanadesikan and Wilk (1969), the lengthy review paper by Gnanadesikan and Kettenring (1972), Gnanadesikan (1973), Cleveland and Kleiner (1975), Friedman and Rafsky (1981), Kleiner and Hartigan (1981), Seaman, Turner, and Young (1987; using polyhedron graphs), Bacon-Shone and Fung (1987; a graphical method based on Wilks's, 1963, statistic in which 'masking' and 'swamping' are easily revealed), Bartkowiak (1989; using angular distances) and Easton and McCulloch (1990; a multivariate generalization of quantile–quantile plots).

Let us consider some possibilities for demonstrating the presence of outliers in (predominantly) bivariate data. The most rudimentary form of representation is the scatter diagram of two, out of the $p$, components. Figure 7.3 shows the scatter diagram for the data on yields of grass at Rothamsted (see page 290). Some observations do seem to 'stick out' and to be separated from others by 'distinct gaps'; notably the observations A and B previously identified as discordant outliers. But in a different respect, the observation C also seems somewhat suspicious. The observation C (and to a lesser degree B) might well have the effect of reducing the apparent correlation between yields on the two plots, whilst A leads one to assume a larger variation in yields than would have appeared plausible in its absence. This effect of A (and to a lesser extent B) is heightened if we consider the projection of the observations on to the dotted line through the data set shown on Figure 7.3 (roughly the regression line of plot 12 yield on plot 3 yield) whilst C and D now appear in no way aberrant. In contrast if we project onto the perpendicular to the dotted line then B and C are particularly extreme, whilst A appears more reasonable.

This example embodies many of the considerations employed in designing graphical methods for exhibiting multivariate outliers. In the first place, the scatter diagram itself may throw up outliers as observations on the periphery of the 'data cloud', distinctly separated from others. The marginal samples may or may not endorse such a declaration (this could be possible for A, not for C). The warning given in Section 7.4.1 is apposite here; there is no reason why outlying behaviour need be manifest in the *marginal* samples. But this is not to deny the importance of certain one-dimensional graphical methods (such as probability plots of linear combinations of the components, or of values of appropriate reduction measures)—see below.

Thus to summarize the remarks above, the perturbation of some aggregate measure, such as the correlation coefficient, from what is anticipated may reveal the presence of outliers (likely here in different respects for A, B, and C. Then again, a change of coordinate basis, and re-representation of the data on the new basis, can reveal outliers not immediately apparent previously. Rotation of the axes of Figure 7.3 in the direction of the dotted line (or of its perpendicular) can help to identify A and B (or C) depending on which of the new axes is considered. An *appropriate* graphical representation of the *ordered* sample values (either marginally in the original data, for particular linear combinations of the components corresponding with a transformations of axes, or expressed in terms of a suitable reduction measure) can on occasions dramatically augment the visual impact of outliers.

We shall consider in more detail in Sections 7.4.3–7.4.7 some work which utilizes such ideas.

Another approach that has been receiving attention is the representation of sample values (or transformed versions of them) in some *pictorial* form where an outlier may have the effect of jolting the intuition by virtue of an occasional odd impression that it creates. It is hard to believe that an outlier need have an obvious effect such as providing the only smiling visage in a group of apparent misanthropes, when we employ Chernoff's faces to represent the observations. But some forms of pictorial display do seem reasonably responsive to outliers. Kleiner and Hartigan (1981) make such a claim for their 'trees' and 'castles' where the matching of relationships between the components to specific features of the pictorial form provides a 'major advantage' over earlier (non-relation-based) proposals such as profiles, stars, glyphs, faces, boxes and Andrew's curves (see Section 7.4.5). See also Seaman, Turner, and Young (1987); they use polyhedron graphs.

### 7.4.3 Principal component analysis method

Several writers have suggested performing a preliminary principal component analysis on the data, and looking at sample values of the projection of the observations on to the principal components of different order. The example above, on Rothamsted grass yields, shows how projecting the observations on the leading or secondary principal component axes (roughly the dotted line and its perpendicular) can highlight different types of outlier. This distinction in the relative utility of the first few, and last few, principal components in outlier detection is basic to the methods described in the literature. Gnanadesikan and Kettenring (1972) discuss this in some detail, remarking how the first few principal components are sensitive to outliers inflating variances or covariances (or correlations, if the principal component analysis has been conducted in terms of the sample correlation matrix, rather than the sample covariance matrix), whilst the last few are sensitive to outliers adding spurious dimensions to the data or obscuring singularities. These distinctions are also discussed by Gnanadesikan (1977, Section 6.4.1).

Suppose we write

$$Z = LX \tag{7.4.1}$$

where $L$ is a $p \times p$ orthogonal matrix whose rows, $l_i'$, are the eigenvectors of the sample variance–covariance matrix $S$ corresponding with its eigenvalues, $c_i$, expressed in descending order of magnitude and $X$ is the $p \times n$ matrix whose $i$th column consists of the transformed observations $x_i - \bar{x}$. Then the $l_i'$ are the principal component coordinates and the $i$th row of $Z$, $z_i'$, gives the projections on to the $i$th principal component coordinate of the deviations of the $n$ original observations about $\bar{x}$. Thus the top few, or lower few, rows of $Z$, provide the means of investigating the presence of outliers affecting the first few, or last few, principal components.

The construction of scatter diagrams for pairs of $z_i$ (among the first few, or last few, principal components) can graphically exhibit outliers. Additionally univariate outlier tests can be applied to individual $z_i$; or the ordered values in $z_i$ can usefully be plotted against an appropriate choice of plotting positions. What is 'appropriate' is not easily assessed in any exact form, especially in the absence of reliable distributional assumptions about the original data. However, if $p$ is reasonably large, it is likely that the linear transformations involved in the principal component analysis may lead (*via* Central Limit Theorem arguments) to the $z_i$ being samples from approximately normal distributions. In such cases *normal probability plotting* in which the $j$th ordered value in $z_i$ is plotted against $\alpha_j$, where

$$\alpha_j = E[U_{(j)}]$$

with $U_{(j)}$ the $j$th order statistic of the normal distribution, $N(0, 1)$, may well reveal outliers as extreme points in the plot lying off the linear relationship exhibited by the mass of points in the plot. Such an informal procedure has been found to be a useful aid to the identification of multivariate outliers.

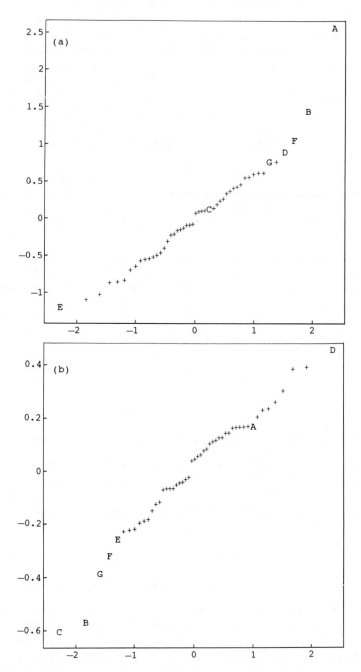

**Figure 7.5** Normal probability plots for principal components of Rothamsted grass yield data. (a) First principal component; (b) Second principal component

To illustrate this we again consider the Rothamsted grass yield data. Figures 7.5(a) and 7.5(b) show normal probability plots for the first and second principal components respectively. The first principal component (see Figure 7.5(a)) shows no marked contra-indication of normality except for the observation A which shows up clearly as an outlier. The outliers E, D, F and B show as extreme values although (apart perhaps from B) they do not lie off the linear relationship in the manner we would expect were they discordant outliers. In the plot (Figure 7.5(b)) of projections onto the second principal component, C, B, G, F, E and D are distinguished as extremes (A is inconspicuous) and there is less indication of normality on this dimension.

Added flexibility of approach is provided by basing principal component analysis on the sample correlation matrix, R, instead of on S, and also by replacing R or S with modified *robust* estimates. The robustness aspect has already been considered in the discussion of the *accommodation* of multivariate outliers in Section 7.2.

Some modifications of approach to outlier detection by principal component analysis are suggested by Hawkins (1974) and by Fellegi (1975). Hawkins (1980a, Section 8.2) is critical of the proposal by Gnanadesikan and Kettenring (1972) that gamma probability plots (see Section 7.4.4) should be carried out on the sums of squares of principal component residuals.

It has been suggested that outliers can be detected as byproducts of other forms of multivariate analysis. For example, Bollen (1987) shows how outliers lead to 'improper' solutions in *factor analysis*; Yenyukov (1988) proposes a *projection pursuit* method for detecting outliers. See also Section 7.4.6 below.

### 7.4.4  Use of reduction measures in the form of generalized distances

Another way in which informal quantitative and graphical procedures may be used to exhibit outliers is to construct reduced univariate measures based on the observations $x_j$ (analogous to the reduction measures more formally considered earlier). Gnanadesikan and Kettenring (1972) consider various possible measures in the classes:

$$\text{I:} \quad (x_j - \bar{x})' S^b (x_j - \bar{x}), \tag{7.4.2}$$

$$\text{II:} \quad (x_j - \bar{x})' S^b (x_j - \bar{x}) / [(x_j - \bar{x})'(x_j - \bar{x})]. \tag{7.4.3}$$

Particularly extreme values of such statistics, possibly demonstrated by graphical display, may reveal outliers of different types. Such measures are of course related to the projections on the principal components and Gnanadesikan and Kettenring (1972) remark that, with class I measures, as $b$ increases above $+1$ more and more emphasis is placed on the first few principal components whereas when $b$ decreases below $-1$ this emphasis progressively shifts to the last few principal components (a similar effect holds for class II measures according as $b \gtrless 0$). Extra flexibility arises by considering $x_j - x_{j'}(j \neq j')$ rather than $x_j - \bar{x}$ in the different measures, or $R$ in place of $S$.

Let us consider some specific examples of the class I measures.

$$(b = 0) \qquad q_j^2 = (x_j - \bar{x})'(x_j - \bar{x}) = \frac{n}{n-1}[\mathrm{tr}(A) - \mathrm{tr}(A^{(j)})]. \qquad (7.4.4)$$

This squared Euclidean distance from $\bar{x}$ is sensitive to outliers 'inflating the overall scale'.

$$(b = 1) \qquad t_j^2 = (x_j - \bar{x})'S(x_j - \bar{x}) = \sum_i c_i[l_i'(x_j - \bar{x})]^2, \qquad (7.4.5)$$

is sensitive to outliers affecting the 'orientation and scale of the first few principal components'.

$$(b = -1) \qquad d_j^2 = (x_j - \bar{x})'S^{-1}(x_j - \bar{x}) = \sum_i c_i^{-1}[l_i'(x_j - \bar{x})]^2 \qquad (7.4.6)$$

is particularly useful for 'uncovering observations which lie far afield from the general scatter of points'.

For graphical display of outliers, the 'gamma-type probability plots' of ordered values, with appropriately estimated shape parameter, are a useful approximate procedure. Essentially the argument is as follows. If the multivariate observation $x_j$ comes from a normal distribution, then the distance measures $R_j(\bar{x}, \Gamma)$ may be regarded as (approximately) independent observations from a gamma distribution, whatever form is taken for $\Gamma$. Thus if we knew, or could reasonably estimate, the parameters in the gamma distribution, a plot of the ordered $R_j(\bar{x}, \Gamma)$ against quantiles of the gamma distribution should be linear with anomalous $R_j(\bar{x}, \Gamma)$ (for example, discordant outliers) showing as extreme values lying markedly off the overall linear relationship. (See Wilk and Gnanadesikan, 1964; Wilk, Gnanadesikan, and Huyett, 1962a, 1962b; Gnanadesikan, 1977; and further comments in Section 7.4.7.) If $X$ is multivariate normal then the exact marginal distribution of the $d_j^2$ is known to be related to a Beta form with parameters $(n - p - 1)/2$ and $p/2$, but the $d_j^2$ are not, of course, independent (see Section 7.3.1). We note again that consideration of the maximum value of $d_j^2$ is equivalent to Wilks's (1963) method; see also (above) the proposals of Healy (1968) for plotting the $d_j^2$ when $X$ is approximately normal.

Rousseeuw and van Zomeren (1990) propose a graphical method for displaying outliers based on (7.4.6) with 'very robust' estimates replacing $\bar{x}$ and $S$. They give examples to show that this has the effect of enhancing the display of outliers (with some, perhaps, which may not be shown up by using (7.4.6)).

Rao (1964) proposes examination of the sums of squares of the lengths of the projections of individual observations on the last few ($q$) principal component coordinates for assessing the propriety of individual observations. Thus outliers may be revealed by particularly large values of

$$\sum_{i=p-q-1}^{p} [\mathbf{l}_i' (\mathbf{x}_j - \bar{\mathbf{x}})]^2$$

The suggestions of Gnanadesikan and Kettenring (1972), and of Gnanadesikan (1977, Section 6.4.1), for informally considering residuals in least-squares fits of structural models, as a means of detecting outliers, are more appropriate to the discussion of outliers in regression models and designed experiments (Chapter 8).

### 7.4.5  Function plots

Gnanadesikan (1977, Section 6.2) reviews an interesting class of methods for displaying multivariate data, in which each observation $\mathbf{x}_j = (x_{1j}, x_{2j}, \ldots, x_{pj})$ is represented by a function $f_{\mathbf{x}}(t) = x_{1j}a_1(t) + x_{2j}a_2(t) + \ldots + x_{pj}a_p(t)$ where $a_1(t), a_2(t), \ldots, a_p(t)$ are orthonormal over some range of values of $t$. Such *function plots* can be useful in revealing outliers. The original idea, due to Andrews (1972), was that $\mathbf{x}_j$ should be represented by the function

$$f_{\mathbf{x}_j}(t) = x_{1j}/\sqrt{2} + x_{2j}\sin t + x_{3j}\cos t + x_{4j}\sin 2t + x_{5j}\cos 2t \ldots \quad (7.4.7)$$

over the range $(-\pi, \pi)$ for $t$. Each sample point in P-space then appears as a curve over such values of $t$. The idea is that this might reveal certain important qualitative features in the data.

Gnanadesikan (1973) shows in an example how (7.4.7) might usefully distinguish outliers. He considers a quadrivariate sample of 50 observations on log-lengths and log-widths of sepals and petals for *Iris setosa*, described by Fisher (1936). He chooses a grid of values of $t$ over $(-\pi, \pi)$, determines $f_{\mathbf{x}_j}(t)$ ($j = 1, 2, \ldots, 50$) over the grid, and at each grid value estimates certain quantiles of $f_{\mathbf{x}}(t)$ from the data. The quantiles chosen are the 10, 25, 50, 75, and 90 percentiles. These are presented graphically along with any individual $f_{\mathbf{x}_j}(t)$ values, at each $t$, *outside* the deciles. The results are shown in Figure 7.6 where we see a very clear indication of the outlying nature of observations number 16 and 42. (The median values are labelled M, quartiles Q, and extreme deciles T.) We would not claim that other techniques, such as residuals based on a principal components analysis, would fail to exhibit these outliers. But this use of Andrews's representation, and of more general types of function plot, may prove to have interesting possibilities for the study of outliers.

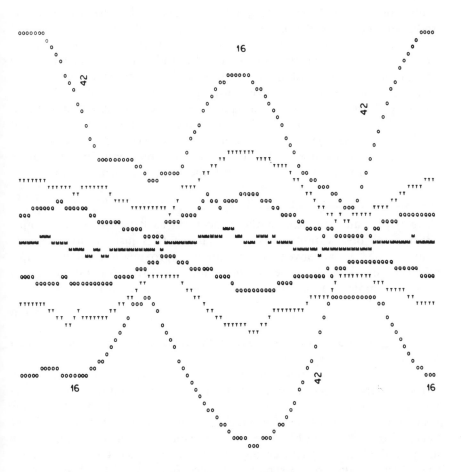

**Figure 7.6** Andrews's plot of the Iris setosa data (reproduced by permission of R. Gnanadesikan and the International Statistical Institute)

Suggested modifications of this approach include previous reordering of the components of $x_j$ in (7.4.7) 'in order of importance', and alternative specific forms for the $a_i(t)$. (See Gnanadesikan, 1977, Section 6.2). Kulkarni and Paranjape (1984) use Andrews's plots for outlier detection in a study of quality control curves for a multivariate process.

### 7.4.6 Correlation methods and influential observations

We have already remarked on the way in which outliers may affect, and be revealed by, the correlation structure in the data. Some proposals for identifying multivariate outliers specifically consider this matter.

Gnanadesikan and Kettenring (1972) suggests that we examine the product-moment correlation coefficients $r_{-j}(s, t)$ relating to the $s$th and $t$th marginal samples after the omission of the single observation $x_j$. As we vary $j$ we can examine, for any choice of $s$ and $t$, the way in which the correlation changes: substantial variations reflecting possible outliers.

Devlin, Gnanadesikan, and Kettenring (1975) make use of the *influence function* of Hampel (1974) to investigate how outliers affect correlation estimates in bivariate data ($p = 2$). Their main interest is in robust estimation of correlation—see Section 7.2. But they are also concerned with the detection of outliers *per se*. They consider a multivariate distribution indexed by a parameter $\theta$ and define in relation to an estimator $\hat{\theta}$ the 'sample influence function'

$$L(x_j; \hat{\theta}) = (n - 1)(\hat{\theta} - \hat{\theta}_{-j}) \quad (j = 1, 2, \ldots, n) \tag{7.4.8}$$

where $\hat{\theta}_{-j}$ is an estimator of the same form as $\hat{\theta}$ based on the sample omitting the observation $x_j$. We see that $\hat{\theta} + L$ is just the $j$th jackknife pseudo-value. As a convenient first-order approximation to the sample influence function of $r$, the product–moment correlation estimate in a bivariate sample, they propose (with an obvious notation)

$$L(x_{1j}, x_{2j}; r) = (n - 1)(r - r_{-j}). \tag{7.4.9}$$

$L(x_{1j}, x_{2j}; r)$ provides an estimate of the influence on $r$ of the omission of the observation $(x_{1j}, x_{2j})$.

Two suggestions are made for presenting graphically how $L(x_{1j}, x_{2j}; r)$ varies over the sample, with a view to identifying as outliers the observations which exhibit a particularly strong influence on $r$. The first amounts to superimposing selected (hyperbolic) contours of $L(x_1, x_2; r)$ on the scatter diagram, thus distinguishing the outliers. Some qualitative comments are made (and illustrated) concerning the choice of which contours to plot. The second relates to the sample influence function of the Fisher transformation $z(r) = \tanh^{-1}(r)$:

$$L(x_{1j}, x_{2j}; z(r)) = (n - 1)[z(r) - z(r_{-j})]. \tag{7.4.10}$$

For large $n$, the distribution of $L(x_{1j}, x_{2j}; z(r))$ is approximately that of the product of two independent standard normal variables, and it is proposed that ordered values of $L$ be plotted against the appropriate quantiles. The distinct $L$ values over the sample are not seriously correlated, and a further normalizing transformation is proposed prior to the probability plotting. Again, it will be extreme values in the plot, lying away from the overall linear relationship, which indicate outliers.

See Gnanadesikan (1977, Section 6.4) for further discussion and practical illustration of such methods.

Influence functions of other statistics (apart from the correlation coefficient) have been proposed as a basis for detecting outliers. Radhakrishnan

(1985) adopts this approach using discriminant analysis, whilst Cleroux, Hebling and Ranger (1986) use a *coefficient of vector correlation*. This method of examining outliers exemplifies a distinction we will be considering in some detail later (see Section 8.7). We have concentrated on the notion of an outlier as a manifestation of a particular type of contamination of the basic model, under which 'extreme' observations are likely to occur. For multivariate data such 'extremes' are to be found (in some appropriately defined sense) on the periphery of the data set. The presence of outliers, defined in this way, *can* grossly distort estimates, and tests, of basic parameters. (This was the motivation for our consideration of methods of accommodating outliers.) But there is also a real prospect, particularly in multivariate data and even more so for regression or designed-experiment data, that estimates of basic parameters can be seriously distorted by observations *which are not outlying* in the sense described above. As we have remarked above, there has been, over recent years, a rapid growth of interest in such 'influential observations' especially in the context of regression analysis. In Section 8.7 we shall review this developing area further and clarify the need to distinguish when appropriate between *outliers* and *influential observations*.

### 7.4.7 A 'gap test' for multivariate outliers

We noted earlier the characterization of multivariate outliers suggested by Rohlf (1975); that they are separated from other observations 'by distinct gaps'. Rohlf has used this idea to develop a *gap test* for multivariate outliers based on minimum spanning trees. Eschewing the nearest neighbour distances as measures of separation in view of the masking effect a cluster of outliers may exert on each other, he considers instead the lengths of edges in the minimum spanning tree MST (or shortest simply connected graph) of the data set as measures of adjacency. He argues that a single isolated point will be connected to only one other point in the MST by a relatively large distance, and that *at least one* edge connection from a cluster of outliers must also be relatively large. Accordingly a gap test for outliers is proposed with the following form. Firstly, examination of the marginal samples yields estimates $s_k$ ($k = 1, 2, \ldots, p$) of the standard deviations. The observations are rescaled as $x'_{ki} = x_{ki}/s_k$ ($k = 1, 2, \ldots, p$; $i = 1, 2, \ldots, n$). Distances between $\mathbf{x}'_i$ and $\mathbf{x}'_j$ in the MST are calculated as

$$d_{ij} = \left\{ \sum_{k=1}^{p} (x'_{ki} - x'_{kj})^2 / p \right\}^{1/2} \qquad (7.4.11)$$

and in particular we denote by $z_i$ the length of the $n - 1$ *edges* of the MST.

The $z_i$ are now examined for homogeneity, either by means of a probability plot of their ordered values or by testing if the ratio of the square of the maximum, $z_{(n-1)}^2$, to the average, $\overline{z^2}$, of the squares is of reasonable value.

The 'gamma-type plot' of the $z_{(i)}$ against quantiles of a gamma distribution has heuristic justification if $X$ comes from a $p$-variate normal distribution, on the following argument.

If the components of $X$ were independent normal (each with unit variance) the inter-point squared Euclidean distances would be independently distributed as $2\chi_p^2$. If the components of $X$ are not independent, these distances will be dependent and may not follow too closely a gamma distribution. However, Rohlf claims that empirical investigations demonstrate that the particular subset of squared *edge* distances, $z_i^2$, do appear to have approximately independent common gamma distributions (on the assumption of homogeneity: that is, absence of contaminants). The relevant shape parameter will need to be estimated either (iteratively) by the maximum likelihood method, or by using the order statistics approach of Wilk, Gnanadesikan, and Huyett (1962a, 1962b). The value of the scale parameter will not need to be estimated since its value effects only the slope of the gamma plot and not its linearity, and it is therefore irrelevant to the detection of markedly anomalous values (here discordant outliers).

Wilk, Gnanadesikan, and Huyett (1962b) consider maximum likelihood estimation of the scale and shape parameters, $\lambda$ and $\eta$, in the gamma distribution, $\Gamma(\eta, \lambda)$, based on an ordered random sample of observations, $y_{(1)}, y_{(2)}, \ldots, y_{(m)}$. They show that $\hat{\eta}$ satisfies

$$\frac{\Gamma'(\hat{\eta})}{\Gamma(\hat{\eta})} - \ln \hat{\eta} = \ln Q \qquad (7.4.12)$$

where $\Gamma(\ )$ denotes the gamma function and $Q$ is the ratio of the geometric and arithmetic means of the $y_{(i)}$. They present useful tabulated aids for determining $\hat{\eta}$. It is interesting to note that $\eta$ may be estimated separately from $\lambda$, an important consideration in that the probability plotting procedure does not require $\lambda$ to be known (it can be arbitrarily assigned).

Rohlf (1975) also remarks on a further advantage of such a means of estimation for current purposes: $\hat{\eta}$ does not depend strongly on the larger values in the ordered sample, so that $\hat{\eta}$ will be reasonably robust against the very outliers we are seeking to identify. In Wilk, Gnanadesikan, and Huyett (1962a) the 'gamma-type' probability plot is described in detail and useful tables of quantiles of the gamma distribution are presented.

In something nearer a formal test of discordancy for a single outlier is required, Rohlf makes the following proposals. If we knew $\lambda$ and $\eta$ in the (approximate) gamma distribution for which the $z_i$ are (approximately) independent observations, then we could compare $z_{(n-1)}/\lambda$ with the upper percentage points for the maximum observation in a sample of size $n - 1$ from a gamma distribution with shape parameter $\eta$. Equivalently

$$z_{(n-1)}^2/[(n-1)\overline{z^2}]$$

where $\bar{z}^2 = (\sum_1^{n-1} z_i^2)/(n - 1)$, has a Beta distribution with parameters $\eta$ and $(n - 1)\eta$ (independent of $\lambda$) and use can be made of existing results on (approximate) upper percentage points of such a Beta distribution (see Rohlf, 1975, for details). Not knowing the value of $\eta$ it is proposed that we should relate $z_{(n-1)}^2/[(n - 1)\bar{z}^2]$ to the approximate upper percentage points of the appropriate Beta distribution for an *estimated* value $\hat{\eta}$. Rohlf presents a table of upper bounds to the upper 5 per cent and 1 per cent points for $n = 10$, 20(20)100, 200 and $\eta = 0.1(0.1)1.0(0.5)5.0$, 6.0(2.0)12.0. Rohlf's proposals are specifically concerned with normal data, and we have of course dealt at length in Section 7.3.1 with other proposals for this case.

The idea of using the MST to reflect outliers is interesting but clearly needs more detailed study and illustration. Although not specific to the outlier problem, the methods (and practical illustrations) for the use of minimal spanning trees in comparing two multivariate samples given by Friedman and Rafsky (1981) reveal new prospects in the use of this approach for multivariate data display which could have useful implications for outlier study. See also Pincus (1984).

# CHAPTER 8

# The Outlier Problem for Structured Data: Regression, the Linear Model and Designed Experiments

We have concentrated so far on the examination of outliers in a single sample of observations from a specific distribution. It has been assumed that the only departure from the null hypothesis of homogeneity of distribution arises in explanation of contaminants, on the basis of one of the contamination models discussed in Chapters 2 and 7. In many statistical analyses, of course, departure from the assumption of homogeneity of distribution need have nothing to do with outliers. Such departure is a natural, and often welcome, manifestation of the appropriateness of some **linear model** explaining how mean values *vary* with different levels of factors of classification or with different values of a set of independent variables; or it may express a time-dependent effect in the generation of the data. This applies, of course, to the whole range of *designed experiments*, *regression situations* and *contingency table* data, etc. Even so, a *further* degree of inhomogeneity may be revealed by the presence of outliers, which express *ad hoc* influences *additional* to linear model effects. Thus it is appropriate to extend our study of outliers to such more highly structured situations.

A crucial distinction must now be recognized in the occurrence of outliers. In a single univariate sample an outlier was identified subjectively as an observation which engenders 'surprise' in its extreme value relative to the other sample members: it 'sticks out at the end' of the sample. Subjective identification is by formal processing either to accommodate the outlier or, following a test of discordancy, to remove the offending observation prior to further processing of the data. For more structured data we would wish to retain the stimulus of 'surprise' but this concept is now far more nebulous.

For example, consider the set of data shown in Table 8.1. The values $u$ and $x$ are respectively the loads (in kg) applied to similar structures and their resulting extensions of length (in cm). From basic physical principles, we must expect some relationship between the loads and extensions and if we plot the data (as in Figure 8.1) we find confirmation of a relationship. Perhaps a linear regression model is appropriate.

**Table 8.1**  Extension of a structure under different loads

| $u_{(i)}$ kg | 11.2 | 21.1 | 29.9 | 34.1 | 43.8 | 53.4 | 59.9 | 61.2 | 68.9 |
|---|---|---|---|---|---|---|---|---|---|
| $x_{(i)}$ cm | 1.6 | 2.1 | 3.4 | 3.3 | 4.2 | 3.1 | 4.9 | 6.2 | 6.3 |

There is no reason now why outliers should be sought merely in the *extremes* of the sample of $x$-values (or of $u$-values). However, in Figure 8.1 an anomaly is apparent; the observation (53.4, 3.1) *disturbs the general pattern* to a degree which is discomforting. It is natural to regard this observation as an *outlier* in a more subtle sense of 'lying outside' the typical relationship between extension and load revealed by the remaining data. An appropriate visual inspection still suffices to detect the outlier in this case, but it is clear that in a more complicated regression situation with many independent variables this may well not be so straightforward.

Thus we now need to refine our notion of an outlier. It is not a simple extreme value, but it has a more general *pattern-disrupting* form. We can,

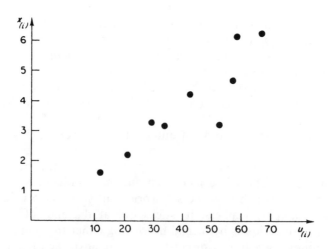

**Figure 8.1**  Relationship between load and extension for data of Table 8.1

however, no longer rely merely on direct subjective impact but may need (as in multivariate samples) to adopt an appropriate outlier detection process before we can even contemplate testing discordancy or refining inference procedures to take specific regard of outliers. (This latter consideration is, of course, the stimulus for developing outlier-robust procedures which provide protection against *possible* outliers, rather than being designed to accommodate pre-identified outliers. Trimming or Winsorization exemplify this approach for univariate samples.)

We should note the **influence** that the outlier in the load/extension data set has on an analysis of the underlying relationship. It can distort the estimates of intercept and slope parameters and will certainly inflate their standard errors. This effect is to be expected of an outlier (or a contaminant). The concept of **influential observations**, as those which have a marked effect on the inference process in a statistical analysis, is one which has attracted a great deal of attention in recent years; see, for example, Huber (1981) and Hampel *et al.* (1986). Belsley, Kuh and Welsch (1980) and Rousseau and Leroy (1987) provide detailed coverage of this topic in the context of regression. See also Cook and Weisberg (1982). But whilst outliers are often influential, not all influential observations need be outliers.

Suppose we decided to take another observation of load and extension, and chose a load of 102 kg under which an extension of 8.7 cm occurred. See Figure 8.2. This might have little effect on the fitted linear relationship but will certainly be *highly influential* in reducing the standard error of the estimate of the slope parameter. So this observation is not outlying in a 'pattern breaking' respect. It yields, of course, highly extreme values of load and extension but we can hardly be 'surprised' at that, since we *designed it to do so*.

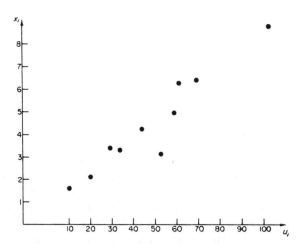

**Figure 8.2**  Augmented load and extension data

Such examples suggest that we need to distinguish carefully between outliers (which are perhaps influential) and a wider notion of influential observations which may not be outlying in the usual sense of the word. Outliers are not assessed *solely* in terms of the influence they exert on *specific* parameter estimates. An extreme ('outlying') value in the *design space* of an experiment lacks the *fortuitous (probabilistic) stimulus* for its extremeness which we have adopted as a characteristic of outlying behaviour, or contamination; it is not an outlier in our terms. Surprisingly, this distinction does not feature widely in published comment on the study of influential data—but see Cook and Weisberg (1980) and Draper and John (1981), also Section 8.7 below. However, we shall adopt it rigidly, considering in this chapter those ideas and methods specific to the conventional view of outliers in structured data sets. The exception is Section 8.7 where we address ourselves to the broader concept of *influential data* with a brief review of key aspects and references.

At that stage we will seek to place in perspective the central concepts that underpin the study of outliers and the application of robust methods in general. In particular, we will show how outlier methods and robustness *interrelate*, and *differ*. But no detailed treatment of robustness will be offered there in view of the proliferation of published work on this topic, and its tangential role in respect of the outlier principles we have been employing. (Some particular results on the influential effect of outliers have also been covered in our treatment of the accommodation of outliers in different contexts.)

A further data set is given in Table 8.2 (extracted from Bross, 1961; earlier presented by Daniel, 1960) which presents hypothetical yields of some product at different levels of two chemicals, A and B.

**Table 8.2**  Yields of a process at different levels of two chemicals

|          | Levels of B |     |     |     | Total |
|----------|------|-----|-----|-----|-------|
|          | 35   | 32  | 37  | 40  | 144   |
| Levels   | 29   | 29  | 34  | 36  | 128   |
| of A     | 25   | 29  | 30  | 20  | 104   |
|          | 19   | 25  | 25  | 35  | 104   |
|          | 22   | 20  | 29  | 29  | 100   |
| Total    | 130  | 135 | 155 | 160 | 580   |

We should not be particularly surprised in such a situation if the means of the underlying distributions differed at different levels of A or B; indeed we would wish to examine the propriety of some linear model for the means. Thus again, the extreme values (19 and 40) in the data set are not the only (or even the predominant) candidates as outliers, and we must employ some basis other than the observed quantitative values of the yields for detecting any outliers.

If we consider the first row of the data (corresponding with the first level of A) the observations are uniformly the largest in their respective columns (i.e. at each level of B). But this would be no basis for suspecting the *integrity* of the data; it is just one rather specific manifestation of the type of effect we are investigating in the analysis of such a two-way experimental design. It could point to a significant effect on yield by chemical A, of a rather more specific style than arises from merely rejecting the null hypothesis of no A-effect.

On the other hand the single observation 20 in the last column of the data stands out in *individual* isolation as being relatively more extreme both within its column, and within its row, than do any other observations in the data. It appears to disrupt seriously the *overall pattern* of results where (roughly speaking) yields decrease with the levels of A and increase with the levels of B. We could formalize this impression by considering the *estimated residuals*, $\tilde{\varepsilon}_{ij}$, in relation to a fitted additive linear model

$$x_{ij} = \mu + \alpha_i + \beta_j + \varepsilon_{ij} \quad (i = 1, \ldots, 5; j = 1, \ldots, 4)$$

where $x_{ij}$ is the yield at levels $i$ and $j$ of A and B, respectively, $\varepsilon_{ij}$ is the corresponding true *residual* and $\sum_1^5 \alpha_i = \sum_1^4 \beta_j = 0$. We can estimate $\varepsilon_{ij}$ by

$$\tilde{\varepsilon}_{ij} = x_{ij} - \bar{x}_{i.} - \bar{x}_{.j} + \bar{x}_.$$

(on the usual dot convention for aggregation over the levels of the factors). The table of estimated residuals is as follows, highlighting the aberrance of the observation 20:

|     |     |     |     |
|----:|----:|----:|----:|
| 2   | −2  | −1  | 1   |
| 0   | −1  | 0   | 1   |
| 2   | 5   | 2   | −9  |
| −4  | 1   | −3  | 6   |
| 0   | −3  | 2   | 1   |

It is this type of *individual* disruption of pattern that we must regard as an expression of the outlying nature of a *single* observation. We need (at least in complex experiments) to develop procedures for formal detection of such outliers and for proper assessment of their statistical significance (discordancy). Additionally, the accommodation aspect is vital. A prime aim is to examine the linear model in a designed experiment (or estimate and test parameters in a regression model) with as little interference as possible from isolated contaminants.

In the various structured models relevant to the different topics of this chapter it is appealing to examine disruption of pattern through the behaviour of **estimated residuals**, and many of the published results on outliers approach the problem in this way. But we must recognize some shortcomings in the use of estimated residuals. They are inevitably inter-correlated and may even have differing variances. Any outliers not only affect their own residuals but have a carry-over effect on others. Thus their aberrance tends to be somewhat smoothed out: they hide behind their neighbours in the data set! Extreme examples of this include two-way experiments with one factor having only two levels, where estimated residuals arise in pairs of identical value and opposite sign, or a $3 \times 3$ Latin square where they take equal values in groups of three and inter-residual correlations are either 1 or $-0.5$. Accordingly other principles for outlier detection, testing, or accommodation are also to be found in the literature and we shall be examining them. Often they involve informal graphical display procedures.

In principle the study of outliers in regression situations, or in designed experiments, can be subsumed in a wider investigation of outliers in general linear models. Much published work has this wider emphasis, and will be discussed later in the chapter. However, we shall start with the simplest situation: namely that of a linear regression model with a single regressor variable.

## 8.1 OUTLIERS IN SIMPLE LINEAR REGRESSION

Consider a set of $n$ observations $x_j$ of independent random variables $X_j$ ($j = 1, \ldots, n$) whose means depend linearly on predetermined values $u_j$ of a variable $U$. Thus

$$x_j = \theta_0 + \theta_1 u_j + \varepsilon_j \tag{8.1.1}$$

where the $\varepsilon_j$ are independent with zero mean. Usually the $\varepsilon_j$ are assumed to come from a common distribution; more specifically we might assume $\varepsilon_j \sim N(0, \sigma^2)$. We shall consider the detection, testing, and accommodation of outliers in this situation. If $\theta_0$ and $\theta_1$ are estimated by least-squares (or equivalently by maximum likelihood on the *normal* error model) as $\tilde{\theta}_0$ and $\tilde{\theta}_1$ we can estimate the residuals $\varepsilon_j$ as

$$\tilde{\varepsilon}_j = x_j - \tilde{\theta}_0 - \tilde{\theta}_1 u_j \tag{8.1.2}$$

and in seeking outliers it is again sensible to examine the relative sizes of the $\tilde{\varepsilon}_j$.

Even for the simple linear regression model (8.1.1) however, we encounter a difficulty. The estimated residuals do not have constant variance. In fact

$$\text{var}(\tilde{\varepsilon}_j) = \sigma^2 \{(1 - 1/n) - (u_j - \bar{u})^2 / \sum_1^n (u_j - \bar{u})^2\}. \tag{8.1.3}$$

Thus the $\tilde{\varepsilon}_j$ are more variable the closer $u_j$ is to $\bar{u}$; see Figure 8.3. This 'ballooning' effect of the residuals (Behnken and Draper, 1972) needs to be taken into account, if it is at all marked, in examining the size of the residuals $\tilde{\varepsilon}_j$ as a reflection of outliers. It is interesting to note the wider implications of 'ballooning' in the screening of regression data. It is common practice to examine estimated residuals for model validation. Clearly any tendency for reduced scatter at the extremes of the set of $u$-values does *not* necessarily imply heteroscedasticity.

The inhomogeneity of the variances of estimated residuals is not restricted to the simple linear regression model (8.1.1). For the general linear model

$$x = A\theta + \epsilon \qquad (8.1.4)$$

we have (in the full rank case)

$$\text{var}(\tilde{\epsilon}) = \sigma^2(I_n - R) \qquad (8.1.5)$$

with

$$R = A(A'A)^{-1}A'$$

so that we will need later to consider the implications for outlier detection of the extent to which the form of the design matrix $A$ induces inhomogeneity of variance in the estimated residuals.

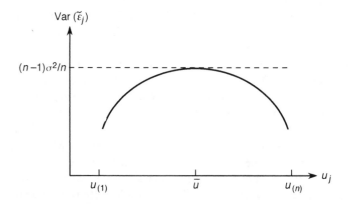

**Figure 8.3** 'Ballooning' effect in linear regression

### 8.1.1 Discordancy tests for simple linear regression

Returning to the simple model (8.1.1), one possible approach *to tests of discordancy* is to examine *appropriately weighted* estimated residuals (**studentized residuals**)

$$e_j = \frac{\tilde{\varepsilon}_j}{s_j} = \tilde{\varepsilon}_j \bigg/ \left\{ s \sqrt{\left(1 - 1/n\right) - \left(u_j - \bar{u}\right)^2 \bigg/ \sum_1^n \left(u_j - \bar{u}\right)^2}\right\} \quad (8.1.6)$$

where $s^2 = \sum \tilde{\varepsilon}_j^2/(n - 2)$ is an unbiased estimate of $\sigma^2$, and $s_j^2$ is thus an unbiased estimate of var $(\tilde{\varepsilon}_j)$.

The weighted or *studentized residuals* provide an intuitively appealing measure of the aberrance of the separate observations. Correspondingly, through the **maximum absolute studentized residual** $e_{max}$, we might now seek to detect and test the discordancy of a single outlier: in terms of the statistic

$$t = \max |\tilde{\varepsilon}_j/s_j|. \quad (8.1.7)$$

*If $t$ is sufficiently large, we adjudge the observation yielding max $|\tilde{\varepsilon}_j/s_j|$ to be a discordant outlier.* Note the implication of this policy. It is no longer necessarily the most extreme estimated residual which indicates the prime candidate for designation as an outlier. Even a modest estimated residual, corresponding with a small variance (8.1.3), can boost an observation into an outlying role. Note also that we are not employing (as some had proposed earlier) the use of the maximum (absolute) *normed* residual (MNR): $\max |\tilde{\varepsilon}_j/(\tilde{\varepsilon}'\tilde{\varepsilon})|$. This also ignores the differing variances of the $\tilde{\varepsilon}_j$.

To conduct the test we need to know the distribution of $t$; its exact form is intractable but much work has been published on approximate forms for the distribution or its percentage points, based on Bonferroni inequalities or large scale simulation studies. We shall examine this work in some detail below when we consider (in Section 8.2.1) discordancy tests for outliers in the general linear model.

For the moment we continue to consider results for the simple linear regression model (8.1.1). Approximate critical values for $t$ under the basic model can be obtained (using an obvious transformation) from tables in Srikantan (1961) or (directly) from tables in Lund (1975). Extracts from the latter are presented as Table XXXVII on pages 522–3. The entries under $q = 2$ are relevant to the model (8.1.1). Tietjen, Moore, and Beckman (1973) employ large-scale simulation to determine approximate critical values (at 10, 5 and 1 per cent levels) for a range of sample sizes similar to those used by Lund (1975). We should note that Srikantan (1961) and Lund (1975) employ a first-order Bonferroni inequality (the right-hand side of (8.2.17) below) to obtain an *upper bound* to the critical values. Tietjen, Moore, and Beckman (1973) present simulation-based *point estimates*; their values tend to be lower than those of Lund by 1 or 2 in the second place of decimals. (See further discussion of Lund's tables in Section 8.2.1).

In our more detailed discussion below, for the general linear model, we will further examine the implications of the use of different orders of the Bonferroni inequality for approximating the critical values of the maximum studentized residual. We will be noting (Srikantan, 1961, and others) that, if the maximum absolute correlation $|\rho_{ij}|$ between the residuals $e_i$ and $e_j$ is not too large, then the approximate critical value given by the first-order Bonferroni upper bound will be the *exact* critical value. Thus the configuration of $u$-values (or the matrix $A$ in the general case), on which the $\rho_{ij}$ depend, is relevant to the accuracy of approximations to the critical values. Tietjen, Moore, and Beckman (1973) show however, by simulation in some simple cases, that such dependence may not be very marked and that the effect may be of limited practical importance.

Prescott (1975) suggests that we ignore the differing variances of the $\tilde{\varepsilon}_j$ and replace $s_j$ in (8.1.7) by $\bar{s}$ where $\bar{s}^2$ is the 'average variance' $\sum_1^n \tilde{\varepsilon}_j^2/n$ introduced by Behnken and Draper (1972). (8.1.7) then reduces to a multiple of the MNR: viz. $n^{1/2} \max |\tilde{\varepsilon}_j|/\sqrt{(\sum_1^n \tilde{\varepsilon}_j^2)}$, and approximate critical values are obtained (invoking first-order Bonferroni bounds for the equi-variance case considered by Stefansky 1971, 1972) on the assumption that the estimated residuals have common variance equal to the population 'average variance' $(n-2)\sigma^2/n$. This yields approximate critical values which are very similar to the simulated values given by Tietjen, Moore, and Beckman (1973).

*Example 8.1 Consider the load/extension data of Table 8.1. Fitting the linear regression model* (8.1.1) *we obtain*

$$\tilde{\theta}_0 = 0.67656 \quad \tilde{\theta}_1 = 0.07565.$$

*The estimated residuals, $\tilde{\varepsilon}_j$, and studentized residuals, $\tilde{\varepsilon}_j/s_j$, are as given in Table 8.3. Thus the observation* (53.4, 3.1) *stands out as an outlier. However, it is not statistically significant (the 5 per cent and 1 per cent critical values being 2.29, 2.44: see Table XXXVII).*

**Table 8.3**

| $y_i$ | $\tilde{\varepsilon}_j$ | $\tilde{\varepsilon}_j/s_j$ |
|---|---|---|
| 1.6 | 0.076 | 0.123 |
| 2.1 | −0.173 | −0.245 |
| 3.4 | 0.462 | 0.616 |
| 3.3 | 0.044 | 0.058 |
| 4.2 | 0.210 | 0.272 |
| 3.1 | −1.616 | −2.137 |
| 4.9 | −0.308 | −0.422 |
| 6.2 | 0.894 | 1.236 |
| 6.3 | 0.411 | 0.614 |

### 8.1.2 Outlier accommodation in simple linear regression

The dual aspect of interest is the *accommodation* of outliers in linear regression. Elashoff (1972) considers the estimation of $\theta_0$ and $\theta_1$ in the special case of contaminants represented by a particular combined mixture-slippage model. She assumes that the $\varepsilon_j$ are uncorrelated and arise from a mixed normal distribution

$$(1 - \gamma)N(0, \sigma^2) + \gamma N(\lambda(u_j), \sigma^2)$$

with $\gamma$ known and $\lambda(u_j) = c(u_j - u_{(1)})^2$, where $c$ is constant. This is a highly specific model and such study of the accommodation of outliers in linear regression is likely to be of limited applicability.

Detection and accommodation methods, based on division of the data into sub-samples, are described by Schweder (1976) and by Hinich and Talwar (1975), respectively. In the former case it is assumed that an uncontaminated sub-sample can be identified, in the latter case trimmed means of the sets of sub-sample estimated regression coefficients are utilized.

Tiku (1978) considers estimation and testing of parameters in (8.1.1) (and in more general models) when several observations are available at each $u$-value and censoring of extremes occurs at some $u$-values, commenting that the results will be relevant when observations are discarded because they are outliers. However, such a situation seems rather restricted in prospect. See also Chen and Dixon (1972) who compare by Monte Carlo methods the effects of Winsorization and trimming at each $u$-level when there is equal replication. They adopt a normal scale-mixture model. Moussa-Hamouda and Leone (1977) also use trimming methods to produce easily computed robust estimators of $\theta_0$, $\theta_1$ and $\sigma$. They tabulate weighting coefficients for sample sizes 5 to 20 and different models, including again a normal scale-mixture one.

Ramsay (1977) presents a Monte Carlo-based comparison of different methods of accommodating outliers in linear regression. Three classes of $M$-estimator are considered; the contamination model is of mixture form with normal components. Carroll (1979) considers the effect of an asymmetric error distribution on the estimation of the variances of robust estimators of $\theta_0$ and $\theta_1$. The robust estimation methods examined are special forms of Huber's Proposal 2 M-estimation for the regression coefficients (see Huber, 1977b, p. 37). Theoretical, and Monte Carlo, results suggest that estimation of the intercept parameter $\theta_0$ and of its variance can be seriously affected by asymmetry (with underestimation of the variance); typically, estimation of the slope parameter $\theta_1$ is little affected. See also Denby and Mallows (1977).

Krasker and Welsch (1982) examine efficient bounded-influence estimation of $\theta_1$ in (8.1.1) with special attention to extremes ('outliers') in the $u$-values. This is more relevant to the discussion of influential observations in Section 8.7, as is Hill (1982). Brown and Kildea (1978) use weighted medians of sums and differences of pairs of observations to produce Hodges–Lehmann type estimates in regression.

Other robust procedures for estimating or testing the regression coefficients are presented by Adichie (1967a, 1967b); they are non-parametric (based on rank tests) and provide *en passant* a degree of protection against outliers.

Concern for the accommodation of outliers in linear regression is a central feature in many studies which advocate methods different from classical least squares to estimate regression parameters, in a manner which is robust against the presence of contamination. Much of this work is presented in the context of the multiparameter linear model and will be more fully discussed below (in Section 8.2.3). Some proposals (even in this wider framework) explore the simple linear regression situation as a special case and require (at least first) mention at this stage.

Rousseeuw (1984) introduces the idea of **least median of squares regression** in the search for outlier robustness. Rather than minimizing the *sum* of squared residuals $\sum \varepsilon_i^2$, he proposes minimizing the *median* of the squared residuals $\varepsilon_i^2$ ($i = 1, 2, \ldots, n$). It was found that the resulting estimator was outlier-robust up to a remarkably high degree of contamination—as much as 50 per cent! For the simple linear regression problem the approach amounts to finding the narrowest linear band covering half the observations, but this is still not a particularly simple computational task, a matter which is further discussed by Steele and Stigler (1986) in the light of the piecewise linear form of the corresponding objective function which needs to be minimized: they propose an improved algorithm. The approach is further discussed by Massart *et al.* (1986). Edelsbrunner and Souvaine (1990) offer a further algorithmic advance for the simple linear regression case with time and space requirements which are $O(n^2)$ and $O(n)$, respectively.

Giltinan, Carroll, and Ruppert (1986) motivate their discussion of outlier-robust estimation of the residual variance, $\sigma^2$, in a heteroscedastic linear model by initially discussing the simple linear regression case (see Section 8.2.3 below for more detail).

All the testing and accommodation methods for the general linear model (described in Section 8.2 below) have obvious application to the special case of simple linear regression.

### 8.1.3 Outliers in linear structural and functional models

Another way of expressing linear relationships is through models which were originally described as representing 'regression with error in both variables'. Suppose that variables $X$ and $Y$ are linearly related:

$$Y = \alpha + \beta X. \qquad . (8.1.8)$$

However, we can observe $X$ and $Y$ *only with superimposed errors*, so that in seeking to sample $(X, Y)$ we in fact obtain values of

$$\left.\begin{array}{l} U = X + \varepsilon \\ V = Y + \eta \end{array}\right\} \qquad (8.1.9)$$

where $(\varepsilon, \eta)$ are random variables.

To fully specify the model we must say more about the basic variable $X$, and the error variables $(\varepsilon, \eta)$. If $X$ is a *random variable*, (8.1.8) and (8.1.9) constitute a **linear structural model**. Frequently we might take $X$ to be $N(\mu, \sigma^2)$ and $\varepsilon, \eta$ to be independent of $X$ and of each other, with $N(0, \sigma_1^2)$ and $N(0, \sigma_2^2)$ distributions. Thus the random sample $(u_1, v_1)$, $\ldots, (u_n, v_n)$ comes from a bivariate normal distribution $N(\mu, \alpha + \beta\mu,$ $\sigma^2 + \sigma_1^2, \beta^2\sigma^2 + \sigma_2^2, \beta\sigma^2)$. There are of course six basic parameters in this linear structural model.

If $X$ is a *mathematical* (not a random) *variable* taking (unobservable) values $x_1, x_2, \ldots, x_n$, we have a **linear functional model**. Again, with independent normal distributions for $\varepsilon, \eta$, the observations $(u_i, v_i)$ are normal: specifically, $N(x_i, \alpha + \beta x_i, \sigma_1^2, \sigma_2^2, 0)$ and we now have $n + 4$ parameters $(x_1, \ldots, x_n, \alpha, \beta, \sigma_1^2, \sigma_2^2)$.

As with any other models there must be a prospect of encountering outliers manifesting some form of contamination. Little study has been made of outliers in linear structural and functional models and progress may not be too straightforward. However, a few contributions have appeared.

Firstly, we consider two papers concerning the accommodation of outliers in linear structural or functional models.

Brown (1982) considers both models and employs a mixture form of contamination of one or other of the error components $\varepsilon$ and $\eta$. Specifically, $\varepsilon$ arises with probability $1 - \gamma$ from $N(0, \sigma_1^2)$ and with probability $\gamma$ from $N(0, k\sigma_1^2)$ where $\gamma$ is small and $k > 1$, whilst $\eta$ arises from $N(0, \sigma_2^2)$. Alternatively, $\eta$ has a similar mixture-type contamination and $\varepsilon$ is uncontaminated. An outlier-robust estimate of $\beta$ is proposed (with $\alpha \equiv 0$) which reduces the weight attached to observations adjudged highly discrepant when assessed in relation to a preliminary estimate of $\beta$. Investigation of some of the properties of the proposed estimator is conducted by simulation in special cases.

Carroll and Gallo (1982) consider a similar mixture model for contamination specifically for the *linear functional case* (again with $\alpha \equiv 0$). Again simulation is employed for initial and limited scrutiny of an *ad hoc* class of 'robust' estimators of $\beta$. The estimators include a modified least-squares form, a modified $M$-estimator and a bounded influence estimator.

A different approach in terms of model and aim is advanced by Barnett (1985). Here we are concerned with the *detection and testing* of outliers in the *linear structural model*. As for unstructured bivariate data we will again need a formal process for highlighting outliers. But the structure now suggests a variety of different forms for contamination. These can be illustrated using *mean-slippage* types of contamination model. Consider a single contaminant, and *four* prospects.

One prospect is slippage in the means of *both* $\varepsilon$ and $\eta$ for the one contaminant. Since $(U, V)$ is bivariate normal, this reverts to the problem of a single contaminant in a sample from such a bivariate distribution—see discussion in Section 7.3.1. In particular we detect and test the outlier by examining

$$\max\begin{pmatrix} u_i - \bar{u} \\ v_i - \bar{v} \end{pmatrix}' S^{-1} \begin{pmatrix} u_i - \bar{u} \\ v_i - \bar{v} \end{pmatrix}$$

where $\bar{u}$, $\bar{v}$ are sample means, and $S$ is the unbiased estimator of the variance–convariance matrix. The test, and its properties, are well known. *The outlier might appear anywhere on the periphery of the data cloud.*

Alternatively, we might have slippage in the mean of *one, or the other,* of $\varepsilon$ and $\eta$. This reverts to the detection and testing of an outlier in a single component of a bivariate normal distribution: see Section 7.4.1. Specifically, it is now reasonable to detect, and test, the outlier as that observation yielding (for slippage in the mean of $\varepsilon$)

$$\max\left\{\begin{pmatrix} u_i - u \\ v_i - \bar{v} \end{pmatrix}' S^{-1} \begin{pmatrix} u_i - u \\ v_i - \bar{v} \end{pmatrix} - \frac{(v_i - \bar{v})^2}{S^2(v)}\right\}$$

where $S^2(v)$ is an appropriate estimator of $\text{var}(V) = \sigma_2^2 + \beta^2\sigma^2$; or (equivalently) yielding

$$\max|(u_i - \bar{u}) - \tilde{q}(v_i - \bar{v})|$$

where $\tilde{q}$ is an estimator of $q = \beta\sigma^2/(\sigma_2^2 + \beta^2\sigma^2)$.

*The outlier now appears as far out as possible, but as close as possible to* $\bar{v}$. (For slippage in the mean of $\eta$ the obvious adjustment is needed).

Finally, we might have slippage in the mean of $X$. Here it turns out that we should detect and test the outlier by examining the observation which yields

$$\max\left\{\begin{pmatrix} u_i - \bar{u} \\ v_i - \bar{v} \end{pmatrix}' S^{-1} \begin{pmatrix} u_i - \bar{u} \\ v_i - \bar{v} \end{pmatrix} - \frac{[(v_i - \bar{v}) - \tilde{\beta}(u_i - \bar{u})]^2}{S^2(v)}\right\}$$

where $\tilde{\beta}$ is an appropriate estimator of $\beta$.

This is interpreted by noting that *the outlier is now as far out as possible, but as close as possible to the linear relationship.*

Barnett (1985) develops and applies these ideas to some biological data (on Pacific squid) and presents tables of critical values for the various null distributions, determined by simulation.

Ketellapper and Ronner (1984) apply robust regression estimation methods to the structural linear relationship—giving particular attention to bounded-influence estimators. Ronner and Steerneman (1985) discuss the estimation of the slope parameter in the linear structural case when outliers are present in the set of $x$-values. Ronner, Steerneman, and Kuper (1985) discuss the performance of moment estimators when there is contamination in the $X$ variable, while Nyquist (1987) examines robust alternatives to the method-of-moments estimators.

Stefanski (1985) presents a unified approach to linear and nonlinear models of functional and structural types which takes passing interest in the problem of contamination.

Abdullah (1989a, 1989b) and Zamar (1989) are concerned with outlier-robust estimation in the linear functional model and consider the properties of various classes of estimator, including $M$-estimators. The work by Zamar includes multivariate generalizations of the linear functional model.

## 8.2 OUTLIERS WITH GENERAL LINEAR MODELS

Many of the general principles for outlier detection, testing discordancy and accommodating outliers in a general linear model have already been indicated or illustrated by the more specific results for simple linear regression. Whilst we shall draw a distinction between methods based exclusively on residuals, and other (including graphical) methods, most attention will be given to the totally residual-based approach. This reflects the emphasis in published work on outliers for the general linear model.

### 8.2.1  Residual-based methods

We are concerned with situations where the observation vector $\mathbf{x} = (x_1, x_2, \ldots, x_n)'$ is represented by a basic general linear model

$$\mathbf{x} = A\mathbf{\theta} + \mathbf{\epsilon} \qquad (8.2.1)$$

with $\mathbf{\theta}$ a $q \times 1$ vector of parameters, $A$ a known $n \times q$ matrix of coefficients (assumed here to be of full rank) and $\mathbf{\epsilon}$ a $n \times 1$ vector of residuals. We assume (in most cases) that $\mathbf{\epsilon}$ has zero mean vector and variance matrix $V(\mathbf{\epsilon}) = \sigma^2 I_n$ where $I_n$ is the $n \times n$ identity matrix, so that the true residuals have common variance and are uncorrelated. (If this latter condition were not satisfied we could often restore it by an appropriate orthogonal transformation.) Any distribution-theoretic results (as needed, for example, in tests of significance) will be based on the assumption that $\mathbf{\epsilon}$ is multivariate normal.

In the absence of any contaminants, requiring a modification of the model (8.2.1) and possibly revealed as outliers, we have the familiar least-squares analysis of the linear model (8.2.1). The least-squares estimator of $\theta$ is

$$\tilde{\theta} = (A'A)^{-1}A'x \qquad (8.2.2)$$

with

$$V(\tilde{\theta}) = (A'A)^{-1}\sigma^2, \qquad (8.2.3)$$

and that of $\epsilon$ is

$$\tilde{\epsilon} = x - A\tilde{\theta} = (I_n - R)x = (I_n - R)\epsilon \qquad (8.2.4)$$

where

$$R = A(A'A)^{-1}A' \qquad (8.2.5)$$

and with

$$V(\tilde{\epsilon}) = (I_n - R)\sigma^2. \qquad (8.2.6)$$

The last term of (8.2.4) shows how the estimated residuals $\tilde{\epsilon}$ relate to the unknown true residuals $\epsilon$, but the determination of $\tilde{\epsilon}$ must be sought in terms of *known* quantities such as $(I_n - R)x$. The estimated residuals $\tilde{\epsilon}_j$ have zero means. From (8.2.6) we see that they are typically *correlated and have differing variances*. Explicitly we can write

$$\text{var}(\tilde{\epsilon}_j) = (1 - A'_j (A'A)^{-1}A_j)\sigma^2 = (1 - r_{jj})\sigma^2 \qquad (8.2.7)$$

(say) where $A'_j$ is the $j$th row of $A$.

The error variance $\sigma^2$ will be unknown. An unbiased estimate is obtained as

$$\tilde{\sigma}^2 = \tilde{\epsilon}'\tilde{\epsilon}/(n - q) = \epsilon'(I_n - R)\epsilon/(n - q) \qquad (8.2.8)$$

in view of the idempotency of $(I_n - R)$; $\tilde{\epsilon}'\tilde{\epsilon}$ is termed the *residual sum of squares* and is denoted $S^2$. $V(\tilde{\epsilon})$ can now be estimated as

$$S^2(\tilde{\epsilon}) = (I_n - R)\tilde{\sigma}^2 \qquad (8.2.9)$$

so that the estimated variance of $\tilde{\epsilon}_j$ is

$$s_j^2 = (1 - r_{jj})\tilde{\sigma}^2 = (1 - r_{jj})(\tilde{\epsilon}'\tilde{\epsilon})/(n - q) = (1 - r_{jj})S^2/(n - q). \qquad (8.2.10)$$

We shall again have reason to consider the *studentized residuals*

$$e_j = \tilde{\epsilon}_j/s_j = \frac{\tilde{\epsilon}_j}{S} \sqrt{\frac{(n - q)}{(1 - r_{jj})}}. \qquad (8.2.11)$$

They have an immediate intuitive appeal in that they constitute weighted versions of the estimated residuals $\tilde{\varepsilon}_j$, where the weights are inversely proportional to estimates of the standard deviations of the $\tilde{\varepsilon}_j$. *The variances of the $e_j$ should thus be more or less constant* (precisely so if $S$ in (8.2.11) were replaced by $\sigma\sqrt{(n-q)}$), avoiding the inconvenience of the disparate variances of the $\tilde{\varepsilon}_j$.

If $\epsilon$ is normally distributed then $\tilde{\theta}$, $\tilde{\epsilon}$, and $\tilde{\sigma}^2$ are of course maximum likelihood estimators, and their distributional behaviour is well known ($\tilde{\theta}_j$ and $\tilde{\varepsilon}_j$ are normally distributed, $S^2$ has a $\sigma^2\chi^2$ distribution with $(n-q)$ degrees of freedom).

A fundamental work on the analysis of least-squares residuals is by Anscombe (1961). He proposes methods using estimated residuals to examine the assumption that the $\varepsilon_j$ are independent and normally distributed with constant variance. The methods include a study of the regression of the estimated residuals on the corresponding fitted values, and take account of the intercorrelation of the estimated residuals. Tukey's test of non-additivity is also considered. Particular attention is given to the case where all the residuals have equal variance, typical of factorial design experiments with equal replication (see Section 8.3). Outliers receive only passing mention.

For a generalized approach to the analysis of residuals (but again with only brief mention of outliers) see Cox and Snell (1968, 1971). See also Behnken and Draper (1972), Draper and Smith (1966), Wooding (1969), Beckman and Trussell (1974), and Wilson (1979). The intercorrelation, and differing variances, of the estimated residuals manifest in (8.2.6) has prompted much interest in transformations which remove the correlation and equalize the variances. Studentization takes us only a part of the way: the $\tilde{\varepsilon}_j$ remain correlated. Theil (1965) proposed a method for eliminating the correlation effect. *Recursive residuals*, which on the basic model are uncorrelated with zero mean and unit variance and where the $j$th recursive residual is a function only of the first $j$ observations, have been considered by Brown, Durbin, and Evans (1975), Hedayat and Robson (1970), and others. These would seem to have potential for the study of outliers, although no progress on this front is evident. There is a major difficulty in that the labelling of the observations is usually done at random, or in relation to some concomitant variable, rather than 'adaptively' in response to the observed sample values (which might be a desirable prospect from the outlier standpoint).

Srikantan (1961), however, is concerned with *using estimated residuals specifically to investigate outliers*. He adopts a mean-slippage alternative model for a single contaminant and assumes a normal error structure. For the *labelled* slippage model (see Section 2.3) where the index of the contaminant is *specified* he shows that for a one-sided (two-sided) test of discordancy the corresponding *studentized* residual is the test statistic of the *uniformly most powerful (unbiased) test of discordancy.* Thus if the alternative hypothesis is

$$\bar{H}: E(\mathbf{X}) = A\boldsymbol{\theta} + \mathbf{a} \qquad (8.2.12)$$

where $\mathbf{a}$ is an $n \times 1$ vector of zeros apart from one possible non-zero value in the $i$th position representing contamination, the tests are based on $d_i = \tilde{\varepsilon}_i / s_i$ in the following way. Suppose

$$t_i = d_i^2$$

$$u_i = \begin{cases} d_i^2 & (d_i \geq 0) \\ 0 & (d_i < 0) \end{cases}$$

$$v_i = \begin{cases} 0 & (d_i \geq 0) \\ d_i^2 & (d < 0) \end{cases} \qquad (8.2.13)$$

The uniformly most powerful (one-sided) tests of $a_i = 0$ against $a_i > 0$ and $a_i < 0$ have rejection regions $u_i > R_\alpha$ and $v_i < R_\alpha$, respectively, where $R_\alpha$ has to be determined to produce a test of size $\alpha$. Against $a_i \neq 0$, the two-sided test with rejection region $t_i \geq R'_\alpha$ is uniformly most powerful unbiased (again $R'_\alpha$ has to be chosen to yield a test of size $\alpha$). We should note that one sided tests in the present context are likely to be of somewhat limited interest (see pages 115–6).

For the (more realistic) *unlabelled* slippage model, where the index $i$ of the contaminant is unspecified, Srikantan recommends similar tests based on $t = \max_j d_j^2$, $u = \max_j u_j$, and $v = \max_j v_j$, but there is no consideration of their optimality properties. We shall return to the thorny issue of the determination of critical values (the corresponding $R'_\alpha$ and $R_\alpha$) of these tests.

A different approach to tests for discordancy of outliers, based on residuals, is described by Andrews (1971). He considers the vector of *normalized residuals* $\mathbf{r} = \tilde{\varepsilon}/(\tilde{\varepsilon}'\tilde{\varepsilon})^{1/2} = \tilde{\varepsilon}/S$ and develops tests of discordancy based on a projection of this vector onto a suitable subspace. Adopting the outlier-generating model (8.2.12) we have

$$S\mathbf{r} = S'\mathbf{r}' + (I_n - R)\mathbf{a} \qquad (8.2.14)$$

where $(S')^2$ and $\mathbf{r}'$ are the residual sum of squares, and vector of estimated residuals, respectively, when the basic (no-outlier) model is true, i.e. $\mathbf{a} = 0$. Thus if $\mathbf{a} \neq 0$, $\mathbf{r}'$ is perturbed by a vector in the direction $(I_n - R)\mathbf{g}$ where $\mathbf{g}$ is an $n \times 1$ vector with $n - 1$ zeros and a unit value at the position corresponding with the contaminant, as indicated by $\mathbf{a}$. Discordancy is revealed by $\mathbf{r}$ being too close to $(I_n - R)\mathbf{g}$ and we can assess this in terms of how small is the norm $||\mathbf{r}*||$ of the orthogonal complement $r*$ of the projection of $\mathbf{r}$ on $(I_n - R)\mathbf{g}$.

Of course, we will not often wish to specify the index of the potential discordant value, but will need to test the *unlabelled* mean-slippage model corresponding with (8.2.12). This leads to a test which attributes discordancy to a single outlier when $\min_j \|\mathbf{r}_j^*\|$ is sufficiently small, where $j$ denotes the index of the potential contaminant. Andrews develops this test in the cases where the error distribution is normal, and exponential. The approach is also extended to multiple outliers. He claims that his approach is essentially different from others which examine merely the absolute values of the residuals (or normed residuals), and he suggests that it is an improvement in view of its ability to take account of the form of $A$. But this facility, as expressed through the use of projection vectors, seems to reduce merely to the use of the individually studentized residuals $e_j$ of (8.2.11) rather than of the undifferentially weighted normed residuals.

In summary, tests of discordancy (and accommodation) of outliers in the general linear model situation make widespread use of the maximum (positive, negative, or absolute) studentized residual as test statistic, with discordancy attributed to the observation yielding the maximum provided that maximum is sufficiently large. This is precisely the Srikantan (1961) prescription. Illustrating it for the two-sided test, we reject the basic hypothesis of (8.2.1) in favour of one which postulates a single contaminant, e.g. (8.2.12), if

$$\tau = \max_j |e_j| = \max_j |\tilde{\varepsilon}_j/s_j| > h_\alpha \qquad (8.2.15)$$

where for a test of size $\alpha$ the critical value $h_\alpha$ needs to be chosen to ensure that, under (8.2.1), (8.2.15) holds with probability $\alpha$. The observation which yields the maximum value of $e_j$ is declared to be the discordant outlier.

Most discussion of such a test centres on the problem of determining the critical values $h_\alpha$. We assume first of all that $\sigma^2$ is unknown and has to be estimated solely from the current data. Srikantan (1961) used the first Bonferroni inequality to determine upper bounds for the critical values of his statistics (discussed above: defined in terms of squared studentized residuals). He was able to present numerical results for some models with $q$ up to 3 in value. We have also noted in Section 8.1 the special case approximations of Prescott (1975), and simulation results of Tietjen, Moore, and Beckman (1973), both concerned particularly with the simple linear regression model (8.1.1) (although Prescott also refers to the general linear model).

So much has been written on the determination of the critical values (and on the joint and marginal distributions of the studentized residuals) that it is necessary to try to put the respective contributions in perspective. Most proposals employ Bonferroni inequalities; let us recall their form in the present context. If $A_1, A_2, \ldots, A_n$ are $n$ events, we have:

$$\sum_{i=1}^{n} P(A_i) - \sum_{i<j} P(A_i A_j) \leq P(U\{A_i\}) \leq \sum_{i=1}^{n} P(A_i). \qquad (8.2.16)$$

If $A_i$ is the event $|e_i| > \xi$, where $|e_i|$ is the absolute value of the $i$th studentized residual, then (8.2.16) becomes

$$\sum_{i=1}^{n} P(|e_i| > \xi) - \sum_{i<j} P(|e_i| > \xi, |e_j| > \xi) \leq P(\max_i |e_i| > \xi) \leq \sum_{i=1}^{n} P(|e_i| > \xi).$$
$$(8.2.17)$$

The right-hand inequality is known as a *first-order Bonferroni inequality*; it is clearly relevant to determining an upper bound to the critical value $h_\alpha$ in (8.2.15) and hence a conservative discordancy test. The two-sided inequality is a *second-order Bonferroni inequality*; it has potential value in exhibiting the accuracy of the upper bound. It is not easy, however, to evaluate the modifying term $\sum_{i<j} P(|e_i| > \xi, |e_j| > \xi)$ and this aspect has been much considered.

Our starting point should be the work of Srikantan (1961) who used the first-order Bonferroni inequality (involving percentage points of the $F$-distribution) to obtain an upper bound for $h_\alpha$. But he went further in a crucial respect, demonstrating that for linear model problems the upper bound can in fact be the exact value, due to the modifying term in the second-order inequality being zero. The exact value is obtained if the bound, $h'_\alpha$ say, satisfies

$$2(h'_\alpha)^2/(n - q) \geq 1 + |\rho_{ij}| \qquad \text{(all } i \neq j) \qquad (8.2.18)$$

with $\rho_{ij}$ the correlation between $e_i$ and $e_j$ i.e.

$$\rho_{ij} = \lambda_{ij}/(\lambda_{ii}\lambda_{jj})^{1/2} \qquad (8.2.19)$$

where the matrix $\{\lambda_{ij}\}$ is just $I_n - R$. (Equivalent conditions, omitting modulus signs, apply to one-sided tests.) For rather special models (e.g. with $q = 2$ and $u_i = i$) the implications of (8.2.18) and (8.2.19) are pursued. For the case mentioned, $h'_\alpha$ is exact for $n \leq 9$ when $\alpha = 0.05$ and for $n \leq 14$ when $\alpha = 0.01$.

An implication of Srikantan's work is that the extreme-tail behaviour of the distribution of the maximum absolute studentized residual can be reflected exactly by the first-order Bonferroni inequality. In view of (8.2.18), and the dependence of the $\rho_{ij}$ on the design matrix $A$, however, it is not easy to determine precisely what constitutes the 'extreme tail'. It depends on $A$, and need not encompass the significance levels (e.g. 5 per cent) conventionally employed. See also Fieller (1976, Chapter 5) on the insensitivity of critical values to the form of $A$ in small samples.

Ellenberg (1973) determined the joint distribution of subsets of the $e_i$ for a general linear model with normal error structure in terms of 'a standardized version of the Inverted–Student Function' and proposed use of the appropriate form of (8.2.16) for putting bounds on the critical value $h_\alpha$ of (8.2.15).

Specifically, if $\eta_i = \sqrt{(n-q)}e_i$ (the so-called standardized residuals, used by some in this context in place of the studentized residuals) it is found that $\eta_i^2$ has a Beta distribution with parameters $(\frac{1}{2}, (n-q-1)/2)$ for all $i = 1, 2, \ldots, n$.

Thus from (8.2.17)

$$P(\max|e_i| > \xi) \leqslant nP(|e_i| > \xi) \qquad (8.2.20)$$

and an upper bound to $h_\alpha$ (the critical value) is given by $h_\alpha'$ where

$$P(|e_i| > h_\alpha') = \alpha/n \qquad (8.2.21)$$

Using the Beta distribution we find that

$$h_\alpha' = \sqrt{(n-q)}[F/(n-q-1+F)]^{1/2} \qquad (8.2.22)$$

where $F$ is the upper $\alpha/n$ point of the $F$-distribution with 1 and $n-q-1$ degrees of freedom. Such percentage points for the $F$-distribution will need to be calculated to employ (8.2.22). Lund (1975) *used this approach and presented the most comprehensive, readily accessible, set of tables* of upper bounds $h_\alpha'$. Values are given for $n = 5(1)10(2)20(5)50(10)100$ and $q = 1(1)6(2)10,15,25$. These are reproduced as Table XXXVII on pages 522–3 but it should be noted that Doornbos (1981) has found some errors of one unit in the last decimal place in Lund's tables. This is consistent with the discrepancy we have noted above between Lund's values and those of Prescott (1975) and Tietjen, Moore, and Beckman (1973) for the case $q = 2$. Doornbos (1981) refers to an extensive tabulation he has produced in a Memorandum of the Eindhoven University of Technology.

Tse (1988) recalculates the Lund (1975) tabulated values, confirms the Doornbos (1981) claim and concludes that the Lund values, while accurate to 0.01, are systematically too large. Tse reports that he has recalculated the critical values for $\alpha = 0.10$, 0.05 and 0.01 over an extended range up to $q < n - 2$ (obtainable on request from the author). Monte Carlo studies of *test size* for the tabulated critical values (Lund, 1975) suggest that the tests are systematically conservative with sizes up to 10 per cent lower than the critical levels (e.g. 9 per cent instead of 10 per cent and 4.5 per cent instead of 5 per cent).

Ellenberg (1976) demonstrates the equivalence of the discordancy test *based on studentized residuals* and others *based on reduction of residual sums of squares* (and associated statistics) *on omission of individual observations* from the sample (so-called *deletion methods*: see below). He also remarks that even if the Srikantan condition (8.2.18) for exactness of the upper bound does not hold, a substantial computational saving in calculating the modification term in (8.2.17) can be achieved from the fact that any $(e_i, e_j)$ for which

$$h'_\alpha \geq |(1 + |\rho_{ij}|)/2|^{1/2} \sqrt{(n - q)} \qquad (8.2.23)$$

will not contribute to the modification. This is illustrated for a particular practical problem considered by Mickey, Dunn, and Clark (1967).

Both Doornbos (1981) and Cook and Prescott (1981) are also concerned with assessing the accuracy of $h'_\alpha$ by consideration of the second-order Bonferroni inequality.

Doornbos (1981) shows that provided

$$\max_{i,j} |\rho_{ij}| \leq g_\alpha \qquad (8.2.24)$$

for suitable $g_\alpha$ (which are functions of $h'_\alpha$ for the current $(n, q)$ and for $(n, q + 1)$ and which he tabulates quite extensively) then use of $h'_\alpha$ leads to a test of size $p_\alpha$ where

$$\alpha - \alpha^2/2 < p_\alpha < \alpha. \qquad (8.2.25)$$

Even if (8.2.24) does not hold, but *most* $\rho_{ij}$ satisfy $|\rho_{ij}| \leq g_\alpha$, then an acceptable lower bound for $p_\alpha$ can be easily determined. This is again illustrated for data from Mickey, Dunn, and Clark (1967). (Note that Doornbos, 1981, points out that the joint distribution of $e_i, e_j$, exhibited by Ellenberg, 1976, was essentially developed by Doornbos and Prins, 1958, and by Joshi, 1972a; also Joshi, 1975, tabulates $h'_\alpha$ for $q \leq 4$.)

Cook and Prescott (1981) also develop a method, based on $F_{1, n - q - 1}$ tail-probabilities for functions of $h'_\alpha$ and $\rho_{ij}$ (not satisfying (8.2.23) ), for determining the second-order Bonferroni-type lower bound for $h_\alpha$. Examples are given including again the ubiquitous Mickey, Dunn, and Clark (1967) data.

All these methods for assessing (and evaluating) the accuracy of $h'_\alpha$ depend only on knowledge of the $\rho_{ij}$ (i.e. on $A$) and thus can be readily implemented.

See also Mirvaliev (1978b) on approximations to critical values; Gentle (1978) on simulation-based calculation of the power of discordancy tests for $q = 1, 2, 3$ (with comment on the multiple outlier problem) and Brown and Kildea (1979) on use of *signs* of residuals for testing (and accommodating) outliers in the linear model.

When we have some knowledge about $\sigma^2$, *external to the data*, the discordancy test needs appropriate modification. Joshi (1972a) considers this for the normal error model where we have either an external estimate $s_\nu^2$ of $\sigma^2$ distributed as $\sigma^2 \chi_\nu^2/\nu$ independent of $\mathbf{X}$, or where we know $\sigma^2$ precisely. The test structure is the same as before except that the (internally) studentized residuals $\tilde{\varepsilon}_i/s_i$ are replaced by *externally studentized, pooled studentized,* or *standardized residuals*:

$$\tilde{\varepsilon}_i/[(1 - r_{ii})s_v^2]^{1/2},$$

$$(v + n - q)^{1/2}\tilde{\varepsilon}_i/[(1 - r_{ii})(vs_v^2 + S^2)]^{1/2},$$

or

$$\tilde{\varepsilon}_i/[(1 - r_{ii})\sigma^2)^{1/2}.$$

A test of discordancy again proceeds in terms of the maximum (positive, negative, or absolute) value of the weighted residuals. For the pooled studentized residuals, $v = 0$ reduces to the original test based on the $\tilde{\varepsilon}_i/s_i$.

Joshi's approach to the determination of the critical values of the tests again yields only upper bounds but they turn out to be intermediate between these given by the first Bonferroni inequality and by the more precise (but less computationally tractable) second Bonferroni inequality. The performance of the tests is considered, and illustrated numerically for the simplest case of a single univariate normal sample. Throughout it is assumed under the alternative hypothesis that there is a single contaminant arising from a constant slippage of the mean; its index is unknown and is assumed to be chosen at random from the set $(1, 2, \ldots, n)$.

We have so far considered a contamination model involving slippage of the mean. Cook, Holschuh, and Weisberg (1982) examine a linear model problem with normal error structure where there is a single contaminant arising from *inflation of the error variance*: specifically, we have $\mathbf{x} = A\theta + \boldsymbol{\epsilon}$ where the variance-covariance matrix of $\boldsymbol{\epsilon}$ is of the form $\sigma^2(I + B)$ where $B$ is a diagonal matrix with only one non-zero term $b(> 0)$. Maximum likelihood estimators are characterized. It is seen that the observation naturally identified as outlying need not necessarily be the one with largest absolute studentized residual, unless the latter also has largest absolute (non-studentized) residual (as is so for balanced designed experiments of appropriate form—see Section 8.3)

The problem of detecting outliers in a linear model (8.2.1) but where the residuals are correlated in the sense that $V(\boldsymbol{\epsilon})$ has a form $\sigma^2 V$ (for prescribed $V$) rather than $\sigma^2 I_n$, is briefly considered by Tamhane (1982).

There is another interesting (if equivalent) way of justifying use of the *individually* studentized residuals as appropriate representations of the data for detecting outliers and for testing discordancy.

We recall that the basic model is (8.2.1): $\mathbf{x} = A\theta + \boldsymbol{\epsilon}$. The prospect of a single contaminant reflecting slippage in the mean can be expressed in terms of the set of alternative hypotheses

$$\bar{H}_j : \mathbf{x} = A\theta + \mathbf{a}_j + \boldsymbol{\epsilon} \qquad (j = 1, 2, \ldots, n) \qquad (8.2.26)$$

where $a_j$ is $n \times 1$ with a value $a$ in the $j$th position and zeros elsewhere. Thus $H_j$ declares that $x_j$ is the contaminant, from the distribution $N(A'_j \theta + a, \sigma^2)$ where $A_j$ is the $j$th row of $A$. As in our study of multivariate outliers, for example, we might choose to detect a single outlier in terms of greatest increase in the maximized likelihood under the set of hypotheses $\bar{H}_j$ relative to the basic model $H: \mathbf{x} = A\theta + \boldsymbol{\epsilon}$.

Under $H$, the likelihood is maximized by putting $\theta = \tilde{\theta}$ and $\sigma^2 = S^2/n = (\boldsymbol{\epsilon}'\boldsymbol{\epsilon})/n$. The maximized log-likelihood is

$$-\frac{n}{2} \log\left(\frac{2\pi S^2}{n}\right) - \frac{n}{2}.$$

Under $\bar{H}_j$, $x_j$ arises from $N(\mu, \sigma^2)$ whilst $\mathbf{x}_{-j}$ (the set of observations excluding $x_{-j}$) arises (independently) from $N(A_{-j}\theta, \sigma^2 I_{n-1})$ where $A_{-j}$ is the reduced matrix $A$ obtained on deletion of the $j$th row. The likelihood is now maximized by putting $\mu = x_j$ and

$$\theta = (A'_{-j} A_{-j})^{-1} A'_{-j} \mathbf{x}_{-j}$$

$$\sigma^2 = S^2_{-j}/n$$

where $S^2_{-j}$ is the sum of squares of the estimated residuals when the reduced data vector $\mathbf{x}_{-j}$ is fitted by least-squares to the model $\mathbf{x}_{-j} = A_{-j}\theta + \boldsymbol{\epsilon}_{-j}$. The maximized log-likelihood now becomes

$$-\frac{n}{2} \log\left(\frac{2\pi S^2_{-j}}{n}\right) - \frac{n}{2}.$$

Thus the increase in the maximized log-likelihood is

$$\frac{n}{2} \log(S^2/S^2_{-j})$$

and so, on the above criterion (with no restriction on the value of $a$), a single outlier is detected as that observation whose omission from the sample effects the greatest reduction in the residual sum of squares. This approach is termed a **single deletion method**. Thus the outlier is the observation yielding

$$\max_j(S^2/S^2_{-j}).$$

It is to be adjudged discordant if this maximum is sufficiently large relative to its null distribution.

In fact we can show that a simple relationship holds:

$$S^2 = S^2_{-j} + \tilde{\epsilon}_j^2/[1 - A'_j (A'A)^{-1} A_j]$$

$$= S^2_{-j} + \tilde{\epsilon}_j^2/(1 - r_{jj}). \qquad (8.2.27)$$

Hence

$$S^2/S^2_{-j} = S^2/[S^2 - \tilde{\varepsilon}_j^2/(1 - r_{jj})]$$

$$= \{1 - \tilde{\varepsilon}_j^2/[S^2(1 - r_{jj})]\}^{-1}$$

$$= [1 - e_j^2/(n - q)]^{-1} \qquad (8.2.28)$$

and we see that maximization of $S^2/S^2_{-j}$ is merely equivalent to maximization of the square of the (individually) *studentized residual*, $e_j$. So the *maximum likelihood ratio* procedure involves examination of the squares (or absolute values) of the studentized residuals, with detection of the outlier as the observation yielding the maximum absolute studentized residual, and declaration of discordancy if it is statistically too large relative to the basic model. This equivalence between reduction in the residual sum of squares (the deletion approach) and the absolute values of the studentized residuals is demonstrated by Ellenberg (1973, 1976). In spite of this, however, the literature continues to draw distinctions between the two approaches as we shall see below.

Thus the preoccupation with studentized residuals as an indication of outliers in the general linear model finds a sound foundation on the maximum likelihood ratio principle, and discordancy tests based on $\max_j(S^2/S^2_{-j})$ and on $\max_j(e_j)$ are equivalent. The maximum likelihood ratio basis for these tests is exhibited by Fieller (1976).

A detailed study by Butler (1983) takes further the use of residuals in outlier discordancy tests and demonstrates some optimality properties of the tests. Butler casts the linear model outlier problem in a multivariate framework; with null model.

$$X = A\Theta + E. \qquad (8.2.29)$$

($X$ is of dimension $n \times k$; $A$ is $n \times p$; $\Theta$ is $p \times k$ and $E$ is $n \times k$). Various cases and contamination models are considered, including normal-error mean-slippage models for the univariate case ($k = 1$) and multivariate case ($k > 1$), with 1 or with $s$ ($> 1$) contaminants. Use of cross-validation techniques, likelihood ratio methods and of the correlation between true and estimated residuals all serve to motivate the use of residual-based procedures for tests of discordancy (demonstrating *en passant* the equivalence with single or, if appropriate, block deletion methods).

Naik (1989) also considers the *multivariate* linear model (8.2.29) and exhibits some local optimality properties for an outlier discordancy test based on the Mardia (1970) measure of multivariate kurtosis.

More recent developments have been particularly concerned with the detection and testing of *multiple outliers* (and often with the attendant difficulties relating to *masking* and *swamping*). This is true of the work of Joshi and Lalitha (1986) and Lalitha and Joshi (1986a and 1986b). In the first-mentioned of these papers, there is detailed consideration of the case of *two outliers in the same direction* for the null model (8.2.1) with normal error structure. If $e_i$ is the studentized residual $(i = 1, 2, \ldots, n)$ they propose a test statistic (for two outliers in the 'positive' direction).

$$U = \max_{1 \leq i < j \leq n} \{(e_i + e_j)/(2 + 2\rho_{ij})^{1/2}\} \qquad (8.2.30)$$

where $\rho_{ij}$ is the correlation between the estimated residuals $\tilde{\varepsilon}_i$ and $\tilde{\varepsilon}_j$.
Specifically if

$$\Lambda = I_n - A(A'A)^{-1}A' = \{\lambda_{ij}\} \qquad (8.2.31)$$

we have

$$\rho_{ij} = \lambda_{ij}/(\lambda_{ii}\lambda_{jj})^{1/2}.$$

For two outliers in the 'negative' direction the proposed test statistic is

$$U^* = -\min_{1 \leq i \leq j \leq n} \{(e_i + e_j)/(2 + 2\rho_{ij})^{1/2}\} \qquad (8.2.32)$$

and for two outliers in opposite directions:

$$V = \max_{1 \leq i \leq j \leq n} \{|(e_i - e_j)/(2 - 2\rho_{ij})^{1/2}|\}. \qquad (8.2.33)$$

Joshi and Lalitha (1986) use the first Bonferroni inequality to approximate the upper percentage points of $U$ in (8.2.30) as the basis of a test of discordancy. Specifically, if the upper $\alpha$-point is $u_\alpha$, we have

$$I_{1 - \mu_\alpha^2}[(p - 1)/2, \ 1/2] = 4\alpha/[n(n - 1)]$$

where $I_q(a, b)$ is the incomplete beta function. For V, we have $v_\alpha = u_\alpha/2$.

For an independent random sample $x_i(i = 1, 2, \ldots, n)$ from $N(\mu, \sigma^2)$ the tests reduce to Murphy's test for two outliers, and the studentized range test, respectively (see **N3** and **N6** in Chapter 6).

The performance of the tests based on $U^*$ (or $U$) and $V$ for a mean-slippage alternative (contamination) models is examined in Lalitha and Joshi (1986a and 1986b, respectively) with tabulated upper and lower bounds given for appropriately defined power characteristics.

Two comparative studies of different discordancy tests for outliers in linear models are reported in Balasooriya and Tse (1986) and Balasooriya, Tse, and Liew (1987). In the former case five tests are compared by Monte Carlo methods and that based on studentized residuals for a single outlier (Srikantan, 1961; see above) is judged to be the best. Balasooriya, Tse, and Liew (1987) apply seven commonly used procedures to six published data sets to examine respective outcomes. The procedures examined include three based on internally studentized or normal residuals (Srikantan, 1961; Prescott, 1975 and Andrews and Pregibon, 1978), and one each on externally studentized residuals and Cook's measure of influence (see (8.2.42) and Section 8.7 below). Their conclusion is that, across the empirical studies, it is only the three procedures based on studentized residuals that agree closely with each other! See also Birch and Fleischer (1984).

Schall and Dunne (1988) derive discordancy tests for multiple outliers in the *extended linear model* (with correlated residuals)

$$X = A\Theta + \varepsilon$$

with

$$V(\varepsilon) = V\sigma^2 \tag{8.2.34}$$

for prescribed $V$. They also examine correspondingly extended measures of influence.

### 8.2.2 Augmented residual-based, and non-residual-based, methods

In the previous section we have considered methods for detecting and testing outliers based specifically on use of estimated residuals relative to the null model, with particular attention to the use of the studentized residuals. We now review an extensive but less homogeneous body of work which *extends, modifies or augments the direct study of estimated residuals*, or *uses criteria not based on residuals*. In this latter category we will include methods based on the reduction in residual sums of squares, but we should recall the duality described above between many residual-based, and reduction-based, procedures. This point needs amplification.

Various methods based on reduction of *residual sums of squares* may *seem* at first sight (and indeed are often claimed by their progenitors) not to involve direct study of the estimated residuals. It is sometimes advanced that this avoids a disadvantageous cloaking (or confounding) effect of an outlier, arising from the fact that *all* the estimated residuals reflect the influence of the outlier. The view is expressed that there is a contradiction in trying to detect an outlier by examining estimated residuals which are 'biased by the presence' of the outlier we are trying to detect.

However, the results of the previous section cast doubt on any advantage in the use of residual sums of squares rather than of estimated (studentized) residuals. Regarding an observation $x_j$ as missing will have the same effect as estimating the parameters under the model $\bar{H}_j$ and the resulting residual sum of squares will be just $S^2_{-j}$. But in view of (8.2.27) the separate residual sums of squares are readily obtained merely by reducing the overall residual sum of $S^2$ by the corresponding weighted squared estimated residual ($\tilde{e}_j^2/(1 - r_{jj})$). Thus three conclusions arise. The calculation is often the same in both approaches, involving determination of the estimated residuals and of $S^2$. The visual impact is equivalent, being the reflection of the values of weighted *squared* estimated residuals (it might *appear* heightened on the residual sum of squares approach merely because we are there looking at a reduction from $S^2$ by the *square* of the estimated residual, rather than at just the absolute value of the residual). Finally, often no basic statistical distinction exists; in particular the method is still as subject to any cloaking, or confounding, influence of the outliers we are seeking to detect as it is using residuals directly. We shall illustrate this later (in Section 8.3) with an example using data arising from a factorial experiment.

Mickey (1974) and Mickey, Dunn, and Clark (1967) also proposed such examination of the separate residual sums of squares after omission of the separate residuals singly, suggesting that a large enough reduction is evidence of a discordant outlier. Their test statistic is

$$\max_{j} \left\{ \frac{S^2 - S^2_{-j}}{S^2_{-j}/(n - q - 1)} \right\}$$

with attribution of discordancy if it is sufficiently large. (They seem to think it necessary, however, to conduct $n$ separate regression analyses, rather than just one yielding $S^2$ and the $\tilde{\varepsilon}_j$.)

Snedecor and Cochran (1967, page 157) discuss a test of discordancy based on the maximum of the studentized differences, i.e. the test statistic is

$$\max_{j} \{ (x_j - \tilde{x}_{-j})/[V(x_j - \tilde{x}_{-j})]^{1/2} \}$$

where $\tilde{x}_{-j}$ is the least-squares estimate of $x_j$ when $x_j$ is regarded as a missing observation. Again the detected outlier is acclaimed discordant if the test statistic is sufficiently large.

Ellenberg (1976) demonstrates that these two methods are equivalent, and both coincide with the test based on the maximum absolute studentized residual. He employs the result (10.2.27) which he derived earlier—Ellenberg (1973).

Where some exceptions to this duality principle do arise, as we shall see, is when *additional criteria are invoked*: using modified or extended notions of 'residual' or recursive multi-stage methods, as is often the case with multiple outliers.

An approach to the detection of outliers in linear models (principally two-way experimental designs) not based exclusively on examining estimated residuals is that of Gentleman and Wilk (1975a). They present a method for detecting the '$k$ *most likely outlier subset*' as that set of $k$ observations warranting attention as outliers prior to a further examination of their discordancy, or to an attempt to analyse the data in a way which is (relatively) insensitive to their presence. The procedure consists of specifying $k$, determining the '$k$ most likely outlier subset', assessing its statistical significance and, if not significant, proceeding to consider successively smaller numbers, $k - 1$, $k - 2$, . . . of outliers until a significant outlier subset is detected (if at all).

The method involves a deal of computational effort. It proceeds as follows.

For given $k$, we consider all $\binom{n}{k}$ partitions of the data set obtained on specifying particular subsets of $k$ observations. If there are truly $k$ contaminants present, then $\binom{n-k}{k}$ of these subsets will *not* contain contaminants, $\binom{n}{k} - \binom{n-k}{k}$ will do so. We suppose (under the contamination model) that the observation vector $\mathbf{x}$ has the form

$$\mathbf{x} = A\mathbf{\theta} + \mathbf{\delta} + \mathbf{\epsilon} \tag{8.2.35}$$

where $A$ is a $n \times q$ design matrix, $\mathbf{\theta}$ a vector of $q$ parameters, $\mathbf{\epsilon}$ the error vector and $\mathbf{\delta}$ a $n \times 1$ vector with $n - k$ zeros and $k$ unknown non-zero values corresponding with the $k$ mean biases of the contaminants.

Suppose that $\tilde{\mathbf{\epsilon}}$, $\hat{\mathbf{\epsilon}}$ are the estimated residuals obtained by fitting the null model $\mathbf{x} = A\mathbf{\theta} + \mathbf{\epsilon}$ and by fitting the contamination model (8.2.35), respectively, when a particular set of $k$ observations are under consideration. Then the difference in the sums of squares

$$Q_k = \tilde{\mathbf{\epsilon}}'\tilde{\mathbf{\epsilon}} - \hat{\mathbf{\epsilon}}'\hat{\mathbf{\epsilon}} \tag{8.2.36}$$

(which is inevitably non-negative) provides a measure of the effect of assuming that the $k$ chosen observations are in fact contaminants. (Note that $\hat{\mathbf{\epsilon}}'\hat{\mathbf{\epsilon}}$ is also the sum of squares of residuals arising from fitting the null model to the reduced data set of $n - k$ observations.)

If we were to evaluate the $Q_k$ for all $\binom{n}{k}$ partitions of the data we could examine their relative sizes as indications of the prospect of the corresponding $k$ observations being contaminants. The *largest* $Q_k$ is used to detect the $k$ *most likely outlier subset*. To determine if it is large enough not to have arisen purely by chance under the null model we would really need to know its null distribution.

This distribution is unknown, and informal methods are proposed to assess the significance of the largest $Q_k$. These include the plotting of some large subset of the larger $Q_k$ against 'typical values' of these obtained (presumably by simulation) under the null model, or by plotting the residuals on the outlier model (8.2.35) corresponding with the set of 'outliers' detected by the largest $Q_k$. The first of these leans very heavily on the assumptions of the model and presumably will be influenced by non-normality, heteroscedasticity, and non-additivity, as well as by contaminated observations.

If the '$k$ most likely outlier subset' is not statistically significant, we go through the same exercise with $k$ reduced to $k - 1$, and so on, until we detect a significant outlier subset.

The whole approach is cumbersome, although some computational simplification is possible, e.g. for two-way designs, see Gentleman and Wilk (1975a). It is interesting in principle, but clearly needs much examination and refinement if it is to be a practical proposition. The philosophy behind starting with a specific $k$ and then considering *smaller* values is that if we obtain significance at any stage we might argue that there is little point in considering even smaller $k$ (even fewer outliers); the reverse policy does not have this apparent advantage. However, we must take care that swamping does not occur. This approach does help, though, to reduce the opposite phenomenon of masking, provided $k$ is initially chosen sufficiently (but not unreasonably) large.

See also John (1978), John and Draper (1978), and Draper and John (1980).

Gentleman (1980) extends the discussion of the '$k$ most likely outliers' approach, considering methods for reducing the otherwise prohibitive computational effort with data in the form of two-way tables, broadening the discussion to the general linear model and comparing the approach with two simpler methods for choosing a subset of $k$ outliers. These two simpler methods amount to choosing the $k$ observations with largest absolute studentised residuals either *en bloc* or in a sequential manner. Such methods provide a starting point for the multiple outlier problem in the linear model, but they do not go very far. In particular the crucial problem of how to choose $k$ (rather than employing a prescribed value for it) requires closer investigation. (See also the Bayesian approach of Dempster and Rosner, 1971, discussed in Section 9.3).

Chambers and Heathcote (1981) consider joint estimation of parameters and detection of outliers in a linear model. (Compare, in this spirit of combining the accommodation and detection aims, the Bayesian methods discussed in Sections 9.2 and 9.3 below.) Their method is developed for large samples. It amounts for a given sample size $n$ to obtaining a family of estimators $\{\boldsymbol{\theta}_n^*(t); -T < t < T\}$ of $\boldsymbol{\theta}$ indexed by a scalar parameter $t$ (which takes values over a finite interval including the origin), through minimization of a loss function $L_n(\boldsymbol{\theta}:t)$. The $\boldsymbol{\theta}_n^*(t)$ are obtained by minimizing a function of the empirical cumulant generating function of the error distribution. Specifically

$$L_n(\boldsymbol{\theta}:t) = t^{-2}\ln\{|n^{-1}\sum_{j=1}^{n} \exp[it(x_j - A_j'\,\boldsymbol{\theta})]|^2\}, \qquad (8.2.37)$$

where $A_j'$ is the $j$th row of $A$. If $t = 0$, the method reverts to the least squares approach. It is shown that under fairly general conditions the asymptotic covariance matrix of $n^{1/2}\boldsymbol{\theta}_n^*(t)$ is of a form $\sigma^2(t)C$, where $C$ is a symmetric non-singular matrix and $\sigma^2(t)$ is determined from the characteristic function of the error distribution.

It is proposed that we should choose $t$ to minimize $\sigma^2(t)$. We obtain $t = 0$ if and only if $\epsilon$ has an uncontaminated normal distribution, and the extent to which the minimizing $t$ differs from zero is used as an indication of the presence of contamination. If the minimum occurs for $t \neq 0$ it is suggested that we delete the observation with largest estimated residual and repeat the process until the minimum occurs at $t = 0$. *A word of caution is needed however!* A minimum at $t \neq 0$ can arise for various reasons. It *might* arise because the error distribution is of contaminated form (and thus relevant to our notion of outliers) but it might merely indicate skewness or long tails.

Another joint detection/accommodation method is proposed by Aitkin and Tunnicliffe Wilson (1980) who use the EM algorithm to estimate the parameters in a normal-mixture contamination model in univariate, *and regression*, situations.

Cook (1979) briefly considers the outlier detection problem in respect of the way in which contaminants (on a mean-slippage model) affect partial *F*-tests, studentized residuals, residual variances, and the convex hull of the data set. For the multiple outlier problem, the Gentleman and Wilk (1975a) '*k* most likely outlier subset' method is discussed and comment is made on the fact that (for example) the 'most likely' *pair* of outliers need not be the observations with largest studentized residuals: the correlations between the residuals 'can combine two seemingly unlikely candidates as the most likely outlier pair'.

**The multiple outlier problem** has been considered by many other writers over recent years.

Hawkins, Bradu, and Kass (1984) consider detection of several outliers in linear model data with normal error structure using the notion of 'elemental sets'. The approach extends the idea of using *tetrads* (e.g. Bradu and Hawkins, 1982; see also Sections 8.4 and 12.1 below.) For any observation $x_i$ we consider all data subsets of size $q$ (the dimensionality of the regression model): each subset is used to estimate the $q$ parameters and each yields a set of estimated residuals (the *elemental residuals*). Summary statistics of the set of elemental residuals are used to assess the outlying candidacy of $x_i$; contamination will shift the mean (with no effect on the variance). To minimize effects of masking and swamping robust location estimators are proposed such as the median. Proposals are made for implementing the method for the multiple outlier problem in a computationally efficient manner. Bradu and Hawkins (1991) apply the elemental set approach to a study of the *least median of squares* principle of Rousseeuw (1984).

A multistage procedure for multiple outliers is proposed by Marasinghe (1985). An initial subset of $k$ candidate outliers is chosen as follows. The first is chosen as that observation with largest absolute studentized residual; the next as that with largest absolute studentized residual in the reduced data set of $(n - 1)$ observations, and so on. If $Q'_k$ is the reduction in the residual sum of squares in fitting the null (normal-error) model to the reduced set of $(n - k)$ observations, we are required to employ a test with test statistic

$$F_k = 1 - Q_k^t / S \qquad (8.2.38)$$

where $S = \sum_1^n e_i^2$ (from fitting the full data set). We note that $F_k$ can be written

$$F_k = \sum_1^k f_i^2$$

where $f_1, f_2, \ldots, f_k$ are 'adjusted residuals' corresponding to the successive fits on the reducing-sized data sets.

Approximate critical values for the test are obtained by Marasinghe from Bonferroni inequalities for given $k$, and presented for $k = 2(1)5$, $n = 2(k + 1)$ to 100 and $\alpha = 0.01$ and 0.05, but only for $q = 2$ (simple linear regression).

If $F_k$ is *less than* the tabulated critical value we reject the no-outlier null hypothesis, reject the 'most extreme' observation (the first-chosen subset member) and conduct a similar procedure for the reduced subset of size $(k - 1)$ from a reduced sample of size $n - 1$ until we no longer reject the null hypothesis; thus determining a set of multiple discordant outliers.

Limitations to the approach are discussed qualitatively, comparisons are made with other methods using Monte Carlo results and the method is applied to the ubiquitously-quoted stack-loss data (Daniel and Wood, 1971).

Computationally the approach is much more manageable than that of Gentleman and Wilk (1975a) discussed above. Fung (1988) warns that starting with $k$ too large (as recommended by Marasinghe) can lead to masking difficulties and illustrates this for the stack-loss data.

Another multiple outlier approach based on *recursive residuals* is described by Kianifard and Swallow (1989). A specified number of *least extreme* observations are identified in terms of some chosen regression diagnostic (such as studentized residuals or values of Cook's D). For the remaining $(n - k)$ we compute their 'recursive residuals'—studentized residuals in the reduced sample set of $(n - k)$ observations—and proceed to test these to identify discordant outliers. Claimed advantages include possible reduction of the masking effect, manageable computational effort and the adaptive principle of the approach (i.e. with the procedure itself determined by the sample we are examining). Some comparative performance characteristics are presented.

Andrews and Pregibon (1978) extend earlier ideas of Andrews (1971) in proposing methods for detecting one or more *important* outliers in the sense that they strongly influence resulting estimates of $\theta$ in the linear model. The motivitation is that such influence accrues if the removal of observations has a large effect on $|A'A|$ or on the residual sum of squares $(\mathbf{x} - A\tilde{\boldsymbol{\theta}})'(\mathbf{x} - A\tilde{\boldsymbol{\theta}})$. The latter consideration is germane to our interest in outliers as reflections of contamination of the response $\mathbf{x}$. Concern for the design structure, expressed through $|A'A|$, carries the interest over to the broader topic of *influence* and the Andrews and Pregibon proposal that we consider the effect of different observations through changes in (effectively) the product

$$|A'A|(\mathbf{x} - A\tilde{\boldsymbol{\theta}})'(\mathbf{x} - A\tilde{\boldsymbol{\theta}})$$

needs to be viewed in this wider context (see Section 8.7).

Motivated by the Andrews and Pregibon approach, Dempster and Gasko-Green (1981) consider a wider class of procedures which remove observations one at a time assessing the order in which they are removed in terms of joint consideration of 'residual size and influence'. In this stepwise process a quantitative (probability) measure is attached to the successive observations which are removed, expressing their relative discrepancies from the linear model. Different methods are compared theoretically and through application to specific data sets. A principal point of interest is the extent to which the stepwise removal process should continue. A method for simultaneous identification of outliers and predictors using dummy variables is described by Peixoto and La Motte (1989). Bassett (1988) re-examines earlier proposals for detecting outliers by use of extreme regression quantiles.

### 8.2.3 Accommodation of outliers for the linear model

We discussed earlier the range of techniques available for accommodating outliers in robust analyses of univariate and multivariate samples.

Included in the discussion was the premium-protection approach of Anscombe (1960a), where in estimating the mean of the underlying distribution a location-slippage contamination model was employed and the mean was estimated either by the overall sample mean for the sample of size $n$, or by the sample mean of $(n - 1)$ observations omitting an extreme sample value if it was sufficiently large (or small) in value.

Such a principle provides a first example of outlier accommodation methods for the linear model. Anscombe (1960a) describes how the same approach can be used in *designed experiments* (or in analysing *general linear models*). Attention is restricted to situations where, in the absence of contaminants, all residuals have *common variance*, and where inter-residual correlation is nowhere 1 (or $-1$). We shall see in Section 8.3 how this applies for a large class of balanced experimental designs. The common variance can be expressed as $l\sigma^2/n$, where $l$ is the number of degrees of freedom associated with the residual sum of squares when we fit the null model.

Only the case where $\sigma$ is known is considered in detail. If $\tilde{\varepsilon}_{max}$ is the estimated residual having greatest absolute value it is proposed that: *if* $|\tilde{\varepsilon}_{max}| > h\sigma$, *we reject the observation yielding* $\tilde{\varepsilon}_{max}$, *treat it as a missing value, and estimate the unknown parameters (means) by a least-squares analysis; if* $|\tilde{\varepsilon}_{max}| \leq h\sigma$ *we retain all observations and conduct a full least-squares analysis.*

The constant $h$ needs to be chosen to produce acceptable premium and protection guarantees. Anscombe shows that the proposed rule has a simple interpretation. If $|\tilde{\varepsilon}_{max}| > h\sigma$, we merely replace $x_m$ (the observation yielding $\tilde{\varepsilon}_{max}$) by $x_m - n\tilde{\varepsilon}_{max}/l$ in the estimating equations for the means. Determination of premium and of protection are less straightforward than in the univariate sample case. One possibility is to consider the way in which application of the rule affects the *determinant* of the corresponding variance matrix (in the absence of contaminants). Anscombe argues that an appropriate notion of premium is in terms of that proportional increase in $\sigma^2$ which would increase the variance-matrix determinant by as much as the proposed rule does, in the absence of contamination. He shows how, on this definition, the premium can be approximately determined.

An approximation to the *protection* provided by the rule, when one of the observations is a contaminant and has bias $\beta\sigma$ in the mean, is also given. Anscombe reviews the numerical properties of the rule by tabulating values of the protection, of the cut-off level $h$ and of the probability of inappropriate rejection of $x_m$ as discordant for premium levels of 2 per cent and 1 per cent, and a range of values of $l/n$.

He also deals briefly with two other matters: multiple outliers and unknown $\sigma$. For the former case he suggests that the rule is applied consecutively to successively smaller samples until we reach the stage that no further rejection of observations takes place. For the (commonly encountered) situation where $\sigma$ is unknown he gives some approximate results for the equivalent 'studentized' rule where $\sigma^2$ is merely replaced by $s^2$, an estimate of $\sigma^2$ based on $l + l_0$ degrees of freedom ($l_0$ corresponding with additional external or prior information about $\sigma^2$). The results are not pursued to the level of useful application, and Anscombe conjectures that the rule will have 'low power' unless $l + l_0$ is reasonably large (say, 30 or so). Some further observations are given by Anscombe and Tukey (1963) in the context of a wider discussion of the analysis of residuals in designed experiments and general linear models. See also Anscombe (1961).

Tiao and Guttman (1967) reconsider the premium-protection approach of Anscombe (1960a) in some detail for the case of a single univariate sample, depending on the extent of knowledge of the residual variance $\sigma^2$. A brief comment on the general linear model is interesting. Recognizing the difficulties arising from the intercorrelation of the estimated residuals, they propose using uncorrelated *modified residuals*, having the form (when $\sigma^2$ is known)

$$\mathbf{z} = \boldsymbol{\epsilon} + \sigma A P \mathbf{u} \qquad (8.2.39)$$

where $P$ is any $q \times q$ matrix satisfying

$$PP' = (A'A)^{-1} \qquad (8.2.40)$$

and $\mathbf{u}$ is $N(\mathbf{0}, I_q)$ independent of $\mathbf{X}$.
From (8.2.5) and (8.2.6) we clearly have

$$V(\mathbf{z}) = [(I_n - A(A'A)^{-1}A') + APP'A']\sigma^2 = I_n\sigma^2. \qquad (8.2.41)$$

Such modified residuals (cf. *adjusted residuals* above) might seem attractive in view of their independence when the error distribution is normal. Various accommodation procedures (including premium-protection rules and Winsorization) utilizing them are discussed in detail by Golub, Guttman, and Dutter (1973) for a location-slippage model for possible contaminants. They remark that for large $n$ the correlation factors in (8.2.39) arising from introducing alien independent perturbations of the residuals to ' "break" the correlation pattern' are small. But there is no detailed consideration of how large $n$ needs to be, nor is there any comparison of their approach with others based on the unadjusted residuals taking proper account of their intercorrelations.

When we considered the accommodation of outliers in univariate samples we explored in some detail the implications of using the simple, intuitively motivated, idea of *trimming*. It is sensible to ask if this idea can be applied to the more structured case of the linear model. Ruppert and Carroll (1980) propose two methods for determining trimmed least squares estimators of $\boldsymbol{\theta}$.

Suppose that we have some preliminary estimator, $\tilde{\boldsymbol{\theta}}_0$, of $\boldsymbol{\theta}$ (for example, the least squares estimator for the complete sample). We could examine the estimated residuals relative to $\tilde{\boldsymbol{\theta}}_0$ and trim (remove) a fraction of the observations with smallest and largest estimated residuals—say, the $[n\alpha]$ smallest and $[n\alpha]$ largest ones. A *trimmed estimator*, $\tilde{\boldsymbol{\theta}}_1(\alpha)$, of $\boldsymbol{\theta}$ is then obtained by least squares estimation using the retained observations.

The proposed alternative approach is due originally to Koenker and Bassett (1978). Suppose $0 < \alpha < 1$ and $\phi_\alpha(x) = \alpha - I(x < 0)$ where $I(\ )$ is the indicator variable. If $q_\alpha(x) = x\phi_\alpha(x)$ then any $\tilde{\boldsymbol{\theta}}(\alpha)$ which minimizes

$$\sum_{i=1}^{n} q_\alpha(x_i - A'_i \tilde{\boldsymbol{\theta}}(\alpha))$$

is called a *regression quantile*. We trim (remove) any observation whose residual relative to $\tilde{\boldsymbol{\theta}}(\alpha)$ is negative or whose residual relative to $\tilde{\boldsymbol{\theta}}(1 - \alpha)$ is positive (for $\alpha < \frac{1}{2}$), and calculate $\tilde{\boldsymbol{\theta}}_2(\alpha)$ as the least squares estimator using the remaining observations.

Ruppert and Carroll examine and compare $\tilde{\theta}_1(\alpha)$ and $\tilde{\theta}_2(\alpha)$. The latter is asymptotically similar in behaviour to a trimmed mean. The former has asymptotic behaviour which depends strongly on the form of the preliminary estimator, $\tilde{\theta}_0$ and can be relatively inefficient even in the null (normal-error) case and even for small trimming factor, $\alpha$. However, $\tilde{\theta}_1(\alpha)$ becomes asymptotically equivalent to $\tilde{\theta}_2(\alpha)$ if we take as preliminary estimator

$$\tilde{\theta}_0 = \tfrac{1}{2}[\tilde{\theta}(\alpha) + \tilde{\theta}(1 - \alpha)]:$$

that is, the average of the two regression quantiles used in determining $\tilde{\theta}_2(\alpha)$. Behaviour in the non-null case is examined principally for long-tailed error distributions, rather than for contamination models.

Moussa-Hamouda and Leone (1977) also consider 'trimmed' estimators of the slope and intercept parameters (and of the error variance) in simple linear regression. The estimators are easy to compute and relevant coefficients are tabulated in some cases (including a normal-mixture contamination model). Denby and Mallows (1977) use $M$-estimates to effect 'trimming' in their proposals for two plotting methods to display outliers in multiple regression. Welsh (1987) proposes a new method for defining trimmed means in the linear model (which extends to Winsorization). See also Maronna and Yohai (1981). Carroll and Ruppert (1982a) consider robust estimation using a Huber 'Proposal 2' type estimator in a heteroscedastic linear model; see also Giltinan, Carroll, and Ruppert (1986).

Denby and Larsen (1977) present an extensive Monte Carlo study of eight different types of robust regression estimator. The alternative models only partially reflect outlier considerations; one is a mixture model with normal components of highly specific form (the minor component has mixing parameter 0.05 and a variance ten times that of the major component). Another major comparative review is given by Chatterjee and Hadi (1986) who examine the interrelationships between different methods of assessing the 'influence' of the various observations on inferences drawn about the fitted linear model. They distinguish measures based on *residuals*, on the *prediction matrix*, on the *volumes of confidence ellipsoids* and on *influence functions* adopting as yardstick the broad definition of influence given by Belsley, Kuh, and Welsch (1980):

*An influential observation is one which, either individually or together with several other observations, has a demonstrably larger impact on the calculated values of various estimates . . . than is the case for . . . the other observations.*

Of course this can well be true of an *outlier* but (as we saw clearly in the load/extension example at the beginning of this chapter) it can also be true of non-outliers, e.g. observations arising at extremal points in the design space. Thus the authors include various forms of estimated residual (and reconfirm their equivalence), prediction matrix measures such as the diagonal terms in $A(A'A)^{-1}A'$, the Andrews and Pregibon (1978) statistic (Section 8.2.2 above) and influence function measures based on Cook's distance measure, which can be written

$$D_i = (\tilde{X} - \tilde{X}_{(i)})'(\tilde{X} - \tilde{X}_{(i)})/q\tilde{\sigma}^2 \qquad (8.2.42)$$

where the predicted values $\tilde{X}$ and $\tilde{X}_{(i)}$ are evaluated for the whole data set, and after removal of the $i^{\text{th}}$ observation, respectively. Tentative recommendations are made about a useful subset of measures to cover various aspects of influence. There is an extensive discussion section with comments from many prominent workers in the field.

We have already referred (in Section 8.1.2 above) to the robust regression procedure of *least median of squares* due to Rousseeuw (1984). This is relevant also to the multiparameter line model (8.2.1). Again, see Steele and Stigler (1986), Massart, *et al.* (1986) and Edelsbrunner and Souvaine (1990) for refinements, including improved computational algorithms.

A unified body of work on outlier-robust regression is concerned with $L_1$ norm estimation, where the fitting principle is minimization of the sum of absolute deviations from the fitted line, rather than minimization of the sum of squares. Dielman (1986) compares forecasts under the two approaches. Bradu (1987) proposes a particular $L_1$ norm estimation procedure with affinities to the method of *median polish*. The work is directed principally to factorial equi-replicated designed experiments (see Section 8.4 below). Another alternative computational approach for $L_1$ norm estimation is advanced by Kim (1987) which essentially combines use of linear programming and reweighted least squares. McKean and Sievers (1987) stress the value of $L_1$ norm estimation for outlier-robust estimation of the linear model and propose two *coefficients of determination* with useful properties as measures of association. A major text on $L_1$ norm methods with many contributions relevant to outlier study is given by Dodge (1987).

Bickel (1984) examines estimators of the regression coefficients based on minimizing asymptotic variance subject to bounded asymptotic bias over a prescribed family of estimators. The resulting estimators prove robust over certain contamination neighbourhoods. Rousseeuw and Yohai (1984) define a *finite-sample breakdown point* in terms of the smallest fraction of contaminated data which can lead to an estimator taking an unlimited range of values. They propose a new class of **S-estimators** with high breakdown point and asymptotically normal distribution.

Martin, Yohai, and Zamar (1989) have proposed a so-called **min-max bias** method of robust regression, based on minimizing the maximum asymptotic bias of robust regression estimators of two types: both based on *M*-estimators. The estimator types are broad enough to include many previously studied such as the *Huber Proposal 2 estimators* (of regression coefficients and residual variance) and the above-mentioned *Rousseeuw-Yohai S-estimators*.

A cross-validatory approach is proposed by Chowdhury and Balasooriya (1988) to accommodate outliers in the estimation of regression parameters in the linear model and to improve prediction of future values.

O'Gorman and Myers (1987) discuss measures of error with regression outliers.

Other work of passing relevance to the outlier problem is presented by Yohai (1974) and Krasker (1980). There are many Bayesian proposals for accommodation of outliers in linear models—see Section 8.7 below.

## 8.3  OUTLIERS IN NON-LINEAR REGRESSION

Some proposals for handling **outliers in non-linear models** are beginning to appear. Manski (1984; with discussion contributions by others) describes some *adaptive* methods of estimation for non-linear regression models. The link with outliers resides in consideration of *contaminated* normal, lognormal or exponential models.

Hardle and Gasser (1985) exhibit an outlier-robust kernel estimator of the derivative of a non-linear regression model.

Naes (1986) considers a **linear mixed model** with fixed and random components and proposes an ordering principle, based on residuals from generalized least squares estimation of the fixed component. The principle leads essentially to the sum of two independent components which distinguish 'different types of outliers'.

Gasser, Sroka, and Jennen-Steinmetz (1986) propose methods for nonparametric estimation of residuals (and of the residual variance) in nonlinear regression models. They employ the idea of a **pseudoresidual** (the residual from fitting a straight line to the two points on either side of the observation of interest) and show how pseudoresiduals can highlight outliers (and heteroscedasticity).

A sample re-use approach (specifically the jackknife) to estimation in non-linear regression is described by Simonoff and Tsai (1986) with attention to outliers and general robustness measures.

A range of non-standard circumstances are reflected in the following works, usually illustrated by interesting fields of study: Cunningham and Heathcote (1989; non-normal errors, multicollinearity; business transactions), McKean and Sievers (1989; use of rank scores, asymmetric error distribution), Welsh (1989; concomitant scale estimation, increasing dimension), Kao and Dutkowsky (1989; non-linear bounded influence regression, bank borrowing) and Parker (1989; Poisson regression, surveillance data).

There has been much attention given to **logistic regression:** a specific example of a non-linear model. In this situation we observe at each design point $a \epsilon A$ an observation $r_i$ from a binomial distribution $B(n, p(a))$ where $p(a)$ depends on $a$ according to the model

$$\log[p/(1 - p)] = A\theta \qquad (8.3.1)$$

This is, of course, just a special case of a **generalized linear model** (GLM; see McCullagh and Nelder 1989) with logit link function and binomial error structure.

Regression diagnostics for this situation were proposed by Pregibon (1981). Drawing parallels with linear model regression diagnostics he proposes use of parameter estimates, standard errors, residuals, etc. from a maximum likelihood model fit, to reflect on possible contamination as evidenced by the presence of outliers to which the fitting of the logistic model is 'extremely sensitive'. He illustrates the results on a data set on blood restriction in the skin of the fingers due to Finney (1947).

Jennings (1986) explores further the problem of detecting outliers in logistic regression and examines the use of estimated residuals. Specifically, he claims that similarities with linear model methods are to an extent delusory and that differences need examining (rather than stressing superficial similarities). A major issue is the form of the distribution of residuals, which depends on the design matrix A (a feature not present in the linear model situation). In particular, the residual from the true model, $x - p(a)$, must take just one of two values so that it thus seems to make little sense to talk about asymptotic normality of 'standardized residuals' as some other workers do. He urges care in carrying over to logistic regression the outlier (and other diagnostic) principles of linear regression.

Follman and Lambert (1989) discuss the use of non-parametric mixing to cope with 'overdispersed' binomial observations at each design point (where the variance is larger than would arise in binomial data) and there is also possible contamination giving rise to outliers.

In Künsch, Stefanski, and Carroll (1989) logistic regression is studied in detail in a broader treatment of bounded-influence estimation for the generalized linear model (GLM). Two interesting examples of data with outliers are discussed: that of Finney (1947) and data from the US federal food-stamp program (Stefanski, Carroll, and Ruppert, 1986).

Tests of discordancy for outliers in logistic regression are discussed in some detail by Bedrick and Hill (1990). They employ exact conditional methods to derive single and multiple outlier tests under a slippage alternative model: conditioning being in respect of the sufficient statistic $A'X$. (For a single contaminant the alternative model is

$$\log[p(1 - p)] = A + \lambda e_i \qquad (8.3.2)$$

where $l_i$ is a vector with 1 in the $i^{th}$ position and 0 elsewhere.) Various test statistics are proposed including (for a single outlier) the score, and reduction in deviance, tests which Pregibon (1981, 1982) had previously proposed in the forms of *unconditional* tests. Bedrick and Hill show how to determine exact critical levels in the *conditional* case. They also examine the multiple outlier problem but show that the critical level evaluations can be prohibitive for more than two or three outliers.

Two interesting practical medical examples are chosen to illustrate the methods; one is concerned with lymph node surgery in carcinoma of the prostate, the other with dose–response data on the tumorigenicity of benzepyrene in mice. Azzalini, Bowman, and Hardle (1989) describe the use of *bootstrap methods* for model checking in logistic regression, with attention to outliers.

A fascinating practical study (including identification of outliers) of logistic regression in the context of the accident to the *Challenger* space shuttle is given by Dalal, Fowlkes, and Hoadley (1989).

Regression diagnostics for *multiple-group* logistic regression are proposed by Lesaffre and Albert (1989). In this situation the regression diagnostics proposed by Pregibon (1981) for logit regression on dichotomous data are extended for use in the polychotomous case, where outcomes may arise in any one of a set of $m$ categories ($m \geq 2$). A feature of the approach is the facility to highlight the corresponding multivariate outliers.

Other work on outliers in binary data (**binary regression**) is described in Zheng (1987, in Chinese) and Copas (1988). Copas defines an outlier in binary data as the realization of a low-probability outcome: a success, when the probability of a success is small. Methods are given for studying the presence and effects of such outliers (under a specified contamination model).

Naes and Martens (1987) discuss regression models in which the coefficients are regarded as random (rather than as fixed parameter values). They show, under independence and normality assumptions, that two independent quadratic forms can be determined for outlier detection, one for the usual response type of outlier, the other for outlying behaviour of the random coefficients themselves. McCabe (1987) also addresses the issue of random-effects outliers (under elliptical symmetry rather than normality of distribution).

Two studies of the **proportional hazards model** of Cox have relevance to outliers. Pettitt and Bin Daud (1989) show that use of various forms of residual have only limited value in reflecting influential data points and suggest alternative more useful forms of diagnostic plots. Sleeper and Harrington (1990) show that model-fitting using natural *splines* can be useful for accommodating outliers.

Nyquist (1988) claims that the variance estimator in *jackknifed* **ridge regression** is outlier-robust.

Inadequacy of linear model fit can sometimes be remedied by transforming the yield data. A particular case is the Box-Cox transformation

$$x^{(\lambda)} = (x^\lambda - 1)/\lambda \quad (\lambda \neq 0)$$
$$= \log x \quad\quad (\lambda = 0)$$

$$(8.3.3)$$

and we proceed to estimate simultaneously the parameter set $(\theta, \sigma, \lambda)$. It is well known that the maximum likelihood estimator of $(\theta, \sigma, \lambda)$ is very sensitive to outliers, and robust procedures have been proposed (Carroll, 1980, 1982a; Bickel and Doksum, 1981). Carroll and Ruppert (1985) extend the earlier work (for outliers and influential design-space points) using generalized bounded-influence $M$-estimators. They illustrate the results for classical data sets including the salinity data discussed by Ruppert and Carroll (1980) and the stack-loss data, and they show that model selection, use of transformations and outlier detection are fundamentally linked.

DeGruttola, Ware, and Louis (1987) present a method for influence analysis of generalized least-squares estimators in the *multivariate linear model* including examination of the effects of outliers. The approach is illustrated using data on the effects of lead on the cognitive development of children. See also Schall and Dunne (1987; outliers and influence) and Koenker and Portnoy (1990; $M$-estimation illustrated for financial data sets).

## 8.4  OUTLIERS IN DESIGNED EXPERIMENTS

Basic problems in the study of outliers in data from designed experiments are that they are difficult to detect (in the respect discussed above) and that their presence influences the analysis of variance of the data set in a way which may cloak significant effects or show up apparent effects which, were it not for the outliers, would not arise.

Most techniques for outlier detection, for testing their discordancy or for minimizing their influences in analysing the underlying linear additive model, again take as a basic measure of the import of individual observations their *residuals about the fitted linear model*. A concern about the intercorrelation, and carry-over influences, between residuals has prompted some use of modified (outlier-robust) residuals or of measures not directly based on residuals. In all cases the aim is to exhibit, and assess, the extent to which individual observations disrupt the overall pattern anticipated in the data (by virtue of the linear model for the means which is implied by the experimental design). In principle all our earlier study of outliers in relation to the general linear model carry over to investigation of data from designed experiments. However some special (simplifying) features, and a substantial amount of literature specific to designed experiments, make it desirable to examine the designed experiment situation separately and in some detail.

Bross (1961) presents what he describes as a 'strategic appraisal' of the problems of handling outliers in 'patterned experiments'. He stresses the special difficulties of their detection: an outlier disrupts an anticipated pattern of interrelationship in the data, the pattern is itself a *non-null* representation and needs to be characterized before it can be 'disrupted'. He develops this theme in terms of isolated departures in the values of observations relative to those of *neighbouring* observations. In describing the influence of outliers he stresses the *combined* effect of the outlier and of the analysis of variance techniques applied to the overall data set. Although Bross proposes no formal set of procedures for coping with outliers he sketches a non-parametric principle which we shall return to below.

A natural starting point for studying discordancy tests for outliers using residuals is found in the work of Daniel (1960). (See also the later proposals in Daniel, 1978).

Into a set of artificial data purporting to be the outcomes of a $4 \times 5$ factorial experiment, he introduces a substantial contamination of a single observation (the data are the same in a reordered form as the data of Table 8.2 above from Bross, 1961). He stresses how the outlier introduces substantial biases in the fitted values and in the residuals not only in its own specific location *but throughout the row and column* in which it appears.

To consider the implications of these results let us start with an unreplicated two-way design where the (crossed) factors A and B have $r$ and $c$ levels, respectively, and we suspect *no* contaminants affecting the resulting data. Under the usual additive linear model for the means, with normal error structure, the observations $x_{ij}(i = 1, 2, \ldots, r; j = 1, 2, \ldots, c)$ can be written

$$x_{ij} = \mu + \alpha_i + \beta_j + \varepsilon_{ij} \qquad (8.4.1)$$

where $\sum \alpha_i = \sum \beta_j = 0$ and the $\varepsilon_{ij}$ are independent $N(0, \sigma^2)$. The estimated means (fitted values) are

$$\tilde{\mu}_{ij} = \bar{x}_{i.} + \bar{x}_{.j} - \bar{x}_{..} \qquad (8.4.2)$$

and the *estimated residuals* are

$$\tilde{\varepsilon}_{ij} = x_{ij} - \tilde{\mu}_{ij}. \qquad (8.4.3)$$

The $\tilde{\varepsilon}_{ij}$ will be $N(0, \nu\sigma^2/(rc))$ where $\nu = (r - 1)(c - 1)$ but they will *not* be independent. The linear constraints on the $\tilde{\varepsilon}_{ij}$ imply correlation between $\tilde{\varepsilon}_{ij}$ and $\tilde{\varepsilon}_{i'j'}$ in the form

$$\rho_{iji'j'} = \begin{cases} -1/(r-) & i = i', j \neq j' \\ -1/(c-1) & i \neq i', j = j' \\ 1/\nu & i \neq i', j \neq j'. \end{cases} \qquad (8.4.4)$$

Formalizing his numerical example Daniel considers the expected bias in the residuals due to a single contaminant at position $(i, j)$ in the two-way layout, reflecting a contamination in the mean value of order $rc$. The expected bias will be

$$\delta_{i'j'} = \begin{cases} rc & i' = i, j' = j \\ -r & i' = i, j' \neq j \\ -c & i' \neq i, j' = j \\ 1 & i' \neq i, j' \neq j. \end{cases} \qquad (8.4.5)$$

He argues that the correlation between the new (contaminated) estimated residuals, $\tilde{\varepsilon}_{ij}$, and their biases will be high, and proposes that the value of this correlation be used as an indication of the presence of the outlier and as a basis for testing its discordancy. The correlation can (after appropriate manipulation) be expressed in the form

$$\tilde{\rho} = \frac{rc}{v} \frac{\tilde{\varepsilon}_{\max}^2}{\sum \tilde{\varepsilon}_{ij}^2} \qquad (8.4.6)$$

where $\tilde{\varepsilon}_{\max}^2 = \max_{i,j} \{\tilde{\varepsilon}_{ij}^2\}$. Thus, in spite of the carry-over effect of the outlier, the indication of its presence resides solely in the *largest* absolute value of the estimated residuals. (This remains true for other designs where the estimated residuals have *common* variance.) To assess if the outlier corresponding with this 'largest residual' is *discordant* we need to compare $\tilde{\varepsilon}_{\max}^2$ appropriately with the true error variance $\sigma^2$.

The residual sum of squares $\sum \tilde{\varepsilon}_{ij}^2$ reflects the influence of the outlier and will be correspondingly inflated (relative to the null situation) to an unknown extent. Thus to assess discordancy where (typically) $\sigma^2$ is unknown we need to replace $\sum \tilde{\varepsilon}_{ij}^2$ by a measure of residual variability which is not influenced by the outlier. To this end it is proposed (implicitly) that we estimate residual variation by removing the outlier, treating it as a missing value and replacing it by the corresponding least-squares estimate. We are led on this argument to a test statistic for discordancy in the form

$$t^2 = \frac{\tilde{\varepsilon}_{\max}^2}{s^2}, \qquad (8.4.7)$$

where

$$s^2 = \left( \sum \tilde{\varepsilon}_{ij}^2 - \frac{rc}{v} \tilde{\varepsilon}_{\max}^2 \right) / (v - 1) = S_M / (v - 1)$$

is an estimate of $\sigma^2$ based on the residual sum of squares $S_M$ in an analysis of variance where the outlier is regarded as a missing observation.

Daniel's arguments concerning the null distribution of $t^2$ are in error (as remarked in a footnote in Daniel, 1960). However, the assessment of discordancy has been taken up by others, as we shall see shortly. But the general principle is not in dispute. We see that a single outlier is again to be detected as *the observation yielding the largest absolute value among the residuals*; its discordancy is to be assessed in terms of its null (uncontaminated) distribution. Note that the common variances of the $\tilde{\varepsilon}_{ij}$ in this situation make it unnecessary to studentize them.

Use of the maximum absolute residual by Daniel (1960) and also, in the context of accommodation, by Anscombe (1960a) (see Section 8.2.3) was based on intuitive arguments with no overt consideration of the outlier model or of any resulting optimality of the proposed procedures. Noting its form as a natural extension of the optimum single sample procedures of Paulson (1952b), Ferguson (1961a) enquired whether the optimality properties also extended to the designed experiment situation and found this to be so.

He adopted a mean-slippage model to explain a single outlier in a general linear model, developed the appropriate rather complex sampling distribution theory and proceeded to investigate the more general linear model formulation in terms of a multiple-decision approach. (For the two-way design the contamination model is

$$x_{ij} = \mu + \alpha_i + \beta_j + \sigma a_{ij} + \varepsilon_{ij} \qquad (8.4.8)$$

where $a_{ij} \neq 0$ for precisely *one* pair of values $(i, j)$ and is zero otherwise.)

Subject to assumptions that no estimated residuals have zero variance, and no two estimated residuals have unit correlation, the outlier test based on maximum absolute *studentized* residual proves to be optimal in the sense of being *invariant admissible*. In the case of designs where all estimated residuals have equal variance (true of the unreplicated two-way design and many others including all equally replicated ordinary factorial designs, Latin squares, and balanced incomplete blocks—see Anscombe, 1960a) the optimality property can be re-expressed in the terms that the outlier test is a Bayes solution with respect to a uniform prior distribution over the set of hypotheses specifying *equal* shifts in the mean for the contaminant.

We must note how the work by Anscombe and by Ferguson extends the range of applicability and propriety of Daniel's proposal—to a wider set of designed experiments and to many general linear model problems.

Later study of this test centres on the determination of critical values for attributing discordancy, and on placing it in the perspective of alternative proposals that have been made for examining the assumptions underlying the analysis of variance (for example, additivity, normality, and homoscedasticity, in addition to non-contamination).

Let us remind ourselves of the basic nature of the Daniel test. It uses as test statistic the ratio of the maximum squared (estimated) residual to the residual sum of squares in a least-squares analysis which regards as *missing* the observation yielding the maximum squared residual. For a general model it can be expressed as

$$t^2 = \tilde{\varepsilon}_{max}^2/[S_M/(l-1)] \tag{8.4.9}$$

where $l(<n)$ is the number of degrees of freedom associated with the residual sum of squares $S_C$ when the model is fitted to the complete data set of $n$ observations; $S_M$ is the corresponding residual sum of squares when the observation yielding $\tilde{\varepsilon}_{max}^2$ is regarded as missing.

Related statistics have been proposed by others, as John and Prescott (1975) point out. Over forty years ago, Quenouille (1953) suggested that we investigate an outlier in a designed experiment by considering a statistic which can be formally expressed as

$$\tau_1^2 = l(S_C - S_M)/S_C \tag{8.4.10}$$

with obvious intuitive appeal. Recognizing the fact that the missing observation is not arbitrary, but corresponds with the largest (absolute) residual, he suggested that the critical level of $\tau_1^2$ is assessed as

$$nP(F_{1,l} > \tau_1^2) \tag{8.4.11}$$

where $F_{1,l}$ has an $F$-distribution with 1 and $l$ degrees of freedom.

In an interesting paper using a simulation method (Goldsmith and Boddy, 1973; discussed in more detail below) the authors suggested using the statistic

$$\tau_2^2 = (l-1)(S_C - S_M)/S_M \tag{8.4.12}$$

and, supported by their simulation study, proposed assessing its critical level as

$$1.25(l-1)P(F_{1,l-1} > \tau_2^2). \tag{8.4.13}$$

Daniel (1960) originally suggested that the critical level of $t^2$ be assessed as

$$P(F_{1,l-1} > nt^2/l^2) \tag{8.4.14}$$

but he pointed out a flaw in his argument (with unexamined implications).

Simulation studies by John and Prescott (1975) suggest that all three distributional proposals, (8.4.11), (8.4.13), and (8.4.14) are, in general, unsatisfactory. Incidentally, the three statistics $t^2$, $\tau_1^2$, and $\tau_2^2$ are all equivalent, and amount to representing the position and import of a single outlier in terms merely of the largest (absolute) residual. (This arises from the relationship between residuals and residual sums of squares discussed at the end of Section 8.2.1. It is illustrated below in relation to the work of Goldsmith and Boddy, 1973.) Specifically we have

$$\tau_2^2 = \frac{n}{l} t^2 = (l - 1)\left[\frac{l}{\tau_1^2} - 1\right]^{-1}. \tag{8.4.15}$$

John and Prescott suggest that the critical level of $\tau_2^2$ should be determined as

$$nP(F_{1, l-1} > \tau_2^2) \tag{8.4.16}$$

and show by simulation how the Quenouille and Daniel proposals are 'far too conservative', those of Goldsmith and Boddy too liberal, whilst their own suggestion appears to be quite accurate for a range of factorial designs with factors at two or three levels. They assess the accuracy of their simulation results by comparison with exact critical values determined by Stefansky (1972), for a range of designs: $3^2, 4^2, 4 \times 3, 5 \times 3, 6 \times 4$ and $8 \times 7$.

Although the John and Prescott proposal is simple and appears reasonable in many cases, the most accurate, and useful, results to date on the null distribution of the test statistic (and hence the best prescription for application of the outlier test of discordancy) derive from the work of Stefansky (1971, 1972).

Stefansky re-expresses $t^2$ in terms of the *maximum normed residual* (MNR). If the estimated residuals are $\tilde{\varepsilon}_i$ ($i = 1, 2, \ldots, n$), the *normed residuals* are

$$z_i = \tilde{\varepsilon}_i \Big/ \sqrt{\sum_1^n \tilde{\varepsilon}_i^2} \tag{8.4.17}$$

and the MNR, denoted $|z|_{(n)}$, is the largest of the absolute values $|z_i|$ ($i = 1, 2, \ldots, n$).

For the two-way design, the statistic (8.4.7) of Daniel can be expressed as

$$t^2 = (v - 1)(|z|_{(n)})^2 / [1 - n(|z|_{(n)})^2 / v], \tag{8.4.18}$$

a strictly increasing function of $|z|_{(n)}$. Thus we can conduct the test in terms of the equivalent test statistic $|z|_{(n)}$, or some simple function of it, provided its null distribution is known. Stefansky addresses herself to determining the null distribution.

The equivalence of $t^2$ and $|z|_{(n)}$ holds for designs where the residuals (in the absence of a contaminant) have equal variances, and the following results hold in such situations. The relationship can be written, in the wider context, as

$$t^2 = (l - 1)(|z|_{(n)})^2 / \{1 - n(|z|_{(n)})^2 / l\} \tag{8.4.19}$$

where $l$ is as defined below equation (8.4.9).

Stefansky (1971) extends an earlier observation of Pearson and Chandra Sekar (1936) that critical values of statistics based on the MNR in the single sample case could be calculated *exactly* from tables of the $t$-distribution for sufficiently large values of the statistic provided we know the largest value that can be taken by the *second largest* of the absolute values of the normed residuals. Specifically, quantities

$$F_i = n(l - 1)z_i^2/(l - nz_i^2) \qquad (8.4.20)$$

are considered. Then $F_{(n)} = \max\{F_i\} = nt^2/l = \tau_2^2$.

The method uses Bonferroni-type inequalities (see Section 8.2.1) to provide lower and upper bounds for the critical values of the MNR (or related quantities). The higher the order of inequality used the sharper the bounds and the more complex the calculations. (But see the discussion in Section 8.2.1).

However, the crucial result is that sufficiently far out in the tail of the distribution *exact* critical values are obtained. The stage at which this facility holds depends on the values of quantities $M_k$: the greatest obtainable values of the $k$th-largest $|z_i|$. Stefansky (1971) shows how to determine $M_k$ for the range of designs with homoscedastic residuals as discussed above. In Stefansky (1972) it is shown that $P(|z|_{(n)} > z)$ can be determined *exactly* from the $r$th Bonferroni upper (lower) bound if $z > M_{2r}$ $(M_{2r+1})$.

It is demonstrated that for the unreplicated two-way design, application of this principle to the first-order inequalities is of little practical interest since it requires $z$ to be unreasonably large. However, the second-order inequalities yield exact critical values beyond about the upper 10 per cent point of the distribution, for two-way designs with up to about nine levels for each factor.

Assessment of the precise range over which exact results can be obtained depends of course on knowing $M_2$, $M_3$, and $M_4$ etc. For the $r \times c$ design with one observation per cell we have

$$M_2 = [m(M - 1)/2]^{1/2}$$

where $m = \min(r, c)$, $M = \max(r, c)$,

$$M_3 = [(r - 1)(c - 1)/(3rc - 2r - 2c - 2)]^{1/2},$$
$$M_4 = 0.5.$$

Correspondingly, exact upper 1 per cent and 5 per cent points of $|z|_{(n)}$ are obtained for $r = 3(1)9$, $c = 3(1)9$.

Prescott (1977) also examines the matter of determining exact critical values for discordancy tests in designed experiments, with particular attention to the second largest estimated residual in designs where estimated residuals have common variance. Galpin and Hawkins (1981) further explore the use of different orders of Bonferroni inequality (including *third* order), and consider in detail two-and three-way designs. They tabulate 1 per cent, 2.5 per cent, 5 per cent and 10 per cent points for the MNR for both types of design, for numbers of levels of each factor ranging from 3 to 10 inclusive. Values are given to five decimal places; the Stefansky values for the two-way design are given to only three decimal places.

Table XXXVIII on page 524 presents upper 5 per cent and 1 per cent points for two-way designs for $r = 3(1)10$, $c = 3(1)10$ (extracted from Stefansky, 1972, with corrections and enhancement from Galpin and Hawkins, 1981). Table XXXIX on pages 525–6 presents (from Galpin and Hawkins, 1981, rounded to three decimal places) upper 5 per cent and 1 per cent points for three-way designs where the numbers of levels $r$, $c$ and $d$ of the three factors each range over 3(1)10.

Multiway designs can be collapsed to two-way structure by sacrificing access to certain interactions. Apart from this prospect, the best prescriptions for factorial designs with more than two factors appear to be those of Galpin and Hawkins (1981), and of John and Prescott (1975) although we must bear in mind the limited range of the John and Prescott simulation studies which encompass only small designs (e.g. $2^3$, $2 \times 3^3$, $2^4$, etc.). Even in this range, the rather poor performance of the John and Prescott critical level proposal when main effects and first order interactions are fitted in the $2 \times 3^2$ design sounds a warning note.

Some interesting practical examples of residual-based methods of testing outliers in designed experiments are given by Goldsmith and Boddy (1973). Twenty-three sets of data, previously discussed in the statistical literature, are re-analysed to detect and test outliers. A 'consecutive' style of analysis is adopted, proceeding from a discordant outlier to the search for further outliers. Computer-based application of the technique is discussed in some detail in relation to any orthogonal design (admitting possible missing values). The authors express dissatisfaction with methods based on the largest residual in that its determination on the basis of fitting the model to the *whole* data set (including any outlier) will produce a test of relatively low power.

Their 'alternative' proposal is to regard each observation in turn as missing and to scan the set of $n$ residual sums of squares to see if one of them is noticeably smaller than the others, indicating an outlier. Concentrating on the minimum residual sum of squares, they employ as test statistic the quantity $\tau_2^2$ (8.4.12) above. However, as Williams (1973) points out, this is equivalent to other proposals based on the largest (absolute) residual, as we have remarked above.

*Example 8.2   To illustrate some of the above ideas we consider a set of data from an unreplicated $2^3$ experiment attributed by Goldsmith and Boddy (1973) to an unpublished lecture by C. Daniel in 1960. The first three columns* *in Table 8.4 below describe (in common notation) the eight treatment combinations, the corresponding yields, and the treatment effect totals respectively.*

*All the effects are of similar order of magnitude and any main-effects analysis will not show up significant treatment effects. There are no immediately obvious outliers in the data. But perhaps outliers are present, masking genuinely significant treatment effects. According to Goldsmith and Boddy (1973), Daniel argued that the largeness of all the interaction terms was in itself suspicious and might indicate one discordant outlier: specifically the yield at 'a', in view of the pattern of signs ($-1, -1, +1, +1$) of the interaction effects. Accordingly he estimated the outlier discrepancy as the mean of the absolute values of the interaction effects (64) and produced 'amended effects' as shown in column 4, in which the interaction terms are now low in comparison with main effects.*

*But let us look at the residuals after fitting a main effects model to the original data. These have values as shown in column 5, with corresponding residual sum of squares 2152.5. The residual at the treatment combination a is indeed the largest, although those at (1), ab and ac are also large. Are the residuals random, or is there contamination at the treatment combination a transmitted (via the correlation between the residuals in the same row or column of the $A \times B \times C$ design) as larger residuals at (1), ab, ac, (and abc)? If we regard the yield at a as missing, its least-squares estimate is 81 (vastly different from 145, by precisely the discrepancy 64 proposed by Daniel). The new residuals are shown in column 6; the residual sum of squares is reduced from 2152.5 (on four degrees of freedom) to 104.5 (on three degrees of freedom).*

**Table 8.4**

| (1) Treatment combination | (2) Yield | (3) Effect total | (4) 'Amended effects' | (5) Residuals | (6) 'New residuals' | (7) Residual Mean Square (corresponding yield missing) |
|---|---|---|---|---|---|---|
| (1) | 121 | 1129 | 1065 | $-15.25$ | 0.75 | 562.5 |
| a | 145 | $-93$ | $-157$ | 32.00 | 0 | 34.8 |
| b | 150 | 53 | 117 | 0.50 | 0.50 | 717.3 |
| ab | 109 | $-45$ | 19 | $-17.25$ | $-1.25$ | 519.1 |
| c | 160 | 79 | 143 | 4.00 | 4.00 | 706.8 |
| ac | 112 | $-59$ | 5 | $-20.75$ | $-4.75$ | 430.5 |
| bc | 180 | 67 | 3 | 10.75 | $-5.25$ | 640.5 |
| abc | 152 | 85 | 21 | 6.00 | 6.00 | 693.5 |

*The outlier test statistic value is*

$$\tau_2^2 = \frac{3 \times 2048}{104.5} = 58.79$$

*with critical level (on the John and Prescott proposal)* 0.04. *Thus at the* 5 *per cent level we would reject the a-yield of* 145 *as a discordant outlier.*

*A major difficulty in handling outliers in designed experiments lies in their initial detection. The residual* 32.00 *in column* (5) *hardly renders the a-yield a compelling candidate. Goldstein and Boddy* (1973) *suggest that clearer initial detection may result from scanning the values of the residual mean squares when each yield in turn is regarded as a missing value. This requires a lot of calculation if the number of observations is at all large. However, for the current data, the residual mean squares (shown in column* 7) *really do highlight the outlier; all but the one corresponding with treatment combination a are of similar order to the original (full data) mean square* 538.1; *the mean square when a is missing is dramatically smaller at* 34.8.

*Finally, we can illustrate the equivalence of using reductions in residual sums of squares, and estimated residuals. In this factorial experiment all the estimated residuals have equal variance. In fact* $r_{jj} = 0.5$ (all $j$). *Now* $S^2 = 2152.5$ *and if we reduce* $S^2$ *separately by the* $\tilde{\varepsilon}_j^2/(1 - r_{jj})$ *using the values for* $\tilde{\varepsilon}_j$ *in column* (5), *we obtain (after appropriate weighting by the degrees of freedom) the residual mean squares corresponding with the entries in column* (7). *Alternatively, if we examine the squares of the estimated residuals we see that the outlier at 'a' shows up just as graphically as in column* (7).

Margolin (1976) briefly and informally considers the detection of outliers in data from a complicated factorial design using interactive computing in APL.

John (1978) gives two detailed practical examples of applying residual-based outlier methods to detect one and two outliers in data from designed experiments. One example is a 1/3 replicate of a $3^4$, the other a confounded $2^5$.

John and Draper (1978) re-examine the use of the Gentleman and Wilk (1975a) $Q_k$ statistic (see 8.2.36) for two-way designs. Instead of plotting the $Q_k$ they propose a two-stage test for 2 or 1 outliers, and tabulate some corresponding (Monte Carlo based) critical values. Draper and John (1980) extend this to tests of three or less outliers. They also briefly consider the general linear model.

A novel approach to outlier detection in designed experiments is described by Bradu and Hawkins (1982). Expressing dissatisfaction with the use of residuals to detect *multiple* outliers, they propose examination of the medians of the *tetrads* of the cells in a two-way design as the basis of a one-step method for detecting multiple outliers. If we consider cells indexed $(i, k)$, $(i, l)$, $(j, k)$, $(j, l)$ the tetrad is defined here by

$$T_{ik;jl} = x_{ik} - x_{jk} - x_{il} + x_{jl} \qquad (8.4.21)$$

(but see Section 12.1).

For a simple additive (crossed) model, $T_{ik;jl}$ has expected value zero. This will not be so if any of the cells are contaminated (that is, depart from the additive model). If $i \neq j$ and $k \neq l$ the tetrad is said to be *proper*. The proposed method consists of determining the *medians* m(i,k) of all proper tetrads containing the cell $(i, k)$, ranking the cells in terms of decreasing values of $|m(i, k)|$, and applying some diagnostic aid to interpretation. The contaminated cells will tend to have the larger values of $|m(i, k)|$ and a graphical procedure (using half-normal plots: see Section 8.5) is suggested for determining the location, and number, of contaminated cells. The method is illustrated for two sets of data: a $7 \times 7$ set of actual data and a simulated $20 \times 20$ set. It is claimed that this approach provides a clearer indication of outliers than does the use of estimated residuals.

See also Hawkins, Bradu, and Kass (1984) and Farebrother (1988).

Mason, Gunst, and Hess (1989; Chapter 24) discuss outlier detection in design models.

Some non-parametric methods have been proposed for detecting outliers in designed experiments. We have already referred to the general proposals by Bross (1961), who seeks to give expression to the idea of outliers as disrupters of anticipated pattern in the data. He sketches out a non-parametric approach in which we detect outliers in terms of pairwise inversions of the observations relative to the anticipated pattern using (*inter alia*) a *sequence sign* test.

Suppose that in a two-way design we expect real effects to show up in a monotone change in the means within each row and each column. We can look for the reflection of such a relationship in the actual data: anomalous inversions in the values of successive observations within a row or a column may indicate outliers.

Appropriate test statistics can be constructed in terms of accumulated numbers of inversions. But there are problems with this approach, arising from the hierarchy of models we have to consider and the intangibility of 'anticipated pattern'. We do not know at the outset what sort of pattern to expect as a reflection of the additive model for the means. We do not even know if the data support an additive model, or if apparent non-additivity reflects interactions or isolated contaminants. More fundamentally, the data may be just a random sample of observations from a common basic distribution. That is to say there are no real effects. But if there are no real effects we have no structured pattern against which to detect outliers; inversions, for example, will be irrelevant in this context.

Of course a single-sample test of discordancy would be appropriate if there are no real effects but we have no way of knowing if effects are present or not. This uncertainty is the stimulus for studying the additive model; we want such study to be safeguarded from outliers—we find ourselves once more in a vicious circle of conflicting aims and indications.

Other non-parametric methods of a more detailed form (but with similar conceptual difficulties) have been proposed. Brown (1975) develops an approximate $\chi^2$ test of discordancy for outliers based on the signs of the estimated residuals in the rows and columns of the data for a two-way design, and considers its extension to more complicated designs. For the two-way design he proposes the statistic

$$c^{-1} \sum_{i=1}^{r} R_i^2 + r^{-1} \sum_{j=1}^{c} C_j^2 - (rc)^{-1} T^2 \qquad (8.4.22)$$

where $R_i$ and $C_j$ are the sums of the signs of the residuals in the $i$th row and $j$th column, respectively, and $T$ the overall sum of the signs of the residuals. This is sensitive to the presence of outliers and will have a null distribution which is approximately a multiple $(1 - 2/\pi)$ of $\chi^2$ with $r + c - 1$ degrees of freedom. Unfortunately the method does not pinpoint the outlier; also a significant result *could* arise from non-null manifestations other than a single contaminant.

Some studies and proposals have been made relating to the *accommodation* of outliers in designed experiments. Herzberg and Andrews (1978) examine the robustness of two specific types of design (chain block and coat-of-mail) to missing values in the data (and by implication to outliers). See also Draper and Herzberg (1979). Carroll (1980) re-examines the two data sets of John (1978) using robust regression methods of the $M$-estimate type. He extends this study of the John data sets in Carroll (1982a) using robust methods to transform the data and protect against outliers.

Outlier-robustness of optimum balanced $2^p$ factorial designs is discussed by Ghosh and Kipngeno (1985). They define an outlier sensitivity factor by which to assess the effects of outliers on different designs. Akhtar and Prescott (1987) review robust response surface designs, whilst Akhtar (1987) examines outlier-robust central composite designs.

The robustness of balanced $2^p$ fractional replicate designs is discussed by Kuwada (1987) in terms of the outlier sensitivity measure of Ghosh and Kipngeno (1985) mentioned above. Broader aspects of the robustness of row-column (crossed factor) designs are considered by Singh, Gupta, and Singh (1987), whilst Welch and Gutierrez (1988) investigate matched-pairs designs in terms of producing (outlier) robust computationally feasible permutation tests.

## 8.5  GRAPHICAL METHODS AND DIAGNOSTICS FOR LINEAR MODEL OUTLIERS

In the spirit of current interest in informal methods of exploratory data analysis there have been many proposals for screening data for outliers using graphical and other diagnostic procedures. Some are specific to our interest in outliers as reflections of contamination of the error component in the linear model. Others take a broader stance: they aim to represent influential observations in the design-space or the response-space.

In reviewing such procedures for linear model outliers we start with a variety of graphical procedures which have been proposed for investigating the validity of the various assumptions underlying the analysis of variance of data arising from a designed experiment. Sometimes these procedures are aimed specifically at exhibiting, or examining, outliers. More often other assumptions such as normality, additivity, or homoscedasticity are under investigation, but the performances of the relevant procedures are sensitive also to the presence of outliers, and *en passant* provide indications of outlying behaviour of data points. Frequently it is difficult to distinguish which specific departure from the underlying assumptions is manifest in an unacceptable graphical plot; sometimes the procedure is more sensitive to one departure than to another. Most of the procedures regard residuals as the natural reflection of the impropriety of the assumptions, although non-residual-based methods have also been advanced.

Again we can conveniently commence with some work by Daniel (1959) on the method of *half-normal plots*. Daniel considers factorial experiments of the form $2^p$: that is, $p$ factors each at two levels. He considers the *ordered* absolute values of the effect totals and remarks that in the absence of any real effects these are *observations of the order statistics from a known distribution*. If the error distribution is normal, the *absolute* effect totals will behave as independent *half-normal* deviates with common variance. That is, their probability density function has the form

$$f(x) = \sqrt{(2/\pi\sigma^2)}\exp[-x^2/2\sigma^2] \quad (x \geqslant 0) \qquad (8.5.1)$$

Plotting the ordered values on appropriately constructed probability paper will produce, in the null case, observations lying close to a straight line through the origin. Departures from linearity will indicate real effects (large values lying off the straight line) or violations of the basic assumptions of the model.

In particular the presence of a single outlier will similarly inflate the absolute values of all effect totals. It will show up, therefore, by the probability plot (although possibly linear apart from a few high values indicating real effects) *being not directed towards the origin* but towards a value similar to the contamination bias reflected by the outlier. Note that we become aware of the outlier in this way; but we do not determine the offending individual observation.

With more than two outliers the plot does not necessarily reveal their presence in a very dramatic form, unless the biases happen to be of the same sign and similar magnitude. We must also recognize that other aberrances such as non-normality, non-constant error variances, etc. can affect the linearity of the plot and might do so, particularly in small data sets, in a way which is indistinguishable from the manifestation of an outlier.

See also Birnbaum (1959) on half-normal plotting methods.

*Example 8.3    Consider again the $2^3$ data discussed in Example 8.2. Figure 8.4(a) shows a half-normal plot of (the absolute values of) the original seven effects totals. We are not made aware of any marked treatment effects, but the failure of the linear pattern to go through the origin strongly suggests the presence of an outlier. If we correct the a-yield as in Example 8.2 we see in Figure 8.4(b) that there are marked A, B and C effects and, apart from these, the line now goes through the origin.*

*Example 8.4    John (1978) gives the yields in a confounded $2^5$ design concerned with coating of metals for aircraft components. If we determine the effects totals and conduct a half-normal plot we obtain the representation of the data shown in Figure 8.5. Notice the implications: a strong effect D and perhaps two linear components one of which goes fairly near the origin. In fact there are two outliers in the data (though these are not quite discordant at the 5 per cent level). We see how the presence of more than one outlier makes it difficult to interpret the half-normal plot with respect to the outliers.*

Within the context of another wide-ranging study of how to assess the validity of assumptions underlying analysis of variance, Anscombe and Tukey (1963) consider graphical display of *residuals*, including probability plots and plots against fitted values or external concomitant variables. They are at pains to emphasize the overlap of influences and indications, remarking that apparent non-normality, non-additivity, different error variances, and isolated contaminants can all show up in similar (and indistinguishable) ways.

One simple graphical presentation is obtained by plotting the ordered values of the residuals on normal probability paper. With normal error structure we expect a straight line relationship. Non-linearity will be indicative of skewness or flatness of the error distribution; outliers will be manifest in marked isolated departures at either end of the plot. Such a procedure (termed FUNOP—full normal plot—by Tukey, 1962) has its limitations, of course, arising from the intercorrelation of residuals, or non-attributability of effects, for the detection of outliers. Tukey (1962) also proposed a procedure entitled FUNOR-FUNOM for compressing the values of larger residuals, in contrast to reducing them to zero as is implied in the outright rejection, as outliers, of the corresponding observations.

One advantage of this simple approach is that it considers extreme values of residuals, not merely extreme *absolute* values. As we shall note below,

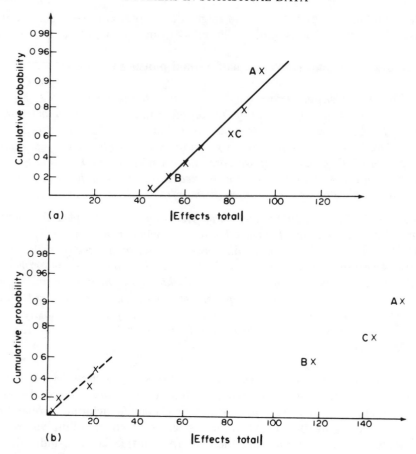

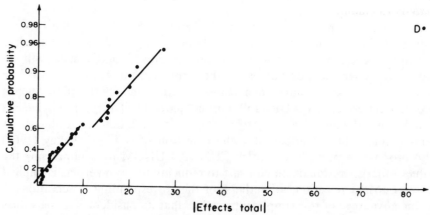

**Figure 8.4**  (a) Half-normal plot for original data of Example 8.2 (b) Half-normal plot for corrected data of Example 8.2

**Figure 8.5**  Half-normal plot for John (1978) data

it can also be of interest to consider the values of residuals relative to the *design matrix*.

Gnanadesikan and Kettenring (1972) also consider the self- camouflaging effect of outliers, due to their influence carrying over to other residuals and thus making their detection problematical. A solution is suggested in the use of 'modified residuals', based on outlier-robust fitted values (e.g. estimating means by medians, trimmed means, etc.) rather than full-sample least-squares estimates. They suggest that probability plots of such ordered *modified residuals* may be more informative about outliers.

Probability plotting methods are also used to augment other methods of studying outliers in the battery of procedures described by Gentleman and Wilk (1975a, 1975b); Gentleman and Wilk (1975b) re-examine full- and half-normal plots of residuals and confirm their usefulness in detecting an outlier when a single contaminant is present in a two-way design. They demonstrate the confusion that can arise from compensating effects when there are two or more contaminants, and suggest that the probability plots have little value in such cases. Regarding the distribution of the residuals they show, in terms of the Shapiro and Wilk (1965) $W$-test for normality (see test **N17** in Chapter 6), that the intercorrelation of the residuals, is not exhibited in apparent non-normality. Indeed, in terms of $W$, the residuals on the null additive model can exhibit a degree of 'super-normality'. This remains so for certain small designs (e.g. $2 \times 3$, $3 \times 4$) even if a contaminant is present. Chen (1971) shows, however, that $W$ is in general sensitive to the presence of outliers. Prescott (1976a, b) compares the effect of outliers on the Shapiro–Wilk $W$-test and on an entropy-based test of normality, by means of *sensitivity contours*.

For the general linear model Larson and McCleary (1972) consider what they term 'partial residual plots', whilst Wood (1973) compares a range of plotting methods for examining the assumptions underlying a linear model analysis (including that of non-contamination). See also Denby and Mallows (1977).

A sequence of papers over the last 10 years or so (including those of Atkinson; 1981, 1982a, 1982b, 1986 and 1987) on the general theme of 'regression diagnostics' place emphasis on different graphical methods for detecting 'outliers' and 'influential observations' in linear model data. For example, Atkinson (1981) contemplates the prospect of transcription, computer data punching, or data manipulation producing gross errors in the response variables $x_i$ or in the explanatory variables $A_i'$ (the $i$th row of $A$). For response outliers he recommends *half normal plots* of suitably modified estimated residuals. But instead of using studentized residuals $e_i$, he suggests using 'cross-validatory' or 'jackknife' residuals (in which the observation under investigation, $x_i$, has been omitted): of the form

$$e_i^* = \frac{x_i - A_i' \, \tilde{\theta}_{(i)}}{S_{(i)}[1 + A_i' \, (A_{(i)}' \, A_{(i)})^{-1} A_i']^{1/2}}$$

where the subscript $(i)$ denotes omission of the $i$th observation. Although it is readily seen that the $e_i^*$ and the $e_i$ are functionally related, as

$$e_i^* = e_i \bigg/ \left( \frac{n-q}{n-q-1} - \frac{e_i^2}{n-q-1} \right)^{1/2},$$

Atkinson claims that $e_i^*$ 'reflects large residuals more dramatically' then $e_i$. He proposes the use of a half-normal plot of the $|e_i^*|$ to detect response outliers.

For influential observations in the design space it is suggested that we use an influence statistic such as that of Cook (1977),

$$D_i = \frac{e_i^2 r_{ii}}{q(1 - r_{ii})},$$

but 'improved' by replacing $e_i$ by $e_i^*$ and standardizing with respect to a D-optimum design with constant $r_{ii} = q/n$. A corresponding form suitable for half-normal plotting is

$$T_i^* = \left\{ \left( \frac{n-q}{q} \right) \left( \frac{r_{ii}}{1 - r_{ii}} \right) \right\}^{1/2} |r_i^*|$$

(cf. Belsley, Kuh, and Welsch, 1980, Section 2.1).

The advantages of the plots are investigated by application to sets of real and simulated data. Use of transformations of the data, and of robust regression methods, are also briefly considered.

This work is greatly expanded and set in the context of a wide range of other proposals in the substantial review paper: Atkinson (1982a). Atkinson (1986) in a paper with the intriguing title 'Masking unmasked' explores diagnostic methods based on single and multiple deletion methods. Masking can be a major problem with single deletion methods when there are multiple outliers. Atkinson proposes use of *elemental sets* to fit *least median of squares* regression as an exploratory tool for highlighting outliers (and influential observations). See also Atkinson (1987).

Williams (1987) discusses diagnostic procedures *for the generalized linear model* based on deviance and single-case deletion.

Other work on regression diagnostics with relevance to outliers includes Chalton and Troskie (1992; Q-plots and singular value decomposition); Simonoff (1988a; on 'non-random missingness'); Gray (1989); McKean, Sheather, and Hettmansperger (1980; recommending rank-based fits to exhibit outliers) and Michels and Trenkler (1990; residual plots using generalized recursive residuals).

A novel approach is suggested by Cerdan (1989) using augmented principal-component plots to yield graphical plots which show up 'outliers and extremes' (the method is related to the use of least-squares residuals).

## 8.6 SOME FURTHER COMMENTS ON INFLUENTIAL OBSERVATIONS

At the beginning of this chapter, attention was drawn to the widespread current interest in *influential observations*: those with the property that 'important features of the analysis are altered substantially when they are deleted' (Cook and Weisberg, 1980). We noted that 'outliers' and 'influential observations' are not coincident concepts. An outlier may substantially alter an estimate of some parameter, or the outcome of a specific test, but this eventuality is seldom the stimulus for the identification of the outlier. In contrast, a clearly manifest outlier (engendering surprise by virtue of its marked extremeness) *need not have* a substantial effect on a *particular* estimate or test—especially if we employ an accommodation procedure.

A particular distinction arises in a field which is rich in proposals for dealing with influential observations: the analysis of linear model data. The idea of an influential observation in the *design space* has no natural interpretation in outlier terms. The idea of influence is part of the broader field of robustness and it is worth contrasting robustness methods with outlier methods. In robust statistics we are concerned with methods of parametric statistical inference in which we seek to devise and examine procedures (for estimation and testing) which **retain desirable statistical properties** even when the assumed parametric model $F$ turns out to be incorrect. We aim to safeguard against serious bias, inefficiency or loss of power when $F$ is, in fact, inappropriate and some other model $G$ is the correct one. Often $G$ (or even $F$) are not fully prescribed or taken into account in proposing such robust procedures. Robustness also considers the status of specific observations in the data set under consideration— through concepts such as **influence** and breakdown.

As an example, we may wish to adopt methods that work well for, say, a **normal** model, but which would remain reasonably good if the actual underlying model turned out to be skew or longer-tailed; or to contain **contamination** in the sense that a small proportion of the sample arises from a different source. For example, the bulk of the data may come from $N(\mu, \sigma^2)$, with a few observations from $N(\mu + a, \sigma^2)$ or $N(\mu, k\sigma^2)$ (with $k > 1$). Note that $F$ and $G$ are not in this case **distinct** parametric families, but $F$ is a limiting form of $G$. Such a prospect can lead to the **contaminants** appearing in the **extremes** of the data set: they may be **outliers** (but they need not necessarily be so).

Thus, the multiplicity of contexts within which robustness is studied can include fully prescribed $G$ (of distinct parametric family form from $F$), partially or unspecified $G$, or even highly specific $G$ with just low-level contamination superimposed on $F$. In this broad context, robustness can include situations in which outliers *might* be present in the data. It is not surprising, therefore, that some workers in the field of robustness have been inclined to view the study of outliers as encompassed with what they see as 'the broader field' of robustness. And yet robust procedures tend **not** to be designed specifically to deal with outliers and contamination *per se*: the alternative prospect $G$ merely happens to sweep in this one specific type of alternative to $F$ often in an unspecified wide range of alternatives.

Thus we have an interesting dichotomy where:

- **robustness** may be seen to include in a much broader compass (of possible alternative models to $F$), **some** concern for the effects of **outliers**, and

- the **study of outliers** may be seen to include, in a much broader compass (of different aims and objectives) **some** concern for the notion of **robustness** of inference methods in the face of outliers.

It is vital that we should recognize the **limited** extent of overlap of the two areas of study and not seek to marginalize the relevance of one or other of the two by tacitly assuming that one encompasses the other ('less important') one!

Such polarization is fortunately less in evidence now than it has been over the last 20 years or so.

As with outliers, rejection is not the only option for influential observations. As Cook and Weisberg (1980) remark:

Cases can be influential because they correspond to outlying responses, remote points in the factor space or, perhaps, a combination of the two. The judgement that a case is influential does not necessarily imply that it should be deleted or down weighted, although this may be an attractive option if the corresponding Studentized residual is large.

If a case is influential because it is remote in the factor space, then it could be the most important case in the data since it may provide the only information in a region where the ability to take observations is limited. Alternatively, such a case might be deleted if it is believed that the model fit to the bulk of the data is not appropriate in a neighborhood of the case in question. Generally, decisions regarding such cases may be difficult since there will be relatively little internal evidence for assessing their validity. The decision to retain them may necessarily be based on faith alone if external evidence of their authenticity is lacking.

It is not part of our brief to present detailed methodology for handling influential observations. The seminal concept of *influence* in robust statistical analysis has been described earlier in our discussion of the accommodation of outliers; see, in particular, Chapters 3 and 5.

We discussed the *influence function* as the basic tool, and associated concepts of *gross-error sensitivity*, the *rejection point*, and the *breakdown point* (or *bound*) and their roles in examining methods for the accommodation of outliers in univariate samples. Some corresponding uses in the study of multivariate data are described in Sections 7.2 and 7.4.6. See also Prescott (1976a), Eplett (1980), and Lambert (1981).

Specific methods for revealing influential observations in linear model data have been reviewed in our discussion of graphical methods in Section 8.5. The greatest attention has been given to the linear model situation and whilst the influence function and its derivative forms have been used for assessing influence, the main thrust has been on the promotion of specific statistics to reflect, or test for, the presence of influential observations. Detailed treatments of this topic have been presented by Belsley, Kuh, and Welsch (1980) and Cook and Weisberg (1982). Pettit (1983), in a wide-ranging consideration of the Bayesian analysis of outliers, presents a useful short review of the different proposed influence measures for the linear model.

In this context we have already noted in Section 8.2.2 the proposal of Andrews and Pregibon (1978); and in Section 8.5 the suggestion of Atkinson (1981) that studentized residuals be replaced with 'cross-validatory' or 'jackknife' residuals to be used directly for revealing response-variable outliers, or substituted into an influence measure due to Cook (1977) for detecting influential points in the design space.

The only further study we shall make of influential observations is to review briefly the range of influence measures that have been proposed in the linear model case.

Cook (1977) argues that the least squares estimates $\tilde{\boldsymbol{\theta}}$ and the estimated residuals $\tilde{\varepsilon}$ give an incomplete picture, sometimes a misleading one. In particular they fail to reflect the implications of the different variances $(1 - r_{ii})\sigma^2$ of the estimated residuals $\tilde{\varepsilon}_i$. The $r_{ii}$ are of course manifestations of the deployment of the design points. Others (including Huber, 1975; Box and Draper, 1975) had previously used the $r_{ii}$ to identify design-space 'outliers' or to assess the vulnerability of the design to such 'outliers'. Cook (1977) proposes using the $\tilde{\varepsilon}_i$ and the $r_{ii}$ to obtain a composite view of the presence and effect of influential data points (either in response-space or design-space). He proposes use of the statistic

$$D_i = [(\tilde{\boldsymbol{\theta}} - \tilde{\boldsymbol{\theta}}_{(i)})' A' A (\tilde{\boldsymbol{\theta}} - \tilde{\boldsymbol{\theta}}_{(i)})]/qs^2 \tag{8.6.1}$$

where $\tilde{\boldsymbol{\theta}}_{(i)}$ is the least squares estimate with the $i$th observation deleted. A large value of $D_i$ indicates strong influence from the associated observation. A test based on $F_{q,n-q}$ is proposed. It can be shown that $D_i$ involves the squares of the studentized residuals $e_i$ and the weight factors $r_{ii}$ in the form

$$D_i = e_i^2 r_{ii}/[q(1 - r_{ii})] \tag{8.6.2}$$

reflecting anomalous behaviour in both the response-space and the design-space. Of course $D_i$ can be large if *either* $e_i$ *or* $r_{ii}$ is large, (or indeed both) but the maximum value of $D_i$ *need not necessarily* coincide with a maximum value of either.

Cook and Weisberg (1980) review alternative influence measures in which $A'A$ or $s^2$ in (8.6.1) are replaced by other similarly motivated quantities. In particular they consider replacing $s^2$ with $s_{(i)}^2$ and $A'A$ with $A_{(i)}' A_{(i)}$ (either singly or jointly), where the subscript $(i)$ represents deletion of the $i$th observation. Other possibilities are to replace $A'A$ with $[\mathrm{diag}(A'A)^{-1}]^{-1}$ or with the identity matrix. The various measures are said to have different geometric representations but to convey similar information to (8.6.1). The proposal of Atkinson (1981) that we should examine graphically the quantities (8.6.2), with $e_i^2$ replaced by the 'cross-validitory' or 'jackknife' residuals $(e_i^*)^2$ and with an additional scale factor $(n - q)$, falls within this same class. Atkinson suggests, however, that the modified form of the $D_i$ should not be considered alone; we should simultaneously examine the 'jackknife' residuals $e_i^*$ (see Section 8.5). Others have proposed similar joint scrutiny, e.g. Obenchain (1977); Hoaglin and Welsch (1978).

A different type of combined influence measure is that of Andrews and Pregibon (1978) in the form of the product of the residual sum of squares and $|A'A|$. Effectively they examine the effect of deleting observations on the value of a quantity $|B'B|$ where $B$ is the conjunction $\{A:\mathbf{x}\}$ of the design matrix and the response vector. Their influence statistic which can be expressed as

$$R_i = |B_{(i)}' \, B_{(i)}| / |B'B| \qquad (8.6.3)$$

(where $(i)$ implies omission of the data relating to the $i$th observation) is reminiscent of Wilks's (1963) *one-outlier scatter ratios* for *unstructured* multivariate data.

For further contributions in the area of influential observations see Cook (1979), Heathcote (1979), Draper and John (1981), Carroll (1982b), Carroll and Ruppert (1982b), Hill (1982), and Welsch (1982) and the many other relevant references throughout this chapter.

# PART IV

# *Special Topics*

# CHAPTER 9

# Bayesian Approaches to Outliers

Passing reference has already been made to the use of Bayesian methods in different aspects of the study of outliers. In this chapter we consider in more detail some of the specific Bayesian proposals that have been made for testing the discordancy of outliers or for coping with their presence in 'contaminated' data. We concentrate mainly on univariate data since most contributions are in this area but cross-reference is made to Bayesian methods described earlier, particularly for multivariate or structured-data situations.

## 9.1 BASIC BAYESIAN CONSIDERATIONS

In the opening chapter we remarked briefly on what might seem to be the somewhat anomalous role of Bayesian methods in the study of outliers. An essential characteristic of an outlier is the degree of 'surprise' it engenders when we examine a set of data. Early informal methods of handling outliers involved the development of procedures for detecting and *rejecting* them as 'foreign influences' reflecting undesirable errors in the data collection process. We must examine whether such an attitude can really fit the Bayesian idiom with its dual regard for the *total* data set as the basic information ingredient from which conditional inferences are to be drawn and with the likelihood as the full statistical expression of the information in the data. Preliminary processing of the data for detection, and possible rejection, of outliers implies (in Bayesian terms) a possibly unwarranted preoccupation with a specific feature of the data, with insufficient regard to its total import. The crucial statement of the likelihood involves a commitment to a fully specified model—which was certainly not a feature of the *ad hoc* studies of outliers.

The Bayesian approach requires an *a priori* statement about the propriety of possible models, or about possible values of the parameters in a parametric family of models. This would have to include a prior assessment of probabilities attaching to the presence and form of outliers as a manifestation of contamination. On the Bayesian idiom this assessment should be made before the data are available, and irrespective of the characteristics of any realized sample. But before we have collected our data how are we to recognize the prospect of outliers? There is nothing to surprise us! There seems to be a degree of conflict here, between a data-keyed response to anomalous observations and a data-independent incorporation of prospective outliers in the likelihood and prior probability assignment.

We have been at pains to stress throughout this book the need to advance beyond the early informal view of outliers to a recognition of the importance of adopting a specific form of outlier-generating model in any development of statistical technique for handling outliers. The inescapable modelling element in the Bayesian approach is thus welcome. Its refinement through an attribution of prior probabilities is also a potentially valuable component in outlier study—provided we really do have some tangible prior information. But to be compelled to produce a prior assessment in any circumstance might be more of an embarrassment than an aid. When all is considered, however, perhaps the major philosophical distinction that remains is found in the irrelevance of the data to the outlier-model specification, and of the sampling rule to the final inference, both of which are essential attitudes in the Bayesian approach. Both conflict with the view we have advanced earlier for recognizing, interpreting, and handling outliers on a more classical approach.

We shall be considering Bayesian methods for **accommodation of outliers**. There is no direct Bayesian analogue of the test of discordancy (although we shall note some general proposals for a Bayesian form of test). Its place is taken in the Bayesian approach by a battery of procedures designed to provide a *posterior probability assessment of the degree of contamination* of the data using a basic model which contains additional parameters reflecting possible contamination.

One of the earliest discussions of the Bayesian approach to outliers is that of de Finetti (1961). He is primarily concerned with exploring basic attitudes rather than developing technique. The discussion is set in the context of some quantity $X$ having an initial (prior) distribution which becomes modified in the light of a sample of observations $x_1, x_2, \ldots, x_n$ to yield a final (posterior) distribution for $X$. There is no estimation or testing problem—the total inference is expressed by the final distribution of $X$. Claiming that *all* inference problems are so represented, de Finetti argues that any reasonable approach to outlier rejection needs to be couched in such terms. He stresses that this raises a fundamental difficulty in that the final distribution (total inference) depends on all the data, an attitude which conflicts with the preliminary rejection of some observations (as outliers). He concludes that if rejection of outliers has any propriety this must hinge on the fact that any observation serving as a candidate for rejection has a 'weak or practically negligible' influence on the final distribution.

This viewpoint opposes much of the *rationale* for outlier processing described throughout this book. There would seem to be little point in, or basis for, *either rejecting or accommodating* outliers if their presence has *negligible* influence on the inferential import of the data!

We have already referred (in Sections 2.3 and 5.5) to the approach to the modelling of outliers employed by Kale, Sinha, Veale, and others (see Kale and Sinha, 1971). Here it is assumed that $n - k$ of the observations $x_1, x_2, \ldots, x_n$ arise from some basic population $F$ whilst the remainder (the outliers) arise from populations $G_1, G_2, \ldots, G_k$ different from $F$. It is assumed that prior to taking the observations there is no way of identifying the anomalous subset of size $k$. Furthermore, such identification does not arise from the observed values. Instead, it is assumed that any subset of $k$ of the $n$ observations is equally likely to be the set of observations arising from $G_1, G_2, \ldots, G_k$. We termed this the *exchangeable* model.

It is a moot point whether the use of the uniform distribution for the indices of the anomalous subset in this model implies that the approach is Bayesian in spirit. For a full Bayesian approach a specification of prior probabilities for the forms of the populations $F$ and $G_i$ $(i = 1, 2, \ldots, k)$ would be required, and inferences would need to be expressed in terms of an appropriate posterior distribution. Kale (1974b) considers this extended prospect in the case where $F$ and $G_i$ $(i = 1, 2, \ldots, k)$ are all members of the single-parameter family of exponential distributions with parameter values $\theta$ and $\theta_i$ $(i = 1, 2, \ldots, k)$, respectively. He shows that with minimal restrictions on the prior distribution $p(\theta, \theta_1, \theta_2, \ldots, \theta_k)$ we obtain the same prescription for identifying the anomalous subset in terms of that set of $k$ indices which has maximum posterior probability of corresponding with $\theta_1, \theta_2, \ldots, \theta_k$: namely that if $\theta_j \leqslant \theta$ $(j = 1, 2, \ldots, k_1)$ and $\theta_j \geqslant \theta$ $(j = k_1 + 1, k_1 + 2, \ldots, k)$ the anomalous observations are $x_{(1)}, x_{(2)}, \ldots, x_{(k_1)}$ and $x_{(n - k + k_1 + 1)}, x_{(n - k + k_1 + 2)}, \ldots, x_{(n)}$. This is of course the intuitively sensible conclusion.

Kale also postulates that since no serious restriction was placed on $p(\theta, \theta_1, \theta_2, \ldots, \theta_k)$ this result will hold on a more classical approach employing no specification of prior attitudes about the parameters. Indeed, Kale (1974a) has proved the corresponding result for the case $\theta_j = \theta'$ $(j = 1, 2, \ldots, k)$ with $\theta' \geqslant \theta$ or $\theta' \leqslant \theta$.

Many of the proposals for Bayesian analysis of outliers which we shall be considering involve a similar form of *exchangeable* model to account for the presence of outliers, but take the discussion further by considering the choice of $k$, other forms of distribution $F$ and $G_i$ (although usually limited to $G_i \equiv G$; $i = 1, 2, \ldots, k$), and the estimation of parameters in the face of outliers. These include the work of Dempster and Rosner (1971), Guttman (1973b), Guttman and Khatri (1975), Dutter and Guttman (1979) on identification and an assessment of the extent of contamination; and Box and Tiao (1968), Sinha (1972, 1973b), Abraham and Box (1978) on accommodation.

## 9.2 BAYESIAN ACCOMMODATION OF OUTLIERS

An interesting sidelight on the notion of *Bayesian accommodation (and modelling) of outliers* is provided by the following simple example. Suppose we have a single observation $x$ of a random variable $X$ with mean $\mu$ and variance $\sigma^2$, and $\mu$ has a prior distribution with mean $\mu_0$ and variance $\sigma_0^2$. Suppose $x$ is very large. How would we expect this to influence the posterior mean of $\mu$? There is intuitive appeal in the idea that $x$ should discredit the prior mean $\mu_0$ and that the posterior mean should be close to $x$! But if the distribution of $X$ and the prior distribution of $\mu$ are normal it is easily shown that the posterior mean of $\mu$ is

$$E(\mu \mid x) = x - (x - \mu_0)\sigma^2/(\sigma^2 + \sigma_0^2) \qquad (9.2.1)$$

Thus $\mu$ does not approach $x$ as $x$ increases. It remains a constant multiple of $x - \mu_0$ below $x$. This rather strange effect is well known; Lindley (1968; reply to Discussion) showed that if the prior distribution of $\mu$ has the Student's $t$ form, rather than being normal, the posterior mean approaches $x$ as $x$ increases. Dawid (1973) generalizes the discussion, remarking that it is possible for both extreme prospects to arise:

that

$$E(\mu \mid x) \to x \text{ as } x \to \infty$$

or

$$E(\mu \mid x) \to \mu_0 \text{ as } x \to \infty.$$

He refers to the latter prospect as one 'where a highly discordant observation discredits itself, and opens up an approach to the treatment of outliers'. There are shades of de Finetti here! An interesting observation on this appears in Geertsema (1989) who quotes Lindley (1956): '. . . the experimental result may reduce the amount of information . . . when a 'surprising' value of $x$ occurs [which] may result in us being less sure about $\theta$ than before the experiment'.

Any undesirable shrinkage of the posterior mean is obviously an effect of the 'thin tails' of the prior distribution. Leonard (1974) proposes a two-part prior distribution with which, essentially, moderately discrepant $x$ are shrunk towards $\mu_0$ whilst highly discrepant $x$ lead us to 'throw away the . . . prior assumptions and let . . . $x$ speak for itself'. See also Meeden and Isaacson (1977).

These results do not take us very far in the matter of outlier accommodation (there being just one observation in the sample) but they open up further thoughts on Bayesian outlier modelling and relate to the notion of *outlier proneness* discussed earlier in Section 2.3.

This aspect is taken up by O'Hagan (1979) who shows, in considering Bayesian estimation of a location parameter using a univariate sample of size $n$, how new definitions of outlier-proneness and outlier-resistance arise with a useful operational interpretation. Although outliers (manifesting contamination) are not formally included in the model used, certain types of basic distribution with 'suitably thick tails' (those that are *outlier-prone*) lead to the effect that a Bayesian analysis essentially ignores any outliers: thus we have a form of *outlier rejection* mechanism in the Bayesian analysis.

The idea of outlier-proneness is expressed through the posterior distribution of the location parameter $\theta$. The basic distribution is said to be **right outlier-prone of order $n - 1$** if as $x_n \to \infty$

$$P(\theta \leqslant c \mid x_1, x_2, \ldots, x_n) \to P(\theta \leqslant c \mid x_1, x_2, \ldots, x_{n-1}) \qquad (9.2.2)$$

for all $c$, $x_1, x_2, \ldots, x_{n-1}$ and any prior distribution for $\theta$. **Left outlier-proneness of order $n - 1$** is defined by changing the condition to $x_n \to -\infty$. If left- and right-outlier-proneness of order $n - 1$ prevail, the distribution is **outlier-prone of order $n - 1$**. O'Hagan shows how outlier-proneness requires a sufficient thickness of tail, e.g. Student's $t$ distribution is outlier-prone of order 1.

A distribution is **outlier-resistant** in O'Hagan's sense if

$$P(\theta \leqslant c \mid x_1, x_2, \ldots, x_n)$$

is a decreasing function of $x_n$ for all $n$, $c$, $x_1, x_2, \ldots, x_n$ and any prior distribution for $\theta$. The normal distribution is outlier-resistant.

Whilst the approach just outlined may be regarded as a mechanism for outlier rejection, it is more reasonably interpreted as a means of *accommodating* outliers. The interest is in estimation of $\theta$; an outlier is either usefully included or effectively ignored *automatically* depending on whether the basic distribution is outlier-resistant or outlier-prone. This is the emphasis of O'Hagan (1981) where, under the intriguing title 'A moment of indecision', he concludes:

For a suitably thick-tailed prior . . . and sufficiently large observation, the sample information dominates the prior. . . . Similarly, a thick-tailed sampling distribution can be dominated by the prior if the observation is sufficiently large.

O'Hagan (1988) extends this principle to show that heavy-tailed models lead to Bayesian methods which automatically 'accommodate' or 'reject' outliers. This is developed further in O'Hagan (1990). See also Goldstein (1982) on the modelling of contamination from this view point, and Rubin (1977) on the general topic of robust Bayesian inference.

We can, of course, adopt a more direct approach to the Bayesian accommodation of outliers, modelling possible contamination explicitly and seeking methods of drawing inferences about non-contamination parameters which are relatively free from the effects of any contaminants.

A major contribution in this category is the work of Box and Tiao (1968) who consider a **Bayesian analysis of the linear model** when outliers may be present in the data. They particularize their results to a linear model where the error terms arise as independent observations from normal distributions with zero mean but where slippage in the variance may have occurred for a limited number of observations.

We start by considering proposals for the general linear model where the observation vector **x** has the form

$$\mathbf{x} = A\mathbf{\theta} + \mathbf{\epsilon} \tag{9.2.3}$$

with $\mathbf{\theta}$ a $p \times 1$ vector of parameters, $A$ an $n \times p$ design matrix and $\mathbf{\epsilon}$ a $n \times 1$ vector of independent random errors. It is supposed that the individual errors may have arisen from one or other of two distributions: a basic distribution $f(\varepsilon | \xi_1)$ or an alternative (outlier generating or contaminating) distribution $g(\varepsilon | \xi_2)$. Interest centres on drawing inferences about $\mathbf{\theta}$, with the parameters $\xi_1$ and $\xi_2$ regarded as nuisance parameters.

Attribution of the individual errors $\varepsilon_i$ to $f(\ )$ or to $g(\ )$ is not triggered by the corresponding observed $x_i$. Indeed, the structure of the model (9.2.3) may render intuitive detection of outliers impossible (see Chapter 7, on the linear model). Instead events $a_{(k)}$ are defined under which a specific $k$ of the $\varepsilon_i$ come from $g(\ )$, the remainder from $f(\ )$, and inferences employ the corresponding likelihood which is made up of $2^n$ components $P(a_{(k)} | \mathbf{\theta}, \xi_1, \xi_2)$ corresponding with all possible $a_{(k)}$. A general theory is developed leading to a formal expression for the posterior distribution of $\mathbf{\theta}$ based on general prior distributions $\{p^{(k)}\}$ for $\{a_{(k)}\}$ and $p(\mathbf{\theta}, \xi_1, \xi_2)$ for $(\mathbf{\theta}, \xi_1, \xi_2)$. Note the rather limiting assumption that the number of contaminants, $k$, is known. We shall return to the question of choice of $k$ later (Section 9.3).

Tangible expression can be given to this structure in terms of the above normal-error model with possible scale-shift. Thus in the work of Box and Tiao (1968) the errors $\varepsilon_i$ arise either from $N(0, \sigma^2)$ or from $N(0, b\sigma^2)$. A particular case is studied where each $\varepsilon_i$ arises with probability $(1 - \lambda)$ from $N(0, \sigma^2)$ or with probability $\lambda$ from $N(0, b\sigma^2)$. It is assumed that $b$ is prescribed (presumably $b > 1$ to make sense of the notion of outliers) and that $(\mathbf{\theta}, \log \sigma)$ is independent, uniform, *a priori*. The posterior distribution of $\mathbf{\theta}$ is exhibited in the form of a $p$-dimensional multivariate $t$-distribution. Marginal distributions of the components $\theta_i$ are also derived.

It is interesting to note that the particular explanation adopted for the way in which the errors arise from $N(0, \sigma^2)$ or $N(0, b\sigma^2)$ implies a mixture-type (rather than slippage-type) model for outlier generation. Box and Tiao point out that a modified approach making formal recognition of the 'mixing' leads to the same results as were obtained under their wider formulation where the likelihood consists of contributions from each of the $2^n$ configurations of error source.

The method is illustrated for estimation of the mean, $\mu$ in an unstructured univariate model. We have $x_1, x_2, \ldots, x_n$ as independent observations each arising with probability $(1 - \lambda)$ from $N(\mu, \sigma^2)$ or with probability $\lambda$ from $N(\mu, b\sigma^2)$, with $\lambda$ and $b$ prescribed and $\sigma^2$ unknown. The posterior distribution of $\mu$ (adopting uniform, independent, prior distributions for $\mu$ and $\log \sigma$) turns out to have the form

$$\pi(\mu \mid x) = \sum_{(k)} w_{(k)} \frac{(n - \phi k)^{1/2}}{s_{(k)}} f_{n-1} \left[ \frac{\mu - \tilde{\mu}_{(k)}}{s_{(k)} / \sqrt{(n - \phi k)}} \right]. \qquad (9.2.4)$$

The summation in (9.2.4) ranges over all events $a_{(k)}$, and the weights $w_{(k)}$ are proportional to

$$\left( \frac{\lambda}{1 - \lambda} \right)^k b^{-k/2} \left( \frac{n}{n - k} \right)^{1/2} \left( \frac{s_{(k)}^2}{s^2} \right)^{-1/2(n-1)}, \qquad (9.2.5)$$

with $\bar{x}$, $s^2$ the sample mean and variance, respectively, $\bar{x}_{(k)}$ the mean of those $x_i$ attributed to $N(\mu, b\sigma^2)$ under $a_{(k)}$ and

$$\left. \begin{aligned} \phi &= 1 - b^{-1} \\ \tilde{\mu}_{(k)} &= \bar{x} - \frac{k\phi}{n - k\phi}(\bar{x}_{(k)} - \bar{x}) \\ (n-1)s_{(k)}^2 &= (n-1)s^2 - \phi \left\{ \sum{}'(x_i - \bar{x})^2 + \frac{\phi k^2}{(n - \phi k)}(\bar{x}_{(k)} - \bar{x})^2 \right\} \end{aligned} \right\} \qquad (9.2.6)$$

where $\sum'$ implies summation over all $x_i$ attributed to $N(\mu, b\sigma^2)$ under $a_{(k)}$. The function $f_{n-1}(\ )$ is the probability density function of the $t$-distribution, so that the posterior distribution of $\mu$ is a weighted average of $2^n$ scaled $t$-distributions with $n - 1$ degrees of freedom.

Determination of (9.2.5) is tedious. Proposals are made by Box and Tiao for easing the load which make the exercise feasible at least for moderate $n$ (up to 20 or so) and small $\lambda$. The method is illustrated on a classical set of data due to Darwin on heights of plants which had been examined earlier by Box and Tiao (1962) from an alternative viewpoint. In the current context the posterior distribution of the mean $\mu$, allowing for possible contaminants, is exhibited in relation to extreme alternatives that there are no contaminants or that there are two contaminants arising from the alternative model $N(\mu, 25\sigma^2)$.

A corresponding Bayesian accommodation approach to the linear model with contamination expressed through *slippage of the mean* is presented by Abraham and Box (1978). They introduce contamination by extending the model (9.2.3) to one of the form

$$x = A\theta + \delta Z + \epsilon \qquad (9.2.7)$$

where $Z$ is a vector of size $n$ with $Z_i = 1$ if the observation is a contaminant and zero otherwise. The number of contaminants is $k$ and there is a common degree of contamination, $\delta$, of the mean for each contaminant. The error terms are independent $N(0, \sigma^2)$. Inferences about $\theta$ are expressed through the posterior distribution of $\theta$ given $x$, adopting independent uniform prior distributions for $\theta$, $\delta$ and $\log \sigma$ and the prior assumption that the components of $Z$ each have (independently) a prescribed small probability $\alpha$ of taking unit value (i.e. representing contamination). Thus no direct inferences need to be drawn about $\delta$ or $k$; they are not prescribed but have been allowed for statistically in the analysis.

The marginal posterior distribution of $\theta$ is seen to have the form of a weighted combination of multivariate $t$-distributions with weights which are the posterior probabilities of particular subsets of observations being contaminants. The computational effort in using the results can be extreme and necessitates assuming that $k$ is small (corresponding to a small value of $\alpha$). The posterior mean and dispersion matrix of $\theta$ for prescribed $k$ are exhibited in forms which have similarity of structure to the (classical) least squares results.

Specializing to the case of a univariate normal random sample the results are again illustrated for data due to Darwin on differences in heights of self- and cross-fertilized plants grown in a split-plot environment (see Box and Tiao, 1973).

This model (9.2.7) is also examined, and the methods illustrated, by Dutter and Guttman (1979) who confirm the above results and present a multivariate generalization. See also Guttman and Dutter (1977).

An interesting aspect of this approach is its facility for separating the accommodation, and assessment of degree of contamination, aspects of the Bayesian study of outliers. Both prospects are provided, the latter in terms of posterior probabilities that individual observations (or subsets) are contaminating. This dual facility (by no means unique to Abraham and Box, 1978) makes it difficult to maintain a clear distinction of principle between the two aspects of study. The present topic is considered here because it would seem to be *primarily* concerned with accommodation. Other treatments which place major (if not exclusive) emphasis on the assessment of the degree of contamination will be considered in Section 9.3; some of these in turn do contain accommodation proposals. The dual approach is examined in general terms (with comments on specific proposals such as those just considered) by Freeman (1980).

An accommodation approach to a more general form of contaminated linear model is described by Guttman, Dutter, and Freeman (1978). With normal error structure they use a contamination model

$$x = A\theta + \delta + \epsilon \qquad (9.2.8)$$

where $\delta$ has a prescribed number, $k$, of unknown non-zero elements, thus generalizing the Abraham and Box (1978) model to *unequal* degrees of slippage of the mean.

When we examine the resulting form of the posterior mean and dispersion matrix we find (not surprisingly) that the effect of unconstrained non-zero elements in δ is that we obtain just the usual least squares expressions *but omitting the suspected contaminants.*

West (1984) considers Bayesian inference for linear models using heavy-tailed error distributions to reflect contamination, in particular using error distributions in the form of scale mixtures of normal distributions. The methods are applied to data on scab disease in potatoes. See also West (1985; for the *generalized linear model*).

Pettit and Smith (1985) provide a review of earlier proposals in the Bayesian mould to the modelling of outliers in linear models, drawing comparisons with non-Bayesian approaches. They indicate that the anticipated 'computational explosion' implicit in many Bayesian methods need not lead to prohibitive effort.

Outlier-robust estimation of θ in the linear model (9.2.3) is approached by Marazzi (1985) on two different principles: restricted Bayes and restricted minimax.

Some recent Bayesian methods for outliers have been more concerned with *detection and testing* than with accommodation (Chaloner and Brant, 1988; Guttman and Pena, 1988; Moser and Marco, 1988). Chaloner and Brant (1988) consider a direct Bayesian approach to outlier detection and the analysis of residuals. In Guttman and Pena (1988) a Bayesian outlier detection procedure is developed in the context of the normal linear model with uninformative priors, based on measuring changes in the marginal posterior density of the error variance. In Moser and Marco (1988) an outlier test procedure is developed based on the predictive distribution of 'suspected' outliers and easily implementable using tables of the F- and t-distributions.

The *exchangeable* model for outliers has been used by Kale, Sinha, and Veale in developing classical methods for estimating or testing the mean of an exponential distribution, where outliers may be present in the data. (Section 5.5 presented some details of this work.) In the same applications context of life-testing and reliability, with exponentially distributed life-times, Sinha (1972, 1973b) has considered corresponding Bayesian methods; see also Bhattacharya and Singh (1988).

Sinha (1972) considers $n$ independent observations $x_1, x_2, \ldots, x_n$ where all but one $(x_i)$ arise from an exponential distribution with p.d.f.

$$f(x, \theta) = \frac{1}{\theta} e^{-x/\theta} \quad (\theta > 0)$$

whilst $x_i$ arises from an exponential distribution with p.d.f. $f(x, \theta/\eta)$ where $0 < \eta \leq 1$. The index $i$ is assumed, *a priori*, to be equally likely to take any of the values $1, 2, \ldots, n$. If $f(x, \theta)$ above is the basic lifetime distribution, whilst $f(x, \theta/\eta)$ is an inconvenient intrusion representing perhaps an unidentified alien component in the sample, a quantity of basic interest in reliability studies is the survivor function

$$R_\theta(\tau) = P(X \geq \tau) = e^{-\tau/\theta}$$

and we might wish to estimate this free from serious influence of $x_i$. In the absence of contamination in the data (i.e. if $\eta = 1$) a desirable estimator is

$$\tilde{R}(\tau) = \begin{cases} [1 - \tau/(n\bar{x})]^{n-1} & (\tau \leq n\bar{x}) \\ 0 & (\text{otherwise}) \end{cases} \qquad (9.2.9)$$

$\tilde{R}(\tau)$ is the uniform minimum variance unbiased estimator of $R(\tau)$. Sinha examines the variance: var $[\tilde{R}(\tau)]$. Clearly this depends on both $\tau$ and $\theta$ but this joint dependence takes a simple form in that var $[\tilde{R}(\tau)]$ is a function of the ratio $\tau/\theta$, and we will denote it by $V(\tau/\theta)$.

One aspect of the influence of an outlier on estimation of $\tilde{R}(\tau)$ is the effect of $x_i$ on $E[\tilde{R}(\tau)]$ and on the mean square error of $\tilde{R}(\tau)$. Both of these are also functions of $\tau/\theta$, and of $\eta$. We will denote these quantities by $\mu_\eta(\tau/\theta)$ and $MSE_\eta(\tau/\theta)$. Their explicit forms are intractable, but Sinha derives lower and upper bounds for each of them.

An alternative approach to investigating $\mu_\eta(\tau/\theta)$ and $MSE_\eta(\tau/\theta)$ for fixed $\eta$ is to determine the distribution of the basic statistic $n\bar{x}/\theta$ arising from some prescribed prior probability distribution for $\eta$ and thence to set bounds on the mean and $MSE$ of $\tilde{R}(\tau)$. No prior distribution is assigned to $\theta$; the approach is accordingly termed 'semi-Bayesian'. For convenience a prior Beta distribution is adopted for $\eta$, with p.d.f.

$$p(\eta) \propto \eta^{p-1}(1 - \eta)^{q-1} \qquad (p, q > 0). \qquad (9.2.10)$$

Denoting the posterior mean and $MSE$ of $\tilde{R}(\tau)$ by $\mu_{p,q}(\tau/\theta)$ and $MSE_{p,q}(\tau/\theta)$ respectively, Sinha shows that

$$\frac{p}{p+q} e^{-\tau/\theta} {}_2F_2(1, q, n, p+q+1, \tau/\theta) \leq \mu_{p,q}(\tau/\theta) \leq e^{-\tau/\theta} \qquad (9.2.11)$$

and

$$\left(\frac{p}{p+q}\right) {}_2F_2(1, q, n, p+q+1, \tau/\theta)k(\tau/\theta) - e^{-2\tau/\theta} \leq MSE_{p,q}(\tau/\theta)$$

$$\leq k(\tau/\theta) - \frac{2p}{p+q} e^{-2\tau/\theta} {}_2F_2(1, q, n, p+q+1, \tau/\theta) + e^{-2\tau/\theta} \qquad (9.2.12)$$

where ${}_2F_2(\ )$ is the hypergeometric function, and

$$k(\tau/\theta) = \{(\tau/\theta)^n e^{-\tau/\theta}/\Gamma(n)\} J \qquad (9.2.13)$$

with

$$J = \int_0^\infty [e^{-u\tau/\theta} u^{2n-2}/(1 + u)^{n-1}] du.$$

In principle these results give some indication of how the sampling behaviour of $\tilde{R}(\tau)$ is affected by the presence of a single contaminant, but their complexity militates against any simple interpretation of this influence; the prior probability structure for $\eta$ is arbitrary and there seems no good reason why we should maintain $\tilde{R}(\tau)$ as an estimator when an outlier is present. Indeed, Kale and Sinha (1971) proposed the use of $s = \sum_1^{n-1} x_{(j)} + x_{(n-1)}$ instead of $n\bar{x}$, when a single spurious observation is present, since this is most likely to correspond with $x_{(n)}$. Accordingly

$$\tilde{R}^*(\tau) = \begin{cases} (1 - \tau/s)^{n-1} & (\tau \leqslant s) \\ 0 & (\text{otherwise}) \end{cases} \qquad (9.2.14)$$

has some appeal as an estimator of $R(\tau)$, but no Bayesian analysis of $\tilde{R}^*(\tau)$ is offered.

Sinha (1973b) considers a fuller Bayesian treatment employing again the prior distribution (9.2.10) for $\eta$, and three possible families of prior distributions for $\theta$ (independent of $\eta$). He derives the Bayes estimators of $\eta$, of $\theta$ (the mean life-time), and of the survivor function $R(\tau)$. The forms are again highly complicated, and specific to the chosen prior structures. No simple qualitative interpretation of the influence of the outlier is offered, nor does it seem feasible. (A 'predictive discordancy test' for exponential data is proposed by Geisser, 1989.)

An alternative basic exponential model with p.d.f.

$$g(x, \mu) = \exp[-(x - \mu)] \quad (x > \mu) \qquad (9.2.15)$$

where the outlier arises from an exponential distribution with p.d.f. $g(x, \mu + \delta)$ $(x > \mu + \delta, \delta > 0)$ is also considered in Sinha (1972) and Sinha (1973b).

We should also note a proposal by Lingappaiah (1976) for estimating the shape parameter in a different wide-ranging family of distributions (including the Weibull and gamma) where several contaminants may be present arising from different members of the same family of distributions. The basic model has p.d.f.

$$f(x) = bx^{\alpha-1}\beta^{\alpha/b}\exp(-\beta x^b)/\Gamma(\alpha/b) \quad (x > 0). \qquad (9.2.16)$$

In a sample of size $n$ we contemplate the prospect that $k$ of the $n$ observations arise from (9.2.16) with $\beta$ replaced by $\theta_i\beta$ $(i = 1, 2, \ldots, k; 0 < \theta_i \leqslant 1)$.

Adopting an exponential prior distribution for $\beta$ and Beta prior distributions for the $\theta_i$, and assuming for fixed $k$ that the set of contaminants is equally likely to be any set of the $k(< n)$ observations, the posterior distribution of $\beta$ is obtained for fixed $(\alpha, b)$. The Bayes estimator of $\beta$ is also derived.

Particular cases are derived for $k = 1$ and where (9.2.16) reduces to a Weibull, gamma, or exponential distribution.

Lingappaiah (1989) considers a Bayes approach to prediction of order statistics in exponential and gamma samples which contain outliers. Lingappaiah (1983) discusses Bayesian estimation of the parameters of a contaminated Burr distribution of specific form: with distribution function

$$F(x) = 1 - (1 + x^c)^{-k}.$$

## 9.3  BAYESIAN ASSESSMENT OF CONTAMINATION

We now turn to the issue of how Bayesian methods have been developed to investigate **the presence of contamination** in a sample—both with respect to *the number k of possible contaminants* and more particularly with respect to *the extent of the contamination.* This latter interest is usually addressed by extending the parameter set of the basic model to include a contamination component and determining the posterior distribution of the contamination parameters (possibly together with that of the basic parameters— hence there might also be an accommodation aspect to the proposals we are about to review).

We start with a contribution due to Dempster and Rosner (1971). No prior distribution is adopted for the set of possible values of *k*. Instead, the choice of *k* involves a combination of the fixed-*k* analyses and classical significance test ideas. This latter aspect will not be considered. We concentrate on the fundamental Bayesian aspects, suggesting minor extensions of interpretation where the authors are not too specific in their proposals.

Suppose that $x_1, x_2, \ldots, x_n$ are independent observations from normal distributions with common unknown variance $\sigma^2$. If no contaminants are present the distributions all have zero mean. Alternatively, a location-shift model prevails with *k* of the means different from zero. A Bayesian analysis for a prescribed value of *k* proceeds as follows.

For given *k*, there are $\binom{n}{k}$ subsets of observations which are candidates for assessment as sets of *k* contaminants. Assuming each subset to be equally likely, *a priori*, the posterior probability $\pi(I)$ for subset $I$ is proportional to

$$\left[ \sum_{i \notin I} x_i^2 \right]^{-n/2}$$

on the assumption of 'relatively innocuous' uniform prior distributions for $\log \sigma$ and for the unknown location parameters.

Clearly $\pi(I)$ is maximized when $I$ consists of the *k* observations with largest absolute values. Such would have been the observations singled out as outlying in a more traditional data-oriented approach where the detection stage proceeds intuitively in terms of the degree of 'surprise'. Here, no such pre-detection is admitted—indeed it is ruled out by the adopted uniform prior distribution of $I$.

It is proposed that the (marginal) posterior probability that $x_i$ is a contaminant is measured by

$$p_i = \sum_{i \in I} \pi(I).$$

The question of whether or not the detected set of $k$ outliers is discordant presumably hinges on how large is

$$\pi_k = \max_{I \in \mathcal{I}} \pi(I) \qquad (9.3.1)$$

where $\mathcal{I}$ is the set of all subsets $I$ of size $k$. We might decide that this needs to exceed, say, 0.95 before we attribute discordancy to the outlier subset.

The question of choice of $k$ is crucial. Dempster and Rosner propose that we consider the sequence of maximized posterior probabilities $\pi_k$ for $k = 1, 2, 3, \ldots$ and that we choose $k$ to yield $\pi_k$ 'large enough to provide reasonable assurance that the $k$ most discrepant data points are outliers', coupling this rather general prescription with informal aids involving significance testing concepts. An alternative might be to choose $k$ to maximize $\pi_k$ and to conclude that the $k$ detected outliers are discordant if $\max_k \pi_k$ is sufficiently large. But clearly any such proposal would need a careful study of its implications.

For prescribed $k$, Dempster and Rosner suggest estimators of $\sigma^2$ and of the anomalous means $\mu_i$, which robustly accommodate the set of $k$ outliers. For the former purpose they propose

$$\tilde{\sigma}_k^2 = \sum_{I \in \mathcal{I}} \pi(I) S_I^2 \qquad (9.3.2)$$

where

$$S_I^2 = \frac{1}{n - k} \sum_{i \notin I} x_i^2. \qquad (9.3.3)$$

For the $\mu_i$, the estimators are

$$\tilde{\mu}_i(I) = \sum_{I \in \mathcal{I}} \pi(I) \tilde{\mu}_i(I) \qquad (9.3.4)$$

where

$$\tilde{\mu}_i(I) = \begin{cases} x_i, & \text{if } i \in I. \\ 0, & \text{otherwise.} \end{cases} \qquad (9.3.5)$$

To illustrate their proposals, Dempster and Rosner reconsider the data discussed in Daniel (1959) in his work on half-normal plots. (See Section 8.4) It is interesting to reproduce some of the results. The Daniel data (the 31 contrasts in a $2^5$ experiment) arranged in ascending order of magnitude appear thus:

| | | | | | | |
|---|---|---|---|---|---|---|
| 0.0000 | 0.0281 | − 0.0561 | − 0.0842 | − 0.0982 | 0.1263 | 0.1684 |
| 0.1964 | 0.2245 | − 0.2526 | 0.2947 | − 0.3087 | 0.3929 | 0.4069 |
| 0.4209 | 0.4350 | 0.4630 | − 0.4771 | 0.5472 | 0.6595 | 0.7437 |
| − 0.7437 | − 0.7577 | − 0.8138 | − 0.8138 | − 0.8980 | 1.080 | − 1.305 |
| 2.147 | − 2.666 | − 3.143 | | | | |

Taking the data at face value as observations from $N(\mu_i, \sigma^2)$ (i.e. ignoring their structured origin in the $2^5$ experiment) Dempster and Rosner seek outliers using the ideas above. They tabulate for $k = 1(1)5$ the $\tilde{\mu}_i$ and $p_i$ and $\tilde{\sigma}_k$ and $\pi_k$. The results are reproduced as Table 9.1 below and provide rather compelling evidence for the three observations with largest absolute value being contaminants. We notice in particular how $\pi_j$ builds up to 0.9247 at $k = 3$, dropping to 0.3311 at $k = 4$, and the maintenance of anomalously high $\mu_i$ and $p_i$ for the last three observations for $k = 3$ and $k = 4$.

A somewhat similar, if more specific, application of Bayesian methods to the 'detection of spuriosity' is described by Guttman (1973b). Adapting a slippage-type alternative hypothesis to describe the occurrence of contaminants, he produces a procedure for determining whether or not a 'spurious observation' has occurred in the data.

Guttman concentrates on a set of independent normal observations $x_1, x_2, \ldots, x_n$ arising, in the absence of 'spuriosity', from a common normal distribution, $N(\mu, \sigma^2)$. Under the alternative model, *one* observation comes from $N(\mu + a, \sigma^2)$. It is assumed that any of the observations is equally likely to be the *one* contaminant and Guttman offers a 'succinct description' of his model in terms of the likelihood

$$P\{\mu, a, \sigma^2 \,|\, \mathbf{x}\} = \frac{1}{n} (2\pi\sigma^2)^{-n/2} \sum_{j=1}^{n} \left\{ \exp\left[ -\frac{1}{2\sigma^2} (x_j - \mu - a)^2 \right] \right.$$

$$\left. \times \exp\left[ -\frac{1}{2\sigma^2} \sum_{i \neq j} (x_i - \mu)^2 \right] \right\}. \tag{9.3.6}$$

Note that this is *not* equivalent to a mixture-type model where each observation has some small probability of arising from $N(\mu + a, \sigma^2)$ and a large complementary probability of arising from $N(\mu, \sigma^2)$.

Adopting a non-informative prior distribution for $\mu$ and $\sigma^2$ with density proportional to $\sigma^{-2}$ the posterior distribution of $(\mu, \sigma^2, a)$ is obtained. Integrating out $\mu$ and $\sigma^2$ the posterior distribution of $a$ is obtained in a form which can be regarded as a weighted combination of densities of the Student's $t$ type (cf. results of Abraham and Box, 1978, above).

This latter characteristic enables the posterior mean and variance of $a$ to be obtained as

**Table 9.1** Bayesian outlier analysis of the Daniel data (reproduced by permission of Academic Press)

| | $k=1$ | | $k=2$ | | $k=3$ | | $k=4$ | | $k=5$ | |
|---|---|---|---|---|---|---|---|---|---|---|
| $x_i$ | $\tilde{\mu}_i$ | $p_i$ | $\tilde{\mu}_i$ | $p_i$ | $\tilde{\mu}_i$ | $p_i$ | $\tilde{\mu}_i$ | $p_i$ | $\tilde{\mu}_i$ | $p_i$ |
| 0.0000 | 0.0000 | 0.0019 | 0.0000 | 0.0015 | 0.0000 | 0.0016 | 0.0000 | 0.0127 | 0.0000 | 0.0160 |
| 0.0281 | 0.0001 | 0.0019 | 0.0000 | 0.0015 | 0.0000 | 0.0016 | 0.0004 | 0.0127 | 0.0004 | 0.0160 |
| 0.0561 | 0.0001 | 0.0019 | 0.0001 | 0.0015 | 0.0001 | 0.0016 | 0.0007 | 0.0128 | 0.0009 | 0.0161 |
| 0.0842 | 0.0002 | 0.0019 | 0.0001 | 0.0015 | 0.0001 | 0.0016 | 0.0011 | 0.0129 | 0.0014 | 0.0162 |
| 0.0982 | 0.0002 | 0.0019 | 0.0001 | 0.0015 | 0.0002 | 0.0016 | 0.0013 | 0.0129 | 0.0016 | 0.0163 |
| 0.1263 | 0.0002 | 0.0019 | 0.0002 | 0.0015 | 0.0002 | 0.0016 | 0.0016 | 0.0131 | 0.0021 | 0.0165 |
| 0.1684 | 0.0003 | 0.0019 | 0.0002 | 0.0015 | 0.0003 | 0.0016 | 0.0022 | 0.0133 | 0.0029 | 0.0169 |
| 0.1964 | 0.0004 | 0.0019 | 0.0003 | 0.0015 | 0.0003 | 0.0017 | 0.0027 | 0.0136 | 0.0034 | 0.0173 |
| 0.2245 | 0.0004 | 0.0019 | 0.0003 | 0.0015 | 0.0004 | 0.0017 | 0.0031 | 0.0139 | 0.0040 | 0.0177 |
| 0.2526 | 0.0005 | 0.0019 | 0.0004 | 0.0015 | 0.0004 | 0.0017 | 0.0036 | 0.0142 | 0.0046 | 0.0183 |
| 0.2947 | 0.0006 | 0.0020 | 0.0005 | 0.0015 | 0.0005 | 0.0018 | 0.0044 | 0.0148 | 0.0057 | 0.0192 |
| 0.3087 | 0.0006 | 0.0020 | 0.0005 | 0.0016 | 0.0005 | 0.0018 | 0.0046 | 0.0150 | 0.0060 | 0.0195 |
| 0.3929 | 0.0008 | 0.0020 | 0.0006 | 0.0016 | 0.0007 | 0.0019 | 0.0065 | 0.0166 | 0.0087 | 0.0222 |
| 0.4069 | 0.0008 | 0.0020 | 0.0007 | 0.0016 | 0.0008 | 0.0019 | 0.0069 | 0.0170 | 0.0092 | 0.0227 |
| 0.4209 | 0.0009 | 0.0020 | 0.0007 | 0.0017 | 0.0008 | 0.0019 | 0.0073 | 0.0173 | 0.0098 | 0.0233 |
| 0.4350 | 0.0009 | 0.0021 | 0.0007 | 0.0017 | 0.0009 | 0.0020 | 0.0077 | 0.0177 | 0.0104 | 0.0239 |
| 0.4630 | 0.0010 | 0.0021 | 0.0008 | 0.0017 | 0.0009 | 0.0020 | 0.0086 | 0.0185 | 0.0117 | 0.0252 |
| 0.4771 | 0.0010 | 0.0021 | 0.0008 | 0.0017 | 0.0010 | 0.0021 | 0.0090 | 0.0189 | 0.0124 | 0.0260 |
| 0.5472 | 0.0012 | 0.0022 | 0.0010 | 0.0018 | 0.0012 | 0.0022 | 0.0118 | 0.0215 | 0.0166 | 0.0304 |
| 0.6595 | 0.0015 | 0.0023 | 0.0013 | 0.0020 | 0.0017 | 0.0026 | 0.0181 | 0.0275 | 0.0270 | 0.0410 |
| 0.7437 | 0.0018 | 0.0025 | 0.0016 | 0.0022 | 0.0022 | 0.0030 | 0.0253 | 0.0341 | 0.0398 | 0.0535 |
| 0.7437 | 0.0018 | 0.0025 | 0.0016 | 0.0022 | 0.0022 | 0.0030 | 0.0253 | 0.0341 | 0.0398 | 0.0535 |
| 0.7577 | 0.0019 | 0.0025 | 0.0017 | 0.0022 | 0.0023 | 0.0031 | 0.0268 | 0.0354 | 0.0425 | 0.0561 |
| 0.8138 | 0.0021 | 0.0026 | 0.0019 | 0.0024 | 0.0028 | 0.0034 | 0.0340 | 0.0417 | 0.0559 | 0.0687 |
| 0.8138 | 0.0021 | 0.0026 | 0.0019 | 0.0024 | 0.0028 | 0.0034 | 0.0340 | 0.0417 | 0.0559 | 0.0687 |
| 0.8980 | 0.0025 | 0.0028 | 0.0024 | 0.0027 | 0.0037 | 0.0041 | 0.0492 | 0.0547 | 0.0865 | 0.0963 |
| 1.0804 | 0.0037 | 0.0034 | 0.0038 | 0.0035 | 0.0068 | 0.0063 | 0.1193 | 0.1104 | 0.4076 | 0.3773 |
| 1.3049 | 0.0059 | 0.0046 | 0.0071 | 0.0054 | 0.0163 | 0.0125 | 0.4414 | 0.3382 | 1.0523 | 0.8064 |
| 2.1468 | 0.0507 | 0.0236 | 0.1482 | 0.0690 | 1.9957 | 0.9296 | 2.1321 | 0.9932 | 2.1443 | 0.9989 |
| 2.6659 | 0.3029 | 0.1136 | 2.3677 | 0.8882 | 2.6537 | 0.9954 | 2.6650 | 0.9997 | 2.6658 | 1.0000 |
| 3.1430 | 2.5128 | 0.7995 | 3.1055 | 0.9881 | 3.1419 | 0.9996 | 3.1429 | 1.0000 | 3.1430 | 1.0000 |
| $\tilde{\sigma}_k$ | 0.8650 | | 0.7102 | | 0.5856 | | 0.5572 | | 0.5221 | |
| $\pi_k$ | 0.7995 | | 0.8762 | | 0.9247 | | 0.3311 | | 0.1825 | |

$$\mu_\alpha = \frac{n}{n-1} \sum_{j=1}^{n} c_j(x_j - \bar{x}) \tag{9.3.7}$$

$$\sigma_a^2 = \left(\frac{n}{n-1}\right)^2 \left\{ \sum_{j=1}^{n} c_j(x_j - \bar{x})^2 - \left[\sum_{j=1}^{n} c_j(x_j - \bar{x})\right]^2 + \frac{(n-1)}{n(n-4)} \sum c_j A_{-j} \right\} \tag{9.3.8}$$

where

$$c_j = (A_{-j})^{-(n-2)/2} \bigg/ \sum_{j=1}^{n} (A_{-j})^{-(n-2)/2},$$

with

$$A_{-j} = \sum_{i \neq j} (x_i - \bar{x}_{-j})^2 \quad \text{and} \quad \bar{x}_{-j} = (n-1)^{-1} \sum_{i \neq j} x_i.$$

The attribution of 'spuriosity' to a member of the sample is approached in terms of the posterior distribution of $a$, and in particular of the values of $\mu_a$ and $\sigma_a^2$. If the weights $c_j$ are roughly equal (to $n^{-1}$) we have little evidence of contamination; a rough argument is given suggesting that contamination should be attributed to an observation $x_j$ whose weight $c_j$ exceeds

$$\frac{1}{n} + \frac{2}{n} \sqrt{\left(\frac{n-1}{n+1}\right)} \quad \text{(for } n \geq 3\text{)}.$$

Alternatively, Bayesian confidence (credibility) intervals for the parameter $a$ yield criteria expressed in terms of the distribution function of the t-distribution with $(n-2)$ degrees of freedom.

Guttman illustrates his recommendations by reference to some sets of simulated data.

The approach can be immediately extended to multivariate data $\mathbf{x}_1$, $\mathbf{x}_2, \ldots, \mathbf{x}_n$ arising from $\mathbf{N}(\boldsymbol{\mu}, V)$ if there is no contamination, and with contamination (if present) manifest in a single observation from $\mathbf{N}(\boldsymbol{\mu} + \mathbf{a}, V)$. Again it is through the posterior distribution of $\mathbf{a}$ that we seek to detect a discordant outlier. The detection criteria again revolve around the values taken by a set of weights attached, in the posterior distribution of $\mathbf{a}$, to the separate observations. It is interesting to note that the implicit concept of extremeness used to detect the outlier is again expressible in terms of the distance metric, or scatter-ratio, discussed in Chapter 7.

Guttman and Khatri (1975) present a direct extension of the Guttman (1973b) proposals to (univariate) situations where there are either 1 or 2 contaminants corresponding to slippage of the variance, or of the mean and the variance, of the basic normal distribution. Specifically, they include the cases where the contamination distributions are as follows:

$$k = 1 \quad \mathbf{N}\left(\mu + \frac{a\sigma}{\delta}, \frac{\sigma^2}{\delta^2}\right) \quad (0 < \delta \leq 1),$$

$$k = 2 \qquad N\left(\mu + a_1, \frac{\sigma^2}{\delta_1^2}\right) \quad \text{and} \quad N\left(\mu + a_2, \frac{\sigma^2}{\delta_2^2}\right),$$

$$\text{or} \qquad N\left(\mu + \frac{a_1\sigma}{\delta_1}, \frac{\sigma^2}{\delta_1^2}\right) \quad \text{and} \quad N\left(\mu + \frac{a_2\sigma}{\delta_2}, \frac{\sigma^2}{\delta_2^2}\right),$$

for both arbitrary $a_1$ and $a_2$, and with $a_1 = -a_2$ ($0 < \delta_j \leqslant 1; j = 1, 2$). (The parameters $\delta$, $\delta_1$, and $\delta_2$ are assumed known.)

Assessment of the presence of contamination *in linear model data*, using a Bayesian approach, features in the work of Chaloner and Brant (1988, using analysis of residuals) and of Guttman and Pena (1988, involving influence measures evaluated in terms of posterior distributions of the parameters).

A novel contribution is made by Ozturk, Patil, and Taillie (1992) who postulate a binomial basic model $B(n, p)$ with possible contamination from a $B(n, kp)$ source (where $p < kp < 1$) and construct a test for 'outliers' (contamination) in this respect.

An interesting and valuable feature of the Bayesian approach to outliers is its facility for drawing inferences about the *extent* of contamination and the *number* of contaminants (although rather limited progress has been made on the latter topic). A general discussion of 'the number of outliers in data from a linear model' is given by Freeman (1980). A major difficulty is a computational one: of determining posterior distributions under the vast number of combinations, $\binom{n}{k}$, of possible sets of contaminants for the range of different values of $k$. Pettit (1983) terms this the 'computational explosion' and has proposed an interesting prospective resolution of this difficulty. He suggests, as an approximation principle, that it should be assumed that any contaminants will manifest themselves as outliers, or at least as sample extremes. For a range of situations the probability of the latter prospect has been estimated and conditions described under which this will be close to unity. In this case, substantial savings of computational effort accrue.

Some proposals for the Bayesian handling of outliers in the specific context of time series are given by Muirhead (1986). The topic is dealt with in detail in Section 10.3.

Pettit (1988) develops a Bayesian approach to modelling outliers in the exponential family of distributions (in particular the exponential distribution). He provides a measure of the prospect of contamination and shows 'that masking and swamping cannot occur'!

A number of studies have been made of the sensitivity of the posterior distribution to the presence of contamination (outliers). These include Berger and Berliner (1986), Spall (1988). Sivaganesan (1988, 1989), and Sivaganesan and Berger (1989). See also Wasserman (1989).

The Bayesian approach to inference is particularly suited to the problem of *prediction* (of 'future' or 'missing' observations). A detailed treatment by Butler (1986) allows for the prospect of contamination and, *en passant*, provides a means of detecting contaminants. This interesting proposal is compared by simulation studies with competing procedures. See also Johnson (1987).

Methods for monitoring Bayesian models which allow for the presence of outliers (and possible contamination) are considered by West (1986) and West and Harrison (1986), again providing means of 'coping with' ('rejecting') outliers. See also Kitagawa (1984), for interesting proposals on how to separate a composite sample into distinct data generation sources, when there is a 'dominant' population and observations which are not from that population are regarded as 'outliers'.

In spite of some conceptual advantages of the Bayesian approach we must acknowledge the problems it raises for the study of outliers. There can be substantial, or even prohibitive, computational demands. Other unsatisfactory or unresolved elements include the frequent unhappy and arbitrary mix of Bayesian and sampling theoretic components, the expedient use of (perhaps improper) uniform prior distributions and the consequent abandonment of an element of 'surprise' in the identification of outliers. We must conclude that much remains to be done to enhance interpretative appeal and the applicability and practical relevance of Bayesian methods in this field.

# CHAPTER 10

# Outliers in Time Series: An Important Area of Outlier Study

We noted in Chapter 1 two examples of time series exhibiting outliers. In the first example, illustrated in Figure 1.6, a realization of a non-stationary series of sales figures showed a distinct disruption of the quarterly cyclic pattern of sales, possibly reflecting an outlier. The second example of moisture content of tobacco (Figure 1.7) exhibited apparently *isolated* outliers which might have reflected malfunction of the recording equipment. The analysis of time series is an important area of statistics and there is a clear need to understand the nature of outliers and to have available methods to detect, or accommodate, them. Consider, for example, the effect an outlier might have on a forecast of a future value.

Again we shall find major difficulties in detecting an outlier. As in general linear model data, it is not necessarily an extreme value and it can be cloaked to some extent by the general structure of the process. In particular, we can experience a smoothing-out effect when we attempt to examine outliers in terms of derived quantities such as the values of estimated residuals about a fitted model. Whilst in the linear model any outlier does not tend to influence adjacent observations *per se*, merely the estimated residuals, etc., the same need not be true for time series data in view of the correlational pattern of the basic process.

We need to consider the prospect of two types of outlier.

In the first case an isolated measurement, or execution, error is *super-imposed* on an otherwise reasonable realization of the process. This will *not* be reflected in the values of adjacent observations and its manifestation can be dramatic and obvious. Such a 'hiccough' effect is possibly what we are noticing in the data of Figure 1.7.

Alternatively, a more *inherent* form of contamination can arise and be reflected through the correlation structure of the process in neighbouring (usually later) observations. The data of Figure 1.6 possibly illustrate this effect. For this type of outlier there is a prospect that the realization itself conspires to conceal the outlier and the detection of outliers becomes even more problematical. On the other hand any smoothing-out effected by autocorrelations can have an intrinsic role in accommodating the outlier. It is possible that its influence on parameter estimation or testing in the basic time series model may be less acute than with independent error structure but this is not always so: outliers in time series can have dramatic effect. We shall examine the possibilities in more detail below.

The two types of outlier are usually referred to as **additive outliers** (AO) and as **innovations outliers** (IO), respectively. See, for example, Martin (1979).

Huber (1972) claims that the 'hiccough' effect is rare; the more usual outlier is of the inherent type revealed more obscurely in 'bumps' and 'quakes'. These are respectively local changes in the mean and variance (requiring corresponding slippage-type alternative models) whose effect extends to influence subsequent observations. For testing of outliers, or for their accommodation, Huber suggests examining coefficients of skewness (presumably of residuals) or kurtosis, or applying a smoothing process, respectively, but offers little by way of detailed prescriptions. Others have expressed the diametrically opposite viewpoint to Huber with regard to the type of outlier likely to be encountered in time series data. For example, Cox (1979) in the discussion of Kleiner, Martin, and Thomson (1979) asks, 'Have the authors ever encountered in applications innovations outliers?' They reply 'although AO outliers are much more frequent, IO type outliers occasionally occur'.

In spite of the clear importance of this topic, published work on outliers in time series was slow to appear, in terms either of frequency-domain or time-domain analyses. In the last decade, however, increasing attention has been given to this challenging area of study. We shall examine the early definitive work, and review the many developments that are currently opening up. Almost all of this work concerns *stationary* time series and can be conveniently divided into the distinct areas of detection and testing, accommodation or robustness, and Bayesian methods (involving a joint approach to detection and accommodation). The subsequent sections deal, respectively, with these different areas.

## 10.1 DETECTION AND TESTING OF OUTLIERS IN TIME SERIES

An early and most detailed examination of the detection and testing of outliers in stationary time series is that of Fox (1972). He distinguishes two types (I and II) of outlier. These are essentially the isolated independent gross execution or recording errors, independent of other observations, and the 'inherent' type of anomalous observation which influences succeeding observations, which we defined above and which, as we have said, are now known as *additive*, and *innovations, outliers* respectively. Four situations are postulated for outliers in a time series data set:

(i) all outliers are additive outliers (AO)
(ii) all outliers are innovations outliers (IO)
(iii) all outliers are of the same type, but we do not know which type, and
(iv) both types of outliers are present.

How we are to assess which of these situations prevails was not considered by Fox apart from remarking that (ii) will be distinguished from (i) by the presence of the carryover effect. (But see below) Only situations (i) and (ii) are examined in some detail, on the assumption that the process is free of trend or seasonal factors; the possible effect of their removal on the examination of outliers is not discussed. Thus the methods presented have obvious limitations, compounded by an additional assumption that there is at most one outlier, but they do provide a starting point in this difficult area of study.

A test for *additive outliers* is developed in relation to the mean-slippage outlier model for a discrete time series

$$x_t = u_t + \delta_{jt} a \qquad (10.1.1)$$

where $\delta_{jt}$ is the Kronecker delta function (i.e. $\delta_{jt} = 1$; if $t = j$; 0 otherwise) and the $u_t$ satisfy an autoregressive scheme of order $p$

$$u_t = \sum_{l=1}^{p} \alpha_l u_{t-l} + z_t \qquad (t = p+1, \ldots, n) \qquad (10.1.2)$$

where the $z_t$ are independent $N(0, \sigma^2)$. Thus we have a set of $n$ observations of a discrete process, further restricted by the assumption that $p$ is known and that $\{u_t\}$ is a stationary process, with a superimposed contaminant at time point $j$. Both the cases of prescribed $j$ and unknown $j$ are considered and *maximum likelihood ratio* tests of $H: a = 0$ against $\bar{H}: a \neq 0$ are developed. For the latter more realistic case ($j$ unknown) the *maximum likelihood ratio* statistic is equivalent to

$$\max_{j=p+1,\ldots,n-p} (k_{j,n})$$

where

$$k_{j,n} = \mathbf{x}' \hat{W}^{-1} \mathbf{x} / (\mathbf{x} - \tilde{\mathbf{a}})' \tilde{W}^{-1} (\mathbf{x} - \tilde{\mathbf{a}}). \qquad (10.1.3)$$

In (10.1.3) $\tilde{\mathbf{a}}$ is $\tilde{a}(0, 0, \ldots, 0, 1, 0, \ldots, 0)'$ where the 1 appears in position $j$ and $\tilde{a}$ is the maximum likelihood estimate of $a$ under (10.1.1) and (10.1.2); $\hat{W}^{-1}$ and $\tilde{W}^{-1}$ are the maximum likelihood estimates of $W^{-1}$ under $H$ and $\bar{H}$, respectively, where the covariance matrix of the process is assumed to have the form

$$V = W\sigma^2 \tag{10.1.4}$$

which depends only on $p$ and the autoregressive coefficients $\alpha_l (l = 1, 2, \ldots, p)$. Note that the elements of $W$ have the form $w_{t,t'} = w_{|t-t'|}$.

The test of discordancy detects the outlier as the observation maximizing $k_{j,n}$ and declares it discordant if the maximum value is sufficiently large. Under $H$ we are involved in determining the distribution of the maximum of a set of $n$ correlated $F$-variates. Significance levels, power calculations, and the behaviour of modified tests are all examined by Fox using simulation methods.

The outlier model used by him for a test of discordancy for an *innovations outlier* has the form

$$x_t = \sum_{l=1}^{p} \alpha_l x_{t-l} + \delta_{jt} a + z_t \tag{10.1.5}$$

with all quantities defined and limited as before. We can see here the carry-over effect of the contaminant. Again the *maximum likelihood ratio* test of $H:a = 0$ against $\bar{H}:a \neq 0$ is developed, and studied by simulation, for the case where $j$ is specified. The more important case of an unspecified value for $j$ is not pursued. Some implications of employing the wrong model $((10.1.5)$ instead of $(10.1.1)$ and $(10.1.2)$ or vice versa) are also examined by simulation.

Chernick, Downing, and Pike (1982) comment that outliers (of unspecified type) can seriously distort estimates of autocorrelations in a stationary time series. They suggest that we should seek to detect outliers by examining the influence function matrix of the estimated autocorrelations. They consider a discrete time series $x_1, x_2, \ldots, x_n$. Choosing a fixed number, $m$, of lags (with $m$ much smaller than $n$) we are advised to consider an $n \times m$ matrix $\{I[H, p(k), (y_j, y_{j+k})]\}$ where $y_i$ is the standardized observation $(x_i - \mu)/\sigma$ (with $\mu$ and $\sigma$ the mean and standard deviation of $x_i$: independent of $i$), $p(k)$ is the autocorrelation of lag $k$, $H$ is the marginal distribution of $y_i$, and $I(\ )$ is the influence function, which the authors show to have the simple form

$$y_j y_{j+k} - p(k)(y_j^2 + y_{j+k}^2)/2$$

(using the result employed by Devlin, Gnanadesikan, and Kettenring, 1975, in their study of the influence function of the correlation coefficient in a bivariate sample). With a transformation of the $y_j, y_{j+k}$ to

and
$$\left.\begin{aligned}
U_{j,k} &= \tfrac{1}{2}[(y_j + y_{j+k})/\sqrt{1 + p(k)} + (y_j - y_{j+k})/\sqrt{1 - p(k)}] \\
V_{j,k} &= \tfrac{1}{2}[(y_j + y_{j+k})/\sqrt{1 + p(k)} - (y_j - y_{j+k})/\sqrt{1 - p(k)}]
\end{aligned}\right\} \tag{10.1.6}$$

we have

$$I_{j,k} = I[H, p(k), (y_j, y_{j+k})] = \sqrt{[1 - p^2(k)]}\, U_{j,k} V_{j,k}. \qquad (10.1.7)$$

For a stationary gaussian process $U_{j,k}$ and $V_{j,k}$ are independent $N(0, 1)$, so that the distribution of $I[H, p(k), (y_j, y_{j+k})]$ has a fairly simple, tractable form and it is possible to determine values which would be unusually large in absolute terms at some specified critical level.

The way in which the presence of an outlier at an isolated time point will influence autocorrelations of different lags prompts the authors to propose a *visual procedure for detecting an outlier*. It is suggested that the entries in the influence function matrix $\{I_{j,k}\}$ should be replaced by zero if they are not unusually large and by + or −, respectively, (depending on their sign) if they exceed the critical value. If the resulting pattern shows a 'clothes pin' effect in the sense that many of the observations in the $t$th row and in the above diagonal [$(t - 1, 1), (t - 2, 2), \ldots$] are large in absolute value, and of similar sign, we conclude that $x_t$ is an outlier. (The authors do not discuss explicitly the relevance of the *signs* of the $I_{j,k}$ which are unusually large in *absolute* value.)

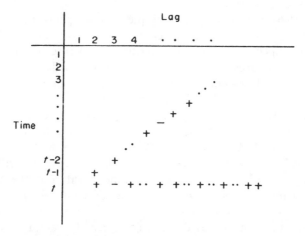

The method is illustrated for various data sets including some power-plant data and some on storage levels of radioactive materials. In practice, of course, we will need to estimate $\mu$, $\sigma$ and $p(k)$. Also the stationary gaussian assumption may not be reasonable, but it is suggested that the method should provide a reasonable basis for indicating outliers.

Stoodley and Mirnia (1979) describe an informal 'automatic' procedure for detecting 'transients, step changes and slope changes' in a time series. Their 'transient' may be interpreted as an outlier (probably of type AO). They are concerned with medical data and concentrate on the autoregressive integrated moving average model: specifically the ARIMA $(0,1,1)$ model. Employing exponentially weighted moving averages and a two-sided backward CUSUM technique they show how to monitor the series for 'out of control' indications; one form of which signifies a 'transient'.

An empirical study of the detection of outliers in time series of body temperatures of cows is described by de Alba and Zartman (1980). They employ a similar model to that used by Stoodley and Mirnia (1979) but incorporating the specific prospect of innovations outliers.

Tiao (1985) describes methods for outlier detection in autoregressive moving average (ARMA) processes. An interesting informal (graphical) approach is described by Embrechts, Herzberg and Ng (1986) who extend the use of Andrews's plots (see section 7.4.5 above) to detect periodic components and outliers in time series.

We noted that Fox (1972) made no methodological proposals to distinguish between the two basic types of outlier (AO or IO). This is remedied in the work of Muirhead (1986) who presents a test of discordancy for a single outlier of unknown type and proceeds to examine the properties of an appealing likelihood-ratio-based rule for distinguishing whether the outlier is of AO or IO type. He also compares this rule with a corresponding Bayesian procedure (see Section 10.3 below). See also Chang, Tiao, and Chen (1988); and Section 10.2 below.

Schmid (1986) considers the multiple outlier problem deriving, and examining asymptotic behavioural characteristics of, tests of discordancy for up to five specifically AO-type outliers occurring at prescribed time points in an autoregressive AR process of known order $p$. Schmid (1989) derives UMPU and asymptotically UMPU tests (where the autoregressive parameters are known and unknown, respectively).

Similar restrictions (of basic process) apply to most of the proposals of Abraham and Yatawara (1988) who present score tests (with derived properties) for ARMA $(p, q)$ and AR $(p)$ processes with a single outlier at a prescribed time point; for the AR $(p)$ process they go on to consider the case of an outlier at an *unknown* time point and the problem of distinguishing outlier type. In Abraham and Chuang (1989) some statistics used to detect outliers in linear model data (related to Cook's statistic) are transferred to the time series context and seem to retain their relevance and capability. Another method for distinguishing an AO outlier from an IO outlier is also presented. There is an interesting four-stage procedure for modelling time series data in the presence of contamination (manifest in the form of outliers) in which an ARMA $(p, q)$ process is approximated by an AR $(p + q)$ process with detection of, and adjustment for, the outliers.

Relationships between outliers and missing values in time series have also been discussed; see, for example, Kanto (1983), Ljung (1989a, 1989b), and Pourahmadi (1989). The problem of estimation of missing values, and of interpolation, can be related to that of outliers. For example, if we have rejected discordant outliers, estimation of such 'missing values' may be necessary.

Some of the Bayesian methods in Section 10.3 are also relevant to outlier *detection*.

## 10.2 ACCOMMODATION OF OUTLIERS IN TIME SERIES

The methods for outlier detection or testing which we have just considered all operate in the time-domain. When we are interested in *outlier-robust inference* procedures for time series we find that some proposals have been made both for time-domain, and for frequency-domain, characteristics.

As in robustness studies for other situations, we shall note that many of the alternative models used in the time series investigations are not specifically of a contamination form: again robustness is sought with respect to broader considerations, such as distributions which are long-tailed, or skew. None the less, increasingly more outlier-specific proposals are appearing in the literature. Sections 10.2.1 and 10.2.2 deal with general matters of outlier-robust estimation and testing for the time-domain and frequency-domain approaches, respectively. In Section 10.2.3 we will consider specific topics such as seasonality, change points, use of the Kalman filter, etc.

### 10.2.1 Time-domain characteristics

We might wish to estimate robustly (particularly in respect of possible forms of contamination) such quantities as the mean, variance and auto-correlations of a stationary time series, or certain parameters of an assumed basic model. Proposals have been made for specific classes of model, such as autoregressive with parameter $p$ as (10.1.2), moving average with parameter $q$, autoregressive-moving average with parameters $p$ and $q$, respectively, or autoregressive-integrated moving average. We will denote these AR ($p$), MA ($q$), ARMA ($p$, $q$), and ARIMA, respectively.

Where it is convenient to do so, we discuss results separately for these processes starting with estimation or testing of trend ($\mu$) and progressing to variance, autocorrelation, and the basic model parameters.

As a simple example consider the *first order* autoregression model AR(1)

$$x_t - \mu = \alpha(x_{t-1} - \mu) + z_t \qquad (10.2.1)$$

for a discrete time series; where the $z_t$ are independent $N(0, \sigma^2)$, $\mu$ is the mean and the autoregression parameter $\alpha$ takes some value in $(-1, 1)$.

Suppose we want to estimate $\mu$ but contemplate the possibility that one of the observations in a set of $n$ observations is contaminated. Specifically we observe $y_1, y_2, \ldots, y_n$ where $y_j \equiv x_j$ for $j \neq i$ and $y_i = x_i + \delta$. Thus our alternative model has an additive outlier at an unknown time point.

Guttman and Tiao (1978) consider this problem and employ a premium-protection approach on the assumption of known $\sigma^2 = 1$. Firstly, for known

$\alpha$, they propose estimating $\mu$ by the maximum likelihood estimator $\hat{\mu}$ if all $n$ estimated residuals $r_i = y_i - \hat{\mu}$ are not too large in absolute value and by a modified value $\hat{\mu} + a_i$ if the largest absolute estimated residual is larger than some prescribed value, where

$$\hat{\mu} = [y_1 + y_n + (1 - \alpha) \sum_{t=2}^{n-1} y_t] / [n - (n - 2)\alpha] \qquad (10.2.2)$$

and $a_i$ is a linear function of $r_i$ and the immediately adjacent estimated residuals.

The premium and protection of the estimator (10.2.2) are examined by simulation for different values of $\alpha$, and possible mis-specification of $\alpha$. The latter prospect is shown to have marked effects, and leads naturally to the need to consider the more realistic prospect of assuming that $\alpha$ is unknown.

This is approached by Guttman and Tiao with the suggestion that we consider the usual estimate of $\alpha$, namely the first serial correlation coefficient,

with

$$\hat{\alpha} = \sum_{t=2}^{n} (y_t - \bar{y}_2)(y_{t-1} - \bar{y}_1) \bigg/ \sum_{t=1}^{n} (y_t - \bar{y})^2$$

$$\bar{y}_1 = \sum_{t=1}^{n-1} y_t / (n - 1), \qquad\qquad (10.2.3)$$

and

$$\bar{y}_2 = \sum_{t=2}^{n} y_t / (n - 1)$$

and proceed initially as above with $\alpha$ replaced by this estimate. They show that a simple contaminant yields a bias of $-\alpha(1 - \alpha^2)\delta^2/(n - 1)$ in the estimate of $\alpha$: quite a marked effect if $|\alpha|$ is not close to zero. Accordingly a new type of estimator of $\mu$ is proposed, of the same basic form as before but with different estimators $\tilde{\alpha}_i$ used depending on which estimated residuals are being considered. The $\tilde{\alpha}_i$ are the first serial correlation estimates (10.2.3) when $y_i$ is removed from the sample and the sample means ($\bar{y}, \bar{y}_1, \bar{y}_2$) appropriately adjusted.

The premium and protection of this estimator of $\mu$ are examined by simulation with mildly encouraging results for the limited range of values of $n$, $\alpha$ and $\delta$ which are considered.

Cressie and Glonek (1984) discuss the known bias of the sample product-moment autocovariance estimator when applied to a stationary time series. They show that replacing the sample mean with the sample median can reduce the bias (even in the case of additive contamination) and explore the conditions for this to be a useful reduction for the AR (1) case. Yoshida,

Kondo, and Inagaki (1984) explore the asymptotic bias of four autocorrelation estimators for a stationary Gaussian process with additive outliers. Two wide-ranging studies of robust estimation of autoregressive parameters are presented by Denby and Martin (1979) and Martin (1979). The former work compares the behavioural properties of least squares estimators and (appropriately defined) $M$-estimators *of the autoregressive parameters* again for the restricted situation of an AR (1) process, but clearly distinguishing the cases of outliers of both types: AO and IO. (Both studies include brief comments on robust estimation of the mean $\mu$ of $x_t$.)

Suppose we observe $y_1, y_2, \ldots, y_n$. The models are as follows, for normal-mixture contamination.

*IO model*

$$\left.\begin{array}{l} x_t = \alpha x_{t-1} + z_t \\ y_t = x_t \end{array}\right\} \tag{10.2.4}$$

where $|\alpha| < 1$ and the $z_t$ are independent $(1 - \gamma)N(0, \sigma^2) + \gamma N(0, \tau^2)$ with $\gamma$ small and $\tau^2 > \sigma^2$.

*AO model*

$$\left.\begin{array}{l} x_t = \alpha x_{t-1} + z_t \\ y_t = x_t + w_t \end{array}\right\} \tag{10.2.5}$$

where $|\alpha| < 1$, the $z_t$ are independent $N(0, \sigma^2)$, the $w_t$ are independent $(1 - \gamma)\omega + \gamma N(0, \tau^2)$ with $y$ small and where $\omega$ has a degenerate distribution at the origin.

Thus in the IO model, occasional innovations $z_t$ have larger variance than the majority and can appear as outliers. In the AO model an additional random component is occasionally superimposed on the underlying process, again with the prospect of exhibiting outlying behaviour. The IO outliers clearly transmit their effect through to later observations; AO outliers do not.

Denby and Martin compare least squares estimators and $M$-estimators of $\alpha$ in respect of the effects of IO and AO types of contamination.

An $M$-estimator $\hat{\alpha}_M$ is obtained as a solution of

$$\sum_{t=1}^{n-1} y_t \psi(y_{t-1} - y_t \tilde{\alpha}_M) = 0 \tag{10.2.6}$$

where $\psi(y)$ is bounded, with $y\psi(y) \geqslant 0$ and $\psi'(0) = 1$.

If we drop the boundedness assumption and put $\psi(y) = y$ we obtain the least squares estimator $\tilde{\alpha}_{LS}$.

A *generalized M-estimator* (*GM-estimator*) $\tilde{\alpha}_{GM}$ is obtained as a solution of

$$\sum_{t=1}^{n-1} g(y_t)\psi(y_{t+1} - y_t\tilde{\alpha}_{GM}) = 0 \qquad (10.2.7)$$

with $g(y)$ bounded and $yg(y) \geq 0$. Note that $\tilde{\alpha}_{GM}$ has a bounded influence function, which is not true of $\tilde{\alpha}_{LS}$.

Particular choices of $\psi(y)$ and $g(y)$ are discussed with the aim of achieving robust estimators. The following results are obtained (from asymptotic calculations and simulations).

Using $\tilde{\alpha}_{LS}$ we encounter no great loss of efficiency if innovations outliers exist, although it would be better to use $\tilde{\alpha}_M$. If additive outliers exist, serious efficiency losses accrue from $\tilde{\alpha}_{LS}$. There is also serious bias in this latter case with $\tilde{\alpha}_{LS}$ and $\tilde{\alpha}_M$. For example, the asymptotic bias of $\tilde{\alpha}_{LS}$ is

$$b_{LS}(\alpha) = -\alpha\gamma\tau^2/(\sigma_x^2 + \gamma\tau^2)$$

which vanishes only if $\alpha = 0$ or $\tau^2/\sigma_x^2$ is negligible. Here $\sigma_x^2$ is the variance of $x_t$; that is $\sigma_x^2 = \sigma^2/(1 - \alpha^2)$. Suppose $\sigma_x^2 = 1$ (for example, $\sigma^2 = 0.75$, $\alpha = 0.5$), $\gamma = 0.1$ and $\tau^2 = 10$: 'a fairly modest' amount of contamination. We have here an astounding degree of asymptotic bias, namely 50 per cent!

For robustness against additive outliers, particular forms of $\tilde{\alpha}_{GM}$ are shown to be highly desirable.

Martin (1979) extends the study to estimation of the $p$ parameters in an AR($p$) model. Space does not permit us to do full justice to this extensive examination but some broad summary conclusions are as follows. Least-squares estimators of the autoregressive parameters in IO models are usually robust; the estimators of $\mu$ and $\sigma$ are not so. For AO models, least-squares estimators of the autoregressive parameters are *not* robust. The same general conclusions apply for naturally defined $M$-estimators and the authors propose use of a form of '*generalized M-estimators*' and of '*conditional mean M-estimators*' for the AO model. They also briefly consider the important question of how to detect whether an IO or an AO model is the more appropriate, as a basis for deciding what forms of robust estimates to employ. One proposal is an exploratory plot of the estimated residuals $r_j$ obtained from a generalized $M$-estimator fit of the data. If we plot $r_j$ against $r_{j+1}$ an IO model will tend to reveal outliers as points lying close to either axis. The greater correlation of the adjacent $r_j$ for an AO model will disrupt this pattern: the outliers will not be close to the axes. However, one must question how useful such a distinction can be if we only contemplate a small number of outliers (perhaps none in fact).

Barthoulot (1983) examines $M$-estimators for AR (1) models with asymmetric contamination.

Martin and Yohai (1984, 1986) provide extended definitions of general robustness measures which they propose as particularly relevant to the AR (1) model with contamination. In the 1984 paper they offer a general definition of *gross-error-sensitivity* based on their earlier proposed '*time series influence curve*'. The approach is illustrated for GM- and RA-estimators (see above). The latter are based on robust estimators of *residual autocovariances* and were proposed by Bustos and Yohai (1986) for ARMA models (see below). Martin and Yohai (1986) is an extensive and detailed study of *influence functionals* (IF) for time series, and their use in studying outlier accommodation issues in time series: principally estimation of the autoregressive, or moving average, parameter in an AR (1), or an MA (1), model. An interesting feature of this work is its use of a *general* outlier-generating model which can encompass many different manifestations (from 'isolated outliers' to 'patches of outliers') including the AO formulation.

**ARMA** Proposals for accommodation have also been made for cases where the basic (uncontaminated) model is of ARMA form. Specifically we mention Pham (1984), Bustos and Yohai (1986), Masarotto (1987), Stockinger and Dutter (1987), and Li and Dickenson (1988).

Bustos and Yohai (1986) examine an ARMA model in the presence of AO outliers. They show that since usual estimators of the model parameters can be markedly affected by such contamination, it is necessary to seek appropriate outlier-robust estimators.

They propose the use of the appropriate form of GM- (Generalized $M$-) estimator or a new estimator: the RA estimator based, as remarked above, on robust versions of *residual autocorrelations estimators*. Specifically, these are defined as follows. The ARMA ($p$, $q$) model can be expressed

$$\phi(B)(x_t - \mu) = \theta(b)Z_t \quad (1 \leqslant t \leqslant T) \tag{10.2.8}$$

where the $Z_t$ are i.i.d. random variables and $\phi(B)$ and $\theta(B)$ are order $p$ and $q$ (respectively) polynomials in the backward-shift operator:

$$Bx_t = x_{t-1}.$$

If $\lambda = (\phi, \theta, \mu)$, the least-squares estimator minimizes

$$\sum_{t=p+1}^{T} r_t^2(\lambda) \tag{10.2.9}$$

where the residuals $r_t(\lambda)$ are given by

$$r_t(\lambda) = \theta(B)^{-1}\phi(B)(x_t - \mu) \tag{10.2.10}$$

where we adopt the convention $x_t = \mu$ for $t \leqslant 0$.

Such estimators are known to be sensitive to just a few additive outliers, and can be inefficient.

To construct the $RA$ estimator, (10.2.9) is effectively expressed in a form that explicitly involves (the usual non-robust) covariance estimators of residuals, and then replaces these with robust versions (in one of various forms). The authors argue that the RA estimator should be robust when $q = 0$, i.e. for an RA($p$) model. When $q > 0$, they suggest a further modification which uses truncated forms of robust residuals—to yield the *truncated residual autocovariances* (TRA) *estimator*.

In Monte Carlo investigations (with additive outliers) the RA and TRA estimators are shown to compare favourably with $M$- and $GM$-estimators. Asymptotic properties of the estimators are exhibited, and they are shown to compare favourably with the conventional least-squares, and $M$-estimators, by means of an extensive Monte Carlo study. Also mentioned is the earlier AM (approximate maximum likelihood) estimator of Martin (1981) but this is discarded in view of its inconsistency—see also Pham (1984).

Masarotto (1987) seeks outlier-robust estimators (for AO contamination) of the ARMA parameters, which are consistent under the basic (no-contamination) model. He proposes *generalized M-estimators of order r*, obtained from the corresponding estimators for AR models by replacing the best infinite memory linear predictor by one of finite length, $r$. A limited-scale simulation study suggests that the proposed estimator might have advantages (in terms of robustness, efficiency and computational facility) over GM- and TRA estimators; but the basic model of the simulation is just MA(1).

Li and Dickinson (1988) construct a $(p + q)$-stage lattice whitening filter for producing consistent estimators of the $p$ autoregressive parameters in an ARMA $(p, q)$ model—with incidental facilities for estimating the $q$ moving average parameters and robustness properties against AO contamination.

A review of robust methods for estimation in univariate ARMA models is given by Stockinger and Dutter (1987). See also Chuang and Abraham (1989).

**ARIMA** The ARIMA model extends the ARMA by allowing for non-stationarity, e.g. by changes in the mean level (shifts) or seasonality. In contrast with (10.2.8), it takes the form

$$\phi(B)\nabla^d x_t = \theta(B)z_t \qquad (10.2.11)$$

where $\nabla$ is the difference operator: $\nabla x_t = x_t - x_{t-1}$.

Essentially, (10.2.11) says that (whilst $x_t$ may not be stationary) the $d$th difference $\nabla^d x_t$ is stationary and can be represented by an ARMA $(p, q)$ model.

Some work on the accommodation of outliers in ARIMA models is beginning to appear. Some of this is described below (Section 10.2.3) under appropriate more specific headings: e.g. diagnostics, forecasting, etc. Directly relevant to the outlier-robust estimation of the basic parameters of the

ARIMA model is the work of Chang, Tiao, and Chen (1988). They are concerned with the comprehensive problem, in an ARIMA model, of detecting outliers, distinguishing AO and IO types and proceeding to the estimation of the basic ARIMA parameters in the presence of contamination. On this latter issue, however, the properties of the proposed estimators are examined by simulation only in the context of AR(1) data.

### 10.2.2 Frequency-domain characteristics

The quantity of principal interest in frequency-domain study of a time series is of course the spectrum. Any second-order stationary process $X_t$ has an autocovariance function

$$c(k) = \text{cov}(X_t, X_{t+k})$$

which describes the covariational structure of the process. Under wide conditions $c(k)$ can be expressed as

$$c(k) = \int_{-\pi}^{\pi} \exp(ik\omega)\,dF(\omega) \qquad (10.2.12)$$

where $F(\omega)$ is bounded and non-decreasing. If $f(\omega) = F'(\omega)$ exists we have

$$c(k) = \int_{-\pi}^{\pi} \exp(ik\omega)f(\omega)\,d\omega \qquad (10.2.13)$$

The functions $F(\omega)$ and $f(\omega)$ are the *spectral distribution function* and *spectral density function* (*spectrum, power spectrum, power spectral density*), respectively.

From (10.2.13), we see that the $c(k)$ are Fourier coefficients of $f(\omega)$. With reasonable conditions on $f(\omega)$ we can equivalently write

$$f(\omega) = \frac{1}{2\pi} \sum_{j=-\infty}^{\infty} c(j)\exp(-ij\omega). \qquad (10.2.14)$$

Since $\text{Var}(X_t) = c(0) = \int f(\omega)d\omega$ we see that the spectrum enables us to express $\text{Var}(X_t)$ as a sum of components associated with all frequencies $\omega$ in a range $(-\pi, \pi)$; a similar interpretation holds for the autocovariances $c(k)$ from (10.2.13). Note that $f(\omega)$ is a symmetric function.

Thus the nature of the process can be described in *frequency terms* using $f(\omega)$, rather than in terms of the time variable, via $c(k)$.

It becomes of interest to estimate $f(\omega)$ and $F(\omega)$ and it is well-known that this is not entirely free from difficulty even for an uncontaminated process. The difficulties are compounded if we anticipate the presence of outliers. The question of robust estimation of the spectrum for IO and AO models has been discussed in some detail by Kleiner, Martin, and Thomson (1979) and by the contributors to the Discussion of that work. Again only AR($p$) autoregressive processes are considered, where for an uncontaminated basic model the $x_t$ have the form

$$x_t = \sum_{j=1}^{p} \alpha_j x_{t-j} + z_j \tag{10.2.15}$$

(where the $z_j$ are uncorrelated with zero mean and constant variance $\sigma^2$, and $\sum_{j=1}^{p} \alpha_j^2 < \infty$). For this process we have

$$F(\omega) = \sigma^2 / |G(\omega)|^2 \tag{10.2.16}$$

where

$$G(\omega) = 1 - \sum_{j=1}^{p} \alpha_j \exp(i\omega j). \tag{10.2.17}$$

Usual estimates of the spectrum employ a smoothed version of the periodogram (in the form of the discrete Fourier transform of modified data $\tilde{x}$, obtained by applying a *window*, or *taper*, to reduce 'leakage' due to the finiteness of any actual realization) or from sample analogues of (10.2.16) and (10.2.17) using, for example, least squares estimates $\tilde{\alpha}$ of $\alpha$.

Kleiner, Martin, and Thomson (1979) show that when innovations outliers can arise, neither form of estimator is seriously influenced in its shape characteristics (although a robust estimate of the scale parameter is required). *With additive outliers, however, both forms are seriously non-robust.* (This distinction of qualitative effect has often been remarked in respect also of time-domain study.) The outliers need not be large relative to the scale of the process (only relative to the $z_t$ process) for them to markedly obscure the nature of the spectrum: see example below.

To overcome the difficulty with additive outliers two methods are proposed both based on robust 'prewhitening' of the data and on applying a weighting factor to the corresponding estimated residuals. Prewhitening is essentially a technique for preliminary filtering of the data to transform the original time series into one with an essentially flat spectrum (as for a 'white noise' process). The prewhitened series is then analysed and the estimate of its spectrum is readjusted for the effects of the prewhitening. The two methods differ only in the manner of prewhitening. They involve use of robust estimates $\tilde{\alpha}$ of the autoregressive parameters $\alpha$, and of prediction residuals designed to be unaffected by any outliers (but to be otherwise relatively unaffected by prewhitening). *GM-estimates* (see Section 10.2.1) are used for $\tilde{\alpha}$. Both methods are shown, by application to various data sets, to be successful in their aim of accommodating outliers.

It is instructive to consider one of the examples used by Kleiner, Martin, and Thomson (1979). Their *Example* 4 concerns 1000 measurements of the distortions in microns along some 4 metres of copper-plated circular section

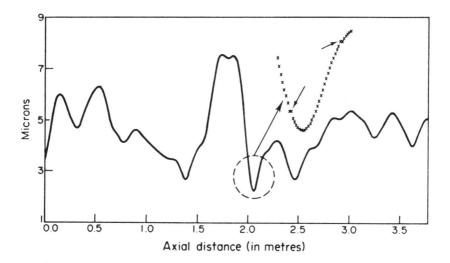

Axial distance (in metres)

**Figure 10.1** Distortion data for steel telephone tubing (reproduced from Kleiner, Martin, and Thomson, 1979, with permission from the Royal Statistical Society)

steel tubing carrying large numbers of telephone calls. Figure 10.1 shows the original data with two essentially unnoticable outliers exhibited in the figure by an enlarged section. These are malfunction errors in the recording equipment and although small in absolute terms they are large in relative terms, since the ratio of prediction variance to process variance, when a seventh order process is fitted, is of order $10^{-6}$. They seriously distort (essentially occlude) the principal features of the power spectrum when this is estimated non-robustly.

Using the authors' robust autoregressive spectrum estimate, the structure of the spectrum is recovered dramatically. Figure 10.2 shows this effect: the broken curve is the non-robust estimate, the continuous curve is the robust estimate. The non-robust estimate is exhibiting (above frequency 0.1) just the effect of the two outliers.

It is interesting to note that little more is to be found in the literature on this important issue of the frequency-domain approach to the handling of outliers in time series.

### 10.2.3  Specific inference problems

It is worth stressing a feature of time-series outliers that has been mentioned already. It is widely observed that outliers of the distinct types IO and AO have different qualitative levels of effect on the estimation of parameters or missing values, fitting of models, forecasting of future values, assessment of seasonality, etc. The distinctions lie in the much lower influence of the

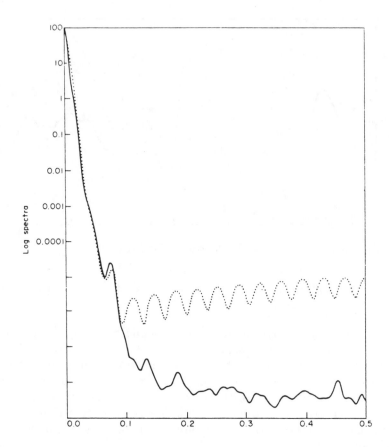

**Figure 10.2** Estimated spectrum for data shown in Figure 10.1 (reproduced from Kleiner, Martin, and Thomson, 1979, with permission from the Royal Statistical Society)

innovations outlier (IO) (whose effect seems to 'massaged' through the time process) compared with that of the additive outlier (AO) (which seems to noticeably perturb the inferences).

We will now consider some specific types of time series investigations and how they might be affected by outliers (or contamination).

**Time series diagnostics** A battery of procedures has developed over recent years for regression diagnostics—screening calculations and displays to explore the fit of a data set to a particular form of *regression* model. Similar facilities have been appearing for time series models and one of the 'symptoms' in any 'diagnosis' is obviously the outlier.

Pena (1987) is concerned with 'measuring the importance of outliers in ARIMA models' in terms of the influence of each observation on parameter estimates in the model. This is examined by missing out each observation (in turn) and noticing the changes in the estimates, employing intervention analysis or Kalman filter techniques to cope with the missing values. Pena claims that his leave-one-out approach is a 'natural generalization' of the influence measures for regression data.

Bruce and Martin (1989) adopt the same philosophical approach, again for the ARIMA model, and include 'leave-$k$-out' diagnostics to deal with 'patches' of outliers and with *masking*. They present an extensive package of diagnostic procedures for ARIMA (and more basic) models. See also Ledolter (1990).

**Model specification and changes** There is bound to be difficulty in determining an appropriate model, or in detecting changes in the model, when outliers are present.

Tsay (1986) presents a composite approach for ARMA ($p$, $q$) models consisting of outlier detection, followed by determination of outlier form (AO or IO), removal of possible effects of contamination, estimation of the order ($p$, $q$) and ultimately estimation of the $p + q$ parameters. Estimation is effected by means of an iterative least squares approach. The methods are illustrated by means of an interesting example on the annual consumption of spirits in the UK from 1870 to 1938. Iterative processing and adjustment for outliers leads to an ARMA (3,2) model and, as a refinement, to an ARMA (3, 3) model. The authors observe that the residuals in the former case (see Figure 10.3a) 'do not behave as a white noise process' whereas in the latter case (see Figure 10.3b) the plot 'fails to show any major model inadequacy'. (Note the difference in scale in the two parts of Figure 10.3.)

Tsay (1988) examines the problem of detecting a shift in the mean level or the residual variance of a time series in the presence of contamination. This theme is also pursued by Bansal and Papantoni-Kazakos (1989).

Li (1988) discusses the testing of the adequacy of fit of a prescribed time series model by means of an outlier-robust generalization of the portmanteau (Box–Pierce) statistic.

Kohn and Ausley (1986) examine the specification of an ARIMA model with missing data. They employ the Kalman filter (see Ausley and Kohn, 1985) for prediction and interpolation. Although not directly concerned with outliers, the methods are interesting and relevant to the outlier theme in the context of methods that examine outliers by successive omission of suspect observations. The Kalman filter has also been used by Kirkendall (1989) for exponential smoothing when detection of outliers and shifts of level are under examination. In de Jong (1988) it is shown that outliers can lead to complete breakdown of the Kalman filter approach, whilst Meinhold and Singpurwalla (1989) discuss a rationale for 'robustification' of Kalman filter models. Abraham and Chuang (1987) employ the EM algorithm for time series estimation in the presence of outliers. Thombs

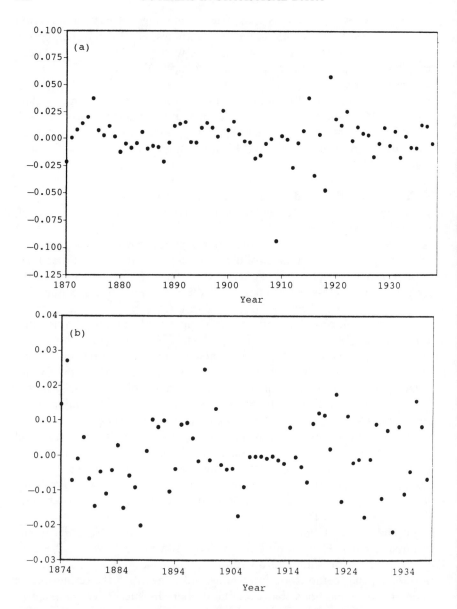

**Figure 10.3** (a) The residual plot of model (3, 2); (b) the residual plot of model (3, 3) (from Tsay, 1986). Reproduced by permission of the American Statistical Association.

and Schucany (1990) present *bootstrap* prediction intervals for autoregression in the presence (*inter alia*) of contaminated normal error distribution.

Forecasting is an important topic in time series study and can be expected to be susceptible to outliers. This theme is examined by Hillmer (1984),

Bretschneider (1986), and Ledolter (1987 and 1989, with particular reference to ARIMA models). The effects of outliers on seasonal adjustment (and estimation of seasonal components) is discussed by Scott (1987), Chiu (1989), and Kitagawa (1989).

An allied topic is outlier-robust use of control charts: see Edgeman (1989) and Rocke (1989).

## 10.3 BAYESIAN METHODS

Abraham and Box (1979) take a Bayesian look at the problem of outliers in time series. They are interested in an AR($p$) model with possible contamination. Again we encounter the IO and AO contamination models. For the IO model we have a process

$$y_t = \sum_{j=1}^{p} \alpha_j y_{t-j} + z_t + \delta u_t \qquad (10.3.1)$$

where the $u_t$ are independent with $P(u_t = 1) = \phi$, $P(u_t = 0) = 1 - \phi$, the $z_t$ are independent $N(0, \sigma^2)$ and $\alpha$, $\delta$ and $\sigma$ are unknown. The posterior distribution of $(\alpha, \delta, \sigma)$ is examined assuming prior independence and ignorance, expressed by a prior density

$$\pi(\alpha, \delta, \sigma) \propto |I(\alpha)|^{1/2} \sigma^{-1} \qquad (10.3.2)$$

where $I(\alpha)$ is the information matrix of $\alpha$. The marginal posterior distributions of $\alpha$ and of $\delta$ are singled out for closer study. It is assumed that $\phi$ is known.

The special case $p = 1$ is considered in some detail, and a simulated time series is analysed.

The AO model is also considered. Here it is $w_t = y_t + \delta u_t$ that is assumed to be an AR($p$) process;

$$w_t = \sum_{j=1}^{p} \alpha_j w_{t-j} + z_t \qquad (10.3.3)$$

and again the $u_t$ are independent with $P(u_t = 1) = \phi$, $P(u_t = 0) = 1 - \phi$. The joint posterior distribution of $(\alpha, \delta)$ is determined using the prior distribution (10.3.2), but the *individual* marginal distributions of $\alpha$ and of $\delta$ prove to be intractable. Muirhead (1986) describes a Bayes rule for distinguishing between AO and IO outliers and shows that it compares well with likelihood ratio-based methods.

West (1981) develops robust approximate Bayesian methods for sequentially updating inferences about the mean (or mean vector for vector processes) of a linear time series. Robustness is sought in relation to nonnormal error distributions in models analogous to the (10.2.5) representation of the AO structure, but where the $w_k$ are not restricted to distributions of the form $(1 - \gamma)w + \gamma N(0, \tau^2)$. Many forms of distribution for $w_k$ are considered including Student's $t$, exponential power, logistic and stable distributions. Note that contamination is not specifically represented; any outlying behaviour arises from long-tailed distributions. See also Masreliez and Martin (1977).

A general approach to robust Bayesian inference, and specific consideration of the outliers problem appears in Smith (1983). This is specialized to time series models by the 'elaboration of standard models' to take account (*inter alia*) of individual aberrant observations (outliers)'.

The study of 'interventions' in time series, by Box and Tiao (1975), whilst not specifically directed to outliers, has obvious links with our theme. The methods seek to represent and analyse the effects of marked changes ('interventions') in the nature of a time series. See also Davis (1979).

Ameen and Harrison (1985) extend Bayesian forecasting methods into time series problems including those based on ARIMA models for situations where 'abrupt changes' occur (e.g. as outliers from contamination sources).

Some work on outliers in multiple time series, and multivariate time series, is beginning to appear. Tsaknakis, Kazakos, and Papantoni-Kazakos (1986) extend outlier-robust prediction and interpolation procedures from the univariate case to contaminated vector-valued second-order time series. Khattree and Naik (1987) propose a method for detecting outliers in bivariate time series data. Gunther (1988) shows that a recursive method for estimating parameters in an autoregressive multiple time series model is highly sensitive (non-robust) in the presence of outliers, whilst Verme (1989) examines the effects of outliers on estimating the cross-correlation function.

## 10.4  COMMENT

It is clear that much still remains to be done in the provision of a comprehensive battery of procedures for handling outliers in time series. Gaps remain at the level of basic modelling (see Maronna, 1979) in the handling of more complex types of time series (moving-average or ARIMA), in the problem of prediction, for non-stationary data (see Fieller, 1979) and for continuous-time processes although, as we have seen, conspicuous advances have been made over the last decade.

It is bound to be difficult to devise methods for non-stationary time series (see, however, Huggins, 1989; Robertson, 1990), but an example of Subba Rao (1979) in the discussion of Kleiner, Martin, and Thomson (1979) provides a cogent tailpiece illustration of how vital it is to get to grips with this problem. He shows the following diagrammatic representation (reproduced by permission of the Royal Statistical Society) of a simulated time series of 600 observations.

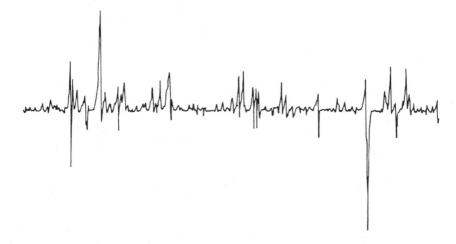

The process looks stationary, with several substantial outliers. *The process is in fact not stationary*: it has the form

$$X_t = 0.4X_{t-1} + 0.8X_{t-1}z_{t-1} + z_t$$

where the $z_t$ are independent $N(0, 1)$. IT CONTAINS NO OUTLIERS.

# CHAPTER 11

# Outliers in Directional Data

Situations often arise where the data are in the form of *angular measurements* giving the orientations of straight lines in the plane ('*circular data*') or in space ('*spherical data*'). It is common practice to refer to such angular measurements as *directional data*, so we use the term here (but see below). To quote from Mardia (1975b),

Directional data analysis is emerging as an important area of statistics. . . . Directional data are often met in astronomy, biology, geology, medicine and meteorology, such as in investigating the origins of comets, solving bird navigational problems, interpreting palaeomagnetic currents, assessing variation in the onset of leukaemia, analysing wind directions, etc.
  The directions are regarded as points on the circumference of a circle in two dimensions or on the surface of a sphere in three dimensions. In general, directions may be visualized as points on the surface of a hypersphere but observed directions are obviously angular measurements. (Mardia, 1975b)

In his authoritative book on this topic, Mardia (1972) observes:

The interest in developing techniques to analyse directional data is as old as the subject of mathematical statistics itself. Indeed, the theory of errors was developed by Gauss primarily to analyse certain directional measurements in astronomy. It is a historical accident that the observational errors involved were sufficiently small to allow Gauss to make a linear approximation and, as a result, he developed a linear rather than a directional theory of errors. In many applications, however, we meet directional data which cannot be treated in this manner. . . . (Mardia, 1972)

In dealing with directional data it is essential to distinguish between situations where the directional sense (as well as the orientation) of a line is meaningful, and situations where it is not; or, as one might say, between directed and undirected lines. Examples of the first are wind directions (a south wind is not the same as a north wind!), and the directions of flight of birds; such observations are called *vectors*. Examples of the second are the orientations of the normals to the orbital planes of comets, and the orientations of the poles (or normals) to fracture planes in structural geology; such observations are called *axes*. Vectorial data and axial data require different methods of analysis. This is of particular importance with spherical data (orientations in space); most circular data situations are vectorial.

Mardia's above-quoted remark that directions may be 'regarded as points on the circumference of a circle in two dimensions or on the surface of a sphere in three dimensions' relates to *vectors*; for *axes* the points are on a semicircle or hemisphere respectively.

For detailed accounts of directional statistics and directional data analysis, see Fisher (1993) for circular data, and Fisher, Lewis, and Embleton (1987, 1993) for spherical data.

Outlying observations arise with angular data just as with data on the line, so the specific methodology required for analysing directional data needs to include techniques for dealing with outliers, whether by accommodation or by discordancy testing with consequent identification or rejection. Until the late 1970s there was little or no published work on such techniques; since then there have been a number of papers, an account of which is given in this chapter. However, much work remains to be done on the treatment of directional outliers.

We first recall that, analogously to the mean $\mu$ and variance $\sigma^2$ of a random variable on the line, a vectorial random variable $\Theta$ on the circle has (in most cases) a *mean direction* $\mu$ given by

$$E\{\sin(\Theta - \mu)\} = 0 \qquad (11.0.1)$$

and a *circular variance* $1 - \rho$, vanishing in the degenerate case $\Theta \equiv \mu$, where

$$\rho = E\{\cos(\Theta - \mu)\}; \qquad (11.0.2)$$

$\rho$ is the *resultant length*. (11.0.1), (11.0.2) can be written

$$E(\cos \Theta) = \rho \cos \mu, \qquad E(\sin \Theta) = \rho \sin \mu. \qquad (11.0.3)$$

The estimates of these parameters from a sample $\theta_i(i = 1, \ldots, n)$ are the *sample mean direction* $\hat{\mu}$ and the *sample mean resultant length* $\hat{\rho}$ given by

$$\frac{1}{n} \sum_{i=1}^{n} \cos \theta_i = \hat{\rho} \cos \hat{\mu}, \qquad \frac{1}{n} \sum_{i=1}^{n} \sin \theta_i = \hat{\rho} \sin \hat{\mu}. \qquad (11.0.4)$$

A *vectorial* random variable on the sphere may be denoted $(\Theta, \Phi)$ where $\Theta$ is the colatitude and $\Phi$ the longitude $(0 \leqslant \Theta \leqslant \pi, 0 \leqslant \Phi < 2\pi)$, or equivalently $(X, Y, Z)$, where the direction cosines $X, Y, Z$ are given by

$$X = \sin \Theta \cos \Phi, \quad Y = \sin \Theta \sin \Phi, \quad Z = \cos \Theta. \qquad (11.0.5)$$

It has as parameters a *mean direction* $(\mu, \nu)$, a *spherical variance* $1 - \rho$ and equivalently a *resultant length*, $\rho$, given by

$$E(X) = E(\sin\Theta\cos\Phi) = \rho \sin\mu \cos\nu,$$
$$E(Y) = E(\sin\Theta\sin\Phi) = \rho \sin\mu \sin\nu,$$
$$E(Z) = E(\cos\Theta) = \rho \cos\mu. \tag{11.0.6}$$

The estimates from a sample $(\theta_i, \phi_i)$ $(i = 1, \ldots, n)$ are the *sample mean direction* $(\hat{\mu}, \hat{\nu})$ and the *sample mean resultant length* $\hat{\rho}$, given by

$$\frac{1}{n}\sum_{i=1}^{n}\sin\theta_i\cos\phi_i = \hat{\rho} \sin\hat{\mu}\cos\hat{\nu}, \quad \frac{1}{n}\sum_{i=1}^{n}\sin\theta_i\cos\phi_i = \hat{\rho} \sin\hat{\mu}\sin\hat{\nu},$$

$$\frac{1}{n}\sum_{i=1}^{n}\cos\theta_i = \hat{\rho} \cos\hat{\mu}, \tag{11.0.7}$$

An *axial* random variable on the sphere may be denoted $(\Theta, \Phi)$ on the hemisphere $(0 \leqslant \Theta \leqslant \frac{\pi}{2}, 0 \leqslant \Phi < 2\pi)$ or equivalently $(X, Y, Z)$ given by (11.0.5). Equally it may be regarded as a *pair* of diametrically opposite points $(\Theta, \Phi)$, $(\Pi - \Theta, \Phi \pm \pi$, or $(X, Y, Z)$, $(-X, -Y, -Z)$ on the whole spherical surface $(0 \leqslant \Theta \leqslant \pi, 0 \leqslant \Phi < 2\pi)$. It has *directions of greatest and least concentration* given by the eigenvectors of the following matrix $\mathbf{T}$ which correspond to the greatest and least eigenvalues of $\mathbf{T}$:

$$\mathbf{T} = \begin{pmatrix} E(X^2) & E(XY) & E(XZ) \\ E(XY) & E(Y^2) & E(YZ) \\ E(XZ) & E(YZ) & E(Z^2) \end{pmatrix}. \tag{11.0.8}$$

The eigenvalues themselves measure the concentration of the distribution about these two directions.

The estimates of these parameters from a sample $(\theta_i, \phi_i)$ $(i = 1, \ldots, n)$ are obtained correspondingly from the sample matrix

$$\hat{\mathbf{T}} = \frac{1}{n}\begin{pmatrix} \sum x_i^2 & \sum x_i y_i & \sum x_i z_i \\ \sum x_i y_i & \sum y_i^2 & \sum y_i z_i \\ \sum x_i z_i & \sum y_i z_i & \sum z_i^2 \end{pmatrix} \tag{11.0.9}$$

where $\quad x_i = \sin\theta_i \cos\phi_i, \; y_i = \sin\theta_i \sin\phi_i, \; z_i = \cos\theta_i. \tag{11.0.10}$

In the main, the basic models assumed in existing results relating to outliers in directional data are the *von Mises distribution* for vectorial data on the circle, the *Fisher distribution* for vectorial data on the sphere, and the *Watson distribution* for axial data on the sphere. In the same way that the normal distribution is used as a kind of all-purpose probability model for data on the line, the von Mises, Fisher and Watson distributions serve as all-purpose models on the circle and sphere. They are the parametric models assumed in this chapter; we shall use the following notation:

$\mathrm{VM}(\mu, \kappa)$     von Mises distribution for an angle $\Theta$ in the plane, with mean direction $\mu$ and concentration parameter $\kappa$.

Density     $f(\theta) = C_{\mathrm{VM}}\exp[\kappa\cos(\theta - \mu)], \; 0 \leqslant \theta < 2\pi \tag{11.0.11}$

where $C_{VM} = 1/[2\pi I_0(\kappa)]$, and $I_0(\kappa)$ is the Bessel function of order zero with imaginary argument.

$\mathbf{F}(\mu, v; \kappa)$    Fisher distribution for a direction $(\Theta, \Phi)$ in space with mean direction $(\mu, v)$ and concentration parameter $\kappa$.

Density    $f(\theta, \phi) = C_F \exp[\kappa\{\cos\theta\cos\mu + \sin\theta\sin\mu\cos(\phi - v)\}]\sin\theta$,
$$0 \leqslant \theta \leqslant \pi, 0 \leqslant \phi < 2\pi \qquad (11.0.12)$$

where $C_F = \kappa/[4\pi \sinh \kappa]$    (11.0.13)

$\mathbf{W}(\mu, v; \kappa)$    Watson distribution for an axis $\{(\Theta, \Phi), (\pi - \Theta, \Phi \pm \pi)\}$ in space, with location parameters $\mu, v$ and concentration parameter $\kappa$. If $\kappa > 0$ the distribution is a *Watson bipolar distribution*, with greatest concentration around the poles $(\Theta = 0, \pi$ if $\mu = 0)$ and least concentration around the equator $(\Theta = \frac{\pi}{2})$. If $\kappa < 0$ the distribution is a *Watson girdle distribution*, with greatest concentration around the equator and least concentration around the poles. (There is no such distinction with the Fisher distribution, since $\mathbf{F}(\mu, v; -\kappa)$ is merely the reflection of $\mathbf{F}(\mu, v; \kappa)$ in the origin.) The extreme case of a girdle distribution is a concentration of points on the equator; the extreme case of a bipolar distribution is a concentration of points at the North (South) pole. In the boundary case $\kappa = 0$ the random variable is, as one might expect, uniformly distributed over the surface of the sphere.

Density    $f(\theta, \phi) = C_W \exp[\kappa\{\cos\theta\cos\mu + \sin\theta\sin\mu\cos(\phi - v)\}^2]\sin\theta$,
$$0 \leqslant \theta \leqslant \pi, 0 \leqslant \phi < 2\pi \qquad (11.0.14)$$

where $1/C_W = 4\pi \int\limits_0^1 \exp(\kappa u^2)\,du.$    (11.0.15)

The von Mises distribution and the Fisher distribution are both unimodal and symmetric; this is no more and no less of a limitation on their usefulness as general-purpose models than is the case for the normal distribution, which is likewise unimodal and symmetric. The same applies to the Watson bipolar and girdle models, which both have rotational symmetry; in the bipolar case about the direction of greatest concentration, the '*principal axis*', and in the girdle case about the direction of least concentration, the '*polar axis*'.

## 11.1 OUTLIERS ON THE CIRCLE

When the data to be analysed consist of directions in the plane one might perhaps expect fewer outlier problems to arise, because on the circle there is only restricted room for an observation to outlie!

On the line, the more extreme the value $x$ of the outlier, the greater its separation from the main data mass. But on the circle, supposing a distribution effectively concentrated on part of the circumference, the outlier—as its value $\theta$ goes on increasing—sneaks back into the main data mass from the other side and passes itself off as a respectable reading. (Lewis, 1975)

However, outliers do arise in circular data. As an illustration we quote the following data relating to the homing ability of the northern cricket frog, *Acris crepitans*, given by Ferguson, Landreth, and McKeown (1967) and later discussed by Collett (1980). The angles are the homing directions taken by 14 frogs on being released after a period of enclosure in a dark environmental chamber.

$$104°, \ 110°, \ 117°, \ 121°, \ 127°, \ 130°, \ 136°,$$
$$145°, \ 152°, \ 178°, \ 184°, \ 192°, \ 200°, \ 316°.$$

Homing directions of 14 frogs in relation to due North

A diagram of the data has already been given as Figure 1.10 on page 22. Clearly the observation 316° has claims to be considered as an outlier in relation to, say, a von Mises basic model.

Discordancy tests for this situation—a single outlier in a sample of circular data—are discussed by Collett (1980). He proposes four test statistics $L$, $C$, $D$ and $M$: two ($L$ and $M$) defined specifically for a von Mises basic model (11.0.11), and two ($C$ and $D$) set up on intuitive grounds and applicable to other models. $L$ is the maximum likelihood ratio test statistic for a location slippage alternative in which $n - 1$ of the observations come from $\mathbf{VM}(\mu, \kappa)$ and one from $\mathbf{VM}(\mu^*, \kappa)$, where $\mu^* \neq \mu$. $M$ and $C$ are constructed in terms of $\hat{\rho}$, the sample mean resultant length (11.0.4) based on all $n$ observations $\theta_1, \ldots, \theta_n$, and $\hat{\rho}_i(i = 1, \ldots, n)$, the sample mean resultant length based on the $n - 1$ observations omitting $\theta_i$. The more remote $\theta_i$ is from the other $n - 1$ observations, the smaller the circular variance $1 - \hat{\rho}_i$ of the reduced sample will be compared with the circular variance $1 - \hat{\rho}$ of the full sample, and the greater will be the increase in mean resultant length from $\hat{\rho}$ to $\hat{\rho}_i$. Based on these considerations we have as possible discordancy test statistics (i) $\min_i[(1 - \hat{\rho}_i)/(1 - \hat{\rho})]$, the relative reduction in the value of the circular variance on omitting the outlier; this statistic, originally suggested by Mardia (1975b), is essentially the same as Collett's $M$; and (ii) $\max_i(\hat{\rho}_i/\hat{\rho})$, the relative increase in the value of the sample mean resultant length on omitting the outlier, which is Collett's $C$. Finally there is a Dixon-type statistic $D$. Taking the observations as points $A_1, A_2, \ldots, A_n$ in order round the circle (it being immaterial which point is chosen as $A_1$), the greatest of the arcs $A_1A_3, A_2A_4, \ldots, A_{n-1}A_1, A_nA_2$ is identified; suppose this is $A_{j-1}A_{j+1}$, containing the single observation $A_j$, which is thus identified as the extreme point in the sample. The evidence for $A_j$ being discordant is less if $A_j$ is near $A_{j-1}$ or $A_{j+1}$ greater if $A_j$ is near the middle of the arc $A_{j-1}A_{j+1}$. Accordingly $D$ is defined in terms of the ratios of the arc lengths $A_{j-1}A_j$, $A_jA_{j+1}$ as

$$D = \min(A_{j-1}A_j/A_jA_{j+1}, \; A_jA_{j+1}/A_{j-1}A_j).$$

Collett investigates the performance of the four discordancy test procedures with reference to the above location-slippage alternative; he uses as criteria the probabilities $P_5$, $P_1$ and $P_1 - P_3$ defined in Section 4.2. For larger samples ($n > 15$) and large $\kappa$ he finds that there is little to choose between the tests; the use of $M$ is then recommended since its asymptotic null distribution is directly related to that of $T_{N1}$ (Chapter 6, page 221), so that the required critical values can be obtained directly from those of $T_{N1}$ (given in this book in Table XIIIa, page 485). For small samples and/or moderate or small $\kappa$, $C$ and $D$ perform better than $M$ or $L$, and either is recommended for practical use. Collett gives tables of simulated 5 per cent and 1 per cent critical values for the $C$- and $D$-tests for $n = 5(1)10(2)20(5)40$ and $\kappa = 2(1)5(2.5)10$.

Mirvaliev (1978a, 1983) considers the treatment of outlying values in data sets consisting of replicated geodetic angular measurements $x_{ijk}$ of angles $A_ioA_j$ between directions $oA_1, \ldots, oA_n$ in the plane. Although the first paper is entitled 'Rejection of outlying results of angular measurements' the methods, based on analysis of variance, are linear not directional, in the spirit of Gauss's work on small angles referred to in the quotation from Mardia (1972) at the beginning of this chapter. In the data sets analysed by Mirvaliev, none of the angular observations exceeds $2°$.

We turn now to *accommodation* of outliers in data on the circle. Collett (1978) proposes an estimator $\tilde{\kappa}$ of the concentration parameter $\kappa$ for a von Mises $\mathbf{VM}(\mu, \kappa)$ sample which is robust against an outlier near the 'anti-mode' $\mu \pm \pi$. He constructs this by assuming that the observations come from a mixture $(1 - \lambda)\mathbf{VM}(\mu, \kappa) + \lambda\mathbf{VM}(\mu \pm \pi, \kappa)$ of two von Mises distributions, i.e. a distribution whose density is, from (11.0.11),

$$f(\theta) = C_{VM}\{(1 - \lambda)\exp[\kappa \cos(\theta - \mu)] + \lambda \exp[-\kappa \cos(\theta - \mu)]\}$$

where $\lambda \in (0, 1)$ is a contamination parameter. This leads to a robust estimator $\tilde{\kappa}$, based on the angles $\theta_i^* \equiv \theta_i(\mathrm{mod}\ \pi)$, which is independent of the contamination parameter $\lambda$. It is evaluated by solving (cf. (11.0.4))

$$\frac{1}{n}\sum_{i=1}^{n}\cos 2\theta_i^* = \tilde{\rho} \cos \tilde{\mu}, \quad \frac{1}{n}\sum_{i=1}^{n}\sin 2\theta_i^* = \tilde{\rho} \sin \tilde{\mu},$$

and

$$I_2(\tilde{\kappa})/I_0(\tilde{\kappa}) = \tilde{\rho} \tag{11.1.1}$$

where $I_0(\kappa)$, $I_2(\kappa)$ are the Bessel functions with imaginary argument of orders zero and two.

Lenth (1981) and Ducharme and Milasevic (1987) investigate procedures for the robust estimation of location, i.e. mean direction $\mu$. Lenth proposes the use of $M$-estimation; corresponding to equation (5.1.12)

$$\sum_{i=1}^{n} \psi(x_j - T_n) = 0$$

and its equivalent formulation

$$\sum_{i=1}^{n} \rho(x_j - T_n) = \text{minimum}$$

for a general $M$-estimator $T_n$ of location on the line, he introduces an $M$-estimator $M_n$ of location on the circle, given by

$$\sum_{i=1}^{n} \rho\{t(\theta_j - M_n)\} = \text{minimum.} \tag{11.1.2}$$

For the function $t(\phi)$, which has to be periodic, he recommends choosing

$$t(\phi) = [2\kappa(1 - \cos\phi)]^{1/2}\text{sgn}(\sin\phi). \tag{11.1.3}$$

If, as in most practical cases, $\kappa$ is not known, a robust estimator $\tilde{\kappa}$ must be used in (11.1.2), (11.1.3), analogously to (5.1.15). For this purpose Lenth suggests a weighted version of the maximum likelihood estimator $\hat{\kappa}$ (see Mardia, 1972); Collett's estimator (11.1.1) could also be used.

Particular $M$-estimators, $M_n$, investigated by Lenth include Huber estimators with $\psi(u)$ given by

$$\psi(\mu) = \begin{cases} u & |u| \leqslant c \\ c\,\text{sgn}\,u & |u| > c \end{cases} \tag{11.1.4}$$

as in (5.1.17), and Andrews estimators with $\psi(u)$ as in (5.1.20). The performances of these estimators, with the Huber $c$ and Andrews $d$ parameters taking values 1.2, 1.5, 1.8, and of the sample mean direction and other estimators, are assessed in relation to various alternatives by a small-scale simulation study based on 250 random samples of size 20 from each distribution; the alternatives include, along with several intrinsic models such as the wrapped Cauchy, a number of contamination alternatives constructed by mixing $\text{VM}(\mu, \kappa)$ with probability $1 - \lambda$ and the circular uniform distribution with probability $\lambda$. Lenth's conclusion from this pilot study is that the Huber and Andrews $M$-estimators of location give good protection against possible outliers, at the cost of a tolerably small premium.

Ducharme and Milasevic (1987) consider three estimators of location: the circular median (see, e.g., Mardia (1972) or Fisher (1993)), the sample mean direction (11.0.4), and an estimator proposed by Watson (1983) based on the eigenvalues of the $2 \times 2$ matrix analogous to (11.0.9). They compare the asymptotic variances of these estimators for uncontaminated and for contaminated von Mises distributions; their contamination model is the mixture distribution

$$(1 - \lambda)\mathrm{VM}(\mu, \kappa) + \lambda\mathrm{VM}(\mu, \beta\kappa).$$

In a pilot study they take $\beta = 0.01$ (so that the contaminating distribution is nearly uniform); $\lambda = 0.1$ and $0.2$; and $\kappa = 0, 0.1, 0.5, 1, 2, 3, 5, 10, 20$. They find that, for these values of the parameters, the circular mean direction is the most efficient of the three estimators for $\kappa \leqslant 5$ but that the circular median (with never less than 80 per cent relative efficiency) rapidly overtakes it with regard to efficiency as $\kappa$ increases (e.g., the relative asymptotic efficiency of median versus mean at $\kappa = 20$ is 1.22, 1.77 for $\lambda = 0.1, 0.2$ respectively). Watson's estimator is less efficient than the circular mean direction in all the cases covered by the pilot study, and less efficient than the circular median in nearly all the cases.

For an investigation of the robustness properties of the circular mean and circular median in terms of their influence curves, see Wehrly and Shine (1981).

To conclude this account of the treatment of outliers on the circle, Bagchi and Guttman (1990) present a Bayesian analysis of data from a von Mises distribution $\mathrm{VM}(\mu, \kappa)$ which may be contaminated by $k( \geqslant 1)$ observations arising from a location-slipped von Mises distribution $\mathrm{VM}(\mu + \nu, \kappa)$. The analysis provides detection of what the authors call *spurious* observations and hence, in particular, provides an estimate of the number $k$. Both for the case of known and unknown concentration parameter $\kappa$, uniform prior distributions for $\mu$, $\nu$ are assumed, with $\mu$, $\nu$ independent; in the case of unknown $\kappa$, a non-informative prior is assumed for $\kappa$. Two data sets are analysed, one being the homing directions of 14 frogs discussed above (and fast becoming a 'classic' data set, like R. A. Fisher's iris versicolor and iris setosa data or Beveridge's wheat price index series!)

## 11.2  OUTLIERS ON THE SPHERE

### 11.2.1  Vectorial data

Table 11.1 shows a sample of 17 directional measurements $(\theta_i, \phi_i)$ of magnetic remanence in red-beds (Embleton and Giddings, 1974). It is not immediately apparent from an inspection of such a table whether any of the observations can be regarded as outliers, and if so which. From this point of view the data are multivariate and thus, as discussed in Section 2.2.3, *graphical methods* of highlighting outliers come into their own. Such a method is given by Lewis and Fisher (1982) for investigating the fit of a Fisher distribution to a spherical sample. In their procedure three quantile plots are drawn, reflecting respectively the marginal distribution of colatitude and longitude with respect to the mean direction and the independence or otherwise of these two variables. They apply their method to the data in Table 11.1; inspection of the quantile plots, separately and in relation to one another, reveals clearly that the two observations $(\theta_6, \phi_6)$ and $(\theta_7, \phi_7)$ are outliers. These can then be formally tested for discordancy by one of the procedures described next.

Six possible discordancy test statistics $C, E, D, G, X, Y$ for testing an outlying observation in a sample $(\theta_i, \phi_i)$ $(i = 1, \ldots, n)$ from a Fisher distribution (11.0.12) are considered by Fisher, Lewis, and Willcox (1981). As in our discussion of discordancy tests for outliers on the circle in Section 11.1, we define $\hat{\rho}_i$ to be the sample mean resultant length (11.0.7) based on the $n - 1$ observed directions omitting $(\theta_i, \phi_i)$, and we note that, the more remote $(\theta_i, \phi_i)$ is from the other $n - 1$ observations, the greater will be the increase in mean resultant length from the value $\hat{\rho}$ (based on the complete sample) to the value $\hat{\rho}_i$ . As in Collett (1980), these considerations motivate the following statistics:

$$C = \max_i(\hat{\rho}_i/\hat{\rho}), \text{ the analogue of Collett's } C; \qquad (11.2.1)$$

$$E = \min_i[(1 - \hat{\rho}_i)/(1 - \hat{\rho})], \text{ the analogue of Collett's } M. \qquad (11.2.2)$$

Fisher, Lewis, and Willcox present these in different forms related to (11.2.1) and (11.2.2) by straightforword transformations. Their $D$ is constructed as the maximum likelihood ratio test statistic for a dispersion slippage alternative in which $n - 1$ of the observations come from $F(\mu, v; \kappa)$ and one from $F(\mu, v; \kappa^*)$ where $\kappa^* < \kappa$. $D$ is somewhat complicated in form, and we do not give details since it is shown to perform poorly relatively to the others.

**Table 11.1**

| $i$ | $\theta_i$ | $\phi_i$ | $i$ | $\theta_i$ | $\phi_i$ | $i$ | $\theta_i$ | $\phi_i$ |
|---|---|---|---|---|---|---|---|---|
| 1 | 79 | 294 | 7 | 76 | 335 | 13 | 69 | 297 |
| 2 | 59 | 287 | 8 | 54 | 318 | 14 | 66 | 297 |
| 3 | 52 | 284 | 9 | 54 | 309 | 15 | 57 | 296 |
| 4 | 64 | 282 | 10 | 66 | 308 | 16 | 60 | 279 |
| 5 | 32 | 277 | 11 | 56 | 304 | 17 | 47 | 279 |
| 6 | 40 | 238 | 12 | 55 | 299 | | | |

The other three test statistics $G, X$ and $Y$ exploit the fact that, if $(\Theta, \Phi)$ is distributed as $F(\mu, v; \kappa)$ and $\Theta'$ is the colatitude measured with respect to the mean direction $(\mu, v)$, then $C' = 1 - \cos \Theta'$ has a truncated

426 OUTLIERS IN STATISTICAL DATA

exponential distribution with parameter $\kappa$, the truncation being negligible unless $\kappa$ is less than say 2. So if $\theta_i'(i = 1, \ldots, n)$ are the sample colatitudes measured with reference to the sample mean direction $(\hat{\mu}, \hat{\nu})$ (11.0.7), the $n$ quantities $c_i' = 1 - \cos \theta_i'$ can be regarded as exponentially distributed with parameter $\kappa$ on the null hypothesis, and discordancy tests for exponential samples (Chapter 6) can be applied. Writing the ordered $\theta_i'$s as $\theta_{(1)}' < \theta_{(2)}' < \ldots < \theta_{(n)}'$ and the ordered $c_i'$s as $c_{(1)}' < c_{(2)}' < \ldots < c_{(n)}'$, they accordingly define the test statistics

$$G = [\pi - \theta_{(n)}']/[\theta_{(n)}' - \theta_{(n-1)}'] \tag{11.2.3}$$

$$X = c_{(n)}' / \sum_{i=1}^{n} c_i' \tag{11.2.4}$$

$$Y = [c_{(n)}' - c_{(n-1)}']/c_{(n)}' . \tag{11.2.5}$$

$X$ and $Y$ are the same as the test statistics $T_{Ea1}$ and $T_{Ea2}$ in Chapter 6 (pages 197 and 198 respectively). $G$ uses the data in a similar way to Collett's $D$ in Section 11.1, i.e. as a ratio of arcs.

Fisher, Lewis, and Willcox evaluate the relative performances of the six tests by a large-scale simulation study, using $P_1$, $P_3$ and $P_5$ (Section 4.2) as performance criteria. They find that $C$ and $G$ are the best of the six tests, and that $Y$ and $D$ are the worst. To apply the $C$- or $G$-tests, however, one needs to know the value of $\kappa$ or to have a good estimate of it; for practical purposes, therefore, they recommend the use of the $E$-test (11.2.2), and give what amounts to the following formula from which the significance probability $SP(t)$ can be calculated to good approximation:

$$SP(t) = P(E < t)$$

$$\simeq [(n - 1)t]^{n-2}/n^{n-3}$$

$$- I[t > n/(2n - 2)](n - 1)\{[(2n - 2)t/n] - 1\}^{n-2} \tag{11.2.6}$$

Here $I$ is the indicator function, taking the values 0 or 1 according as $t \le$ or $> n/(2n - 2)$.

They also show how to adapt their single-outlier methods to the *block* testing of multiple outliers in a sample from a Fisher distribution, and they confirm by means of a block test based on a generalized form of $E$ in (11.2.2) that the observations $(\theta_6, \phi_6)$ and $(\theta_7, \phi_7)$ in Table 11.1 do in fact give evidence of discordancy: the significance probability being about $2\frac{1}{2}$ per cent. Kimber (1985) compares the use of block testing and consecutive testing of multiple outliers in Fisher samples, using the same data set (Table 11.1) for purposes of illustration.

We turn next to an accommodation problem, viz. the estimation of the concentration parameter $\kappa$ for a sample from a Fisher distribution (11.0.12) by procedures robust against outliers. This is discussed by Fisher (1982), who proposes two robust estimators. These are based on the ordered quantities $c'_{(i)}$ defined above, where $c'_i = 1 - \cos \theta'_i$ and $\theta'_i$ is the colatitude of $(\theta_i, \phi_i)$ measured with reference to the sample mean direction $(\hat{\mu}, \hat{\nu})$. If there are no contaminants in the sample (and assuming $\kappa$ is not very small), $\{c'_i\}$ $(i = 1, \ldots, n)$ can, as stated above, be regarded as a sample from an exponential distribution with mean $1/\kappa$. Fisher accordingly uses Chikka-goudar and Kunchur's (1980) robust $L$-estimator (5.5.6), (5.5.7) for an exponential mean and obtains for $\kappa$ the robust estimator

$$\hat{\kappa}_R = 1/\sum_{j=1}^{n} a_j c'_{(j)} \qquad (11.2.7)$$

where the $a_j$ are given by (5.5.7).

If it is decided to regard $s$ upper outliers in the sample $\{c'_i\}$ as discordant, he suggests the use of the one-sided Winsorized mean $\overset{w}{c}_{0,s}$ defined in (3.2.2), to give a robust estimator

$$\hat{\kappa}_{W,s} = (n - s + 1)/(n\overset{w}{c}_{0,s}). \qquad (11.2.8)$$

To exemplify the methods, estimates of $\kappa$ are given by Fisher for three data sets from Fisher, Lewis, and Willcox (1981). These results suggest that $\hat{\kappa}_R$ is preferable to $\hat{\kappa}_{W,1}$, which has a pronounced positive bias; $\hat{\kappa}_R$ also has the advantage of being free from any *post hoc* choice of the Winsorizing parameter $s$.

Reviewing Fisher's proposals, Kimber (1985) investigates the performance of $\hat{\kappa}_R$, $\hat{\kappa}_{W,1}$, $\hat{\kappa}_{W,2}$, and of corresponding approximately unbiased versions; in particular

$$\tilde{\kappa}_{W,s} = \{(n - s - 1)/(n - s + 1)\}\hat{\kappa}_{W,s} \qquad (11.2.9)$$

which has expected value $\kappa$ *when the distribution of the $c'_i$ is exactly exponential*. For each estimator, and for samples of sizes $n = 10, 20, 100$ from $F(0, 0; \kappa)$ with various values of $\kappa \geq 5$, the mean, median and MSE of the estimator are estimated in each case from 1000 simulated samples, (a) when the samples are uncontaminated, (b) when they are contaminated, either 1 or 2 of the $n$ observations being replaced by observations with slipped values of $\theta$ representing both mild and gross contamination. Kimber finds that $\hat{\kappa}_R$ performs poorly in the presence of gross errors, and no better than the Winsorized estimators in the presence of mild contamination. He notes that the estimator $(n - 1)^{3/2}/(n^{1/2}\sum c'_i)$, which is also considered by Fisher (1982), performs well provided that contamination is only mild; his conclusion is, however, that

if large outliers may be present, then only the Winsorized estimate, for suitably chosen [$s$], gives adequate protection. In the routine approach, a sensible choice is to use this with [$s$] = $k$, the maximum number of outliers envisaged . . . .

Turning to a regression context (see Chapter 8), Rivest (1989) discusses the treatment of outliers in regression on the sphere, when the random error attaching to each dependent observation $(\theta_i, \phi_i)$ is a value from a Fisher distribution (cf. ordinary regression on the line, when the random errors are assumed normally distributed). He gives a test for discordancy of a single outlying pair of directions, and presents an analysis of a geophysical data set (Chang, 1986) of 11 points (i.e. paired directions), two of which are possible outliers.

## 11.2.2 Axial data

As with Fisher samples (Section 11.2.1), outliers in Watson samples can be highlighted by graphical methods. In a procedure given by Best and Fisher (1986) for investigating the fit of a Watson distribution to a spherical axial sample, two quantile plots are drawn, reflecting respectively the marginal distributions of colatitude and longitude with respect to the axis of rotational symmetry. Suppose that

$$\Theta'\left(0 \leqslant \Theta' \leqslant \frac{\pi}{2}\right)$$

is the angle between $(\Theta, \Phi)$ and the axis of rotational symmetry, i.e. the principal axis for a bipolar distribution and the polar axis for a girdle distribution.

For a Watson bipolar distribution $(\kappa > 0)$ the distribution of $B' = 1 - \cos^2\Theta'$ is approximately exponential with parameter $\kappa$ provided $\kappa$ exceeds, say, 8. Given, therefore, a Watson bipolar sample $(\theta_i, \phi_i)$ $(i = 1, \ldots, n)$ for which $\kappa$ is not too small, one calculates the angles $\theta_i'$ between $(\theta_i, \phi_i)$ and the estimated principal axis; the corresponding longitudes $\phi_i'$; and $b'_{(1)} < b'_{(2)} < \ldots b'_{(n)}$, the ordered values of $b_i' = 1 - \cos^2\theta_i'$. An exponential quantile plot of the $b'_{(i)}$ is the required colatitude plot, and a uniform quantile plot of the ordered $\phi'_{(i)}$ is the longitude plot.

For a Watson girdle distribution $(\kappa < 0)$ the distribution of $\Theta'$ is, of course, different from the bipolar case. We now have that $G' = \cos^2\Theta'$ is distributed approximately as $(1/2|\kappa|)\chi_1^2$, provided that $|\kappa| (= -\kappa)$ exceeds, say, 3. Given a Watson girdle sample $(\theta_i, \phi_i)$ $(i = 1, \ldots, n)$, the procedure is to calculate the angles $\theta_i'$ between $(\theta_i, \phi_i)$ and the estimated polar axis; the corresponding longitudes $\phi_i'$; and $g'_{(1)} < g'_{(2)} < \ldots < g'_{(n)}$, the ordered values of $g_i' = \cos^2\theta_i'$. The colatitude plot is now a plot of the $g'_{(i)}$ against the quantiles of $\chi_1^2$, while the longitude plot is as before a plot of the ordered values $\phi'_{(i)}$ against uniform quantiles. Best and Fisher (1986), by way of illustration, show these plots for a sample of obviously girdle form, consisting of 40 poles to axial-plane cleavage surfaces; an outlier is manifestly visible.

Discordancy tests of a single outlier in a Watson bipolar sample and in a Watson girdle sample are described in Best and Fisher (1986) and in

Fisher, Lewis and Embleton (1987, 1993); likewise block discordancy tests of more than one outlier. Denote the eigenvalues of the matrix $\hat{T}$ in (11.0.9), based on all the $n$ axes $(\theta_i, \phi_i)$ or $(x_i, y_i, z_i)$, by $\hat{\tau}_1 \leqslant \hat{\tau}_2 \leqslant \hat{\tau}_3$. Suppose $k(\geqslant 1)$ outlying axes are under test; let $\hat{T}^*$ be the corresponding matrix computed from $n - k$ axes after omitting the $k$ axes under test, and denote its eigenvalues by $\hat{\tau}_1^* \leqslant \hat{\tau}_2^* \leqslant \hat{\tau}_3^*$. The discordancy test statistics are functions of $\hat{\tau}_3$ and $\hat{\tau}_3^*$ in the bipolar case, and of $\hat{\tau}_1$ and $\hat{\tau}_1^*$ in the girdle case.

As regards accommodation of outliers in Watson samples, the estimate of location (principal axis or polar axis) given above (see 11.0.9) is likely to be reasonably robust against the presence of one or two outliers. On the other hand, the effect of outliers on the estimation of $\kappa$ may be substantial. The maximum likelihood estimator of $\kappa$ for an uncontaminated sample, $\hat{\kappa}$ say, is given by

$$D(\hat{\kappa}) = \hat{\tau}_3 \text{ (bipolar case)}, \quad \hat{\tau}_1 \text{ (girdle case)}, \qquad (11.2.10)$$

where $D(x)$ is a tabulated function defined by

$$D(x) = \int_0^1 u^2 \exp(xu^2) du / \int_0^1 \exp(xu^2) du.$$

A robust procedure for estimating $\kappa$ is presented in Fisher, Lewis, and Embleton (1987, 1993). The number, $k$, of outlying observations which it is desired to accommodate is first chosen. In the bipolar case, $\hat{\tau}_3$ in (11.2.10) is then replaced by the $(0, k)$ Winsorized mean of $b'_{(1)}, b'_{(2)}, \ldots, b'_{(n)}$; in the girdle case, $\hat{\tau}_1$ in (11.2.10) is replaced by the $(0, k)$ Winsorized mean of $g'_{(1)}, g'_{(2)}, \ldots, g'_{(n)}$.

# CHAPTER 12

# Some Little-explored Areas: Contingency Tables and Sample Surveys

## 12.1 OUTLIERS IN CONTINGENCY TABLES

The problem of outliers in cross-classified data can often be serious. Nelder (1971) [actually 1972] points out that 1 per cent of gross errors in a complex cross-classification can result in a false inference . . . [The] major problem is detection of such outliers rather than the subsequent outlier-resistant analysis. (Bradu and Hawkins, 1982)

In the particular context of contingency tables,

. . . we want to detect cell entries that deviate extremely from results based on independence. This requires a scale for measuring deviance. . . . [We] can use a variety of scales. (Mosteller and Parunak, 1985)

**Table 12.1**

| Factor $A$ (rows) | | Factor $B$ (columns) | |
|---|---|---|---|
| | | $B_1 \ldots B_j \ldots B_c$ | |
| | $A_1$ | $x_{11} \ldots x_{1j} \ldots x_{1c}$ | $R_1$ |
| | $\cdot\,\cdot$ | $\ldots\ldots\ldots\ldots\ldots$ | $\cdot\,\cdot$ |
| | $A_i$ | $x_{i1} \ldots x_{ij} \ldots x_{ic}$ | $R_i$ |
| | $\cdot\,\cdot$ | $\ldots\ldots\ldots\ldots\ldots$ | $\cdot\,\cdot$ |
| | $A_r$ | $x_{r1} \ldots x_{rj} \ldots x_{rc}$ | $R_r$ |
| | | $C_1 \ldots C_j \ldots C_c$ | $N$ |

Suppose, to fix ideas, that we have a two-way $r \times c$ contingency table giving the frequencies $x_{ij}$ ($i = 1, \ldots, r; j = 1, \ldots, c$) with which the different levels of two factors are encountered in a set of data. We use the notation shown in Table 12.1. The row factor $A$ has levels $A_1, \ldots, A_r$ and the column factor

$B$ has levels $B_1, \ldots, B_c$; $x_{ij}$ is the number of observations in a sample of size $N$ at joint factor level $(A_i, B_j)$, and is shown in cell $(i, j)$; $R_i (i = 1, \ldots, r)$ and $C_j = (j = 1, \ldots, c)$ are the marginal totals, each summing to $N$.

If the factor levels were ordered, either in terms of quantitative measures or qualitative (e.g. 'good', 'passable', 'poor'), the idea of a multivariate outlier (see Chapter 7) as an observation on the periphery of a data cloud would carry over to contingency table data in a natural way. We might then think of a contingency table as being a grouped version of a scatter diagram. From this point of view we could expect outliers to be reflected in enhancement of the frequencies in sparse cells around the edges of the table. But these are likely to be difficult to detect, being exhibited typically in increases of perhaps only one in frequencies within cells where we might expect frequencies to be very small. Indeed, few proposals seem to have been made to deal with outliers in this spirit; some comments are made by Barnett (1992):

It seems to be by no means a trivial problem to devise methods for identifying and testing the discordancy of outliers within such *grouped* data sets (whether of one, two or more dimensions) and perhaps others will find this a topic worthy of fuller study.

Published methods for investigating outliers in contingency tables proceed in the main from a different standpoint, exemplified by the above-quoted comment by Mosteller and Parunak (1985). In the general context, an outlier in a set of data is an observation which appears inconsistent with an assumed model for the data. In the context of a contingency table, such as Table 12.1, the usual assumed null model is of independence between the row and column classifications; and an *outlier* is a particular frequency which deviates markedly from the corresponding expected frequency on the null model. (It is often convenient to think of the relevant cell as the outlier, and we shall use the terms 'outlying frequency $x_{ij}$' and 'outlying cell $(i, j)$' as equivalent.)

Denote by $S$ the entire set of $rc$ cells in the contingency table; by $Q$ a subset of $k$ cells $(k \geq 1)$ to be examined for possible discordancy; and by $R$ the subset of $rc - k$ cells not under question, i.e. which exclude $Q$. Write $p_{ij}$ for the probability that an observation falls in the joint category $(A_i, B_j)$. The *independence model* for the data is

$$H_0: \quad p_{ij} = \alpha_i \beta_j \quad \left( \sum_i \alpha_i = \sum_j \beta_j = 1 \right) \quad \text{for all } (i, j) \in S. \quad (12.1.1)$$

The alternative model is

$$H_1: \quad p_{ij} = \alpha_i \beta_j \quad \left( \sum_{i,j} \alpha_i \beta_j = 1 \right) \quad \text{for all } (i, j) \in R; \quad (12.1.2)$$

(this is called a *quasi-independence* model; a term dating back at least to Goodman, 1968).

Suppose that an initial test indicates rejection of the independence hypothesis $H_0$. The problem is then to identify a 'most outlying' cell or set of cells whose removal will lead to acceptance of the corresponding quasi-independence model $H_1$. As with data situations surveyed in earlier chapters (univariate and multivariate samples, linear models, etc.), multiple outliers can be considered *en bloc* or *consecutively*. We first discuss detection procedures for a single cell, and then the consecutive application of these procedures to the identification of a possible discordant subset of $k$ cells ($k \geqslant 1$).

There are various criteria for identifying a most outlying single frequency, say $x_{IJ}$ in cell ($I$, $J$), and assessing its discordancy. We consider the five following criteria in particular:

- the standardized residual $r_{IJ}$;
- the adjusted residual $\tilde{r}_{IJ}$;
- the deleted residual $r_{IJ}^*$;
- the reduction in the Pearson goodness of fit statistic on omitting the queried cell, $X^2 - X_{IJ}^2$;
- the reduction in the likelihood ratio goodness of fit statistic on omitting the queried cell, $G^2 - G_{IJ}^2$.

The *standardized residual* is

$$r_{IJ} = (x_{IJ} - e_{IJ})/\sqrt{(e_{IJ})}, \qquad (12.1.3)$$

where $e_{IJ}$ is the maximum likelihood estimator under $H_0$ of $Np_{IJ} = N\alpha_I\beta_J$, the expected value of cell entry ($I$, $J$). With a multinomial sampling model, we have

$$e_{IJ} = R_I C_J / N \qquad (12.1.4)$$

The estimated asymptotic variance of $r_{IJ}$ is

$$\hat{v}_{IJ} = (1 - R_I/N)(1 - C_J/N) \qquad (12.1.5)$$

(see, e.g. Haberman, 1973). This motivates the use, as our second criterion, of the *adjusted residual*

$$\tilde{r}_{IJ} = r_{IJ}/\sqrt{(\hat{v}_{IJ})}$$

$$= (x_{IJ} - R_I C_J/N)/\sqrt{(R_I C_J(N - R_I)(N - C_J)/N^3)} \qquad (12.1.6)$$

Consider now the quasi-independence model (12.1.2) with a single cell ($I$, $J$) excluded. We write the $I^{\text{th}}$ row total, $J^{\text{th}}$ column total and grand total for the *retained* data (subset $R$) as

$$R_I^* = R_I - x_{IJ}, \; C_J^* = C_J - x_{IJ}, \; \text{and} \; N^* = N - x_{IJ}.$$

The log likelihood is

$$\sum_{i \neq I} R_i \log \alpha_i + R_I^* \log \alpha_I + \sum_{j \neq J} C_j \log \beta_j + C_J^* \log \beta_J. \qquad (12.1.7)$$

Write $e_{ij}^*$ for the maximum likelihood estimator under $H_1$ of $N^* \alpha_i \beta_j$, the expected value of cell entry $(i, j)$. Maximizing (12.1.7) with respect to $\alpha$ and $\beta$, subject to

$$\sum_{i=1}^{r} \alpha_i \sum_{j=1}^{c} \beta_j - \alpha_I \beta_J = 1,$$

we find the following expressions for $e_{ij}^*$ (Brown, 1974):

$$e_{ij}^*(i \neq I, j \neq J) = R_i C_j (N^* - R_I^* - C_J^*)/(N^* - R_I^*)(N^* - C_J^*)$$
$$e_{iJ}^*(i \neq I) = R_i C_J^*/(N^* - R_I^*) \qquad (12.1.8)$$
$$e_{Ij}^*(j \neq J) = R_I^* C_j/(N^* - C_J^*)$$

and finally

$$e_{IJ}^* = R_I^* C_J^*/(N^* - R_I^* - C_J^*), \qquad (12.1.9)$$

which can be viewed as the fitted value for cell $(I, J)$ if the observation were missing. Its difference from the observation $x_{IJ}$ provides an obvious basis for a detection criterion. For this purpose we define, analogously to $r_{IJ}$ in (12.1.3), the so-called (Simonoff, 1988b) *deleted residual*

$$r_{IJ}^* = (x_{IJ} - e_{IJ}^*)/\sqrt{(e_{IJ}^*)} \qquad (12.1.10)$$

(rather a misnomer, actually, since it is the observation which is deleted, not the residual). From (12.1.9),

$$e_{IJ}^* = (R_I - x_{IJ})(C_J - x_{IJ})/(N - R_I - C_J + x_{IJ}). \qquad (12.1.11)$$

The usual $\chi^2$ statistics (in their Pearson and likelihood ratio forms) for testing independence provide two more criteria. The Pearson statistic under the independence model (12.1.1) can be written, from (12.1.3),

$$X^2 = \sum_{i,j} r_{ij}^2 \quad (i = 1, \ldots, r; \; j = 1, \ldots, c). \qquad (12.1.12)$$

Under the quasi-independence model (12.1.2) with exclusion of the single cell $(I, J)$, its value is reduced to

$$X_{IJ}^2 = \sum_{i,j} (x_{ij} - e_{ij}^*)^2/e_{ij}^* \quad (i = 1, \ldots, r; \; j = 1, \ldots, c; \; (i, j) \neq (I, J)). \qquad (12.1.13)$$

This can be calculated from (12.1.8) and (12.1.9), and hence we obtain the *reduction* $X^2 - X_{IJ}^2$ on omitting *cell* $(I, J)$. An explicit expression for this reduction is given by Brown (1974; equation (6), p. 407).

Alternatively we can base the reduction on the likelihood ratio statistic, whose value under the independence model (12.1.1) is

$$G^2 = 2 \sum_{i,j} x_{ij} \log(x_{ij}/e_{ij}) \quad (i = 1, \ldots, r; \; j = 1, \ldots, c). \quad (12.1.14)$$

Under the quasi-independence model with cell $(I, J)$ excluded, its value is

$$G_{IJ}^2 = 2 \sum_{i,j} x_{ij} \log(x_{ij}/e_{ij}^*) \quad ((i = 1, \ldots, r; j = 1, \ldots, c); \; (i,j) \neq (I, J)). \quad (12.1.15)$$

From (12.1.8), (12.1.9), (12.1.14) and (12.1.15), the *reduction* $G^2 - G_{IJ}^2$ on *omitting cell* $(I, J)$ can be calculated.

Any of the five above criteria $r_{IJ}$, $\tilde{r}_{IJ}$, $r_{IJ}^*$, $X^2 - X_{IJ}^2$, $G^2 - G_{IJ}^2$ can be used to *identify* the most outlying cell. To *test* this cell for discordancy, the use of the adjusted residual $\tilde{r}_{IJ}$, the reduction $X^2 - X_{IJ}^2$, or the reduction $G^2 - G_{IJ}^2$ is recommended. The approximate null distributions of *random* values of each of these three statistics are respectively $N(0, 1)$, $\chi_1^2$, $\chi_1^2$. The null distribution of the deleted residual $r_{IJ}^*$ does not appear to be known (Simonoff, 1988b). We require here to test the *greatest* out of $rc$ values, and accordingly use Bonferroni bounds as described in Section 4.1.5. See Fuchs and Kenett (1980) for a detailed discussion of the test using adjusted residuals with Bonferroni-based critical values.

We now generalize to the situation in which an identified subset $Q$ of $k$ cells has been declared discordant, and a further subset of $k'$ cells in the subset $R$ of $rc - k$ cells is to be examined. This provides the basis for the location and testing of multiple outliers, both *consecutively* $(k \geq 1, k' = 1)$ and *en bloc* $(k = 0, k' > 1)$. Our detailed discussion above relates to the case $k = 0, k' = 1$.

The null model of quasi-independence is now

$$H_0' : \; p_{ij} = \alpha_i \beta_j \left( \sum_{i,j} \alpha_i \beta_j = 1 \right) \text{ for all } (i, j) \in R. \quad (12.1.16)$$

Write $N'$ for the total frequency of cells in $R$. We require the maximum likelihood estimator under $H_0'$, $e_{ij}'$ say, of $N' \alpha_i \beta_j$, the expected value of cell entry $(i, j)$. A similar calculation will also give us $e_{ij}^{*'}$, the corresponding estimator based on the $rc - k - k'$ cells remaining when the subset of $k'$ is excluded from $R$. The values of the standardized, adjusted and deleted residuals can then be obtained for consecutive testing in the case $k' = 1$; and the values of the reductions in $X^2$ and $G^2$ can be obtained for consecutive testing $(k' = 1)$ or block testing $(k' > 1)$.

The quasi-independence model $H_0'$ in (12.1.16) can be written

$$p_{ij} = \delta_{ij}\alpha_i\beta_j \left( \sum_{i,j}\delta_{ij}\alpha_i\beta_j = 1; \ i = 1, \ldots, r; \ j = 1, \ldots, c \right) \quad (12.1.17)$$

where $\delta_{ij} = 0$ for each of the $k$ cells in $Q$ and 1 for each of the $rc - k$ cells in $R$. If these $rc - k$ cells have row totals $R_i'$, column totals $C_j'$ and grand total $N'$, maximization of the log likelihood $\sum_{i=1}^{r} R_i' \log \alpha_i + \sum_{i=1}^{c} C_j' \log \beta_j$ subject to $\sum_{i,j}\delta_{ij}\alpha_i\beta_j = 1$ leads to the following equations for the estimators $\hat{\alpha}_i$, $\hat{\beta}_j$ :

$$\hat{\alpha}_i \sum_j \delta_{ij}\hat{\beta}_j = R_i' / N' \quad (12.1.18)$$

$$\hat{\beta}_j \sum_i \delta_{ij}\hat{\alpha}_i = C_j' / N' \quad (12.1.19)$$

These are solved iteratively, starting say with $\hat{\beta}_j^{(0)} = 1/c$, all $j$, whence successively we obtain $\hat{\alpha}_i^{(1)}$ from (12.1.18), $\hat{\beta}_j^{(1)}$ from (12.1.19), $\hat{\alpha}_i^{(2)}$ from (12.1.18), and so on until convergence. The values of the maximum likelihood estimators $e_{ij}' = N'\hat{\alpha}_i\hat{\beta}_j$ now follow.

Other iterative procedures for fitting the quasi-independence model (12.1.16) are given by Goodman (1968, pp. 1118–1120), Fienberg (1969, pp. 167–168) and Brown (1974, p. 412). See also Plackett (1981).

We have now reviewed a choice of statistics for *detecting* extreme (outlying) cells and/or for *testing* them for discordancy. What procedures are to be recommended?

Brown (1974) applies $|r_{IJ}|$, $|\tilde{r}_{IJ}|$, $|r_{IJ}^*|$ and $X^2 - X_{IJ}^2$ to the reanalysis of a $14 \times 14$ table earlier analysed by Fienberg (1969). (This data set, published originally by Pearson (1904), gives the frequencies of different combinations of father's occupation and son's occupation for 775 father/son pairs. It would be natural to expect some outlying frequencies along the diagonal.) Brown identifies cells consecutively, of course on an *inward* basis (see page 128 above), i.e. from most extreme to less extreme. Identification continues until a subset $R$ is reached for which the value of $X_Q^2$ is non-significant; here $X_Q^2$ denotes the Pearson $\chi^2$ statistic which tests the quasi-independence model (12.1.16). As Simonoff (1988b) observes,

this stopping rule increases the chance that an outlier will not be tested, since the omnibus nature of the overall chi-squared test for quasi-independence means that it has lower power to detect an outlier cell than a test specifically designed for that purpose.

Brown (1974) states that, in the case of a single discordant frequency ($k' = 1$), it is always correctly identified by criteria $|\tilde{r}_{IJ}|$ and $X^2 - X_{IJ}^2$, but not infallibly by $|r_{IJ}|$ or $|r_{IJ}^*|$. See, however, the discussion below of the multiple outlier case.

Quoting further from Brown (1974),

The correct identification of a single cell, when it is the only cell that does not fit a quasi-independent model, is a far simpler problem than identifying a subset of cells, all of which deviate from a model fitted to the other cells. . . . As in stepwise regression, the sequential procedure described here does not necessarily identify that subset whose exclusion will reduce the $\chi^2$ by the maximum amount. In examples where some of the selected cells have a common property, such as lying on the diagonal or in the same row, we have found it helpful to first eliminate all the cells with the same property and then do the analysis again.

The above-mentioned $14 \times 14$ table of father/son occupations would appear to be a case in point.

In an important paper, Simonoff (1988b) draws attention to the danger of *masking* when the consecutive testing of cells for discordancy is done on an *inward* (or *forward selection*) basis (again see page 128 above). From the evidence provided by simulation studies, described below, Simonoff recommends the use of inward procedures for *identifying* outliers, combined with *outward* (or *backward selection*) procedures for *testing* them once they are identified.

The simulation studies investigate the performance of ten different combinations of identification and test methods, shown in Table 12.2. Detailed results of some of the studies are presented in the paper, viz. of simulations of sizes 1600 carried out on three types of $5 \times 5$ table containing contaminant frequencies in three of their cells. The frequency in each contaminant cell $(I, J)$ is generated according to what one might call a probability slippage model in which the cell probability under independence, $p_{IJ} = \alpha_I \beta_J$, is replaced by $p_{IJ}(1 + \Delta_{IJ})$ for a specified slippage constant $\Delta_{IJ}$ (with subsequent norming of all probabilities so that they add to 1).

**Table 12.2**

| Identification criterion | Discordancy test criterion | Direction of consecutive testing |
|---|---|---|
| Adjusted residual $\tilde{r}_{IJ}$ | $\tilde{r}_{IJ}$ $X^2 - X_{IJ}^2$ $G^2 - G_{IJ}^2$ | Outward |
| Deleted residual $r_{IJ}^*$ | $X^2 - X_{IJ}^2$ $G^2 - G_{IJ}^2$ | Outward |
| Adjusted residual $\tilde{r}_{IJ}$ | $\tilde{r}_{IJ}$ $X^2 - X_{IJ}^2$ $G^2 - G_{IJ}^2$ | Inward |
| Deleted residual $r_{IJ}^*$ | $X^2 - X_{IJ}^2$ $G^2 - G_{IJ}^2$ | Inward |

In the simulations, the five following performance measures are estimated:

$\beta_1$ = the probability of detecting any contaminant cell, i.e. the power.

$\beta_2$ = the probability of identifying as discordant all the contaminant cells and only those cells.

$\beta_3$ = the probability of identifying as discordant at least one of the contaminant cells.

$N_C$ = the average number of contaminant cells (correctly) identified.

$N_I$ = the average number of non-contaminant cells (incorrectly) identified.

From these and other simulations (involving $3 \times 3$ and $7 \times 7$ tables) carried out by Simonoff, the main conclusions are as follows.

1. *For identification, it is more effective to use deleted residuals $r_{IJ}^*$ than adjusted residuals $\tilde{r}_{IJ}$*. Adjusted residuals give higher values of $N_I$, and are thus more vulnerable to *swamping*.

   Simonoff (1988b) illustrates this graphically by the invented $5 \times 5$ contingency table shown in Table 12.3.

**Table 12.3**

|          | $j = 1$ | 2  | 3  | 4  | 5  |
|----------|---------|----|----|----|----|
| $i = 1$  | 18      | 41 | 41 | 20 | 21 |
| 2        | 39      | 20 | 20 | 22 | 22 |
| 3        | 24      | 20 | 20 | 16 | 18 |
| 4        | 20      | 20 | 19 | 19 | 19 |
| 5        | 23      | 19 | 20 | 17 | 20 |

The table evidently contains three outlying cells, namely $(1, 2)$, $(1, 3)$ and $(2, 1)$. Using *deleted* residuals, the first three cells identified are $(2, 1)$, $(1, 3)$, $(1, 2)$, a correct result. But using *adjusted* residuals the first three cells identified are $(1, 1)$, $(2, 1)$ and $(1, 5)$. The non-outlying cells $(1, 1)$ and $(1, 5)$ have been wrongly identified as outliers due to *swamping*; the outlying cells $(1, 2)$ and $(1, 3)$ have been wrongly ignored due to *masking*.

   Note that the comparison made here between $r_{IJ}^*$ and $\tilde{r}_{IJ}$ is not inconsistent with the previously mentioned preference of Brown (1974) for $\tilde{r}_{IJ}$, since Brown's comments relate only to the case of a single outlying cell.

2. *For consecutive testing of a number of outlying cells, outward testing is more effective than inward testing*. Outward testing is less vulnerable to *masking* than inward testing; this is shown by the low, often very low, values of $\beta_2$ found for inward testing, and also the higher values of power ($\beta_1$) found for outward testing.

3. *The best choice of discordancy test statistic is $G^2 - G_{IJ}^2$*. Compared with $X^2 - X_{IJ}^2$ and $\tilde{r}_{IJ}$, it gives the highest values of $\beta_2$, $\beta_3$ and $N_C$. (For a general discussion of the comparison between the $X^2$ and $G^2$ statistics, see Fienberg (1979).)

Simonoff (1988b) concludes (our italics inserted for emphasis):

The detection of outlying cells in contingency tables fits well in the general framework of outlier detection. A backwards- stepping [i.e., outward] approach, identifying cells by largest deleted residual and testing them by difference in $G^2$, results in a generally slightly conservative procedure that has good power to detect single and multiple outliers, while being resistant to swamping and masking.

A major strength of this procedure is that it can be applied easily when fitting complicated models. *It is also easily applied to multidimensional contingency tables* when fitting (for instance) log-linear models or special models for tables with ordered categories (see, e.g., Agresti, 1983; Clogg, 1982).

...The focus here has been on the detection of outlier cells, but *notions of "influential" cells akin to those in regression models are also appropriate in this context.* These are cells that have a strong effect on the fit, even if they are not outlying themselves. The methods discussed here could be adapted to detect such cells by using the change in $G^2$, ... instead of the residuals in the identification portion of stepping.

For related work using adjusted residuals, see Haberman (1973) and Muñoz-Garcia, Moreno-Rebollo, and Pascual-Acosta (1987). Perli, Hommel, and Lehmacher (1985) propose 'sequentially rejective test procedures' based on the calculation of Pearson's $\chi^2$ for successively reduced tables.

Mosteller and Parunak (1985), in an interesting paper entitled 'Identifying extreme cells in a sizable contingency table: probabilistic and exploratory approaches', present three methods for identifying outliers. The first of these is a simulation procedure. The significance of a residual in any particular cell is estimated by simulating its null distribution from a large number of contingency tables generated under the assumption of independence with row and column totals the same as the observed values. A modified form of residual is used by Mosteller and Parunak (though the method can be applied with any of the other forms), viz.

$$\overset{MP}{r}_{IJ} = (\tilde{r}_{IJ}/A_{IJ})\sqrt{(1 - (1/N))} \tag{12.1.20}$$

where $A_{IJ}$ is a standardization factor given by

$$A_{IJ} = 1 + \exp(- R_I C_J/N) \tag{12.1.21}$$

and $\tilde{r}_{IJ}$ is given by (12.1.6).

The second and third methods use an exploratory approach. Cells are identified as outliers when their residuals (however defined) exceed the upper quartile of the set of $N$ residuals by more than 1.5 times the interquartile range. In the second method the standardized residual $r_{IJ} = (x_{IJ} - e_{IJ})/\sqrt{(e_{IJ})}$ of (12.1.3) is used, with $e_{IJ} = R_I C_J/N$. A similar procedure is employed in the third method, but with a different set of expected frequencies $e_{IJ}$. A multiplicative null model is assumed for these, and the parameters in the model are fitted by applying median polish to the two-way table of $\log x_{ij}$.

Mosteller and Parunak illustrate these methods by analysing a $19 \times 6$ contingency table of archaeological data due to Casjens (1974). This gives the counts of 19 different kinds of artefact found on sites at different distances (six distance categories) from permanent water in an area of Nevada. The aim is to investigate the association between the types of site used by the prehistoric hunter-gatherers who inhabited the area, as characterized by the kinds of artefact present, and the proximity to water. Two 'discordant' outliers are found, and given an archaeological interpretation.

A different approach to the identification of outliers in two-way contingency tables, the *tetrad* approach, was proposed by Fienberg (1969) and later developed by Kotze and Hawkins (1984); Bradu and Hawkins (1982) applied similar methods to the identification of outliers in two-way factorial layouts. Suppose a particular cell $(I, J)$ in a complete $r \times c$ table is under examination. There are $(r - 1)(c - 1)$ $2 \times 2$ subtables containing $(I, J)$, typically consisting of the frequencies

$$x_{IJ} \quad x_{Ih}$$

$$x_{gJ} \quad x_{gh}$$

$$(g \neq I, h \neq J).$$

The sample contrast $\log (x_{IJ}x_{gh}/x_{gJ}x_{Ih})$ is called a *tetrad*. For details of the identification procedure, based on median tetrads, and examples of its application to particular data sets, see Kotze and Hawkins (1984). See also Paulson and Thornton (1975).

## 12.2 OUTLIERS IN SAMPLE SURVEYS

Earlier chapters in this book have borne witness to the vast interest over many years in the treatment of outliers, and to the extensive literature on outlier problems in a wide range of data contexts: univariate and multivariate samples, regression, designed experiments, time series, directional data, and so on. Yet, strangely, there has up to now been very little published work on outliers in *sample surveys*. This is extraordinary, since outliers are just as liable to arise in sample survey data as in any other data context. In a survey of the numbers of people in households in the United Kingdom, it would be surprising to encounter a household with 8 members (the proportion of such households in 1991 was about 1 in 1000; see *Family Spending: A Report on the 1991 Family Expenditure Survey*, HMSO, 1992). Or again, in a survey including heights of adults it would be surprising (but not impossible) to find in the sample an individual with height 208 cm. These values, which cause *surprise*, are surely *outlying values*.

It is true, of course, that problems of *missing* values in sample surveys have received considerable attention. Such values are due typically to non-response or to deletion in the editing process (e.g. of a clearly unacceptable value such as a height of 508 cm). Much work has been done on methods for filling in suitable quantities—*imputed* values—in replacement of missing values; see, e.g., Rubin (1987). There is an obvious link between *imputation* and the treatment of outlying values. For an example of the treatment of outliers by imputation procedures, see Little and Smith (1987).

Barnett (1992, 1993) has recently sought to draw attention to the dearth of work on outliers in sample surveys and to some of the problems which urgently need addressing. Three of the basic aspects are as follows:

- the modelling of contamination in sample surveys;
- the construction of outlier-robust procedures;
- the detection and testing of outliers.

What models can be postulated to explain the occurrence of outliers in a sample survey? We can leave aside instances of wrong measurement or misrecording, i.e. *deterministic* outliers (such as the man's height of 508 cm); these are dealt with by editing. To fix ideas, take the case of a univariate sample. If this was from an *infinite* population,

an outlier might arise due to contamination. Thus $x_{(n)}$ might *not* be an observation from the basic model **F**, but from a mixture $(1 - \lambda)\mathbf{F} + \lambda\mathbf{G}$ where $\lambda$ is small and **G** is a shifted form of **F** ... The model $(1 - \lambda)\mathbf{F} + \lambda\mathbf{G}$ is the outlier-generating model, implying (infrequent) random mixing of two random generating mechanisms for the observations. In a sample survey we are dealing with a different situation. Here we have a fixed (predetermined) finite population $X_1, X_2, \ldots, X_N$ being the values taken by a variable $X$ for a specific set of $N$ individuals. The random element enters only when we *specify and apply a probabilistic procedure* to select $n$ of these fixed values. Thus a *simple random (s.r.) sample* $x_1, x_2, \ldots, x_n$ is obtained by successively choosing from $X_1, X_2, \ldots, X_N$ at *random and without replacement*.

A particular $x_i$, say the largest one $x_{(n)}$, may be surprisingly large and be declared an outlier, but how are we now to model this in terms of contamination or an outlier-generating model? If s.r. sampling has been truly applied, $x_{(n)}$ must in fact be one of the $X_i (i = 1, 2, \ldots, N)$ so cannot reflect contamination. It is merely a facet of the finite population. . . . So if $x_{(n)}$ is a surprisingly large value of $X_i (i = 1, 2, \ldots, N)$ this merely means that the population contains values larger than we expected. . . .

How can $x_{(n)}$ reflect a value untypical of (not a member of) the underlying fixed population? This is the fundamental dilemma in trying to develop an approach to outliers in finite population ('survey') sampling, and hitherto undefined principles will have to be set up. (Barnett, 1992)

One possibility is that not all the $N$ values in the finite population are values of the variable $X$, but a small number, $k$, of them are values of a different variable $Y$ which have been inadvertently included. That is, the finite population from which we are sampling is not $X_1, X_2, \ldots, X_N$, but is in fact

$$X_1, X_2, \ldots, X_{N-k}; \quad Y_1, Y_2, \ldots, Y_k.$$

We have here a finite-population version of the *mixture model* for contamination.

A second possibility is to assume a *super-population model*, in which $X_1, X_2, \ldots, X_N$ is regarded as being itself a sample of size $N$ from some distribution $\mathbf{F}$. Contamination of the infinite population $\mathbf{F}$, which can be modelled in one of the usual ways, will then feed through into the sample of size $n$ and provide a straightforward model for outliers in that sample.

Thus, as Barnett (1993) says,

in developing outlier methods for survey sampling we have to resolve the basic anomaly of what non-null model should be used and there would seem to be much potential in such a study.

Some further proposals for modelling contamination in finite populations are made in Barnett (1993).

We next review the very limited amount of published work on outliers in sample surveys. This provides several examples of the construction of outlier-robust procedures.

Hidiroglou and Srinath (1981) consider outlier-robust methods for estimating the total, $T = Y_1 + \ldots + Y_N$, of a finite population of $N$ positive values from a sample of size $n$. They describe the problem as

... the estimation of the population total of some characteristic from a simple random sample containing a few large or extreme observations. These observations [subsequently called *outliers* in the paper] are assumed to be true observations belonging to the population being sampled. ... It is important to deflate the weights for such units at the estimation stage once they have been sampled and identified.

They define an outlier as a value $Y$ which exceeds some known or chosen value $c$, where $c$ is a boundary point or cutoff point (whose value might be determined, for example, from the results of previous related surveys). Denote by $k(\geqslant 1)$ the number of outliers, as just defined, in the sample. The estimators of $T$ proposed by Hidiroglou and Srinath are all of the form of weighted combinations of the non-outlier and outlier values in the sample, viz.

$$\hat{T}_{\mathrm{HS}r} = [(N - rk)/(n - k)] \sum_{y \leqslant c} y + r \sum_{y > c} y. \qquad (12.2.1)$$

They make a detailed study of three particular cases, viz. $r = 1$; $r = N(n + k)/2n^2$ (no specific reason is given for this choice of $r$); and $r$ chosen to minimize $\mathrm{MSE}(\hat{T}_{\mathrm{HS}r})$. They also consider the case when the number of outliers in the whole population is known. The relative efficiencies of the proposed estimators are investigated for various values of six parameters involved ($N$, $n/N$, the ratio of the means of the outlier and non-outlier units in the population, etc.); no overall optimal choice emerges.

Gross *et al.* (1986) address the same problem of estimating a finite population total $T$ in the presence of outliers, and they define an outlier in the same way as a value of $Y$ for which $Y > c$. In their formulation the $k$ outliers in the sample are regarded as forming a separate stratum of size $k$, a so-called 'surprise stratum', and they set up the outlier-robust 'surprise estimator'

$$\hat{T}_S = [(N - k)/(n - k)] \sum_{y \leqslant c} y + \sum_{y > c} y. \tag{12.2.2}$$

This is in fact Hidiroglou and Srinath's estimator (12.2.1) with $r = 1$. Gross *et al.* (1986) consider two further estimators in their paper, viz. the Winsorized estimator

$$\hat{T}_{W1} = N \overset{W}{\bar{y}}_{0,k} = (N/n)\left( \sum_{y \leqslant c} y + kc \right), \tag{12.2.3}$$

and a modified Winsorized estimator

$$\hat{T}_{W2} = \hat{T}_{W1} + n(\bar{y} - \overset{W}{\bar{y}}_{0,k}) \tag{12.2.4}$$

where $\bar{y}$ is the sample mean $(\sum y)/n$.

They apply these estimators on a stratified basis to a simulated population which models the quarterly survey of capital expenditure conducted by the Australian Bureau of Statistics. The broad conclusion is drawn that $\hat{T}_{W2}$ performs marginally better, in terms of bias and MSE, than $\hat{T}_{W1}$; and that both are superior to $\hat{T}_S$.

Like the above-cited authors, Chambers (1986) considers the problem of robust estimation of a finite population total $T = Y_1 + \ldots + Y_N$ in a sample survey, given a sample of size $n$ possibly containing outliers. He makes the further assumption, however, that there is a concomitant variable $X$ available associated with the survey variable (typically its 'size'), and that all $N$ values $X_1, \ldots, X_N$ are known. He distinguishes between two types of outlier, which he calls 'non-representative' (incorrect readings, etc.) and 'representative' (surprisingly extreme but genuinely arising from the population). Any outliers in the sample are taken to be of representative type (non-representative outliers having been dealt with by survey editing and imputation procedures). A superpopulation model is assumed with, for $i = 1, \ldots, N$,

$$E(y_i) = \beta x_i;$$

$$V(y_i) \equiv \sigma_i^2 = \sigma^2 g(x_i), \quad \text{where } g(\cdot) \text{ is a known function;} \tag{12.2.5}$$

$$\text{and} \quad (y_i - \beta x_i)/\sigma_i \quad i.i.d. \ \ N(0,1).$$

Write $\sum_1$ for summation over the $n$ values in the sample and $\sum_2$ for summation over the other $N - n$ values. Then $\sum_1 y$, $\sum_1 x$ and $\sum_2 x$ are known, but not $\sum_2 y$. The least-squares estimator of $\beta$ is

$$\hat{\beta}_{LS} = \sum_1 \frac{x_i y_i}{\sigma_i^2} \Big/ \sum_1 \frac{x_i^2}{\sigma_i^2} \qquad (12.2.6)$$

which can be written

$$\sum_1 \frac{x_i}{\sigma_i} \left( \frac{y_i - \hat{\beta}_{LS} x_i}{\sigma_i} \right) \sum_2 x \Big/ \sum_1 \frac{x_i^2}{\sigma_i^2} = 0 \qquad (12.2.7)$$

The estimator $\hat{\beta}_{LS}$ is non-robust to outliers, and so therefore is the straightforward estimator of $T$:

$$\hat{T}_{LS} = \sum_1 y + \hat{\beta}_{LS} \sum_2 x \qquad (12.2.8)$$

Chambers replaces it by the outlier-robust estimator

$$\hat{T}_{OR} = \sum_1 y + \hat{\beta}_{OR} \sum_2 x + A \qquad (12.2.9)$$

where $\qquad A = \sum_1 \frac{x_i}{\sigma_i} \psi \left( \frac{y_i - \hat{\beta}_{OR} x_i}{\sigma_i} \right) \sum_2 x \Big/ \sum_1 \frac{x_i^2}{\sigma_i^2} \qquad (12.2.10)$

Here $\hat{\beta}_{OR}$ is some outlier-robust estimator of $\beta$, and $\psi$ is a suitably chosen function. The term

$$\frac{y_i - \hat{\beta}_{LS} x_i}{\sigma_i}$$

in (12.2.7) is replaced in (12.2.10) by

$$\psi \left( \frac{y_i - \hat{\beta}_{OR} x_i}{\sigma_i} \right),$$

in the spirit of $M$-estimation.

Census data from Baltimore, Maryland are used as the basis for a simulation study to compare the performance of a number of estimators of $T$. The survey variable $Y$ is the 1970 census count of population in a census block, for $N = 557$ blocks; the associated variable $X$ is the corresponding 1960 census count of the number of occupied dwellings in each block. The scatter plot of $(x, y)$ shows a number of outlying points. In the simulations, 10 per cent samples ($n = 55$) are drawn, by six different sampling methods. Nine different estimators of $T$ are compared; these include four versions of $\hat{T}_{OR}$, with the variance function in (12.2.5) taken as $g(x) = x^2$. The estimator which performs best, with respect to bias and MSE jointly, is of form $\hat{T}_{OR}$ with $\psi(\cdot)$ given by

$$\psi(t) = t \exp[- \tfrac{1}{2} a(|t| - b)^2] \qquad (12.2.11)$$

and $a$, $b$ chosen adaptively.

Tambay (1988) discusses the treatment of outliers in sub-annual economic surveys designed to estimate the levels and trends of economic activity, and applies several strategies to the robust estimation of monthly levels and trends from the monthly Shipments, Inventories and Orders survey at *Statistics Canada*. As with the three papers cited above the problem is essentially to accommodate outliers in the estimation of a finite population total $T = Y_1 + \ldots + Y_N$. (Here each $Y_i$ for *levels* of activity is equal to $X_i W_i$, where $X_i$ is the directly observed value of the economic variable and $W_i$ is the sample design weight; for *trends*, $Y_i = W_i D_i$ where $D_i$ is the difference $X_i - X_i'$ between the values of the economic variable for the current and the previous month.)

Tambay refers to outliers as 'valid sample observations which have a large, undue influence on the estimates', cf. the 'representative outliers' of Chambers (1986). However, he defines them for practical purposes as values falling outside intervals of the form $(-\infty, (1 - c_1)m + c_1 q_3)$ for levels or $[(1 - c_2)m + c_2 q_1, (1 - c_2)m + c_2 q_3]$ for trends, where $m$ is the median and $q_1$, $q_3$ are the lower and upper quartiles of the relevant sample; cf. the 'outsiders' of Kimber (1990). This resembles the approach of Hidiroglou and Srinath (1981) and Gross et al. (1986). To accommodate such outliers he uses Winsorization and also a modified Winsorization similar to, though not the same as, (12.2.4). An account is given of a detailed study of the performance of these methods for various values of the outlier detection parameters $c_1$, $c_2$, and for various procedures for interrelating the data for successive months of the 14-month period analysed.

Little and Smith (1987) present a three-stage procedure for dealing with missing and outlying values in multivariate survey data. The first stage is the detection of outlying response vectors or 'cases'; the next is the detection of outlying values within outlying cases; finally comes the fitting of likely values to replace missing or outlying values (imputation). The methods are illustrated by the analysis of a subset of multiresponse data from the *Annual Survey of Manufactures* (U.S. Department of Commerce, Bureau of the Census, 1981). The data set is in effect a finite population of $N = 152$ cases for a particular industry, each case being a vector of 14 variables relating to numbers, wages, conditions, etc. of the workforce. Multivariate distance methods (see Chapter 7) are used to identify outlying cases and outlying values within cases. As regards robust estimation, discordant values are replaced by their conditional means given the non-discordant values in the case vector.

Waternaux, Laird, and Ware (1989) discuss the analysis of *longitudinal* survey data, with particular reference to a longitudinal study of the effects of prenatal and postnatal lead exposure on the cognitive development of 214 children at 6, 12, and 18 months of age. Multivariate distance measures are used to detect outlying *individuals* (children), and regression diagnostic analyses are used to detect outlying *observations*. A small number of individuals are also identified, by regression methods, as *influential*. Conclusions are drawn regarding the effects of lead exposure which, the authors say, 'do not seem to be driven by influential observations or outliers'; accommodation procedures are not used.

Cressie (1989) applies empirical Bayes methods to the estimation of population 'undercount' for 51 states in the United States 1980 Census, i.e. the difference between the true count and the census count, expressed as a percentage of the true count. Probability plots based on empirical Bayes estimates *identify*, for separate consideration, an outlier *State*, namely New York.

For other published work, see Scott, Brewer, and Ho (1978), and Vardeman and Meeden (1983).

Finally, how can outliers in sample surveys be *identified and tested*? This seems on the face of it to be a straightforward matter. Suppose, to simplify the discussion, that the survey data are univariate. In the few literature references reviewed above, Hidiroglou and Srinath (1981), Gross *et al.* (1986) and Tambay (1988), dealing with univariate data, use simple cutoff rules to define 'outliers'. Why not make use of the rich variety of discordancy test procedures (see Chapter 6) which are available for outliers in samples from *infinite* populations? For example, if testing an upper outlier $x_{(n)}$, one might use a normal sample statistic such as

$$\tau_{N1} = (x_{(n)} - \bar{x})/s \qquad (12.2.12)$$

or

$$\tau_{N7} = (x_{(n)} - x_{(n-1)})/(x_{(n)} - x_{(1)}) \qquad (12.2.13)$$

described respectively on pages 221–3, 226–7; or again an exponential sample statistic such as

$$\tau_{Ea1} = x_{(n)}/\sum x \qquad (12.2.14)$$

or

$$\tau_{E2} = (x_{(n)} - x_{(n-1)})/(x_{(n)} - x_{(1)}) \qquad (12.2.15)$$

described respectively on pages 197–8, 199. (We use here the notation $\tau_{N1}$, $\tau_{N7}$, etc., instead of $T_{N1}$, $T_{N7}$ etc., to avoid confusion with population totals $T$.) The expressions (12.2.13) and (12.2.15) are, of course, statistics of the same form but with different distributions and different critical values. Knowledge of the general form of the finite population of $N$ values should give guidance as to a suitable choice of discordancy test statistic (normal-based, exponential-based, or whatever).

The question then is, how does such a test statistic behave when applied to a sample from a finite population, what are its critical values (say for testing at 5 per cent and 1 per cent), and how do they relate to the corresponding critical values for the familiar infinite population situation (as in Chapter 6)?

To investigate this, Barnett and Roberts (1993) have recently carried out an extensive simulation study. Taking first the case of the normal sample statistic $\tau_{N1}$ in (12.2.12), they generate a finite population $X_1, \ldots, X_N$ of size $N = 1000$ by random sampling from $N(0, 1)$; call this finite population FP1. From it they select, for each of various sample sizes $n$ in the range 4 to 200, 5000 random samples. These give, for each $n$, 5000 values of $\tau_{N1}$ and hence an estimate of the 5 per cent critical value for use when testing $x_{(n)}$ by means of this statistic. By smoothing against $n$, a sequence of 5 per cent critical values for different $n$ is obtained which are accurate to at least three significant figures. These turn out to be markedly different from the corresponding critical values for the infinite population case. Barnett and Roberts then repeat the process with further finite populations FP2, FP3, . . . of size 1000 sampled from $N(0, 1)$. *Each FP gives a different sequence of critical values!* Thus there is **no unique critical value** of $\tau_{N1}$ for a level-$\alpha$ test of an upper outlier in a sample of size $n$ from a finite 'normal' population.

Similar calculations are carried out for the $\tau_{N7}$ statistic (12.2.13) and for other forms of infinite population (exponential and uniform) with various appropriate discordancy test statistics ($\tau_{Ea1}$, $\tau_{E2}$, etc.). In every case the same general pattern of behaviour is found.

The simulation results . . . lead to the following strange conclusion. *Even if we are sure of the distributional form of the finite population* (e.g. it is 'normal' in the sense that it is as if it has arisen as a large random sample of size $N$ from a true normal distribution) *there is no common set of critical values which can be used in a test of discordancy.*

Thus there seems to be no test of discordancy available for testing our upper outlier in a finite population of known shape—or by extension for testing any outlier manifestation (lower outlier, outlier pair, etc.). (Barnett and Roberts, 1993)

Barnett and Roberts (1993) discuss the implications of these results in the final section of their paper, entitled 'A way forward'. It would seem that a new approach is needed, very possibly based on sample quantiles.

# CHAPTER 13

# Important Strands: Computer Software, Data Studies, Standards and Regulations

Throughout this book we have concentrated on basic conceptual matters and detailed methodological developments in our efforts to present a comprehensive coverage of the current state of knowledge about outliers. We have also illustrated many of the topics by worked practical examples of the use of outlier methods.

There are nonetheless other aspects of the treatment of outliers which, while less well developed, merit consideration.

In this brief chapter we signal a few of these matters and provide a brief introductory commentary on them.

There are three particular topics that will be highlighted:

- **statistical computing**—particularly the level of provision of outlier methods within statistical software systems, and the implications of automated and repeated statistical analysis for the handling of outliers;
- **data analyses**—with further illustrations from the wide range of practical fields in which extensive case study material involving the handling of outliers can provide useful reinforcement of outlier methodology;
- **standards and regulations**—covering the quality, performance and reliability of systems involving statistical variability—especially the requirements for handling outliers in the burgeoning and conflicting official national regulations governing the manufacture of products, control of administrative structures, and so on.

### 13.1 OUTLIERS AND STATISTICAL COMPUTING

**The statistical software package** is a central tool in modern statistical analysis, whether conducted by the professional statistician in a high-tech, window based, multifaceted graphical environment, or by the 'do-it-your-self' non-statistician in a user-friendly minimum intervention mode. There are two principal considerations in respect of the provision of facilities provided for handling outliers. These are the following.

(i) What outlier methods does a package offer for one-off application: are they well-chosen, appropriate, efficient? *Are* there any such facilities?
(ii) What methods for outlier analysis should we expect from a computer-ised system which provides routine, repetitive (possibly on-line), data analysis of the process it is monitoring?

**We start with outlier methods in some of the major statistical software packages.** The overall situation is not encouraging and suggests that *none of the packages offers a comprehensive procedure-based or integrated approach to outlier handling.* Instead we find a range of *optional* data-diag-nostic facilities which reveal a degree of confusion between the need for measures of robustness and for the more specific (albeit overlapping) outlier methods. This is typified by the provision of measures of 'influence' or 'leverage' of individual observations in unstructured data sets or in linear model (regression) data—the latter showing the confusion that we have already commented on between outliers reflecting surprising degrees of extremeness in response variables, and lack of robustness indicated by high values of measures such as Cook's $D^2$ which can arise as much from isolated fringe points in the design space. There is a great contrast between the rich and sophisticated panoply of outlier methodology (as reflected throughout this book) and the paucity of coverage in the statistical packages—which constitute the most general means of application of modern statistical methods. It is hard to find even a single (albeit univariate, normal distribution based) test of discordancy, let alone any attempt to reflect wider interests (e.g. accommodation) or more complex problems (e.g. outliers in the linear model or multivariate framework) to any marked degree. Frequently, even the inevitable distributional dependence of the limited facilities offered is neither pointed out nor safeguarded!

It is not our purpose to offer a comprehensive outlier-consumers' guide to statistical packages or a 'best-buy' in this regard. We merely seek to demonstrate some of the apparent shortcomings by commenting on some of the packages at different stages of their development and refinement over the last fifteen years or so.

The Biomedical Computer Programs of the BMDP system as described in BMDP (1978) do in fact make several *mentions* of 'outliers'. A prominent general feature is the facility to set bounds on what are to be regarded as 'acceptable' values (what amounts to a pre-specified truncation principle) with facilities to replace such values with 're-estimated' values. Within the 'acceptable' limits we are able to flag values which still exhibit extreme (standardized) deviations or residuals. For multivariate or structured data, we can examine various forms of Mahalanobis distances with large values (in the case of *normal* distributions) indicating 'outliers'. Ten years later (BMDP, 1988a, Vol. 1), slight refinements of detail are evident but the approach is in principle unchanged. There is greater emphasis on residual-based diagnostics for regression (and other linear model) data, and increasing attention to robustness (non-outlier-specific) assessments. Trimming is offered as an option to examine the effects of outliers. BMDP (1988b, Vol. 2) includes related facilities for 'missing data analysis'.

In the *SAS System for Linear Models* (SAS, 1986) reference to outliers is made under the heading of diagnostic measures for regression data, where some such measures are provided to cope with 'model misspecification, outliers, influential observations, and collinearity' . . . 'terms that have no precise definition . . . [and whose] meanings vary from one author to another'. We are offered general descriptions of these terms, including that for an *outlier*, which it is said:

refers generally to an observation whose value of the dependent variable is either much larger or much smaller than the values of the dependent variable for other observations with similar values of the independent variables (SAS, 1986).

More detailed (but still relatively sparse) coverage is given in the *SAS Users' Guide* (SAS, 1988). In the *SAS/STAT® Users' Guide* for release 6.03, two procedures mention outliers. These are FASTCLUS (for cluster analysis) where it is remarked that an initial specification of an over-large number of clusters (20–100) can have the advantage of showing up outliers as 'clusters with only one member', and PRINCOMP (principal components) where comment is made on the use of this approach *per se* (*qua* Gnanadesikan, 1977) for exhibiting multivariate outliers. In the SAS Applications Programming guide, outliers are also discussed under regression (signalling special attention if a studentized residual exceeds 2.5 in value).

The SPSS^X system (SPSS, 1983) mentions outliers briefly under coverage of the analysis of residuals in regression problems. The user is able to identify the '10 worst outliers' in terms of chosen criteria, including use of unstandardized, studentized, studentized deleted, or deleted residuals, Mahalanobis distances or Cook's distances: the default option is standardized residuals. Additionally, the user can exhibit 'outliers' as observations whose standardized residuals have absolute values in excess of a specified value: the default value is 3! In the SPSS^X *Advanced Statistics Guide* (SPSS, 1985), presumably intended for the more statistically sophisticated user, the facilities appear to be unchanged.

In the *Reference Manual* for MINITAB (see, for example, MINITAB, 1989) there is no mention of 'outlier' in the index. Residual-based diagnostics are offered. Standardized residuals with (absolute) value in excess of 2 are marked (R)—a label (X) is also placed against observations of high leverage.

The S-PLUS system offers similar facilities for marking observations with high (absolute) levels of studentized residual, or high leverage. In *The New S Language* (Becker, Chambers, and Wilks, 1988) an exclusion facility provides 'an easy way to exclude outliers'.

None of the procedures so far described relate the reference level to the sample size (shades of Chauvenet!). Genstat tries to remedy this but in rather a bizarre way.

In the *Reference Manual* for the Genstat system (Genstat, 1987) we find, under the topic of 'simple linear regression' what appears to be the only treatment for outliers. To warn about 'potential "outliers" in your analysis' a list of particularly large standardized residuals is provided. The notion 'particularly large' is defined in a specific way, namely as exceeding in absolute terms a 'value $c$ corresponding to probability $1/d$ of being exceeded by a standard Normal deviate, where $d$ is the number of residual degrees of freedom', provided such a value $c$ exceeds 2 and is less than 4. If not (and the text says this will happen if $d$ is less than 20, or greater than 15 773, respectively) $c$ takes the boundary values of 2 and 4. Thus the reference values are double-tailed probabilities in $N(0, 1)$ (not single-tailed as the text suggests) and it is interesting to note how far they depart from, say, the critical values for a 5 per cent test of discordancy for a linear regression model with normal errors as given in Table XXXV (p. 520, using $q = 1$). We have the following values.

| $d$ | 10 | 15 | 24 | 40 | 80 | 100 |
|---|---|---|---|---|---|---|
| Genstat | 2.00 | 2.00 | 2.04 | 2.24 | 2.50 | 2.58 |
| Table XXXV | 2.47 | 2.61 | 2.89 | 3.07 | 3.33 | 3.41 |

As remarked above, this is not intended as a comprehensive review of statistical packages. They are clearly not conspicuously attentive to the outlier problem.

There are of course many more, and more specific, packages available. In the latter category is a robust statistics system ISP due to Huber—one might hope that such a specialist facility would adopt a much more sophisticated approach to outliers! In fact in the *PC-ISP Users' Guide* the only mention of outliers seems to be in relation to an interactive dynamic graphics facility which reveals clusters and helps to 'identify outliers'.

In a review of STATGRAPHICS (version 2.6), Hunt (1989) observes that the regression procedures include 'an interactive facility to see the effect of outliers' with the option 'to remove outliers' so that 'the effect of any outliers can be easily seen'.

Beyond the basic provisions of statistical packages we do find some consideration of the processing of outliers in statistical computing, for

instance in source-language algorithms and procedures or tailored programs designed for outlier study. Examples are found in Simonoff (1984a,
a routine for four outlier detection statistics); Young *et al.*
(1985, PROC ELPRINT, an intriguing joint facility for handling 'footnotes and outlier
detection'); Leroy and Rousseeuw (1985, robust regression using 'progres');
Raouf and Chaudhary (1987, a modular outlier testing program); Rock
(1987, ROBUST: an interactive FORTRAN-77 package), and Sadler and
Smith (1987, variance function estimation in immunoassay). Hopkins and
Morse (1990) give an index of the algorithms from the journal *Applied
Statistics* which indicate some (e.g. numbers 30, 126, 190) with outlier
relevance.

**Consider a different aspect of computerized data analysis.** There are
applications where similar analyses need to be carried out routinely and
regularly (perhaps in industrial process monitoring or even in established
sample surveys). There is a clear advantage in adopting an automated (but
properly monitored) approach to such data handling. But what of outliers?
We cannot expect our automatic process to 'show surprise' at the extremeness of some observations. However, it should do precisely the automated
equivalent of that—not only in respect of outliers but of process change
generally, as in changes of level (e.g. perhaps of quality of a product) or
of variability. There does not seem to be a refined approach to the outlier
aspect of this problem (see also the comments on Standards and Regulations in Section 13.3 below). What is needed is a facility in the automated
statistical processing system to satisfactorily distinguish between indications
of change in the *overall system* (e.g. how the mean level, or variance,
change with time) and of *isolated disruptions* (e.g. outliers in an otherwise
independent sequence of identically distributed observations, or even in a
time-series). This would require the most up-to-date use of outlier methods,
for rejection or (perhaps more reasonably) for accommodation, based
securely on the proper distributional and structural considerations. It is
precisely in situations of this type that we might expect to have a good
knowledge of the distributional/structural base and yet there is obviously
much that needs to be done to begin to introduce the appropriate outlier
methods.

An interesting methodological feature arises. Consider a long-term process which continues to yield observations of what is expected to be a
controlled variable. We will rapidly be able to infer the form of the
distribution as an appropriate base model for outlier procedures. Perhaps
it is *normal*, perhaps not! Why not act *adaptively*, assessing outliers in
relation to our developing *empirical* knowledge as the process continues,
and even as it goes through changes (e.g. shifts of level or variability).
There seem to be most interesting prospects here for developing new and
useful outlier methods, possibly employing the up-to-date sample reuse
methods of jackknife or bootstrap.

## 13.2 OUTLIERS IN DETAILED DATA STUDIES

Practical examples from many fields of application have been used throughout this text to illustrate various outlier methods. There has recently been a marked increase in published case studies which feature prominently the need to detect and assess the import of outliers. These cover a wide range of practical situations, although there is particular prominence in the fields of health and medicine. Such case studies provide a rich source of material on which to illustrate and contrast different methods of outlier processing, either by noting the specific techniques used in the publications themselves or for didactic purposes through the process of trying out alternative approaches.

It is therefore valuable to provide references to, and brief commentary on, some publications of this type and this is done in the present section.

A substantial study of *economics* data is presented by Krämer *et al.* (1985) under the title 'Diagnostic checking in practice'. They examine more than ten data sets that have previously been the subject of detailed published analyses. Among various diagnostic measures (of model adequacy, error structure, etc.) for the linear model, they include an outlier test described in the following terms:

This test is based on successively adding an outlier dummy in each period and rejecting the model whenever the maximum value of the t-statistics of the corresponding parameter estimates is too large.

attributing the test to Cook and Weisberg (1982, pp. 20–32). In fact, this employs a location-slippage contamination model and it is interesting to observe multiple applications of it to various data sets.

In a different field, Seaver and Triantis (1986) examine (*inter alia*) the testing of outliers in data from an *industrial* manufacturing process. They place major emphasis on outliers in an examination of data concerned with plant output and technical efficiency measures in paper pulp and linerboard production. Various methods (Gray and Ling, 1984; Cook and Weisberg, 1980 and 1982; Andrews and Pregibon, 1978; etc.) are employed for single or multiple outlier detection in a major data set, with detailed discussion of the empirical implications.

A large number of case studies involving outliers have been carried out in the fields of *health* and *medicine*. Tango (1986) considers clinical laboratory data and outliers in the context of using Akaike's information criterion, while Bartkowiak *et al.* (1987) are concerned with locating outliers in large epidemiological data sets. The combining of event rates in clinical trials is the topic of Berlin *et al.* (1989) who show the effects of an outlier on the data analysis.

Chow (1989) is concerned specifically with detecting outliers in bioavailability/bioequivalence trials; see also Chow and Tse (1990). Lesaffre and

Albert (1989) give interesting examples from medicine in relation to multiple-group logistic regression diagnostics. Stroup, Williamson and Herndon (1989) examine outliers in respect to surveillance data for notifiable diseases—see also Parker (1989). AIDS and HIV infection are the focus of the work by Taylor (1989) which is concerned (among other considerations) with aberrant values; see also Taylor *et al.* (1990).

Thompson and Zucchini (1989) refer to outliers in the context of statistically analysing ROC curves. Peters (1990) mentions outliers in his discussion of the teaching of statistics to medical students. Gleser (1989) refers to possible 'outliers' in his commentary on a detailed data study of indoor air pollution and pulmonary performance.

Examples from *biochemistry* are to be found in Cologne, Mendelman, and Chafflin (1989) who exhibit an outlier in the analysis of data on penicillin-binding proteins; and in Healy (1989) who refers to unusually high rates of occurrence of 'blunders' in UK hospital biochemistry data. In *biology* (specifically *entomology*) we have already referred in some detail (**Example 4.4** above) to the intriguing study of uneven sex ratios in the light-brown apple moth, *Epiphyas postvittana*, by Geier, Briese, and Lewis (1978).

A fascinating *physical* example of empirical data analysis is the examination by Dalal, Fowlkes, and Hoadley (1989) of different launches of the space shuttle *Challenger* prior to the catastrophic failure that took place during the launch on 28 January 1986. Binomial and binary logistic regression models are fitted to the data on O-ring thermal distress and show temperature dependence. Interesting aspects of the data study relate to the effects of removing individual data points in turn; residual analysis revealed clear outliers which were discussed.

## 13.3 OUTLIERS IN STANDARDS REQUIREMENTS

Different countries have for a long time laid down standards (either legislative or advisory) specifying how items should be manufactured, how processes should be controlled or even how systems should be operated. Thus in the UK we have the BSI (British Standards Institute) which specifies standards for anything from staples in surgical implants (BS3531(11))) to 'stays'. In the US, the ANSI (American National Standards Institute) provides a similar service. The 'standards' are laid out in painstaking detail and cover many thousands of products and processes. Many countries have their own standards organizations, and efforts at harmonization are evident in many of them conforming to (or contributing to) the standards of an umbrella organization: the ISO (International Organization for Standardization). The ISO Technical Committee 69 (ISO/TC 69) is responsible for 'applications of statistical methods' and its proceedings are observed by nearly 50 countries (with 10 of them as active participants); see Merle (1983).

It is inevitable that statistical analysis and interpretation must play a central role in such a framework—indeed it would be worrying otherwise and much effort has been exerted to ensure that proper statistical considerations prevail. What, for example, is one to mean by saying that a sausage or a steam-roller satisfies the quality requirements if this is not expressed in some statistical form?

In the past there have been pressures to incorporate within particularly appropriate standards (e.g. for quality control through batch inspection) the relevant statistical procedural details: how to sample the process and how to test and interpret the outcomes. With the increasing recognition of the ubiquitous need for a statistical approach, the standard-specific recommendation has been augmented by more general prescriptions for statistical probity.

For example, in the wide-ranging standard BS5750 (covering all aspects of quality management and quality systems) which is presently preoccupying many organizations from general factors to government ministries, we read in the introductory pages (Part 0: Section 0.2)

**Use of statistical methods** . . . Correct application of modern statistical methods is an important element at all stages in the quality loop and is not limited to the post-production (or, inspection) stages. Applications may be for purposes such as. . . .'

Relevant broad areas of statistical methodology are described (e.g. design of experiments, regression, sampling inspection) and attention is drawn to sources of comprehensive coverage within the standards documentation: e.g. ISO Standards Handbook 3, *Statistical Methods*; unfortunately Handbook 3 has no index in which to seek the entry 'outlier'.

It is encouraging to see what a central role statistical methods now play in standards and regulations. The index to the *BSI Standards Catalogue 1990* (BSI, 1990), for example, has many statistical references—including estimation, interpretation of data, reliability, statistical methods, tests for departure from normality, etc.

Individual and lengthy publications of the BSI (and of the ISO) cover a wide range of special fields including guidance on presenting data on reliability of electronic components, on statistical interpretation of data, on methods of testing (of fatigue, or performance of rubber, for instance), on statistical terminology, on quality control methods, on glossaries of statistical terms (even with definitions in English and French), and so on.

A similarly wide coverage is evident in the publications of the ISO stemming from the work of the committee ISO/TC 69 and of its six major subcommittees, covering terminology, interpretation of statistical data, applications of statistical methods, quality control, acceptance sampling, and application of precision data.

It is interesting to examine to what extent this sophisticated statistical framework takes specific regard of outliers, which are clearly most relevant and important in this field of endeavour.

A welcome example is found in American National Standard E178–75 entitled 'Standard Recommended Practice for Dealing with Outlying Observations' published in 1975 under the jurisdiction of the ASTM Committee E-11 on Statistical Methods. This was developed from even earlier recommendations in 1961 and 1968, and presents principles and techniques for handling outliers. In E178–75, distinctions are drawn between extreme manifestations of random variation and gross errors of deterministic origin. The declared purpose is to offer statistical procedures which help to determine which of these two origins is the more plausible, and which manifest themselves in effect in tests of discordancy for assumed univariate normal data. Warnings about the normal distribution basis and the need for tests for normality, and qualitative comments about the inappropriateness of the procedures for non-normal data, are given. The specific tests that are proposed include the extreme studentised deviate test (N1, see p. 485 above) and various tests using the Dixon statistics (e.g. N10, see p. 498 above). Tables of critical levels are given for implementation of the tests, and several illustrative examples. A Grubbs-type test is also covered, as are some tests for multiple outliers and separate procedures to distinguish cases where the variance of the basic distribution is, or is not, known. Brief discussion is also given on accommodation, e.g. 'Winsorization' is mentioned.

Although the (1975) E178–75 coverage is limited (and presumably a later version might exist) it is none the less a most impressive fact to note that, within the public domain and more than thirty years ago in its original form, recommendations are made for specific actions in regard to the treatment of outliers, with good reason to expect (by virtue of the authority of the ANS standards) that the recommendations would be followed.

What of the specific situation with regard to outliers in the UK under the BSI requirements, or in the international arena under ISO regulations? We have noted the wide coverage of statistical methods per se, but in contrast the term 'outlier' does not in fact feature within the index of the BSI Standards Catalogue. We have thus to look within the pages of the statistical publications themselves to find any specific recommendations on outliers.

The BSI Handbook 25 (BSI, 1985) is entitled Quality Management Systems. Statistical interpretation of data. It consists of six separate standards. One of these, BS 5497, is Precision of test methods, Part 1 prepared as a guide to examining the repeatability and reproducibility of a test method for inter-laboratory tests. It does, in fact, contain specific mention of outliers. In the 1987 version of BS 5497 (BSI, 1987), outliers are defined as observations that 'deviate so far from the comparable entries in the same table that they are considered as irreconcilable with the other data'. It is suggested that Cochran's test (Cochran, 1941) should be applied to a set of samples (e.g. from different laboratories), followed for a discordant sample by a Dixon test applied to the observations in the identified sample (in two-sided form, such as test **N8** in Chapter 6 above, or the corresponding version with implicit protection for masking if the sample size $n$ exceeds 7). The proposed scheme requires repeated applications of such tests until no further discordant samples or observations are found. Warnings are given about dependence on normality and the dangers in repeated applications of the inability to prescribe an overall significance level. An interesting distinction is drawn between critical levels between 5 per cent and 1 per cent (revealing 'stragglers') and less than 1 per cent (revealing 'statistical outliers'). Thus in one particular field we find at least a nod in the right direction!

It is understood that standard BS 5497 is about to be replaced, and will coincide with ISO 5725–2, a new international standard. These new British and International standards will contain proposals for handling outliers in a similar spirit to those of the earlier BS 5497, but with the Dixon tests replaced by Grubbs tests for one or two outliers.

It would seem that a more organized approach to the outlier problem *per se* was under consideration by the BSI until recent years. Under the work of their committee QMS 12 on the Statistical Interpretation of Data, there had been an intention to publish material on 'the determination and handling of outliers'. Later, however, it was agreed by the committee that this work should be cancelled on various grounds, notably that the proliferation of method made it unlikely that a neat set of proposals could be made. It is to be hoped that these problems will be overcome and that the BSI will be able to produce the more comprehensive guidance on the handling of outliers that is so vitally needed.

CHAPTER 14

# Perspective

We have covered a lot of ground in this study of outliers. We hope that we have helped both the experimental scientist with a range of useful techniques and provided food for thought for the professional statistician in clarifying basic principles and indicating some of the methodological gaps.

The outlier problem continues to arouse more interest year by year, in spite of its long history. The pages of the international statistical journals contain ever more contributions to this fascinating and useful field of study, as is evidenced by the substantial list of references at the end of the book: unavoidably about twice as long as in the first edition of the book.

Current research activity continues to centre largely on informal methods for processing outliers in highly structured situations (designed experiment, contingency table and regression data, with an emphasis on computer application and graphical display) and on refinements of understanding of familiar univariate and multivariate single-sample procedures. On both fronts useful results abound, but evermore remains to be done. Some of the directions for the continuing study of univariate single-sample procedures are clear: to find yet more about performance criteria of tests of discordancy which embodies inturn the need for a greater understanding of distributional theory elements. Many of the outlier-generating models warrant further investigation; the interactions between, and respective relevances of, the different models have not been fully examined. Many useful contributions are being made on basic distribution-theory results which will provide a springboard for more detailed investigation of test performance, behaviour of accommodation methods etc. Procedures for 'recursive' (consecutive) testing are continuing to attract attention.

We have noted new developments in handling outliers in directional data, contingency tables, sample surveys and time series. These important practical and conceptual matters require even more study and are also likely to involve some challenging mathematical work.

But when all is said and done, the major problem in outlier study remains the one that faced the very earliest workers in the subject—*What is an outlier and how should we deal with it?* We have taken the view that the stimulus lies in the subjective concept of surprise engendered by one, or a few, observations in a set of data: that this surprise initiates an investigation of the statistical propriety (or influence) of the detected outliers. We have noted that surprise is not always immediate; for example, in multivariate or highly structured data an explicit detection process may be required to reveal the outliers (the 'surprising' observations). However, the chain of operations remains the same. We find an observation surprising, we then proceed to investigate it from the statistical viewpoint, and conclude by rejecting it, welcoming it, or accommodating it. If accommodation is the aim, the analysis proceeds a stage further to a treatment of the overall data set with allowance for the presence of the outlier. There continues to be major emphasis on the accommodation aspect of outlier study. The recent thrust on identifying 'influential observations' (particularly in regression and designed experiment data) is interesting. It bears a close relationship to the accommodation problem in the central role it gives to examining the extent to which tests or estimates of parameters are influenced by individual observations. This is at the heart of robust inference. There is beginning to be more study of interrelationships of principle and objective between the associated ideas of handling outliers and of robust methods generally.

We have drawn attention to the extent to which statistical computational software, and international standards and regulations, reflect a concern for and interest in outliers. Much more needs to be done in these fields, which do not reflect the ever-increasing attention to the outlier problem within statistical methodology at large. The idea of an outlier is even getting through to the school level, as is evidenced by the paper by Huck, Cross, and Clark in the journal *Teaching Statistics* (Huck, Cross, and Clark, 1986).

At the conceptual level various viewpoints exist. Using a Bayesian approach, there are philosophical obstacles to the pre-processing of selected observations from a larger data set (though less serious objections to the subjectivity of the outlier detection process, at least if formalized). The statistician adopting a more classical approach may find the subjective element not entirely acceptable. Whilst prepared to countenance the rejection, for example, of individual members of a sample, it may be felt that this should be done 'objectively' by a routine procedure carried out regularly and indiscriminately on every similar set of data that arises. However, this carries a risk of an unreasonable degree of 'sanitization' of data sets (and consequent loss of fine detail in the data) if carried too far. We return to this point below.

Concerning such fundamental objections, it is indeed not clear that the implications of the subjective detection of outliers are, or even can be, appropriately measured and reflected in the fuller statistical analysis of a set of data. Of course, proper concern for the construction of the outlier model can go a long way to resolve this problem.

Specific difficulties which we have to keep firmly in view are those of *masking* and *swamping*. Their manifestation in single univariate samples is straightforward. In more complex situations these difficulties need to be viewed in a much wider context. In a designed experiment it remains true that one outlier may not be declared discordant because of the masking effect of another. In another respect the very presence of outliers may be masked by particularly strong real effects or by breakdowns in the conventional assumptions, such as normality, homoscedasticity, additivity, and so on. In reverse we may falsely attribute idiosyncratic facets of the behaviour of the data to such breakdowns, whilst in reality they truly reflect the presence of individual contaminants.

Multivariate or highly structured data further highlight the 'subjectivity' versus 'objectivity' argument. We have remarked above on the more nebulous nature of 'surprise' in such situations. Explicit detection procedures are needed for outliers here, and almost inevitably data analysis methods will be computerized, perhaps augmented by graphical display. We will not be surprised at pleas for outlier packages which relieve the analyst of any responsibility. But computers are not easily taught to be surprised. The concept is a human one. It can only to a limited extent be translated into a mechanized form. Further, we have seen how little attention statistical packages pay to the outlier problem.

Retaining the viewpoint that the element of surprise must be central to the study of outliers, so, it would follow, must the individual continue to shoulder the major burden of outlier detection through personal and regular intervention in any data-screening procedure. Of course we can, and should, react to summarized or graphically presented data from the computer to assist in the detection of outliers. Of course we can build into the computer formal procedures for testing discordancy. But we do this to a large extent at the sacrifice of the subjective 'surprise' stimulus.

Even on an 'objective' approach to outlier processing the computer may have its disadvantages to be set against its indispensability as a digester and presenter of large-scale data. In a highly structured situation such as a designed experiment the notion of an outlier remains so primitive (perhaps an extreme residual, perhaps some other sort of unexpected break in pattern) that any total replacement of situation-specific analysis by depersonalized routine computer processing could inhibit the development of clearer understanding of outliers in such areas.

What of the future? Outliers continue to occupy a centre-stage position in the development of statistical methods. They exist as part of the experimentalist's reality; they are part of the analyst's inescapable responsibility. Surely the professional statistician needs to have a thorough modern understanding of how to handle outliers in statistical data.

# Statistical Tables

# Contents List of Tables

(*Contd.*)

## (*Contd.*)

$$L = \sum_{i=1}^{[n/2]} a_{n, n-i+1}[x_{(n-i+1)} - x_{(i)}]$$

and

$$S^2 = \sum_{i=1}^{n} (x_i - \bar{x})^2$$

(*Contd.*)

(*Contd.*)

**Table I** Optimal values of $D_1$, $D_2$ and $M$ for From's estimator of the exponential parameter

| $\alpha$ | | 0.1 | | | 0.2 | | | 0.3 | | | 0.4 | | | 0.5 | |
| $n$ | $M$ | $D_1$ | $D_2$ | $M$ | $D_1$ | $D_2$ | $M$ | $D_1$ | $D_2$ | $M$ | $D_1$ | $D_2$ | $M$ | $D_1$ | $D_2$ |
|---|---|---|---|---|---|---|---|---|---|---|---|---|---|---|---|
| 4 | 3 | 0.2140 | 0.0185 | 3 | 0.2215 | 0.0373 | 3 | 0.2226 | 0.0595 | 3 | 0.2181 | 0.0858 | 3 | 0.2102 | 0.1144 |
| 6 | 5 | 0.1527 | 0.0125 | 5 | 0.1579 | 0.0250 | 5 | 0.1585 | 0.0409 | 5 | 0.1552 | 0.0612 | 5 | 0.1498 | 0.0838 |
| 8 | 7 | 0.1184 | 0.0092 | 7 | 0.1221 | 0.0183 | 7 | 0.1224 | 0.0308 | 7 | 0.1199 | 0.0476 | 7 | 0.1159 | 0.0664 |
| 10 | 9 | 0.0965 | 0.0072 | 9 | 0.0993 | 0.0143 | 9 | 0.0995 | 0.0246 | 9 | 0.0975 | 0.0389 | 9 | 0.0944 | 0.0551 |
| 12 | 11 | 0.0814 | 0.0058 | 11 | 0.0835 | 0.0115 | 11 | 0.0836 | 0.0203 | 11 | 0.0820 | 0.0330 | 11 | 0.0795 | 0.0473 |
| 15 | 14 | 0.0658 | 0.0045 | 14 | 0.0674 | 0.0089 | 14 | 0.0674 | 0.0161 | 14 | 0.0661 | 0.0269 | 14 | 0.0643 | 0.0390 |
| 20 | 19 | 0.0498 | 0.0032 | 19 | 0.0509 | 0.0062 | 19 | 0.0508 | 0.0119 | 19 | 0.0499 | 0.0207 | 19 | 0.0487 | 0.0304 |
| 25 | 24 | 0.0401 | 0.0024 | 24 | 0.0408 | 0.0047 | 24 | 0.0408 | 0.0094 | 24 | 0.0401 | 0.0168 | 24 | 0.0392 | 0.0250 |
| 30 | 29 | 0.0335 | 0.0019 | 29 | 0.0340 | 0.0037 | 29 | 0.0304 | 0.0077 | 29 | 0.0334 | 0.0143 | 29 | 0.0328 | 0.0213 |
| 40 | 39 | 0.0252 | 0.0013 | 39 | 0.0255 | 0.0025 | 39 | 0.0255 | 0.0057 | 39 | 0.0251 | 0.0110 | 39 | 0.0247 | 0.0165 |
| 50 | 49 | 0.0202 | 0.0009 | 49 | 0.0204 | 0.0019 | 49 | 0.0204 | 0.0045 | 49 | 0.0201 | 0.0090 | 49 | 0.0198 | 0.0136 |

**Table I** (*Contd.*)

| α | | 0.6 | | | 0.7 | | | 0.8 | | | 0.9 | |
|---|---|---|---|---|---|---|---|---|---|---|---|---|
| n | M | $D_1$ | $D_2$ | M | $D_1$ | $D_2$ | M | $D_1$ | $D_2$ | M | $D_1$ | $D_2$ |
| 4 | 3 | 0.2021 | 0.1418 | 3 | 0.1964 | 0.1648 | 3 | 0.1945 | 0.1819 | 3 | 0.1959 | 0.1932 |
| 6 | 5 | 0.1445 | 0.1049 | 5 | 0.1411 | 0.1214 | 5 | 0.1400 | 0.1326 | 5 | 0.1409 | 0.1393 |
| 8 | 7 | 0.1122 | 0.0835 | 7 | 0.1100 | 0.0963 | 7 | 0.1094 | 0.1044 | 7 | 0.1099 | 0.1089 |
| 10 | 9 | 0.0917 | 0.0695 | 9 | 0.0901 | 0.0798 | 9 | 0.0897 | 0.0861 | 9 | 0.0901 | 0.0894 |
| 12 | 11 | 0.0775 | 0.0596 | 11 | 0.0763 | 0.0682 | 11 | 0.0761 | 0.0732 | 11 | 0.0764 | 0.0758 |
| 15 | 14 | 0.0628 | 0.0493 | 14 | 0.0621 | 0.0561 | 14 | 0.0619 | 0.0599 | 14 | 0.0621 | 0.0617 |
| 20 | 19 | 0.0478 | 0.0383 | 19 | 0.0473 | 0.0433 | 19 | 0.0473 | 0.0459 | 19 | 0.0474 | 0.0471 |
| 25 | 24 | 0.0386 | 0.0314 | 24 | 0.0383 | 0.0353 | 24 | 0.0382 | 0.0373 | 24 | 0.0383 | 0.0381 |
| 30 | 29 | 0.0323 | 0.0267 | 29 | 0.0321 | 0.0298 | 29 | 0.0321 | 0.0314 | 29 | 0.0322 | 0.0320 |
| 40 | 39 | 0.0244 | 0.0205 | 39 | 0.0243 | 0.0228 | 38 | 0.0243 | 0.0238 | 38 | 0.0242 | 0.0242 |
| 50 | 49 | 0.0196 | 0.0167 | 49 | 0.0195 | 0.0184 | 48 | 0.0195 | 0.0192 | 47 | 0.0196 | 0.0195 |

**Table II** Optimal choice of the number, $m^*$, of lower ordered sample values out of $n$ used in estimating the scale $\theta$ of an exponential distribution, and associated relative efficiency $e_{m^*}$, when one observation has slipped in scale to $\theta/h$ (reproduced by permission of the author and the American Statistical Association)

| h | 0.05 | | 0.10 | | 0.15 | | 0.20 | | 0.25 | |
|---|---|---|---|---|---|---|---|---|---|---|
| n | $m^*$ | $e_{m^*}$ | $m^*$ | $e_{m^*}$ | $m^*$ | $e_{m^*}$ | $m^*$ | $e_{m^*}$ | $m^*$ | $e_{m^*}$ |
| 2 | 1 | 88.59 | 1 | 21.96 | 1 | 9.66 | 1 | 5.38 | 1 | 3.43 |
| 3 | 1 | 74.71 | 2 | 17.58 | 2 | 7.88 | 2 | 4.49 | 2 | 2.93 |
| 4 | 2 | 67.10 | 2 | 15.97 | 2 | 6.83 | 3 | 3.95 | 3 | 2.63 |
| 5 | 3 | 60.42 | 3 | 14.51 | 3 | 6.29 | 4 | 3.57 | 4 | 2.41 |
| 6 | 4 | 54.84 | 4 | 13.27 | 4 | 5.81 | 4 | 3.29 | 5 | 2.25 |
| 7 | 5 | 50.19 | 5 | 12.22 | 5 | 5.40 | 5 | 3.10 | 6 | 2.12 |
| 8 | 6 | 46.28 | 6 | 11.33 | 6 | 5.05 | 6 | 2.93 | 7 | 2.01 |
| 9 | 7 | 42.96 | 7 | 10.57 | 7 | 4.75 | 7 | 2.79 | 8 | 1.93 |
| 10 | 8 | 40.10 | 8 | 9.91 | 8 | 4.49 | 8 | 2.66 | 9 | 1.85 |
| 15 | 12 | 30.73 | 13 | 7.65 | 13 | 3.59 | 13 | 2.22 | 13 | 1.62 |
| 20 | 17 | 25.06 | 17 | 6.36 | 18 | 3.06 | 18 | 1.96 | 18 | 1.49 |
| 30 | 27 | 18.45 | 27 | 4.87 | 27 | 2.46 | 28 | 1.68 | 28 | 1.34 |
| 40 | 36 | 14.77 | 37 | 4.03 | 37 | 2.14 | 38 | 1.52 | 38 | 1.25 |
| 50 | 46 | 12.39 | 47 | 3.49 | 47 | 1.93 | 48 | 1.42 | 48 | 1.20 |

| h | 0.30 | | 0.35 | | 0.40 | | 0.45 | | 0.50 | |
|---|---|---|---|---|---|---|---|---|---|---|
| n | $m^*$ | $e_{m^*}$ | $m^*$ | $e_{m^*}$ | $m^*$ | $e_{m^*}$ | $m^*$ | $e_{m^*}$ | $m^*$ | $e_{m^*}$ |
| 2 | 1 | 2.39 | 1 | 1.79 | 1 | 1.41 | 1 | 1.16 | 1 | 1.00 |
| 3 | 2 | 2.11 | 2 | 1.62 | 2 | 1.32 | 2 | 1.13 | 2 | 1.00 |
| 4 | 3 | 1.93 | 3 | 1.52 | 3 | 1.27 | 3 | 1.11 | 3 | 1.00 |
| 5 | 4 | 1.80 | 4 | 1.45 | 4 | 1.23 | 4 | 1.09 | 4 | 1.00 |
| 6 | 5 | 1.70 | 5 | 1.39 | 5 | 1.20 | 5 | 1.08 | 6 | 1.00 |
| 7 | 6 | 1.63 | 6 | 1.35 | 6 | 1.17 | 6 | 1.07 | 7 | 1.00 |
| 8 | 7 | 1.56 | 7 | 1.31 | 7 | 1.16 | 7 | 1.06 | 8 | 1.00 |
| 9 | 8 | 1.51 | 8 | 1.28 | 8 | 1.14 | 8 | 1.05 | 9 | 1.00 |
| 10 | 9 | 1.47 | 9 | 1.26 | 9 | 1.13 | 9 | 1.05 | 10 | 1.00 |
| 15 | 14 | 1.33 | 14 | 1.17 | 14 | 1.08 | 14 | 1.03 | 14 | 1.00 |
| 20 | 19 | 1.25 | 19 | 1.13 | 19 | 1.06 | 19 | 1.02 | 20 | 1.00 |
| 30 | 29 | 1.17 | 29 | 1.08 | 29 | 1.04 | 29 | 1.01 | 30 | 1.00 |
| 40 | 39 | 1.12 | 39 | 1.06 | 39 | 1.03 | 39 | 1.01 | 40 | 1.00 |
| 50 | 49 | 1.10 | 49 | 1.05 | 49 | 1.02 | 49 | 1.00 | 50 | 1.00 |

**Table III** Critical values for 5% and 1% tests of discordancy for an upper outlier in a gamma sample, using the ratio $x_{(n)}/\sum x_i$ as test statistic. This table is reproduced, with permission from McGraw-Hill Book Company, from Eisenhart, Hastay, and Wallis (1947), Tables 15.1 and 15.2, with appropriate change of notation

**Gal(Ea1)**

*5% critical values*

| r \ n | 0.5 | 1 | 1.5 | 2 | 2.5 | 3 | 3.5 | 4 | 4.5 | 5 | 8 | 18 | 72 | ∞ |
|---|---|---|---|---|---|---|---|---|---|---|---|---|---|---|
| 2 | 0.9985 | 0.9750 | 0.9392 | 0.9057 | 0.8772 | 0.8534 | 0.8332 | 0.8159 | 0.8010 | 0.7880 | 0.7341 | 0.6602 | 0.5813 | 0.5000 |
| 3 | 0.9669 | 0.8709 | 0.7977 | 0.7457 | 0.7071 | 0.6771 | 0.6530 | 0.6333 | 0.6167 | 0.6025 | 0.5466 | 0.4748 | 0.4031 | 0.3333 |
| 4 | 0.9065 | 0.7679 | 0.6841 | 0.6287 | 0.5895 | 0.5598 | 0.5365 | 0.5175 | 0.5017 | 0.4884 | 0.4366 | 0.3720 | 0.3093 | 0.2500 |
| 5 | 0.8412 | 0.6838 | 0.5981 | 0.5441 | 0.5065 | 0.4783 | 0.4564 | 0.4387 | 0.4241 | 0.4118 | 0.3645 | 0.3066 | 0.2513 | 0.2000 |
| 6 | 0.7808 | 0.6161 | 0.5321 | 0.4803 | 0.4447 | 0.4184 | 0.3980 | 0.3817 | 0.3682 | 0.3568 | 0.3135 | 0.2612 | 0.2119 | 0.1667 |
| 7 | 0.7271 | 0.5612 | 0.4800 | 0.4307 | 0.3974 | 0.3726 | 0.3535 | 0.3384 | 0.3259 | 0.3154 | 0.2756 | 0.2278 | 0.1833 | 0.1429 |
| 8 | 0.6798 | 0.5157 | 0.4377 | 0.3910 | 0.3595 | 0.3362 | 0.3185 | 0.3043 | 0.2926 | 0.2829 | 0.2462 | 0.2022 | 0.1616 | 0.1250 |
| 9 | 0.6385 | 0.4775 | 0.4027 | 0.3584 | 0.3286 | 0.3067 | 0.2901 | 0.2768 | 0.2659 | 0.2568 | 0.2226 | 0.1820 | 0.1446 | 0.1111 |
| 10 | 0.6020 | 0.4450 | 0.3733 | 0.3311 | 0.3029 | 0.2823 | 0.2666 | 0.2541 | 0.2439 | 0.2353 | 0.2032 | 0.1655 | 0.1308 | 0.1000 |
| 12 | 0.5410 | 0.3924 | 0.3264 | 0.2880 | 0.2624 | 0.2439 | 0.2299 | 0.2187 | 0.2098 | 0.2020 | 0.1737 | 0.1403 | 0.1100 | 0.0833 |
| 15 | 0.4709 | 0.3346 | 0.2758 | 0.2419 | 0.2195 | 0.2034 | 0.1911 | 0.1815 | 0.1736 | 0.1671 | 0.1429 | 0.1144 | 0.0889 | 0.0667 |
| 20 | 0.3894 | 0.2705 | 0.2205 | 0.1921 | 0.1735 | 0.1602 | 0.1501 | 0.1422 | 0.1357 | 0.1303 | 0.1108 | 0.0879 | 0.0675 | 0.0500 |
| 24 | 0.3434 | 0.2354 | 0.1907 | 0.1656 | 0.1493 | 0.1374 | 0.1286 | 0.1216 | 0.1160 | 0.1113 | 0.0942 | 0.0743 | 0.0567 | 0.0417 |
| 30 | 0.2929 | 0.1980 | 0.1593 | 0.1377 | 0.1237 | 0.1137 | 0.1061 | 0.1002 | 0.0958 | 0.0921 | 0.0771 | 0.0604 | 0.0457 | 0.0333 |
| 40 | 0.2370 | 0.1576 | 0.1259 | 0.1082 | 0.0968 | 0.0887 | 0.0827 | 0.0780 | 0.0745 | 0.0713 | 0.0595 | 0.0462 | 0.0347 | 0.0250 |
| 60 | 0.1737 | 0.1131 | 0.0895 | 0.0765 | 0.0682 | 0.0623 | 0.0583 | 0.0552 | 0.0520 | 0.0497 | 0.0411 | 0.0316 | 0.0234 | 0.0167 |
| 120 | 0.0998 | 0.0632 | 0.0495 | 0.0419 | 0.0371 | 0.0337 | 0.0312 | 0.0292 | 0.0279 | 0.0266 | 0.0218 | 0.0165 | 0.0120 | 0.0083 |
| ∞ | 0 | 0 | 0 | 0 | 0 | 0 | 0 | 0 | 0 | 0 | 0 | 0 | 0 | 0 |

**Table III** (*Contd.*)

1% *critical values*

| r / n | 0.5 | 1 | 1.5 | 2 | 2.5 | 3 | 3.5 | 4 | 4.5 | 5 | 8 | 18 | 72 | ∞ |
|---|---|---|---|---|---|---|---|---|---|---|---|---|---|---|
| 2 | 0.9999 | 0.9950 | 0.9794 | 0.9586 | 0.9373 | 0.9172 | 0.8988 | 0.8823 | 0.8674 | 0.8539 | 0.7949 | 0.7067 | 0.6062 | 0.5000 |
| 3 | 0.9933 | 0.9423 | 0.8831 | 0.8335 | 0.7933 | 0.7606 | 0.7335 | 0.7107 | 0.6912 | 0.6743 | 0.6059 | 0.5153 | 0.4230 | 0.3333 |
| 4 | 0.9676 | 0.8643 | 0.7814 | 0.7212 | 0.6761 | 0.6410 | 0.6129 | 0.5897 | 0.5702 | 0.5536 | 0.4884 | 0.4057 | 0.3251 | 0.2500 |
| 5 | 0.9279 | 0.7885 | 0.6957 | 0.6329 | 0.5875 | 0.5531 | 0.5259 | 0.5037 | 0.4854 | 0.4697 | 0.4094 | 0.3351 | 0.2644 | 0.2000 |
| 6 | 0.8828 | 0.7218 | 0.6258 | 0.5635 | 0.5195 | 0.4866 | 0.4608 | 0.4401 | 0.4229 | 0.4084 | 0.3529 | 0.2858 | 0.2229 | 0.1667 |
| 7 | 0.8376 | 0.6644 | 0.5685 | 0.5080 | 0.4659 | 0.4347 | 0.4105 | 0.3911 | 0.3751 | 0.3616 | 0.3105 | 0.2494 | 0.1929 | 0.1429 |
| 8 | 0.7945 | 0.6152 | 0.5209 | 0.4627 | 0.4226 | 0.3932 | 0.3704 | 0.3522 | 0.3373 | 0.3248 | 0.2779 | 0.2214 | 0.1700 | 0.1250 |
| 9 | 0.7544 | 0.5727 | 0.4810 | 0.4251 | 0.3870 | 0.3592 | 0.3378 | 0.3207 | 0.3067 | 0.2950 | 0.2514 | 0.1992 | 0.1521 | 0.1111 |
| 10 | 0.7175 | 0.5358 | 0.4469 | 0.3934 | 0.3572 | 0.3308 | 0.3106 | 0.2945 | 0.2813 | 0.2704 | 0.2297 | 0.1811 | 0.1376 | 0.1000 |
| 12 | 0.6528 | 0.4751 | 0.3919 | 0.3428 | 0.3099 | 0.2861 | 0.2680 | 0.2535 | 0.2419 | 0.2320 | 0.1961 | 0.1535 | 0.1157 | 0.0833 |
| 15 | 0.5747 | 0.4069 | 0.3317 | 0.2882 | 0.2593 | 0.2386 | 0.2228 | 0.2104 | 0.2002 | 0.1918 | 0.1612 | 0.1251 | 0.0934 | 0.0667 |
| 20 | 0.4799 | 0.3297 | 0.2654 | 0.2288 | 0.2048 | 0.1877 | 0.1748 | 0.1646 | 0.1567 | 0.1501 | 0.1248 | 0.0960 | 0.0709 | 0.0500 |
| 24 | 0.4247 | 0.2871 | 0.2295 | 0.1970 | 0.1759 | 0.1608 | 0.1495 | 0.1406 | 0.1338 | 0.1283 | 0.1060 | 0.0810 | 0.0595 | 0.0417 |
| 30 | 0.3632 | 0.2412 | 0.1913 | 0.1635 | 0.1454 | 0.1327 | 0.1232 | 0.1157 | 0.1100 | 0.1054 | 0.0867 | 0.0658 | 0.0480 | 0.0333 |
| 40 | 0.2940 | 0.1915 | 0.1508 | 0.1281 | 0.1135 | 0.1033 | 0.0957 | 0.0898 | 0.0853 | 0.0816 | 0.0668 | 0.0503 | 0.0363 | 0.0250 |
| 60 | 0.2151 | 0.1371 | 0.1069 | 0.0902 | 0.0796 | 0.0722 | 0.0668 | 0.0625 | 0.0594 | 0.0567 | 0.0461 | 0.0344 | 0.0245 | 0.0167 |
| 120 | 0.1225 | 0.0759 | 0.0585 | 0.0489 | 0.0429 | 0.0387 | 0.0357 | 0.0334 | 0.0316 | 0.0302 | 0.0242 | 0.0178 | 0.0125 | 0.0083 |
| ∞ | 0 | 0 | 0 | 0 | 0 | 0 | 0 | 0 | 0 | 0 | 0 | 0 | 0 | 0 |

$n$ = number of observations.
$r$ = shape parameter of the gamma distribution ($r = 1$ for exponential distribution).

**Table IV** Critical values for 5% and 1% Dixon-type tests of discordancy for an upper outlier in an exponential sample, using $[x_{(n)} - x_{(n-1)}]/x_{(n)}$ or $[x_{(n)} - x_{(n-1)}]/[x_{(n)} - x_{(1)}]$ as test statistic

| For testing | For testing | Ea2, E2 | |
|---|---|---|---|
| $T_{Ea2} = \dfrac{x_{(n)} - x_{(n-1)}}{x_{(n)}}$ | $T_{E2} = \dfrac{x_{(n)} - x_{(n-1)}}{x_{(n)} - x_{(1)}}$ | | |
| $n$ | $n$ | 5% | 1% |
| 2 | 3 | 0.974 | 0.995 |
| 3 | 4 | 0.894 | 0.957 |
| 4 | 5 | 0.830 | 0.912 |
| 5 | 6 | 0.782 | 0.875 |
| 6 | 7 | 0.746 | 0.845 |
| 7 | 8 | 0.717 | 0.821 |
| 8 | 9 | 0.694 | 0.800 |
| 9 | 10 | 0.675 | 0.783 |
| 10 | 11 | 0.658 | 0.768 |
| 11 | 12 | 0.644 | 0.755 |
| 12 | 13 | 0.631 | 0.743 |
| 13 | 14 | 0.620 | 0.733 |
| 14 | 15 | 0.610 | 0.724 |
| 15 | 16 | 0.601 | 0.715 |
| 16 | 17 | 0.593 | 0.707 |
| 17 | 18 | 0.586 | 0.700 |
| 18 | 19 | 0.579 | 0.694 |
| 19 | 20 | 0.573 | 0.687 |
| 20 | 21 | 0.567 | 0.682 |

$n$ = number of observations.

**Table V** Critical values for 5% and 1% tests of discordancy for a lower outlier in an exponential sample, using $x_{(1)}/\sum x_j$ as test statistic. Values of the statistic *lower* than the critical value are significant

**Ea3**

| $n$ | 5% | 1% |
|-----|------|------|
| 3 | 0.00844 | 0.00167 |
| 4 | 0.00424 | $0.0^3836$ |
| 5 | 0.00255 | $0.0^3502$ |
| 6 | 0.00170 | $0.0^3335$ |
| 7 | 0.00122 | $0.0^3239$ |
| 8 | $0.0^3913$ | $0.0^3179$ |
| 9 | $0.0^3710$ | $0.0^3140$ |
| 10 | $0.0^3568$ | $0.0^3112$ |
| 12 | $0.0^3388$ | $0.0^4761$ |
| 14 | $0.0^3281$ | $0.0^4552$ |
| 16 | $0.0^3213$ | $0.0^4419$ |
| 18 | $0.0^3167$ | $0.0^4328$ |
| 20 | $0.0^3135$ | $0.0^4264$ |
| 30 | $0.0^4589$ | $0.0^4116$ |
| 40 | $0.0^4329$ | $0.0^5644$ |
| 50 | $0.0^4209$ | $0.0^5410$ |
| 100 | $0.0^5518$ | $0.0^5102$ |

$n$ = number of observations.

**Table VI** Critical values for 5% and 1% Dixon-type tests of discordancy for a lower outlier in an exponential sample, using $[x_{(2)} - x_{(1)}]/[x_{(n)} - x_{(1)}]$ as test statistic

### E4

| $n$ | 5% | 1% |
|-----|-------|-------|
| 3   | 0.905 | 0.980 |
| 4   | 0.618 | 0.808 |
| 5   | 0.429 | 0.618 |
| 6   | 0.316 | 0.479 |
| 7   | 0.246 | 0.381 |
| 8   | 0.198 | 0.312 |
| 9   | 0.165 | 0.262 |
| 10  | 0.140 | 0.224 |
| 11  | 0.121 | 0.194 |
| 12  | 0.106 | 0.171 |
| 13  | 0.094 | 0.152 |
| 14  | 0.085 | 0.136 |
| 15  | 0.077 | 0.124 |
| 16  | 0.070 | 0.113 |
| 17  | 0.064 | 0.103 |
| 18  | 0.059 | 0.095 |
| 19  | 0.055 | 0.088 |
| 20  | 0.051 | 0.082 |

$n$ = number of observations.

**Table VII** Critical values for 5% and 1% tests for two upper outliers in an exponential sample, using $[x_{(n-1)} + x_{(n)}]/\sum x_j$ as test statistic

### Ea5

| Sample size $n$ | Significance level 5% | 1% | Sample size $n$ | Significance level 5% | 1% |
|-----|-------|-------|-----|-------|-------|
| 5   | 0.884 | 0.935 | 20  | 0.418 | 0.476 |
| 6   | 0.825 | 0.887 | 22  | 0.391 | 0.447 |
| 7   | 0.771 | 0.840 | 24  | 0.368 | 0.421 |
| 8   | 0.723 | 0.795 | 26  | 0.348 | 0.398 |
| 9   | 0.681 | 0.753 | 28  | 0.330 | 0.378 |
| 10  | 0.643 | 0.716 | 30  | 0.314 | 0.360 |
| 11  | 0.609 | 0.681 | 35  | 0.281 | 0.321 |
| 12  | 0.579 | 0.650 | 40  | 0.254 | 0.291 |
| 13  | 0.552 | 0.621 | 45  | 0.232 | 0.266 |
| 14  | 0.527 | 0.595 | 50  | 0.214 | 0.245 |
| 15  | 0.505 | 0.571 | 60  | 0.186 | 0.213 |
| 16  | 0.484 | 0.549 | 70  | 0.165 | 0.188 |
| 17  | 0.466 | 0.528 | 80  | 0.148 | 0.169 |
| 18  | 0.448 | 0.510 | 90  | 0.135 | 0.154 |
| 19  | 0.433 | 0.492 | 100 | 0.124 | 0.141 |

**Table VIII** Critical values for 5% and 1% tests of discordancy for a lower and upper outlier-pair in a gamma sample, using the ratio $x_{(m)}/x_{(1)}$ as test statistic. This table is reproduced, with permission of the Biometrika Trustees, from Pearson and Hartley (1966), Table 31, after changing the notation where appropriate

## Ga7(Ea7)

*Upper 5% points*

| r \ n | 2 | 3 | 4 | 5 | 6 | 7 | 8 | 9 | 10 | 11 | 12 |
|---|---|---|---|---|---|---|---|---|---|---|---|
| 1 | 39.0 | 87.5 | 142 | 202 | 266 | 333 | 403 | 475 | 550 | 626 | 704 |
| 1.5 | 15.4 | 27.8 | 39.2 | 50.7 | 62.0 | 72.9 | 83.5 | 93.9 | 104 | 114 | 124 |
| 2 | 9.60 | 15.5 | 20.6 | 25.2 | 29.5 | 33.6 | 37.5 | 41.1 | 44.6 | 48.0 | 51.4 |
| 2.5 | 7.15 | 10.8 | 13.7 | 16.3 | 18.7 | 20.8 | 22.9 | 24.7 | 26.5 | 28.2 | 29.9 |
| 3 | 5.82 | 8.38 | 10.4 | 12.1 | 13.7 | 15.0 | 16.3 | 17.5 | 18.6 | 19.7 | 20.7 |
| 3.5 | 4.99 | 6.94 | 8.44 | 9.70 | 10.8 | 11.8 | 12.7 | 13.5 | 14.3 | 15.1 | 15.8 |
| 4 | 4.43 | 6.00 | 7.18 | 8.12 | 9.03 | 9.78 | 10.5 | 11.1 | 11.7 | 12.2 | 12.7 |
| 4.5 | 4.03 | 5.34 | 6.31 | 7.11 | 7.80 | 8.41 | 8.95 | 9.45 | 9.91 | 10.3 | 10.7 |
| 5 | 3.72 | 4.85 | 5.67 | 6.34 | 6.92 | 7.42 | 7.87 | 8.28 | 8.66 | 9.01 | 9.34 |
| 6 | 3.28 | 4.16 | 4.79 | 5.30 | 5.72 | 6.09 | 6.42 | 6.72 | 7.00 | 7.25 | 7.48 |
| 7.5 | 2.86 | 3.54 | 4.01 | 4.37 | 4.68 | 4.95 | 5.19 | 5.40 | 5.59 | 5.77 | 5.93 |
| 10 | 2.46 | 2.95 | 3.29 | 3.54 | 3.76 | 3.94 | 4.10 | 4.24 | 4.37 | 4.49 | 4.59 |
| 15 | 2.07 | 2.40 | 2.61 | 2.78 | 2.91 | 3.02 | 3.12 | 3.21 | 3.29 | 3.36 | 3.39 |
| 30 | 1.67 | 1.85 | 1.96 | 2.04 | 2.11 | 2.17 | 2.22 | 2.26 | 2.30 | 2.33 | 2.36 |
| $\infty$ | 1.00 | 1.00 | 1.00 | 1.00 | 1.00 | 1.00 | 1.00 | 1.00 | 1.00 | 1.00 | 1.00 |

**Table VIII** (*Contd.*)

*Upper 1% points*

| n / r | 2 | 3 | 4 | 5 | 6 | 7 | 8 | 9 | 10 | 11 | 12 |
|---|---|---|---|---|---|---|---|---|---|---|---|
| 1 | 199 | 448 | 729 | 1036 | 1362 | 1705 | 2063 | 2432 | 2813 | 3204 | 3605 |
| 1.5 | 47.5 | 85 | 120 | 151 | 184 | 21(6) | 24(9) | 28(1) | 31(0) | 33(7) | 36(1) |
| 2 | 23.2 | 37 | 49 | 59 | 69 | 79 | 89 | 97 | 106 | 113 | 120 |
| 2.5 | 14.9 | 22 | 28 | 33 | 38 | 42 | 46 | 50 | 54 | 57 | 60 |
| 3 | 11.1 | 15.5 | 19.1 | 22 | 25 | 27 | 30 | 32 | 34 | 36 | 37 |
| 3.5 | 8.89 | 12.1 | 14.5 | 16.5 | 18.4 | 20 | 22 | 23 | 24 | 26 | 27 |
| 4 | 7.50 | 9.9 | 11.7 | 13.2 | 14.5 | 15.8 | 16.9 | 17.9 | 18.9 | 19.8 | 21 |
| 4.5 | 6.34 | 8.5 | 9.9 | 11.1 | 12.1 | 13.1 | 13.9 | 14.7 | 15.3 | 16.0 | 16.6 |
| 5 | 5.85 | 7.4 | 8.6 | 9.6 | 10.4 | 11.1 | 11.8 | 12.4 | 12.9 | 13.4 | 13.9 |
| 6 | 4.91 | 6.1 | 6.9 | 7.6 | 8.2 | 8.7 | 9.1 | 9.5 | 9.9 | 10.2 | 10.6 |
| 7.5 | 4.07 | 4.9 | 5.5 | 6.0 | 6.4 | 6.7 | 7.1 | 7.3 | 7.5 | 7.8 | 8.0 |
| 10 | 3.32 | 3.8 | 4.3 | 4.6 | 4.9 | 5.1 | 5.3 | 5.5 | 5.6 | 5.8 | 5.9 |
| 15 | 2.63 | 3.0 | 3.3 | 3.4 | 3.6 | 3.7 | 3.8 | 3.9 | 4.0 | 4.1 | 4.2 |
| 30 | 1.96 | 2.2 | 2.3 | 2.4 | 2.4 | 2.5 | 2.5 | 2.6 | 2.6 | 2.7 | 2.7 |
| ∞ | 1.00 | 1.0 | 1.0 | 1.0 | 1.0 | 1.0 | 1.0 | 1.0 | 1.0 | 1.0 | 1.0 |

Values in the column $n = 2$ and in the rows $r = 1$ and $\infty$ are exact. Elsewhere the third digit may be in error by a few units for the 5% points and several units for the 1% points. The third digit figures in parentheses for $r = 1.5$ are the most uncertain.

$n$ = number of observations.

$r$ = shape parameter of the gamma distribution.

**Table IX** Critical values for 5% and 1% tests of discordancy for a lower and upper outlier-pair in a gamma sample, using Kimber's (1988) statistic

**Ga9 (Ea9)**

| | $n$ \ $r$ | $\frac{1}{2}$ | 1 | $1\frac{1}{2}$ | 2 | $2\frac{1}{2}$ | 3 | 4 | 5 | 7 | 10 |
|---|---|---|---|---|---|---|---|---|---|---|---|
| | 5 | 0.768 | 0.521 | 0.396 | 0.315 | 0.261 | 0.221 | 0.173 | 0.137 | 0.101 | 0.071 |
| | 6 | 0.714 | 0.475 | 0.359 | 0.285 | 0.238 | 0.199 | 0.154 | 0.127 | 0.092 | 0.066 |
| | 7 | 0.676 | 0.442 | 0.330 | 0.262 | 0.218 | 0.184 | 0.142 | 0.117 | 0.084 | 0.060 |
| | 8 | 0.632 | 0.413 | 0.306 | 0.243 | 0.202 | 0.171 | 0.132 | 0.109 | 0.078 | 0.056 |
| 5 per cent | 10 | 0.567 | 0.367 | 0.268 | 0.211 | 0.175 | 0.151 | 0.117 | 0.094 | 0.069 | 0.048 |
| critical | 12 | 0.513 | 0.329 | 0.239 | 0.188 | 0.155 | 0.135 | 0.104 | 0.084 | 0.061 | 0.043 |
| values | 14 | 0.469 | 0.299 | 0.217 | 0.170 | 0.140 | 0.122 | 0.094 | 0.076 | 0.055 | 0.039 |
| | 16 | 0.433 | 0.273 | 0.199 | 0.156 | 0.128 | 0.111 | 0.085 | 0.069 | 0.050 | 0.035 |
| | 18 | 0.402 | 0.252 | 0.184 | 0.145 | 0.119 | 0.101 | 0.078 | 0.064 | 0.046 | 0.033 |
| | 20 | 0.376 | 0.234 | 0.172 | 0.136 | 0.111 | 0.094 | 0.072 | 0.060 | 0.043 | 0.031 |
| | 5 | 0.876 | 0.647 | 0.497 | 0.405 | 0.356 | 0.297 | 0.242 | 0.190 | 0.144 | 0.099 |
| | 6 | 0.835 | 0.595 | 0.465 | 0.373 | 0.317 | 0.271 | 0.209 | 0.173 | 0.125 | 0.088 |
| | 7 | 0.796 | 0.553 | 0.429 | 0.342 | 0.286 | 0.247 | 0.188 | 0.157 | 0.113 | 0.080 |
| | 8 | 0.757 | 0.516 | 0.396 | 0.314 | 0.261 | 0.227 | 0.172 | 0.144 | 0.103 | 0.073 |
| 1 per cent | 10 | 0.686 | 0.455 | 0.342 | 0.270 | 0.224 | 0.195 | 0.149 | 0.123 | 0.089 | 0.063 |
| critical | 12 | 0.626 | 0.407 | 0.302 | 0.238 | 0.198 | 0.171 | 0.132 | 0.108 | 0.079 | 0.056 |
| values | 14 | 0.575 | 0.368 | 0.272 | 0.214 | 0.178 | 0.153 | 0.118 | 0.096 | 0.071 | 0.050 |
| | 16 | 0.532 | 0.336 | 0.249 | 0.196 | 0.163 | 0.138 | 0.108 | 0.087 | 0.064 | 0.045 |
| | 18 | 0.496 | 0.310 | 0.231 | 0.182 | 0.151 | 0.126 | 0.099 | 0.080 | 0.058 | 0.041 |
| | 20 | 0.465 | 0.287 | 0.216 | 0.170 | 0.141 | 0.116 | 0.092 | 0.075 | 0.053 | 0.038 |

**Table X** Critical values for 5% and 1% tests for the presence of an undefined number of discordant values in an exponential sample, using Shapiro and Wilk's 'W-Exponential' statistic

**E14**

| n | Lower 1% | Lower 5% | Upper 5% | Upper 1% |
|---|---|---|---|---|
| 3 | 0.254 | 0.270 | 0.993 | 0.9997 |
| 4 | 0.310 | 0.160 | 0.858 | 0.968 |
| 5 | 0.0905 | 0.119 | 0.668 | 0.860 |
| 6 | 0.0665 | 0.0956 | 0.509 | 0.678 |
| 7 | 0.0591 | 0.0810 | 0.416 | 0.571 |
| 8 | 0.0512 | 0.0710 | 0.350 | 0.485 |
| 9 | 0.0442 | 0.0633 | 0.300 | 0.401 |
| 10 | 0.0404 | 0.0568 | 0.253 | 0.339 |
| 12 | 0.0358 | 0.0494 | 0.202 | 0.272 |
| 14 | 0.0317 | 0.0428 | 0.165 | 0.213 |
| 16 | 0.0280 | 0.0374 | 0.136 | 0.177 |
| 18 | 0.0250 | 0.0332 | 0.116 | 0.148 |
| 20 | 0.0227 | 0.0302 | 0.100 | 0.129 |
| 30 | 0.0164 | 0.0213 | 0.0593 | 0.0719 |
| 40 | 0.0131 | 0.0164 | 0.0414 | 0.0499 |
| 50 | 0.0111 | 0.0137 | 0.0317 | 0.0360 |
| 60 | 0.0095 | 0.0117 | 0.0252 | 0.0291 |
| 70 | 0.0084 | 0.0103 | 0.0209 | 0.0241 |
| 80 | 0.0075 | 0.0091 | 0.0177 | 0.0205 |
| 90 | 0.0069 | 0.0082 | 0.0156 | 0.0176 |
| 100 | 0.0063 | 0.0074 | 0.0139 | 0.0153 |

$n$ = number of observations.

**Table XI** Critical values for 5% and 1% consecutive tests of up to $k$ upper outliers in an exponential sample. See details of test **Ea18** on pages 213–4.

**Ea18**

$k = 2$

| $n$ | 5% $s_2$ | 5% $s_1$ | 1% $s_2$ | 1% $s_1$ |
|---|---|---|---|---|
| 10 | 0.434 | 0.483 | 0.514 | 0.570 |
| 11 | 0.401 | 0.453 | 0.475 | 0.536 |
| 12 | 0.372 | 0.427 | 0.441 | 0.506 |
| 13 | 0.348 | 0.403 | 0.412 | 0.480 |
| 14 | 0.327 | 0.383 | 0.387 | 0.456 |
| 15 | 0.308 | 0.364 | 0.364 | 0.435 |
| 16 | 0.292 | 0.347 | 0.345 | 0.415 |
| 17 | 0.277 | 0.332 | 0.327 | 0.398 |
| 18 | 0.264 | 0.318 | 0.312 | 0.381 |
| 19 | 0.252 | 0.306 | 0.297 | 0.366 |
| 20 | 0.241 | 0.294 | 0.284 | 0.353 |
| 22 | 0.222 | 0.273 | 0.262 | 0.328 |

| $n$ | 5% $s_2$ | 5% $s_1$ | 1% $s_2$ | 1% $s_1$ |
|---|---|---|---|---|
| 24 | 0.206 | 0.256 | 0.242 | 0.307 |
| 26 | 0.193 | 0.240 | 0.226 | 0.289 |
| 28 | 0.181 | 0.227 | 0.212 | 0.273 |
| 30 | 0.171 | 0.215 | 0.200 | 0.258 |
| 35 | 0.150 | 0.190 | 0.175 | 0.229 |
| 40 | 0.134 | 0.171 | 0.156 | 0.205 |
| 45 | 0.121 | 0.155 | 0.140 | 0.186 |
| 50 | 0.111 | 0.142 | 0.128 | 0.171 |
| 60 | 0.0947 | 0.122 | 0.109 | 0.147 |
| 70 | 0.0830 | 0.107 | 0.0955 | 0.129 |
| 80 | 0.0740 | 0.0960 | 0.0850 | 0.115 |
| 90 | 0.0670 | 0.0869 | 0.0767 | 0.104 |
| 100 | 0.0612 | 0.0794 | 0.0699 | 0.0948 |
| 120 | 0.0523 | 0.0679 | 0.0597 | 0.0809 |
| 140 | 0.0459 | 0.0595 | 0.0521 | 0.0707 |

**Table XI** (*Contd.*)

k = 3

| n | 5% | | | 1% | | |
|---|---|---|---|---|---|---|
| | $s_3$ | $s_2$ | $s_1$ | $s_3$ | $s_2$ | $s_1$ |
| 15 | 0.306 | 0.321 | 0.380 | 0.359 | 0.379 | 0.452 |
| 16 | 0.287 | 0.304 | 0.363 | 0.337 | 0.359 | 0.432 |
| 17 | 0.271 | 0.288 | 0.347 | 0.316 | 0.339 | 0.412 |
| 18 | 0.257 | 0.274 | 0.333 | 0.299 | 0.322 | 0.395 |
| 19 | 0.248 | 0.266 | 0.325 | 0.284 | 0.307 | 0.380 |
| 20 | 0.236 | 0.255 | 0.313 | 0.270 | 0.294 | 0.366 |
| 22 | 0.216 | 0.235 | 0.291 | 0.248 | 0.272 | 0.343 |
| 24 | 0.196 | 0.214 | 0.269 | 0.228 | 0.252 | 0.321 |
| 26 | 0.183 | 0.200 | 0.251 | 0.212 | 0.235 | 0.302 |
| 28 | 0.170 | 0.188 | 0.236 | 0.196 | 0.219 | 0.283 |
| 30 | 0.163 | 0.180 | 0.229 | 0.184 | 0.206 | 0.268 |

| n | 5% | | | 1% | | |
|---|---|---|---|---|---|---|
| | $s_3$ | $s_2$ | $s_1$ | $s_3$ | $s_2$ | $s_1$ |
| 35 | 0.139 | 0.155 | 0.198 | 0.160 | 0.182 | 0.239 |
| 40 | 0.125 | 0.140 | 0.179 | 0.141 | 0.160 | 0.213 |
| 45 | 0.113 | 0.128 | 0.166 | 0.126 | 0.145 | 0.193 |
| 50 | 0.101 | 0.114 | 0.147 | 0.116 | 0.133 | 0.179 |
| 60 | 0.0860 | 0.0973 | 0.127 | 0.0983 | 0.113 | 0.154 |
| 70 | 0.0751 | 0.0852 | 0.111 | 0.0849 | 0.0981 | 0.133 |
| 80 | 0.0685 | 0.0781 | 0.103 | 0.0754 | 0.0873 | 0.119 |
| 90 | 0.0619 | 0.0706 | 0.0932 | 0.0685 | 0.0796 | 0.109 |
| 100 | 0.0559 | 0.0636 | 0.0853 | 0.0624 | 0.0726 | 0.0994 |
| 120 | 0.0469 | 0.0534 | 0.0698 | 0.0532 | 0.0618 | 0.0849 |
| 140 | 0.0417 | 0.0476 | 0.0626 | 0.0458 | 0.0532 | 0.0727 |

**Table XI** (*Contd.*)

| | | | | k = 4 | | | | |
|---|---|---|---|---|---|---|---|---|
| | | 5% | | | | 1% | | |
| n | $s_4$ | $s_3$ | $s_2$ | $s_1$ | $s_4$ | $s_3$ | $s_2$ | $s_1$ |
| 20 | 0.235 | 0.241 | 0.261 | 0.321 | 0.269 | 0.277 | 0.302 | 0.377 |
| 22 | 0.213 | 0.221 | 0.240 | 0.299 | 0.243 | 0.253 | 0.278 | 0.351 |
| 24 | 0.195 | 0.203 | 0.223 | 0.280 | 0.222 | 0.233 | 0.258 | 0.329 |
| 26 | 0.180 | 0.189 | 0.208 | 0.263 | 0.205 | 0.216 | 0.240 | 0.309 |
| 28 | 0.168 | 0.176 | 0.195 | 0.248 | 0.190 | 0.201 | 0.225 | 0.292 |
| 30 | 0.157 | 0.166 | 0.184 | 0.235 | 0.177 | 0.189 | 0.212 | 0.277 |
| 35 | 0.136 | 0.144 | 0.161 | 0.208 | 0.152 | 0.163 | 0.185 | 0.245 |
| 40 | 0.120 | 0.128 | 0.144 | 0.187 | 0.134 | 0.144 | 0.165 | 0.220 |
| 45 | 0.107 | 0.115 | 0.130 | 0.169 | 0.120 | 0.130 | 0.149 | 0.200 |
| 50 | 0.0974 | 0.105 | 0.119 | 0.155 | 0.108 | 0.118 | 0.136 | 0.183 |
| 60 | 0.0826 | 0.0892 | 0.101 | 0.134 | 0.0915 | 0.0998 | 0.115 | 0.157 |
| 70 | 0.0719 | 0.0779 | 0.0888 | 0.117 | 0.0795 | 0.0869 | 0.101 | 0.138 |
| 80 | 0.0639 | 0.0693 | 0.0792 | 0.105 | 0.0704 | 0.0772 | 0.0897 | 0.123 |
| 90 | 0.0576 | 0.0625 | 0.0715 | 0.0948 | 0.0633 | 0.0695 | 0.0809 | 0.111 |
| 100 | 0.0525 | 0.0571 | 0.0653 | 0.0866 | 0.0576 | 0.0633 | 0.0737 | 0.101 |
| 120 | 0.0447 | 0.0487 | 0.0558 | 0.0740 | 0.0489 | 0.0538 | 0.0628 | 0.0866 |
| 140 | 0.0391 | 0.0426 | 0.0488 | 0.0647 | 0.0427 | 0.0470 | 0.0548 | 0.0756 |

**Table XII** Critical values for 5% and 1% consecutive tests of up to $k$ lower outliers in an exponential sample. See details of test **Ea19** on pages 214–5.

**Ea19**

| | | k = 2 | | |
|---|---|---|---|---|
| | | 5% | | 1% |
| n | $t_2$ | $t_1$ | $t_2$ | $t_1$ |
| 10 | 0.837 | 0.977 | 0.921 | 0.996 |
| 15 | 0.829 | 0.977 | 0.917 | 0.995 |
| 20 | 0.825 | 0.976 | 0.915 | 0.995 |
| 50 | 0.819 | 0.975 | 0.911 | 0.995 |
| 100 | 0.817 | 0.975 | 0.910 | 0.995 |
| 200 | 0.816 | 0.975 | 0.910 | 0.995 |

| | | | k = 3 | | | |
|---|---|---|---|---|---|---|
| | | 5% | | | 1% | |
| n | $t_3$ | $t_2$ | $t_1$ | $t_3$ | $t_2$ | $t_1$ |
| 15 | 0.705 | 0.855 | 0.984 | 0.808 | 0.931 | 0.997 |
| 20 | 0.698 | 0.852 | 0.984 | 0.802 | 0.929 | 0.997 |
| 50 | 0.687 | 0.846 | 0.983 | 0.792 | 0.926 | 0.997 |
| 100 | 0.683 | 0.845 | 0.983 | 0.789 | 0.925 | 0.997 |
| 200 | 0.681 | 0.843 | 0.983 | 0.788 | 0.925 | 0.997 |

**Table XII** (*Contd.*)

### k = 4

| | | 5% | | | | 1% | | |
|---|---|---|---|---|---|---|---|---|
| n | $t_4$ | $t_3$ | $t_2$ | $t_1$ | $t_4$ | $t_3$ | $t_2$ | $t_1$ |
| 20 | 0.596 | 0.717 | 0.868 | 0.988 | 0.693 | 0.817 | 0.938 | 0.998 |
| 50 | 0.580 | 0.706 | 0.863 | 0.987 | 0.677 | 0.807 | 0.935 | 0.998 |
| 100 | 0.575 | 0.702 | 0.861 | 0.987 | 0.672 | 0.804 | 0.934 | 0.997 |
| 200 | 0.573 | 0.700 | 0.860 | 0.987 | 0.671 | 0.803 | 0.934 | 0.997 |

**Table XIII** Critical values for 5% and 1% tests of discordancy for a single outlier in a normal sample, using the deviation from the sample mean or population mean, studentized or standardized, as test statistic

| | Table XIIIa N1 | | Table XIIIb N2 | | Table XIIIc Nμ1 | | Table XIIId Nμ2 | |
|---|---|---|---|---|---|---|---|---|
| n | 5% | 1% | 5% | 1% | 5% | 1% | 5% | 1% |
| 3 | 1.15 | 1.15 | 1.15 | 1.15 | 1.68 | 1.72 | 1.70 | 1.73 |
| 4 | 1.46 | 1.49 | 1.48 | 1.50 | 1.85 | 1.95 | 1.90 | 1.97 |
| 5 | 1.67 | 1.75 | 1.71 | 1.76 | 1.98 | 2.12 | 2.05 | 2.15 |
| 6 | 1.82 | 1.94 | 1.89 | 1.97 | 2.07 | 2.25 | 2.16 | 2.30 |
| 7 | 1.94 | 2.10 | 2.02 | 2.14 | 2.15 | 2.36 | 2.26 | 2.42 |
| 8 | 2.03 | 2.22 | 2.13 | 2.28 | 2.21 | 2.45 | 2.33 | 2.52 |
| 9 | 2.11 | 2.32 | 2.21 | 2.38 | 2.26 | 2.52 | 2.40 | 2.61 |
| 10 | 2.18 | 2.41 | 2.29 | 2.48 | 2.31 | 2.59 | 2.45 | 2.68 |
| 12 | 2.29 | 2.55 | 2.41 | 2.63 | 2.40 | 2.70 | 2.55 | 2.80 |
| 14 | 2.37 | 2.66 | | | 2.47 | 2.79 | | |
| 15 | 2.41 | 2.71 | 2.55 | 2.81 | 2.50 | 2.83 | 2.66 | 2.94 |
| 16 | 2.44 | 2.75 | | | 2.53 | 2.86 | | |
| 18 | 2.50 | 2.82 | | | 2.58 | 2.91 | | |
| 20 | 2.56 | 2.88 | 2.71 | 3.00 | 2.62 | 2.95 | 2.79 | 3.10 |
| 30 | 2.74 | 3.10 | | | 2.80 | 3.17 | 2.96 | 3.30 |
| 40 | 2.87 | 3.24 | | | 2.92 | 3.30 | 3.08 | 3.43 |
| 50 | 2.96 | 3.34 | | | 2.98 | 3.39 | | |
| 60 | 3.03 | 3.41 | | | | | 3.23 | 3.59 |
| 100 | 3.21 | 3.60 | | | 3.23 | 3.61 | | |
| 120 | 3.27 | 3.66 | | | | | 3.46 | 3.83 |

**Table XIII** (*Contd.*)

| | Table XIIIe N$\sigma$1 | | Table XIIIf N$\sigma$2 | | Table XIIIg N$\mu\sigma$1 | |
|---|---|---|---|---|---|---|
| n | 5% | 1% | 5% | 1% | 5% | 1% |
| 3 | 1.74 | 2.22 | 1.93 | 2.39 | 2.12 | 2.71 |
| 4 | 1.94 | 2.43 | 2.15 | 2.61 | 2.23 | 2.81 |
| 5 | 2.08 | 2.57 | 2.29 | 2.76 | 2.32 | 2.88 |
| 6 | 2.18 | 2.68 | 2.40 | 2.87 | 2.39 | 2.93 |
| 7 | 2.27 | 2.76 | 2.48 | 2.95 | 2.44 | 2.98 |
| 8 | 2.33 | 2.83 | 2.55 | 3.02 | 2.49 | 3.02 |
| 9 | 2.39 | 2.88 | 2.60 | 3.07 | 2.53 | 3.06 |
| 10 | 2.44 | 2.93 | 2.65 | 3.12 | 2.57 | 3.09 |
| 12 | 2.52 | 3.01 | | | 2.63 | 3.14 |
| 14 | 2.59 | 3.07 | | | 2.68 | 3.19 |
| 15 | 2.62 | 3.10 | 2.82 | 3.29 | 2.71 | 3.21 |
| 16 | 2.64 | 3.12 | | | 2.73 | 3.23 |
| 18 | 2.69 | 3.17 | | | 2.77 | 3.26 |
| 20 | 2.73 | 3.21 | 2.94 | 3.39 | 2.80 | 3.29 |
| 30 | 2.88 | 3.38 | 3.08 | 3.53 | 2.93 | 3.40 |
| 40 | 2.99 | 3.44 | 3.18 | 3.62 | 3.02 | 3.48 |
| 50 | 3.05 | 3.53 | | | 3.08 | 3.54 |
| 60 | | | 3.30 | 3.73 | 3.14 | 3.59 |
| 100 | 3.27 | 3.67 | | | 3.28 | 3.72 |
| 120 | | | | | 3.33 | 3.76 |
| 200 | | | | | 3.47 | 3.89 |
| 500 | | | | | 3.71 | 4.11 |
| 1000 | | | | | 3.88 | 4.26 |

n = number of observations.

**Table XIVa** Critical values for 5% and 1% tests of discordancy for a single outlier in a normal sample, using the externally studentized deviation from the mean as test statistic

**Nv1**

*5% critical values*

| $v$ \ $n$ | 5 | 6 | 8 | 10 | 15 | 20 | 30 | 40 | 60 | ∞ |
|---|---|---|---|---|---|---|---|---|---|---|
| 3 | 2.37 | 2.24 | 2.09 | 2.01 | 1.91 | 1.87 | 1.82 | 1.80 | 1.78 | 1.74 |
| 4 | 2.71 | 2.55 | 2.37 | 2.27 | 2.15 | 2.10 | 2.04 | 2.02 | 1.99 | 1.94 |
| 5 | 2.95 | 2.78 | 2.57 | 2.46 | 2.32 | 2.26 | 2.20 | 2.17 | 2.14 | 2.08 |
| 6 | 3.15 | 2.95 | 2.72 | 2.60 | 2.45 | 2.38 | 2.31 | 2.28 | 2.25 | 2.18 |
| 7 | 3.30 | 3.09 | 2.85 | 2.72 | 2.55 | 2.47 | 2.40 | 2.37 | 2.33 | 2.27 |
| 8 | 3.43 | 3.21 | 2.95 | 2.81 | 2.64 | 2.56 | 2.48 | 2.44 | 2.41 | 2.33 |
| 9 | 3.54 | 3.31 | 3.04 | 2.89 | 2.71 | 2.63 | 2.54 | 2.50 | 2.47 | 2.39 |
| 10 | 3.64 | 3.39 | 3.12 | 2.96 | 2.77 | 2.68 | 2.60 | 2.56 | 2.52 | 2.44 |
| 12 | 3.80 | 3.54 | 3.25 | 3.08 | 2.88 | 2.78 | 2.69 | 2.65 | 2.61 | 2.52 |

*1% critical values*

| $v$ \ $n$ | 5 | 6 | 8 | 10 | 15 | 20 | 30 | 40 | 60 | ∞ |
|---|---|---|---|---|---|---|---|---|---|---|
| 3 | 3.65 | 3.32 | 2.96 | 2.78 | 2.57 | 2.47 | 2.38 | 2.34 | 2.29 | 2.22 |
| 4 | 4.11 | 3.72 | 3.31 | 3.10 | 2.84 | 2.73 | 2.62 | 2.57 | 2.52 | 2.43 |
| 5 | 4.45 | 4.02 | 3.56 | 3.32 | 3.03 | 2.91 | 2.79 | 2.73 | 2.68 | 2.57 |
| 6 | 4.70 | 4.24 | 3.74 | 3.48 | 3.17 | 3.04 | 2.91 | 2.85 | 2.79 | 2.68 |
| 7 | 4.93 | 4.43 | 3.89 | 3.62 | 3.29 | 3.14 | 3.01 | 2.94 | 2.88 | 2.76 |
| 8 | 5.11 | 4.58 | 4.02 | 3.73 | 3.38 | 3.23 | 3.08 | 3.02 | 2.95 | 2.83 |
| 9 | 5.26 | 4.71 | 4.13 | 3.82 | 3.46 | 3.30 | 3.15 | 3.08 | 3.01 | 2.88 |
| 10 | 5.39 | 4.82 | 4.22 | 3.90 | 3.53 | 3.37 | 3.21 | 3.13 | 3.06 | 2.93 |
| 12 | 5.62 | 5.01 | 4.38 | 4.04 | 3.65 | 3.47 | 3.30 | 3.22 | 3.15 | 3.01 |

$n$ = number of observations.
$v$ = degrees of freedom of independent estimate of $\sigma^2$.

**Table XIVb** Critical values for 5% and 1% tests of discordancy for a single outlier in a normal sample, using the externally and internally studentized deviation from the mean as test statistic

**Nv2**

*5% critical values*

| $v$ / $n$ | 1 | 2 | 3 | 4 | 6 | 12 | 50 |
|---|---|---|---|---|---|---|---|
| 3 | 1.37 | 1.48 | 1.55 | 1.59 | 1.63 | 1.68 | 1.72 |
| 4 | 1.60 | 1.68 | 1.73 | 1.77 | 1.81 | 1.87 | 1.92 |
| 5 | 1.76 | 1.82 | 1.87 | 1.90 | 1.94 | 2.00 | 2.06 |
| 6 | 1.89 | 1.94 | 1.97 | 2.00 | 2.04 | 2.09 | 2.16 |
| 7 | 1.99 | 2.03 | 2.06 | 2.08 | 2.11 | 2.17 | 2.24 |
| 8 | 2.07 | 2.10 | 2.13 | 2.15 | 2.18 | 2.23 | 2.30 |
| 10 | 2.20 | 2.23 | 2.24 | 2.26 | 2.29 | 2.33 | 2.40 |
| 12 | 2.31 | 2.32 | 2.34 | 2.35 | 2.37 | 2.41 | 2.48 |
| 15 | 2.42 | 2.44 | 2.45 | 2.46 | 2.47 | 2.51 | 2.58 |
| 20 | 2.57 | 2.58 | 2.58 | 2.59 | 2.60 | 2.63 | 2.68 |

*1% critical values*

| $v$ / $n$ | 1 | 2 | 3 | 4 | 6 | 12 | 50 |
|---|---|---|---|---|---|---|---|
| 3 | 1.40 | 1.58 | 1.70 | 1.79 | 1.90 | 2.04 | 2.17 |
| 4 | 1.69 | 1.82 | 1.92 | 1.99 | 2.09 | 2.22 | 2.30 |
| 5 | 1.90 | 2.00 | 2.08 | 2.14 | 2.23 | 2.36 | 2.51 |
| 6 | 2.06 | 2.14 | 2.21 | 2.26 | 2.33 | 2.46 | 2.61 |
| 7 | 2.19 | 2.25 | 2.31 | 2.35 | 2.42 | 2.53 | 2.69 |
| 8 | 2.29 | 2.35 | 2.40 | 2.43 | 2.49 | 2.60 | 2.75 |
| 10 | 2.46 | 2.50 | 2.54 | 2.57 | 2.61 | 2.70 | 2.85 |
| 12 | 2.59 | 2.62 | 2.65 | 2.67 | 2.70 | 2.79 | 2.92 |
| 15 | 2.73 | 2.75 | 2.77 | 2.79 | 2.82 | 2.88 | 3.01 |
| 20 | 2.90 | 2.91 | 2.93 | 2.94 | 2.96 | 3.01 | 3.12 |

$n$ = number of observations.
$v$ = degrees of freedom of independent estimate of $\sigma^2$.

**Table XIVc** Critical values for 5% and 1% tests of discordancy for a single outlier in a normal sample, using the greatest externally studentized deviation from the mean as test statistic

$Nv3$

*5% critical values*

| $v$ / $n$ | 6 | 8 | 10 | 15 | 20 | 30 | 40 | 60 | $\infty$ |
|---|---|---|---|---|---|---|---|---|---|
| 3 | 2.6 | 2.4 | 2.3 | 2.2 | 2.1 | 2.0 | 2.0 | 2.0 | 1.9 |
| 4 | | 2.7 | 2.6 | 2.4 | 2.3 | 2.3 | 2.2 | 2.2 | 2.1 |
| 5 | | 2.9 | 2.8 | 2.6 | 2.5 | 2.4 | 2.4 | 2.4 | 2.3 |
| 6 | | | 2.9 | 2.7 | 2.6 | 2.5 | 2.5 | 2.5 | 2.4 |
| 8 | | | | 2.9 | 2.8 | 2.7 | 2.7 | 2.6 | 2.6 |
| 10 | | | | 3.1 | 3.0 | 2.9 | 2.8 | 2.8 | 2.7 |
| 15 | | | | 3.3 | 3.2 | 3.1 | 3.0 | 2.9 | 2.8 |
| 20 | | | | | 3.3 | 3.2 | 3.1 | 3.0 | 2.9 |
| 30 | | | | | | 3.4 | 3.3 | 3.2 | 3.1 |
| 40 | | | | | | | 3.4 | 3.3 | 3.2 |
| 60 | | | | | | | | 3.5 | 3.3 |

*1% critical values*

| $v$ / $n$ | 8 | 10 | 15 | 20 | 30 | 40 | 60 | $\infty$ |
|---|---|---|---|---|---|---|---|---|
| 3 | 3.3 | 3.1 | 2.8 | 2.7 | 2.6 | 2.5 | 2.5 | 2.4 |
| 4 | 3.7 | 3.4 | 3.1 | 3.0 | 2.9 | 2.8 | 2.7 | 2.6 |
| 5 | 4.0 | 3.7 | 3.3 | 3.2 | 3.0 | 2.9 | 2.9 | 2.8 |
| 6 | | 3.9 | 3.5 | 3.3 | 3.2 | 3.1 | 3.0 | 2.9 |
| 8 | | 4.1 | 3.7 | 3.5 | 3.3 | 3.2 | 3.2 | 3.0 |
| 10 | | 4.3 | 3.8 | 3.6 | 3.5 | 3.4 | 3.3 | 3.1 |
| 15 | | | 4.1 | 3.9 | 3.7 | 3.6 | 3.5 | 3.3 |
| 20 | | | | 4.0 | 3.9 | 3.7 | 3.6 | 3.4 |
| 30 | | | | | 4.0 | 3.8 | 3.7 | 3.5 |
| 40 | | | | | | 4.0 | 3.8 | 3.6 |
| 60 | | | | | | | 4.0 | 3.7 |

$n$ = number of observations.
$v$ = degrees of freedom of independent estimate of $\sigma^2$.

**Table XIVd** Critical values for 5% and 1% tests of discordancy for a single outlier in a normal sample, using the greatest externally and internally studentized deviation from the mean as test statistic

Nv4

*5% critical values*

| $v$ $n$ | 1 | 2 | 3 | 4 | 6 | 12 | 50 |
|---|---|---|---|---|---|---|---|
| 3 | 1.39 | 1.54 | 1.63 | 1.69 | 1.76 | 1.8 | 1.9 |
| 4 | 1.65 | 1.76 | 1.83 | 1.88 | 1.95 | 2.03 | 2.1 |
| 5 | 1.83 | 1.92 | 1.97 | 2.02 | 2.08 | 2.16 | 2.2 |
| 6 | 1.98 | 2.04 | 2.09 | 2.12 | 2.18 | 2.26 | 2.35 |
| 7 | 2.09 | 2.14 | 2.18 | 2.21 | 2.26 | 2.34 | 2.43 |
| 8 | 2.18 | 2.22 | 2.26 | 2.29 | 2.33 | 2.40 | 2.49 |
| 10 | 2.33 | 2.36 | 2.38 | 2.40 | 2.44 | 2.50 | 2.59 |
| 12 | 2.44 | 2.46 | 2.48 | 2.50 | 2.53 | 2.58 | 2.67 |
| 15 | 2.57 | 2.58 | 2.60 | 2.61 | 2.63 | 2.68 | 2.77 |
| 20 | 2.72 | 2.73 | 2.74 | 2.75 | 2.77 | 2.80 | 2.87 |

*1% critical values*

| $v$ $n$ | 1 | 2 | 3 | 4 | 6 | 12 | 50 |
|---|---|---|---|---|---|---|---|
| 3 | 1.41 | 1.60 | 1.74 | 1.84 | 1.97 | 2.15 | 2.3 |
| 4 | 1.70 | 1.86 | 1.97 | 2.06 | 2.18 | 2.35 | 2.53 |
| 5 | 1.93 | 2.05 | 2.14 | 2.21 | 2.32 | 2.48 | 2.67 |
| 6 | 2.10 | 2.20 | 2.28 | 2.34 | 2.43 | 2.58 | 2.77 |
| 7 | 2.24 | 2.32 | 2.39 | 2.44 | 2.52 | 2.66 | 2.85 |
| 8 | 2.36 | 2.42 | 2.48 | 2.53 | 2.60 | 2.73 | 2.92 |
| 10 | 2.54 | 2.59 | 2.63 | 2.67 | 2.73 | 2.84 | 3.01 |
| 12 | 2.68 | 2.71 | 2.75 | 2.78 | 2.82 | 2.92 | 3.09 |
| 15 | 2.84 | 2.86 | 2.89 | 2.91 | 2.94 | 3.02 | 3.18 |
| 20 | 3.01 | 3.03 | 3.05 | 3.06 | 3.09 | 3.15 | 3.28 |

$n$ = number of observations.
$v$ = degrees of freedom of independent estimate of $\sigma^2$.

**Table XV** Critical values for 5% and 1% tests of discordancy for k upper outliers in a normal sample (part of Table XV is reproduced by permission of the American Statistical Association and the American Society for Quality Control)

Table XVa
N3

| $n$ | $k=2$ 5% | $k=2$ 1% | $k=3$ 5% | $k=3$ 1% | $k=4$ 5% | $k=4$ 1% |
|-----|------|------|------|------|------|------|
| 5   | 2.10 | 2.16 |      |      |      |      |
| 6   | 2.41 | 2.50 |      |      |      |      |
| 7   | 2.66 | 2.79 | 2.97 | 3.08 |      |      |
| 8   | 2.87 | 3.02 | 3.29 | 3.42 |      |      |
| 9   | 3.04 | 3.22 | 3.58 | 3.73 | 3.82 | 3.98 |
| 10  | 3.18 | 3.40 | 3.82 | 4.00 | 4.17 | 4.34 |
| 12  | 3.44 | 3.70 | 4.24 | 4.44 | 4.72 | 4.92 |
| 14  | 3.66 | 3.92 | 4.57 | 4.83 | 5.20 | 5.42 |
| 16  | 3.83 | 4.10 | 4.85 | 5.14 | 5.60 | 5.85 |
| 18  | 3.96 | 4.25 | 5.08 | 5.38 | 5.91 | 6.20 |
| 20  | 4.11 | 4.41 | 5.30 | 5.60 | 6.22 | 6.54 |
| 30  | 4.56 | 4.92 | 6.03 | 6.41 | 7.26 | 7.64 |
| 40  | 4.84 | 5.29 | 6.49 | 6.98 | 7.93 | 8.38 |
| 50  | 5.06 | 5.51 | 6.82 | 7.34 | 8.38 | 8.88 |
| 100 | 5.62 | 6.06 | 7.77 | 8.27 | 9.71 | 10.3 |

Table XVb
N4

| $n$ | $k=2$ 5% | $k=2$ 1% | $k=3$ 5% | $k=3$ 1% | $k=4$ 5% | $k=4$ 1% |
|-----|-------|--------|-------|-------|-------|-------|
| 5   | 0.019 | 0.0034 |       |       |       |       |
| 6   | 0.056 | 0.019  |       |       |       |       |
| 7   | 0.102 | 0.044  | 0.032 | 0.010 |       |       |
| 8   | 0.148 | 0.075  | 0.064 | 0.028 |       |       |
| 9   | 0.191 | 0.108  | 0.099 | 0.048 | 0.045 | 0.018 |
| 10  | 0.231 | 0.141  | 0.129 | 0.070 | 0.070 | 0.032 |
| 12  | 0.300 | 0.204  | 0.196 | 0.120 | 0.125 | 0.070 |
| 14  | 0.357 | 0.261  | 0.250 | 0.172 | 0.174 | 0.113 |
| 16  | 0.405 | 0.310  | 0.300 | 0.219 | 0.219 | 0.151 |
| 18  | 0.446 | 0.353  | 0.337 | 0.260 | 0.259 | 0.192 |
| 20  | 0.480 | 0.391  | 0.377 | 0.300 | 0.299 | 0.231 |
| 30  | 0.601 | 0.527  | 0.506 | 0.434 | 0.434 | 0.369 |
| 40  | 0.672 | 0.610  | 0.588 | 0.522 | 0.523 | 0.460 |
| 50  | 0.720 | 0.667  | 0.646 | 0.592 | 0.588 | 0.531 |
| 100 | 0.833 | 0.802  |       |       |       |       |

Table XVc
Nμ3

| | k = 2 | | k = 3 | | k = 4 | |
|---|---|---|---|---|---|---|
| n | 5% | 1% | 5% | 1% | 5% | 1% |
| 4 | 2.68 | 2.77 | | | | |
| 5 | 2.90 | 3.05 | | | | |
| 6 | 3.10 | 3.27 | 3.85 | 4.04 | | |
| 7 | 3.27 | 3.47 | 4.07 | 4.29 | | |
| 8 | 3.41 | 3.63 | 4.30 | 4.52 | 4.98 | 5.25 |
| 9 | 3.53 | 3.77 | 4.50 | 4.72 | 5.23 | 5.52 |
| 10 | 3.63 | 3.89 | 4.66 | 4.90 | 5.44 | 5.76 |
| 12 | 3.83 | 4.12 | 4.94 | 5.24 | 5.84 | 6.17 |
| 14 | 3.98 | 4.30 | 5.18 | 5.53 | 6.18 | 6.55 |
| 16 | 4.10 | 4.45 | 5.39 | 5.75 | 6.47 | 6.84 |
| 18 | 4.22 | 4.59 | 5.57 | 5.95 | 6.72 | 7.10 |
| 20 | 4.34 | 4.70 | 5.75 | 6.13 | 6.94 | 7.35 |
| 30 | 4.71 | 5.13 | 6.32 | 6.78 | 7.73 | 8.17 |
| 40 | 4.96 | 5.44 | 6.74 | 7.26 | 8.32 | 8.83 |
| 50 | 5.12 | 5.65 | 6.99 | 7.52 | 8.65 | 9.19 |
| 100 | 5.68 | 6.13 | 7.87 | 8.38 | 9.88 | 10.5 |

Table XVd
Nμ4

| | k = 2 | |
|---|---|---|
| | 5% | 1% |
| | 0.034 | 0.0072 |
| | 0.086 | 0.027 |
| | 0.128 | 0.052 |
| | 0.169 | 0.091 |
| | 0.215 | 0.125 |
| | 0.255 | 0.157 |
| | 0.286 | 0.186 |
| | 0.349 | 0.244 |
| | 0.400 | 0.298 |
| | 0.448 | 0.346 |
| | 0.484 | 0.386 |
| | 0.510 | 0.416 |
| | 0.616 | 0.543 |
| | 0.683 | 0.618 |
| | 0.729 | 0.670 |
| | 0.836 | 0.808 |

Table XVe
Nσ3

| | k = 2 | | k = 3 | | k = 4 | |
|---|---|---|---|---|---|---|
| n | 5% | 1% | 5% | 1% | 5% | 1% |
| 4 | 2.39 | 2.93 | | | | |
| 5 | 2.81 | 3.38 | | | | |
| 6 | 3.10 | 3.69 | 3.37 | 3.97 | | |
| 7 | 3.33 | 3.93 | 3.81 | 4.45 | | |
| 8 | 3.51 | 4.12 | 4.13 | 4.80 | 4.35 | 5.02 |
| 9 | 3.66 | 4.28 | 4.45 | 5.14 | 4.83 | 5.61 |
| 10 | 3.79 | 4.41 | 4.66 | 5.40 | 5.16 | 5.93 |
| 12 | 4.00 | 4.63 | 5.03 | 5.81 | 5.74 | 6.56 |
| 14 | 4.17 | 4.79 | 5.31 | 6.09 | 6.18 | 7.01 |
| 16 | 4.31 | 4.93 | 5.60 | 6.34 | 6.56 | 7.38 |
| 18 | 4.43 | 5.05 | 5.80 | 6.53 | 6.86 | 7.67 |
| 20 | 4.53 | 5.14 | 5.96 | 6.66 | 7.10 | 7.90 |
| 30 | 4.89 | 5.52 | 6.57 | 7.31 | 7.97 | 8.83 |
| 40 | 5.13 | 5.76 | 6.95 | 7.67 | 8.55 | 9.36 |
| 50 | 5.28 | 5.94 | 7.18 | 7.94 | 8.89 | 9.74 |
| 100 | 5.77 | 6.32 | 8.02 | 8.73 | 10.1 | 10.9 |

Table XVf
Nσ4

| | k = 2 | |
|---|---|---|
| | 5% | 1% |
| | 0.0012 | 0.00004 |
| | 0.037 | 0.0077 |
| | 0.162 | 0.049 |
| | 0.351 | 0.137 |
| | 0.585 | 0.268 |
| | 0.869 | 0.438 |
| | 1.22 | 0.631 |
| | 2.03 | 1.21 |
| | 2.94 | 1.95 |
| | 4.03 | 2.77 |
| | 5.11 | 3.69 |
| | 6.21 | 4.42 |
| | 12.4 | 9.8 |
| | 19.5 | 16.2 |
| | 26.8 | 22.9 |
| | 67.7 | 60.0 |

| | Table XVg Nμσ3 | | | | | | Table XVh Nμσ4 | |
|---|---|---|---|---|---|---|---|---|
| | $k = 2$ | | $k = 3$ | | $k = 4$ | | $k = 2$ | |
| $n$ | 5% | 1% | 5% | 1% | 5% | 1% | 5% | 1% |
| 4 | 3.27 | 4.14 | | | | | 0.071 | 0.015 |
| 5 | 3.44 | 4.26 | | | | | 0.232 | 0.067 |
| 6 | 3.65 | 4.46 | 4.48 | 5.47 | | | 0.450 | 0.166 |
| 7 | 3.82 | 4.58 | 4.72 | 5.69 | | | 0.683 | 0.335 |
| 8 | 3.95 | 4.70 | 4.91 | 5.91 | 5.57 | 6.72 | 1.05 | 0.539 |
| 9 | 4.04 | 4.78 | 5.10 | 6.08 | 5.88 | 6.99 | 1.41 | 0.741 |
| 10 | 4.12 | 4.87 | 5.25 | 6.23 | 6.12 | 7.23 | 1.82 | 0.984 |
| 12 | 4.28 | 5.00 | 5.54 | 6.51 | 6.54 | 7.63 | 2.66 | 1.57 |
| 14 | 4.42 | 5.14 | 5.81 | 6.70 | 6.89 | 7.94 | 3.63 | 2.40 |
| 16 | 4.54 | 5.25 | 6.02 | 6.87 | 7.18 | 8.21 | 4.73 | 3.24 |
| 18 | 4.64 | 5.35 | 6.18 | 7.03 | 7.40 | 8.43 | 5.89 | 4.24 |
| 20 | 4.73 | 5.44 | 6.30 | 7.16 | 7.60 | 8.63 | 7.12 | 5.28 |
| 30 | 5.01 | 5.68 | 6.77 | 7.61 | 8.31 | 9.26 | 13.3 | 10.7 |
| 40 | 5.24 | 5.85 | 7.15 | 7.90 | 8.81 | 9.73 | 20.4 | 16.9 |
| 50 | 5.36 | 6.03 | 7.34 | 8.17 | 9.12 | 10.1 | 27.9 | 23.8 |
| 100 | 5.81 | 6.39 | 8.11 | 8.83 | 10.2 | 11.1 | 68.6 | 61.4 |

$n$ = number of observations.
$k$ = number of outliers.

**Table XVI** Critical values for 5% and 1% tests of discordancy for a lower and upper outlier-pair in a normal sample, using as statistic the ratio of the reduced sum of squares to either the total sum of squares or the population variance

| | Table XVIa N5 | | Table XVIb Nμ5 | | Table XVIc Nσ5 | | Table XVId Nμσ5 | |
|---|---|---|---|---|---|---|---|---|
| $n$ | 5% | 1% | 5% | 1% | 5% | 1% | 5% | 1% |
| 4 | 0.00044 | 0.00001 | 0.019 | 0.0030 | 0.00009 | 0.00002 | 0.038 | 0.0062 |
| 5 | 0.011 | 0.002 | 0.049 | 0.017 | 0.025 | 0.0049 | 0.131 | 0.042 |
| 6 | 0.044 | 0.014 | 0.092 | 0.040 | 0.123 | 0.036 | 0.308 | 0.126 |
| 7 | 0.078 | 0.033 | 0.127 | 0.062 | 0.265 | 0.100 | 0.511 | 0.224 |
| 8 | 0.120 | 0.060 | 0.169 | 0.093 | 0.492 | 0.239 | 0.801 | 0.400 |
| 9 | 0.159 | 0.093 | 0.205 | 0.127 | 0.727 | 0.370 | 1.09 | 0.583 |
| 10 | 0.195 | 0.122 | 0.239 | 0.158 | 1.01 | 0.562 | 1.46 | 0.839 |
| 12 | 0.266 | 0.181 | 0.299 | 0.214 | 1.80 | 1.09 | 2.24 | 1.43 |
| 14 | 0.320 | 0.236 | 0.352 | 0.265 | 2.66 | 1.75 | 3.15 | 2.11 |
| 16 | 0.369 | 0.288 | 0.393 | 0.308 | 3.62 | 2.54 | 4.20 | 2.88 |
| 18 | 0.411 | 0.325 | 0.432 | 0.347 | 4.69 | 3.35 | 5.22 | 3.74 |
| 20 | 0.448 | 0.363 | 0.468 | 0.381 | 5.78 | 4.16 | 6.43 | 4.71 |
| 30 | 0.571 | 0.509 | 0.581 | 0.517 | 11.7 | 9.45 | 12.5 | 10.2 |
| 40 | 0.644 | 0.584 | 0.651 | 0.591 | 18.8 | 15.6 | 19.4 | 16.3 |
| 50 | 0.699 | 0.648 | 0.703 | 0.654 | 26.0 | 22.2 | 26.8 | 22.8 |
| 100 | 0.821 | 0.794 | 0.823 | 0.796 | 66.6 | 59.2 | 67.4 | 60.0 |

$n$ = number of observations.

**Table XVIIa, b** Critical values for 5% and 1% tests of discordancy for a lower and upper outlier-pair in a normal sample, using the studentized range (XVIIa) or the standardized range (XVIIb) as test statistic

| n | Table XVIIa N6 5% | Table XVIIa N6 1% | Table XVIIb Nσ6(Nμσ6) 5% | Table XVIIb Nσ6(Nμσ6) 1% |
|---|---|---|---|---|
| 3 | 2.00 | 2.00 | 3.31 | 4.12 |
| 4 | 2.43 | 2.45 | 3.63 | 4.40 |
| 5 | 2.75 | 2.80 | 3.86 | 4.60 |
| 6 | 3.01 | 3.10 | 4.03 | 4.76 |
| 7 | 3.22 | 3.34 | 4.17 | 4.88 |
| 8 | 3.40 | 3.54 | 4.29 | 4.99 |
| 9 | 3.55 | 3.72 | 4.39 | 5.08 |
| 10 | 3.69 | 3.87 | 4.47 | 5.16 |
| 12 | 3.91 | 4.13 | 4.62 | 5.29 |
| 14 | 4.09 | 4.34 | 4.74 | 5.40 |
| 16 | 4.24 | 4.52 | 4.85 | 5.49 |
| 18 | 4.37 | 4.67 | 4.93 | 5.57 |
| 20 | 4.49 | 4.80 | 5.01 | 5.65 |
| 30 | 4.89 | 5.26 | 5.30 | 5.91 |
| 40 | 5.16 | 5.56 | 5.50 | 6.09 |
| 50 | 5.35 | 5.77 | 5.65 | 6.23 |
| 60 | 5.51 | 5.94 | 5.76 | 6.34 |
| 100 | 5.90 | 6.36 | 6.08 | 6.64 |
| 200 | 6.39 | 6.84 | | |
| 500 | 6.94 | 7.42 | | |
| 1000 | 7.33 | 7.80 | | |

$n$ = number of observations.

**Table XVIIc** Critical values for 5% and 1% tests of discordancy for a lower and upper outlier-pair in a normal sample, using the externally studentized range as test statistic

**Nv6**

*5% critical values*

| v \ n | 1 | 2 | 3 | 4 | 5 | 6 | 8 | 10 | 12 | 15 | 20 | 30 | 60 |
|---|---|---|---|---|---|---|---|---|---|---|---|---|---|
| 3 | 27.0 | 8.33 | 5.91 | 5.04 | 4.60 | 4.34 | 4.04 | 3.88 | 3.77 | 3.67 | 3.58 | 3.49 | 3.40 |
| 4 | 32.8 | 9.80 | 6.82 | 5.76 | 5.22 | 4.90 | 4.53 | 4.33 | 4.20 | 4.08 | 3.96 | 3.85 | 3.74 |
| 5 | 37.1 | 10.9 | 7.50 | 6.29 | 5.67 | 5.30 | 4.89 | 4.65 | 4.51 | 4.37 | 4.23 | 4.10 | 3.98 |
| 6 | 40.4 | 11.7 | 8.04 | 6.71 | 6.03 | 5.63 | 5.17 | 4.91 | 4.75 | 4.59 | 4.45 | 4.30 | 4.16 |
| 7 | 43.1 | 12.4 | 8.48 | 7.05 | 6.33 | 5.90 | 5.40 | 5.12 | 4.95 | 4.78 | 4.62 | 4.46 | 4.31 |
| 8 | 45.4 | 13.0 | 8.85 | 7.35 | 6.58 | 6.12 | 5.60 | 5.30 | 5.12 | 4.94 | 4.77 | 4.60 | 4.44 |
| 9 | 47.4 | 13.5 | 9.18 | 7.60 | 6.80 | 6.32 | 5.77 | 5.46 | 5.27 | 5.08 | 4.90 | 4.72 | 4.55 |
| 10 | 49.1 | 14.0 | 9.46 | 7.83 | 6.99 | 6.49 | 5.92 | 5.60 | 5.39 | 5.20 | 5.01 | 4.82 | 4.65 |
| 12 | 52.0 | 14.8 | 9.95 | 8.21 | 7.32 | 6.79 | 6.18 | 5.83 | 5.61 | 5.40 | 5.20 | 5.00 | 4.81 |
| 14 | 54.3 | 15.4 | 10.3 | 8.52 | 7.60 | 7.03 | 6.39 | 6.03 | 5.80 | 5.57 | 5.36 | 5.15 | 4.94 |
| 16 | 56.3 | 15.9 | 10.7 | 8.79 | 7.83 | 7.24 | 6.57 | 6.19 | 5.95 | 5.72 | 5.49 | 5.27 | 5.06 |
| 18 | 58.0 | 16.4 | 11.0 | 9.03 | 8.03 | 7.43 | 6.73 | 6.34 | 6.09 | 5.85 | 5.61 | 5.38 | 5.15 |
| 20 | 59.6 | 16.8 | 11.2 | 9.23 | 8.21 | 7.59 | 6.87 | 6.47 | 6.21 | 5.96 | 5.71 | 5.47 | 5.24 |

**Table XVIIc** (*Contd.*)

*1% critical values*

| ν \ n | 1 | 2 | 3 | 4 | 5 | 6 | 8 | 10 | 12 | 15 | 20 | 30 | 60 |
|---|---|---|---|---|---|---|---|---|---|---|---|---|---|
| 3 | 135.0 | 19.0 | 10.6 | 8.12 | 6.98 | 6.33 | 5.64 | 5.27 | 5.05 | 4.84 | 4.64 | 4.45 | 4.28 |
| 4 | 164.3 | 22.3 | 12.2 | 9.17 | 7.80 | 7.03 | 6.20 | 5.77 | 5.50 | 5.25 | 5.02 | 4.80 | 4.59 |
| 5 | 185.6 | 24.7 | 13.3 | 9.96 | 8.42 | 7.56 | 6.62 | 6.14 | 5.84 | 5.56 | 5.29 | 5.05 | 4.82 |
| 6 | 202.2 | 26.6 | 14.2 | 10.6 | 8.91 | 7.97 | 6.96 | 6.43 | 6.10 | 5.80 | 5.51 | 5.24 | 4.99 |
| 7 | 215.8 | 28.2 | 15.0 | 11.1 | 9.32 | 8.32 | 7.24 | 6.67 | 6.32 | 5.99 | 5.69 | 5.40 | 5.13 |
| 8 | 227.2 | 29.5 | 15.6 | 11.5 | 9.67 | 8.61 | 7.47 | 6.87 | 6.51 | 6.16 | 5.84 | 5.54 | 5.25 |
| 9 | 237.0 | 30.7 | 16.2 | 11.9 | 9.97 | 8.87 | 7.68 | 7.05 | 6.67 | 6.31 | 5.97 | 5.65 | 5.36 |
| 10 | 245.6 | 31.7 | 16.7 | 12.3 | 10.2 | 9.10 | 7.86 | 7.21 | 6.81 | 6.44 | 6.09 | 5.76 | 5.45 |
| 12 | 260.0 | 33.4 | 17.5 | 12.8 | 10.7 | 9.48 | 8.18 | 7.49 | 7.06 | 6.66 | 6.28 | 5.93 | 5.60 |
| 14 | 271.8 | 34.8 | 18.2 | 13.3 | 11.1 | 9.81 | 8.44 | 7.71 | 7.26 | 6.84 | 6.45 | 6.08 | 5.73 |
| 16 | 281.8 | 36.0 | 18.8 | 13.7 | 11.4 | 10.1 | 8.66 | 7.91 | 7.44 | 7.00 | 6.59 | 6.20 | 5.84 |
| 18 | 290.4 | 37.0 | 19.3 | 14.1 | 11.7 | 10.3 | 8.85 | 8.08 | 7.59 | 7.14 | 6.71 | 6.31 | 5.93 |
| 20 | 298.0 | 37.9 | 19.8 | 14.4 | 11.9 | 10.5 | 9.03 | 8.23 | 7.73 | 7.26 | 6.82 | 6.41 | 6.01 |

$n$ = number of observations.

$\nu$ = degrees of freedom of independent estimate of $\sigma^2$. The critical values for $\nu = \infty$ are given by Table XVIIb.

**Table XVIII** Critical values for 5% and 1% tests of discordancy for $k$ lower outliers and $k$ upper outliers in a normal sample, using the standardized $(k-1)$th quasirange as test statistic

$$N\sigma 9(N\mu\sigma 9)$$

| | $k = 2$ | | $k = 3$ | | $k = 4$ | |
|---|---|---|---|---|---|---|
| $n$ | 5% | 1% | 5% | 1% | 5% | 1% |
| 4 | 1.58 | 2.17 | | | | |
| 5 | 2.05 | 2.61 | | | | |
| 6 | 2.35 | 2.89 | 1.11 | 1.56 | | |
| 7 | 2.58 | 3.10 | 1.51 | 1.95 | | |
| 8 | 2.75 | 3.26 | 1.78 | 2.21 | 0.859 | 1.23 |
| 9 | 2.90 | 3.40 | 1.99 | 2.41 | 1.20 | 1.56 |
| 10 | 3.03 | 3.52 | 2.15 | 2.57 | 1.44 | 1.80 |
| 12 | 3.23 | 3.71 | 2.42 | 2.82 | 1.79 | 2.14 |
| 14 | 3.39 | 3.86 | 2.62 | 3.01 | 2.04 | 2.38 |
| 16 | 3.53 | 3.98 | 2.79 | 3.17 | 2.24 | 2.57 |
| 18 | 3.64 | 4.09 | 2.93 | 3.30 | 2.40 | 2.73 |
| 20 | 3.74 | 4.18 | 3.05 | 3.41 | 2.54 | 2.86 |
| 30 | 4.11 | 4.52 | 3.47 | 3.81 | 3.02 | 3.32 |
| 40 | 4.35 | 4.75 | 3.74 | 4.07 | 3.32 | 3.61 |
| 50 | 4.53 | 4.92 | 3.94 | 4.26 | 3.54 | 3.82 |
| 60 | 4.67 | 5.05 | 4.10 | 4.41 | 3.71 | 3.98 |
| 70 | 4.79 | 5.16 | 4.23 | 4.54 | 3.85 | 4.12 |
| 80 | 4.88 | 5.26 | 4.34 | 4.64 | 3.97 | 4.23 |
| 90 | 4.97 | 5.34 | 4.43 | 4.73 | 4.07 | 4.33 |
| 100 | 5.05 | 5.41 | 4.52 | 4.81 | 4.16 | 4.41 |

$n$ = number of observations.
$2k$ = number of outliers ($k$ upper and $k$ lower).

**Table XIXa, b, c, d, e, f, g** Critical values for 5% and 1% Dixon-type tests of discordancy for one or more outliers in a normal sample

| | Table XIXa N7(N$\mu$7) | | Table XIXb N8(N$\mu$8) | | Table XIXc N9(N$\mu$9) | | Table XIXd N10(N$\mu$10) | |
|---|---|---|---|---|---|---|---|---|
| $n$ | 5% | 1% | 5% | 1% | 5% | 1% | 5% | 1% |
| 3 | 0.941 | 0.988 | | | | | | |
| 4 | 0.765 | 0.889 | 0.831 | 0.922 | 0.955 | 0.991 | | |
| 5 | 0.642 | 0.780 | 0.717 | 0.831 | 0.807 | 0.916 | 0.960 | 0.992 |
| 6 | 0.560 | 0.698 | 0.621 | 0.737 | 0.689 | 0.805 | 0.824 | 0.925 |
| 7 | 0.507 | 0.637 | 0.570 | 0.694 | 0.610 | 0.740 | 0.712 | 0.836 |
| 8 | 0.468 | 0.590 | 0.524 | 0.638 | 0.554 | 0.683 | 0.632 | 0.760 |
| 9 | 0.437 | 0.555 | 0.492 | 0.594 | 0.512 | 0.635 | 0.580 | 0.701 |
| 10 | 0.412 | 0.527 | 0.464 | 0.564 | 0.477 | 0.597 | 0.537 | 0.655 |
| 12 | 0.376 | 0.482 | 0.429 | 0.520 | 0.428 | 0.541 | 0.473 | 0.590 |
| 14 | 0.349 | 0.450 | 0.397 | 0.485 | 0.395 | 0.502 | 0.432 | 0.542 |
| 16 | 0.329 | 0.426 | 0.376 | 0.461 | 0.369 | 0.472 | 0.401 | 0.508 |
| 18 | 0.313 | 0.407 | 0.354 | 0.438 | 0.349 | 0.449 | 0.377 | 0.480 |
| 20 | 0.300 | 0.391 | 0.340 | 0.417 | 0.334 | 0.430 | 0.358 | 0.458 |
| 25 | 0.277 | 0.362 | 0.316 | 0.386 | 0.304 | 0.394 | 0.324 | 0.417 |
| 30 | 0.260 | 0.341 | 0.300 | 0.368 | 0.283 | 0.369 | 0.301 | 0.389 |

| | Table XIXe N11(N$\mu$11) | | Table XIXf N12(N$\mu$12) | | Table XIXg N13(N$\mu$13) | |
|---|---|---|---|---|---|---|
| $n$ | 5% | 1% | 5% | 1% | 5% | 1% |
| 4 | 0.967 | 0.992 | | | | |
| 5 | 0.845 | 0.929 | 0.976 | 0.995 | | |
| 6 | 0.736 | 0.836 | 0.872 | 0.951 | 0.983 | 0.995 |
| 7 | 0.661 | 0.778 | 0.780 | 0.885 | 0.881 | 0.945 |
| 8 | 0.607 | 0.710 | 0.710 | 0.829 | 0.803 | 0.890 |
| 9 | 0.565 | 0.667 | 0.657 | 0.776 | 0.737 | 0.840 |
| 10 | 0.531 | 0.632 | 0.612 | 0.726 | 0.682 | 0.791 |
| 12 | 0.481 | 0.579 | 0.546 | 0.642 | 0.600 | 0.704 |
| 14 | 0.445 | 0.538 | 0.501 | 0.593 | 0.546 | 0.641 |
| 16 | 0.418 | 0.508 | 0.467 | 0.557 | 0.507 | 0.595 |
| 18 | 0.397 | 0.484 | 0.440 | 0.529 | 0.475 | 0.561 |
| 20 | 0.372 | 0.464 | 0.419 | 0.506 | 0.450 | 0.535 |
| 25 | 0.343 | 0.428 | 0.382 | 0.464 | 0.406 | 0.489 |
| 30 | 0.322 | 0.402 | 0.355 | 0.433 | 0.376 | 0.457 |

$n$ = number of observations.

**Table XIXh, i** Critical values for 5% and 1% tests of discordancy for one, or two, upper outliers in a normal sample, using $[x_{(n)} - x_{(n-1)}]/\sigma$ (XIXh) or $[x_{(n-1)} - x_{(-2)}]/\sigma$ (XIXi) as test statistic

| | Table XIXh<br>$N\sigma7(N\mu\sigma7)$ | | Table XIXi<br>$N\sigma8(N\mu\sigma8)$ | |
|---|---|---|---|---|
| $n$ | 5% | 1% | 5% | 1% |
| 3 | 2.17 | 2.90 | | |
| 10 | 1.46 | 2.03 | 0.96 | 1.38 |
| 20 | 1.28 | 1.80 | 0.79 | 1.14 |
| 30 | 1.20 | 1.70 | 0.73 | 1.05 |
| 40 | 1.14 | 1.63 | 0.68 | 1.00 |
| 60 | 1.08 | 1.56 | 0.63 | 0.93 |
| 80 | 1.04 | 1.50 | 0.61 | 0.90 |
| 100 | 1.02 | 1.47 | 0.58 | 0.86 |
| 200 | 0.95 | 1.38 | 0.54 | 0.81 |
| 500 | 0.87 | 1.28 | 0.48 | 0.73 |
| 1000 | 0.83 | 1.22 | 0.45 | 0.67 |

$n$ = number of observations.

**Table XX** Critical values for 5% and 1% tests of discordancy for one or more outliers in a normal sample, using as test statistic the sample skewness (XXa), the sample kurtosis (XXb), or the sample kurtosis based on deviations from the population mean (XXc)

| | Table XXa<br>N14 | | Table XXb<br>N15 | |
|---|---|---|---|---|
| $n$ | 5% | 1% | 5% | 1% |
| 5 | 1.0 | 1.3 | 2.9 | 3.1 |
| 7 | | | 3.55 | 4.23 |
| 8 | | | 3.70 | 4.53 |
| 9 | | | 3.86 | 4.82 |
| 10 | 0.9 | 1.3 | 3.95 | 5.00 |
| 12 | | | 4.05 | 5.20 |
| 15 | 0.8 | 1.2 | 4.13 | 5.30 |
| 20 | 0.8 | 1.1 | 4.17 | 5.36 |
| 25 | 0.71 | 1.06 | 4.16 | 5.30 |
| 30 | 0.66 | 0.99 | 4.11 | 5.21 |
| 40 | 0.59 | 0.87 | 4.06 | 5.04 |
| 50 | 0.53 | 0.79 | 3.99 | 4.88 |
| 60 | 0.49 | 0.72 | | |
| 70 | 0.46 | 0.67 | | |
| 75 | | | 3.87 | 4.59 |
| 80 | 0.43 | 0.63 | | |
| 90 | 0.41 | 0.60 | | |
| 100 | 0.39 | 0.57 | 3.77 | 4.39 |
| 200 | 0.28 | 0.40 | 3.57 | 3.98 |
| 500 | 0.18 | 0.26 | 3.37 | 3.60 |
| 1000 | 0.13 | 0.18 | 3.26 | 3.41 |

Table XXc
**Nμ6**

| n | 5% | 1% |
|---|-----|-----|
| 3 | 2.8 | 3.0 |
| 4 | 3.3 | 3.7 |
| 5 | 3.6 | 4.3 |
| 6 | 3.8 | 4.7 |
| 7 | 3.9 | 4.9 |
| 8 | 4.0 | 5.1 |
| 9 | 4.0 | 5.3 |
| 10 | 4.1 | 5.4 |
| 12 | 4.2 | 5.5 |
| 14 | 4.2 | 5.5 |
| 16 | 4.2 | 5.5 |
| 18 | 4.2 | 5.5 |
| 20 | 4.2 | 5.4 |
| 30 | 4.2 | 5.2 |
| 40 | 4.1 | 5.0 |
| 50 | 4.0 | 4.9 |
| 100 | 3.8 | 4.3 |

$n$ = number of observations.

**Table XXI** Critical values for 5% and 1% tests of discordancy for $k$ outliers in a normal sample, using Tietjen and Moore's $E_k$-statistic
**N16**

| | $k = 2$ | | $k = 3$ | | $k = 4$ | |
|---|-----|-----|-----|-----|-----|-----|
| n | 5% | 1% | 5% | 1% | 5% | 1% |
| 5 | 0.010 | 0.002 | | | | |
| 6 | 0.034 | 0.012 | | | | |
| 7 | 0.065 | 0.028 | 0.016 | 0.006 | | |
| 8 | 0.099 | 0.050 | 0.034 | 0.014 | | |
| 9 | 0.137 | 0.078 | 0.057 | 0.026 | 0.021 | 0.009 |
| 10 | 0.172 | 0.101 | 0.083 | 0.044 | 0.037 | 0.018 |
| 12 | 0.234 | 0.159 | 0.133 | 0.083 | 0.073 | 0.042 |
| 14 | 0.293 | 0.207 | 0.179 | 0.123 | 0.112 | 0.072 |
| 16 | 0.340 | 0.263 | 0.227 | 0.166 | 0.153 | 0.107 |
| 18 | 0.382 | 0.306 | 0.267 | 0.206 | 0.187 | 0.141 |
| 20 | 0.416 | 0.339 | 0.302 | 0.236 | 0.221 | 0.170 |
| 30 | 0.549 | 0.482 | 0.443 | 0.386 | 0.364 | 0.308 |
| 40 | 0.629 | 0.574 | 0.534 | 0.480 | 0.458 | 0.408 |
| 50 | 0.684 | 0.636 | 0.599 | 0.550 | 0.529 | 0.482 |

$n$ = number of observations.
$k$ = number of outliers.

**Table XXIIa** Critical values for 5% and 1% tests for the presence of an undefined number of contaminants in a normal sample, using Shapiro and Wilk's $W$-statistic
**N17**

| $n$ | 5% | 1% |
|---|---|---|
| 3 | 0.767 | 0.753 |
| 4 | 0.748 | 0.687 |
| 5 | 0.762 | 0.686 |
| 6 | 0.788 | 0.713 |
| 7 | 0.803 | 0.730 |
| 8 | 0.818 | 0.749 |
| 9 | 0.829 | 0.764 |
| 10 | 0.842 | 0.781 |
| 12 | 0.859 | 0.805 |
| 14 | 0.874 | 0.825 |
| 16 | 0.887 | 0.844 |
| 18 | 0.897 | 0.858 |
| 20 | 0.905 | 0.868 |
| 25 | 0.918 | 0.888 |
| 30 | 0.927 | 0.900 |
| 35 | 0.934 | 0.910 |
| 40 | 0.940 | 0.919 |
| 45 | 0.945 | 0.926 |
| 50 | 0.947 | 0.930 |

$n$ = number of observations.

**Table XXIIb** Values of the constants $a_{n, n-i+1}$ required for calculating Shapiro and Wilk's $W$-statistic $T_{N17} = L^2/S^2$, where

$$L = \sum_{i=1}^{[n/2]} a_{n, n-i+1} [x_{(n-i+1)} - x_{(i)}] \text{ and } S^2 = \sum_{i=1}^{n} (x_i - \bar{x})^2$$

| i \ n | 1 | 2 | 3 | 4 | 5 | 6 | 7 | 8 | 9 | 10 |
|---|---|---|---|---|---|---|---|---|---|---|
| 3 | 0.7071 | | | | | | | | | |
| 4 | 0.6872 | 0.1677 | | | | | | | | |
| 5 | 0.6646 | 0.2413 | | | | | | | | |
| 6 | 0.6431 | 0.2806 | 0.0875 | | | | | | | |
| 7 | 0.6233 | 0.3031 | 0.1401 | | | | | | | |
| 8 | 0.6052 | 0.3164 | 0.1743 | 0.0561 | | | | | | |
| 9 | 0.5888 | 0.3244 | 0.1976 | 0.0947 | | | | | | |
| 10 | 0.5739 | 0.3291 | 0.2141 | 0.1224 | 0.0399 | | | | | |
| 12 | 0.5475 | 0.3325 | 0.2347 | 0.1586 | 0.0922 | 0.0303 | | | | |
| 14 | 0.5251 | 0.3318 | 0.2460 | 0.1802 | 0.1240 | 0.0727 | 0.0240 | | | |
| 16 | 0.5056 | 0.3290 | 0.2521 | 0.1939 | 0.1447 | 0.1005 | 0.0593 | 0.0196 | | |
| 18 | 0.4886 | 0.3253 | 0.2553 | 0.2027 | 0.1587 | 0.1197 | 0.0837 | 0.0496 | 0.0163 | |
| 20 | 0.4734 | 0.3211 | 0.2565 | 0.2085 | 0.1686 | 0.1334 | 0.1013 | 0.0711 | 0.0422 | 0.0140 |
| 25 | 0.4450 | 0.3069 | 0.2543 | 0.2148 | 0.1822 | 0.1539 | 0.1283 | 0.1046 | 0.0823 | 0.0610 |
| 30 | 0.4254 | 0.2944 | 0.2487 | 0.2148 | 0.1870 | 0.1630 | 0.1415 | 0.1219 | 0.1036 | 0.0862 |
| 35 | 0.4096 | 0.2834 | 0.2427 | 0.2127 | 0.1883 | 0.1673 | 0.1487 | 0.1317 | 0.1160 | 0.1013 |
| 40 | 0.3964 | 0.2737 | 0.2368 | 0.2098 | 0.1878 | 0.1691 | 0.1526 | 0.1376 | 0.1237 | 0.1108 |
| 45 | 0.3850 | 0.2651 | 0.2313 | 0.2065 | 0.1865 | 0.1695 | 0.1545 | 0.1410 | 0.1286 | 0.1170 |
| 50 | 0.3751 | 0.2574 | 0.2260 | 0.2032 | 0.1847 | 0.1691 | 0.1554 | 0.1430 | 0.1317 | 0.1212 |

| i \ n | 11 | 12 | 13 | 14 | 15 | 16 | 17 | 18 | 19 | 20 |
|---|---|---|---|---|---|---|---|---|---|---|
| 25 | 0.0403 | 0.0200 | | | | | | | | |
| 30 | 0.0697 | 0.0537 | 0.0381 | 0.0227 | 0.0076 | | | | | |
| 35 | 0.0873 | 0.0739 | 0.0610 | 0.0484 | 0.0361 | 0.0239 | 0.0119 | | | |
| 40 | 0.0986 | 0.0870 | 0.0759 | 0.0651 | 0.0546 | 0.0444 | 0.0343 | 0.0244 | 0.0146 | 0.0049 |
| 45 | 0.1062 | 0.0959 | 0.0860 | 0.0765 | 0.0673 | 0.0584 | 0.0497 | 0.0412 | 0.0328 | 0.0245 |
| 50 | 0.1113 | 0.1020 | 0.0932 | 0.0846 | 0.0764 | 0.0685 | 0.0608 | 0.0532 | 0.0459 | 0.0386 |

| i \ n | 21 | 22 | 23 | 24 | 25 |
|---|---|---|---|---|---|
| 45 | 0.0163 | 0.0082 | | | |
| 50 | 0.0314 | 0.0244 | 0.0174 | 0.0104 | 0.0035 |

$n$ = number of observations.

Table XXIII Critical values for 5% and 1% consecutive tests of types ESD, KUR and RST for up to $k$ outliers in a normal sample with $\mu$ and $\sigma^2$ unknown. For each test, level, sample size and value of $k$ (2, 3, 4, or 5) the required $k$ critical values are given. See details of test **N19** on pages 235–6

**N19**

|  |  | ESD | | KUR | | RST | |
|---|---|---|---|---|---|---|---|
|  |  | 5% | 1% | 5% | 1% | 5% | 1% |
| $k = 2$ |  |  |  |  |  |  |  |
|  | $n$ |  |  |  |  |  |  |
|  | 20 | 2.83 | 3.09 | 4.56 | 5.77 | 5.17 | 6.29 |
|  |  | 2.52 | 2.76 | 3.66 | 4.45 | 3.86 | 4.64 |
|  | 30 | 3.05 | 3.35 | 4.48 | 5.64 | 4.63 | 5.44 |
|  |  | 2.67 | 2.92 | 3.56 | 4.25 | 3.55 | 4.08 |
|  | 40 | 3.17 | 3.52 | 4.31 | 5.43 | 4.37 | 5.10 |
|  |  | 2.77 | 2.98 | 3.50 | 4.07 | 3.44 | 3.83 |
|  | 50 | 3.27 | 3.61 | 4.26 | 5.17 | 4.28 | 4.89 |
|  |  | 2.85 | 3.08 | 3.49 | 4.01 | 3.42 | 3.81 |
|  | 60 | 3.34 | 3.70 | 4.18 | 5.05 | 4.21 | 4.81 |
|  |  | 2.90 | 3.17 | 3.46 | 3.93 | 3.38 | 3.77 |
|  | 80 | 3.45 | 3.80 | 4.04 | 4.78 | 4.13 | 4.66 |
|  |  | 2.97 | 3.23 | 3.39 | 3.80 | 3.36 | 3.71 |
|  | 100 | 3.52 | 3.87 | 3.91 | 4.57 | 4.08 | 4.59 |
|  |  | 3.03 | 3.28 | 3.37 | 3.74 | 3.35 | 3.66 |
| $k = 3$ |  |  |  |  |  |  |  |
|  | $n$ |  |  |  |  |  |  |
|  | 20 | 2.88 | 3.13 | 4.76 | 6.04 | 6.62 | 8.33 |
|  |  | 2.60 | 2.83 | 3.83 | 4.72 | 5.05 | 6.25 |
|  |  | 2.45 | 2.68 | 3.48 | 4.23 | 4.11 | 5.01 |
|  | 30 | 3.12 | 3.41 | 4.70 | 5.91 | 5.47 | 6.50 |
|  |  | 2.73 | 3.01 | 3.69 | 4.43 | 4.25 | 4.93 |
|  |  | 2.56 | 2.75 | 3.31 | 3.87 | 3.54 | 4.10 |
|  | 40 | 3.22 | 3.58 | 4.42 | 5.61 | 4.95 | 5.75 |
|  |  | 2.81 | 3.03 | 3.58 | 4.16 | 3.92 | 4.42 |
|  |  | 2.62 | 2.82 | 3.28 | 3.73 | 3.39 | 3.80 |
|  | 50 | 3.34 | 3.68 | 4.37 | 5.43 | 4.76 | 5.51 |
|  |  | 2.89 | 3.15 | 3.56 | 4.14 | 3.79 | 4.30 |
|  |  | 2.68 | 2.89 | 3.23 | 3.69 | 3.31 | 3.70 |

Table XXIII (*Contd.*)

|  |  | ESD | | KUR | | RST | |
|---|---|---|---|---|---|---|---|
|  |  | 5% | 1% | 5% | 1% | 5% | 1% |
|  | 60 | 3.42 | 3.75 | 4.27 | 5.19 | 4.63 | 5.28 |
|  |  | 2.95 | 3.20 | 3.51 | 4.04 | 3.73 | 4.14 |
|  |  | 2.73 | 2.95 | 3.23 | 3.60 | 3.27 | 3.60 |
|  | 80 | 3.49 | 3.85 | 4.06 | 4.92 | 4.43 | 5.04 |
|  |  | 3.03 | 3.27 | 3.46 | 3.88 | 3.63 | 4.02 |
|  |  | 2.81 | 3.01 | 3.19 | 3.53 | 3.22 | 3.51 |
|  | 100 | 3.60 | 3.97 | 4.00 | 4.66 | 4.39 | 4.94 |
|  |  | 3.10 | 3.34 | 3.41 | 3.76 | 3.59 | 3.95 |
|  |  | 2.86 | 3.06 | 3.17 | 3.48 | 3.21 | 3.46 |
| $k = 4$ |  |  |  |  |  |  |  |
| $n$ |  |  |  |  |  |  |  |
|  | 20 | 2.95 | 3.20 | 4.97 | 6.37 | 8.54 | 11.09 |
|  |  | 2.63 | 2.83 | 3.92 | 4.73 | 6.54 | 8.08 |
|  |  | 2.49 | 2.68 | 3.57 | 4.26 | 5.42 | 6.65 |
|  |  | 2.39 | 2.58 | 3.39 | 4.13 | 4.60 | 5.71 |
|  | 30 | 3.16 | 3.48 | 4.78 | 6.11 | 6.33 | 7.66 |
|  |  | 2.77 | 3.02 | 3.75 | 4.46 | 4.97 | 5.81 |
|  |  | 2.59 | 2.79 | 3.40 | 3.97 | 4.23 | 4.88 |
|  |  | 2.49 | 2.70 | 3.22 | 3.77 | 3.73 | 4.30 |
|  | 40 | 3.32 | 3.64 | 4.61 | 5.81 | 5.60 | 6.56 |
|  |  | 2.86 | 3.10 | 3.64 | 4.24 | 4.44 | 5.04 |
|  |  | 2.67 | 2.87 | 3.30 | 3.76 | 3.84 | 4.30 |
|  |  | 2.55 | 2.74 | 3.13 | 3.55 | 3.43 | 3.83 |
|  | 50 | 3.40 | 3.74 | 4.44 | 5.57 | 5.24 | 6.06 |
|  |  | 2.93 | 3.18 | 3.59 | 4.15 | 4.18 | 4.69 |
|  |  | 2.72 | 2.92 | 3.28 | 3.73 | 3.66 | 4.08 |
|  |  | 2.59 | 2.78 | 3.10 | 3.49 | 3.30 | 3.65 |
|  | 60 | 3.48 | 3.82 | 4.36 | 5.32 | 5.05 | 5.76 |
|  |  | 2.98 | 3.20 | 3.54 | 4.00 | 4.03 | 4.46 |
|  |  | 2.77 | 2.97 | 3.25 | 3.63 | 3.56 | 3.91 |
|  |  | 2.63 | 2.82 | 3.08 | 3.42 | 3.23 | 3.52 |
|  | 80 | 3.57 | 3.91 | 4.14 | 4.87 | 4.76 | 5.33 |
|  |  | 3.05 | 3.31 | 3.47 | 3.90 | 3.88 | 4.29 |
|  |  | 2.84 | 3.04 | 3.21 | 3.54 | 3.46 | 3.76 |
|  |  | 2.69 | 2.87 | 3.05 | 3.35 | 3.16 | 3.41 |
|  | 100 | 3.64 | 3.96 | 4.03 | 4.69 | 4.63 | 5.16 |
|  |  | 3.13 | 3.34 | 3.44 | 3.81 | 3.82 | 4.15 |
|  |  | 2.89 | 3.06 | 3.20 | 3.49 | 3.41 | 3.64 |
|  |  | 2.74 | 2.90 | 3.05 | 3.33 | 3.13 | 3.35 |

**Table XXIII** (*Contd.*)

| k = 5 | n | ESD 5% | ESD 1% | KUR 5% | KUR 1% | RST 5% | RST 1% |
|---|---|---|---|---|---|---|---|
| | 20 | 2.97 | 3.18 | 5.10 | 6.40 | 11.31 | 14.91 |
| | | 2.65 | 2.89 | 4.02 | 5.00 | 8.75 | 11.39 |
| | | 2.51 | 2.69 | 3.62 | 4.36 | 7.32 | 9.43 |
| | | 2.42 | 2.61 | 3.46 | 4.09 | 6.28 | 8.09 |
| | | 2.37 | 2.57 | 3.37 | 4.01 | 5.46 | 7.05 |
| | 30 | 3.19 | 3.48 | 4.80 | 6.14 | 7.33 | 8.91 |
| | | 2.78 | 3.03 | 3.78 | 4.48 | 5.76 | 6.74 |
| | | 2.60 | 2.80 | 3.42 | 3.98 | 4.95 | 5.85 |
| | | 2.51 | 2.74 | 3.25 | 3.75 | 4.38 | 5.16 |
| | | 2.45 | 2.62 | 3.14 | 3.62 | 3.91 | 4.58 |
| | 40 | 3.31 | 3.63 | 4.61 | 5.80 | 6.20 | 7.47 |
| | | 2.88 | 3.13 | 3.68 | 4.27 | 4.95 | 5.69 |
| | | 2.69 | 2.89 | 3.34 | 3.78 | 4.29 | 4.84 |
| | | 2.55 | 2.74 | 3.14 | 3.56 | 3.85 | 4.37 |
| | | 2.47 | 2.65 | 3.02 | 3.44 | 3.50 | 3.92 |
| | 50 | 3.45 | 3.77 | 4.51 | 5.67 | 5.71 | 6.65 |
| | | 2.96 | 3.21 | 3.61 | 4.16 | 4.56 | 5.14 |
| | | 2.74 | 2.94 | 3.31 | 3.72 | 4.02 | 4.46 |
| | | 2.61 | 2.79 | 3.12 | 3.48 | 3.62 | 4.00 |
| | | 2.52 | 2.70 | 2.99 | 3.31 | 3.33 | 3.65 |
| | 60 | 3.51 | 3.81 | 4.37 | 5.28 | 5.40 | 6.10 |
| | | 3.01 | 3.24 | 3.57 | 4.07 | 4.34 | 4.87 |
| | | 2.77 | 2.96 | 3.27 | 3.66 | 3.83 | 4.22 |
| | | 2.65 | 2.83 | 3.09 | 3.45 | 3.51 | 3.86 |
| | | 2.56 | 2.72 | 2.98 | 3.29 | 3.22 | 3.54 |
| | 80 | 3.61 | 3.93 | 4.22 | 5.05 | 5.06 | 5.71 |
| | | 3.11 | 3.36 | 3.53 | 3.96 | 4.13 | 4.58 |
| | | 2.86 | 3.08 | 3.26 | 3.57 | 3.68 | 4.01 |
| | | 2.72 | 2.89 | 3.09 | 3.36 | 3.37 | 3.62 |
| | | 2.62 | 2.76 | 2.97 | 3.25 | 3.14 | 3.37 |
| | 100 | 3.70 | 4.01 | 4.11 | 4.80 | 4.90 | 5.47 |
| | | 3.16 | 3.42 | 3.48 | 3.83 | 4.02 | 4.41 |
| | | 2.91 | 3.10 | 3.22 | 3.51 | 3.58 | 3.85 |
| | | 2.77 | 2.93 | 3.07 | 3.33 | 3.30 | 3.54 |
| | | 2.67 | 2.84 | 2.97 | 3.20 | 3.09 | 3.33 |

**Table XXIV** Critical values for 5% and 1% Dixon-type tests of discordancy for one or more outliers in a uniform sample

U

| | $\dfrac{x_{(n)} - x_{(n-1)}}{x_{(n)} - x_{(1)}}$ | | $\dfrac{x_{(n)} - x_{(n-2)}}{x_{(n)} - x_{(1)}}$ | | $\dfrac{x_{(n)} - x_{(n-2)}}{x_{(n)} - a}$ | |
|------|--------|--------|--------|--------|--------|--------|
| $n$ | 5% | 1% | 5% | 1% | 5% | 1% |
| 4 | 0.7763 | 0.9000 | 0.9747 | 0.9950 | 0.8637 | 0.9411 |
| 6 | 0.5271 | 0.6838 | 0.7514 | 0.8591 | 0.6574 | 0.7779 |
| 8 | 0.3930 | 0.5359 | 0.5818 | 0.7057 | 0.5207 | 0.6661 |
| 10 | 0.3124 | 0.4376 | 0.4707 | 0.5899 | 0.4291 | 0.5441 |
| 12 | 0.2589 | 0.3690 | 0.3942 | 0.5044 | 0.3644 | 0.4698 |
| 16 | 0.1926 | 0.2803 | 0.2968 | 0.3891 | 0.2794 | 0.3679 |
| 20 | 0.1533 | 0.2257 | 0.2377 | 0.3160 | 0.2264 | 0.3018 |
| 25 | 0.1221 | 0.1815 | 0.1902 | 0.2557 | 0.1829 | 0.2462 |
| 30 | 0.1015 | 0.1517 | 0.1585 | 0.2137 | 0.1534 | 0.2079 |
| 35 | 0.0868 | 0.1303 | 0.1358 | 0.1848 | 0.1321 | 0.1798 |
| 40 | 0.0758 | 0.1141 | 0.1188 | 0.1623 | 0.1159 | 0.1584 |
| 45 | 0.0673 | 0.1016 | 0.1056 | 0.1446 | 0.1033 | 0.1415 |
| 50 | 0.0605 | 0.0915 | 0.0951 | 0.1304 | 0.0932 | 0.1279 |
| 60 | 0.0503 | 0.0763 | 0.0792 | 0.1090 | 0.0779 | 0.1073 |
| 75 | 0.0402 | 0.0611 | 0.0633 | 0.0875 | 0.0625 | 0.0864 |
| 100 | 0.0301 | 0.0459 | 0.0475 | 0.0657 | 0.0470 | 0.0652 |
| 200 | 0.0150 | 0.0230 | 0.0237 | 0.0331 | 0.0236 | 0.0329 |

**Table XXV** Critical values for 5% and 1% tests of discordancy for $k = 1, 2$, or 3 outliers in a Gumbel least-value sample, using a Dixon-type statistic (**GU1, GU4**) or reduction in sum of squares (**GU2, GU5**).

| | | Table XXVa GU1 | | Table XXVb GU2 | | Table XXVc GU4 | | Table XXVd GU5 | |
|---|---|---|---|---|---|---|---|---|---|
| | | k upper outliers | | | | k lower outliers | | | |
| | n | 5% | 1% | 5% | 1% | 5% | 1% | 5% | 1% |
| k = 1 | 5 | 0.55 | 0.71 | 0.20 | 0.08 | 0.74 | 0.85 | 0.07 | 0.02 |
| | 6 | 0.46 | 0.62 | 0.31 | 0.16 | 0.68 | 0.79 | 0.10 | 0.04 |
| | 7 | 0.38 | 0.53 | 0.41 | 0.25 | 0.64 | 0.76 | 0.14 | 0.06 |
| | 8 | 0.34 | 0.46 | 0.48 | 0.32 | 0.60 | 0.72 | 0.18 | 0.08 |
| | 9 | 0.32 | 0.42 | 0.53 | 0.39 | 0.58 | 0.70 | 0.20 | 0.10 |
| | 10 | 0.29 | 0.39 | 0.58 | 0.46 | 0.56 | 0.68 | 0.24 | 0.13 |
| | 11 | 0.27 | 0.38 | 0.61 | 0.49 | 0.55 | 0.67 | 0.26 | 0.15 |
| | 12 | 0.25 | 0.34 | 0.65 | 0.53 | 0.53 | 0.65 | 0.28 | 0.16 |
| | 13 | 0.24 | 0.33 | 0.67 | 0.56 | 0.52 | 0.64 | 0.30 | 0.18 |
| | 14 | 0.22 | 0.31 | 0.69 | 0.60 | 0.51 | 0.63 | 0.32 | 0.19 |
| | 15 | 0.21 | 0.29 | 0.71 | 0.62 | 0.50 | 0.62 | 0.34 | 0.21 |
| | 16 | 0.21 | 0.28 | 0.73 | 0.66 | 0.50 | 0.61 | 0.35 | 0.22 |
| | 17 | 0.20 | 0.27 | 0.75 | 0.67 | 0.49 | 0.60 | 0.37 | 0.24 |
| | 18 | 0.19 | 0.26 | 0.76 | 0.69 | 0.48 | 0.59 | 0.38 | 0.26 |
| | 19 | 0.18 | 0.25 | 0.77 | 0.71 | 0.47 | 0.58 | 0.40 | 0.26 |
| | 20 | 0.18 | 0.24 | 0.78 | 0.72 | 0.46 | 0.58 | 0.41 | 0.27 |
| k = 2 | 8 | 0.49 | 0.61 | 0.25 | 0.14 | 0.73 | 0.81 | 0.07 | 0.03 |
| | 9 | 0.44 | 0.57 | 0.31 | 0.19 | 0.70 | 0.79 | 0.09 | 0.04 |
| | 10 | 0.40 | 0.52 | 0.37 | 0.25 | 0.68 | 0.77 | 0.11 | 0.06 |
| | 11 | 0.37 | 0.48 | 0.41 | 0.29 | 0.65 | 0.74 | 0.13 | 0.07 |
| | 12 | 0.34 | 0.44 | 0.45 | 0.35 | 0.64 | 0.74 | 0.15 | 0.09 |
| | 13 | 0.32 | 0.41 | 0.48 | 0.38 | 0.62 | 0.72 | 0.17 | 0.10 |
| | 14 | 0.31 | 0.40 | 0.51 | 0.40 | 0.61 | 0.70 | 0.18 | 0.11 |
| | 15 | 0.29 | 0.37 | 0.55 | 0.45 | 0.60 | 0.69 | 0.20 | 0.13 |
| | 16 | 0.28 | 0.36 | 0.57 | 0.49 | 0.59 | 0.68 | 0.22 | 0.13 |
| | 17 | 0.27 | 0.34 | 0.59 | 0.50 | 0.58 | 0.67 | 0.23 | 0.14 |
| | 18 | 0.26 | 0.33 | 0.62 | 0.54 | 0.57 | 0.67 | 0.25 | 0.16 |
| | 19 | 0.25 | 0.32 | 0.63 | 0.55 | 0.57 | 0.67 | 0.26 | 0.17 |
| | 20 | 0.24 | 0.31 | 0.65 | 0.57 | 0.56 | 0.66 | 0.27 | 0.18 |
| k = 3 | 10 | 0.50 | 0.61 | 0.24 | 0.15 | 0.75 | 0.82 | 0.06 | 0.03 |
| | 12 | 0.42 | 0.52 | 0.32 | 0.22 | 0.71 | 0.78 | 0.09 | 0.05 |
| | 14 | 0.37 | 0.46 | 0.39 | 0.30 | 0.68 | 0.76 | 0.12 | 0.07 |
| | 16 | 0.34 | 0.42 | 0.45 | 0.36 | 0.65 | 0.73 | 0.14 | 0.09 |
| | 18 | 0.31 | 0.38 | 0.50 | 0.42 | 0.63 | 0.71 | 0.17 | 0.11 |
| | 20 | 0.29 | 0.36 | 0.54 | 0.46 | 0.61 | 0.69 | 0.19 | 0.13 |

**Table XXVI** Values of the constants $c_{n,i}$ required for calculating the discordancy test statistics $T_{GU3}$ and $T_{GU6}$ for one or more outliers in a Gumbel least-value sample with $a$ and $b$ unknown. See pages 258 and 260.

GU3 and GU6

| $i$ \ $n$ | 1 | 2 | 3 | 4 | 5 | 6 | 7 | 8 | 9 | 10 | 11 | 12 |
|---|---|---|---|---|---|---|---|---|---|---|---|---|
| 3 | 0.822 | 1.159 | | | | | | | | | | |
| 4 | 0.869 | 1.415 | 1.471 | | | | | | | | | |
| 5 | 0.896 | 1.549 | 1.878 | 1.714 | | | | | | | | |
| 6 | 0.914 | 1.633 | 2.108 | 2.258 | 1.913 | | | | | | | |
| 7 | 0.927 | 1.690 | 2.258 | 2.582 | 2.579 | 2.081 | | | | | | |
| 8 | 0.936 | 1.732 | 2.365 | 2.801 | 2.993 | 2.858 | 2.225 | | | | | |
| 9 | 0.943 | 1.764 | 2.444 | 2.961 | 3.281 | 3.356 | 3.104 | 2.353 | | | | |
| 10 | 0.949 | 1.789 | 2.506 | 3.082 | 3.495 | 3.711 | 3.681 | 3.324 | 2.467 | | | |
| 11 | 0.954 | 1.809 | 2.555 | 3.178 | 3.660 | 3.978 | 4.100 | 3.975 | 3.523 | 2.570 | | |
| 12 | 0.958 | 1.826 | 2.595 | 3.255 | 3.792 | 4.188 | 4.420 | 4.455 | 4.244 | 3.704 | 2.664 | |
| 13 | 0.961 | 1.840 | 2.629 | 3.319 | 3.900 | 4.357 | 4.674 | 4.826 | 4.782 | 4.491 | 3.871 | 2.750 |
| 14 | 0.964 | 1.852 | 2.657 | 3.373 | 3.990 | 4.497 | 4.881 | 5.124 | 5.203 | 5.084 | 4.720 | 4.026 |
| 15 | 0.966 | 1.862 | 2.681 | 3.419 | 4.066 | 4.614 | 5.053 | 5.369 | 5.543 | 5.553 | 5.366 | 4.933 |
| 16 | 0.968 | 1.871 | 2.703 | 3.458 | 4.131 | 4.715 | 5.199 | 5.574 | 5.825 | 5.936 | 5.881 | 5.630 |
| 17 | 0.970 | 1.879 | 2.721 | 3.493 | 4.188 | 4.801 | 5.324 | 5.749 | 6.063 | 6.254 | 6.304 | 6.190 |
| 18 | 0.972 | 1.886 | 2.737 | 3.523 | 4.238 | 4.877 | 5.433 | 5.900 | 6.267 | 6.525 | 6.659 | 6.652 |
| 19 | 0.973 | 1.892 | 2.752 | 3.550 | 4.282 | 4.943 | 5.528 | 6.031 | 6.444 | 6.757 | 6.961 | 7.042 |
| 20 | 0.975 | 1.897 | 2.765 | 3.574 | 4.321 | 5.002 | 5.613 | 6.147 | 6.599 | 6.960 | 7.223 | 7.376 |
| 21 | 0.976 | 1.902 | 2.777 | 3.596 | 4.356 | 5.055 | 5.688 | 6.250 | 6.735 | 7.138 | 7.452 | 7.666 |
| 22 | 0.977 | 1.907 | 2.787 | 3.615 | 4.388 | 5.102 | 5.755 | 6.341 | 6.857 | 7.297 | 7.653 | 7.920 |
| 23 | 0.978 | 1.911 | 2.797 | 3.633 | 4.417 | 5.145 | 5.816 | 6.424 | 6.966 | 7.438 | 7.833 | 8.145 |
| 24 | 0.979 | 1.915 | 2.806 | 3.649 | 4.443 | 5.184 | 5.871 | 6.499 | 7.065 | 7.565 | 7.994 | 8.346 |
| 25 | 0.980 | 1.918 | 2.814 | 3.664 | 4.467 | 5.220 | 5.921 | 6.567 | 7.154 | 7.679 | 8.139 | 8.527 |

**Table XXVI** (*Contd.*)

| $n$ \ $i$ | 13 | 14 | 15 | 16 | 17 | 18 | 19 | 20 | 21 | 22 | 23 | 24 |
|---|---|---|---|---|---|---|---|---|---|---|---|---|
| 14 | 2.830 | | | | | | | | | | | |
| 15 | 4.169 | 2.904 | | | | | | | | | | |
| 16 | 5.132 | 4.304 | 2.974 | | | | | | | | | |
| 17 | 5.878 | 5.319 | 4.430 | 3.039 | | | | | | | | |
| 18 | 6.480 | 6.111 | 5.495 | 4.549 | 3.100 | | | | | | | |
| 19 | 6.981 | 6.755 | 6.332 | 5.662 | 4.662 | 3.158 | | | | | | |
| 20 | 7.405 | 7.293 | 7.016 | 6.542 | 5.820 | 4.768 | 3.213 | | | | | |
| 21 | 7.770 | 7.750 | 7.590 | 7.264 | 6.741 | 5.971 | 4.870 | 3.265 | | | | |
| 22 | 8.088 | 8.146 | 8.080 | 7.873 | 7.501 | 6.931 | 6.114 | 4.966 | 3.315 | | | |
| 23 | 8.368 | 8.492 | 8.505 | 8.395 | 8.143 | 7.727 | 7.113 | 6.251 | 5.059 | 3.362 | | |
| 24 | 8.617 | 8.797 | 8.878 | 8.849 | 8.696 | 8.402 | 7.944 | 7.287 | 6.382 | 5.147 | 3.407 | |
| 25 | 8.839 | 9.069 | 9.208 | 9.249 | 9.179 | 8.986 | 8.651 | 8.151 | 7.454 | 6.508 | 5.232 | 3.451 |

$n$ = number of observations.

**Table XXVIIa**   Critical values for 5% and 1% tests of discordancy for an upper outlier $x_{(n)}$ in a Poisson sample, using $x_{(n)}$ conditional on $\sum x_j$ as test statistic

*5% critical values*                        **P1**

| $\sum x_j - x_{(n)}$ / $n$ | 0 | 1 | 2 | 3 | 4 | 5 | 6 | 8 | 10 | 12 | 14 | 16 | 18 | 20 | 22 | 24 |
|---|---|---|---|---|---|---|---|---|---|---|---|---|---|---|---|---|
| 3 | 4 | 7 | 9 | 11 | 13 | 15 | 16 | 20 | 24 | 27 | 31 | 34 | 38 | 41 | 44 | 48 |
| 4 | 4 | 6 | 8 | 9 | 11 | 13 | 14 | 17 | 21 | 24 | 27 | 30 | 33 | 36 | 39 | 42 |
| 5 | 3 | 5 | 7 | 9 | 10 | 12 | 13 | 16 | 19 | 22 | 25 | 28 | 30 | 33 | 36 | 39 |
| 6 | 3 | 5 | 7 | 8 | 10 | 11 | 12 | 15 | 18 | 21 | 24 | 26 | 29 | 32 | 34 | 37 |
| 8 | 3 | 5 | 6 | 7 | 9 | 10 | 11 | 14 | 17 | 19 | 22 | 25 | 27 | 30 | 32 | 35 |
| 10 | 3 | 4 | 6 | 7 | 8 | 10 | 11 | 14 | 16 | 19 | 21 | 24 | 26 | 28 | 31 | 33 |
| 12 | 3 | 4 | 6 | 7 | 8 | 10 | 11 | 13 | 16 | 18 | 21 | 23 | 25 | 28 | 30 | 32 |
| 16 | 3 | 4 | 5 | 7 | 8 | 9 | 10 | 13 | 15 | 17 | 20 | 22 | 24 | 27 | 29 | 31 |
| 20 | 2 | 4 | 5 | 6 | 8 | 9 | 10 | 12 | 15 | 17 | 19 | 22 | 24 | 26 | 28 | 31 |
| 25 | 2 | 4 | 5 | 6 | 7 | 9 | 10 | 12 | 14 | 17 | 19 | 21 | 23 | 26 | 28 | 30 |
| 50 | 2 | 4 | 5 | 6 | 7 | 8 | 9 | 12 | 14 | 16 | 18 | 20 | 22 | 25 | 27 | 29 |
| 100 | 2 | 3 | 5 | 6 | 7 | 8 | 9 | 11 | 13 | 15 | 18 | 20 | 22 | 24 | 26 | 28 |

The outlier $x_{(n)}$ is significant at 5% if $\sum x_j$ is greater than or equal to the tabulated value in the column corresponding to the observed value of $\sum x_j - x_{(n)}$.

*1% critical values*

| $\sum x_j - x_{(n)}$ / $n$ | 0 | 1 | 2 | 3 | 4 | 5 | 6 | 8 | 10 | 12 | 14 | 16 | 18 | 20 | 22 | 24 |
|---|---|---|---|---|---|---|---|---|---|---|---|---|---|---|---|---|
| 3 | 6 | 8 | 11 | 13 | 15 | 17 | 19 | 23 | 26 | 30 | 34 | 37 | 41 | 44 | 48 | 51 |
| 4 | 5 | 7 | 9 | 11 | 13 | 14 | 16 | 19 | 23 | 26 | 29 | 32 | 35 | 38 | 41 | 45 |
| 5 | 4 | 6 | 8 | 10 | 11 | 13 | 15 | 18 | 21 | 24 | 27 | 30 | 32 | 35 | 38 | 41 |
| 6 | 4 | 6 | 8 | 9 | 11 | 12 | 14 | 17 | 20 | 22 | 25 | 28 | 31 | 33 | 36 | 39 |
| 8 | 4 | 5 | 7 | 8 | 10 | 11 | 13 | 15 | 18 | 21 | 23 | 26 | 29 | 31 | 34 | 36 |
| 10 | 3 | 5 | 7 | 8 | 9 | 11 | 12 | 15 | 17 | 20 | 22 | 25 | 27 | 30 | 32 | 35 |
| 12 | 3 | 5 | 6 | 8 | 9 | 10 | 12 | 14 | 17 | 19 | 22 | 24 | 27 | 29 | 31 | 34 |
| 16 | 3 | 5 | 6 | 7 | 9 | 10 | 11 | 14 | 16 | 18 | 21 | 23 | 26 | 28 | 30 | 32 |
| 20 | 3 | 4 | 6 | 7 | 8 | 10 | 11 | 13 | 16 | 18 | 20 | 23 | 25 | 27 | 29 | 32 |
| 25 | 3 | 4 | 6 | 7 | 8 | 9 | 11 | 13 | 15 | 18 | 20 | 22 | 24 | 27 | 29 | 31 |
| 50 | 3 | 4 | 5 | 6 | 8 | 9 | 10 | 12 | 14 | 17 | 19 | 21 | 23 | 25 | 27 | 30 |
| 100 | 2 | 4 | 5 | 6 | 7 | 8 | 9 | 12 | 14 | 16 | 18 | 20 | 22 | 24 | 27 | 29 |

The outlier $x_{(n)}$ is significant at 1% if $\sum x_j$ is greater than or equal to the tabulated value in the column corresponding to the observed value of $\sum x_j - x_{(n)}$.

$n$ = number of observations.
$x_{(n)}$ = greatest observation.
$\sum x_j$ = sum of observations.

**Table XXVIIb**  Critical values for 5% and 1% tests of discordancy for a lower outlier $x_{(1)}$ in a Poisson sample, using $x_{(1)}$ conditional on $\sum x_j$ as test statistic

**P2**

*5% critical values*

| $x_{(1)}$ / $n$ | 0 | 1 | 2 | 3 | 4 | 5 | 6 | 7 | 8 | 9 | 10 | 11 | 12 | 13 | 14 | 15 |
|---|---|---|---|---|---|---|---|---|---|---|---|---|---|---|---|---|
| 3 | 11 | 16 | 21 | 25 | 29 | 33 | 37 | 41 | 45 | 49 | 53 | 57 | 61 | 64 | 68 | 72 |
| 4 | 16 | 23 | 30 | 36 | 42 | 47 | 53 | 58 | 64 | 69 | 74 | 80 | 85 | 90 | 95 | |
| 5 | 21 | 31 | 39 | 47 | 55 | 62 | 69 | 76 | 83 | 89 | 96 | | | | | |
| 6 | 27 | 38 | 49 | 58 | 68 | 76 | 85 | 93 | | | | | | | | |

The outlier $x_{(1)}$ is significant at 5% if $\sum x_j$ is greater than or equal to the tabulated value in the column corresponding to the observed value of $x_{(1)}$.

*1% critical values*

| $x_{(1)}$ / $n$ | 0 | 1 | 2 | 3 | 4 | 5 | 6 | 7 | 8 | 9 | 10 | 11 | 12 | 13 | 14 | 15 |
|---|---|---|---|---|---|---|---|---|---|---|---|---|---|---|---|---|
| 3 | 15 | 20 | 26 | 30 | 35 | 39 | 44 | 48 | 52 | 56 | 60 | 64 | 68 | 72 | 76 | 80 |
| 4 | 21 | 30 | 37 | 43 | 50 | 56 | 62 | 67 | 73 | 79 | 84 | 90 | 95 | 100 | | |
| 5 | 28 | 39 | 48 | 57 | 65 | 72 | 80 | 87 | 95 | | | | | | | |
| 6 | 35 | 48 | 60 | 70 | 80 | 89 | 98 | | | | | | | | | |

The outlier $x_{(1)}$ is significant at 1% if $\sum x_j$ is greater than or equal to the tabulated value in the column corresponding to the observed value of $x_{(1)}$.

$n$ = number of observations.
$x_{(1)}$ = smallest observation.
$\sum x_j$ = sum of observations.

**Table XXVIIIa**   Critical values for 5% and 1% tests of discordancy for an upper outlier-pair $x_{(n-1)}$, $x_{(n)}$ in a Poisson sample,, using $x_{(n-1)} + x_{(n)}$ conditional on $\sum x_j$ as test statistic

**P3**

*5% critical values*

| $\sum x_j - x_{(n-1)} - x_{(n)}$ /  $n$ | 0 | 1 | 2 | 3 | 4 | 5 | 6 | 8 | 10 | 12 | 14 |
|---|---|---|---|---|---|---|---|---|---|---|---|
| 4 | 7 | 11 | 14 | 17 | 20 | 23 | 25 | 30 | 35 | 40 | 45 |
| 5 | 6 | 9 | 12 | 14 | 16 | 19 | 21 | 25 | 29 | 34 | 38 |
| 6 | 6 | 8 | 11 | 13 | 15 | 17 | 19 | 23 | 26 | 30 | 34 |
| 8 | 5 | 7 | 9 | 11 | 13 | 15 | 16 | 20 | 23 | 26 | 29 |
| 10 | 5 | 7 | 9 | 10 | 12 | 14 | 15 | 18 | 21 | 24 | 27 |
| 12 | 4 | 6 | 8 | 10 | 11 | 13 | 14 | 17 | 20 | 23 | 26 |
| 16 | 4 | 6 | 8 | 9 | 11 | 12 | 13 | 16 | 19 | 22 | 24 |
| 20 | 4 | 6 | 7 | 9 | 10 | 12 | 13 | 16 | 18 | 21 | 23 |

The outlier-pair $x_{(n-1)}$, $x_{(n)}$ is significant at 5% if $\sum x_j$ is greater than or equal to the tabulated value in the column corresponding to the observed value of $\sum x_j - x_{(n-1)} - x_{(n)}$.

*1% critical values*

| $\sum x_j - x_{(n-1)} - x_{(n)}$ /  $n$ | 0 | 1 | 2 | 3 | 4 | 5 | 6 | 8 | 10 | 12 | 14 |
|---|---|---|---|---|---|---|---|---|---|---|---|
| 4 | 10 | 14 | 17 | 20 | 23 | 26 | 29 | 34 | 39 | 45 | 50 |
| 5 | 8 | 11 | 14 | 16 | 19 | 21 | 24 | 28 | 32 | 37 | 41 |
| 6 | 7 | 10 | 12 | 15 | 17 | 19 | 21 | 25 | 29 | 33 | 36 |
| 8 | 6 | 8 | 11 | 13 | 14 | 16 | 18 | 21 | 25 | 28 | 31 |
| 10 | 6 | 8 | 10 | 12 | 13 | 15 | 17 | 20 | 23 | 26 | 29 |
| 12 | 5 | 7 | 9 | 11 | 12 | 14 | 16 | 19 | 22 | 25 | 27 |
| 16 | 5 | 7 | 8 | 10 | 12 | 13 | 14 | 17 | 20 | 23 | 26 |
| 20 | 5 | 7 | 8 | 10 | 11 | 12 | 14 | 17 | 19 | 21 | 24 |

The outlier-pair $x_{(n-1)}$, $x_{(n)}$ is significant at 1% if $\sum x_j$ is greater than or equal to the tabulated value in the column corresponding to the observed value of $\sum x_j - x_{(n-1)} - x_{(n)}$

$n$ = number of observations.
$x_{(1)}$ = greatest observation.
$x_{(n-1)}$ = second greatest observation.
$\sum x_j$ = sum of observations.

**Table XXVIIIb** Critical values for 5% and 1% tests of discordancy for a lower outlier-pair $x_{(1)}$, $x_{(2)}$ in a Poisson sample, using $x_{(1)} + x_{(2)}$ conditional on $\sum x_j$ as test statistic

*5% critical values*                                                      **P4**

| $x_{(1)} + x_{(2)}$ / $n$ | 0 | 1 | 2 | 3 | 4 | 5 | 6 | 7 | 8 |
|---|---|---|---|---|---|---|---|---|---|
| 4 | 7 | 11 | 14 | 17 | 20 | 23 | 25 | 28 | 30 |
| 5 | 11 | 16 | 20 | 24 | 27 | 31 | 34 | 38 | 41 |
| 6 | 15 | 20 | 26 | 30 | 35 | 39 | 44 | 48 | 52 |
| 8 | 22 | 31 | 38 | 45 | 51 | 57 | 63 | 69 | 75 |
| 10 | 31 | 42 | 51 | 60 | 68 | 76 | 84 | 91 | 99 |

The outlier-pair $x_{(1)}$, $x_{(2)}$ is significant at 5% if $\sum x_j$ is greater than or equal to the tabulated value in the column corresponding to the observed value of $x_{(1)} + x_{(2)}$.

*1% critical values*

| $x_{(1)} + x_{(2)}$ / $n$ | 0 | 1 | 2 | 3 | 4 | 5 | 6 | 7 | 8 |
|---|---|---|---|---|---|---|---|---|---|
| 4 | 10 | 14 | 17 | 20 | 23 | 26 | 29 | 31 | 34 |
| 5 | 14 | 19 | 23 | 28 | 32 | 35 | 39 | 43 | 46 |
| 6 | 19 | 25 | 30 | 35 | 40 | 45 | 50 | 54 | 58 |
| 8 | 28 | 37 | 45 | 52 | 59 | 65 | 71 | 78 | 84 |
| 10 | 38 | 50 | 60 | 69 | 78 | 86 | 94 | | |

The outlier-pair $x_{(1)}$, $x_{(2)}$ is significant at 1% if $\sum x_j$ is greater than or equal to the tabulated value in the column corresponding to the observed value of $x_{(1)} + x_{(2)}$.

$n$ = number of observations.
$x_{(1)}$ = smallest observation.
$x_{(1)}$ = next smallest observation.
$\sum x_j$ = sum of observations.

**Table XXIX** Critical values for 5% and 1% tests of discordancy for an upper outlier $x_{(n)}$ in a binomial sample, using $x_{(n)}$ conditional on $\sum x_j$ as test statistic

**B1, B2**

*5% critical values*

$x_{(n)} = m$

| $n$ \ $m$ | 3 | 4 | 5 | 6 | 7 | 8 | 9 | 10 |
|---|---|---|---|---|---|---|---|---|
| 3 | 3 | 5 | 7 | 10 | 12 | 15 | 18 | 21 |
| 4 | 3 | 6 | 9 | 12 | 16 | 20 | 23 | 27 |
| 5 | 4 | 7 | 11 | 15 | 19 | 24 | 28 | 33 |
| 6 | 4 | 8 | 12 | 17 | 22 | 28 | 33 | 38 |
| 7 | 4 | 9 | 14 | 19 | 25 | 31 | 37 | 44 |
| 8 | 5 | 10 | 15 | 22 | 28 | 35 | 42 | 50 |
| 9 | 5 | 11 | 17 | 24 | 31 | 39 | 47 | 55 |
| 10 | 6 | 11 | 18 | 26 | 34 | 43 | 51 | 60 |

An outlier $x_{(n)}$ equal to $m$ is judged discordant at level 5% if $\sum x_j$ is less than or equal to the tabulated value in the column corresponding to $m$.

$x_{(n)} = m - 1$

| $n$ \ $m$ | 3 | 4 | 5 | 6 | 7 | 8 | 9 | 10 |
|---|---|---|---|---|---|---|---|---|
| 3 | — | — | 4 | 7 | 9 | 12 | 14 | 17 |
| 4 | — | 3 | 5 | 8 | 11 | 15 | 18 | 22 |
| 5 | — | 3 | 6 | 10 | 14 | 18 | 22 | 26 |
| 6 | — | 4 | 7 | 11 | 16 | 20 | 25 | 31 |
| 7 | — | 4 | 8 | 12 | 18 | 23 | 29 | 35 |
| 8 | — | 4 | 9 | 14 | 20 | 26 | 32 | 39 |
| 9 | — | 5 | 9 | 15 | 21 | 28 | 36 | 43 |
| 10 | — | 5 | 10 | 16 | 23 | 31 | 39 | 47 |

An outlier $x_{(n)}$ equal to $m - 1$ is judged discordant at level 5% if $\sum x_j$ is less than or equal to the tabulated value in the column corresponding to $m$.

$x_{(n)} = m - 2$

| $n$ \ $m$ | 3 | 4 | 5 | 6 | 7 | 8 | 9 | 10 |
|---|---|---|---|---|---|---|---|---|
| 3 | — | — | — | 4 | 6 | 9 | 11 | 14 |
| 4 | — | — | 3 | 5 | 8 | 11 | 14 | 17 |
| 5 | — | — | 3 | 6 | 9 | 13 | 17 | 21 |
| 6 | — | — | 3 | 7 | 11 | 15 | 19 | 24 |
| 7 | — | — | 4 | 7 | 12 | 17 | 22 | 27 |
| 8 | — | — | 4 | 8 | 13 | 19 | 25 | 31 |
| 9 | — | — | 4 | 9 | 14 | 20 | 27 | 34 |
| 10 | — | — | 5 | 9 | 15 | 22 | 29 | 37 |

An outlier $x_{(n)}$ equal to $m - 2$ is judged discordant at level 5% if $\sum x_j$ is less than or equal to the tabulated value in the column corresponding to $m$.

## Table XXIX (*Contd.*)

1% critical values

$x_{(n)} = m$

| $n$ \ $m$ | 3 | 4 | 5 | 6 | 7 | 8 | 9 | 10 |
|---|---|---|---|---|---|---|---|---|
| 3 | — | 4 | 6 | 8 | 10 | 13 | 16 | 19 |
| 4 | — | 4 | 7 | 10 | 13 | 17 | 20 | 24 |
| 5 | — | 5 | 8 | 12 | 16 | 20 | 24 | 29 |
| 6 | 3 | 6 | 9 | 14 | 18 | 23 | 27 | 33 |
| 7 | 3 | 6 | 11 | 15 | 21 | 26 | 31 | 38 |
| 8 | 3 | 7 | 12 | 17 | 23 | 29 | 36 | 43 |
| 9 | 3 | 7 | 13 | 19 | 25 | 33 | 40 | 47 |
| 10 | 4 | 8 | 14 | 20 | 28 | 35 | 44 | 52 |

An outlier $x_{(n)}$ equal to $m$ is judged discordant at level 1% if $\sum x_j$ is less than or equal to the tabulated value in the column corresponding to $m$.

$x_{(n)} = m - 1$

| $n$ \ $m$ | 3 | 4 | 5 | 6 | 7 | 8 | 9 | 10 |
|---|---|---|---|---|---|---|---|---|
| 3 | — | — | — | 5 | 8 | 10 | 12 | 15 |
| 4 | — | — | 4 | 6 | 9 | 12 | 15 | 19 |
| 5 | — | — | 5 | 7 | 11 | 14 | 18 | 22 |
| 6 | — | — | 5 | 8 | 12 | 17 | 21 | 26 |
| 7 | — | 3 | 6 | 9 | 14 | 19 | 24 | 29 |
| 8 | — | 3 | 6 | 10 | 15 | 21 | 27 | 33 |
| 9 | — | 3 | 7 | 11 | 17 | 23 | 29 | 36 |
| 10 | — | 3 | 7 | 12 | 18 | 25 | 32 | 40 |

An outlier $x_{(n)}$ equal to $m - 1$ is judged discordant at level 1% if $\sum x_j$ is less than or equal to the tabulated value in the column corresponding to $m$.

$x_{(n)} = m - 2$

| $n$ \ $m$ | 3 | 4 | 5 | 6 | 7 | 8 | 9 | 10 |
|---|---|---|---|---|---|---|---|---|
| 3 | — | — | — | — | 5 | 7 | 9 | 12 |
| 4 | — | — | — | 4 | 6 | 9 | 12 | 15 |
| 5 | — | — | — | 4 | 7 | 10 | 14 | 17 |
| 6 | — | — | — | 5 | 8 | 12 | 16 | 20 |
| 7 | — | — | — | 5 | 9 | 13 | 18 | 22 |
| 8 | — | — | 3 | 6 | 10 | 14 | 19 | 25 |
| 9 | — | — | 3 | 6 | 10 | 16 | 21 | 28 |
| 10 | — | — | 3 | 7 | 11 | 17 | 23 | 30 |

An outlier $x_{(n)}$ equal to $m - 2$ is judged discordant at level 1% if $\sum x_j$ is less than or equal to the tabulated value in the column corresponding to $m$.

$n$ = number of observations.
$x_{(n)}$ = greatest observation.
$\sum x_j$ = sum of observations.
$m$ = parameter of binomial distribution.

**Table XXX** Critical values for 5% and 1% tests of discordancy of a single outlier in a multivariate normal sample where $V$ is known and the test statistic is $\max_{j=1, 2, \ldots, n}(x_j - \bar{x})' V^{-1}(x_j - \bar{x})$ (reproduced by permission of The Institute of Statistical Mathematics)

| $n$ | $p = 2$ | | $p = 3$ | | $p = 4$ | |
|---|---|---|---|---|---|---|
| | 5% | 1% | 5% | 1% | 5% | 1% |
| 3 | 5.32 | 7.53 | 6.69 | 9.07 | 7.92 | 10.45 |
| 4 | 6.48 | 8.95 | 8.05 | 10.70 | 9.47 | 12.28 |
| 5 | 7.29 | 9.92 | 9.00 | 11.81 | 10.54 | 13.51 |
| 6 | 7.91 | 10.64 | 9.72 | 12.63 | 11.34 | 14.41 |
| 7 | 8.41 | 11.21 | 10.28 | 13.28 | 11.97 | 15.12 |
| 8 | 8.82 | 11.68 | 10.74 | 13.80 | 12.49 | 15.70 |
| 9 | 9.18 | 12.08 | 11.15 | 14.24 | 12.93 | 16.19 |
| 10 | 9.48 | 12.42 | 11.49 | 14.62 | 13.31 | 16.61 |
| 12 | 9.99 | 12.98 | 12.05 | 15.26 | 13.94 | 17.29 |
| 14 | 10.40 | 13.44 | 12.53 | 15.76 | 14.45 | 17.83 |
| 16 | 10.77 | 13.88 | 12.93 | 16.18 | 14.87 | 18.28 |
| 18 | 11.06 | 14.13 | 13.26 | 16.53 | 15.23 | 18.66 |
| 20 | 11.32 | 14.42 | 13.55 | 16.84 | 15.55 | 18.99 |
| 25 | 11.88 | 15.02 | 14.15 | 17.47 | 16.19 | 19.67 |
| 30 | 12.31 | 15.49 | 14.63 | 17.96 | 16.70 | 20.21 |

$n$ = number of observations; $p$ = dimension.

**Table XXXI** Critical values for 5% and 1% tests of discordancy of a single outlier in a multivariate normal sample where $\mu$ and $V$ are known, and the test statistic is

$$R_{(n)}(\mu, V) = \max_{j = 1, 2, \ldots, n} (x_j - \mu)' V^{-1}(x_j - \mu)$$

| $n$ | $p = 2$ | | $p = 4$ | | $p = 6$ | | $p = 8$ | | $p = 10$ | |
|---|---|---|---|---|---|---|---|---|---|---|
| | 5% | 1% | 5% | 1% | 5% | 1% | 5% | 1% | 5% | 1% |
| 3 | 8.15 | 11.40 | 12.05 | 15.77 | 15.46 | 19.54 | 18.63 | 23.02 | 21.66 | 26.31 |
| 4 | 8.73 | 11.98 | 12.72 | 16.42 | 16.20 | 20.24 | 19.43 | 23.76 | 22.50 | 27.10 |
| 5 | 9.17 | 12.42 | 13.23 | 16.91 | 16.76 | 20.78 | 20.03 | 24.34 | 23.15 | 27.71 |
| 6 | 9.53 | 12.79 | 13.65 | 17.32 | 17.22 | 21.22 | 20.53 | 24.81 | 23.67 | 28.21 |
| 7 | 9.84 | 13.09 | 14.00 | 17.66 | 17.60 | 21.59 | 20.94 | 25.20 | 24.12 | 28.62 |
| 8 | 10.11 | 13.36 | 14.30 | 17.96 | 17.94 | 21.91 | 21.30 | 25.54 | 24.49 | 28.98 |
| 9 | 10.34 | 13.61 | 14.57 | 18.24 | 18.23 | 22.21 | 21.61 | 25.86 | 24.83 | 29.31 |
| 10 | 10.55 | 13.81 | 14.81 | 18.46 | 18.49 | 22.45 | 21.89 | 26.11 | 25.12 | 29.58 |
| 25 | 12.38 | 15.64 | 16.87 | 20.48 | 20.73 | 24.62 | 24.29 | 28.41 | 27.65 | 31.99 |
| 50 | 13.77 | 17.02 | 18.41 | 21.99 | 22.40 | 26.24 | 26.06 | 30.12 | 29.52 | 33.78 |
| 100 | 15.15 | 18.41 | 19.94 | 23.50 | 24.04 | 27.84 | 27.80 | 31.82 | 31.35 | 35.55 |
| 200 | 16.54 | 19.80 | 21.46 | 25.00 | 25.67 | 29.44 | 29.52 | 33.49 | 33.16 | 37.30 |
| 500 | 18.37 | 21.63 | 23.46 | 26.98 | 27.80 | 31.95 | 31.77 | 35.68 | 35.50 | 39.58 |
| 1000 | 19.76 | 23.03 | 24.96 | 28.47 | 29.39 | 33.11 | 33.44 | 39.33 | 37.25 | 41.30 |

$n$ = number of observations; $p$ = dimension.

**Table XXXII** Critical values for 5% and 1% tests of discordancy of a single outlier in a multivariate normal sample where $\mu$ and $V$ are unknown and the test statistic is

$$R_{(n)}(\bar{\mathbf{x}}, S) = \max_{j = 1, 2, \ldots, n} (\mathbf{x}_j - \bar{\mathbf{x}})' S^{-1} (\mathbf{x}_j - \bar{\mathbf{x}})$$

| $n$ | $p = 2$ | | $p = 3$ | | $p = 4$ | | $p = 5$ | |
|---|---|---|---|---|---|---|---|---|
| | 5% | 1% | 5% | 1% | 5% | 1% | 5% | 1% |
| 5 | 3.17 | 3.19 | | | | | | |
| 6 | 4.00 | 4.11 | 4.14 | 4.16 | | | | |
| 7 | 4.71 | 4.95 | 5.01 | 5.10 | 5.12 | 5.14 | | |
| 8 | 5.32 | 5.70 | 5.77 | 5.97 | 6.01 | 6.09 | 6.11 | 6.12 |
| 9 | 5.85 | 6.37 | 6.43 | 6.76 | 6.80 | 6.97 | 7.01 | 7.08 |
| 10 | 6.32 | 6.97 | 7.01 | 7.47 | 7.50 | 7.79 | 7.82 | 7.98 |
| 12 | 7.10 | 8.00 | 7.99 | 8.70 | 8.67 | 9.20 | 9.19 | 9.57 |
| 14 | 7.74 | 8.84 | 8.78 | 9.71 | 9.61 | 10.37 | 10.29 | 10.90 |
| 16 | 8.27 | 9.54 | 9.44 | 10.56 | 10.39 | 11.36 | 11.20 | 12.02 |
| 18 | 8.73 | 10.15 | 10.00 | 11.28 | 11.06 | 12.20 | 11.96 | 12.98 |
| 20 | 9.13 | 10.67 | 10.49 | 11.91 | 11.63 | 12.93 | 12.62 | 13.81 |
| 25 | 9.94 | 11.73 | 11.48 | 13.18 | 12.78 | 14.40 | 13.94 | 15.47 |
| 30 | 10.58 | 12.54 | 12.24 | 14.14 | 13.67 | 15.51 | 14.95 | 16.73 |
| 35 | 11.10 | 13.20 | 12.85 | 14.92 | 14.37 | 16.40 | 15.75 | 17.73 |
| 40 | 11.53 | 13.74 | 13.36 | 15.56 | 14.96 | 17.13 | 16.41 | 18.55 |
| 45 | 11.90 | 14.20 | 13.80 | 16.10 | 15.46 | 17.74 | 16.97 | 19.24 |
| 50 | 12.23 | 14.60 | 14.18 | 16.56 | 15.89 | 18.27 | 17.45 | 19.83 |
| 100 | 14.22 | 16.95 | 16.45 | 19.26 | 18.43 | 21.30 | 20.26 | 23.17 |
| 200 | 15.99 | 18.94 | 18.42 | 21.47 | 20.59 | 23.72 | 22.59 | 25.82 |
| 500 | 18.12 | 21.22 | 20.75 | 23.95 | 23.06 | 26.37 | 25.21 | 28.62 |

$n$ = number of observations; $p$ = dimension.

**Table XXXIII** Critical values of $\sqrt{r_2}$ for 5% and 1% tests of discordancy for a pair of outliers in a multivariate normal sample where $\mu$ and $V$ are unknown, using the test statistic

$$r_2 = \min_{j_1, j_2} \mathscr{R}_{j_1, j_2}$$

| n | p = 2 | | p = 3 | | p = 4 | | p = 5 | |
|---|---|---|---|---|---|---|---|---|
| | 5% | 1% | 5% | 1% | 5% | 1% | 5% | 1% |
| 5 | 0.0025 | 0.0005 | 0.0000 | | | | | |
| 6 | 0.0337 | 0.0150 | 0.0011 | 0.0002 | | | | |
| 7 | 0.0860 | 0.0498 | 0.0202 | 0.0090 | 0.0006 | 0.0001 | | |
| 8 | 0.1417 | 0.0937 | 0.0580 | 0.0335 | 0.0136 | 0.0060 | 0.0004 | 0.0001 |
| 9 | 0.1942 | 0.1393 | 0.1024 | 0.0674 | 0.0425 | 0.0245 | 0.0098 | 0.0043 |
| 10 | 0.2419 | 0.1831 | 0.1470 | 0.1049 | 0.0788 | 0.0518 | 0.0327 | 0.0189 |
| 12 | 0.3229 | 0.2616 | 0.2288 | 0.1791 | 0.1549 | 0.1163 | 0.0966 | 0.0686 |
| 14 | 0.3879 | 0.3276 | 0.2982 | 0.2460 | 0.2246 | 0.1804 | 0.1631 | 0.1270 |
| 16 | 0.4410 | 0.3828 | 0.3563 | 0.3040 | 0.2853 | 0.2389 | 0.2242 | 0.1838 |
| 18 | 0.4850 | 0.4295 | 0.4054 | 0.3542 | 0.3376 | 0.2908 | 0.2782 | 0.2360 |
| 20 | 0.5221 | 0.4694 | 0.4472 | 0.3976 | 0.3828 | 0.3366 | 0.3257 | 0.2830 |
| 25 | 0.5935 | 0.5472 | 0.5288 | 0.4839 | 0.4722 | 0.4290 | 0.4211 | 0.3798 |
| 30 | 0.6451 | 0.6041 | 0.5882 | 0.5478 | 0.5380 | 0.4984 | 0.4923 | 0.4537 |
| 35 | 0.6842 | 0.6475 | 0.6335 | 0.5969 | 0.5885 | 0.5523 | 0.5473 | 0.5116 |
| 40 | 0.7150 | 0.6818 | 0.6693 | 0.6360 | 0.6285 | 0.5953 | 0.5911 | 0.5580 |
| 45 | 0.7399 | 0.7097 | 0.6982 | 0.6677 | 0.6610 | 0.6304 | 0.6267 | 0.5961 |
| 50 | 0.7605 | 0.7328 | 0.7222 | 0.6941 | 0.6880 | 0.6596 | 0.6564 | 0.6279 |
| 100 | 0.8629 | 0.8477 | 0.8417 | 0.8260 | 0.8225 | 0.8065 | 0.8047 | 0.7883 |
| 200 | 0.9232 | 0.9152 | 0.9118 | 0.9035 | 0.9015 | 0.8929 | 0.8918 | 0.8830 |
| 500 | 0.9650 | 0.9618 | 0.9602 | 0.9568 | 0.9558 | 0.9523 | 0.9517 | 0.9480 |

$n$ = number of observations; $p$ = dimension.

**Table XXXIV**   Critical values for 5% and 1% tests of discordancy of a single outlier in a bivariate normal sample where $\mu$ and $V$ are unknown, and the test statistic is

$$R_{(n)}(\bar{\mathbf{x}}, S_v) = \max_{j=1,2,\ldots,n} (\mathbf{x}_j - \bar{\mathbf{x}})'S_v^{-1}(\mathbf{x}_j - \bar{\mathbf{x}})$$

where $S_v$ is an 'external' estimate of $V$ (reproduced by permission of The Institute of Statistical Mathematics)

*5% test*

| $n$ $\backslash$ $v$ | 3 | 4 | 5 | 6 | 7 | 8 | 9 | 10 | 11 | 12 | 14 |
|---|---|---|---|---|---|---|---|---|---|---|---|
| 20 | 6.88 | 8.53 | 9.72 | 10.64 | 11.44 | 12.10 | 12.67 | 13.18 | 13.56 | 14.04 | 14.76 |
| 22 | 6.72 | 8.30 | 9.45 | 10.36 | 11.10 | 11.73 | 12.28 | 12.76 | 13.19 | 13.58 | 14.26 |
| 24 | 6.58 | 8.13 | 9.24 | 10.12 | 10.83 | 11.44 | 11.97 | 12.43 | 12.84 | 13.21 | 13.86 |
| 26 | 6.47 | 7.98 | 9.06 | 9.92 | 10.61 | 11.20 | 11.71 | 12.16 | 12.55 | 12.91 | 13.54 |
| 28 | 6.37 | 7.86 | 8.92 | 9.75 | 10.42 | 11.00 | 11.49 | 11.93 | 12.31 | 12.66 | 13.28 |
| 30 | 6.29 | 7.75 | 8.79 | 9.61 | 10.27 | 10.83 | 11.31 | 11.74 | 12.11 | 12.45 | 13.05 |
| 32 | 6.23 | 7.66 | 8.69 | 9.49 | 10.14 | 10.69 | 11.16 | 11.57 | 11.94 | 12.27 | 12.86 |
| 34 | 6.17 | 7.58 | 8.60 | 9.38 | 10.02 | 10.56 | 11.02 | 11.43 | 11.79 | 12.12 | 12.69 |
| 36 | 6.12 | 7.51 | 8.51 | 9.29 | 9.92 | 10.45 | 10.91 | 11.31 | 11.67 | 11.98 | 12.55 |
| 38 | 6.07 | 7.45 | 8.44 | 9.21 | 9.83 | 10.35 | 10.81 | 11.20 | 11.55 | 11.87 | 12.42 |
| 40 | 6.03 | 7.40 | 8.38 | 9.14 | 9.75 | 10.27 | 10.71 | 11.10 | 11.45 | 11.76 | 12.30 |
| 45 | 5.94 | 7.29 | 8.25 | 8.99 | 9.59 | 10.09 | 10.53 | 10.91 | 11.24 | 11.54 | 12.07 |
| 50 | 5.88 | 7.20 | 8.14 | 8.87 | 9.46 | 9.95 | 10.38 | 10.75 | 11.08 | 11.37 | 11.89 |
| 55 | 5.82 | 7.13 | 8.06 | 8.78 | 9.36 | 9.84 | 10.26 | 10.63 | 10.95 | 11.24 | 11.74 |
| 60 | 5.78 | 7.07 | 7.99 | 8.70 | 9.27 | 9.75 | 10.16 | 10.52 | 10.84 | 11.13 | 11.62 |
| 100 | 5.59 | 6.82 | 7.70 | 8.37 | 8.91 | 9.36 | 9.75 | 10.09 | 10.39 | 10.65 | 11.12 |
| 150 | 5.50 | 6.71 | 7.56 | 8.21 | 8.74 | 9.18 | 9.55 | 9.88 | 10.17 | 10.43 | 10.88 |
| 200 | 5.45 | 6.65 | 7.49 | 8.13 | 8.64 | 9.09 | 9.46 | 9.78 | 10.06 | 10.32 | 10.76 |

**Table XXXIV** (*Contd.*)

*1% test*

| n v | 3 | 4 | 5 | 6 | 7 | 8 | 9 | 10 | 11 | 12 | 14 |
|---|---|---|---|---|---|---|---|---|---|---|---|
| 20 | 10.72 | 12.99 | 14.61 | 15.85 | 16.86 | 17.71 | 18.44 | 19.08 | 19.66 | 20.17 | 21.07 |
| 22 | 10.36 | 12.53 | 14.07 | 15.24 | 16.20 | 17.00 | 17.69 | 18.29 | 18.83 | 19.31 | 20.16 |
| 24 | 10.07 | 12.16 | 13.63 | 14.76 | 15.68 | 16.44 | 17.10 | 17.67 | 18.18 | 18.64 | 19.44 |
| 26 | 9.84 | 11.86 | 13.28 | 14.37 | 15.25 | 15.98 | 16.62 | 17.16 | 17.66 | 18.09 | 18.86 |
| 28 | 9.64 | 11.63 | 12.99 | 14.05 | 14.90 | 15.62 | 16.22 | 16.75 | 17.22 | 17.64 | 18.37 |
| 30 | 9.47 | 11.40 | 12.74 | 13.77 | 14.60 | 15.30 | 15.88 | 16.40 | 16.85 | 17.26 | 17.97 |
| 32 | 9.33 | 11.22 | 12.54 | 13.54 | 14.35 | 15.02 | 15.60 | 16.10 | 16.54 | 16.94 | 17.63 |
| 34 | 9.21 | 11.06 | 12.36 | 13.34 | 14.13 | 14.79 | 15.35 | 15.84 | 16.28 | 16.66 | 17.34 |
| 36 | 9.10 | 10.93 | 12.20 | 13.17 | 13.95 | 14.59 | 15.14 | 15.62 | 16.04 | 16.42 | 17.08 |
| 38 | 9.01 | 10.81 | 12.06 | 13.01 | 13.78 | 14.41 | 14.95 | 15.42 | 15.84 | 16.21 | 16.86 |
| 40 | 8.93 | 10.70 | 11.94 | 12.88 | 13.63 | 14.26 | 14.79 | 15.25 | 15.66 | 16.03 | 16.66 |
| 45 | 8.75 | 10.48 | 11.69 | 12.60 | 13.33 | 13.93 | 14.45 | 14.89 | 15.29 | 15.64 | 16.25 |
| 50 | 8.62 | 10.31 | 11.49 | 12.38 | 13.09 | 13.68 | 14.18 | 14.62 | 15.00 | 15.34 | 15.93 |
| 55 | 8.51 | 10.18 | 11.33 | 12.21 | 12.90 | 13.48 | 13.97 | 14.39 | 14.77 | 15.10 | 15.68 |
| 60 | 8.42 | 10.07 | 11.20 | 12.06 | 12.75 | 13.32 | 13.80 | 14.21 | 14.58 | 14.91 | 15.48 |
| 100 | 8.05 | 9.59 | 10.66 | 11.46 | 12.10 | 12.63 | 13.07 | 13.45 | 13.79 | 14.09 | 14.61 |
| 150 | 7.87 | 9.37 | 10.40 | 11.18 | 11.79 | 12.30 | 12.73 | 13.10 | 13.42 | 13.71 | 14.21 |
| 200 | 7.79 | 9.26 | 10.28 | 11.04 | 11.62 | 12.14 | 12.57 | 12.92 | 13.24 | 13.52 | 14.01 |

$n$ = number of observations; $v$ = number of degrees of freedom.

**Table XXXV**   5% and 1% critical values for testing discordancy of an 'upper' outlier in a bivariate exponential sample ($\theta$ known)

*5% critical values*

| n θ | 0.1 | 0.2 | 0.3 | 0.4 | 0.5 | 0.6 | 0.7 | 0.8 | 0.9 | 1.0 |
|---|---|---|---|---|---|---|---|---|---|---|
| 4 | 6.61 | 6.77 | 6.88 | 6.97 | 7.05 | 7.11 | 7.16 | 7.21 | 7.25 | 7.28 |
| 5 | 6.87 | 7.03 | 7.15 | 7.24 | 7.31 | 7.38 | 7.43 | 7.47 | 7.51 | 7.55 |
| 6 | 7.08 | 7.25 | 7.37 | 7.46 | 7.53 | 7.59 | 7.64 | 7.69 | 7.73 | 7.76 |
| 7 | 7.26 | 7.43 | 7.55 | 7.64 | 7.71 | 7.77 | 7.83 | 7.87 | 7.91 | 7.95 |
| 8 | 7.42 | 7.59 | 7.71 | 7.80 | 7.87 | 7.93 | 7.98 | 8.03 | 8.07 | 8.10 |
| 9 | 7.56 | 7.73 | 7.85 | 7.94 | 8.01 | 8.07 | 8.12 | 8.17 | 8.21 | 8.24 |
| 10 | 7.68 | 7.85 | 7.97 | 8.06 | 8.13 | 8.19 | 8.25 | 8.29 | 8.33 | 8.36 |
| 12 | 7.89 | 8.06 | 8.18 | 8.28 | 8.35 | 8.41 | 8.46 | 8.50 | 8.54 | 8.58 |
| 15 | 8.15 | 8.32 | 8.44 | 8.54 | 8.61 | 8.67 | 8.72 | 8.76 | 8.80 | 8.84 |
| 20 | 8.48 | 8.66 | 8.78 | 8.87 | 8.94 | 9.00 | 9.05 | 9.10 | 9.14 | 9.17 |
| 30 | 8.95 | 9.12 | 9.25 | 9.34 | 9.41 | 9.47 | 9.52 | 9.56 | 9.60 | 9.64 |
| 40 | 9.27 | 9.45 | 9.58 | 9.67 | 9.74 | 9.80 | 9.85 | 9.89 | 9.93 | 9.97 |
| 60 | 9.73 | 9.92 | 10.04 | 10.13 | 10.20 | 10.26 | 10.31 | 10.36 | 10.39 | 10.43 |
| 100 | 10.31 | 10.49 | 10.62 | 10.71 | 10.78 | 10.84 | 10.89 | 10.93 | 10.97 | 11.00 |

**Table XXXV** (*Contd.*)

*1% critical values*

| θ \ n | 0.1 | 0.2 | 0.3 | 0.4 | 0.5 | 0.6 | 0.7 | 0.8 | 0.9 | 1.0 |
|---|---|---|---|---|---|---|---|---|---|---|
| 4 | 8.50 | 8.68 | 8.80 | 8.89 | 8.97 | 9.03 | 9.08 | 9.12 | 9.16 | 9.19 |
| 5 | 8.76 | 8.94 | 9.06 | 9.15 | 9.22 | 9.28 | 9.33 | 9.38 | 9.42 | 9.45 |
| 6 | 8.97 | 9.15 | 9.27 | 9.36 | 9.43 | 9.49 | 9.54 | 9.59 | 9.63 | 9.66 |
| 7 | 9.14 | 9.32 | 9.45 | 9.54 | 9.61 | 9.67 | 9.72 | 9.76 | 9.80 | 9.84 |
| 8 | 9.30 | 9.48 | 9.60 | 9.69 | 9.76 | 9.82 | 9.87 | 9.92 | 9.95 | 9.99 |
| 9 | 9.43 | 9.61 | 9.73 | 9.83 | 9.90 | 9.96 | 10.01 | 10.05 | 10.09 | 10.12 |
| 10 | 9.55 | 9.73 | 9.85 | 9.95 | 10.02 | 10.08 | 10.13 | 10.17 | 10.21 | 10.24 |
| 12 | 9.76 | 9.94 | 10.06 | 10.15 | 10.23 | 10.29 | 10.34 | 10.38 | 10.42 | 10.45 |
| 15 | 10.01 | 10.19 | 10.32 | 10.41 | 10.48 | 10.54 | 10.59 | 10.63 | 10.67 | 10.70 |
| 20 | 10.33 | 10.52 | 10.64 | 10.73 | 10.81 | 10.87 | 10.91 | 10.96 | 10.99 | 11.03 |
| 30 | 10.79 | 10.97 | 11.10 | 11.19 | 11.26 | 11.32 | 11.37 | 11.41 | 11.45 | 11.48 |
| 40 | 11.11 | 11.30 | 11.42 | 11.51 | 11.59 | 11.64 | 11.69 | 11.74 | 11.77 | 11.81 |
| 60 | 11.56 | 11.75 | 11.88 | 11.97 | 12.04 | 12.10 | 12.15 | 12.19 | 12.23 | 12.26 |
| 100 | 12.13 | 12.32 | 12.45 | 12.54 | 12.61 | 12.67 | 12.71 | 12.76 | 12.79 | 12.82 |

**Table XXXVI**  5% and 1% critical values for testing the discordancy of an upper outlier in a bivariate Pareto distribution ($\theta_1$, $\theta_2$, $a$ known) (reproduced from Barnett, 1979, by permission of the South African Statistical Association)

| n | α | ρ = 0.4<br>a = 2.5 | 0.33<br>3 | 0.29<br>3.5 | 0.25<br>4 | 0.20<br>5 | 0.14<br>7 | 0.10<br>10 | 0.05<br>20 | 0.025<br>40 |
|---|---|---|---|---|---|---|---|---|---|---|
| 4 | 5% | 9.15 | 6.53 | 5.10 | 4.22 | 3.23 | 2.35 | 1.84 | 1.36 | 1.17 |
|   | 1% | 17.81 | 11.42 | 8.26 | 6.46 | 4.55 | 3.01 | 2.19 | 1.49 | 1.22 |
| 5 | 5% | 10.03 | 7.05 | 5.45 | 4.48 | 3.38 | 2.43 | 1.88 | 1.38 | 1.18 |
|   | 1% | 19.50 | 12.32 | 8.82 | 6.85 | 4.76 | 3.12 | 2.25 | 1.51 | 1.23 |
| 6 | 5% | 10.81 | 7.51 | 5.75 | 4.70 | 3.51 | 2.50 | 1.92 | 1.40 | 1.18 |
|   | 1% | 21.00 | 13.11 | 9.31 | 7.18 | 4.95 | 3.21 | 2.29 | 1.53 | 1.24 |
| 8 | 5% | 12.16 | 8.29 | 6.27 | 5.07 | 3.74 | 2.61 | 1.98 | 1.42 | 1.19 |
|   | 1% | 23.59 | 14.46 | 10.13 | 7.73 | 5.26 | 3.35 | 2.36 | 1.55 | 1.25 |
| 10 | 5% | 13.32 | 8.95 | 6.70 | 5.37 | 3.92 | 2.70 | 2.03 | 1.44 | 1.20 |
|   | 1% | 25.82 | 15.59 | 10.81 | 8.19 | 5.51 | 3.46 | 2.42 | 1.57 | 1.26 |
| 12 | 5% | 14.35 | 9.52 | 7.06 | 5.63 | 4.07 | 2.78 | 2.07 | 1.45 | 1.21 |
|   | 1% | 27.80 | 16.58 | 11.40 | 8.58 | 5.72 | 3.56 | 2.47 | 1.58 | 1.26 |
| 15 | 5% | 15.71 | 10.28 | 7.55 | 5.97 | 4.27 | 2.88 | 2.12 | 1.47 | 1.21 |
|   | 1% | 30.42 | 17.89 | 12.17 | 9.08 | 5.99 | 3.68 | 2.52 | 1.60 | 1.27 |
| 20 | 5% | 17.66 | 11.34 | 8.22 | 6.43 | 4.53 | 3.01 | 2.19 | 1.49 | 1.22 |
|   | 1% | 34.16 | 19.71 | 13.23 | 9.77 | 6.38 | 3.84 | 2.60 | 1.63 | 1.28 |
| 30 | 5% | 20.82 | 13.02 | 9.26 | 7.14 | 4.93 | 3.20 | 2.28 | 1.52 | 1.24 |
|   | 1% | 40.23 | 22.60 | 14.89 | 10.84 | 6.91 | 4.08 | 2.72 | 1.67 | 1.29 |
| 40 | 5% | 23.40 | 14.36 | 10.07 | 7.69 | 5.24 | 3.34 | 2.36 | 1.55 | 1.25 |
|   | 1% | 45.17 | 24.90 | 16.18 | 11.67 | 7.33 | 4.26 | 2.80 | 1.69 | 1.30 |

**Table XXXVII** Critical values for 5% and 1% tests of discordancy for a single outlier in a general linear model with normal error structure, using the studentized residual as test statistic (reproduced by permission of the American Statistical Association and the American Society for Quality Control)

*5% critical values*

| n \ q | 1 | 2 | 3 | 4 | 5 | 6 | 8 | 10 | 15 | 25 |
|---|---|---|---|---|---|---|---|---|---|---|
| 5 | 1.92 | | | | | | | | | |
| 6 | 2.07 | 1.93 | | | | | | | | |
| 7 | 2.19 | 2.08 | 1.94 | | | | | | | |
| 8 | 2.28 | 2.20 | 2.10 | 1.94 | | | | | | |
| 9 | 2.35 | 2.29 | 2.21 | 2.10 | 1.95 | | | | | |
| 10 | 2.42 | 2.37 | 2.31 | 2.22 | 2.11 | 1.95 | | | | |
| 12 | 2.52 | 2.49 | 2.45 | 2.39 | 2.33 | 2.24 | 1.96 | | | |
| 14 | 2.61 | 2.58 | 2.55 | 2.51 | 2.47 | 2.41 | 2.25 | 1.96 | | |
| 16 | 2.68 | 2.66 | 2.63 | 2.60 | 2.57 | 2.53 | 2.43 | 2.26 | | |
| 18 | 2.73 | 2.72 | 2.70 | 2.68 | 2.65 | 2.62 | 2.55 | 2.44 | | |
| 20 | 2.78 | 2.77 | 2.76 | 2.74 | 2.72 | 2.70 | 2.64 | 2.57 | 2.15 | |
| 25 | 2.89 | 2.88 | 2.87 | 2.86 | 2.84 | 2.83 | 2.80 | 2.76 | 2.60 | |
| 30 | 2.96 | 2.96 | 2.95 | 2.94 | 2.93 | 2.93 | 2.90 | 2.88 | 2.70 | 2.17 |
| 35 | 3.03 | 3.02 | 3.02 | 3.01 | 3.00 | 3.00 | 2.93 | 2.97 | 2.91 | 2.64 |
| 40 | 3.08 | 3.08 | 3.07 | 3.07 | 3.06 | 3.06 | 3.05 | 3.03 | 3.00 | 2.84 |
| 45 | 3.13 | 3.12 | 3.12 | 3.12 | 3.11 | 3.11 | 3.10 | 3.09 | 3.06 | 2.96 |
| 50 | 3.17 | 3.16 | 3.16 | 3.16 | 3.15 | 3.15 | 3.14 | 3.14 | 3.11 | 3.04 |
| 60 | 3.23 | 3.23 | 3.23 | 3.23 | 3.22 | 3.22 | 3.22 | 3.21 | 3.20 | 3.15 |
| 70 | 3.29 | 3.29 | 3.28 | 3.28 | 3.28 | 3.28 | 3.27 | 3.27 | 3.26 | 3.23 |
| 80 | 3.33 | 3.33 | 3.33 | 3.33 | 3.33 | 3.33 | 3.32 | 3.32 | 3.31 | 3.29 |
| 90 | 3.37 | 3.37 | 3.37 | 3.37 | 3.37 | 3.37 | 3.36 | 3.36 | 3.36 | 3.34 |
| 100 | 3.41 | 3.41 | 3.40 | 3.40 | 3.40 | 3.40 | 3.40 | 3.40 | 3.39 | 3.38 |

## Table **XXXVII** (*Contd.*)

*1% critical values*

| q \ n | 1 | 2 | 3 | 4 | 5 | 6 | 8 | 10 | 15 | 25 |
|---|---|---|---|---|---|---|---|---|---|---|
| 5 | 1.98 | | | | | | | | | |
| 6 | 2.17 | 1.98 | | | | | | | | |
| 7 | 2.32 | 2.17 | 1.98 | | | | | | | |
| 8 | 2.44 | 2.32 | 2.18 | 1.98 | | | | | | |
| 9 | 2.54 | 2.44 | 2.33 | 2.18 | 1.99 | | | | | |
| 10 | 2.62 | 2.55 | 2.45 | 2.33 | 2.18 | 1.99 | | | | |
| 12 | 2.76 | 2.70 | 2.64 | 2.56 | 2.46 | 2.34 | 1.99 | | | |
| 14 | 2.86 | 2.82 | 2.78 | 2.72 | 2.65 | 2.57 | 2.35 | 1.99 | | |
| 16 | 2.95 | 2.92 | 2.88 | 2.84 | 2.79 | 2.73 | 2.58 | 2.35 | | |
| 18 | 3.02 | 3.00 | 2.97 | 2.94 | 2.90 | 2.85 | 2.75 | 2.59 | | |
| 20 | 3.08 | 3.06 | 3.04 | 3.01 | 2.98 | 2.95 | 2.87 | 2.76 | 2.20 | |
| 25 | 3.21 | 3.19 | 3.18 | 3.16 | 3.14 | 3.12 | 3.07 | 3.01 | 2.75 | |
| 30 | 3.30 | 3.29 | 3.28 | 3.26 | 3.25 | 3.24 | 3.21 | 3.17 | 3.04 | 2.21 |
| 35 | 3.37 | 3.36 | 3.35 | 3.34 | 3.34 | 3.33 | 3.30 | 3.25 | 3.19 | 2.81 |
| 40 | 3.43 | 3.42 | 3.42 | 3.41 | 3.40 | 3.40 | 3.38 | 3.36 | 3.30 | 3.05 |
| 45 | 3.48 | 3.47 | 3.47 | 3.46 | 3.46 | 3.45 | 3.44 | 3.43 | 3.38 | 3.23 |
| 50 | 3.52 | 3.52 | 3.51 | 3.51 | 3.51 | 3.50 | 3.49 | 3.48 | 3.45 | 3.34 |
| 60 | 3.60 | 3.59 | 3.59 | 3.59 | 3.58 | 3.58 | 3.57 | 3.56 | 3.54 | 3.48 |
| 70 | 3.65 | 3.65 | 3.65 | 3.65 | 3.64 | 3.64 | 3.64 | 3.63 | 3.61 | 3.57 |
| 80 | 3.70 | 3.70 | 3.70 | 3.70 | 3.69 | 3.69 | 3.69 | 3.68 | 3.67 | 3.64 |
| 90 | 3.74 | 3.74 | 3.74 | 3.74 | 3.74 | 3.74 | 3.73 | 3.73 | 3.72 | 3.70 |
| 100 | 3.78 | 3.78 | 3.78 | 3.77 | 3.77 | 3.77 | 3.77 | 3.77 | 3.76 | 3.74 |

$n$ = number of observations.
$q$ = number of independent variables (including count for intercept if fitted).

**Table XXXVIII** 5% and 1% critical values for the maximum normed residual, for testing the discordancy of a single outlier in a $r \times c$ factorial experiment (reproduced by permission of the American Statistical Association and the American Society for Quality Control)

*5% critical values*

| r / c | 3 | 4 | 5 | 6 | 7 | 8 | 9 | 10 |
|---|---|---|---|---|---|---|---|---|
| 3 | 0.648 | 0.645 | 0.624 | 0.600 | 0.577 | 0.555 | 0.535 | 0.518 |
| 4 | | 0.621 | 0.590 | 0.561 | 0.535 | 0.513 | 0.493 | 0.475 |
| 5 | | | 0.555 | 0.525 | 0.499 | 0.477 | 0.457 | 0.440 |
| 6 | | | | 0.495 | 0.469 | 0.447 | 0.428 | 0.412 |
| 7 | | | | | 0.444 | 0.423 | 0.405 | 0.389 |
| 8 | | | | | | 0.402 | 0.385 | 0.369 |
| 9 | | | | | | | 0.368 | 0.353 |
| 10 | | | | | | | | 0.338 |

*1% critical values*

| r / c | 3 | 4 | 5 | 6 | 7 | 8 | 9 | 10 |
|---|---|---|---|---|---|---|---|---|
| 3 | 0.660 | 0.675 | 0.664 | 0.646 | 0.626 | 0.606 | 0.587 | 0.569 |
| 4 | | 0.665 | 0.640 | 0.613 | 0.588 | 0.565 | 0.544 | 0.525 |
| 5 | | | 0.608 | 0.578 | 0.551 | 0.527 | 0.506 | 0.488 |
| 6 | | | | 0.546 | 0.519 | 0.495 | 0.475 | 0.457 |
| 7 | | | | | 0.492 | 0.469 | 0.449 | 0.431 |
| 8 | | | | | | 0.446 | 0.426 | 0.409 |
| 9 | | | | | | | 0.407 | 0.391 |
| 10 | | | | | | | | 0.375 |

**Table XXXIX** 5% and 1% critical values for the maximum normed residual, for testing the discordancy of a single outlier in a $r \times c \times d$ factorial experiment (reproduced by permission of the American Statistical Association)

*5% critical values*

| $r$ | $c$ | 3 | 4 | 5 | $d$<br>6 | 7 | 8 | 9 | 10 |
|---|---|---|---|---|---|---|---|---|---|
| 3 | 3 | 0.478 | | | | | | | |
| 4 | 3 | 0.455 | | | | | | | |
|   | 4 | 0.423 | 0.388 | | | | | | |
| 5 | 3 | 0.429 | | | | | | | |
|   | 4 | 0.395 | 0.359 | | | | | | |
|   | 5 | 0.367 | 0.331 | 0.305 | | | | | |
| 6 | 3 | 0.406 | | | | | | | |
|   | 4 | 0.371 | 0.336 | | | | | | |
|   | 5 | 0.343 | 0.309 | 0.284 | | | | | |
|   | 6 | 0.321 | 0.288 | 0.264 | 0.245 | | | | |
| 7 | 3 | 0.386 | | | | | | | |
|   | 4 | 0.352 | 0.317 | | | | | | |
|   | 5 | 0.324 | 0.291 | 0.267 | | | | | |
|   | 6 | 0.303 | 0.271 | 0.248 | 0.230 | | | | |
|   | 7 | 0.285 | 0.255 | 0.233 | 0.216 | 0.202 | | | |
| 8 | 3 | 0.369 | | | | | | | |
|   | 4 | 0.335 | 0.301 | | | | | | |
|   | 5 | 0.308 | 0.276 | 0.253 | | | | | |
|   | 6 | 0.287 | 0.257 | 0.235 | 0.218 | | | | |
|   | 7 | 0.270 | 0.241 | 0.220 | 0.204 | 0.191 | | | |
|   | 8 | 0.256 | 0.228 | 0.208 | 0.193 | 0.180 | 0.170 | | |
| 9 | 3 | 0.354 | | | | | | | |
|   | 4 | 0.320 | 0.288 | | | | | | |
|   | 5 | 0.294 | 0.263 | 0.241 | | | | | |
|   | 6 | 0.274 | 0.245 | 0.223 | 0.207 | | | | |
|   | 7 | 0.258 | 0.230 | 0.209 | 0.194 | 0.182 | | | |
|   | 8 | 0.244 | 0.217 | 0.198 | 0.183 | 0.172 | 0.162 | | |
|   | 9 | 0.233 | 0.207 | 0.188 | 0.174 | 0.163 | 0.154 | 0.146 | |
| 10 | 3 | 0.341 | | | | | | | |
|   | 4 | 0.308 | 0.276 | | | | | | |
|   | 5 | 0.282 | 0.252 | 0.231 | | | | | |
|   | 6 | 0.263 | 0.234 | 0.214 | 0.198 | | | | |
|   | 7 | 0.244 | 0.220 | 0.200 | 0.186 | 0.174 | | | |
|   | 8 | 0.237 | 0.208 | 0.189 | 0.175 | 0.164 | 0.155 | | |
|   | 9 | 0.223 | 0.198 | 0.180 | 0.167 | 0.156 | 0.147 | 0.140 | |
|   | 10 | 0.213 | 0.189 | 0.172 | 0.159 | 0.149 | 0.141 | 0.134 | 0.128 |

OUTLIERS IN STATISTICAL DATA

**Table XXXIX** (*Contd.*)

*1% critical values*

| r | c | 3 | 4 | 5 | d 6 | 7 | 8 | 9 | 10 |
|---|---|---|---|---|---|---|---|---|---|
| 3 | 3 | 0.503 | | | | | | | |
| 4 | 3 | 0.488 | | | | | | | |
|   | 4 | 0.460 | 0.425 | | | | | | |
| 5 | 3 | 0.465 | | | | | | | |
|   | 4 | 0.432 | 0.395 | | | | | | |
|   | 5 | 0.403 | 0.365 | 0.336 | | | | | |
| 6 | 3 | 0.443 | | | | | | | |
|   | 4 | 0.408 | 0.370 | | | | | | |
|   | 5 | 0.378 | 0.341 | 0.313 | | | | | |
|   | 6 | 0.354 | 0.318 | 0.291 | 0.270 | | | | |
| 7 | 3 | 0.423 | | | | | | | |
|   | 4 | 0.386 | 0.350 | | | | | | |
|   | 5 | 0.357 | 0.321 | 0.294 | | | | | |
|   | 6 | 0.333 | 0.299 | 0.273 | 0.253 | | | | |
|   | 7 | 0.314 | 0.280 | 0.256 | 0.237 | 0.222 | | | |
| 8 | 3 | 0.405 | | | | | | | |
|   | 4 | 0.368 | 0.332 | | | | | | |
|   | 5 | 0.339 | 0.304 | 0.278 | | | | | |
|   | 6 | 0.316 | 0.283 | 0.258 | 0.239 | | | | |
|   | 7 | 0.298 | 0.265 | 0.242 | 0.224 | 0.210 | | | |
|   | 8 | 0.282 | 0.251 | 0.229 | 0.211 | 0.198 | 0.187 | | |
| 9 | 3 | 0.389 | | | | | | | |
|   | 4 | 0.353 | 0.317 | | | | | | |
|   | 5 | 0.324 | 0.290 | 0.265 | | | | | |
|   | 6 | 0.302 | 0.269 | 0.246 | 0.227 | | | | |
|   | 7 | 0.284 | 0.253 | 0.230 | 0.213 | 0.199 | | | |
|   | 8 | 0.269 | 0.239 | 0.217 | 0.201 | 0.188 | 0.177 | | |
|   | 9 | 0.256 | 0.227 | 0.207 | 0.191 | 0.178 | 0.168 | 0.160 | |
| 10 | 3 | 0.375 | | | | | | | |
|   | 4 | 0.339 | 0.304 | | | | | | |
|   | 5 | 0.311 | 0.278 | 0.254 | | | | | |
|   | 6 | 0.289 | 0.258 | 0.235 | 0.217 | | | | |
|   | 7 | 0.272 | 0.242 | 0.220 | 0.203 | 0.190 | | | |
|   | 8 | 0.257 | 0.228 | 0.208 | 0.192 | 0.179 | 0.169 | | |
|   | 9 | 0.245 | 0.217 | 0.197 | 0.182 | 0.170 | 0.161 | 0.153 | |
|   | 10 | 0.234 | 0.208 | 0.188 | 0.174 | 0.163 | 0.153 | 0.146 | 0.139 |

# References and
# Bibliography

Most of the works listed here have been referred to in the text; the pages on which principal or substantial mention is made of any work are shown in parentheses at the end of the reference. An example is:

Behnken, D. W., and Draper, N. R. (1972). 'Residuals and their variance patterns'. *Technometrics*, **14**, 101–111. (321, 323, 330)

Additional works which have not been specifically mentioned in the text, but which are likely to assist with further study of the history of outlier problems are also listed. We have indicated the area of relevance by means of a chapter reference, accompanied by the symbol *H* to show that the work is of particular historical interest, for example:

Glaisher, J. W. L. (1874). Note on a paper by Mr. Stone 'On the rejection of discordant observations'. *Monthly Notices Roy. Astr. Soc.*, **34**, 251. (Ch. 2, *H*)

---

Abdullah, M. B. (1989a). 'Robust estimation of a linear functional relationship'. *J. Statist. Comp. Sim.*, **33**, 101–123. (328)

Abdullah, M. B. (1989b). 'On W-estimators of a linear functional relationship'. *Commun. Statist. Theor. Meth.*, **18**, 287–314. (328)

Abraham, B., and Box, G. E. P. (1978). 'Linear models and spurious observations'. *Applied Statistics*, **27**, 131–138. (380, 383, 384, 391)

Abraham, B., and Box, G. E. P. (1979). 'Bayesian analysis of some outlier problems in time series'. *Biometrika*, **66**, 229–236. (413)

Abraham, B., and Chuang, A. (1987). 'EM algorithm and the estimation of time series models in the presence of outliers'. *Amer. Statist. Assoc. Proc. Bus. Econ. Statist. Sect.*, **1987**, 459–464. (411)

Abraham, B., and Chuang, A. (1989). 'Outlier detection and time series modeling'. *Technometrics*, **31**, 241–248. (400)

Abraham, B., and Yatawara, N. (1988). 'A score test for detection of time series outliers'. *J. Time Ser. Anal.*, **9**, 109–119. (400)

Adichie, J. N. (1967a). 'Asymptotic efficiency of a class of non-parametric tests for regression parameters'. *Ann. Math. Statist.*, **38**, 884–893. (325)

Adichie, J. N. (1967b). 'Estimates of regression parameters based on rank tests'. *Ann. Math. Statist.*, **38**, 894–903. (325)

Afifi, A. A., and Azen, S. P. (1979). *Statistical Analysis: A Computer oriented approach*, 2nd edn. Academic Press, New York. (21)

Agresti, A. (1983). 'A survey of strategies for modeling cross-classifications having ordinal variables'. *J. Amer. Statist. Assoc.*, **78**, 184–198. (439)

Airy, G. B. (1856). Letter from Professor Airy, Astronomer Royal, to the Editor. *Astr. J.*, **4**, 137–138. (Ch. 2, *H*)

Airy, G. B. (1861). *On the Algebraical and Numerical Theory of Errors of Observations and the Combination of Observations*. Macmillan, London. (Ch. 2, *H*)

Aitkin, M., and Tunnicliffe Wilson, G. (1980). 'Mixture models, outliers and the EM algorithm'. *Technometrics*, **22**, 325–331. (47, 344)

Akhtar, M. (1987). 'Central composite designs robust to outliers'. *Pakistan J. Statist.*, **3**, 41–48. (365)

Akhtar, M., and Prescott, P. (1987). 'A review of robust response surface designs'. *Pakistan J. Statist.*, **3**, 11–26. (365)

de Alba, E., and Zartman, D. L. (1980). 'Testing outliers in time series: an application to remotely sensed temperatures in cattle'. Pp. 127–138 of Anderson, O. D. (Ed.) (1980). *Analysing Time Series*. North-Holland, Amsterdam. (400)

Allen, G. C. (1961). See Bernoulli, D. (1777). (27)

Ameen, J. R. M. and Harrison, P. J. (1985). 'Normal discount Bayesian models'. In *Bayesian Statistics, 2*. North-Holland, Amsterdam. (414)

American National Standard (1975). *E178–75 Standard Recommended Practice for Dealing with Outlying Observations*. ANSI, Washington. (457)

Andrews, D. F. (1971). 'Significance tests based on residuals'. *Biometrika*, **58**, 139–148. (289, 331, 345)

Andrews, D. F. (1972). 'Plots of high-dimensional data'. *Biometrics*, **28**, 125–136. (309)

Andrews, D. F., and Pregibon, D. (1978). 'Finding the outliers that matter'. *J. Roy. Statist. Soc. B*, **40**, 85–93. (339, 345, 350, 373, 374)

Andrews, D. F., Bickel, P. J., Hampel, F. R., Huber, P. J., Rogers, W. H., and Tukey, J. W. (1972). *Robust Estimates of Location: Survey and Advances*. Princeton University Press, Princeton, N. J. (65, 147–152, 156, 158, 161–163)

Anonymous (1821). 'Dissertation sur la recherche du milieu le plus probable, entre les résultats de plusieurs observations ou expériences'. *Annales de Mathématiques Pure et Appliquées*, **12**, 181–204. (Ch. 2, *H*)

Anscombe, F. J. (1960a). 'Rejection of outliers'. *Technometrics*, **2**, 123–147. (33, 34, 50, 61, 143, 273, 346, 357)

Anscombe, F. J. (1961). 'Examination of residuals'. *Proceedings of the Fourth Berkeley Symposium on Mathematical Statistics and Probability*, Vol. 1, pp. 1–36. (330, 347)

Anscombe, F. J., and Barron, B. A. (1966). 'Treatment of outliers in samples of size three'. *J. Res. Nat. Bur. Standards, B.* **70**, 141–147. (55, 62, 65)

Anscombe, F. J., and Tukey, J. W. (1963). 'The examination and analysis of residuals'. *Technometrics*, **5**, 141–160. (347, 367)

Ansley, C. F., and Kohn, R. (1985). 'Estimation, filtering and smoothing in state space models with incompletely specified initial conditions.' *Ann. Statist.*, **13**, 1286–1316. (411)

Arley, N. (1940). 'On the distribution of relative errors from a normal population of errors. A discussion of some problems in the theory of errors'. *Mathematisk-Fysiske Meddelelser udgivet af det Kgl. Danske Videnskabernes Selskab*, **18**. (Ch. 2, *H*)

Arnold, B. C., and Balakrishnan, N. (1989). *Relations, Bounds and Approximations for Order Statistics*, Lecture Notes in Statistics, **53**, Springer-Verlag, New York. (67, 164, 165)

Atkinson, A. C. (1981). 'Two graphical displays for outlying and influential observations in regression'. *Biometrika*, **68**, 13–20. (367, 373, 374)

Atkinson, A. C. (1982a). 'Regression diagnostics, transformations and constructed variables' (with Discussion). *J. Roy. Statist. Soc. B*, **44**, 1–36. (369, 370)

Atkinson, A. C. (1982b). 'Robust and diagnostic regression analyses'. *Commun. Statist. Theor. Meth.*, **11**, 2559–2571. (369)

Atkinson, A. C. (1986a). 'Comment on Chatterjee and Hadi'. (1986). *Statist. Sci.*, **1**, 397–402. (369)

Atkinson, A. C. (1986b). 'Masking unmasked'. *Biometrika*, **73**, 533–541. (369, 370)

Atkinson, A. C. (1987). 'Robust regression and unmasking transformations'. In Pukkila, T., and Puntanen, S. (Eds). *Proceedings of the Second International Tampere Conference in Statistics*. University of Tampere, Finland, **99–112**. (369, 370)

Azzalini, A., Bowman, A. W., and Hardle, W. (1989). 'On the use of nonparametric regression for model checking'. *Biometrika*, **76**, 1–11. (353)

Bacon-Shone, J., and Fung, W. K. (1987). 'A new graphical method for detecting single and multiple outliers in univariate and multivariate data'. *Applied Statistics*, **36**, 153–162. (302)

Bagchi, P., and Guttman, I. (1990). 'Spuriosity and outliers in directional data'. *J. Appl. Statist.*, **17**, 341–350. (424)

Bailey, B. J. R. (1977). 'Tables of the Bonferroni statistics'. *J. Amer. Statist. Assn.*, **72**, 469–478. (121)

Balakrishnan, N. (1988). 'Recurrence relations among moments of order statistics from two related outlier models'. *Biometrical J.*, **30**, 741–746. (189)

Balakrishnan, N. (1992a). 'Outlier and robustness of estimators'. In Balakrishnan, N. (ed). *Handbook of the Logistic Distribution*. Marcel Dekker, New York. **263–290**. (68, 186–189)

Balakrishnan, N. (1992b). 'Relationships between single moments of order statistics from non-identically distributed variables'. In Sen. P. K. and Salama, I. A. (eds). *Order Statistics and Nonparametrics: Theory and Applications*, Elsevier Science Publishers BV. 65–78. (68)

Balakrishnan, N., and Ambagaspitiya, R. S. (1988). 'Relationships among moments of order statistics in samples from two related outlier models and some applications.' *Commun. Statist. Theor. Meth.*, **17**, 2327–2341. (68, 186, 188–190)

Balakrishnan, N., and Ambagaspitiya, R. S. (1989). 'An empirical power comparison of three tests of exponentiality under mixture—and outlier-models.' *Biometrical J.*, **31**, 49–66. (125)

Balakrishnan, N., Chan, P. S., Ho, K. L., and Lo, K. K. (1992). 'Means, variances and covariances of logistic order statistics in the presence of an outlier'. *Selected Tables in Math. Statist.*, **1992**. (68)

Balasooriya, U. (1989). 'Detection of outliers in the exponential distribution based on prediction'. *Commun. Statist. Theor. Meth.*, **18**, 711–720. (137)

Balasooriya, U., and Chan, L. K. (1981). 'Robust estimation of Stigler's data sets by cross validation'. *Applied Statistics*, **30**, 170–177. (56)

Balasooriya, U., and Tse, Y. K. (1986). 'Outlier detection in linear models: a comparative study in simple linear regression'. *Commun. Statist. Theor. Meth.*, **15**, 3589–3597. (339)

Balasooriya, U., Tse, Y. K., and Liew, Y. S. (1987). 'An empirical comparison of some statistics for identifying outliers and influential observations in linear regression models'. *J. Appl. Statist.*, **14**, 177–184. (339)

Bansal, R. K., and Papantoni-Kazakos, P. (1989). 'Outlier-resistant algorithms for detecting a change in a stochastic process'. *Trans. Info. Theor.*, **35**, 521–535. (411)

Barnett, V. (1966). 'Order statistics estimators of the location of the Cauchy distribution'. *J. Amer. Statist. Assn.*, **61**, 1205–1218. (82)

Barnett, V. (1976a). 'Convenient probability plotting positions for the normal distribution'. *Applied Statistics*, **25**, 47–50. (43)

Barnett, V. (1976b). 'The ordering of multivariate data' (with Discussion). *J. Roy. Statist. Soc. A*, **139**, 318–354. (270, 281)

Barnett, V. (1978a). 'The study of outliers: purpose and model'. *Applied Statistics*, **27**, 242–250. (4, 32, 33)

Barnett, V. (1978b). 'Multivariate outliers: Wilk's test and distance measures'. *Bull. Int. Statist. Inst.*, **47**(4), 37–40. (293)

Barnett, V. (1979). 'Some outlier tests for multivariate samples'. *South African Statist. J.*, **13**, 29–52. (272, 294, 295, 296)

Barnett, V. (Ed.) (1981). *Interpreting Multivariate Data*. Wiley, Chichester. (302)

Barnett, V. (1983a). 'Reduced distance measures and transformation in processing multivariate outliers'. *Austral. J. Statist.*, **25**, 1–12. (296, 297)

Barnett, V. (1983b). 'Principles and methods for handling outliers in data sets'. Pp. 131–166 of Wright, T. (Ed.) (1983). *Statistical Methods and the Improvement of Data Quality* Academic Press, Orlando, Florida. (33, 286)

Barnett, V. (1983c). 'Marginal outliers in the bivariate normal distribution'. *Bull. Int. Statist. Inst.*, **50**(4), 579–583. (300)

Barnett, V. (1985). 'Detection and testing of different types of outlier in linear structural relationships'. *Australian. J. Statist.*, **27**, 151–162. (326, 327)

Barnett, V. (1988). 'Outliers and order statistics'. *Commun. Statist. Theor. Meth.*, **17**, 2109–2118. (9)

Barnett, V. (1992). 'Unusual outliers'. In Schach, S. and Trenkler, G. (Eds). (1992). *Data Analysis and Statistical Inference*, Verlag Josel Eul, Köln. (432, 441)

Barnett, V. (1993). 'Outliers in Sample Surveys'. *J. Appl. Statist.* (In press). (441, 442)

Barnett, V., and Lewis, T. (1967). 'A study of low-temperature probabilities in the context of an industrial problem' (with Discussion). *J. Roy. Statist. Soc. A*, **130**, 177–206. (10)

Barnett, V., and Lewis, T. (1978). *Outliers in Statistical Data*, 1st edn. John Wiley, Chichester. (205)

Barnett, V. and Roberts, D. (1993). 'The problem of outlier tests in sample surveys'. *Commun. Statist. Theor. Meth.*, **22**, 1993. (253, 447)

Bartholomew, D. J. (1957). 'Testing for departure from the exponential distribution'. *Biometrika*, **44**, 253–257. (212)

Barthoulot, J. (1983). *Robust estimation in asymmetric contamination and application to autoregressive processes*. European Univ. Studies, Series V: Economics and Management. Lang, Bern. (403)

Bartkowiak, A. (1989). 'Displaying multivariate outliers using the concept of angular distance'. *Biometrie-Praximetrie*, **29**, 1–16. (302)

Bartkowiak, A., Lukasik, S., Chwistecki, K., Mrukowicz, M., and Morgenstern, W. (1987). 'Location of outliers in large epidemiological data'. *EDV in Med. Bio.*, **18**, 108–114. (454)

Bassett, G. W. Jr. (1988). 'A property of the observations fit by the extreme regression quantiles'. *Comp. Statist. Data Anal.*, **6**, 353–359. (346)

Basu, A. P. (1965). 'On some tests of hypotheses relating to the exponential distribution when some outliers are present'. *J. Amer. Statist. Assn.*, **60**, 548–559. Corr. *J. Amer. Statist. Assn.*, **60**, 1249. (193, 210)

Bebbington, A. C. (1978). 'A method of bivariate trimming for robust estimation of the correlation coefficient'. *Applied Statistics*, **27**, 221–226. (81, 281)

Becker, R. A., Chamber, J. M., and Wilks, A. R. (1988). *The New S Language*. Wadsworth & Brooks, Pacific Grove, California. (452)

Beckman, R. J. and Cook, R. D. (1983). 'Outlier . . . . . . s' (with Discussion). *Technometrics*, **25**, 119–163. (33, 135)

Beckman, R. J., and Trussell, H. J. (1974). 'The distribution of an arbitrary studentized residual and the effects of updating in multiple regression'. *J. Amer. Statist. Assn.*, **69**, 199–201. (330)

Bedrick, E. J., and Hill, J. R. (1990). 'Outlier tests for logistic regression: A conditional approach'. *Biometrika*, **77**, 815–827. (352)

Begg, T. B., Preston, S. R., and Healy, M. J. R. (1966), 'The dietary habits of patients with occlusive arterial disease'. *Atti V Conv. internat. Asp. diet. Inf. Senesc.*, pp. 66–79. (298)

Behnken, D. W., and Draper, N. R. (1972). 'Residuals and their variance patterns'. *Technometrics*, **14**, 101–111. (321, 323, 330)

Belsley, D. A., Kuh, E., and Welsch, R. E. (1980). *Regression Diagnostics: Identifying Influential Data and Sources of Collinearity*. Wiley, New York. (317, 349, 370, 373)

Bendre, S. M. (1989). 'Masking and swamping effects on tests for multiple outliers in normal sample.' *Commun. Statist. Theor. Meth.*, **18**, 697–710. (110, 113)

Bendre, S. M., and Kale, B. K. (1985). 'Masking effect on tests for outliers in exponential models'. *J. Amer. Statist. Assoc.*, **80**, 1020–1025. 'Correction' (1986). *J. Amer. Statist. Assoc.*, **81**, 1132.

Bendre, S. M., and Kale, B. K. (1987). 'Masking effect on tests for outliers in normal samples'. *Biometrika*, **74**, 891–896.

Beran, R. (1974). 'Asymptotically efficient adaptive rank estimates in location models'. *Ann. Statist.*, **2**, 63–74. (87)

Beran, R. (1977). 'Minimum Hellinger distance estimates for parametric models'. *Ann. Statist.*, **5**, 445–463. (154)

Beran, R. (1978). 'An efficient and robust adaptive estimator of location'. *Ann. Statist.*, **6**, 292–313. (87, 154)

Berger, J., and Berliner, L. M. (1986). 'Robust Bayes and empirical Bayes analysis with epsilon-contaminated priors'. *Ann. Statist.*, **14**, 461–486. (393)

Berkane, M., and Bentler, P. M. (1987). 'Characterizing parameters of multivariate elliptical distributions'. *Commun. Statist. Sim. Comp.*, **16**, 193–198. (296)

Berkane, M., and Bentler, P. M. (1988). 'Estimation of contamination parameters and identification of outliers in multivariate data'. *Soc. Meth. Res.*, **17**, 55–64. (296)

Berlin, J. A., Laird, N. M., Sacks, H. S., and Chalmers, T. C. (1989). 'A comparison of statistical methods for combining event rates from clinical trials'. *Statist. Med.*, **8**, 141–151. (454)

Bernoulli, D. (1777). 'Dijudicatio maxime probabilis plurium observationum discrepantium atque verisimillima inductio inde formanda'. *Acta Academiae Scientiorum Petropolitanae*, **1**, 3–33. English translation by C. G. Allen (1961), *Biometrika*, **48**, 3–13. (Ch. 2, *H*)

Bernoulli, J. III (1785). 'Milieu'. *Encyclopédie Méthodique*, II, 404–409. (Ch. 2, *H*)

Bertrand, J. (1888a). 'Sur la loi de probabilité des erreurs d'observation'. *C. R. Acad. Sci. Paris*, **106**, 153–156. (Ch. 2, *H*)

Bertrand, J. (1888b). 'Sur la combination des mesures d'une même grandeur'. *C. R. Acad. Sci. Paris*, **106**, 701–704. (Ch. 2, *H*)

Bertrand, J. (1889). *Calcul des Probabilités*. Gauthier–Villars, Paris. (Ch. 2, *H*)

Bessel, F. W., and Baeuer, J. J. (1838). *Gradmessung in Ostpreussen und ihre Verbindung mit Preussischen und Russischen Dreiecksketten*. Berlin. Reprinted in *Abhendlungen von F. W. Bessel* (ed. R. Engleman), Leipzig, 1876. (Ch. 2, *H*)

Best, D. J., and Fisher, N. I. (1986). 'Goodness of fit and discordancy tests for samples from the Watson distribution on the sphere'. *Australian J. Statist.*, **28**, 13–31. (428)

Bhattacharya, S. K., and Singh, N. (1988). 'Bayesian analysis of an outlier model in life testing'. *S. African Statist. J.*, **22**, 55–67. (385)

Bickel, P. J. (1965). 'On some robust estimates of location'. *Ann. Math. Statist.*, **36**, 847–858. (35, 65, 151)

Bickel, P. J. (1967). 'Some contributions to the theory of order statistics'. *Proc. Fifth Berkeley Symp. Math. Statist. Prob.* Vol. 1, 575–591. (146)

Bickel, P. J. (1976). 'Another look at robustness: A review of reviews and some new developments'. *Scand. J. Statist.*, **3**, 145–168. (35)

Bickel, P. J. (1984). 'Robust regression based on infinitesimal neighbourhoods'. *Ann. Statist.*, **12**, 1349–1368. (350)

Bickel, P. J., and Doksum, K. A. (1981). 'An analysis of transformation revisited', *J. Amer. Statist. Assoc.*, **76**, 296–311. (354)

Bickel, P. J., and Hodges, J. L. Jr. (1967). 'The asymptotic theory of Galton's test and a related simple estimate of location'. *Ann. Math. Statist.*, **38**, 73–89. (163)

Birch, J. B. and Fleischer, S. J. (1984). 'Diagnostic methods for detecting outliers in regression analysis'. *Envir. Entomology*, **13**, 19–25. (340)

Birch, J. B., and Myers, R. H. (1982). 'Robust analysis of covariance'. *Biometrics*, **38**, 699–713. (283)

Birnbaum, A. (1959). 'On the analysis of factorial experiments without replication'. *Technometrics*, **1**, 343–357. (367)

Birnbaum, A., and Laska, E. M. (1967). 'Optimal robustness: A general method with applications to linear estimators of location'. *J. Amer. Statist. Assn.*, **62**, 1230–1240. (65)

Birnbaum, A., and Miké, V. (1970). 'Asymptotically robust estimators of location'. *J. Amer. Statist. Assn.*, **65**, 1265–1282. (148)

Birnbaum, A., Laska, E. M., and Meisner, M. (1971). 'Optimally robust linear estimators of location'. *J. Amer. Statist. Assn.*, **66**, 302–310.

Bliss, C. I., Cochran, W. G., and Tukey, J. W. (1956). 'A rejection criterion based upon the range'. *Biometrika*, **43**, 418–422. (65)

BMDP (1978). *Biomedical Computational Programs P-Series.* Univ. Calif. Press, Berkeley, Calif. (450)

BMDP (1988a). *BMDP Statistical Software Manual.* 1. Univ. Calif. Press, Berkeley, Calif. (451)

BMDP (1988b). *BMDP Statistical Software Manual.* 2. Univ. Calif. Press, Berkeley, Calif. (451)

Bollen, K. A. (1987). 'Outliers and improper solutions: A confirmatory factor analysis example'. *Soc. Meth. Res.,* **15**, 375–384. (307)

Boos, D. D. (1980). 'A new method for constructing approximate confidence intervals from *M*-estimates'. *J. Amer. Statist. Assn.,* **75**, 142–145. (86)

Boos, D. D. (1981). 'Minimum distance estimators for location and goodness-of-fit'. *J. Amer. Statist. Assn.,* **76**, 663–670. (154)

Boscovich, R. J. (1757). 'De litteraria expeditione per pontificiam ditionem, et synopsis amplioris operis, ac habentur plura ejus ex exemplaria etiam sensorum impressa'. *Bononiensi Scientiarum et Artum Instuto Atque Academia Commentarii,* **4**, 353–396. (Ch. 2, *H*)

Bowley, A. L. (1928). *F. Y. Edgeworth's Contributions to Mathematical Statistics.* Royal Statistical Society, London. (Ch. 2, *H*)

Box, G. E. P., and Draper, N. R. (1975). 'Robust designs'. *Biometrika,* **62**, 347–352. (373)

Box, G. E. P., and Tiao, G. C. (1962). 'A further look at robustness via Bayes' theorem'. *Biometrika,* **49**, 419–432. (383)

Box, G. E. P., and Tiao, G. C. (1968). 'A Bayesian approach to some outlier problems'. *Biometrika,* **55**, 119–129. (47, 48, 380, 382)

Box, G. E. P., and Tiao, G. C. (1973). *Bayesian Inference in Statistical Analysis.* Addison-Wesley, London. (384)

Box, G. E. P., and Tiao, G. C. (1975). 'Intervention analysis with applications to economic and environmental problems'. *J. Amer. Statist. Assn.,* **70**, 70–79. (414)

Bradley, R. C. (1984). 'Some mixing properties of Tukey's 3R smoother'. *Stochastics,* **11**, 249–264. (153)

Bradu, D. (1987). 'An epsilon-median polish algorithm, outlier location in linear models'. *Comp. Statist. Data Anal.,* **5**, 327–336. (350)

Bradu, D., and Hawkins, D. M. (1982). 'Location of multiple outliers in two-way tables, using tetrads'. *Technometrics,* **24**, 103–108. (344, 363, 431, 440)

Bradu, D., and Hawkins, D. M. (1991). 'Sample size requirements for multiple outlier location techniques based on elemental sets'. *Research Report,* **1991**, 1–17. (344)

Brant, R. (1986). 'Comment on Chatterjee and Hadi'. *Statist. Sci.,* **1**, 405–407. (349)

Brant, R. (1990). 'Comparing classical and resistant outlier rules'. *J. Amer. Statist. Assoc.,* **85**, 1083–1090. (53)

Bretschneider, S. (1986). 'Estimating forecast variance with exponential smoothing: Some new results'. *Int. J. Forecasting,* **2**, 349–355. (413)

Bross, I. D. J. (1961). 'Outliers in patterned experiments: a strategic re-appraisal'. *Technometrics,* **3**, 91–102. (318, 355, 364)

Brown, B. M. (1975). 'A short-cut test for outliers using residuals'. *Biometrika,* **62**, 623–629. (365)

Brown, B. M., and Hettmansperger, T. P. (1989). 'An affine invariant bivariate version of the sign test'. *J. Roy. Statist. Soc. B,* **51**, 117–125. (271)

Brown, B. M., and Kildea, D. G. (1978). 'Asymptotically equivalent forms of Hodges-Lehmann-type estimates in regression'. *Austral. J. Statist.*, **20**, 153–160. (324)

Brown, B. M., and Kildea, D. G. (1979). 'Outlier-detection tests and robust estimators based on signs of residuals'. *Commun. Statist. Theor. Meth.*, **8**, 257–269. (335)

Brown, M. B. (1974). 'Identification of the sources of significance in two-way contingency tables'. *Applied Statistics*, **23**, 405–413. (434–438)

Brown, M. L. (1982). 'Robust line estimation with errors in both variables'. *J. Amer. Statist. Assn.* **77**, 71–79. (326)

Brown, R. L., Durbin, J., and Evans, J. M. (1975). 'Techniques for testing the constancy of regression relationships over time' (with Discussion). *J. Roy. Statist. Soc. B*, **37**, 149–192. (330)

Bruce, A. G., and Martin, R. D. (1989). 'Leave-k-out diagnostics for time series'. *J. Roy. Statist. Soc. B.*, **51**, 363–424. (411)

Brunt, D. (1917). *The Combination of Observations.* University Press, Cambridge. (2nd edn. 1931). (Ch. 2, *H*)

BSI (1985). *Handbook 25 Quality Management Systems. Statistical Interpretation. of Data.* BSI, Milton Keynes. (457)

BSI (1987). *BS5497 Precision of Test Methods (Part 1).* BSI, Milton Keynes. (457)

BSI (1990). *BSI Standards Catalogue 1990.* BSI, Milton Keynes. (456)

Burrows, P. M. (1979). 'Selected percentage points of Greenwood's statistic'. *J. Roy. Statist. Soc. A.*, **142**, 256–258. (211, 212)

Bustos, O. H., and Yohai, V. J. (1986). 'Robust estimates for ARMA models'. *J. Amer. Statist. Assoc.*, **81**, 155–168. (405)

Butler, R. W. (1982). 'Nonparametric interval and point prediction using data trimmed by a Grubbs-type outlier rule'. *Ann. Statist.*, **10**, 197–204. (87)

Butler, R. (1983). 'Outlier discordancy tests in the normal linear model'. *J. Roy. Statist. Soc. B*, **45**, 120–132. (338)

Butler, R. W. (1986). 'Predictive likelihood inference with applications'. *J. Roy. Statist. Soc. B.*, **48**, 1–38. (393)

Calvin, M., Heidelberger, C., Reid, J. C., Tolbert, B. M., and Yankwich, P. F. (1949). *Isotopic Carbon: Techniques in its measurement and chemical manipulation.* Wiley, New York. (4)

Campbell, N. A. (1978). 'The influence function as an aid in outlier detection in discriminant analysis'. *Applied Statistics*, **27**, 251–258. (282)

Campbell, N. A. (1980). 'Robust procedures in multivariate analysis I: Robust covariance estimation'. *Applied Statistics*, **29**, 231–237. (276, 278, 282)

Campbell, N. A. (1982). 'Robust procedures in multivariate analysis II. Robust canonical variate analysis'. *Applied Statistics*, **31**, 1–8. (282)

Caroni, C., and Prescott, P. (1992). 'Sequential application of Wilks's multivariate outlier test'. *Applied Statistics*, **41**, 355–364. (291)

Carroll, R. J. (1979). 'On estimating variances of robust estimators when the errors are asymmetric'. *J. Amer. Statist. Assn.*, **74**, 674–679. (65, 146, 324)

Carroll, R. J. (1980). 'Robust methods for factorial experiments with outliers'. *Applied Statistics*, **29**, 246–251. (354, 365)

Carroll, R. J. (1982a). 'Two examples of transformation when there are possible outliers'. *Applied Statistics*, **31**, 149–152. (354, 365)

Carroll, R. J. (1982b). 'Robust estimation in certain heteroscedastic linear models when there are many parameters'. *J. Statist. Plan. Inf.*, **7**, 1–12. (374)

Carroll, R. J., and Gallo, P. P. (1982). 'Some aspects of robustness in the functional errors-in-variables model'. *Commun. Statist. Theor. Meth.*, **11**, 2573–2585. (326)

Carroll, R. J., and Ruppert, D. (1982a). 'Robust estimation in heteroscedastic linear models'. *Ann. Statist.*, **10**, 429–441. (349)

Carroll, R. J., and Ruppert, D. (1982b). 'Weak convergence of bounded influence regression estimates with applications to repeated significance testing'. *J. Statist. Plan. Inf.*, **7**, 117–129. (374)

Carroll, R. J., and Ruppert, D. (1985). 'Transformations in regression: a robust analysis'. *Technometrics*, **27**, 1–12. (354)

Casjens, L. (1974). '*The Prehistoric Human Ecology of Southern Ruby Valley, Nevada*'. Doctoral Dissertation, Harvard University. (440)

Central Statistical Office (1992). *Family Spending: a Report on the 1991 Family Expenditure Survey*. HMSO, London.

Cerdan, S. V. (1989). 'A note on the behaviour of augmented principal-component plots in regression'. *Commun. Statist. Theor. Meth.*, **18**, 331–342. (371)

Chaloner, K., and Brant, R. (1988). 'A Bayesian approach to outlier detection and residual analysis'. *Biometrika*, **75**, 651–659. (385, 393)

Chalton, D. O., and Troskie, C. G. (1992). 'Q Plots, a graphical aid for regression analysis'. *Commun. Statist. Theor. Meth.*, **21**, 625–636. (370)

Chamberlain, R. (1975). In *British Births 1970. Vol. 1: The First Week of Life.* Heinemann, London. (5)

Chambers, R. L. (1986). 'Outlier robust finite population estimation'. *J. Amer. Statist. Assoc.*, **81**, 1063–1069. (443, 445)

Chambers, R. L., and Heathcote, C. R. (1981). 'On the estimation of slope and the identification of outliers in linear regression'. *Biometrika*, **68**, 21–33. (343)

Chang, T. (1986). 'Spherical regression'. *Ann. Statist.*, **14**, 907–924. (428)

Chang, I. H., Tiao, G. C., and Chen, C. (1988). 'Estimation of time series parameters in the presence of outliers'. *Technometrics*, **30**, 193–204. (400, 407)

Chatfield, C. (1975). *The Analysis of Time Series: Theory and Practice*. Chapman and Hall, London (p. 102). (18)

Chatfield, C. (1978). 'The Holt-Winters forecasting procedure'. *Applied Statistics*, **27**, 264–279. (Ch. 10)

Chatterjee, S., and Hadi, A. S. (1986). 'Influential observations, high leverage points, and outliers in linear regression'. *Statist. Sci.*, **1**, 415–416. (349)

Chauvenet, W. (1863). 'Method of least squares'. Appendix to *Manual of Spherical and Practical Astronomy*, Vol. 2, Lippincott, Philadelphia, pp. 469–566; tables 593–599. Reprinted (1960) 5th edn. Dover, New York. (4, 28, 38)

Chen, E. H. (1971). 'The power of the Shapiro-Wilk W-test for normality in samples from contaminated distributions'. *J. Amer. Statist. Assn.*, **66**, 760–762. (233, 234, 369)

Chen, E. H., and Dixon, W. J. (1972). 'Estimates of parameters of a censored regression sample'. *J. Amer. Statist. Assn.*, **67**, 664–671. (324)

Cheng, K. F. (1991). 'M-estimator using jack-knife pseudovalues'. *Scand. J. Statist.*, **18**, 51–61. (150)

Chernick, M. R., and Murthy, V. K. (1983). 'The use of influence functions for outlier detection and data editing'. *Amer. J. Math. Manag. Sci.*, **3**, 47–61. (170)

Chernick, M. R., Downing, D. J. and Pike, D. H. (1982). 'Detecting outliers in time series data'. *J. Amer. Statist. Assn.*, **77**, 743–747. (398)

Chernoff, H., Gastwirth, J. L., and Johns, M. V. Jr. (1967). 'Asymptotic distribution of linear combinations of order statistics'. *Ann. Math. Statist.*, **31**, 52–72. (146)

Chhikara, R. S., and Feiveson, A. L. (1980). 'Extended critical values of extreme studentized deviate test statistics for detecting multiple outliers'. *Commun. Statist. Sim. Comp.*, **9**, 155–166. (132, 236)

Chikkagoudar, M. S., and Kunchur, S. H. (1980). 'Estimation of the mean of an exponential distribution in the presence of an outlier'. *Canadian J. Statist.*, **8**, 59–63. (68, 177, 178, 183, 427)

Chikkagoudar, M. S., and Kunchur, S. M. (1983). 'Distributions of test statistics for multiple outliers in exponential samples'. *Commun. Statist. Theor. Meth.*, **12**, 2127–2142. (202, 203, 204, 216)

Chikkagoudar, M. S., and Kunchur, S. H. (1987). 'Comparison of many outlier procedures for exponential samples.' *Commun. Statist. Theor. Meth.*, **16**, 627–645. (98, 197, 198, 214–216)

Chiu, S. T. (1989). 'Detecting periodic components in a white Gaussian time series'. *J. Roy. Statist. Soc. B.*, **51**, 249–259. (413)

Chow, S. C. (1989). 'Detection of outliers in bioavailability/bioequivalence trials'. *J. Amer. Statist. Assoc.*, **84**, 254–259. (454)

Chow, S. C., and Tse, S. K. (1990). 'Outlier detection in bioavailability/bioequivalence studies'. *Statist. Med.*, **9**, 549–558. (454)

Chowdhury, A., and Balasooriya, U. (1988). 'A note on a typical observation in linear models. *J. Appl. Statist.*, **15**, 355–362. (351)

Chuang, A., and Abraham, B. (1989). 'Comparison of parameter estimation methods in time series with outliers: A simulation study'. *Proc. Bus. Econ. Statist. Sect. Amer. Statist. Assoc.*, 83–92. (406)

Cleroux, R., Helbling, J. M., and Ranger, N. (1986). 'Some methods of detecting multivariate outliers'. *Comp. Statist. Quart.*, **3**, 177–195. (310)

Cleveland, W. S. (1979). 'Robust locally weighted regression and smoothing scatter plots'. *J. Amer. Statist. Assn.*, **74**, 829–836. (280)

Cleveland, W. S., and Kleiner, B. (1975). 'A graphical technique for enhancing scatter plots with moving statistics'. *Technometrics*, **17**, 447–454. (302)

Clogg, C. C. (1982). 'Some models for the analysis of association in multiway cross-classifications having ordered categories'. *J. Amer. Statist. Assoc.*, **77**, 803–815. (439)

Cochran, W. G. (1941). 'The distribution of the largest of a set of estimated variances as a fraction of their total'. *Ann. Eugen.*, **11**, 47–52. (197, 242, 458)

Collett, D. (1978). *Some problems related to outliers on the line and on the circle*. Ph.D. thesis, University of Hull. (422)

Collett, D. (1980). 'Outliers in circular data'. *Applied Statistics*, **29**, 50–57. (22, 421, 425)

Collett, D., and Lewis, T. (1976). 'The subjective nature of outlier rejection procedures'. *Applied Statistics*, **25**, 228–237. (121)

Collings, B. J., Margolin, B. H., and Oehlert, G. W. (1981). 'Analyses for binomial data, with application to the fluctuation test for mutagenicity'. *Biometrics*, **37**, 775–794. (266)

Collins, J. R. (1986). 'Maximum asymptotic variances of trimmed means under asymmetric contamination'. *Ann. Statist.*, **14**, 348–354. (81)

Collins, J. R., and Wiens, D. P. (1985). 'Minimax variance M-estimators in epsilon-contamination models'. *Ann. Statist.*, **13**, 1078–1096. (65)

Cologne, J. B., Mendelman, P. M., and Chaffin, D. O. (1989). 'Statistical comparison of ligand-binding kinetics.' *Statist. Med.*, **8**, 871–881. (455)

Cook, R. D. (1977). 'Detection of influential observation in linear regression'. *Technometrics*, **19**, 15–18. (370, 373)

Cook, R. D. (1979). 'Influential observations in linear regression'. *J. Amer Statist. Assn.*, **74**, 169–174. (344, 374)

Cook, R. D. (1982). *Residuals and Influence in Regression*. Chapman and Hall, New York. (317)

Cook, R. D., and Hawkins, D. M (1990). 'Comments on Rousseeuw and Van Zomeren. (1990). (278)

Cook, R. D., Holschuh, N., and Weisberg, S. (1982). 'A note on an alternative outlier model'. *J. Roy. Statist. Soc. B*, **44**, 370–376. (49, 336)

Cook, R. D., and Prescott, P. (1981). 'On the accuracy of Bonferroni significance levels for detecting outliers in linear models'. *Technometrics*, **23**, 59–63. (335)

Cook, R. D., and Weisberg, S. (1980). 'Characterisations of an empirical influence function for detecting influential cases in regression'. *Technometrics*, **22**, 495–508. (318, 371, 372, 374, 454)

Cook, R. D., and Weisberg, S. (1982). *Residuals and Influence in Regression*. Chapman and Hall, New York. (317, 373, 454)

Copas, J. B. (1988). 'Binary regression models for contaminated data (with Discussion)'. *J. Roy. Statist. Soc.*, **50**, 225–265. (353)

Cox, D. R. (1979). Discussion of Kleiner, B., Martin, R. D., and Thomson, D. J. (1979). *J. Roy. Statist. Soc. B*, **41**, 343. (396)

Cox, D. R., and Snell, E. J. (1968). 'A general definition of residuals' (with Discussion). *J. Roy. Statist. Soc. B*, **30**, 248–275. (330)

Cox, D. R., and Snell, E. J. (1971). 'On test statistics calculated from residuals'. *Biometrika*, **58**, 589–594. (330)

Cressie, N. (1989). 'Empirical Bayes estimation of undercount in the decennial census'. *J. Amer. Statist. Assoc.*, **84**, 1033–1044. (446)

Cressie, N., and Glonek, G. (1984). 'Median based covariogram estimators reduce bias'. *Statist. Prob. Lett.*, **2**, 299–304. (402)

Critchley, F. (1985). 'Influence in principal components analysis'. *Biometrika*, **72**, 627–636. (282)

Critchley, F., and Vitiello, C. (1991). 'The influence of observations on misclassification probably estimates in linear discriminant analysis'. *Biometrika*, **78**, 677–690. (282)

Crow, E. L., and Siddiqui, M. M. (1967). 'Robust estimation of location'. *J. Amer. Statist. Assn.*, **62**, 353–389. (66)

Cunningham, R. B., and Heathcote, C. R. (1989). 'Estimating a non-gaussian regression model with multicollinearity'. *Australian. J. Statist.*, **31**, 12–17. (351)

Currie, I. D. (1981). 'Further percentage points of Greenwood's statistic'. *J. Roy. Statist. Soc. A.*, **144**, 360–363. (211, 212)

Czuber, E. (1891). *Theorie der Beobachtungsfehler*. Teubner, Leipzig. (Ch. 2, *H*)

D'Agostino, R. B., and Tietjen, G. L. (1971). 'Simulation probability points of $b_2$ for small samples'. *Biometrika*, **58**, 669–672. (231)

Dalal, S. R., Fowlkes, E. B., and Hoadley, B. (1989). 'Risk analysis of the space shuttle: Pre-challenger prediction of failure'. *J. Amer. Statist. Assoc.*, **84**, 945–957. (353, 455)

Dallal, G. E., and Hartigan, J. A. (1980). 'Note on a test of monotone association insensitive to outliers'. *J. Amer. Statist. Assn.*, **75**, 722–725. (283)

Daniel, C. (1959). 'Use of half-normal plots in interpreting factorial two-level experiments'. *Technometrics*, **1**, 311–341. (40, 366, 389)

Daniel, C. (1960). 'Locating outliers in factorial experiments'. *Technometrics*, **2**, 149–156. (318, 355–358)

Daniel, C. (1978). 'Patterns in residuals in the two-way layout'. *Technometrics*, **20**, 385–395. (355)

Daniel, C., and Wood, F. S. (1971). *Fitting Equations to Data.* Wiley, New York. (345)

Daniell, P. J. (1920). 'Observations weighted according to order'. *Amer. J. Math.*, **42**, 222–236. (Ch. 2, *H*)

Das, R., and Sinha, B. K. (1986). 'Detection of multivariate outliers with dispersion slippage in elliptically symmetric distributions'. *Ann. Statist.*, **14**, 1619–1624. (296)

David, H. A. (1952). 'Upper 5 and 1% points of the maximum *F*-ratio'. *Biometrika*, **39**, 422–424. (204, 205)

David, H. A. (1956a). 'On the application to statistics of an elementary theorem in probability'. *Biometrika*, **43**, 85–91. (237)

David, H. A. (1956b). 'Revised upper percentage points of the extreme studentized deviate from the sample mean'. *Biometrica*, **43**, 449–451. (237, 245)

David, H. A. (1962). 'Order statistics in short-cut tests'. In Sarhan and Greenberg (1962). (240)

David, H. A. (1979). 'Robust estimation in the presence of outliers'. Pp. 61–74 in Launer, R. L., and Wilkinson, G. N. (Eds) (1979). *Robustness in Statistics.* Academic Press, New York. (157, 168)

David, H. A. (1981). *Order Statistics*, 2nd edn. Wiley, New York. (104, 115, 122, 123, 146, 157, 168)

David, H. A., and Ghosh, J. K. (1985). 'The effect of an outlier on L-estimators of location in symmetric distributions'. *Biometrika*, **72**, 216–218. (82)

David, H. A., and Paulson, A. S. (1965). 'The performance of several tests for outliers'. *Biometrika*, **52**, 429–436. (222, 237, 238)

David, H. A., and Shu, V. S. (1978). 'Robustness of location estimators in the presence of an outlier'. Pp. 235–250 in David, H. A. (Ed.) (1978). *Contributions to Survey Sampling and Applied Statistics, in honour of H. O. Hartley.* Academic Press, New York. (67, 68, 147, 168)

David, H. A., Hartley, H. O., and Pearson, E. S. (1954). 'The distribution of the ratio, in a single normal sample, of range to standard deviation'. *Biometrika*, **41**, 482–493. (95, 226)

David, H. A., Kennedy, W. J., and Knight, R. D. (1977). 'Means, variances and covariances of the normal order statistics in the presence of an outlier'. In Harter, H. L., and Owen, D. B. (Eds) (1977). *Selected Tables in Mathematical Statistics.* Vol. V. American Mathematical Society, Providence, RI. (82, 169)

David, F. N., Barton, D. E., Ganeshalingam, S., Harter, H. L., Kim, P. J., Merrington, M., and Walley, D. (1968). *Normal Centroids, Medians and Scores for Ordinal Data*, Tracts for computers No. XXIX, Cambridge University Press, London. (72)

Davies, P. L., and Gather, U. (1993). 'The identification of multiple outliers'. *Technometrics*, **35**, 1–20. (53)

Davis, W. W. (1979). 'Robust methods for detection of shifts of the innovation variance of a time series'. *Technometrics*, **21**, 313–320. (414)

Dawid, A. P. (1973). 'Posterior expectations for large observations'. *Biometrika*, **60**, 664–667. (380)

De Gruttola, V., Ware, J. H., and Louis, T. A. (1987). 'Influence analysis of generalized least squares estimators'. *J. Amer. Statist. Assoc.*, **82**, 911–917. (354)

De Jong, P. (1988). 'A cross-validation filter for time series model'. *Biometrika*, **75**, 594–600. (411)

Dempster, A. P., and Gasko-Green, M. (1981). 'New tools for residual analysis'. *Ann. Statist.*, **9**, 945–959. (346)

Dempster, A. P., and Rosner, B. (1971). 'Detection of outliers'. In Gupta, S. S. and Yackel, J. (Eds) (1971) *Statistical Decision Theory and Related Topics*. Academic Press, New York. (343, 379, 388)

Denby, L., and Larsen, W. A. (1977). 'Robust regression estimators compared via Monte Carlo'. *Commun. Statist. Theor. Meth.*, **6**, 335–362. (349)

Denby, L., and Mallows, C. L. (1977). 'Two diagnostic displays for robust regression analysis'. *Technometrics*, **19**, 1–13. (324, 349, 369)

Denby, L., and Martin, R. D. (1979). 'Robust estimation of the first-order autoregressive parameter'. *J. Amer. Statist. Assn.*, **74**, 140–146. (403)

Desu, M. M., Gehan, E. A., and Severo, N. C. (1974). 'A two-stage estimation procedure when there may be spurious observations'. *Biometrika*, **61**, 593–599. (62)

Devlin, S. J., Gnanadesikan, R., and Kettenring, J. R. (1975). 'Robust estimation and outlier detection with correlation coefficients'. *Biometrika*, **62**, 531–545. (280, 310, 398)

Devlin. S. J., Gnanadesikan, R., and Kettenring, J. R. (1981). 'Robust estimation of dispersion matrices and principal components'. *J. Amer. Statist. Assn.*, **76**, 354–362. (276, 277, 282)

De Wet, T., and van Wyk, J. W. J. (1979a). 'Some large sample properties of Hogg's adaptive trimmed means'. *S. African Statist. J.*, **13**, 53–69.

De Wet, T., and van Wyk, J. W. J. (1979b). 'Efficiency and robustness of Hogg's adaptive trimmed means'. *Commun. Statist. Theor. Meth.*, **8**, 117–128. (87, 145)

Dielman, T. E. (1986). 'A comparison of forecasts from least absolute value and least squares regression'. *J. Forecasting*, **5**, 189–195. (350)

Dixit, U. J. (1989). 'Estimation of parameters of the gamma distribution in the presence of outliers'. *Commun. Statist. Theor. Meth.*, **18**, 3071–3085. (185)

Dixit, U. J. (1991). 'On the estimation of the power of the scale parameter of the gamma distribution in the presence of outliers'. *Commun. Statist. Theor. Meth.*, **20**, 1315–1328. (185)

Dixon, W. J. (1950). 'Analysis of extreme values'. *Ann. Math. Statist.*, **21**, 488–506. (49, 91, 123, 126, 210, 222, 228, 230, 240, 245, 248)

Dixon, W. J. (1951). 'Ratios involvig extreme values'. *Ann. Math. Statist.*, **22**, 68–78. (91, 95, 210, 227–230)

Dixon, W. J. (1953). 'Processing data for outliers'. *Biometrics*, **9**, 74–89. (49, 137)

Dixon, W. J. (1960). 'Simplified estimation from censored normal samples'. *Ann. Math. Statist.*, **31**, 385–391. (65, 222)

Dixon, W. J. (1962). 'Rejection of observations'. In Sarhan and Greenberg (1962). (222, 245, 246, 248)

Dixon, W. J., and Tukey, J. W. (1968). 'Approximate behaviour of the distribution of Winsorized $t$ (Trimming/Winsorization 2)'. *Technometrics*, **10**, 83–98. (74, 75, 84, 170)

Dodge, Y. (1987). *Statistical Data Analysis Based on the L1-norm and Related Methods*. North-Holland, Amsterdam. (350)

Donoho, D. L. (1982). *'Breakdown Properties of Multivariate Location Estimators'*, qualifying paper, Harvard University. (278)

Donoho, D., and Huber, P. J. (1983). 'The notion of breakdown point'. *Wadsworth Statist. Prob. Ser.*, **1983**, 157–184.

Donoho, D. L., and Liu, R. C. (1988). 'The "automatic" robustness of minimum distance functionals'. *Ann. Statist.*, **16**, 552–586. (78, 154)

Doolittle, H. M. (1884). 'The rejection of doubtful observations' (Abstract). *Bulletin of the Philosophical Society of Washington, (Math. Soc.)*, **6**, 153–156. (Ch. 2, *H*)

Doornbos, R. (1976). *Slippage Tests*. 2nd edn. Mathematical Centre Tracts, No. 15, Mathematisch Centrum, Amsterdam. (119, 261–263, 265)

Doornbos, R. (1981). 'Testing for a single outlier in a linear model'. *Biometrics*, **37**, 705–712. (334, 335)

Doornbos, R., and Prins, H. J. (1958). 'On slippage tests'. *Indag. Math.*, **20**, 38–55, 438–447. (335)

Draper, N. R., and Herzberg, A. M. (1979). 'Designs to guard against outliers in the presence or absence of model bias'. *Canadian J. Statist.*, **7**, 127–135. (365)

Draper, N. R., and John, J. A. (1980). 'Testing for three or fewer outliers in two-way tables'. *Technometrics*, **22**, 9–15. (343, 363)

Draper, N. R., and John, J. A. (1981). 'Influential observations and outliers in regression'. *Technometrics*, **23**, 21–26. (318, 374)

Draper, N. R., and Smith, H. (1966). *Applied Regression Analysis*. Wiley, New York. (330)

Ducharme, G. R., and Milasevic, P. (1987). 'Some asymptotic properties of the circular median'. *Commun. Statist. Theor. Meth.*, **16**, 659–664. (423, 424)

Dutter, R., and Guttman, I. (1979). 'On estimation in the linear model when spurious observations are present—a Bayesian approach'. *Commun. Statist. Theor. Meth.*, **8**, 611–635. (379, 384)

Easton, G. S., and McCulloch, R. E. (1990). 'A multivariate generalization of quantile-quantile plots'. *J. Amer. Statist. Assoc.*, **85**, 376–386. (302)

Edelsbrunner, H., and Souvaine, D. L. (1990). 'Computing least median of squares regression lines and guided topological sweep'. *J. Amer. Statist. Assoc.*, **85**, 115–119. (325, 350)

Edgeman, R. L. (1989). 'Inverse Gaussian control charts'. *Australian. J. Statist.*, **31**, 78–84. (413)

Edgeworth, F. Y. (1883). 'The method of least squares'. *Philosophical Magazine*, **16**, 360–375. (30, 36)

Edgeworth, F. Y. (1887). 'On discordant observations'. *Philosophical Magazine*, **23**, 364–375. (30, 36)

Eisenhart, C., Hastay, M. W. and Wallis, W. A. (Eds) (1947). *Selected Techniques of Statistical Analysis*, McGraw-Hill, New York. (197, 242, 470)

Elashoff, J. D. (1972). 'A model for quadratic outliers in linear regression'. *J. Amer Statist. Assn.*, **67**, 478–485. (323)

Ellenberg, J. H. (1973). 'The joint distribution of the standardized least squares residuals from a general linear regression'. *J. Amer. Statist. Assn.*, **68**, 941–943. (334, 338, 341)

Ellenberg, J. H. (1976). 'Testing for a single outlier from a general linear regression'. *Biometrics*, **32**, 637–645. (334, 335, 338, 341)

Embleton, B. J. J., and Giddings, J. W. (1974). 'Late Precambrian and Lower Palaeozoic palaeomagnetic results from South Australia and Western Australia'. *Earth Planet. Sci. Lett.*, **22**, 355–365. (424)

Embrechts, P., Herzberg, A. M., and Ng, A. C. K. (1986). 'An investigation of Andrew's plots to detect period and outliers in time series data'. *Commun. Statist. Sim. Comp.*, **15**, 1027–1051. (400)

Eplett, W. J. R. (1980). 'An influence curve for two-sample rank tests'. *J. Roy. Statist. Soc., B*, **42**, 64–70. (158, 373)

Epstein, B. (1960a). 'Tests for the validity of the assumption that the underlying distribution of life is exponential: Part I'. *Technometrics*, **2**, 83–101. (96, 193)

Epstein, B. (1960b). 'Tests for the validity of the assumption that the underlying distribution of life is exponential: Part II'. *Technometrics*, **2**, 167–183. (96, 193, 208)

Escoffier, B., and Le Roux B. (1976). 'Factor's stability in correspondence analysis. How to control the influence of outlying data (in French)'. *Cahiers Anal. Donnees*, **1**, 297–318 (Dunod, Paris). (283)

Farebrother, R. W. (1988). 'Elemental location shift estimators in the constrained linear model'. *Commun. Statist. Theor. Meth.*, **17**, 79–85. (364)

Faye, H. E. (1888). 'Sur certain points de la théorie des erreurs accidentelles'. *C. R. Acad. Sci. Paris*, **106**, 783–786. (Ch. 5, *H*)

Feigin, P. D., and Cohen, A. (1978). 'On a model for concordance between judges'. *J. R. Statist. Soc., B*, **40**, 203–213. (283)

Fellegi, I. P. (1975). 'Automatic editing and imputation of quantitative data' (Summary). *Bull. Int. Statist. Inst.*, **XLVI**, 249–253. (301, 306)

Fenton, R. (1975). Personal correspondence. (19)

Ferguson, D. E., Landreth, H. F., and McKeown, J. P. (1967). 'Sun compass orientation of the northern cricket frog, *Acris crepitans*'. *Anim. Behav.*, **15**, 45–53. (421)

Ferguson, T. S. (1961a). 'On the rejection of outliers'. *Proceedings of the Fourth Berkeley Symposium on Mathematical Statistics and Probability*, Vol. 1, pp. 253–287. (49, 96, 101, 102, 222–224, 227, 230, 243, 284, 292, 357)

Ferguson, T. S. (1961b). 'Rules for rejection of outliers'. *Rev. Inst. Int. de Statist.*, **29**, 29–43. (137, 222, 224, 226, 227, 230, 245)

Fieller, N. R. J. (1976). *Some Problems related to the Rejection of Outlying Observations*. Ph.D. Thesis, University of Hull. (110, 202, 208, 224–226, 240–242, 244, 245, 247, 333, 338)

Fieller, N. R. J. (1979). Discussion of Kleiner, B., Martin, R. D., and Thomson, D. J. (1979). *J. Roy. Statist. Soc., B*, **41**, 340–341. (414)

Fienberg, S. E. (1969). 'Preliminary graphical analysis and quasi- independence for two-way contingency tables'. *Applied Statistics*, **18**, 153–168. (436, 440)

Fienberg, S. E. (1979). 'The use of chi-squared statistics for categorical data problems'. *J. Roy. Statist. Soc. B.*, **41**, 54–64. (438)

de Finetti, B. (1961). 'The Bayesian approach to the rejection of outliers'. *Proceedings of the Fourth Berkeley Symposium on Mathematical Statistics and Probability*, Vol. 1, pp. 199–210. (378)

Finney, D. J. (1947). 'The estimation from individual records of the relationship between dose and quantal response.' *Biometrika*, **34**, 320–334. (352)

Finney, D. J. (1974). 'Problems, data and inference: The Address of the President' (with Proceedings). *J. Roy. Statist. Soc. A*, **137**, 1–23. (10)

Fisher, N. I. (1982). 'Robust estimation of the concentration parameter of Fisher's distribution on the sphere'. *Applied Statistics*, **31**, 152–154. (426, 427)

Fisher, N. I. (1993). *Statistical Analysis of Circular Data*. Cambridge University Press, Cambridge. (418, 424)

Fisher, N. I., Lewis, T., and Embleton, B. J. J. (1987); 1st paperback Edition (with corrections) (1993). *Statistical Analysis of Spherical Data*, Cambridge University Press, Cambridge. (418, 429)

Fisher, N. I., Lewis, T., and Willcox, M. E. (1981). 'Tests of discordancy for samples from Fisher's distribution on the sphere'. *Applied Statistics*, **30**, 230–237. (425, 427)

Fisher, R. A. (1929). 'Tests of significance in harmonic analysis'. *Proc. Roy. Soc. A*, **125**, 54–59. (119, 197)

Fisher, R. A. (1936). 'The use of multiple measurements in taxonomic problems'. *Ann. Eugen*, **7**, 179–188. (309)

Fisher, R. A., Corbet, A. S., and Williams, C. B. (1943). 'The relation between the number of species and the number of individuals in a random sample of an animal population'. *J. Animal Ecol.*, **12**, 42–57. (15)

Follmann, D. A., and Lambert, D. (1989). 'Generalizing logistic regression by nonparametric mixing'. *J. Amer. Statist. Assoc.*, **84**, 295–300. (352)

Fox, A. J. (1972). 'Outliers in time series'. *J. Roy. Statist. Soc. B*, **43**, 350–363. (396, 400)

Freeman, P. R. (1980). 'On the number of outliers in data from a linear model' (with Discussion). Pp. 349–365 of Bernardo, J. M., DeGroot, M. H., Lindley, D. V., and Smith, A. F. M. (Eds) (1980). *Bayesian Statistics*. University Press, Valencia, Spain. (384, 393)

Friedman, J. H., and Rafsky, L. C. (1981). 'Graphics for the multivariate two-sample problem' (with Discussion). *J. Amer. Statist. Assn.*, **76**, 277–295. (302, 313)

From, S. G. (1991). 'Mean square error efficient estimation of an exponential mean under an exchangeable single outlier model'. *Commun. Statist. Sim.*, **20**, 1073–1084. (181–183)

Fuchs, C., and Kenett, R. (1980). 'A test for outlying cells in the multinomial distribution and two-way contingency tables'. *J. Amer. Statist. Assn.*, **75**, 395–398. (435)

Fung, K. Y., and Paul, S. R. (1985). 'Comparisons of outlier detection procedures in Weibull or extreme-value distribution'. *Commun. Statist. Sim. Comp.*, **14**, 895–917. (96, 135, 136, 256–259)

Fung, W. K. (1988). 'Critical values for testing in multivariate statistical outliers'. *J. Statist. Comp. Sim.*, **30**, 195–212. (289, 345)

Galpin, J. S., and Hawkins, D. M. (1981). 'Rejection of a single outlier in two- or three-way layouts'. *Technometrics*, **23**, 65–70. (360–361)

Ganeshanandam, S., and Krzanowski, W. J. (1989). 'On selecting variables and assessing their performance in linear discriminant analysis'. *Australian J. Statist.*, **31**, 433–447. (283)

Gasser, T., Sroka, L., and Jennen-Steinmetz, C. (1986). 'Residual variance and residual pattern in nonlinear regression and for the detection of outliers'. *Biometrika*, **73**, 625–633. (351)

Gastwirth, J. L. (1966). 'On robust procedures'. *J. Amer. Statist. Assn.*, **61**, 929–948. (65, 82, 146, 162)

Gastwirth, J. L., and Cohen, M. L. (1970). 'Small sample behavior of some robust linear estimators of location'. *J. Amer. Statist. Assn.*, **65**, 946–973. (66, 146, 162)

Gather, U. (1985). 'The influence of outlier-proneness on the tail- behaviour of some location estimators'. *Stat. Dec.*, **2**, 165–171. (49, 53)

Gather, U. (1986). 'Robust estimation of the mean of the exponential distribution in outlier situations'. *Commun. Statist. Theor. Meth.*, **15**, 2323–2345. (178, 180)

Gather, U. (1989). 'Testing for multisource contamination in location/scale families'. *Commun. Statist. Theor. Meth.*, **18**, 1–34. (50, 133)

Gather, U., and Kale, B. K. (1988). 'Maximum likelihood estimation in the presence of outliers'. *Commun. Statist. Theor. Meth.*, **17**, 3767–3784. (104, 133, 183)

Gather, U., and Rauhut, B. (1990). 'The outlier behaviour of probability distributions'. *J. Statist. Plan. Inf.*, **26**, 237–252. (53)

Geertsema, J. C. (1987). 'The behaviour of sequential confidence intervals under contamination'. *Sequential Anal.*, **6**, 71–91. (75)

Geertsema, J. C. (1989). 'Discussion of De Waal, D. J. and Groznewald, P. C. N. (1989). On measuring the amount of information from the data in a Bayesian analysis'. *S. African Statist. J.*, **23**, 23–54. (380)

Geier, P. W., and Briese, D. T. (1977). 'Predominantly female progeny in the light-brown apple moth'. *Search*, **8**, 83–85. (128)

Geier, P. W., Briese, D. T., and Lewis, T. (1978). 'The light-brown apple moth'. *Australian J. Ecology*, **3**, 467–488. (128, 455)

Geisser, S. (1989). 'Predictive discordancy tests for exponential observations'. *Canadian J. Statist.*, **17**, 19–26. (387)

Genstat (1987). *Genstat 5 Reference Manual.* OUP, Oxford. (452)

Gentle, J. E. (1978). 'Testing for outliers in linear regression'. Pp. 223–234 of David, H. A. (Ed.) (1978). *Contributions to Survey Sampling and Applied Statistics, in honor of H. O. Hartley.* Academic Press, New York. (335)

Gentleman, J. F. (1980). 'Finding the *k* most likely outliers in two-way tables'. *Technometrics*, **22**, 591–600. (343)

Gentleman, J. F., and Wilk, M. B. (1975a). 'Detecting Outliers: II Supplementing the direct analysis of residuals'. *Biometrics*, **31**, 387–410. (341, 343–345, 363, 369)

Gentleman, J. F., and Wilk, M. B. (1975b). 'Detecting outliers in a two-way table. I. Statistical behaviour of residuals'. *Technometrics*, **17**, 1–14. (369)

George, E. O., and Rousseau, C. C. (1987). 'On the logistic midrange'. *Ann. Inst. Statist. Math.*, **39**, 627–635. (106)

Ghosh, S., and Kipngeno, W. A. K. (1985). 'On the robustness of the optimum balanced 2m factorial designs of resolution V (given by Srivastava and Chopra) in the presence of outliers'. **11**, 119–129. (365)

Gideon, R. A., and Hollister, R. A. (1987). 'A rank correlation coefficient resistant to outliers'. *J. Amer. Statist. Assoc.*, **82**, 656–666. (280)

Giltinan, D. M., Carroll, R. J., and Ruppert, D. (1986). 'Some new estimation methods for weighted regression when there are possible outliers'. *Technometrics*, **3**, 219–230. (325, 349)

Glaisher, J. W. L. (1872). 'On the law of facility of errors of observations and on the method of least squares'. *Mem. Roy. Astr. Soc.*, **39**, 75–124. (36)

Glaisher, J. W. L. (1872–73). 'On the rejection of discordant observations'. *Monthly Notices Roy. Astr. Soc.*, **33**, 391–402. (29)

Glaisher, J. W. L. (1874). Note on a paper by Mr. Stone 'On the rejection of discordant observations'. *Monthly Notices Roy. Astr. Soc.*, **34**, 251. (Ch. 2, *H*)

Gleser, L. J. (1989). 'Commentary on "Indoor air pollution and pulmonary performance: investigating errors in exposure assessment".' *Statist. Med.*, **8**, 1127–1131. (455)

Gnanadesikan, R. (1973). 'Graphical methods for informal inference in multivariate data analysis'. *Bull. Int. Statist. Inst.*, **45**, Book 4, 195–206. (302, 309)

Gnanadesikan, R. (1977). *Methods for Statistical Data Analysis of Multivariate Observations*. Wiley, New York. (279, 302, 304, 307–310, 451)

Gnanadesikan, R., and Kettenring, J. R. (1972). 'Robust estimates, residuals and outlier detection with multiresponse data'. *Biometrics*, **28**, 81–124. (269, 274, 278, 279, 298, 302, 304, 308, 309, 369)

Gnanadesikan, R., and Wilk, M. B. (1969). 'Data analytic methods in multivariate statistical analysis'. In Krishnaiah, P. R. (Ed.) (1969). *Multivariate Analysis*, Vol. II. Academic Press, New York. (302)

Goldsmith, P. L., and Boddy, R. (1973). 'Critical analysis of factorial experiments and orthogonal fractions'. *Applied Statistics*, **22**, 141–160. (358, 361–363)

Goldstein, M. (1982). 'Contamination distributions'. *Ann. Statist.*, **10**, 174–183. (52, 381)

Goldstein, M. (1983). 'Outlier resistant distributions: where does the probability go?' *J. Roy. Statist. Soc. B*, **45**, 355–357. (53)

Golub, G. H., Guttman, I., and Dutter, R. (1973). 'Examination of pseudo-residuals of outliers for detecting spuriosity in the general univariate linear model'. In Kabe, D. G. and Gupta, P. R. (Eds) (1973). *Multivariate Statistical Inference*. North-Holland, Amsterdam. (274, 348)

Goodman, L. A. (1968). 'The analysis of cross-classified data: independence, quasi-independence, and interactions in contingency tables with or without missing entries'. *J. Amer. Statist. Assoc.*, **63**, 1091–1131. (433, 436)

Goodwin, H. M. (1913). *Elements of the Precision of Measurements and Graphical Methods*. McGraw-Hill, New York. (30)

Gould, B. A. Jr. (1855). 'On Peirce's criterion for the rejection of doubtful observations, with tables for facilitating its application'. *Astr. J.*, **4**, 81–87. (Ch. 2, *H*)

Govindarajalu, Z. (1966). 'Best linear estimates under symmetric censoring of the parameters of a double exponential population'. *J. Amer. Statist. Assoc.*, **61**, 248–258. (186)

Gray, J. B. (1989). 'On the use of regression diagnostics'. *The Statistician*, **38**, 97–105. (370)

Gray, J. B., and Ling, R. F. (1984). 'K-clustering as a Detection Tool for Influential Subsets in Regression'. *Technometrics*, **26**, 305–318. (454)

Green, R. F. (1974). 'A note on outlier-prone families of distributions'. *Ann. Statist.*, **2**, 1293–1295. (52, 53)

Green, R. F. (1976). 'Outlier-prone and outlier-resistant distributions'. *J. Amer. Statist. Assn.*, **71**, 502–505. (Ch. 2)

Greenwood, M. (1946). 'The statistical study of infectious diseases'. *J. Roy. Statist. Soc. A.*, **109**, 85–109. (212)

Griffin, P. S. (1988). 'Asymptotic normality of Winsorized means'. *Stochastic Processes and their Applications*, **29**, 107–127. (81)

Griffin, P. S., and Pruitt, W. E. (1987). 'The central limit problem for trimmed sums'. *Math. Proc. Camb. Phil. Soc.*, **102**, 329–349. (81)

Gross, A. M. (1976). 'Confidence interval robustness with long-tailed symmetric distributions'. *J. Amer. Statist. Assn.*, **71**, 409–416. (86, 171–174)

Gross, A. M. (1977). 'Confidence intervals for bisquare regression estimates'. *J. Amer. Statist. Assn*, **72**, 341–354. (86)

Gross, W. F., Bode, G., Taylor, J. M., and Lloyd-Smith, C. W. (1986). 'Some finite population estimators which reduce the contribution of outliers'. In Francis, I. S., Manly, B. F. J., and Lam, F. C. (ed.). (1986). pp 386–390. *Pacific Statistical Congress: Proceedings of the Congress, Auckland, New Zealand, 20–24 May, 1985*. Elsevier Science Publishers BV (North-Holland). (443, 445, 446)

Grubbs, F. E. (1950). 'Sample criteria for testing outlying observations'. *Ann. Math. Statist.*, **21**, 27–58. (38, 49, 92, 95–97, 136, 222, 225, 226, 245)

Grubbs, F. E. (1969). 'Procedures for detecting outlying observations in samples'. *Technometrics*, **11**, 1–21. (32, 33, 97, 222, 225, 245)

Grubbs, F. E., and Beck, G. (1972). 'Extension of sample sizes and percentage points for significance tests of outlying observations'. *Technometrics*, **14**, 847–854. (222, 225)

Gumbel, E. J. (1960). 'Bivariate exponential distributions'. *J. Amer. Statist. Assn.*, **55**, 698–707. (293)

Gunther, R. (1988). 'Adaptive estimation and prediction for an extension of the autoregressive multiple time series model'. *J. Theor. Appl. Statist.*, **19**, 557–564. (414)

Gupta, A. K. (1952). 'Estimation of the mean and standard deviation of a normal population from a censored sample'. *Biometrika*, **39**, 260–273. (168)

Gupta, S. S. (1960). 'Order statistics from the gamma distribution'. *Technometrics*, **2**, 243–262. Correction *Technometrics*, **2**, 523. (285)

Guttman, I. (1973a). 'Premium and protection of several procedures for dealing with outliers when sample sizes are moderate to large'. *Technometrics*, **15**, 385–404. (62, 66, 165, 166)

Guttman, I. (1973b). 'Care and handling of univariate or multivariate outliers in detecting spuriosity—a Bayesian approach'. *Technometrics*, **15**, 723–738. (47–49, 51, 272, 274, 379, 391)

Guttman, I., and Dutter, R. (1977). 'A Bayesian approach to the detection of spuriousness and estimation in the general univariate linear model when outliers are present'. *Trans. 7th Prague Conference on Information Theory, Statistical Decision Function, Random Process and of the 1974 European Meeting of Statisticians*. Acad. Publishing House, Czechoslovak Academy of Sciences, Prague. (384)

Guttman, I., and Khatri, C. G. (1975). 'A Bayesian approach to some problems involving the detection of spuriosity' Pp. 111–145 of Gupta, R. P. (Ed.) (1975). *Applied Statistics*. North-Holland, Amsterdam. (379, 392)

Guttman, I., and Kraft, C. H. (1980). 'Robustness to spurious observations of linearized Hodges—Lehmann estimators and Anscombe estimators'. *Technometrics*, **22**, 55–63. (62, 84, 144)

Guttman, I., and Pena, D. (1988). 'Outliers and influence: evaluation by posteriors of parameters in the linear model'. In *Bayesian Statistics, 3*. Oxford University Press, New York. (385, 393)

Guttman, I., and Smith, D. E. (1969). 'Investigation of rules for dealing with outliers in small samples from the normal distribution I: Estimation of the mean'. *Technometrics*, **11**, 527–550. (62, 84, 144, 145, 165, 166)

Guttman, I., and Smith, D. E. (1971). 'Investigation of rules for dealing with outliers in small samples from the normal distribution II; Estimation of the variance'. *Technometrics*, **13**, 101–111. (62, 84, 165, 170)

Guttman, I., and Tiao, G. C. (1978). 'Effect of correlation on estimation of a mean in the presence of spurious observations'. *Canadian J. Statist.*, **6**, 229–248. (144, 401)

Guttman, I., Dutter, R., and Freeman, P. R. (1978). 'Care and handling of univariate outliers in the general linear model to detect spuriosity—a Bayesian approach'. *Technometrics*, **20**, 187–193. (384)

Haberman, S. J. (1973). 'The analysis of residuals in cross-classified tables'. *Biometrics*, **29**, 205–220. (433, 439)

Halperin. M., Greenhouse, S. W., Cornfield, J., and Zalokar, J. (1955). 'Tables of percentage points for the studentized maximum absolute deviate in normal samples'. *J. Amer. Statist. Assn.*, **50**, 185–195. (96, 239, 246)

Hampel, F. R. (1968). 'Contributions to the theory of robust estimation'. *Ph D thesis, Univ. California, Berkeley.* (69, 74)

Hampel, F. R. (1971). 'A generalized qualitative definition of robustness'. *Ann. Math. Statist.*, **42**, 1887–1896. (69, 74)

Hampel, F. R. (1973). 'Robust estimation: a condensed partial survey'. *Z. Wahrscheinlichkeitstheorie und Verw. Gebiete*, **27**, 87–104. (275)

Hampel, F. R. (1974). 'The influence curve and its role in robust estimation'. *J. Amer. Statist. Assn.*, **69**, 383–393. (35, 36, 72, 80, 81, 84, 149, 151, 156, 164, 169, 277, 310)

Hampel, F. R., Ronchetti, E. M., Rousseeuw, P. J., and Stahel, W. A. (1986). *Robust Statistics: The Approach Based on Influence Functions.* Wiley, New York. (35, 78, 148, 151, 155, 158, 317)

Hara, T. (1988). 'Detection of multivariate outliers with location slippage or scale inflation in left orthogonally invariant or elliptically contoured distributions'. *Ann. Inst. Statist. Math.*, **40**, 395–406. (296)

Hardle, W., and Gasser, T. (1985). 'On robust kernel estimation of derivatives of regression functions'. *Scandinavian J. Statist. Theor. Applic.*, **12**, 233–240. (351)

Harter, H. L. (1969a). *Order Statistics and their Use in Testing and Estimation*, Vol. 1: *tests Based on Range and Studentized Range of Samples from a Normal Population*. U.S. Air Force, Aerospace Research Laboratories, Washington, D.C. (240)

Harter, H. L. (1969b). *Order Statistics and their Use in Testing and Estimation*, Vol. 2: *Estimates Based on Order Statistics of Samples from Various Populations*. U.S. Air Force. Aerospace Research Laboratories, Washington, D.C. (181, 250)

Harter, H. L. (1974–1976). 'The method of least squares and some alternatives Parts I–VI'. *Rev. Int. Inst. de Statist.*, **42**, 147–174 (Part I); **42**, 235–264 (Part II); **43**, 1–44 (Part III); **43**, 125–190 (Part IV); **43**, 269–278 (Part V); **44**, 113–159 (Part VI). (30, 31)

Hartley, H. O. (1950). 'The maximum $F$-ratio as a short-cut test for heterogeneity of variance'. *Biometrika*, **37**, 308–312. (204, 205)

Hasabelnaby, N. A., Ware, J. H., and Fuller, W. A. (1989). 'Indoor air pollution and pulmonary performance: investigating errors in exposure assessment'. *Statist. Med.*, **8**, 1109–1126.

Hawkins, D. M. (1972). 'Analysis of a slippage test for the chi-squared distribution'. *S. African Statist. J.*, **6**, 11–17. (96)

Hawkins, D. M. (1973). 'Repeated testing for outliers'. *Statistica Neerlandica*, **27**, 1–10. (98, 131)

Hawkins, D. M. (1974). 'The detection of errors in multivariate data using principal components'. *J. Amer. Statist. Assn.*, **69**, 340–344. (306)

Hawkins, D. M. (1977). 'Comment on "A new statistic for testing suspected outliers" '. *Commun. Statist. Theor. Meth.*, **6**, 435–438. (136, 223)

Hawkins, D. M. (1978). 'Analysis of three tests for one or two outliers'. *Statistica Neerlandica*, **32**, 137–148. (98, 123, 131, 136, 222)

Hawkins, D. M. (1979). 'Fractiles of an extended multiple outlier test'. *J. Statist. Comp. Sim.*, **8**, 227–236. (136, 232, 236, 241)

Hawkins, D. M. (1980a). *Identification of Outliers*. Chapman and Hall, London. (33, 96, 98, 121, 126, 128, 131, 136, 306)

Hawkins, D. M. (1980b). 'Critical values for identifying outliers'. Letter to *Applied Statistics*, **29**, 95–96. (131, 236)

Hawkins, D. M., and Perold, A. F. (1977). 'On the joint distribution of left- and right-sided outlier statistics'. *Utilitas Math.*, **12**, 129–143. (232)

Hawkins, D. M., Bradu, D., and Kass, G. V. (1984). 'Location of several outliers in multiple-regression data using elemental sets'. *Technometrics*, **26**, 197–208. (344, 364)

He, X., Simpson, D. L., and Portnoy, S. L. (1990). 'Breakdown robustness of tests'. *J. Amer. Statist. Assoc.*, **85**, 446–452. (77)

Healy, M. J. R. (1968). 'Multivariate normal plotting'. *Applied Statistics*, **17**, 157–161. (287, 298, 300, 307)

Healy, M. J. R. (1989). 'Measuring measuring-errors'. *Statist. Med.*, **8**, 893–906. (455)

Heathcote, C. R. (1977). 'The integrated squared error estimation of parameters'. *Biometrika*, **64**, 255–264. (154)

Heathcote, C. R. (1979). 'Bounded influence curve estimation and consistency'. Pp. 103–110 of McNeil, D. R. (Ed.) (1979). *Interactive Statistics*. North-Holland, New York. (374)

Hedayat, A., and Robson, D. S. (1970). 'Independent stepwise residuals for testing homoscedasticity'. *J. Amer. Statist. Assn.*, **65**, 1573–1581. (330)

Hertsgaard, D. (1979). 'Distribution of asymmetric trimmed means'. *Commun. Statist. Theor. Meth.*, **8**, 359–367. (79, 143)

Herzberg, A. M., and Andrews, D. F. (1978). 'The robustness of chain block designs and coat-of-mail designs'. *Commun. Statist. Theor. Meth.*, **7**, 479–485. (365)

Hidiroglou, M. H., and Srinath, K. P. (1981). 'Some estimators of a population total from simple random samples containing large units'. *J. Amer. Statist. Assoc.*, **76**, 690–695. (442, 445, 446)

Hill, M. A., and Dixon, W. J. (1982). 'Robustness in real life: a study of clinical laboratory data'. *Biometrics*, **38**, 377–396. (56)

Hill, R. W. (1982). 'Robust regression when there are outliers in the carriers: the univariate case'. *Commun. Statist. Theor. Meth.*, **11**, 849–868. (374)

Hillmer, S. (1984). 'Monitoring and adjusting forecasts in the presence of additive outliers'. *J. Forecasting*, **3**, 205–215. (412)

Hinich, M. J., and Talwar, P. P. (1975). 'A simple method for robust estimation'. *J. Amer. Statist. Assn.*, **70**, 113–119. (324)

Hirsch, R. F., Wu, G. L., and Tway, P. C. (1987). 'Reliability of factor analysis in the presence of random noise or outlying data'. *Chemometrics and Intelligent Laboratory Systems*, **1**, 265–272. (283)

Hoaglin, D. C., Iglewicz, B., and Tukey, J. W. (1986). 'Outlier labelling'. *J. Amer. Statist. Assoc.*, **81**, 991–999. (53)

Hoaglin, D. C., and Iglewicz, B. (1987). 'Fine-tuning some resistant rules for outlier labelling'. *J. Amer. Statist. Assoc.*, **82**, 1147–1149. (53)

Hoaglin, D. C., and Welsch, R. E. (1978). 'The hat matrix in regression and Anova'. *American Statistician*, **32**, 17–22. (374)

Hodges, J. L. Jr. (1967). 'Efficiency in normal samples and tolerance of extreme values for some estimates of location'. *Proceedings of the 5th Berkeley Symposium on Mathematical Statistics and Probability*, University of California, Berkeley, Calif. (74, 163)

Hodges, J. L. Jr., and Lehmann, E. L. (1963). 'Estimates of location based on rank tests'. *Ann. Math. Statist.*, **34**, 598–611. (82, 152)

Hoenig, J., and Crotty, I. M. (1958). *International J. Social Psychiatry*, **3**, 260–277. (92)

Hogg, R. V. (1967). 'Some observations on robust estimation'. *J. Amer. Statist. Assn.*, **62**, 1179–1186. (65)

Hogg, R. V. (1974). 'Adaptive robust procedures: a partial review and some suggestions for future applications and theory (with comments)'. *J. Amer. Statist. Assn.*, **69**, 909–927. (35, 87, 88, 145, 149, 151)

Hogg, R. V., Fisher, D. M., and Randles, R. H. (1975). 'A two-sample adaptive distribution-free test'. *J. Amer. Statist. Assn.*, **70**, 656–661. (158)

Homan, S. M., and Lachenbruch, P. A. (1986). 'Robust estimation of the exponential mean parameter for small samples: Complete and censored data'. *Commun. Statist. Sim. Comp.*, **15**, 1087–1108. (62, 181)

Hopkins, T., and Morse, D. (1990). 'Index of Statistical Algorothims'. *Applied Statistics*, **39**, 177–187. (453)

Huber, P. J. (1964). 'Robust estimation of a location parameter'. *Ann. Math. Statist.*, **35**, 73–101. (35, 63, 65, 148, 149, 151, 158, 161)

Huber, P. J. (1965). 'A robust version of the probability ratio test'. *Ann. Math. Statist.*, **36**, 1753–1758. (87)

Huber, P. J. (1967). 'The behaviour of maximum likelihood estimates under nonstandard conditions'. *Proc. Fifth Berkeley Symp. Math. Statist. Prob.*, Vol. I, pp. 221–233. (148)

Huber, P. J. (1968). 'Robust estimation'. *Mathematical Centre Tracts, Selected Statistical Papers*, **37**, 3–25. Mathematisch Centrum Amsterdam. (75)

Huber, P. J. (1970). 'Studentizing robust estimates'. In Puri, M. L. (Ed.) (1970) *Nonparametric Techniques in Statistical Inference*. Cambridge University Press, London. (75, 86)

Huber, P. J. (1972). 'Robust statistics: a review (The 1972 Wald Lecture)'. *Ann. Math. Statist.*, **43**, 1041–1067. (35, 41, 55, 74, 82, 86, 88, 162, 396)

Huber, P. J. (1975). 'Robustness and designs'. In Srivastava, J. N. (Ed.) (1975). *A Survey of Statistical Design and Linear Models*. North-Holland, Amsterdam. (373)

Huber, P. J. (1977a). 'Robust covariances'. Pp. 165–191 of Gupta, S. S., and Moore, D. S. (Eds) (1977). *Statistical Decision Theory and Related Topics II*. Academic Press, New York. (275, 277)

Huber, P. J. (1977b). *Robust Statistical Procedures*. SIAM, Philadelphia. (324)

Huber, P. J. (1981). *Robust Statistics*. Wiley, New York. (35, 55, 63, 65, 72, 82, 84, 85, 87, 275, 279, 280, 317)

Huber, P. J. (1983). 'Minimax aspects of bounded-influence regression'. *J. Amer. Statist. Assn.*, **78**, 66–72. Discussion and reply, 72–80. (Krasker, W. S. and Welsch, R. E., 72–73; Cook, R. D., and Weisberg, S., 74–75; Bickel, P. J. 75–77; Mallows, C. L., 77; Carroll, R. J., 78–79; Huber, P. J., 80. (65)

Huck, S. W., Cross, T. L., and Clark, S. B. (1986). 'Overcoming misconceptions about Z-scores'. *Teaching Statistics*, **8**, 38–40. (460)

Huggins, R. M. (1989). 'The sign test for stochastic processes'. *Australian. J. Statist.*, **31**, 153–165. (414)

Hulsmann, J. (1987). 'L-norm estimation of the distribution function'. *Springer, Berlin*, **1987**, 297–301. (154)

Hunt, R. (1989). 'Statistical Software Review of STATGRAPHICS, Version 2–6'. *Applied Statistics*, **38**, 158–160. (452)

Iglewicz, B., and Martinez, J. (1982). 'Outlier detection using robust measures of scale'. *J. Statist. Comp. Sim.*, **15**, 285–294. (234, 235)

Iman, R. L., and Conover, W. J. (1977). 'On the power of the *t*-test and some rank tests when outliers may be present'. *Canadian J. Statist.*, **5**, 187–193. (77, 158)

Irwin, J. O. (1925). 'On a criterion for the rejection of outlying observations'. *Biometrika*, **17**, 238–250. (31, 95, 249, 250)

Jaeckel, L. A. (1969). *Robust Estimates of Location*. Ph.D. dissertation, University of California-Berkeley, University Microfilms Inc., Ann Arbor, Mich.

Jaeckel, L. A. (1971a). 'Robust estimates of location: Symmetry and asymmetric contamination'. *Ann. Math. Statist*, **42**, 1020–1034. (35, 62, 63, 65, 66, 146, 148, 151)

Jaeckel, L. A. (1971b). 'Some flexible estimates of location'. *Ann. Math. Statist.*, **42**, 1540–1552. (65, 87, 145)

Jain, R. B. (1981a). 'Detecting outliers: power and some other considerations'. *Commun. Statist. Theor. Meth.*, **10**, 2299–2314. (49, 126, 236)

Jain, R. B. (1981b). 'Percentage points of many-outlier detection procedures'. *Technometrics*, **23**, 71–76. (236)

Jain, R. B., and Pingel, L. A. (1981a). 'A procedure for estimating the number of outliers'. *Commun. Statist. Theor. Meth.*, **10**, 1029–1041. (132)

Jain, R. B., and Pingel, L. A. (1981b). 'On the robustness of recursive outlier detection procedures to nonnormality'. *Commun. Statist. Theor. Meth.*, **10**, 1323–1334. (236)

Jeffreys, H. (1932). 'An alternative to the rejection of observations'. *Proc. Roy. Soc. London, A*, **137**, 78–87. (36)

Jeffreys, H. (1938). 'The law of error and the combination of observations'. *Phil. Trans. Roy. Soc. London, A*, **237**, 231–271. (Ch. 2, *H*)

Jennings, D. E. (1986). 'Outliers and residual distributions in logistic regression'. *J. Amer. Statist. Assoc.*, **81**, 987–990. (352)

Jennings, L. W., and Young, D. M. (1988). 'Extended critical values of the multivariate extreme deviate test for detecting a single spurious observation'. *Commun. Statist. Sim. Comp.*, **17**, 1359–1373. (289)

Jevons, W. S. (1874). *The Principles of Science*. Macmillan, London (latest edn. 1958). (Ch. 2, *H*)

John, J. A. (1978). 'Outliers in factorial experiments'. *Applied Statistics*, **27**, 111–119. (20, 343, 363, 365, 367)

John, J. A., and Draper, N. R. (1978). 'On testing for two outliers or one outlier in two-way tables'. *Technometrics*, **20**, 69–78. (343, 363)

John, J. A., and Prescott, P. (1975). 'Critical values of a test to detect outliers in factorial experiments'. *Applied Statistics*, **24**, 56–59. (358, 361)

Johns, M. V. (1979). 'Robust Pitman-like estimators'. Pp. 49–60 of Launer, R. L., and Wilkinson, G. N. (Eds) (1979). *Robustness in Statistics*. Academic Press, New York. (155)

Johnson, D. E., McGuire, S. A., and Milliken, G. A. (1978). 'Estimating $\sigma^2$ in the presence of outliers'. *Technometrics*, **20**, 441–456. (84, 167)

Johnson, N. J. (1978). 'Modified $t$ tests and confidence intervals for asymmetrical populations'. *J. Amer. Statist. Assn.*, **73**, 536–544. (157)

Johnson, W. (1987). 'The detection of influential observations for allocation, separation, and the determination of the probabilities in a Bayesian framework'. *J. Bus. Econ. Statist.*, **5**, 369–381. (394)

Joiner, B. L., and Hall, D. L. (1983). 'The ubiquitous role of f'/f in efficient estimation of location'. *Amer. Statist.*, **37**, 128–133. (82, 83)

Jolliffe, I. T. (1989). 'Rotation of ill-defined principal components'. *Applied Statistics*, **38**, 139–147. (282)

Jones, D. H. (1979). 'An efficient adaptive distribution-free test for location'. *J. Amer. Statist. Assn.*, **74**, 822–828. (158)

Joshi, P. C. (1972a). 'Some slippage tests of mean for a single outlier in linear regression'. *Biometrika*, **59**, 109–120. (335)

Joshi, P. C. (1972b). 'Efficient estimation of a mean of an exponential distribution when an outlier is present'. *Technometrics*, **14**, 137–144. (51, 68, 175)

Joshi, P. C. (1975). 'Some distribution theory results for a regression model'. *Ann. Inst. Statist. Math., Tokyo*, **27**, 309–317. (335)

Joshi, P. C. (1988). 'Estimation and testing under exchangeable exponential model with a single outlier'. *Commun. Statist. Theor. Meth.*, **17**, 2315–2326. (181, 182)

Joshi, P. C., and Lalitha, S. (1986). 'Tests for two outliers in a linear model'. *Biometrika*, **73**, 236–239. (338, 339)

Jurečková, J. (1985). 'Robust estimators of location and their second-order asymptotic relations'. In Atkinson, A. C. and Fienberg, S. E. (eds). (1985). *A Celebration of Statistics. The ISI Centenary Volume*. Springer-Verlag, 377–392. (83)

Kabe, D. G. (1970). 'Testing outliers from an exponential population'. *Metrika*, **15**, 15–18. (193, 199, 200, 207, 209, 210)

Kale, B. K. (1974). 'Detection of outliers—a semi-Bayesian approach (preliminary report) (abstract)'. *Inst. Math. Statist. Bull.*, **3**, 153. (379)

Kale, B. K. (1975a). 'A note on outlier-resistant families and mixtures of distributions'. *Technical Report No.* 66, Department of Statistics, University of Manitoba, Winnipeg, Canada. (53)

Kale, B. K. (1975b). 'On outlier-proneness of some families of distributions'. *Technical Report No.* 68, Department of Statistics, University of Manitoba, Winnipeg, Canada. (53)

Kale, B. K. (1975c). 'Trimmed means and the method of maximum likelihood when spurious observations are present'. In Gupta, S. S. (Ed.) (1975) *Applied Statistics.* North-Holland, Amsterdam. (68, 176)

Kale, B. K. (1976). 'Detection of outliers'. *Sankhyā B*, **38**, 356–363. (132)

Kale, B. K. (1977). 'A note on outlier resistant families and mixtures of distributions'. *J. Ind. Statist. Assn.*, **15**, 119. (53)

Kale, B. K., and Sinha, S. K. (1971). 'Estimation of expected life in the presence of an outlier observation'. *Technometrics*, **13**, 755–759. (51, 68, 175, 176, 193, 379, 387)

Kanto, A. (1983). 'On the interpolation of missing value in a time series'. *Acta Univ. Tampere. A*, **153**, 55–61. (400)

Kao, C., and Dutkowsky, D. H. (1989). 'An application of nonlinear bounded influence estimation to aggregate bank borrowing from the federal reserve'. *J. Amer. Statist. Assoc.*, **84**, 700–709. (351)

Kapur, M. N. (1957). 'A property of the optimum solution suggested by Paulson for the $k$-sample slippage problem for the normal distribution'. *Ind. Soc. Agric. Statist.*, **9**, 179–190. (104)

Katz, B. P., and Brown, M. B. (1988). 'Detection of a random alteration in a multivariate observation when knowing probable direction'. *Comp. Statist. Data Anal.*, **6**, 145–155. (301)

Keller-McNulty, S., and Higgins, J. J. (1987). 'Effect of tail weight and outliers on power and type-I error of robust permutation tests for location'. *Commun. Statist. Sim. Comp.*, **16**, 17–35. (87)

Kempthorne, P. J., and Mendel, M. B. (1990). 'Comments on Rousseeuw & Van Zomeren. (1990). (269, 279)

Ketellapper, R. H., and Ronner, A. E. (1984). 'Are robust estimation methods useful in the structural errors-in-variables model?' *Metrika*, **31**, 33–41. (328)

Khalfina, N. M. (1986). 'Detection of outliers in results of observations by means of the Chauvenet test. *Zapiski Nauchnykh Seminarov Leningradskogo Otdeleniya Matematicheskogo, Instituta imeni V. A. Steklova Akademii Nauk SSSR (LOMI)*, **153**, 153–159, 176, 180–181. (29)

Khalfina, N. M. (1989). 'Detection of outliers by the Chauvenet method in observations connected in a homogeneous Markov chain'. *Zapiski Nauchnykh Seminarov Leningradskogo Otdeleniya Matematicheskogo, Instituta imeni V. A. Steklova Akademii Nauk SSSR (LOMI)*, **177**, 163–169, 192. (29)

Khattree, R., and Naik, D. N. (1987). 'Detection of outliers in bivariate time series data'. *Commun. Statist. Theor. Meth.*, **16**, 3701–3714. (414)

Kianifard, F., and Swallow, W. H. (1989). 'Using recursive residuals, calculated on adaptively-ordered observations, to identify outliers in linear regression'. *Biometrics*, **45**, 571–585. (345)

Kim, B. Y. (1987). 'A robust estimation procedure for the linear regression model'. *J. Korean Statist. Soc.*, **16**, 80–91. (350)

Kimber, A. C. (1979). 'Tests for a single outlier in a gamma sample with unknown shape and scale parameters'. *Applied Statistics*, **28**, 243–250. (49, 194, 213)

Kimber, A. C. (1982). 'Tests for many outliers in an exponential sample'. *Applied Statistics*, **31**, 263–271. (49, 96, 98, 115, 131, 139, 197, 214, 215)

Kimber, A. C. (1983a). 'Trimming in Gamma samples'. *Applied Statistics*, **32**, 7–14. (79, 184)

Kimber, A. C. (1983b). 'Discordancy testing in gamma samples with both parameters unknown'. *Applied Statistics*, **32**, 304–310. (115)

Kimber, A. C. (1983c). 'Comparison of some robust estimators of scale in gamma samples with known shape'. *J. Statist. Comp. Sim.*, **12**. (185)

Kimber, A. C. (1985). 'A note on the detection and accommodation of outliers relative to Fisher's distribution on the sphere'. *Appl. Statist.*, **34**, 169–172. (426, 427)

Kimber, A. C. (1988). 'Testing upper and lower outlier pairs in gamma samples'. *Commun. Statist. Sim. Comp.*, **17**, 1055–1072. (102, 206, 207, 479)

Kimber, A. C. (1989). 'On Rosado's objective criterion for outliers'. *Metron*, **47**, 5–11. (204, 205)

Kimber, A. C. (1990). 'Exploratory data analysis for possibly censored data from skewed distributions'. *Applied Statistics*, **39**, 21–30. (53, 445)

Kimber, A. C., and Stevens, H. J. (1981). 'The null distribution of a test for two upper outliers in an exponential sample'. *Applied Statistics*, **30**, 153–157. (72, 96, 136, 202, 203)

King, E. P. (1953). 'On some procedures for the rejection of suspected data'. *J. Amer. Statist. Assn.*, **48**, 531–533. (227)

Kirkendall, N. J. (1989). 'Evaluation of Kalman filter implementations of exponential smoothing with outlier and level shift detection'. *Amer. Statist. Assoc. Proc. Bus. Econ. Statist. Sect.*, **1989**, 408–413. (411)

Kitagawa, G. (1979). 'On the use of AIC for the detection of outliers'. *Technometrics*, **21**, 193–199. (105)

Kitagawa, G. (1984). 'Bayesian analysis of outliers via Akaike's predictive likelihood of a model'. *Commun. Statist. Sim. Comp.*, **13**, 107–126. (394)

Kitagawa, G. (1989). 'Non-Gaussian seasonal adjustment'. *Comp. Math. with Applic.*, **18**, 503–514. (413)

Kleiner, B., and Hartigan, J. A. (1981). 'Representing points in many dimensions by trees and castles'. (with Discussion). *J. Amer. Statist. Assn.*, **76**, 260–276. (302, 303)

Kleiner, B., Martin, R. D., and Thomson, D. J. (1979). 'Robust estimation of power spectra' (with Discussion). *J. Roy. Statist. Soc., B*, **41**, 313–351. (396, 408, 414)

Koenker. R., and Bassett, G. (1978). 'Regression quantiles'. *Econometrika*, **46**, 33–50. (348)

Koenker, R., and Portnoy, S. (1990). 'M estimation of multivariate regression'. *J. Amer. Statist. Assoc.*, **85**, 1060–1068. (354)

Kohn, R., and Ansley, C. F. (1986). 'Estimation, prediction and interpolation for ARIMA models with missing data'. *J. Amer. Statist. Assoc.*, **81**, 751–761. (411)

Kotze, T. J. W., and Hawkins, D. M. (1984). 'The identification of outliers in two-way contingency tables using $2 \times 2$ subtables'. *Applied Statistics*, **33**, 215–223. (440)

Kramer, W., Sonnberger, H., Maurer, J., and Havlik, P. (1985). 'Diagnostic checking in practice'. *Rev. Econ. Statist.*, **67**, 118–23. (454)

Krasker, W. S. (1980). 'Estimation in linear regression models with disparate data points'. *Econometrika*, **48**, 1333–1346. (351)

Krasker, W. S., and Welsch, R. E. (1982). 'Efficient bounded-influence regression estimation'. *J. Amer. Statist. Assn.*, **77**, 595–604. (324)

Kruskal, W. H. (1960b). 'Discussion of the papers of Messrs. Anscombe and Daniel'. *Technometrics*, **2**, 157–158. (39, 52)

Kudo, A. (1956a). 'On the testing of outlying observations'. *Sankhyā*, **17**, 67–76. (104, 222, 224, 238, 239, 245–247)

Kudo, A. (1956b). 'On the invariant multiple decision procedures'. *Bull. Math. Statist.*, **6**, 57–68. (104)

Kudo, A. (1957). 'The extreme value in a multiple normal sample'. *Mem. Fac. Sci. Kyushu Univ.*, *A*, **11**, 143–156. (292)

Kulkarni, S. R., and Paranjape, S. R. (1984). 'Use of Andrews' function plot technique to construct control curves for multivariate process'. *Commun. Statist. Theor. Meth.*, **13**, 2511–2533. (309)

Künsch, H. R., Stefanski, L. A., and Carroll, R. J. (1989). 'Conditionally unbiased bounded-influence estimation in general regression models with applications to generalized linear models'. *J. Amer. Statist. Assoc.*, **84**, 460–466. (352)

Kuwada, M. (1987). 'On the robustness of balanced fractional $2/sp m$ factorial designs of resolution $21 + 1$ in the presence of outliers. *Ann. Inst. Statist. Math.*, **39**, 649–659. (365)

Lachenbruch, P. A., and Mickey, M. R. (1968). 'Estimation of error rates in discriminant analysis'. *Technometrics*, **10**, 1–11. (283)

Lalitha, S., and Joshi, P. C. (1986a). 'Performance of Murphy's test for two outliers in a linear model'. *Quart. J. Netherlands Soc. Statist. Oper. Res.*, **40**, 99–107. (338, 339)

Lalitha, S., and Joshi, P. C. (1986b). 'Performance of studentized range statistic for two outliers in a linear model'. *Statistica Neerlandica*, **40**, 157–167. (338, 339)

Lambert, D. (1981). 'Influence functions for testing'. *J. Amer. Statist. Assn.*, **76**, 649–657. (157, 373)

Lambert, D. (1982). 'Qualitative robustness of tests'. *J. Amer. Statist. Assn.*, **77**, 352–357. (74, 77)

Lambert, D. (1985). 'Robust two-sample permutation tests'. *Ann. Statist.*, **13**, 606–625. (87)

Larson, W. A., and McCleary, S. J. (1972). 'The use of partial residuals in regression analysis'. *Technometrics*, **14**, 781–790. (369)

Launer, R. L., and Wilkinson, G. N. (Eds) (1979). *Robustness in Statistics.* Academic Press, New York. (35)

Laurent, A. G. (1963). 'Conditional distribution of order statistics and distribution of the reduced $i$th order statistic of the exponential model'. *Ann. Math. Statist.*, **34**, 652–657. (193)

Ledolter, J. (1987). 'The effects of outliers on the estimates in and the forecasts from ARIMA time series models'. *Amer. Statist. Assoc. Proc. Bus. Econ. Statist. Sect.*, **1987**, 453–458. (413)

Ledolter, J. (1989). 'The effect of additive outliers on the forecasts from ARIMA models'. *Int. J. Forecasting*, **5**, 231–240. (413)

Ledolter, J. (1990). 'Outlier diagnostics in time series analysis'. *J. Time Ser. Anal.*, **11**, 317–324. (Ref). Ledolter, J. (1989). *Int. J. Forecast*, **5**, 231–240. (411)

Legendre, A. M. (1805). *Nouvelles Methods pour la Determination des Orbites des Cometes.* Courcier, Paris (especially 'Appendice sur la methode des moindres quarrés, pp. 72–80). (Ch. 2, *H*)

Legendre, A. M. (1814). 'Methode des moindres quarrés, pour trouver le milieu le plus probable entre les résultats de différentes observations'. *Mémoires de la Classe des Sciences Mathématiques et Physiques de l'Institut de France*, **ANNEE 1810**, 149–154. (Ch. 2, *H*)

Lehmann, E. L. (1975). *Nonparametrics: Statistical Methods based on Ranks.* McGraw-Hill, New York. (87)

Lenth, R. V. (1981). 'Robust measures of location for directional data'. *Technometrics*, **23**, 77–81.(423)

Leonard, T. (1974). 'A modification to the Bayes estimate for the mean of a normal distribution'. *Biometrika*, **61**, 627–628. (380)

Leone, F. C., Jayachandran, T., and Eisenstat, S. (1967). 'A study of robust estimators'. *Technometrics*, **9**, 652–660. (151, 171)

Leroy, A., and Rousseeuw, P. (1985). 'Computing robust regression estimators with "progres" and some simultation results'. *Statist. Dec. Supp. Issue*, **1985**, 321–325. (453)

Lesaffre, E., and Albert, A. (1989). 'Multiple-group logistic regression diagnostics'. *Applied Statistics*, **38**, 425–440. (353, 454)

Lewis, T. (1975). In Discussion on Mardia, K. V. (1975). 'Statistics of directional data'. *J. Roy. Statist. Soc. B*, **37**, 375–376. (421)

Lewis, T. (1987). 'Uneven sex ratios in the light brown apple moth: a problem in outlier allocation'. Chapter 9 in Hand. D. J. and Everitt, B. S. (Eds). (1987). *The Statistical Consultant in Action. Camb. Univ. Press, Cambridge.* (130)

Lewis, T., and Fieller, N. R. J. (1979). 'A recursive algorithm for null distributions for outliers: I. Gamma samples'. *Technometrics*, **21**, 371–376. (96, 136, 200, 208, 242, 244)

Lewis, T., and Fisher, N. I. (1982). 'Graphical methods for investigating the fit of a Fisher distribution to spherical data'. *Geophys. J. Roy. Astr. Soc.*, **69**, 1–13. (425)

Li, W. K. (1988). 'A goodness-of-fit test in robust time series modelling'. *Biometrika*, **75**, 355–361. (411)

Li, S., and Dickinson, B. W. (1988). 'Application of the lattice filter to robust estimation of AR and ARMA models'. *Trans. Acous., Speech Signal Proc.*, **36**, 502–512. (405, 406)

Likeš, J. (1966). 'Distribution of Dixon's statistics in the case of an exponential population'. *Metrika*, **11**, 46–54. (91, 96, 136, 193, 198–200, 204, 209, 210)

Likeš, J. (1987). 'Some tests for $k = 2$ upper outliers in an exponential sample'. *Biometrical J.*, **29**, 313–321. (136)

Lindley, D. V. (1956). 'On a measure of information provided by an experiment'. *Ann Math. Statist.*, **27**, 986–1005. (380)

Lindley, D. V. (1968). 'The choice of variables in multiple regression' (with Discussion). *J. Roy. Statist. Soc. B*, **30**, 31–66. (380)

Lingappaiah, G. S. (1976). 'Effect of outliers in the estimation of parameters'. *Metrika*, **23**, 27–30. (387)

Lingappaiah, G. S. (1983). 'Bayesian approach to the estimation of parameters in the Burr's distribution with outliers'. *J. Orissa Math. Soc.*, **2**, 55–59. (388)

Lingappaiah, G. S. (1989). 'Bayes approach to prediction in samples from gamma population when outliers are present'. *Indian J. Pure Appl. Math.*, **20**, 858–870. (388)

Little, R. J. A. and (1988). 'Robust estimation of the mean and covariance matrix from data with missing values'. *Applied Statistics*, **37**, 23–38. (279)

Little, R. J. A., and Smith, P. J. (1987). 'Editing and imputation for quantitative survey data'. *J. Amer. Statist. Assoc.*, **82**, 58–68. (441, 445)

Ljung, G. M. (1989a). 'A note on the estimation of missing values in time series'. *Commun. Statist. B. Sim. Comp.*, **18**, 459–465. (400)

Ljung, G. M. (1989b). 'Outliers and missing observations in time series'. *Amer. Statist. Assoc. Proc. Bus. Econ. Statist. Sect.*, **1989**, 397–401. (400)

Lopuhaä, H. P. (1989). 'On the relation between S-estimators and M-estimators of multivariate location and covariance'. *Ann. Statist.*, **17**, 1662–1683. (279)

Lovie, P. (1986). 'Identifying outliers'. In Lovie, A. D. (Ed.). (1986). *New Developments in Statistics for Psychology and the Social Sciences. British Psychological Society*, **1986**, pp 44–69. (52)

Lovie, A. D. (Ed.). (1986). *New Developments in Statistics for Psychology and the Social Sciences*. Methuen, London.

Lund, R. E. (1975). 'Tables for an approximate test for outliers in linear models'. *Technometrics*, **17**, 473–476. (322, 334)

McCabe, B. P. M. (1987). 'Testing regression models for random effects outliers under elliptical symmetry'. *Econ. Lett.*, **25**, 47–49. (353)

McCullagh, P., and Nelder, J. A. (1989). *Generalized Linear Models*. 2nd edn. Chapman and Hall, London. (352)

McKay, A. T. (1935). 'The distribution of the difference between the extreme observation and the sample mean in samples of $n$ from a normal universe'. *Biometrika*, **27**, 466–471. (245)

McKean, J. W., and Sievers, G. L. (1987). 'Coefficients of determination for least absolute deviation analysis'. *Statist. Prob. Lett.*, **5**, 49–54. (350)

McKean, J. W., and Sievers, G. L. (1989). 'Rank scores suitable for the analysis of linear models under asymmetric error distributions'. *Technometrics*, **31**, 207–218. (351)

McKean, J. W., Sheather, S. J., and Hettmansperger, T. P. (1990). 'Regression diagnostics for rank-based methods'. *J. Amer. Statist. Assoc.*, **85**, 1018–1028. (370)

McMillan, R. G. (1971). 'Tests for one or two outliers in normal samples with unknown variance'. *Technometrics*, **13**, 87–100. (49, 97, 98, 131, 137, 138, 222, 224, 225, 237, 238, 240)

McMillan, R. G., and David, H. A. (1971). 'Tests for one of two outliers in normal samples with known variance'. *Technometrics*, **13**, 75–85. (49, 97, 98, 131, 137, 138, 245, 247)

Main, P. (1987). 'Outlier-neutral, outlier-prone and outlier-resistant distributions'. *Trabajos de Estadistica y de Investigacion Operativa*, **2**, 91–101. (53)

Main, P. (1988). 'Asymptotic behaviour of reliability functions'. *Statist. Prob. Lett.*, **7**, 259–263. (53)

Mann, N. R., Scheuer, E. M., and Fertig, K. W. (1973). 'A new goodness-of-fit test for the two-parameter Weibull or extreme-value distribution with unknown parameters'. *Commun. Statist.*, **2**, 383–400. (258)

Manski, C. F. (1984). 'Adaptive estimation of nonlinear regression models'. *Econ. Rev.*, **3**, 145–210. (351)

Marasinghe, M. G. (1985). 'A multistage procedure for detecting several outliers in linear regression'. *Technometrics*, **27**, 395–399. (344)

Marazzi, A. (1985). 'On constrained minimization of the Bayes risk for the linear model'. *Statist. Dec.*, **3**, 277–296. (385)

Marco, V. R., Young, D. M., and Turner, D. W. (1987). 'A note on the effect of simple equicorrelation in detecting a spurious multivariate observation'. *Commun. Statist. Theor. Meth.*, **16**, 1027–1036. (285)

Mardia, K. V. (1962). 'Multivariate Pareto distributions'. *Ann. Statist.*, **33**, 1008–1015. (295)

Mardia, K. V. (1970). 'Measures of multivariate skewness and kurtosis with applications'. *Biometrika*, **57**, 519–530. (293, 296, 338)

Mardia, K. V. (1972). *Statistics of Directional Data*. Academic Press. London. (417, 422–424)

Mardia, K. V. (1974). 'Applications of some measures of multivariate skewness and kurtosis in testing normality and robustness studies'. *Sankhyā, B*, **36**, 115–128. (293)

Mardia, K. V. (1975a). 'Assessment of multinormality and the robustness of Hotelling's $T^2$ test'. *Applied Statistics*, **24**, 163–171. (293)

Mardia, K. V. (1975b). 'Statistics of directional data' (with Discussion). *J. Roy. Statist. Soc. B*, **37**, 349–393. (417, 421)

Mardia, K. V., and Kanazawa, M. (1983). 'The null distribution of multivariate kurtosis'. *Commun. Statist. Sim. Comp.*, **12**, 569–576. (293)

Margolin, B. H. (1976). 'Design and analysis of factorial experiments via interactive computing in APL'. *Technometrics*, **18**, 135–150. (363)

Marks, R. G., and Rao, P. V. (1978). 'A modified Tiao-Guttman rule for multiple outliers'. *Commun. Statist. Theor. Meth. A*, **7**, 113–126. (47, 62, 84, 167)

Marks, R. G., and Rao, P. V. (1979). 'An estimation procedure for data containing outliers with a one-directional shift in the mean'. *J. Amer. Statist. Assn.*, **74**, 614–620. (47, 62, 84, 167)

Maronna, R. A. (1976). 'Robust *M*-estimators of multivariate location and scatter'. *Ann. Statist.*, **4**, 51–67. (275, 277)

Maronna, R. A. (1979). Discussion of Kleiner, B., Martin R. D., and Thomson, D. J. (1979). *J. Roy. Statist Soc., B*, **41**, 344–345. (414)

Maronna, R. A., and Yohai, V. J. (1981). 'Asymptotic behaviour of general *M*-estimates for regression and scale with random carriers'. *Zeitschrift fur Wahrscheinlichkeittheorie*, **58**, 7–20. (349)

Martin, R. D. (1979). 'Robust estimation for time series'. In Laumer, R. L., and Wilkinson, G. N. (Eds) (1979). *Robustness in Statistics*. Academic Press, New York. (396, 403, 404)

Martin, R. D. (1981). 'Robust methods for time series'. In Findley, D. F. (Ed.). (1981). *Applied Time Series Analysis, 2.* (406)

Martin, R. D., and Yohai, V. J. (1984). 'Gross-error sensitivities of GM and RA-estimates. In *Robust and Nonlinear time Series Analysis*. Lecture Notes in Statistics, **26**, Springer-Verlag, New York. (405)

Martin, R. D., and Yohai, V. J. (1986). 'Influence functionals for time series'. *Ann. Statist.*, **14**, 781–855. (405)

Martin, R. D., and Zamar, R. H. (1989). 'Asymptotically min-max bias robust M-estimates of scale for positive random variables'. *J. Amer. Statist. Assoc.*, **84**, 494–501. (65)

Martin, R. D., Yohai, V. J., and Zamar, R. H. (1989). 'Min-max bias robust regression'. *Ann. Statist.*, **17**, 1608–1630. (350)

Martinsek, A. T. (1988). 'Negative regret, optional stopping and the elimination of outliers'. *J. Amer. Statist. Assoc.*, **83**, 160–163. (81)

Masarotto, G. (1987). 'Robust and consistent estimates of autoregressive moving average parameters'. *Biometrika*, **74**, 791–797. (405, 406)

Mason, R. L., Gunst, R. F., and Hess, J. L. (1989). *Statistical Design and Analysis of Experiments with Applications to Engineering and Science*. Wiley, New York. (364)

Masreliez, C. J., and Martin, R. D. (1977). 'Robust Bayesian estimation for the linear model and robustifying the Kalman filter'. *I. E. E. E. Trans. Aut. Control*, **AC-22**, 361–371. (414)

Massart, D. L., Kaufman, L., Rousseeuw, P. J., and Leroy, A. (1986). 'Least median of squares: A robust method for outlier and model error detection in regression and calibration'. *Analytica Chimica Acta*, **187**, 171–179. (325, 350)

Mathar, R. (1985). 'Outlier-prone and outlier-resistant multidimensional distributions'. *Statistics*, **16**, 451–456. (273)

Medvedev, A. G., and Kharin, YU.S. (1988). 'Robustness of decision rules in multivariate classification problems'. *J. Sov. Math.*, **40**, 533–539. (283)

Meeden, G., and Isaacson, D. (1977). 'Approximate behaviour of posterior distribution for a large observation'. *Ann. Statist.*, **5**, 899–908. (380)

Meinhold, R. J., and Singpurwalla, N. D. (1989). 'Robustification of Kalman filter model'. *J. Amer. Statist. Assoc.*, **84**, 479–486. (411)

Mendeleev, D. I. (1895). 'Course of work on the renewal of prototypes or standard measures of lengths and weights' (Russian). *Vremennik Glavnoi Palaty Mer i Vesov*, **2**, 157–185. (Reprinted 1950; *Collected Writings (Socheneniya)*, **22**, 175–213, izdat, Akad. Nauk, SSSR, Leningrad—Moscow). (30, 36)

Merle, C. (1983). 'International standardization: application of statistics'. In *Encyclopaedia of Statistical Sciences*, **4**, Wiley, New York. (455)

Merriman, M. (1877). 'List of writings relating to the method of least squares with historical and critical notes'. *Transactions of the Connecticut Academy of Arts and Sciences*, **4**, 151–232. (Ch. 2, *H*)

Merriman, M. (1884). *A Textbook on the Method of Least Squares*. Wiley, New York. (Ch. 2, *H*)

Michels, P., and Trenkler, G. (1990). 'Testing the stability of regression coefficients using generalized recursive residuals'. *Australian J. Statist.*, **32**, 293–312. (371)

Mickey, M. R. (1974). 'Detecting outliers with stepwise regression'. *Communications—UCLA Health Sciences Facility*, **1**, 1. (341)

Mickey, M. R., Dunn, O. J., and Clark, V. (1967). 'Note on use of stepwise regression in detecting outliers'. *Computers & Biomed. Res.*, **1**, 105–111. (335, 341)

Miller, R. (1974). 'The jackknife—a review'. *Biometrika*, **61**, 1–15. (173)

Minitab (1988). *Minitab Reference Manual, Release 6*. Minitab Inc., State College, PA. (451)

Minitab (1989). *Minitab Reference Manual, Release 7*. Minitab Inc., State College, PA. (451)

Mirvaliev, M. (Translated by Kashper, A.) (1978a). 'Rejection of outlying results of angular measurements'. *Theor. Prob. Ap.*, **23**, 814–819. (422)

Mirvaliev, M. (Translated by Kashper, A.) (1978b). 'The rejection of outlying observarious in regression analysis'. *Theor. Prob. Ap.*, **23**, 598–602. (335)

Mirvaliev, M. (1983). (Translated by Kraiman, A. R.) 'Smoothing of the results of angular measurements made according to Tomilin's scheme'. *Theor. Prob. Appl.*, **28**, 455–462. (422)

Moberg, T. F., Ramberg, J. S., and Randles, R. H. (1978). 'An adaptive *M*-estimator and its application to a selection problem'. *'Technometrics*, **20**, 225–263. (151)

Moore, P. G. (1957). 'The two-sample *t*-test based on range'. *Biometrika*, **44**, 482–485. (240)

Moran, M. A., and McMillan, R. G. (1973). 'Tests for one or two outliers in normal samples with unknown variance: a correction'. *Technometrics*, **15**, 637–640. (98, 222, 237, 238)

Moran, P. A. P. (1953). 'The random division of an interval—Part III'. *J. Roy. Statist. Soc. B.*, **15**, 77–80. (212)

Moser, B. K., and Marco, V. R. (1988). 'Bayesian outlier testing using the predictive distribution for a linear model of constant intraclass form'. *Commun. Statist. Theor. Meth.*, **17**, 849–860. (385)

Mosteller, F. (1948). 'A *k*-sample slippage test for an extreme population'. *Ann. Math. Statist.*, **19**, 58–65. (107)

Mosteller, F., and Parunak, A. (1985). 'Identifying extreme cells in a sizable contigency table: probabilistic and exploratory approaches'. In Hoaglin, D. C., Mosteller, F. and Tukey, J. W. (Eds). (1985). *Exploring Data Tables, Trends and Shapes*, Wiley, New York, pp. 189–224. (431, 432, 439)

Mosteller, F., and Tukey, J. W. (1977). *Data Analysis and Linear Regression.* Addison-Wesley, Reading, Mass. (235, 280)

Mount, K. S., and Kale, B. K. (1973). 'On selecting a spurious observation'. *Can. Math. Bull.*, **16**, 75–78. (51)

Moussa-Hamouda, E., and Leone, F. C. (1977). 'The robustness of efficiency of adjusted trimmed estimators in linear regression'. *Technometrics*, **19**, 19–34. (324, 349)

Muirhead, C. R. (1986). 'Distinguishing outlier types in time series'. *J. Roy. Statist. Soc. B.*, **48**, 39–47. (393, 400, 413)

Muncke, G. W. (1825). 'Beobachtung'. Gehler's *Physikalisches Worterbuch*, 2nd edn. Leipzig, Vol. 1, pp. 884–912. (Ch. 2, *H*)

Muñoz-Garcia, J., Moreno-Rebollo, J. L., and Pascual-Acosta, A. (1987). 'Detecting outliers in s-multinomial distributions through adjusted residuals'. *J. Appl. Statist.*, **14**, 171–176. (493)

Muñoz-Garcia, J., Moreno-Rebollo, J. L., and Pascual-Acosta, A. (1990). 'Outliers: A formal approach'. *Int. Statist. Rev.*, **58**, 215–226. (53)

Murphy, R. B. (1951). *On Tests for Outlying Observations.* Ph. D. thesis, Princeton University, University Microfilms Inc., Ann Arbor, Mich. (109, 224)

Naes, T. (1986). 'Detection of multivariate outliers in linear mixed models'. *Commun. Statist. Theor. Meth.*, **15** 33–47. (351)

Naes, T., and Martens, H. (1987). 'Testing adequacy of linear random models'. *J. Theor. Appl. Statist.*, **18**, 323–331. (353)

Naik, D. N. (1989). 'Detection of outliers in the multivariate linear regression model'. *Commun. Statist. Theor. Meth.*, **18**, 2225–2232. (338)

Naik, U. D. (1972). 'A Bayesian analysis of certain contaminated samples'. *Research Report No.* 104, Department of Probability and Statistics, Sheffield University. (Ch. 9)

Nair, K. R. (1948). 'The distribution of the extreme deviate from the sample mean and its studentized form'. *Biometrika*, **35**, 118–144. (237, 245)

Nair, K. R. (1952). 'Tables of percentage points of the "Studentized" extreme deviate from the sample mean'. *Biometrika*, **39**, 189–191. (237)

Nelder, J. A. (1972). 'A statistician's point of view'. In Jeffers, J. N. R. (Ed.). *Mathematical Models in Ecology. The 12th Symposium of the British Ecological Society, Grange-over-Sands, Lancashire, March 1971.* pp 367–373. Blackwell Scientific Publications, Oxford. (431)

Newcomb, S. (1886). 'A generalized theory of the combination of observations so as to obtain the best result'. *Amer. J. Math.*, **8**, 343–366. (30, 36)

Newcomb, S. (1912). 'Researches on the motion of the moon, Part II. The mean motion of the moon and other astronomical elements derived from observations of eclipses and occultations extending from the period of the Babylonians until A. D. 1908'. *Astronomical papers*, **9**, 1–249, U.S. Government Printing Office, Washington. (Ch. 2, *H*)

Neyman, J., (1979). 'Outlier proneness and resistance. Developments in probability and mathematical statistics generated by studies in meteorology and weather'. *Commun. Statist. Sim. Comp.*, **8**, 1097–1110. (53)

Neyman, J., and Scott, E. L. (1971). 'Outlier proneness of phenomena and of related distribution. In Rustagi, J. (Ed.) (1971). *Optimising Method in Statistics.* Academic Press, New York. (52)

Noether, G. E. (1967). 'Wilcoxon confidence intervals for location parameters in the discrete case'. *J. Amer. Statist. Assn.*, **62**, 184–188. (87)

Noether, G. E. (1973). 'Some simple distribution-free confidence intervals for the center of a symmetric distribution'. *J. Amer. Statist. Assn.*, **68**, 716–719. (87)

Noether, G. E. (1974). 'Distribution-free confidence intervals based on linear rank statistics'. In Williams, E. J. (Ed.) (1974). *Studies in Probability and Statistics.* Jerusalem Academic Press, Jerusalem. (87)

Nyquist, H. (1984). 'A model for robust analysis of multivariate data'. *Metron*, **42**, 15–23. (277)

Nyquist, H. (1987). 'Robust estimation of the structural errors-in-variables model'. *Metrika*, **34**, 177–183. (328)

Nyquist, H. (1988). 'Applications of the jackknife procedure in ridge regression'. *Comp. Statist. Data Anal.*, **6**, 177–183. (353)

Obenchain, R. L. (1977). Letter to the Editor. *Technometrics*, **19**, 348–349. (374)

O'Brien, P. C. (1984). 'Procedures for comparing samples with multiple endpoints'. *Biometrics*, **40**, 1079–1087.

O'Gorman, M. A., and Myers, R. H. (1987). 'Measures of error with outliers in regression'. *Commun. Statist. Sim. Comp.*, **16**, 771–789. (351)

Ogrodnikoff, K. (1928). 'On the occurrence of discordant observations and a new method of treating them'. *Monthly Notices Roy. Astr. Soc.*, **88**, 523–532. (Ch. 5, *H*)

O'Hagan, A. (1979). 'On outlier rejection phenomena in Bayes inference'. *J. Roy. Statist. Soc., B*, **41**, 358–367. (53, 381)

O'Hagan, A. (1981). 'A moment of indecision'. *Biometrika*, **1**, 329–330. (381)

O'Hagan, A. (1988). 'Modelling with heavy tails'. In Bernardo, J. M., DeGroot, M. H., Lindley, D. V., and Smith, A. F. M. (Eds). *Bayesian Statistics 3*, 345–359. (52, 381)

O'Hagan, A. (1990). 'Outliers and credence for location parameter inference'. *J. Amer. Statist. Assoc.*, **85**, 172–176. (381)

Oja, H. (1983). 'Descriptive statistics for multivariate distributions'. *Statist. Prob. Lett.*, **1**, 327–332. (271)

Ozturk, O., Patil, G. P., and Taillie, C. (1992). 'Detection of contaminated observations in binomial sampling: a Bayesian approach'. *Tech. Report*, 1992. (393)

Pagurova, V. I. (1985). 'The Chauvenet test for detecting several outliers'. *Akademiya Nauk SSSR. Teoriya Veroyatnostei i ee Primeneniya*, **30**, 558–561. (29)

Parker, R. A. (1989). 'Analysis of surveillance data with Poisson regression: A Case Study'. *Statist. Med.*, **8**, 285–294. (351, 455)

Parr, W. C. (1981). 'Minimum distance estimation: a bibliography'. *Commun. Statist. Theor. Meth.*, **10**, 1205–1224. (154)

Parr, W. C., and De Wet, T. (1981). 'On minimum Cramér-von-Mises-norm parameter estimation'. *Commun. Statist. Theor. Meth.*, **10**, 1149–1166. (154)

Parr, W. C., and Schucany, W. R. (1980). 'Minimum distance and robust estimation'. *J. Amer. Statist. Assn.*, **75**, 616–624. (153)

Pasman. V.R., and Shevlyakov, G. L. (1987). 'Robust methods for estimating the correlation coefficient.' *Akademiya Nauk SSSR. Avtomatika i Telemekhanika*, **1987**, 70–80. (280)

Patel, K. R., Mudholkar, G. S., and Fernando, J. L. I. (1988). 'Student's *t* approximations for three simple robust estimators'. *J. Amer. Statist. Assoc.*, **83**, 1203–1210. (147, 174)

Patil, S. A. (1985). 'On the mean square errors and biases of estimators based on three observations.' *J. Indian. Statist. Assoc.*, **23**, 139–157. (65)

Patil, S. A., Kovner, J. L., and King, R. M. (1977). 'Tables of percentage points of ratios of linear combinations of order statistics of samples from exponential distributions'. *Commun. Statist. B*, **6**, 115–136. (210)

Paul, S. R., and Barnwal, R. K. (1987). 'Detection of outliers in Poisson samples'. *Commun. Statist. Theor. Meth.*, **16**, 2391–2403. (262, 264)

Paul, S. R., and Fung, K. Y. (1986). 'Critical values for Dixon type test statistics for testing outliers in Weibull or extreme value distributions'. *Commun. Statist. Sim. Comp.*, **15**, 277–283. (257, 259)

Paulson, A. S., and Thornton, J. C. (1975). 'A new approach to goodness of fit and outliers'. In Vogt, W. G., and Mickle, M. H. (Eds) (1975). *Proc. 6th Pittsburg Conf.: Modelling and Simulation*. Instruments Soc. America, Pittsburg, PA. (440)

Paulson, E. (1952b). 'A optimum solution to the *k*-sample slippage problem for the normal distribution'. *Ann. Math. Statist.*, **23**, 610–616. (103, 104, 357)

PC-ISP (1991). *PC-ISP Users' Guide, Version 3.1*. Datavision AG, Switzerland. (452)

Pearson, E. S. (1926). 'A further note on the distribution of range in samples taken from a normal population'. *Biometrika*, **18**, 173–194. (248)

Pearson, E. S. (1932). 'The percentage limits for the distribution of range in samples from a normal population ($n < 100$)'. *Biometrika*, **24**, 404–417. (248)

Pearson, E. S., and Chandra Sekar, C. (1936). 'The efficiency of statistical tools and a criterion for the rejection of outlying observations'. *Biometrika*, **28**, 308–320. (31, 92, 110, 136, 222, 360)

Pearson, E. S., and Hartley, H. O. (1942). 'The probability integral of the range in samples of *n* observations from a normal population'. *Biometrika*, **32**, 301–310. (248)

Pearson, E. S., and Hartley, H. O. (Eds) (1966). *Biometrika Tables for Statisticians*, Vol. 1, 3rd edn., Cambridge University Press, London. (204, 223, 226, 230, 231, 237, 240, 245, 246)

Pearson, E. S., and Stephens, M. A. (1964). 'The ratio of range to standard deviation in the same normal sample'. *Biometrika*, **51**, 484–487. (95, 226)

Pearson, K. (1904). 'On the theory of contingency and its relation to association and normal correlation'. Reprinted in 1948 in *Karl Pearson's Early Statistical Papers*, Cambridge University Press, Cambridge. pp. 443–475. (436)

Pearson, K. (Ed.) (1931). *Tables for Statisticians and Biometricians*. Biometric Lab., University College, London. (40)

Peirce, B. (1852). 'Criterion for the rejection of doubtful observations'. *Astr. J.*, **2**, 161–163. (3, 28, 38)

Peirce, B. (1878). 'On Peirce's criterion' (with remarks by Scott, C. A.). *Proceedings of the American Academy of Arts and Sciences*, **13**, 348–351. (Ch. 2, *H*)

Peirce, C. S. (1873). 'On the theory of errors of observations'. *Report of the Superintendent of the United States Coast Survey*, (for the year ending 1 November 1870) U.S. Government Printing Office, Washington. (Ch. 2, *H*)

Peixoto, J. L., and LaMotte, L. R. (1989). 'Simultaneous identification of outliers and predictors using variable selection techniques'. *J. Statist. Plan. Inf.*, **23**, 327–343. (346)

Pena, D. (1987). 'Measuring the importance of outliers in ARIMA models'. In Vilaplana, J. P., Wertz, W. and Puri, M. L. (Eds), *New Perspectives in Theoretical and Applied Statistics*. John Wiley & Sons, Inc., New York, pp 109–118. (410)

Perli, H.-G., Hommel, G., and Lehmacher, W. (1985). 'Sequentially rejective test procedures for detecting outlying calls in one-and two-sample multinomial experiments'. *Biometrical J.*, **27**, 885–893. (439)

Peters, T. J. (1990). 'Comment on Appleton, D. R. (1990). What statistics should we teach medical undergraduates and graduates?' *Statist. Med.*, **9**, 1013–1027. (455)

Pettit, L. I. (1983). *Bayesian approaches to outliers*. Ph.D thesis. University of Nottingham. (373, 393)

Pettit, L. I. (1988). 'Bayes methods for outliers in exponential samples'. *J. Roy. Statist. Soc. B.*, **50**, 371–380. (393)

Pettit, L. I., and Smith, A.F.M. (1983). 'Bayesian model comparisons in the presence of outliers'. *Bull. Inst. Int. Statist.*, **50**, 292–306. (52)

Pettit, L. I., and Smith, A.F.M. (1985). Outliers and influential observations in linear models. *Bayesian Statist.*, **1985**, 473–494. (358)

Pettitt, A. N., and Bin Daud, I. (1989). 'Case-weighted measures of influence for proportional hazards regressions'. *Applied Statistics*, **38**, 51–67. (353)

Pham, T. D. (1984). 'On robust estimation of parameters for autoregressive moving average models'. *Lec. Notes Statist.*, **26**, 273–286. (405, 406)

Pincus, R. (1984). 'Distribution of the maximal gap in a sample and its application for outlier detection'. *Theor. Decis. Lib. B. Math. Statist.*, **1984**, 90–91. (313)

Plackett, R. L. (1981). *The Analysis of Categorical Data. 2nd Edn.* Griffin, London. (436)

Pollak, M. (1979). A class of robust estimators'. *Commun. Statist. Theor. Meth.*, **8**, 509–531. (155)

Portnoy, S. L. (1977). 'Robust estimation in dependent situation'. *Ann. Statist.*, **5**, 22–43. (144)

Pourahmadi, M. (1989). 'Estimation and interpolation of missing values of a stationary time series'. *J. Time. Ser. Anal.*, **10**, 149–169. (400)

Pregibon, D. (1981). 'Logistic regression diagnostics'. *Ann. Statist.*, **9**, 705–724. (352, 353)

Pregibon, D. (1982). 'Resistant fits for some commonly used logistic models with medical applications'. *Biometrics*, **38**, 485–498. (353)

Prescott, P. (1975). 'An approximate test for outliers in linear models'. *Technometrics*, **17**, 129–132. (323, 332, 334, 339)

Prescott, P. (1976a). 'On a test for normality based on sample entropy'. *J. Roy. Statist. Soc. B*, **38**, 254–256. (369, 373)

Prescott, P. (1976b). 'Comparison of tests for normality using stylized sensitivity surfaces'. *Biometrika*, **63**, 285–289. (72)

Prescott, P. (1977). 'An upper bound for any linear function of normal residuals'. *Commun. Statist. Sim. Comp.*, **6**, 83–88. (360)

Prescott, P. (1978). 'Examination of the behaviour of tests for outliers when more than one outlier is present'. *Applied Statistics.* **27**, 10–25. (47, 72, 131, 136, 139, 222)

Prescott, P. (1979). 'Critical values for a sequential test for many outliers'. *Applied Statistics*, **28**, 36–39. (126, 223, 224, 236)

Pyke, R. (1965). 'Spacings'. *J. R. Statist. Soc. B.*, **27**, 395–449. (258)

Quenouille, M. H. (1953). *The Design and Analysis of Experiment.* Griffin, London. (358)

Quesenberry, C. P. (1986). 'Screening outliers in normal process control data with uniform residuals'. *J. Quality Technology*, **18**, 226–233. (126)

Quesenberry, C. P., and David, H. A. (1961). 'Some tests for outliers'. *Biometrika*, **48**, 379–387. (222, 224, 238, 239)

Raab, G. M. (1981). 'Estimation of a variance function, with application to immunoassay'. *Applied Statistics*, **30**, 32–40. (170)

Radhakrishnan, R. (1983). 'Influence functions for certain parameters in discriminant analysis'. *Metron*, **41**, 183–194. (283)

Radhakrishnan, R. (1985). 'Influence functions for certain parameters in discriminant analysis when a single discriminant function is not adequate'. *Commun. Statist. Theor. Meth.*, **14**, 535–549. (310)

Radhakrishnan, R., and Kshirsagar, A. M. (1981). 'Influence functions for certain parameters in multivariate analysis'. *Commun. Statist. Theor. Meth.*, **10**, 515–529. (280, 282)

Raghunandanan, K., and Srinivasan, R. (1970). 'Simplified estimation of parameters in a logistic distribution'. *Biometrika*, **57**, 677–679. (186, 187)

Raghunandanan, K., and Srinivasan, R. (1971). 'Simplified estimation of parameters in a double exponential distribution'. *Technometrics*, **13**, 689–691. (186, 187)

Ramachandran, K. V., and Khatri, C. G. (1957). 'On a decision procedure based on the Tukey statistic'. *Ann. Math. Statist.*, **28**, 802–806. (104)

Ramsay, J. O. (1977). 'A comparative study of several robust estimators of slope, intercept and scale in linear regression'. *J. Amer. Statist. Assn.*, **72**, 608–615. (324)

Ranganathan, J., and Kale, B. K. (1983). 'Outlier-resistant tolerance intervals for exponential distributions'. *Amer. J. Math. Manag. Sci.*, **3**, 5–24. (184)

Rao, C. R. (1964). 'The use and interpretation of principal component analysis in applied research'. *Sankhyā, A*, **26**, 329–358. (308)

Raouf, A., and Chaudhary, K. M. (1987). 'Development of a modular computer program for testing outliers'. *Pakistan J. Statist.*, **3**, 115–134. (453)

Rasmussen, J. L. (1988). 'Evaluating outlier identification tests: Mahalanobis D squared and Comrey D$k$'. *Multivar. Behav. Res.*, **23**, 189–202. (285)

Rauhut, B. O. (1982). 'Estimation of the mean of the exponential distribution with an outlying observation'. *Commun. Statist. Theor. Meth.*, **11**, 1439–1452. (179, 180)

Rauhut, B. (1987). 'The modelling of outlier situations'. Springer, Berlin–New York, 317–324. (52)

Relles, D. A., and Rogers, W. H. (1977). 'Statisticians are fairly robust estimators of location'. *J. Amer. Statist. Assn.*, **72**, 107–111. (56)

Rider, P. R. (1933). 'Criteria for rejection of observations'. *Washington University Studies—New Series, Science and Technology*, **8**, 3–23. (29)

Rieder, H. (1978). 'A robust asymptotic testing model'. *Ann. Statist.*, **6**, 1080–1094. (87)

Rieder, H. (1982). 'Qualitative robustness of rank tests'. *Ann. Statist.*, **10**, 205–211. (74)

Rivest, L.-P. (1989). 'Spherical regression for concentrated Fisher–von Mises distributions'. *Ann. Statist.*, **17**, 307–317. (428)

Robertson, C. (1990). 'A matrix regression model for the transition probabilities in a finite state stochastic process'. *Applied Statistics*, **39**, 1–19. (414)

Rock, N. M. S. (1987). 'ROBUST: An interactive FORTRAN-77 package for exploratory data analysis'. *Comp. & Geosci*, **13**, 463–494. (453)

Rocke, D. M. (1989). 'Robust control charts'. *Technometrics*, **31**, 173–184. (413)

Rocke, D. M., Downs, G. W., and Rocke, A. J. (1982). 'Are robust estimators really necessary?' *Technometrics*, **24**, 95–101. (56, 62)

Rohlf, F. J. (1975). 'Generalisation of the gap test for the detection of multivariate outliers'. *Biometrics*, **31**, 93–101. (301, 311–313)

Ronchetti, E. (1985). 'Robust model selection in regression'. *Statist. Prob. Lett.*, **3**, 21–23. (35)

Ronner, A. E., and Steerneman, A. G. M. (1985). 'The occurrence of outliers in the explanatory variable considered in an errors-in-variable framework'. *Metrika*, **32**, 97–107. (328)

Ronner, A., Steerneman, T., and Kuper, G. (1985). 'On the performance of moment estimators in a structural regression model with outliers in the explanatory variable'. In *Methods of Operations Research*. 55. Verlagsgruppe Athenaum, Hain. (328)

Rosado, F. F. (1987). 'Outliers in exponential populations'. *Metron*, **45**, 85–91. (204, 205)

Rosner, B. (1975). 'On the detection of many outliers'. *Technometrics*, **17**, 221–227. (132, 136, 232, 235, 236, 292)

Rosner, B. (1977). 'Percentage points for the RST many outlier procedure'. *Technometrics*, **19**, 307–312. (132, 136, 235, 236, 292)

Rousseeuw, P. J. (1984). 'Least median of squares regression'. *J. Amer. Statist. Assoc.*, **79**, 871–880. (325, 344, 350)

Rousseeuw, P. J. (1985). 'Multivariate estimation with high breakdown point'. In *Math. Statist. Applic.*, (Vol, B). (Ed). Grossmann, W., Pflug, G., Vincze, I., and Wertz, W. Riedel Publishing, pp. 283–297. (278)

Rousseeuw, P. J., and Bassett, G. W. (1990). 'The Remedian: a robust averaging method for large data sets'. *J. Amer. Statist. Assoc.*, **85**, 97–104. (36, 152)

Rousseeuw, P. J., and Leroy, A. M. (1987). *Robust Regression and Outlier Detection.* Wiley, New York. (35, 317)

Rousseeuw, P. J., and Ronchetti, E. (1981). 'Influence curves for general statistics'. *J. Comp. Appl. Math.*, **7**, 161–166. (158)

Rousseeuw, P. J., and Yohai, V. (1984). 'Robust regression by means of S-estimators'. *Lec. Notes Statist.*, **26**, 256–272. (279, 350)

Rousseeuw, P. J., and Van Zomeren, B. C. (1990). 'Unmasking multivariate outliers and leverage points'. *J. Amer. Statist. Assoc.*, **85**, 633–639. (278, 307)

Royston, J. P. (1982a). 'An extension of Shapiro and Wilk's W test for normality to large samples'. *Appl. Statist.*, **31**, 115–124. (233, 234)

Royston, J. P. (1982b). 'Algorithm AS181. The W test for normality'. *Appl. Statist.*, **31**, 176–180. (234)

Royston, J. P. (1986). 'A remark on AS181. The W test for normality'. *Appl. Statist.*, **35**, 232–234. (234)

Royston, J. P. (1989). 'Correcting the Shapiro-Wilk W for ties'. *J. Statist. Comp. Sim.*, **31**, 237–249. (234)

Rubin, D. B. (1987). *Multiple Imputation for Nonresponse in Surveys.* Wiley, New York. (441)

Rubin, H. (1977). 'Robust Bayesian inference'. Pp. 351–356 of Gupta, S. S., and Moore, D. S. (Eds) (1977). *Statistical Decision Theory and Related Topics II.* Academic Press, New York. (381)

Ruppert, D., and Carroll, R. J. (1980). 'Trimmed least squares estimation in the linear model'. *J. Amer. Statist. Assn.*, **75**, 828–838. (348, 354)

Ryland, A. (1841). 'Income of scientific and literary societies in England'. *J. Statist. Soc.*, **4**, 264–267. (11)

Sadler, W. A., and Smith, M. H. (1987). 'A computer program for variance function estimation, with particular reference to immunoassay data'. *Comp. Biomed. Res.*, **20**, 1–11. (453)

Sarhan, A. E., and Greenberg, B. G. (Eds) (1962). *Contributions to Order Statistics.* Wiley, New York. (157, 168)

SAS (1986). *SAS System for Linear Models.* SAS Institute Inc., Cary, NC. (451)

SAS (1988). *SAS/STAT® Users' Guide, Release 6.03 Edition.* SAS Institute Inc., Cary, NC. (451)

Saunder, S. A. (1903). 'Note on the use of Peirce's criterion for the rejection of doubtful observations'. *Monthly Notices Roy. Astr. Soc.*, **63**, 432–436. (28)

Schall, R., and Dunne, T. T. (1987). 'On outliers and influence in the general multivariate normal linear model'. (Contr) Pukkila, T. and Puntanen, S. *Proc. Second Int. Tampere Confer. Statist.*, **1987**, 665–678. (354)

Schall, R., and Dunne, T. T. (1988). 'A unified approach to outliers in the general linear model'. *Indian J. Statist. B.*, **50**, 157–167. (340)

Schmid, W. (1986). 'The multiple outlier problem in time series analysis'. *Australian. J. Statist.*, **28**, 400–413. (400)

Schmid, W. (1989). 'Identification of a type $ rm I $ outlier in an autoregressive model'. *J. Theor. Appl. Statist.*, **20**, 531–545. (400)

Schwager, S. J. (1979). *Detection of multivariate outliers*. Ph.D. thesis, Yale University. (Ch. 7)

Schwager, S. J. and Margolin, B. (1982). 'Detection of multivariate normal outliers'. *Ann. Statist.*, **10**, 943–954. (293, 296)

Schweder, T. (1976). 'Some "optimal" methods to detect structural shift or outliers in regression'. *J. Amer. Statist. Assn.*, **71**, 491–501. (324)

Scott, A. J., Brewer, K. R. W., and Ho, E. W. H. (1978). 'Finite population sampling and robust estimation'. *J. Amer. Statist. Assoc.*, **73**, 359–361. (446)

Scott, S. (1987). 'On the impact of outliers on seasonal adjustment'. *Amer. Statist. Proc. Bus. Econ. Statist. Sect.*, **1987**, 469–474. (413)

Seaman, J. W. Jr., Turner, D. W., and Young, D. M. (1987). 'Polyhedron graphs for displaying multivariate data'. *Int. J.*, **14**, 269–277. (302, 303)

Seaver, B. L., and Triantis, K. P. (1986). 'A statistical perspective on efficiency measurement using multiple outlier diagnostics'. *Amer. Statist. Assoc. Proc., Bus. Econ. Statist. Sect.*, **1986**, 451–456. (454)

Senn, S. J. (1979). 'A sixty year old "medical record" '. *Medical Record*, **20**, 528–531. (5)

Shapiro, S. S., and Wilk, M. B. (1965). 'An analysis of variance test for normality (complete samples)'. *Biometrika*, **52**, 591–611. (46, 96, 233, 234, 369)

Shapiro, S. S., and Wilk, M. B. (1972). 'An analysis of variance test for the exponential distribution (complete samples). *Technometrics*, **14**, 355–370. (46, 96, 210)

Shapiro, S. S., Wilk, M. B., and Chen, M. J. (1968). 'A comparative study of various tests for normality'. *J. Amer. Statist. Assn.*, **63**, 1343–1372. (46, 96, 226, 230, 231, 233, 234)

Sheyin, O. B. (1966a). 'Origin of the theory of errors'. *Nature*, **211**, 1003–1004. (Ch. 2, *H*)

Sheyin, O. B. (1966b). 'On selection and adjustment of direct observations' (Russian). *Izvestiia Vysshikh Uchebnykh Zavedenii. Geodeziia i Aerofotos'emka*, **1966**. English translation: *Geodesy and Aero-photography*, **1966** (1967), 114–117. (Ch. 2, *H*)

Sheyin, O. B. (1971). 'J. H. Lambert's work on probability'. *Archive for History of Exact Sciences*, **7**, 244–256. (Ch. 2, *H*)

Shiffler, R. E. (1988). 'Maximum Z scores and outliers'. *The American Statistician*, **42**, 79–80. (223)

Shoemaker, L. H. (1984). 'Robustness properties for a class of scale estimators'. *Commun. Statist. Theor. Meth.*, **13**, 15–28. (36)

Shorack, G. R. (1976). 'Robust studentization of location estimates'. *Statistica Neerlandica*, **30**, 119–142. (86)

Shu, V. S. (1978). 'Robust estimation of a location parameter in the presence of outliers'. *Ph D Thesis, Iowa State Univ. Ames, Iowa.* (67)

Siddiqui, M. M., and Raghunandanan, K. (1967). 'Asymptotically robust estimators of location'. *J. Amer. Statist. Assn.*, **62**, 950–953. (65, 161)

Simar, L. (1983). 'Protecting against gross errors: the aid of Bayesian methods. In Florens, J. P., Mouchart, M., Raoult, J. P., Simar, L. and Smith, A. F. M. (Eds). *Specifying Statistical Models*. Springer–Verlag, New York-Berlin, pp. 1–12. (53)

Simonoff, J. S. (1984a). 'The calculation of outlier detection statistics'. *Commun. Statist. Sim. Comp.*, **13(2)**, 275–285. (136, 236, 453)

Simonoff, J. S. (1984b). 'A comparison of robust methods and detection of outliers techniques when estimating a location parameter'. *Commun. Statist. Theor. Meth.*, **13**, 813–842. (98, 136, 236)

Simonoff, J. S. (1987a). 'Outlier detection and robust estimation of scale'. *J. Statist. Comp. Sim.*, **27**, 79–92. (36, 53, 84, 136)

Simonoff, J. S. (1987b). 'Comment on Hoaglin, Iglewicz and Tukey (letter)'. *J. Amer. Statist. Assoc.*, **82**, 703–704. (53, 136)

Simonoff, J. S. (1987c). 'The breakdown and influence properties of outlier rejection-plus-mean procedures'. *Commun. Statist. Theor. Meth.*, **16**, 1749–1760. (136)

Simonoff, J. S. (1988a). 'Regression diagnostics to detect nonrandom missingness in linear regression'. *Technometrics*, **30**, 205–214. (370)

Simonoff, J. S. (1988b). 'Detecting outlying cells in two-way contingency tables via backwards-stepping'. *Technometrics*, **30**, 339–345. (434–439)

Simonoff, J. S. and Tsai, C. H. (1986). 'Jackknife-based estimators and confidence regions in nonlinear regression'. *Technometrics*, **28**, 103–112. (351)

Simpson, D. G. (1987). 'Minimum Hellinger distance estimation for the analysis of count data'. *J. Amer. Statist. Assoc.*, **82**, 802–807. (154)

Simpson, D. G. (1989). 'Hellinger deviance tests: efficiency, breakdown points and examples'. *J. Amer. Statist. Assoc.*, **84**, 107–113. (78, 154)

Singh, G., Gupta, S., and Singh, M. (1987). 'Robustness of row-column designs'. *Statist. Prob. Lett.*, **5**, 421–424. (365)

Sinha, B. K. (1984). 'Detection of multivariate outliers in elliptically symmetric distributions'. *Anal. Statist.*, **12**, 1558–1565. (296)

Sinha, S. K. (1972). 'Reliability estimation in life testing in the presence of an outlier observation'. *Op. Res.*, **20**, 888–894. (380, 385, 387)

Sinha, S. K. (1973a). 'Distributions of order statistics and estimation of mean life when an outlier may be present'. *Canad. J. Statist.*, **1**, 119–121. (176)

Sinha, S. K. (1973b). 'Lifetesting and reliability estimation for non-homogeneous data—a Bayesian approach'. *Comm. Statist.*, **2**, 235–243. (380, 385, 387)

Sinha, S. K. (1973c). 'Estimation of the parameters of a two-parameter exponential distribution when an outlier may be present'. *Utilitas Mathematica*, **3**, 75–82. (146) Correction (1974), *Utilitas Mathematica*, **4**, 333–334. (176)

Sinha, S. K. (1973d). 'Some distributions relevant in life testing when an outlier may be present'. *Technical Report No.* 42, Department of Statistics, University of Manitoba, Winnipeg, Canada. (176)

Siotani, M. (1959). 'The extreme value of the generalised distances of the individual points in the multivariate normal sample'. *Ann. Inst. Statist. Math. Tokyo*, **10**, 183–208. (285, 293)

Sivaganesan, S. (1988). 'Range of posterior measures for priors with arbitrary contaminations'. *Commun. Statist. Theor. Meth.*, **17**, 1591–1612. (393)

Sivaganesan, S. (1989). 'Sensitivity of posterior mean to unimodality preserving contaminations'. *Int. Math. J. Stochastic Meth. Mod.*, **7**, 77–93. (393)

Sivaganesan, S., and Berger, J. O. (1989). 'Ranges of posterior measures for priors with unimodal contaminations'. *Ann. Statist.*, **17**, 868–889. (393)

Sleeper, L. A., and Harrington, D. P. (1990). 'Regression splines in the Cox model with application to covariate effects in liver disease'. *J. Amer. Statist. Assoc.*, **85**, 941–949. (353)

Small, C. G. (1990). 'A survey of multidimensional medians'. *Int. Statist. Rev.*, **58**, 263–277. (271)

Smith, A. F. M. (1983). 'Bayesian approaches to outliers and robustness'. *Lec. notes Statist.*, **16**, 13–35. (52, 414)

Snedecor, G. W. and Cochran, W. G. (1967). *Statistical Methods*, 6th edn. The Iowa State University Press, Ames, Iowa. (341)

Spall, J. C. (1988). 'Effect of the sample on the posterior probability in Bayesian analysis'. *Commun. Statist. Theor. Meth.*, **17**, 1811–1827. (393)

Spence, I., and Lewandowsky, S. (1989). 'Robust multidimensional scaling'. *Psychometrika*, **54**, 501–513. (283)

SPSSX (1983). *SPSSX Users' Guide*. McGraw-Hill, New York. (451)

SPSSX (1985). *SPSSX Advanced Statistics Guide*. McGraw-Hill, New York. (451)

Srikantan, K. S. (1961). 'Testing for the single outlier in a regression model'. *Sankhyā, A*, **23**, 251–260. (322, 323, 330, 332, 333, 339)

Srivastava, M. S. (1980). 'Effect of equicorrelation in detecting a spurious observation'. *Canadian J. Statist.*, **8**, 249–251. (223)

Srivastava, M. S. and Lee, G. C. (1983). 'On the choice of transformations of the correlation coefficient with or without an outlier'. *Commun. Statist. Theor. Meth.*, **12**, 2533–2547. (280)

Srivastava, M. S. and Lee, G. C. (1984). 'On the distribution of the correlation coefficient when sampling from a mixture of two bivariate normal densities: robustness and the effect of outliers'. *Canadian J. Statist.*, **12**, 119–133. (280)

Srivastava, M. S., and Lee, G. C. (1985). 'On the robustness of tests of correlation coefficient in the presence of an outlier'. *Commun. Statist. Theor. Meth.*, **14**, 25–40. (280)

Stahel, W. A. (1981). *'Robuste Schätzungen: Infinitesimale Optimalität und Schätzungen von Kovarianzmatrizen'*. Ph.D thesis, Eidgenossiche Technische Hochschule, Zürich. (278)

Stampfer, S. (1839). 'Ueber das Verhältniss der Wiener Klafter zum Meter'. *Jahrbucher des K. K. Polytechnisches Institutes (Vienna)*, **20**, 145–176. (Ch. 2, *H*)

Stapanian, M. A., Garner, F. C., Fitzgerald, K. E., Flatman, G. T., and Englund, E. J. (1991). 'Properties of two tests for outliers in multivariate data'. *Commun. Statist. Sim.*, **20**, 667–687. (49, 293)

Steele, J. M., and Steiger, W. L. (1986). 'Algorithms and complexity for least median of squares regression'. *Discrete Appl. Math.*, **14**, 93–100. (325, 350)

Stefanski, L. A. (1985). 'The effects of measurement error on parameter estimation'. *Biometrika*, **72**, 583–592. (328)

Stefanski, L. A., and Meredith, M. (1986). 'Robust estimation of location in samples of size three'. *Commun. Statist. Theor. Meth.*, **15**, 2921–2933. (65)

Stefanski, L. A., Carroll, R. J., and Ruppert, D. (1986). 'Optimally bounded score functions for generalized linear models with applications to logistic regression'. *Biometrika*, **73**, 413–425. (352)

Stefansky, W. (1971). 'Rejecting outliers by maximum normal residual'. *Ann. Math. Statist.*, **42**, 35–45. (222, 323, 359, 360)

Stefansky, W. (1972). 'Rejecting outliers in factorial designs'. *Technometrics*, **14**, 469–479. (323, 359–361)

Stephens, M. A. (1978). 'On the *W*-test for exponentiality with origin known'. *Technometrics*, **20**, 33–35. (97, 210)

Stephens, M. A. (1981). 'Further percentage points for Greenwood's statistic'. *J. Roy. Statist. Soc. A.*, **144**, 364–366. (211, 212)

Stewart, R. M. (1920a). 'Peirce's criterion'. *Popular Astronomy*, **28**, 2–3. (Ch. 2, *H*)

Stewart, R. M. (1920b). 'The treatment of discordant observations'. *Popular Astronomy*, **28**, 4–6. (Ch. 2, *H*)

Stigler, S. M. (1973). 'Simon Newcomb, Percy Daniell, and the history of robust estimation 1885–1920'. *J. Amer. Statist. Assn.*, **68**, 872–879. (30, 156)

Stigler, S. M. (1977). 'Do robust estimates work with real data?' *Ann. Statist.*, **5**, 1055–1078. (56, 57)

Stigler, S. M. (1980). 'Studies in the history of probability and statistics XXXVIII. R. H. Smith, a Victorian interested in robustness'. *Biometrika*, **67**, 217–221. (57)

Stockinger, N., and Dutter, R. (1987). 'Robust time series analysis: a survey'. *Kybernetika. Supplement*, **23**, 90 pp. (405, 406)

Stone, C. J. (1975). 'Adaptive maximum likelihood estimators of a location parameter'. *Ann. Statist.*, **3**, 267–284. (87)

Stone, E. J. (1868). 'On the rejection of discordant observations'. *Monthly Notices Roy. Astr. Soc.*, **28**, 165–168. (28, 29)

Stone, E. J. (1873). 'On the rejection of discordant observations'. *Monthly Notices Roy. Astr. Soc.*, **34**, 9–15. (29, 36)

Stone, E. J. (1874). 'Note on a discussion relating to the rejection of discordant observations'. *Monthly Notices Roy. Astr. Soc.*, **35**, 107–108. (Ch. 2, *H*)

Stoodley, K. D. C., and Mirnia, M. (1979). 'The automatic detection of transients, step changes and slope changes in the monitoring of medical time series'. *The Statistician*, **28**, 163–170. (399, 400)

Stroup, D. F., Williamson, G. D., and Herndon, J. L. (1989). 'Detection of aberrations in the occurrence of notifiable diseases surveillance data'. *Statist. Med.*, **8**, 323–329. (455)

Student (1927). 'Errors of routine analysis'. *Biometrika*, **19**, 151–164. (36)

Subba Rao, T. (1979). Discussion of Kleiner, B., Martin, R. D., and Thomson, D. J. (1979). *J. Roy. Statist. Soc. B*, **41**, 346–347. (414)

Sugiura, N., and Sasamoto, H. (1989). 'Locally best invariant test for outliers in a gamma type distribution'. *Commun. Statist. Sim.*, **18**, 415–427. (102, 212)

Sweeting, T. J. (1983). 'Independent scale-free spacings for the exponential and uniform distributions.' *Statist. Prob. Lett.*, **1**, 115–119. (98, 136, 137)

Sweeting, T. J. (1986). 'Asymptotically independent scale-free spacings with applications to discordancy testing'. *Ann. Statist.*, **14**, 1485–1496. (98, 136, 137)

Takahashi, R. (1987). 'A note on outlier-prone and outlier-resistant distributions'. *J. Japan Statist. Soc.*, **17**, 107–112. (53)

Takeuchi, K. (1971). 'A uniformly asymptotically efficient estimator of a location parameter'. *J. Amer. Statist. Assn.*, **66**, 292–301. (75, 148)

Tambay, J.-L. (1988). 'An integrated approach for the treatment of outliers in sub-annual economic surveys'. *American Statistical Association Proceedings of the Survey Research Methods*, American Statistical Association, Alexandria, Va. pp. 229–234. (445, 446)

Tamhane, A. C. (1982). 'A note on the use of residuals for detecting an outlier in linear regression'. *Biometrika*, **69**, 488–489. (336)

Tamura, R. N., and Boos, D. D. (1986). 'Minimum Hellinger distance estimation for multivariate location and covariance'. *J. Amer. Statist. Assoc.*, **81**, 223–229. (154)

Tanaka, Y., and Odaka, Y. (1989). 'Influential observations in principal factor analysis'. *Psychometrika*, **54**, 475–485. (283)

Tango, T. (1986). 'Estimation of normal ranges of clinical laboratory data'. *Statist. Med.*, **5**, 335–346. (454)

Taylor, J. M. G. (1989). 'Models for the HIV infection and AIDS epidemic in the United States'. *Statist. Med.*, **8**, 45–58. (455)

Taylor, J. M. G., Muñoz, A., Bass, S. M., Saah, A. J., Chmiel, J. S., Kingsley, L. A., and The Multicentre Aids Cohort Study (1990). 'Estimating the distribution of times from HIV seroconversion to Aids using multiple imputation'. *Statist. Med.*, **9**, 505–514. (455)

Thall, P. F. (1979). 'Huber-sense robust *M*-estimation of a scale parameter, with application to the exponential distribution'. *J. Amer. Statist. Assn.*, **74**, 147–152. (156, 157)

Theil, H. (1965). 'The analysis of disturbances in regression analysis'. *J. Amer. Statist. Assn.*, **60**, 1067–1079. (330)

Thombs, L. A., and Schucany, W. R. (1990). 'Bootstrap prediction intervals for autogression'. *J. Amer. Statist. Assoc.*, **85**, 486–492. (411)

Thompson, G. W. (1955). 'Bounds for the ratio of range to standard deviation'. *Biometrika*, **42**, 268–269. (240)

Thompson, M. L., and Zucchini, W. (1989). 'On the statistical analysis of ROC curves.' *Statist. Med.*, **8**, 1277–1290. (455)

Thompson, W. A. Jr., and Willke, T. A. (1963). 'On an extreme rank sum test for outliers'. *Biometrika*, **50**, 375–383. (Ch. 8)

Thompson, W. R. (1935). 'On a criterion for the rejection of observations and the distribution of the ratio of the deviation to the sample standard deviation'. *Ann. Math. Statist.*, **6**, 214–219. (31)

Tiao, G. C. (1985). 'Autoregressive moving average models intervention problems and outlier detection in time series'. In Hannan, E. J., Krishnaiah, P. R., and Rao, M. M. (Eds). *Time Series in the Time Domain*. North-Holland, Amsterdam, pp. 85–118. (400)

Tiao, G. C., and Guttman, I. (1967). 'Analysis of outliers with adjusted residuals'. *Technometrics*, **9**, 541–559. (62, 167, 347)

Tietjen, G. L., and Moore, R. H. (1972). 'Some Grubbs-type statistics for the detection of several outliers'. *Technometrics*, **14**, 583–597. (94, 97, 132, 136, 137, 224, 225, 232)

Tietjen, G. L., Moore, R. H., and Beckman, R. J. (1973). 'Testing for a single outlier in simple linear regression'. *Technometrics*, **15**, 717–721. (322, 323, 332, 334)

Tiku, M. L. (1975). 'A new statistic for testing suspected outliers'. *Commun. Statist. A*, **4**, 737–752. (132, 136, 222)

Tiku, M. L. (1977). Rejoinder to 'Comment on "A new statistic for testing suspected outliers" '. *Commun. Statist. Theor. Meth.*, **6**, 1417–1422. (223)

Tiku, M. L. (1978). 'Linear regression model with censored observations'. *Commun. Statist. Theor. Meth.*, **7**, 1219–1232. (324)

Tiku, M. L. (1980). 'Robustness of MML estimators based on censored samples and robust test statistics'. *J. Statist. Plan. Inf.*, **4**, 123–143. (77, 164, 187, 188)

Tiku, M. L. (1987). A robust procedure for testing an assumed value of the population correlation coefficient. *Commun. Statist. Sim. Comp.*, **16**, 907–924. (280)

Tiku, M. L. (1988). 'Modified maximum likelihood estimation for the bivariate normal'. *Commun. Statist. Theor. Meth.*, **17**, 893–910. (280)

Tiku, M. L., and Balakrishnan, N. (1986). 'A robust test for testing the correlation coefficient'. *Commun. Statist. Sim. Comp.*, **15**, 945–971. (280)

Tiku, M. L., and Balakrishnan, N. (1988a). 'Generalization of the robust bivariate T\$ 2\$ statistic to multivariate populations'. *Commun. Statist. Theor. Meth.*, **17**, 3899–3911. (279)

Tiku, M. L., and Balakrishnan, N. (1988b). 'Robust Hotelling-type T\$ 2\$ statistics based on the modified maximum likelihood estimators'. *Commun. Statist. Theor. Meth.*, **17**, 1789–10. (279)

Tiku, M. L. and Singh, M. (1981). 'Robust test for means when population variances are unequal'. *Commun. Statist. Theor. Meth.*, **10**, 2057–2071. (77)

Tiku, M. L., Tan, W. Y., and Balakrishnan, N. (1986). *Robust Inference.* Marcel Dekker, Inc., New York. (164, 187, 188)

Tingley, M. and Field, C. (1990). 'Small-sample confidence intervals'. *J. Amer. Statist. Assoc.*, **85**, 427–434. (16)

Tippett, L. H. C. (1925). 'On the extreme individuals and the range of samples taken from a normal population'. *Biometrika*, **17**, 364–387. (248)

Titterington, D. M. (1978). 'Estimation of correlation coefficients by ellipsoidal trimming'. *Applied Statistics*, **27**, 227–234. (281)

Truax, D. R. (1953). 'An optimum slippage test for the variances of $k$, normal distributions'. *Ann. Math. Statist.*, **24**, 669–674. (104)

Tsaknakis, H., Kazakos, D., and Papantoni-Kazakos, P. (1986). 'Robust prediction and interpolation for vector stationary processes'. *Prob. Theor. Rel. Fields*, **72**, 589–602. (414)

Tsay, R. S. (1986). 'Time series model specification in the presence of outliers'. *J. Amer. Statist. Assoc.*, **81**, 132–141. (411)

Tsay, R. S. (1988). 'Outliers, level shifts and variance changes in time series'. *J. Forecasting*, **7**, 1–20. (411)

Tse, Y. K. (1988). 'Assessing Lund's critical values for testing for outliers in linear regression models'. *J. Appl. Statist.*, **15**, 363–366. (334)

Tukey, J. W. (1960). 'A survey of sampling from contaminated distributions'. In Olkin, I. (Ed.) (1960). *Contributions to Probability and Statistics.* University Press, Stanford, California. (35, 49, 57, 59, 156)

Tukey, J. W. (1962). 'The future of data analysis'. *Ann. Math. Statist.*, **3**, 1–67. (367)

Tukey, J. W. (1977). *Exploratory Data Analysis*, Vol. 1. Addison–Wesley, Reading, Mass. (72, 147)

Tukey, J. W. (1978). 'The ninther, a technique for low-effort robust (resistant) location in large samples. Pp. 251–257 of David, H. A. (Ed.) (1978) *Contributions to Survey Sampling and Applied Statistics, in honour of H. O. Hartley.* Academic Press, New York. (152)

Tukey, J. W., and McLaughlin, D. M. (1963). 'Less vulnerable confidence and significance procedures for location based on a single sample: Trimming/Winsorization'. *Sankhyā, A*, **25**, 331–352. (75, 85, 86, 145)

United states Department of Commerce, Bureau of the Census (1981). *Annual Survey of Manufactures*. U.S. Government Printing Office, Washington, DC. (445)

Vardeman, S., and Meeden, G. (1983). 'Admissible estimators of the population total using trimming and Winsorization'. *Statist. Prob. Letts.*, **1**, 317–321. (446)

Vaughan, R. J., and Venables, W. N. (1972). 'Permanent expressions for order statistic densities'. *J. Roy. Statist. Soc. B.*, **34**, 308–310. (67, 68)

Veale, J. R. (1975). 'Improved estimation of expected life when one identified spurious observation may be present'. *J. Amer. Statist. Assn.*, **70**, 398–401. (68, 84, 178, 179)

Veale, J. R., and Huntsberger, D. V. (1969). 'Estimation of a mean when one observation may be spurious'. *Technometrics*, **11**, 331–339. (Ch. 3)

Veale, J. R., and Kale, B. K. (1972). 'Tests of hypotheses for expected life in the presence of a spurious observation'. *Utilitas Mathematica*, **2**, 9–23. (51, 68, 77, 175, 176)

Verme, D. A. (1989). 'Effects of outliers on the cross correlation function in transfer function models'. *Amer. Statist. Assoc. Proc. Bus. Econ. Statist. Sect.*, **1989**, 414–417. (414)

Walsh, J. E. (1950). 'Some nonparametric tests of whether the largest observations of a set are too large or too small'. *Ann. Math. Statist.*, **21**, 583–592. Correction (1953), *Ann. Math. Statist.*, **24**, 134–135. (107)

Walsh, J. E. (1959). 'Large sample non-parametric rejections of outlying observations'. *Ann. Inst. Statist. Math. Tokyo*, **10**, 223–232. (107)

Walsh, J. E. (1965). *Handbook of Non-parametric Statistics, II*. Van Nostrand, Princeton, N. J. (107)

Walsh, J. E., and Kelleher, G. J. (1974). 'Nonparametric estimation of mean and variance when a few "sample" values possibly outliers'. *Ann. Inst. Statist. Math. Tokyo*, **25**, 87–90. (107)

Wang, P. C. C. (1981). 'Robust asymptotic tests of statistical hypotheses involving nuisance parameters'. *Ann. Statist.*, **9**, 1096–1106. (77, 87)

Wani, J. K., and Kabe, D. G. (1971). 'Distributions of Dixon's statistics for the truncated exponential, rectangular, and random intervals population'. *Metron*, **29**, 151–160. (251, 252)

Wasserman, L. A. (1989). 'A robust Bayesian interpretation of likelihood regions'. *Ann. Statist.*, **17**, 1387–1393. (393)

Waternaux, C., Laird, N. M., and Ware, J. H. (1989). 'Methods for analysis of longitudinal data: blood-lead concentrations and cognitive development'. *J. Amer. Statist. Assoc.*, **84**, 33–41. (445)

Watson, G. S. (1983). *Statistics on Spheres*. Wiley, New York. (424)

Wegman, E. J., and Carroll, R. J. (1977). 'A Monte Carlo study of robust estimators of location'. *Commun. Statist. Theor. Meth.*, **6**, 795–812. (164)

Wehrly, T. E., and Shine, E. P. (1981). 'Influence curves of estimators for directional data.' *Biometrika*, **68**, 334–335. (424)

Wei, L. J. (1981). 'Estimation of location difference for fragmentary samples'. *Biometrika*, **68**, 471–476. (158)

Weisberg, S. (1986). 'Comment on Chatterjee & Hadi. (1986).' *Statist. Sci.*, **1**, 414–415. (349)

Welch, W. J., and Gutierrez, L. G. (1988). 'Robust permutation tests for matched-pairs designs'. *J. Amer. Statist. Assoc.*, **83**, 450–455. (365)

Welsch, R. E. (1982). 'Influence functions and regression diagnostics'. Pp. 149–169 of Launer, R. L., and Siegel, A. F. (Eds) (1982). *Modern Data Analysis*. Academic Press, New York. (374)

Welsh, A. H., (1987). 'The trimmed mean in the linear model. (Contr) DeJong, P. J., DeWet, T., and Koenker, R.' *Ann. Statist.*, **15**, 20–45. (349)

Welsh, A. H. (1989). 'Concomitant scale estimation in regression problems with increasing dimension'. *Australian. J. Statist.*, **31**, 215–227. (351)

West, M. (1981). 'Robust sequential approximate Bayesian estimation'. *J. Roy. Statist. Soc. B*, **43**, 157–166. (413)

West, M. (1984). 'Outlier models and prior distributions in Bayesian linear regression'. *J. Roy. Statist. Soc. B.*, **46**, 431–439. (385)

West, M. (1985). 'Generalized linear models: scale parameters, outlier accommodation and prior distributions'. *Bayesian Statist.*, **1985**, 531–557. (385)

West, M. (1986). 'Bayesian model monitoring'. *J. Roy. Statist. Soc. B.*, **48**, 70–78. (394)

West, M., and Harrison, P. J. (1986). 'Monitoring and adaptation in Bayesian forecasting models.' *J. Amer. Statist. Assoc.*, **81**, 741–750. (394)

West, S. A. (1975). 'Bias in the estimator of Kendall's rank correlation when extreme pairs are removed from the sample'. *J. Amer. Statist. Assn.*, **70**, 439–442. (Ch. 7)

Wilk, M. B., and Gnanadesikan, R. (1964). 'Graphical methods for internal comparisons in multiresponse experiments'. *Ann. Math. Statist.*, **35**, 613–631. (302, 307)

Wilk, M. B., Gnanadesikan, R., and Huyett, M. J. (1962a). 'Probability plots for the gamma distribution'. *Technometrics*, **4**, 1–20. (307, 312)

Wilk, M. B., Gnanadesikan, R., and Huyett, M. J. (1962b). 'Estimation of parameters of the gamma distribution using order statistics'. *Biometrika*, **49**, 525–545. (307, 312)

Wilks, S. S. (1962). *Mathematical Statistics*. Wiley, New York. (288)

Wilks, S. S. (1963). 'Multivariate statistical outliers' *Sankhyā, A*, **25**, 407–426. (287, 288, 302, 307, 374)

Williams, D. A. (1973). Letter to the Editor. *Applied Statistics*, **22**, 407–408. (361)

Williams, D. A. (1987). 'Generalized linear model diagnostics using the deviance and single case deletions'. *Applied Statistics*, **36**, 181–191. (370)

Willke, T. A. (1966). 'A note on contaminated samples of size three'. *Journal of Research of the National Bureau of Standards, B*, **70**, 149–151. (65)

Wilson, S. R. (1979). 'Examination of regression residuals'. *Austral. J. Statist.*, **21**, 18–29. (330)

Winlock, J. (1856). 'On Professor Airy's objections to Peirce's criterion'. *Astr. J.*, **4**, 145–147. (Ch. 2, *H*)

Wood, F. S. (1973). 'The use of individual effects and residuals in fitting equations to data'. *Technometrics*, **15**, 677–695. (369)

Wooding, W. M. (1969). 'The computation and use of residuals in the analysis of experiment data'. *J. Quality Technology*, **1**, 175–188. Correction, **1**, 294. (330)

Worsley, K. J. (1982). 'An improved Bonferroni inequality and applications'. *Biometrika*, **69**, 297–302. (121)

Wright, T. W. (1884). *A Treatise on the Adjustment of Observations by the Method of Least Squares*. Van Nostrand, New York. (30)

Wright, T. W., and Hayford, J. F. (1906). *Adjustment of Observations*. Van Nostrand, New York. (30)

Yenyukov, I. S. (1988). 'Detecting structures by means of projection pursuit'. *Physica, Heidelberg*, **1988**, 47–58. (306)

Ylvisaker, D. (1977). 'Test resistance'. *J. Amer. Statist. Assn.*, **72**, 551–556. (74)

Yohai, V. J. (1974). 'Robust estimation in the linear model'. *Ann. Statist.*, **2**, 562–567. (351)

Yoshida, M., Kondo, M., and Inagaki, N. (1984). 'Asymptotic properties of several estimators of the autocorrelation based on limited estimating functions for a stationary Gaussian process with additive outliers'. *J. Japan Statist. Soc.*, **14**, 157–168. (402)

Youden, W. J. (1949). 'The fallacy of the best two out of three'. *National Bureau of Standards Technical Bulletin*, **33**, 77–78. (Ch. 2)

Young, D. M., Pavur, R., and Marco, V. R. (1989). 'On the effect of correlation and unequal variances in detecting a spurious observation.' *Canadian J. Statist.*, **17**, 103–105. (285)

Young, S. S., Fraction, G. F., Skrzynecki, R. E., and Shotts, J. B. (1985). 'PROC ELPRINT: A print precedure that supports footnotes and outlier detection'. *Proc. SAS Users Group Int. Conference*, **1985**, 787–792. (453)

Von Zach, F. X. (1805). 'Versuch einer auf Erfahrung gegrundeten Bestimmung terrestrischer Refractionen'. *Monatsliche Correspondenz zur Beförderung der Erdund Himmels-Kunde*, **11**, 389–415. (Ch. 2, *H*)

Zamar, R. H. (1989). 'Robust estimation in the errors-in-variables model'. *Biometrika*, **76**, 149–160. (328)

Zheng, Z. O. (1987). 'The treatment of errors in the independent variables in binary regression models'. *J. Math. Res. Expo.*, **7**, 305–310. (353)

# Index

* Now available in a lower priced paperback edition in the Wiley Classics Library.